SBE BROADCAST ENGINEERING HANDBOOK

ABOUT THE AUTHORS

Jerry C. Whitaker, Editor-in-Chief, is Vice President of Standards Development for the Advanced Television Systems Committee (ATSC) in Washington, D.C. He is the author or editor of more than 40 technical books, including *The DTV Handbook, The Standard Handbook of Video and Television Engineering, The Standard Handbook of Audio and Radio Engineering,* and *Communications Receivers.* Mr. Whitaker is a Fellow of the Society of Broadcast Engineers.

The Society of Broadcast Engineers (SBE) is the only organization devoted to the advancement of all levels and types of broadcast engineering. With more than 5100 members and 115 local chapters, the SBE provides a forum for the exchange of ideas and the sharing of information to help members keep pace with a rapidly changing industry. The SBE amplifies the voices of broadcast engineers by validating their skills with professional certification, by offering educational opportunities to maintain and expand those skills, and by speaking out on technical regulatory issues that affect how members work.

SBE BROADCAST ENGINEERING HANDBOOK

Hands-On Guide to Station Design and Maintenance

Jerry C. Whitaker

and

Society of Broadcast Engineers

New York   Chicago   San Francisco   Athens   London
Madrid   Mexico City   Milan   New Delhi
Singapore   Sydney   Toronto

Cataloging-in-Publication Data is on file with the Library of Congress.

McGraw-Hill Education books are available at special quantity discounts to use as premiums and sales promotions, or for use in corporate training programs. To contact a representative please visit the Contact Us page at www.mhprofessional.com.

SBE Broadcast Engineering Handbook: Hands-On Guide to Station Design and Maintenance

Copyright ©2016 by McGraw-Hill Education. All rights reserved. Printed in the United States of America. Except as permitted under the United States Copyright Act of 1976, no part of this publication may be reproduced or distributed in any form or by any means, or stored in a data base or retrieval system, without the prior written permission of the publisher.

1 2 3 4 5 6 7 8 9 0   DOC/DOC   1 2 1 0 9 8 7 6

ISBN 978-0-07-182626-6
MHID 0-07-182626-2

The pages within this book were printed on acid-free paper.

Sponsoring Editor
Robert Argentieri

Editorial Supervisor
Donna M. Martone

Acquisitions Coordinator
Lauren Rogers

Project Manager
Hardik Popli

Copy Editor
Exsecant

Proofreader
Alekha Jena, Cenveo Publisher Services

Indexer
Jack Lewis

Production Supervisor
Pamela A. Pelton

Composition
Cenveo Publisher Services

Art Director, Cover
Jeff Weeks

Information contained in this work has been obtained by McGraw-Hill Education from sources believed to be reliable. However, neither McGraw-Hill Education nor its authors guarantee the accuracy or completeness of any information published herein, and neither McGraw-Hill Education nor its authors shall be responsible for any errors, omissions, or damages arising out of use of this information. This work is published with the understanding that McGraw-Hill Education and its authors are supplying information but are not attempting to render engineering or other professional services. If such services are required, the assistance of an appropriate professional should be sought.

Front cover artwork credit:
  Upper right—Entercom Indianapolis, IN; photo by John L. Poray

  Bottom Right—WISH-TV Indianapolis, IN; photo by John L. Poray

Back cover artwork credit:
  Upper photo—Entercom Indianapolis, IN; photo by John L. Poray

  Bottom photo—WISH-TV Indianapolis, IN; photo by John L. Poray

CONTENTS

SBE Editorial Advisory Board   xvii
Authors   xix
Preface   xxi

Section 1   Regulatory Issues   *Ralph Hogan*   1.1

Chapter 1.1   FCC Licensing and Administrative Basics for the Technically Minded
Ernie Sanchez   1.3

1.1   Introduction / 1.3
1.2   Licensing Basics / 1.4
1.3   Description of FCC Reference Resources / 1.5

Chapter 1.2   Chief Operator Requirements   *Dennis Baldridge*   1.11

1.1   Introduction / 1.11
1.2   The Chief Operator Selection / 1.11
1.3   Duties Required by the Chief Operator / 1.12
1.4   Summary / 1.15

Chapter 1.3   The Alternative Broadcast Inspection Program (ABIP)   *Larry Wilkins*   1.17

1.1   Introduction / 1.17
1.2   About the Program / 1.17
1.3   Final Comments / 1.20

Chapter 1.4   Broadcast Accessibility Requirements   *Mike Starling*   1.21

1.1   Introduction / 1.21
1.2   Radio Reading Services for the Print Handicapped / 1.22
1.3   TV for the Visually Disabled—Video Description / 1.23
1.4   Broadcast Captioning / 1.24
1.5   Conclusion / 1.30

Chapter 1.5   The Emergency Alert System   *Larry Wilkins*   1.31

1.1   Introduction / 1.31
1.2   Emergency Alert System / 1.33
1.3   For More Information / 1.35

Section 2 RF Transmission *Douglas Garlinger and Gary Sgrignoli* 2.1

Chapter 2.1 AM and FM Transmitters *Scott Marchand and Alex Morash* 2.3

2.1 Introduction / 2.3
2.2 Theory / 2.3
2.3 Transmitter Overview / 2.15
2.4 Factors Affecting Performance / 2.19
2.5 Reducing the Total Cost of Ownership and Extending Transmitter Life / 2.26
References / 2.32

Chapter 2.2 Coaxial Transmission Lines *Derek Small, Nicholas Paulin, Philip Young, and Bill Harland* 2.33

2.1 Introduction / 2.33
2.2 Transmission Line Types / 2.33
2.3 Electrical and Operational Parameters / 2.35
2.4 Power Handling / 2.39
2.5 Differential Expansion / 2.44
2.6 Semiflexible Transmission Line Systems / 2.48
2.7 Rigid Transmission Line / 2.52
2.8 Pressurization / 2.56
2.9 Maintenance and Inspection / 2.57

Chapter 2.3 FM Channel Combiners *Derek J. Small* 2.59

2.1 Introduction / 2.59
2.2 Combiner Types / 2.59
2.3 Frequency Response / 2.61
2.4 Cross-Coupled Filters / 2.66
2.5 Delay/Loss Correction / 2.68
Reference / 2.68

Chapter 2.4 Transmitting Antennas for FM and TV Broadcasting *Kerry W. Cozad* 2.69

2.1 Introduction / 2.69
2.2 General Antenna Characteristics / 2.70
2.3 Installation / 2.77
2.4 Maintenance / 2.80
2.5 Summary / 2.86
Bibliography / 2.86

Chapter 2.5 Practical Aspects of Maintaining Medium-Wave Antenna Systems in AM Transmission *Phil Alexander* 2.87

2.1 Introduction / 2.87
2.2 Elements of the Antenna System / 2.88
2.3 Technical Principles / 2.90
2.4 Troubleshooting a Single Radiator Problem / 2.97
2.5 Troubleshooting Directional Arrays / 2.98

Chapter 2.6 International Shortwave Broadcasting *Douglas Garlinger* 2.101

2.1 Introduction / *2.101*
2.2 FCC Regulation / *2.101*
2.3 International Shortwave Bands / *2.102*
2.4 Frequency Management / *2.104*
2.5 Frequency Requests / *2.106*
2.6 Shortwave Transmitters / *2.106*
2.7 Antenna Types / *2.107*
2.8 Single Sideband / *2.109*
2.9 DRM® Digital Radio Mondale™ / *2.109*
2.10 Ionosphere / *2.109*
2.11 Smoothed Sunspot Number / *2.111*
2.12 Interval Signals / *2.112*
2.13 Reception Reports / *2.113*
2.14 Defining Terms / *2.114*
Bibliography / *2.114*
For Further Information / *2.114*

Chapter 2.7 Evaluation of TV Coverage and Interference *Bill Meintel* 2.115

2.1 Introduction / *2.115*
2.2 The Need for a More Sophisticated Model / *2.116*
2.3 The Longley–Rice Model / *2.129*
2.4 Final Thoughts / *2.132*

Chapter 2.8 DTV RF Considerations *Douglas Garlinger* 2.133

2.1 Introduction / *2.133*
2.2 8-VSB Signal / *2.133*
2.3 FCC Spectral Mask / *2.141*
2.4 8-VSB Transmission Monitoring / *2.143*
2.5 8-VSB Power / *2.146*
2.6 High Power Amplifier Devices / *2.148*
2.7 8-VSB Specialist Certification / *2.151*
2.8 Defining Terms / *2.151*
References / *2.151*

Chapter 2.9 ATSC DTV Transmission System *Gary Sgrignoli* 2.153

2.1 Introduction / *2.153*
2.2 System Description / *2.154*
2.3 VSB Baseband Description / *2.166*
2.4 VSB Baseband Spectral Description / *2.169*
2.5 VSB RF Spectrum Description / *2.172*
2.6 ATSC DTV Transmission System Parameters / *2.179*

Chapter 2.10 Television Transmitters *John Tremblay* 2.183

2.1 Introduction / *2.183*
2.2 The Amplifier / *2.184*
2.3 Power Supplies / *2.197*
2.4 Control and Metering / *2.197*

2.5 Mask Filters / 2.198
2.6 Cooling / 2.199
Reference / 2.200

Chapter 2.11 DTV Mask Filters   Daniel S. Fallon   2.201

2.1 Introduction / 2.201
2.2 Filters and Emission Limits / 2.201
2.3 Types of Filters / 2.203
2.4 Thermal Stability / 2.208
2.5 Installation Considerations / 2.209
2.6 Maintenance / 2.210
2.7 Filter Retuning / 2.210

Chapter 2.12 DTV Television RF Measurements   Linley Gumm   2.211

2.1 Introduction / 2.211
2.2 Measurements Using General Purpose Test Equipment / 2.211
2.3 Determining Signal Quality / 2.223
2.4 Signal Quality Measurements / 2.227
Reference / 2.230

Chapter 2.13 Hybrid Microwave IP and Cellular Data for Newsgathering
Nuraj Lal Pradhan and John Wood   2.231

2.1 Introduction / 2.231
2.2 Migration from Analog FM Modulation to Digital Modulation / 2.231
2.3 COFDM Overview / 2.234
2.4 Cellular News Gathering (CNG) / 2.236
2.5 Migration from Ku to Ka Band IP Satellite Systems / 2.240
2.6 Migration to Hybrid Microwave Solutions / 2.241
2.7 Hybrid Aggregation / 2.241
2.8 Summary / 2.243
List of Acronyms / 2.245
References / 2.246

Section 3   DTV Transport   *Dr. Richard Chernock*   3.1

Chapter 3.1 MPEG-2 Transport   John R. Mick Jr   3.3

3.1 Introduction / 3.3
3.2 MPEG-2 Systems Specification Introduction / 3.3
3.3 The MPEG-2 Transport Stream / 3.6
3.4 Tables, Sections, and Descriptors / 3.11
3.5 MPEG-2 Program Specific Information (PSI) / 3.15
3.6 TS_Program_Map_Section Syntax / 3.18
3.7 MPEG-2 Packetized Elementary Stream (PES) Packets / 3.19
3.8 MPEG-2 System Timing / 3.22
References / 3.27

Chapter 3.2 Program and System Information Protocol: PSIP *Dr. Richard Chernock* 3.29

3.1 Introduction / *3.29*
3.2 Virtual Channels / *3.30*
3.3 PSIP Tables / *3.30*
3.4 PSIP Descriptors / *3.36*
3.5 The Big Picture / *3.38*

Chapter 3.3 IP Transport for Mobile DTV *Gomer Thomas* 3.43

3.1 Introduction / *3.43*
3.2 Content Delivery Framework / *3.43*
3.3 Services / *3.44*
3.4 Signaling / *3.45*
3.5 Timing and Buffer Model / *3.46*
3.6 Announcements / *3.47*
3.7 Terms / *3.47*
References / *3.47*

Chapter 3.4 Mobile Emergency Alert System *Wayne C. Luplow, Wayne Bretl, and Jay C. Adrick* 3.49

3.1 Introduction / *3.49*
3.2 M-EAS as Part of ATSC Mobile DTV / *3.49*
3.3 M-EAS Relationship to National Alerting Infrastructure / *3.50*
3.4 M-EAS Input Sources / *3.50*
3.5 Use Scenario / *3.52*
3.6 M-EAS Advantages / *3.52*
3.7 Implementation / *3.52*
3.8 Emergency Alerting as Part of ATSC 3.0 / *3.53*
Bibliography / *3.56*

Chapter 3.5 ATSC Mobile DTV System *Jerry C. Whitaker* 3.57

3.1 Introduction / *3.57*
3.2 ATSC Mobile DTV, A/153 / *3.58*
3.3 Supporting Recommended Practice / *3.66*
3.4 Transmission Infrastructure / *3.66*
References / *3.67*

Section 4 Information Technology Systems *Wayne M. Pecena* 4.1

Chapter 4.1 Information Technology and the Broadcast Plant *Wayne M. Pecena* 4.3

4.1 Introduction / *4.3*
4.2 The IP Network—A Technology Review / *4.4*
4.3 Networking Standards / *4.5*
4.4 The OSI Model / *4.6*
4.5 Encapsulation and De-Encapsulation / *4.9*
4.6 The Data-Flow Layers / *4.11*

4.7 Conclusion / 4.46
A4.1 Appendix / 4.48
Suggested Further Reading / 4.53

Chapter 4.2 Network Systems   *Gary Olson*   4.55

4.1 Introduction / 4.55
4.2 Network Infrastructure / 4.55
4.3 Network Topology / 4.56
4.4 File-Based Workflow Architecture / 4.62

Chapter 4.3 Time and Frequency Transfer over Ethernet Using NTP and PTP
Nikolaus Kerö   4.69

4.1 Introduction / 4.69
4.2 PTP—Precision Time Protocol / 4.69
4.3 IEEE 1588—What the Standard Specifies / 4.74
4.4 SyncE—Synchronous Ethernet: A Solution to All Problems? / 4.77
4.5 Genlock over IP / 4.77
4.6 Conclusions / 4.77
References / 4.78

Chapter 4.4 Standards for Video Transport over an IP Network   *John Mailhot*   4.79

4.1 Introduction / 4.79
4.2 Historical View of Television Transport over Carrier Networks / 4.79
4.3 MPEG-2 Transport Streams over IP Networks / 4.81

Section 5   Production Systems   *Andrea Cummis*   5.1

Chapter 5.1 Production Facility Design   *Richard G. Cann,*
Anthony Hoover, Frederic M. Remley, and Ernst-Joachim Voelker   5.3

5.1 Introduction / 5.3
5.2 Studio Design Considerations / 5.3
References / 5.17
Bibliography / 5.17

Chapter 5.2 Audio System Interconnections   *Greg Shay and Martin Sacks*   5.19

5.1 Introduction / 5.19
5.2 Audio over IP—A Primer / 5.20
5.3 What Can You Do with AOIP? / 5.25
5.4 Network Requirements / 5.31
5.5 Network Engineering for Audio Engineers / 5.32
Resources / 5.35

Chapter 5.3 Audio Monitoring Systems   *Martin Dyster*   5.37

5.1 Introduction / *5.37*
5.2 Audio Monitoring in Broadcast / *5.38*
5.3 Connectivity—Signal Types / *5.42*
5.4 The Future of Audio Monitoring / *5.43*

Chapter 5.4 Remote Audio Broadcasting   *Martin Dyster*   5.45

5.1 Introduction / *5.45*
5.2 News Remote Broadcasting / *5.46*
5.3 The Future of Remote Broadcasting / *5.48*

Chapter 5.5 Master Control and Centralized Facilities   *John Luff*   5.49

5.1 Introduction / *5.49*
5.2 The Function of Master Control / *5.49*
5.3 Centralizing Broadcast Operations / *5.57*

Chapter 5.6 Video Switchers   *Brian J. Isaacson*   5.63

5.1 Introduction / *5.63*
5.2 Switcher Features / *5.63*
5.3 The Big Picture / *5.69*

Chapter 5.7 Automation Systems   *Gary Olson*   5.71

5.1 Introduction / *5.71*
5.2 But Enough History / *5.73*
5.3 Orchestration—The Next Generation of Automation / *5.75*
5.4 Summary / *5.76*

Chapter 5.8 Media Asset Management   *Sam Bogoch*   5.77

5.1 Introduction / *5.77*
5.2 A Brief History of Modern MAM / *5.78*
5.3 Four Ways to Categorize MAM / *5.79*
5.4 XML—A Key Interchange Format / *5.80*
5.5 Workflow Automation and Process Orchestration—A Paradox / *5.81*
5.6 Conclusions / *5.81*

Chapter 5.9 Production Intercom Systems   *Vinnie Macri*   5.83

5.1 Introduction / *5.83*
5.2 Analog Party-Line/TW Intercoms, Wired / *5.84*
5.3 Digital Partyline Systems / *5.92*

5.4 Wireless Production Intercoms / 5.93
5.5 IFB / 5.96
5.6 Matrix Intercoms / 5.97
5.7 Virtual Intercoms / 5.104

Chapter 5.10 Broadcast Studio Lighting   *Frank Marsico*   5.105

5.1 Introduction / 5.105
5.2 Three Point Lighting / 5.105
5.3 Light Source Information / 5.106
5.4 The Lighting System / 5.110
5.5 Lighting Instruments / 5.130
5.6 Accessory Hardware / 5.138
5.7 Summary / 5.141
Bibliography / 5.141
References / 5.141
Resources / 5.141
5.8 Organizations / 5.142

Chapter 5.11 Cellular/IP ENG Systems   *Joseph J. Giardina and Herbert Squire*   5.145

5.1 Introduction / 5.145
5.2 The Cellular Revolution / 5.146
5.3 The Ideal ENG Device / 5.147
5.4 IP-Based ENG Systems / 5.149
5.5 Getting the Story / 5.157
5.6 Finding a Practical Solution / 5.158
5.7 No More "Film at Eleven!" / 5.161

Section 6 Facility Issues   *Jerry C. Whitaker*   6.1

Chapter 6.1 Broadcast Facility Design   *Gene DeSantis*   6.3

6.1 Introduction / 6.3
6.2 Construction Considerations / 6.3
Bibliography / 6.9

Chapter 6.2 Wire Management   *Fred Baumgartner*   6.11

6.1 Introduction / 6.11
6.2 Labels / 6.12
6.3 Documentation / 6.12
6.4 Physical Layer / 6.15
6.5 Ringing out the Plant / 6.20
6.6 Cleaning and Removing / 6.21
6.7 Bottom Line / 6.23
6.8 For Further Information / 6.24

Chapter 6.3 Equipment Rack Enclosures and Devices *Jerry C. Whitaker* 6.25

    6.1 Introduction / *6.25*
    6.2 Rack/Equipment Layout / *6.25*
    6.3 Industry Standard Equipment Enclosures / *6.25*
    6.4 Rack Grounding / *6.35*
    6.5 Computer Floors / *6.37*
Reference / *6.39*
Bibliography / *6.39*

Chapter 6.4 Broadcast Systems Cooling and Environmental Management *Fred Baumgartner* 6.41

    6.1 Introduction / *6.41*
    6.2 Cooling System Design Considerations / *6.41*
    6.3 Transmitters / *6.46*
    6.4 Satellite Antennas / *6.48*
    6.5 Operating Parameters / *6.49*

Chapter 6.5 Facility Ground System *Jerry C. Whitaker* 6.53

    6.1 Introduction / *6.53*
    6.2 The Grounding Electrode / *6.53*
    6.3 Ground System Options / *6.62*
References / *6.67*
Bibliography / *6.67*

Chapter 6.6 AC Power Systems *Jerry C. Whitaker* 6.69

    6.1 Introduction / *6.69*
    6.2 Designing for Fault-Tolerance / *6.69*
    6.3 Plant Maintenance / *6.72*
    6.4 Standby Power Systems / *6.73*
References / *6.85*
Bibliography / *6.85*

Chapter 6.7 Transmission System Maintenance *Steve Fluker* 6.87

    6.1 Introduction / *6.87*
    6.2 Transmitter Site Visits / *6.87*
    6.3 Transmitters / *6.90*
    6.4 Transmission Line / *6.96*
    6.5 Generator Maintenance / *6.99*
    6.6 Supporting Equipment Inspection / *6.105*
    6.7 HVAC and Electrical Systems / *6.115*
    6.8 Tower Maintenance / *6.119*

Section 7 Broadcast Management   *Wayne M. Pecena*   7.1

Chapter 7.1 Management and Leadership   *Mike Seaver*   7.3

7.1 Introduction / 7.3
7.2 Scope of Management and Leadership / 7.3
7.3 Summary / 7.15

Chapter 7.2 Project and Systems Management   *Jerry C. Whitaker and Robert Mancini*   7.17

7.1 Introduction / 7.17
7.2 Plan for Success / 7.17
7.3 Elements of Process / 7.19
7.4 Conclusion / 7.22
Bibliography / 7.22

Chapter 7.3 Systems Engineering   *Gene DeSantis*   7.23

7.1 Introduction / 7.23
7.2 Electronic System Design / 7.24
7.3 Program Management / 7.29
Bibliography / 7.34

Chapter 7.4 Disaster Planning and Recovery for Broadcast Facilities
Thomas G. Osenkowsky   7.35

7.1 Introduction / 7.35
7.2 Anticipation / 7.35
7.3 Preparation / 7.36
7.4 Assessment / 7.36
7.5 Who Is in Charge? / 7.38
7.6 Personnel Matters / 7.38
7.7 Common Precautions / 7.38
7.8 Conclusion / 7.39

Chapter 7.5 Safety Considerations   *Jerry C. Whitaker*   7.41

7.1 Introduction / 7.41
7.2 Electric Shock / 7.41
7.3 Operating Hazards / 7.46
7.4 Nonionizing Radiation / 7.51
References / 7.59
Bibliography / 7.59

Chapter 7.6 About the Society of Broadcast Engineers   *Ralph Hogan, Wayne M. Pecena, and John L. Poray*   7.61

7.1 Introduction / 7.61
7.2 The SBE Certification Program / 7.62
7.3 SBE Education Programs / 7.63
7.4 SBE's History Is People / 7.63

Chapter 7.7 Looking Toward the Future—ATSC 3.0 *Jerry C. Whitaker and Dr. Richard Chernock*   **7.65**

7.1 Introduction / 7.65
7.2 About the ATSC 3.0 Process / 7.66
7.3 Moving Forward / 7.71
References / 7.72

Annex
Jerry C. Whitaker   **A.1**

Appendix A. Reference Data and Tables *Jerry C. Whitaker*   **A.3**

A.1 Introduction / A.3
Reference / A.35

Appendix B. The Electromagnetic Spectrum *John Norgard*   **A.37**

B.1 Introduction / A.37
B.2 Spectral Subregions / A.38
B.3 Frequency Assignment and Allocations / A.45
Bibliography / A.47

Appendix C. Standards Organizations and SI Units *Jerry C. Whitaker and Robert Mancini*   **A.49**

C.1 Introduction / A.49
C.2 The Standards Development Organization / A.50
C.3 Principal Standards Organizations / A.53
C.4 Acquiring Reference Documents / A.57
C.5 Tabular Data / A.58
Reference / A.60
Bibliography / A.60

Index   I.1

SBE EDITORIAL ADVISORY BOARD

Jerry C. Whitaker, CPBE, 8-VSB, Advanced Television Systems Committee, Washington, DC; Editor-in-Chief

Timothy Carrol, Linear Acoustic, Lancaster, PA

Dr. Richard Chernock, Triveni Digital, Princeton Junction, NJ

Andrea Cummis, CBT, CTO, AC Video Solutions, Roseland, NJ

Douglas Garlinger, CPBE, 8-VSB, CBNT, WISH-TV, Indianapolis, IN

Ralph Hogan, CPBE, CBNE, DRB, KJZZ-FM/KBAQ-FM, Tempe, AZ

John A. Luff, HD Consulting, Pittsburgh, PA

Wayne M. Pecena, CPBE, 8-VSB, AMD, DRB, CBNE, Texas A&M University, College Station, TX

John L. Poray, CAE, Society of Broadcast Engineers, Indianapolis, IN

Gary Sgrignoli, Meintel, Sgrignoli, & Wallace, Mr. Prospect, IL

Special thanks to Chriss Scherer, CPBE, CBNT, of the Society of Broadcast Engineers, for editorial support.

AUTHORS

Jay C. Adrick, GatesAir, Mason, OH

Phil Alexander, CSRE, AMD Broadcast Engineering Service and Technology, Indianapolis, IN

Dennis Baldridge, CPBE, 8-VSB, AMD, DRB, CBNT, Consultant, Hillsboro, WI

Fred Baumgartner, CPBE, CBNT, KMGH TV, Scripps, Denver, CO/Nautel Television

Sam Bogoch, Axle Video LLC, Boston, MA

Wayne Bretl, Zenith R&D Labs, Lincolnshire, IL

Richard G. Cann

Dr. Richard Chernock, Triveni Digital, Princeton Junction, NJ

Kerry W. Cozad, Continental Electronics, Dallas, TX

Andrea Cummis, CBT, CTO, AC Video Solutions, Roseland, NJ

Gene DeSantis

Martin Dyster, The Telos Alliance, Lancaster, PA

Daniel S. Fallon, Dielectric, Raymond, ME

Steve Fluker, CBT, WFTV/WRDQ, Cox Media Group, Orlando, FL

Douglas Garlinger, CPBE, 8-VSB, CBNT, WISH-TV, Indianapolis, IN

Joseph J. Giardina, DSI RF Systems, Somerset, NJ

Linley Gumm, Consultant, Beaverton, OR

Bill Harland, Electronic Research, Inc., Chandler, IN

Ralph Hogan, CPBE, DRB, CBNE, KJZZ-FM/KBAQ-FM, Tempe, AZ

Anthony Hoover

Brian J. Isaacson, Communitek Video Systems, New York, NY

Nikolaus Kerö, Oregano Systems, Vienna, Austria

John Luff, HD Consulting, Pittsburgh, PA

Wayne C. Luplow, Zenith R&D Labs, Lincolnshire, IL

Vinnie Macri, Communications Specialist, Millington, NJ

John Mailhot, Imagine Communications, Frisco, TX

Robert Mancini, Mancini Enterprises, Whittier, CA

Scott Marchand, Nautel, Hackett's Cove, NS

Frank Marsico, Shadowstone, Inc., Clifton, NJ

Bill Meintel, Meintel, Sgrignoli, & Wallace, Warrenton, VA

John R. Mick Jr., Mick Ventures, Inc., Denver, CO

Alex Morash, Nautel, Hackett's Cove, NS

John Norgard

Gary Olson, GHO Group, New York, NY

Thomas G. Osenkowsky, CPBE, Radio Engineering Consultant, Brookfield, CT

Nicholaus Paulin, Electronic Research, Inc., Chandler, IN

Nuraj Lal Pradhan, VISLINK, Inc., North Billerica, MA

Wayne M. Pecena, CPBE, 8-VSB, AMD, DRB, CBNE, Texas A&M University—KAMU, College Station, TX

John L. Poray, CAE, Society of Broadcast Engineers, Indianapolis, IN

Frederic M. Remley

Martin Sacks, The Telos Alliance, Cleveland, OH

Ernie Sanchez, The Sanchez Law Firm P.C., Washington, DC

Mike Seaver, CBT, Seaver Management and Consulting, Quincy, IL

Gary Sgrignoli, Meintel, Sgrignoli, & Wallace, Mt. Prospect, IL

Greg Shay, The Telos Alliance, Cleveland, OH

Derek Small, Dielectric, Raymond, ME

Herbert Squire, CSRE DSI RF Systems, Somerset, NJ

Mike Starling, Cambridge, MD

Gomer Thomas, Gomer Thomas Consulting, LLC, Arlington, WA

John Tremblay, UBS/LARCAN, Toronto, ON

Ernst-Joachim Voelker

Larry Wilkins, CPBE, AMD, CBNT, Alabama Broadcasters Association, Hoover, AL

Jerry C. Whitaker, CPBE, 8-VSB, ATSC, Washington, DC

John Wood, VISLINK, Inc., North Billerica, MA

Philip Young, Electronic Research, Inc., Chandler, IN

PREFACE

This is an exciting time for broadcasters and for consumers. Digital radio and television systems—the products of decades of work by engineers around the world—are now commonplace. In the studio, the transition from analog to digital systems continues to accelerate. Numerous other advancements relating to the capture, processing, storage, transmission, and reception of audio and video programs are rolling out at a record pace.

Within the last few years, the options available to consumers for audio and video content have grown exponentially. Dramatic advancements in information technology (IT) systems, imaging, display, and compression schemes have all vastly reshaped the technical landscape of radio and television. These changes give rise to a new handbook focused on practical aspects of radio and television broadcasting. The *SBE Broadcast Engineering Handbook* is offered as a hands-on guide to station design and maintenance. This handbook is the latest in a series of books offered by the Society of Broadcast Engineers (SBE) that are focused on broadcast technologies, station operation, and professional certification.

About the SBE

The SBE is a nonprofit professional organization formed in 1964 to serve the needs of broadcast engineers and media professionals. From the studio operator to the maintenance engineer and the chief engineer to the vice president of engineering, SBE members come from commercial and noncommercial radio and television stations and cable facilities. A growing segment of members are engaging the industry on their own as consultants and contractors. Field and sales engineers, and engineers from recording studios, schools, production houses, closed-circuit television, corporate audiovisual departments, the military, government, and other facilities are also members of the SBE.

The SBE is the only organization devoted to the advancement of all levels and types of broadcast engineering. With more than 5100 members and 115 local chapters, the SBE provides educational opportunities, certification of experience and knowledge, a forum for the exchange of ideas, and the sharing of information to help members keep pace with our rapidly changing industry.

About the Handbook

The broadcast industry has embarked on perhaps the most significant transition of technologies and business models in the history of modern radio and television. Nowhere is this transition more evident than in the studio, where IT technologies are becoming the norm. With these and other changes in mind, the SBE undertook a project to produce a comprehensive handbook with McGraw-Hill that covers key areas of interest to broadcast engineers. This book is divided into seven major sections, plus an extensive annex. Each section focuses on a particular area of expertise:

Section 1: Regulatory Issues
Section 2: RF Systems
Section 3: DTV Transport
Section 4: Information Technology Systems
Section 5: Production Systems
Section 6: Facility Issues
Section 7: Broadcast Management

Annex: Appendix A—Reference Data and Tables, Appendix B—The Electromagnetic Spectrum, Appendix C—Standards Organizations and SI Units

The SBE Editorial Advisory Board, which led this project, has made every effort to cover the subject of broadcast technology in a comprehensive manner. References are provided at the end of most chapters to direct readers to sources of additional information. Over 50 experts contributed their time and knowledge to this handbook.

Within the limits of a practical page count, there are always more items that could be examined in greater detail. Excellent books on the subject of broadcast engineering are available that cover areas that may not be addressed in this handbook. Indeed, entire books have been written on each of the sections noted just previously. The goal of this handbook is to provide a concise overview of technologies, principles, and practices of importance to radio and television broadcast engineers, and to provide pointers to additional detailed information.

The field of technology encompassed by radio and television engineering is broad and exciting. It is an area of considerable importance to market segments of all types and—of course—to the public. It is the intent of the *SBE Broadcast Engineering Handbook* to bring these important concepts and technologies together in an understandable form.

Jerry C. Whitaker
Editor-in-Chief

SBE has established a web site to support the SBE Broadcast Engineering Handbook:
http://sbe.org/handbook

Visit the web site for additional information, errata, updates, and links to educational material.

SECTION 1
REGULATORY ISSUES

Section Leader—Ralph Hogan
CPBE, CBNE, DRB, Tempe, Arizona

Over the years, deregulation has reduced or eliminated much of the tedious record keeping and measurements that previously went with being a broadcast engineer and placed a greater emphasis on the responsibilities of the licensee. There are not as many regulatory requirements for the broadcast media engineer to be knowledgeable of as there used to be. However, even with deregulation, there are still many important requirements contained within the Code of Federal Regulations (CFR) that individuals responsible for stations and facilities need to know in today's world.

This section covers some of the important information areas not only to keep the station legal but also serve to give the reader a quick reference resource when the need arises.

Chapters follow on basic Federal Communication Commission (FCC) licensing, requirements of a Chief Operator, and the Alternative Broadcast Inspection Program, along with an important chapter on accessibility requirements for both television and radio. With an increased interest in being able to notify the public in the event of emergencies by utilizing newer technologies, there is a chapter on the Emergency Alert System.

While this section is not an all-inclusive review of FCC rules, it provides the reader with important information on some of the major areas that are necessary to keep the engineer and licensee out of trouble.

The following chapters are included in Section 1:

Chapter 1.1: FCC Licensing and Administrative Basics for the Technically Minded
Chapter 1.2: Chief Operator Requirements
Chapter 1.3: The Alternative Broadcast Inspection Program (ABIP)
Chapter 1.4: Broadcast Accessibility Requirements
Chapter 1.5: The Emergency Alert System

CHAPTER 1.1
FCC LICENSING AND ADMINISTRATIVE BASICS FOR THE TECHNICALLY MINDED

Ernie Sanchez
The Sanchez Law Firm P.C., Washington, DC

1.1 INTRODUCTION

Perhaps you remember the old riddle: "How do you eat an elephant?" The answer, of course, was "One bite at a time." Maybe surprisingly, the same answer applies to starting to understand and find your way through, around, and under the Federal Communication Commission (FCC). You simply cannot let yourself be intimidated by this almost 80-year-old federal agency. It has more than 1700 employees, a budget in excess of $350 million dollars, and amazingly complex regulatory responsibilities that encompass a vast range of old and modern communication technologies. If you try to absorb the totality of the FCC at once, you can expect to be confused, intimated, and overwhelmed. So do not do that. Instead, learn about the FCC and its operations a bit at time, as you need to learn a specific piece of knowledge.

Most of the FCC's broadcast-related activities, however, occur in the Mass Media Bureau, which has fewer than 200 employees. According to the FCC, the Media Bureau's main activities are as follows:

> The Media Bureau develops, recommends, and administers the policy and licensing programs for the regulation of media, including cable television, broadcast television and radio, and satellite services in the United States and its territories. The Bureau advises and recommends to the Commission, or acts for the Commission under delegated authority, in matters pertaining to multichannel video programming distribution, broadcast radio and television, direct broadcast satellite service policy, and associated matters. The Bureau will, among other things:
>
> - Conduct rulemaking proceedings concerning the legal, engineering, and economic aspects of electronic media services.
> - Conduct comprehensive studies and analyses concerning the legal, engineering, and economic aspects of electronic media services.
> - Resolve waiver petitions, declaratory rulings, and adjudications related to electronic media services.
> - Process applications for authorization, assignment, transfer, and renewal of media services, including AM, FM, TV, the cable TV relay service, and related matters.

Some of you may be primarily interested in how some of the main FCC activities function. Your attitude may be "just tell me the basic rules and keep it simple." Others of you may want not only the basic information but also some help in starting to understand why the FCC initiates some of the seemingly obscure, odd, and unexpected things it sometimes does.

For both categories of students, please consider this as only a starting map for some of the key things a broadcast engineer should know about dealing with (and working with) the FCC. We will start with the basics, and then as we work our way through the subject, offer suggestions and more detailed references for those of you who want to dig deeper into the history and policy of how and why the FCC functions.

Happily, there are some wonderful tools available for you to learn about the FCC. Best of all, it is not considered cheating if you use these tools. In fact, to not use them may be considered wasteful and foolish.

1.2 LICENSING BASICS

Licensing basics can be summarized as follows:

- *Existing broadcast facilities expiration*: All FCC broadcast-related licenses are granted for set periods of time. Broadcast stations licenses are currently granted for 8-year periods, with renewals due on a rolling basis depending on the state where the broadcast facility is located. All radio and television facilities in the same state will share the same renewal dates. The FCC periodically publishes radio and TV renewal calendars, with the full range of applicable dates, which should be studied carefully and kept for reference.

- *New broadcast facilities expiration*: When new broadcast facilities are authorized by the FCC, they are assigned expiration dates that conform to the expiration dates of the radio or television stations in the same state.

- *Windows for applying for new facilities*: Any new broadcast facility requires a specific advance authorization from the FCC. In general, new facilities can only be submitted to the FCC when a specific window has been opened that permits applying for a specific category of new facility. While these windows are announced periodically, they do not generally follow a specific pattern or timetable, and in some instances, many years can pass between the opening of windows for specific facilities. If you are interested in applying for a specific category of new broadcast facility, it is essential that you pay very close attention to the opening of such windows. In general, the FCC takes a very inflexible attitude toward application due dates. Failure to meet window requirements is almost always fatal for the noncomplying application.

- *Procedures for applying for new facilities*: The process has two steps. First, one submits an application for a construction permit that spells out the proposed legal and engineering parameters of the proposed facility. Once granted, that construction permit grants 3 years for the construction of the requested facility. Next, a license application is requested from the FCC to "cover" the construction permit originally granted. The second step of the process is designed to insure that the facility has been constructed in accordance with the approvals granted by the FCC. When a construction permit is granted by the FCC, there are sometimes "special operating conditions" specified that must be met by the applicant when the facility is being constructed and before the new station can be tested. These conditions are considered very important to honor and must be carefully followed to avoid potentially severe FCC legal sanctions.

- *Modification of existing broadcast facilities*: Most modifications of broadcast facilities require advance authorization from the FCC. Depending on the nature of the proposed modification, the change may be considered "minor" or "major." Minor modifications are typically processed by the FCC staff in a matter of a few months with a minimum of bureaucracy, usually on a rolling basis.

- *Waiver requests*: The FCC is legally required to give a "hard look" at any requests for a waiver of the normal commission rules where the underlying purpose of the original rule would be better served by grant of the waiver. The FCC may not refuse to consider a waiver just because the requested waiver would violate the original Commission rule.
- *Length of construction permits*: Construction permits are generally for 3 years from the date of issuance. In general, it is against FCC policy to grant extensions of time beyond the initial 3-year period. In certain narrow instances, however, it may be possible to convince the FCC to "toll" an existing construction permit, which means to stop the clock on the running of the construction deadline while certain unforeseeable issues are resolved. The FCC requirements to "toll" a construction permit are very difficult to meet and should not be attempted without experienced professional help.
- *Special Temporary Authority*: Every licensed broadcast facility is required to operate within the technical and legal requirements of its license. If for some reason the facility cannot be operated in accordance with the license, the FCC must be notified promptly and permission be sought in the form of a "Special Temporary Authority" to operate at variance from the existing license.
- *Due process*: The administrative process of the FCC and its decision-making is subject to a Federal Statute called "The Administrative Procedure Act." That statute requires that all FCC actions and procedures are subject to advance public notice and participation by the public in creation of the rules based on evidence-based standards. Furthermore, all FCC actions and determinations are subject to reconsideration and appeal whenever a final action has been taken. Essentially, this means that FCC actions provide for an extraordinary amount of due process and appeals at every level of internal FCC decision-making. In addition, any ultimate FCC decision is subject to potential review by the U.S. Court of Appeals and the U.S. Supreme Court. While not common, hotly contested cases can result in multiple reviews at different internal levels of the FCC and the Federal Court System. In some instances, contested FCC actions can take years—and sometimes decades—to resolve. Critics of the system say that the FCC appeals process heavily favors parties that have the substantial financial resources to pursue extended—and sometimes multiple—legal appeals.

1.3   DESCRIPTION OF FCC REFERENCE RESOURCES

While the FCC is a large complex bureaucracy, there are a number of extremely useful concise references available to guide you through the key issues important to broadcasters and their engineers. A selection of those resources follows:

- *The Public and Broadcasting* is a very concise, clear, and helpful summary of FCC broadcasting rules and regulations. While most of this content is not very technical, it still represents a good starting point for research on topics that might be unfamiliar to you. A hard copy can be secured by calling the FCC toll-free at 1-888-225-5322 (1-888-CALL FCC). This document can also be found on the Commission's web site at http://www.fcc.gov/guides/public-and-broadcasting-july-2008. The current edition was published in July 2008. The Condensed Table of Contents of the Current Edition is listed in Table 1.1.1.

TABLE 1.1.1 Condensed Table of Contents of *The Public and Broadcasting*

Introduction
The FCC and Its Regulatory Authority
   The Communications Act
   How the FCC Adopts Rules
   The FCC and the Media Bureau
   FCC Regulation of Broadcast Radio and Television
The Licensing of TV and Radio Stations
   Commercial and Noncommercial Educational Stations
   Applications to Build New Stations, Length of License Period
   Applications for License Renewal
   Digital Television
   Digital Radio
   Public Participation in the Licensing Process
   Renewal Applications
   Other Types of Applications
Broadcast Programming: Basic Law and Policy
   The FCC and Freedom of Speech
   Licensee Discretion
   Criticism, Ridicule, and Humor Concerning Individuals, Groups, and Institutions
   Programming Access
Broadcast Programming: Law and Policy on Specific Kinds of Programming
   Broadcast Journalism
   Introduction
   Hoaxes
   News Distortion
   Political Broadcasting: Candidates for Public Office
   Objectionable Programming
   Programming Inciting "Imminent Lawless Action"
   Obscene, Indecent, or Profane Programming
   How to File an Obscenity, Indecency, or Profanity Complaint
   Violent Programming
   The V-Chip and TV Program Ratings
   Other Broadcast Content Regulation
   Station Identification
   Children's Television Programming
   Station Conducted Contests
   Lotteries
   Soliciting Funds
   Broadcast of Telephone Conversations
Access to Broadcast Material by People with Disabilities
   Closed Captioning
   Access to Emergency Information

TABLE 1.1.1 Condensed Table of Contents of *The Public and Broadcasting (Continued)*

Business Practices and Advertising
  Business Practices, Advertising Rates, and Profits
  Employment Discrimination and Equal Employment Opportunity ("EEO")
  Sponsorship Identification
  Underwriting Announcements on Noncommercial Educational Stations
  Loud Commercials
  False or Misleading Advertising
  Offensive Advertising
  Tobacco and Alcohol Advertising
  Subliminal Programming
Blanketing Interference
  Rules
  How to Resolve Blanketing Interference Problems
Other Interference Issues
The Local Public Inspection File
  Requirement to Maintain a Public Inspection File
  Purpose of the File
  Viewing the Public Inspection File
  Contents of the File
  The License
  Applications and Related Materials
  Citizen Agreements
  Contour Maps
  Rules
  Material Relating to an FCC Investigation or Complaint
  Ownership Reports and Related Material
  List of Contracts Required to Be Filed with the FCC
  Political File
  EEO Materials
  "The Public and Broadcasting"
  Letters and E-Mails from the Public
  Quarterly Programming Reports
  Children's Television Programming Reports
  Records Regarding Children's Programming Commercial Limits
  Time Brokerage Agreements
  Lists of Donors
  Local Public Notice Announcements
  Must-Carry or Retransmission Consent Election
  DTV Transition Consumer Education Activity Reports
Comments or Complaints About a Station

1.8 REGULATORY ISSUES

- *The Broadcaster's BIGBOOK Project* is a very practical, hands-on, two-volume loose leaf, updated periodically and designed for broadcast station management and technical personnel. These volumes contain basic tutorial information in workbook form with emphasis on the ultrapractical, with special attention to helping identify and organize FCC-related documentation related to regulatory compliance. Volume 1 covers Control Room Operational Issues, and Volume 2 covers Public File requirements and compliance. These volumes are not strong on history or policy explanations but they are very helpful for the person who needs a quick, basic explanation on the FCC rules and how to comply with them. See http://www.windriverbroadcast.com/. Subjects covered by Volume 1 are listed in Table 1.1.2.

TABLE 1.1.2 Contents of Volume 1 of *The Broadcaster's BIGBOOK Project*

Introduction. Instructional pages, regulatory concerns, current trend line information

Announcements/Calendar. Important station announcements, calendar of important dates with respect to compliance matters

Antenna and Tower. Technical information, locations, procedures, FCC information, tower lighting, registration

Emergency Alert System (EAS). General tutorial

Equipment Performance Measurement (EPM). Annual Report's for AM stations; new transmitting equipment installation, FCC-ordered EPMs as required

FCC information, General. Compliance warnings for operators. How to handle an FCC visit. Self-Inspection checklists for your facility

Licenses and Postings. Station license, auxiliary licenses (translators, boosters, STL, RPU equipment). Chief Operator appointment, Public File location

Logs, Operating. Information on transmitter readings as required by station policy, local tech inspection notes and logs of EAS tests, and tower light inspections required by FCC. Also logs of equipment problems, maintenance, and repairs

Power and Pattern Adjustments, AM stations. Nighttime power changes, antenna pattern change instructions, and any presunrise/postsunset or critical hours information, if applicable. Information on directional AM antenna readings, proofs, and instructions

Preventive Maintenance. General purpose PM information is provided

Remote Control/ATS. Brief instructions for operations—sign-on, sign-off (if used), power changes, transmitter reading steps, adjustments, troubleshooting techniques

Transmitter Data, AM, FM, TV. Brief technical information, operating parameters, nominal meter readings. System efficiency calculations to derive radiated power from input power and system losses. Power tables for indirect AM reading conversions

Forms and Copy Masters. Log forms, Chief Operator forms, technical consultant/engineer's reports, EAS/tower lamp log forms, discrepancy/failure report

FM Translators, Auxiliary Licenses, and Records. Information regarding FM translators for FM stations and for AM stations, Studio Transmitter Links (STLs), Remote Pickup Unit (RPU) licenses

On-air Programming Hazards. Obscenity, profanity, indecency. Broadcast hoaxes. Pranks and jokes. Improper EAS messages. Slanderous material. Political broadcasting references

Station Web Site(s). For radio stations, general information. For TV stations, include references to the Public File requirements online as prescribed by the FCC on FCC-hosted web site

Class A TV Cross Reference. Relates Class A Part 74 Rules to Part 73 Rules

LPTV, TV Translator Cross Reference. References to Part 74 Rules

LPFM Cross Reference. 73.800 Series Rules; references to other rule parts

- The FCC web site (www.fcc.gov) contains possibly more than you ever wanted to know about the minute inner workings of the FCC.
- *The National Association of Broadcasters Legal Guide to Broadcast Law and Regulation* (6th edition) is a well-respected, 990-page reference work, used by lawyers and broadcasters to get a basic overview of broadcast legal topics, geared to provide practical, useable information in a clear manner without heavy academic or scholarly emphasis. See http://www.nabstore.com/ProductDetails.asp?ProductCode=9780240811178.
- The Georgetown University Law Center has compiled a detailed guide to FCC legal research. This is extremely useful to anyone who wants to explore in detail the history of FCC legal decision-making in their areas of jurisdiction and related court decisions. See http://www.law.georgetown.edu/library/research/guides/communications.cfm.

CHAPTER 1.2
CHIEF OPERATOR REQUIREMENTS

Dennis Baldridge
Consultant, Hillsboro, WI

1.1 INTRODUCTION

Telecommunications falls under Title 47 of the *Code of Federal Regulations*. In Part 73, the Federal Communications Commission requires the licensee of every radio and television station to appoint a Chief Operator (CO). The designated person is responsible for specific duties as outlined in Section 73.1870 of the *Rules* and regulations. In the past years, this position required the appointed person to hold a specific license in order to be qualified. With the advent of stable modern technology and automated systems, this requirement has been eliminated. The assigned CO needs only to be able to accurately perform the required duties; no special license is required. The CO is the person responsible for the proper operation of the station, according to the FCC Rules and the terms of the station's authorization. If any questions arise, he/she is the go-to person for the FCC. The CO is not a revenue-generating position but can be a revenue-saving position since many fines imposed by the FCC on broadcast facilities are directly or indirectly related to these responsibilities. Thus, the CO is an important position.

1.2 THE CHIEF OPERATOR SELECTION

The person selected to perform the duties of CO must meet certain qualifications as outlined in Section 73.1870(b). For all AM directional stations, AM stations over 10 kW, and TV stations, the CO must be an employee of the station [73.1870(b)(1)]. The CO for AM nondirectional stations not exceeding 10 kW and FM stations may be either an employee of the station or engaged to serve on a contract basis [73.1870(b)(2)]. See Table 1.2.1. In both cases, the CO is to work whatever number of hours each week the licensee determines is necessary to keep the station's technical operation in compliance with the FCC Rules and the terms of the station authorization. Agreements with chief operators serving on a contract basis must be in writing, with a copy kept in the station files [73.1870(b)(3)].

TABLE 1.2.1 Employee Versus Contract

Station	Employee	Contract
AMD or AM > 10 kW	X	
TV	X	
AM ≤ 10 kW	X	X
FM	X	X

1.2.1 The Chief Operator Designation

The actual designation need not be lengthy or complicated. However, it must be in writing with a copy of the designation posted with the station license. It is recommended that an alternate person be named to perform the required duties in the event of illness or vacation. For example, "Radio station W - - - designates John Doe as its Chief Operator. In his absence, Jane Doe will serve as acting Chief Operator." This designation should be on station letterhead, dated and signed by an authorized representative of the licensee (see Fig. 1.2.1).

WSDG-FM
"The Best Blend"

January 12, 20XX

To whom it may concern:

Radio station WSDG designates John Doe as Chief Operator.

In his absence, Jane Doe will serve as acting Chief Operator.

Signed by

Signature

Authorized Licensee Signature, WSDG

FIGURE 1.2.1   Example Chief Operator designation.

1.3 DUTIES REQUIRED BY THE CHIEF OPERATOR

The CO is responsible for completion of the duties specified in Section 73.1870(c). The CO must be familiar with these obligations and reference all other related FCC Rules necessary to verify compliance with these Rules and the terms of the station's authorization. The CO may be the station's engineer, but need not be. The primary focus is that the CO is responsible for the proper technical operation of the broadcast facility and should be competent in these areas. However, when these duties are delegated to other persons, the CO must maintain supervisory oversight sufficient to know that each requirement has been fulfilled in a timely and correct manner. COs continually certify compliance.

The CO's responsibilities can be summarized in four main areas: (1) inspections and calibrations, (2) field measurements as required, (3) station logs, and (4) any other required entries.

1.3.1 Inspections and Calibrations

Section 73.1580 of the FCC's Rules requires that each AM, FM, TV, and Class A TV station licensee or permittee conduct periodic complete calibration and inspections of the transmitting system and all required monitors to ensure proper station operation. This includes not only frequency and power measurements but also any required equipment performance measurements necessary to maintain and verify compliance. The CO is responsible to make sure that station is in compliance with the Rules for the entire transmission system, any required monitors, all metering and control systems, any special conditions listed in the station's authorization, and any necessary repairs or adjustments as needed. Calibrations and adjustments should be documented in the station's logs.

The CO should be sufficiently familiar with the operation of the entire broadcast system that any anomalies can easily be detected and corrected. Although in many cases the Rules do not specify a precise interval for analysis of the system's performance, it should occur at intervals sufficient to maintain compliance. Some aspects, such as power output levels and modulation (where applicable), are relatively easy to evaluate because many stations retain the necessary equipment in operation to evaluate these parameters. Other criterion, such as frequency tests or harmonic measurements, may require additional equipment. It is to the operator's advantage to notate in the logs whenever technical parameters are verified. For example, let us say that the FCC were to receive a complaint against your station that it was off frequency. The FCC field engineer would undoubtedly investigate the situation. If the field engineer were to substantiate the situation, he or she might very well ask to view the station's logs to ascertain how long this variance had been present. If it had been 5 years since you checked the frequency or documented your measurements in the station's logs, the FCC could conclude that you had been out of tolerance for a long time. However, if you had just checked the frequency within the past month, found it to be compliant, and documented the results in your logs, the FCC might conclude that you had only been out for a short time, potentially reducing any apparent liability.

One of the easiest aspects to overlook when inspecting or calibrating the station's systems is that of documenting your findings. In reality, it does not take very much additional time to make appropriate notations that could prove later to be of great benefit. The best practice calls for complete records.

1.3.2 Field Measurements

Sections 73.61 and 73.154 outline conditions that require certain measurements for AM directional stations. Unless specified otherwise in the station's authorization, each directional AM station is to conduct measurements of the field strength at each monitoring point location specified in the station authorization. These measurements are to be taken as often as necessary to ensure that the radiated field at those points does not exceed the values specified in the authorization. Corrections, calibrations, and certification should be included in the station's logs.

All AM stations are required to make annual equipment performance measurements at not more than 14-month intervals [see 73.1590(a)(6)]. These tests should include measurements to verify compliance with the occupied bandwidth mask and harmonic measurements. Some stations retain the necessary equipment to perform accurate tests. Others hire these tests out to contract engineers. Section 73.44 details the measuring equipment requirements if using a spectrum analyzer for these tests. Some operators do not realize that not all spectrum analyzers are capable of 300-Hz resolution bandwidth as specified in the Rules. If utilizing this method, then be certain that your equipment meets the standards necessary for accurate measurements.

One often-overlooked test for FM stations is the requirement for FM equipment performance measurements whenever a subcarrier is added. For example, the CO of an FM station might have performed the required equipment performance measurement when the station commenced broadcasting or upgraded to stereo. However, later on they may have added another subcarrier for an auxiliary service or have added RBDS. Section 73.1590(a)(4) requires new equipment performance measurements upon the installation of an FM subcarrier or stereophonic transmission equipment.

Other times that equipment performance measurements are required include, but are not limited to, the following:

- Initial installation or replacement of a new main transmitter.
- Modification of an existing transmitter made under the provisions of Section 73.1690.
- Modification of transmission systems.
- When required by other provisions of the Rules or the station license.

1.3.3 Station Logs

Section 73.1820 of the FCC Rules details the requirements for the station's log. At least once each week, the CO should review the logs to verify that all required entries are being made correctly and verify that the station has been operated as required by the rules or station authorization. When the review is complete, the CO should date and sign the log. The CO should initiate any necessary corrective actions to maintain compliance. All such actions should be noted in the logs.

Included in the logs are the (Emergency Alert System) EAS records. Since EAS violations represent one of the highest areas of violations issued by the FCC, the CO should verify that the station has received and sent all required tests for each week. Any discrepancies and corrective actions taken should also be noted in the logs.

Whenever a noncompliance condition exists, the station's logs are to contain this information and details of corrective actions taken. The CO is required to advise the station licensee of any condition which is repetitive.

The form of the station logs is not specified in the Rules, only the content. Thus, the layout, design, and amount of information included are left up to the CO's preferences. A good rule of thumb is to include sufficient information to verify full compliance with the FCC's Rules and the terms of the station's authorization. If the station is running under attended operation (see 73.1400), more detailed logs will be required due to the many operators going on and off duty. Attended operation consists of ongoing supervision of the transmission facilities by a station employee or other person designated by the licensee.

Section 73.1400(b) also allows the option of unattended operation.

> Unattended operation is either the absence of human supervision or the substitution of automated supervision of a station's transmission system for human supervision. In the former case, equipment is employed which is expected to operate within assigned tolerances for extended periods of time. The latter consists of the use of a self-monitoring or ATS-monitored and controlled transmission system that, in lieu of contacting a person designated by the licensee, automatically takes the station off the air within three hours of any technical malfunction which is capable of causing interference.

Logs for unattended operation can be minimal, provided they contain sufficient information to verify compliance with the FCC's Rules. A one-page log, reviewed each week by the CO, could provide adequate documentation. It should contain the information about sent and received EAS tests and notations for any calibrations, out of tolerance conditions, or any other required information.

Documentation of proper operation is essential. The station's logs are the proof of compliance, calibration, and maintenance of the broadcast system and should be considered of utmost importance.

1.3.4 Other Entries

Section 73.1870(c)(4) of the FCC Rules mentions other "entries which may be required in the station records." These may include, but are not limited to, the following:

- Quarterly tower light maintenance checks (see Section 17.47)
- Maintenance of the most recent equipment performance measurements [see Section 73.1590(a)]
- Entries specified in 73.1820, and any other entries required by the terms of the station's authorization

1.4 SUMMARY

The CO position is one of great importance. He/she certifies compliance with the FCC's Rules, and the logs are the station's proof that requirements are being fulfilled. The CO must be familiar enough with the FCC's Rules to verify compliance and know when corrective actions are needed. In any technical investigation by the FCC, the logs, specifically maintained by the CO, will most certainly play a significant role in determining the station's liability. The CO is the person specifically chosen by the licensee to certify compliance with the FCC Rules and the terms of the station's authorization. The CO may delegate some of these responsibilities but must maintain oversight.

CHAPTER 1.3
THE ALTERNATIVE BROADCAST INSPECTION PROGRAM (ABIP)

Larry Wilkins
CPBE, Alabama Broadcasters Association, Hoover, Alabama

1.1 INTRODUCTION

The Federal Communication Commission (FCC) is an independent agency of the U.S. government, created by Congressional statute to regulate interstate communications by radio, television, wire, satellite, and cable in all 50 states, the District of Columbia, and U.S. territories. One of the main tasks of the FCC covers the approval and issuance of licenses to operate all forms of communications that use the radio frequency (RF) spectrum. This includes Radio and Television Broadcast.

A set of rules and regulations are in place to aid in creating an "even field" for all users. The FCC has a program of random inspections to insure that users are complying with these rules and regulations. If, during an inspection, the FCC finds areas of noncompliance, it can issue a notice of liability which can result in a substantial monetary fine. Over the years, this has become a burdensome task for the FCC, given the number of broadcast stations and the limited staff. It was also discovered that a number of the noncompliance issues were not intentional but due to operator oversight or misinterpretation of the rules.

1.2 ABOUT THE PROGRAM

The Alternative Broadcast Inspection Program (ABIP) was created in around 1989 to aid the Commission in making sure all broadcast operations are in compliance and to help stations understand the rules in a one-on-one setting. In the beginning, the ABIP inspectors were retired FCC engineers. As the program began to expand around the country, other engineers were recruited and trained by regional FCC inspectors.

Under the program, the FCC enters into a contract with each state broadcasters association to oversee the ABIP. Each state association then hires a qualified engineer to conduct the inspections. The chosen engineer must be approved by the FCC field office serving the area.

The inspections are entirely voluntary on the part of the stations. Those choosing to have their facilities inspected under the ABIP will fill out a request form with their state association. The inspector will then contact the station and set up a date for the inspection. Once the date has been set, the inspector will notify (with the approval of the station) the FCC field office for that state. This initiates a 150-day grace period that will eliminate a potential visit by the FCC before the ABIP inspection is completed and any issues corrected.

1.18 REGULATORY ISSUES

Once the inspection is completed, the ABIP engineer will meet with the station(s) Chief Engineer and Manager to go over any issues that may need to be addressed to insure the operation is in full compliance. The inspector will create a written report, which will outline any items that need to be addressed, along with the procedure for correction. This report is then sent to the station General Manager and/or the Chief Engineer.

The results of the inspection are completely confidential in that no one will see the report except station management and engineering. Neither the FCC nor the state broadcast association receives a copy. Once the inspector is satisfied with any corrective measures to achieve compliance, a Certificate of Compliance is issued by the state broadcasters association and signed by the inspector, state association President, and the Director of the FCC field office. Once the certificate is issued, the FCC will not conduct random inspections of the station for a period of 3 years. Upon expiration of the 3-year Certification of Compliance, the station must be reinspected if it wishes to have the Certification renewed.

Note, however, the Commission reserves the right to conduct an inspection if it receives a complaint or report concerning safety items (e.g., inadequate tower lights or noncompliant painting). The station may also be randomly selected for an Equal Employment Opportunity (EEO) audit. This is normally done by mail and does not result in a personal visit by the FCC inspector.

If the inspection reveals issues that need to be addressed, the station(s) has until the end of the 150-day grace period to bring the station into compliance. The inspector may request photos, copies of documents, or even a return visit to insure compliance.

The ABIP inspection can normally be completed in one day. However, some licensees may have more stations than can be completed, due to logistics, in that time period.

1.2.1 Scope of the ABIP Inspection

The station engineer should be available to accompany the inspector during inspection. Items that are inspected include the following.

- **Public File** (FCC 73.7526): Ascertaining that all documents required by the Commission are in the file, the file is located at the correct address, and can be made available for review by the public during regular business hours. The inspector will have a written outline of required documents and proper retention times to aid in the inspection. Normally, a copy of this document is left with the station personnel that maintain the pubic file. During this part of the inspection, all station authorizations (licenses) are reviewed to assure the information is correct (main station location, main station staffing, ownership, Part 74 transmitter locations, etc.). As of January 2012, television stations are required to post their public file material on the FCC Station Profiles & Public Inspection web site. Radio stations are still required to maintain a paper or electronic file at their main studio location.

- **Technical Documents**: Review of technical documents that are required to be retained (not part of public file). These include Chief Operator designation, NRSC measurements (AM stations only), quarterly tower light inspection, Antenna Structure Registrations (ASR), and equipment performance measurements.

    Note: Tower owners that have "robust, continuous" remote monitoring systems connected to a network operations center staffed 24/7 can apply for a waiver of quarterly physical inspections of tower marking and lighting systems.

- **EAS/CAP Equipment** (FCC Part 11): Inspecting the **Emergency Alert System/Common Alerting Protocol** (EAS/CAP) equipment as to proper operation and monitor assignments and reviewing with the engineer the procedure in place for issuing a Required Weekly Test. Part of this review includes checking the monitor sources and making sure the required EAS/CAP alerts and test can be issued even when the station is unattended. Stations are also required to maintain a copy of the National EAS handbook along with a copy of their State EAS plan.

- **Review of the Station Log** (FCC 73.1820): This log should have listed all dates and times of required EAS/CAP activations. It should also contain information as to proper operation of the tower lights if applicable. The Chief Operator or the alternate Chief Operator of the station(s) is required to review the station log once each week to assure that the required entries have been made. Once reviewed, the Chief Operator is required to sign and date the log. Station logs are required to be retained for a period of 2 years.

- **Review of the Station Web Site** (FCC 73.2080): If a radio station has a web site, a link to the most recent EEO file must be posted on the site. Television stations, as of January 2012, are required to post their entire public file (except letters and emails from the public) on the Station Profile web site hosted by the FCC. A link to that site is required to be posted on the station web site.

- **Remote Control Operation** (FCC 73.1350): It is required that the personnel designated by the licensee to control the transmitter must have the capability to turn the transmitter off at all times, or include an alternate method of taking control of the transmitter which can terminate the station's operation within 3 minutes.

  In general, the licensee or permittee must correct any malfunction that could cause interference or turn the transmitter off within 3 hours of the malfunction. Some malfunctions, however, must be corrected within 3 minutes.

- **Transmitter Power Output** (FCC 73.1580): The transmitter power output (TPO) is checked and compared to that listed on the station authorization. AM stations will require reading the antenna current meter at the base of the tower. Alternatively, stations may use a calibrated line meter at a known impedance point (as is the case with directional AM operation).

  FM stations will typically use the indirect method to determine power output. Total power = plate voltage times plate current times transmitter efficiency.

  Television stations are required to have a means to verify correct TPO, such as an RF power meter.

  At an AM transmitter site, the condition of the fence and gate surrounding the tower and/or property is examined. All AM towers are required to have a "locked" fence around the tower structure.

  At FM transmitter sites, the inspector should attempt to view the antenna to compare the number of bays to the station's authorization.

- **Tower Inspection** (FCC 73.1213): The location of the tower structure should be verified against the station's authorization. A GPS device may be used to do so. Observe the paint condition if applicable. The inspector may use a chip chart to check the fading of the color. Proper tower light operation will be verified, along with a review of the monitoring procedures used by the tower owner.

- **Antenna Structure Registration**: All tower structures over 199 feet above ground should be registered with the Federal Aviation Administration (FAA). Once approved, the tower structure will receive an ASR number. This number should be placed on or near the tower itself in a place where it can be easily observed. Inspector will review the ASR to check tower location and ownership.

- **Maintenance Logs** (FCC 73.1350): While the FCC does not mention station maintenance logs by name in the rules and regulations, it does require the licensee established procedures and schedules for insuring the station is operating technically in full compliance with FCC rules and regulations. A maintenance log will be a written record of the required inspections and monitoring.

- **Directional AM Operation**: If the AM station being inspected operates with directional properties, the inspector will visit each monitor point to verify proper operation. Although most directional AM stations are no longer required to make monthly field measurements, they must have or be able to secure a calibrated field strength meter to insure proper pattern parameters as often as necessary to insure compliance.

  Note: The FCC has approved method of moments (MoM) modeling of an AM directional array. This virtually eliminates monitor points; however, the precision of the sampling system under the MoM rules is required to be checked every 2 years.

- **AM Power and/or Pattern Change**: The inspector will review the procedure for any power or pattern changes that are required by the station authorization. Station(s) must provide a method for verifying that the required changes have taken place as scheduled.

1.3 FINAL COMMENTS

The ABIP should not be viewed as just an insurance policy against a visit by the FCC inspector, but as an educational tool. ABIP inspectors are well versed in the interpretation of current and newly enacted rules by the Commission. Should a question arise about a certain rule, the inspector will contact the FCC field office for an updated interpretation.

CHAPTER 1.4
BROADCAST ACCESSIBILITY REQUIREMENTS

Mike Starling
Esq., Cambridge, Maryland

1.1 INTRODUCTION

One in five Americans, over 50 million, has a sensory or physical disability affecting their ability to use broadcast media.[1] Disabilities are disproportionally concentrated in the elderly, and thus these numbers will increase as the number of elderly "baby boomers" increases.

The United States is not alone, with the United Nations reporting that there are a total of 650 million worldwide who have a disability that limits their daily living and social participation. This fact led the United Nations to adopt the first Human Rights Convention of the 21st century in 2006, the Rights of Persons with Disabilities (CRPD). The CRPD has to date been adopted by 147 countries out of 192 recognized countries and contains a number of recommendations for insuring "universal access" to mass media.[2]

Among the communications technology recommendations are that signatories

> shall take appropriate measures to promote access for persons with disabilities to new information and communications technologies and systems, including the Internet, and to promote the design, development, production and distribution of accessible information and communications technologies and systems at an early stage, so that these technologies and systems become accessible at minimum cost.[3]

U.S. broadcasters have been at the forefront of offering inclusive media services as early as the late 1960s. Thanks to new Internet Protocol (IP)-based technologies and computer-driven support systems, broadcast media outlets are increasingly adopting accessibility technologies to better serve those with disabilities. Specialized broadcast services for those with serious sensory disabilities like blindness and deafness have been in operation for decades, but new services for the deaf-blind and physically disabled are being piloted and launched each year. Ensuring such services are functioning properly for the intended communities is a modern obligation of technical staff.

---

[1] US Census Bureau, press release coinciding with the 22nd Anniversary of the Americans with Disabilities Act, July 25, 2012; "There are approximately 36 million Americans with hearing loss and 25 million with a significant vision loss." https://www.census.gov/newsroom/releases/archives/miscellaneous/cb12-134.html

[2] See the Global Initiative for Inclusive Information and Communications Technologies (G3ICT) publications at: http://g3ict.com/resource_center/G3ict_Publications

[3] CPRD, Article 9, Section 2 (g) and (h): http://www.un.org/disabilities/default.asp?id=269

1.2 RADIO READING SERVICES FOR THE PRINT HANDICAPPED

Not surprisingly, such innovative strategies are being produced by both host stations and broadcast networks—and in many cases such specialized efforts are being "passed through" from third party service providers.

Mass media support for consumers without ready access to the broadcast content began in 1969 as C. Stanley Potter launched the first radio reading service over the 67-kHz subcarrier of KSJR-FM, in Collegeville-St. Cloud, MN—one of the flagship stations of the Minnesota Public Radio network. Shortly thereafter, other services began to sprout up around the country reading today's books, newspapers, and magazines for hundreds of thousands of print-disabled listeners. Every day there are hundreds of volunteers providing readings for hundreds of thousands who cannot read for themselves.

As the term *print-disabled* implies, anyone who cannot see, hold, or comprehend the printed word is eligible to receive radio reading services. Radio reading services have evolved into "Audio Information Services" because they have not been *exclusively for the blind* for decades, and now commonly include those with physical and intellectual disabilities.

The International Association of Audio Information Services (IAAIS) estimates there are currently over 75 radio reading services in the United States, as well as a handful of others in countries around the world.[4]

Among the 75 radio reading services, there are a total of over 120 FM stations carrying the services, with a projected signal coverage reach of 180 million Americans.[5] This represents approximately 58% of the total number of eligible persons with print disabilities. For this reason, the IAAIS, among others, has been branching out with telephone information systems (for interactivity) and Internet distribution to improve reach.[6]

Although most radio reading services operate on 67-kHz subcarriers, nearly two dozen operate on 92 kHz, which was adopted by the Federal Communication Commission (FCC) in the early 1980s based on technical studies and proposals produced by J. C. Kean of National Public Radio (NPR). Preemphasis is set to 150 *us* with modulation set to ±5 kHz for a 67-kHz subcarrier or typically ±6 kHz for a 92-kHz subcarrier. Special receivers are available for such transmissions, costing between $25 and $150.

FM stations that volunteer to host a radio reading service are allowed to recoup the actual costs incurred for providing the service. This is a classic "but-for" analysis that requires counting added costs, which "but-for" the reading service activity would not have been incurred. Most FM hosts do not charge the reading services for providing the subcarrier capacity. In fact, some commercial stations provide such services as evidence of their commitment to public service and building goodwill. The Talking Information Center, housed at WATD-FM in Marshfield, Massachusetts, is one example of a commercial station that has set aside not only subcarrier capacity but also substantial studio allocations for the reading service.

Why should reading services have to operate on somewhat noisy FM subcarriers? And why are not the newspapers and book publishers concerned about loss of sales from having books read over the radio? The answers are related—FM subcarriers are regulated as a "point-to-point" nonbroadcast service, where each receiver is tracked and made available only to eligible recipients. This restricted audience, only for those who need it, is why Congress granted an exemption to the Copyright Act of 1976 for distribution of readings for the blind. The Act provides a copyright exemption for:

> performance of a nondramatic literary work, by or in the course of a transmission *specifically designed for and primarily directed to blind* or other handicapped persons who are unable to read normal printed material as a result of their handicap, or deaf or other handicapped persons who are unable to hear the

---

[4] See www.iaais.org
[5] See www.iaais.org/2015RRS_SignalReachStudy.pdf
[6] The National Federation of the Blind's NEWSLINE is a popular dial in service that offers scores of publications with "text-to-speech" audio, which can be sped up for quick information consumption.

aural signals accompanying a transmission of visual signals, if the performance is made without any purpose of direct or indirect commercial advantage and its transmission is made through the facilities of:

(i) a governmental body; or

(ii) a noncommercial educational broadcast station (as defined in section 397 of title 47); or

(iii) a radio subcarrier authorization (as defined in 47 CFR 73.293–73.295 and 73.593–73.595); or

(iv) a cable system (as defined in section 111 (f))

Radio reading services, by being provided over an FM subcarrier, whose receivers are made available only to the print-handicapped who qualify for this exemption, have enjoyed this exemption since enacted.

In recent years, NPR, along with iBiquity Digital, and other technology partners have demonstrated making such services available on HD Radio multicast channels using conditional access techniques to allow only eligible listeners to tune the service. To date, only Sun Sounds of Arizona, the statewide Audio Information Service operating on flagship station KBAQ-FM in Phoenix, operates on an HD multicast channel. Additionally, a New Orleans radio reading service exists on 88.3 FM and operates on an "open channel." Since these services are administered by nonprofit organizations and staffed by volunteers, and there is no subscription required, these transmissions are also exempt under the "specifically designed for and primarily directed to blind or other handicapped persons."

The audio quality improvement over 67-kHz subcarriers is substantial for those consumers, and HD radios are modernly no more expensive than traditional FM subcarrier radios. In fact, Sun Sounds of Arizona staffers led the IAAIS initiative to define the production criteria for "Talking Radios" through their Standards for Accessible Radios—a recommended standard. That effort led to the first "Vision Free" radio, the DICE ITR 100A. Other models by other manufacturers, such as the Insignia Narrator, continue in production. IAAIS received a Stevie Wonder "WonderVision" award for producing the StAR publication. The StAR was authored and championed by David Noble of Sun Sounds of Arizona, who served as the Government Relations Chair for IAAIS.[7]

1.3 TV FOR THE VISUALLY DISABLED—VIDEO DESCRIPTION

Beginning in the 1990s, television broadcasters have increasingly embraced specialized services for the visually impaired by providing an increasing number of "video described" broadcasts.

One of the earliest efforts at Video Description for TV was produced by a collaboration in the early 1980s between CBS and the KPBS Radio Reading Service in San Diego. Under that multiyear effort, every January 1, the CBS Rose Parade coverage was narrated for the blind and distributed to scores of radio reading services via the Public Radio Satellite System, by a blind producer—Doug Wakefield.

Wakefield would fly in for the broadcast several days in advance while the floats were being assembled. He climbed over the floats and interviewed the teams sponsoring each one about their efforts and kept copious notes on his Perkins Brailler. During the broadcast, Doug would work from his notes, describing flower petals "as big as medicine balls, on stems as large as utility poles" and synchronizing the coverage to the float currently being discussed on the CBS feed.

CBS shared their "international sound" feed with KPBS, which used it as background for the broadcast. Families at home with "blind" family members were instructed to turn down the CBS TV audio and listen to the special feed being carried on the local radio reading service. For the first time Blind family members got to "watch" TV, with the narration being provided by experts, rather than whispers about what is going on from family members.

By 1985, PBS station WGBH-TV began investigating uses for the new "secondary audio program" component of the new "multichannel sound." Margaret Pfanstiehl, head of the Metropolitan Washington Ear radio reading service, conducted tests of descriptive video in 1986. The Corporation

---

[7] See the StAR Project link at: http://iaais.org/memberservices.html

for Public Broadcasting provided a grant in 1988 to establish a permanent DVS organization. DVS subsequently became a feature of many PBS programs starting in 1990. The FCC promulgated rules concerning DVS carriage requirements in the top 25 markets in 2000, but the rules were struck down as exceeding the FCC's authority to Act by a federal court in 2002.[8] However, the 21st Century Communications and Video Act of 2010 reinstated the FCC's involvement in providing rules for video description.

Under the rules, affiliates in the top 25 markets and top 5 rated cable networks have to provide 50 hours per week of video described programming per quarter. Significantly, the rules do not apply to syndicated programming (where the SAP channel is often utilized for Spanish language dubbing). Currently, dozens of TV series are distributed by the major TV networks and many cable-only networks as well.[9]

1.4 BROADCAST CAPTIONING

Open-captioned broadcasts, superimposed on the video image, are credited to have commenced with PBS's "The French Chef" in 1972. Shortly thereafter, WGBH in Boston launched ongoing captioning services for both commercial and PBS network TV programs.

Broadcast captioning is the creation of text for a near-verbatim transcription of the audio being broadcast. We use the term "near-verbatim" as broadcast captioning typically includes "nonspeech" elements, such as speaker identification [Lester Holt], and [laughter] and [applause], to provide the fullest understanding of the overall audio content being captioned. And in some instances, a verbatim transcription is virtually impossible, such as when speakers are talking over each other, or the content is being captioned live and is [inaudible].

In 1990, the Americans with Disabilities Act (ADA) was passed to ensure equal opportunity for persons with disabilities. The ADA prohibits discrimination against persons with disabilities in public accommodations or commercial facilities. Title III of the ADA requires that public facilities, such as hospitals, bars, shopping centers, and museums (but not movie theaters), provide access to verbal information on televisions, films, or slide shows.

Television Decoder Circuitry Act of 1990 required all analog television receivers with screens of at least 13 inches or greater, either sold or manufactured, to have the ability to display closed captioning by July 1, 1993. The Telecommunications Act of 1996 expanded on the Decoder Circuitry Act to place the same requirements on digital television receivers. At that time of NTSC programming, captions were "encoded" into line 21 of the vertical blanking interval. Additionally, TV programming distributors in the United States were required to provide closed captions for Spanish language video programming.

1.4.1 Challenges with the DTV Transition

The final broadcast TV full-power cutover date to digital television broadcasting occurred on June 12, 2009. Throughout the DTV transition, which began by early TV station adopters in 2008, problems surfaced largely with incompatibilities across the set-top box equipment marketplace.

Most prerecorded syndicated TV programming includes captioned content, which is typically passed through by most broadcasters as EIA-608 captions along with a transcoded CEA-708 version encapsulated within CEA-708 packets.

Thus, the numerous benefits of CEA-708 improvements such as enhanced character set with more accented letters and non-Latin letters, and more special symbols, viewer-adjustable text size (often referred to as the "caption volume control"), allowing individuals to adjust their TVs to display small, normal, or large captions, text and background color choices, including both transparent

---

[8] See http://transition.fcc.gov/ogc/documents/opinions/2002/01-1149.html
[9] A partial list can be found at: http://en.wikipedia.org/wiki/Descriptive_Video_Service#Regular_U.S._series_with_DVS_available

and translucent backgrounds in lieu of the "big black block," a full range of typefaces and fonts, and additional language channels are largely unfulfilled promises.

Congress, the entity with plenary power over telecommunications, as an instrument of commerce, has failed to act on mandatory CEA-708 compliance for set-top boxes. Ironically, the Consumer Electronics Association (CEA), through its CEAPAC, has a longstanding lobbying policy opposed to mandates in consumer electronics manufacturing[10] as inimical to fostering disruptive innovation. The innovation of DTV has certainly been disruptive to deaf users, who have been vocal in their dismay over loss of captioning that was previously passed through analog NTSC and local cable set-top boxes.

As a result of vocal discontent with captioning services in DTV, advocacy groups became active and achieved some attention through the Communications and Video Accessibility Act of 2010 (CVAA). The CVAA is a dramatic expansion of obligations on the part of broadcasters and other video program distributors that for the first time addresses implementation of quality standards for closed captioning, on matters of accuracy, timing, completeness, and placement. This is the first time the FCC has addressed quality issues in captions. Additionally, the CVAA requires ATSC-decoding set-top box remotes to have a button to turn on or off the closed captioning in the video output. It also requires broadcasters to provide the pass-through captioning for all television programs redistributed on the Internet.

After various delays, the new rules took place in March 2015. Table 1.4.1 is a summary, provided by Ben Henson of the Georgia Association of Broadcasters, reprinted and modified for length, by permission.

TABLE 1.4.1 Summary of New Captioning Requirements

I. New monitoring and maintenance record retention requirements for closed captioning

As of March 16, 2015, each broadcast TV station is required to maintain records of its closed captioning monitoring and maintenance activities, which shall include, without limitation, information about the monitoring and maintenance of equipment and signal transmissions, to ensure the pass through and delivery of closed captioning to viewers, and technical equipment checks and other activities, to ensure that captioning equipment and other related equipment are maintained in good working order. Each TV station and TV network shall maintain such records for a minimum of 2 years and shall submit such records to the FCC upon request. The FCC did not mandate any specific format for keeping records and thus provided flexibility to entities to establish their own internal procedures for creating and maintaining records that demonstrate compliance efforts and allow for prompt response to complaints and inquiries. At a minimum, it would appear to be reasonable for a TV station to log its closed captioning monitoring and maintenance activities once or twice per month, although no specific timeframe was provided by the FCC.

II. TV stations must obtain one of three types of closed captioning certifications from programming suppliers or report noncertifying suppliers to the FCC

The FCC currently places primary responsibility for compliance with the new closed captioning quality rules with video programming providers (VPDs), such as individual TV stations, because they are ultimately responsible for ensuring the delivery of programming to consumers.

Under the new closed captioning quality rules, the FCC requires VPDs to exercise best efforts to obtain a certification from each programmer that supplies it with programming attesting that the programmer (1) complies with the new captioning quality standards (as described further below); (2) adheres to the "Best Practices" for video programmers (as described further below), or (3) is exempt from the closed captioning rules under one or more preexisting exemptions, in which case such certification must identify the specific exemption claimed. Each TV station needs to obtain a certification from each programming source. Locally produced programming by the station owner is not exempt—a station should prepare a certification for its own produced programming.

A TV station may satisfy its best efforts obligation by locating a programmer's certification on the programmer's web site (or elsewhere). If a TV station is unable to locate such certification, the station must inform the video programmer in writing that the programmer must make such certification widely available within 30 days. Sample forms for requesting a certification from a video program supplier for this purpose are available through various communications attorneys and broadcast trade associations. VPDs that fail to exercise best efforts to obtain the certification may be subject to FCC enforcement action.

(Continued)

---

[10] See, for example, http://www.appliancedesign.com/articles/83374-cea-opposes-government-mandates-on-dtv-products; http://www.telecompaper.com/news/cea-aims-to-block-fcc-digital-tv-mandate--321397

TABLE 1.4.1   Summary of New Captioning Requirements (*Continued*)

If a VPD carries the programming of a video programmer that does not provide the certification described above, the VPD is obligated to report the noncertifying programmer to the FCC. At this writing that process has not been identified. The FCC will compile a list of such programmers that will become available in a public database maintained by the Commission. If a VPD uses its best efforts to obtain one of these certifications from each of its programmers, and the VPD reports to the Commission the identity of any programmer who has refused to provide the requested certification, no sanctions will be imposed on the VPD as a result of any captioning violations that are outside the control of the VPD.

Certification option 1: meeting closed captioning quality standards
Under new section 79.1(j)(2) of the FCC's rules, closed captioning must convey the aural content of video programming in the original language (i.e., English or Spanish) to individuals who are deaf and hard of hearing to the same extent that the audio track conveys such content to individuals who are able to hear. Captioning must be accurate, synchronous, complete, and appropriately placed as those terms are defined below.

Accuracy. Captioning must match the spoken words (or song lyrics when provided on the audio track) in their original language (English or Spanish), in the order spoken, without substituting words for proper names and places, and without paraphrasing, except to the extent that paraphrasing is necessary to resolve any time constraints. Captions must contain proper spelling (including appropriate homophones), appropriate punctuation and capitalization, correct tense and use of singular or plural forms, and accurate representation of numbers with appropriate symbols or words. If slang or grammatical errors are intentionally used in a program's dialogue, they must be mirrored in the captions. Captioning must provide nonverbal information that is not observable, such as the identity of speakers, the existence of music (whether or not there are also lyrics to be captioned), sound effects, and audience reaction, to the greatest extent possible, given the nature of the program. Captions must be legible, with appropriate spacing between words for readability.

Synchronicity. Captioning must coincide with the corresponding spoken words and sounds to the greatest extent possible, given the type of the programming. Captions must begin to appear at the time that the corresponding speech or sounds begin and end approximately when the speech or sounds end. Captions must be displayed on the screen at a speed that permits them to be read by viewers.

Completeness. Captioning must run from the beginning to the end of the program, to the fullest extent possible.

Placement. Captioning must be viewable and not blocked by other important visual content on the screen, including, but not limited to, character faces, featured text (e.g., weather or other news updates, graphics, and credits), and other information that is essential to understanding a program's content when the closed captioning feature is activated. Caption font must be sized appropriately for legibility. Lines of caption may not overlap one another and captions must be adequately positioned so that they do not run off the edge of the video screen.

The FCC recognizes that the standards discussed above will vary for different types of programming: prerecorded, live, and near-live.

a. **Prerecorded programming**

   Prerecorded programming is programming that is produced, recorded, and edited in advance of its first airing on television. Captioning that is added after prerecorded programming is produced but before it airs is known as off-line captioning. Because of the greater opportunity to review and edit off-line captioning, prerecorded programming is expected to achieve full compliance with these standards, except for de minimis errors.

   At times, captioning is added to prerecorded programming as it airs to the public. Because there is no opportunity to proofread real-time captions, the FCC believes that real-time captioning for prerecorded programming will not be expected to achieve full compliance. However, the FCC also noted that it expects that the use of real-time captioning for prerecorded programming will be limited only to those situations when it is necessary.

b. **Live programming**

   Live programming is video programming that is shown on television substantially simultaneously with its performance, such as news, sports, and awards programs. Real-time captioning is used for live programming. Given the lack of time to review and correct real-time captioning, the FCC recognizes that full compliance with the accuracy standards may not be achievable, and will review such complaints on a case-by-case basis. The FCC encourages contractual provisions (i) allowing programmers to provide captioners advance notice of vocabulary likely to be used, (ii) requiring captioners to have access to reliable, high-speed Internet to minimize interruptions or malfunctions, (iii) requiring programmers to provide captioners with high-quality audio program signals to improve accuracy, and (iv) requiring captioners to have certain skills and training.

   The FCC also recognizes that a slight delay between the dialogue and the appearance of captions on live programming is inevitable. The FCC encourages industry participants to minimize such delays, such as by including contractual provisions (i) allowing programmers to provide captioners with advance materials, (ii) requiring programmers to provide captioners with high-quality audio program signals, and (iii) requiring captioners to have certain skills and training.

TABLE 1.4.1   Summary of New Captioning Requirements (*Continued*)

The FCC further recognizes the challenges inherent in ensuring that captioning for live programming is captioned up to its very last second. The FCC encourages the following measures: (i) entities that send audio feed should alert the captioner that the program's end is imminent, (ii) a fadeout of the last scene to add a few seconds for the transition to the next program, (iii) advance delivery of the audio to the captioner by a few seconds, and (iv) allowing captions appearing toward the end of the program to be placed on the screen during the subsequent advertisement or program provided such placement would not interfere with the advertisement or program.

The FCC recognizes that certain live programming, such as sports programs, make extensive use of graphics and scrawls, for which it can be challenging to avoid having captions block graphics and scrawls. Accordingly, the Commission will take such matters into consideration when reviewing complaints regarding violations of the placement standard.

c. Near-live programming

Near-live programming is defined as video programming recorded less than 24 hours prior to the time it was first aired on television. Examples include late-night talk and comedy shows and some public affairs programming. For purposes of caption quality standards, the FCC will treat near-live programming as if it were live programming. The FCC encourages the adoption of either of the following industry practices: (i) in advance of air time, programmers deliver a script or partial script to the captioner, allowing the captioner to create a caption file that can be combined simultaneously with the near-live program when it is aired or (ii) providing captioners with access to the live feed of the taping, allowing the captioner to then improve caption quality prior to the airing of the near-live program.

Certification option 2: Best Practices by video programmers

To satisfy its obligation to exercise its best efforts to obtain certification from its programmers regarding closed caption quality, a VPD also may seek certification from its video programmers that they will adhere to the following Best Practices codified under new Section 79.1(k):

Agreements with captioning services. Video programmers complying with the Best Practices will take the following actions to promote the provision of high-quality television closed captions through new or renewed agreements with captioning vendors.

Performance requirements. Include performance requirements designed to promote the creation of high-quality closed captions for video programming comparable to the Captioning Vendor Best Practices set forth in Section 79.1(k)(2)-(4) of the FCC's rules.

Verification. Include a means of verifying compliance with the above performance requirements such as through periodic spot checks of captioned programming.

Training. Include provisions designed to ensure that captioning vendors' employees and contractors who provide caption services have received appropriate training and that there is oversight of individual captioner's performance.

Operational best practices. Video programmers complying with the Best Practices will take the following actions to promote delivery of high-quality television captions through improved operations:

Preparation materials. To the extent available, provide captioning vendors with advance access to preparation materials such as show scripts, lists of proper names (people and places), and song lyrics used in the program, as well as to any dress rehearsal or rundown that is available and relevant.

Quality audio. Make commercially reasonable efforts to provide captioning vendors with access to a high-quality program audio signal to promote accurate transcription and minimize latency.

Captioning for prerecorded programming

The presumption is that prerecorded programs, excluding programs that initially aired with real-time captions, will be captioned offline before air except when, in the exercise of a programmer's commercially reasonable judgment, circumstances require real-time or live display captioning. Examples of commercially reasonable exceptions may include instances when (1) a programmer's production is completed too close to initial air time be captioned offline or may require editorial changes up to air time (e.g., news content, reality shows), (2) a program is delivered late, (3) there are technical problems with the caption file, (4) last minute changes must be made to later network feeds (e.g., when shown in a later time zone) due to unforeseen circumstances, (5) there are proprietary or confidentiality considerations, or video programming networks or channels with a high proportion of live or topical time-sensitive programming, but also some prerecorded programs, use real-time captioning for all content (including prerecorded programs) to allow for immediate captioning of events or breaking news stories that interrupt scheduled programming.

Make reasonable efforts to employ live display captioning instead of real-time captioning for prerecorded programs if the complete program can be delivered to the caption service provider in sufficient time prior to airing.

(*Continued*)

TABLE 1.4.1 Summary of New Captioning Requirements (*Continued*)

Monitoring and remedial Best Practices. Video programmers complying with the Best Practices will take the following actions aimed at improving prompt identification and remediation of captioning errors as they occur:

Preair monitoring of offline captions. As part of the overall preair quality control process for television programs, conduct periodic checks of offline captions on prerecorded programs to determine the presence of captions.

Real-time monitoring of captions. Monitor television program streams at point of origination (e.g., monitors located at the network master control point or electronic monitoring) to determine presence of captions.

Programmer and captioning vendor contacts. Provide to captioning vendors appropriate staff contacts who can assist in resolving captioning issues. Make captioning vendor contact information readily available in master control or other centralized location, and contact captioning vendor promptly if there is a caption loss or obvious compromise of captions.

Recording of captioning issues. Maintain a log of reported captioning issues, including date, time of day, program title, and description of the issue. Beginning one year after the effective date of the captioning quality standards (i.e., March 16, 2016), such log shall reflect reported captioning issues from the prior year.

Troubleshooting protocol. Develop procedures for troubleshooting consumer captioning complaints within the distribution chain, including identifying relevant points of contact, and work to promptly resolve captioning issues, if possible.

Accuracy spot checks. Within 30 days following notification of a pattern or trend of complaints from the FCC, conduct spot checks of television program captions to assess caption quality and address any ongoing concerns.

Certification option 3: certification of exemption
A VPD's third alternative is to obtain a certification from the programmer (or itself) that the video program in question is exempt from closed captioning requirements. The self-implementing exemptions in the FCC's rules have not changed for many years, although the FCC has solicited comment on whether to alter these exemptions. Self-implementing exemptions apply, for example, to programs broadcast in a language other than English or Spanish, programs aired between 2 a.m. and 6 a.m., and programs aired on digital subchannels that had less than $3 million in annual revenues the prior calendar year. A video programmer also may have applied for an exemption based on the economically burdensome standard. In the latter case, the program in question is considered exempt while the application for exemption remains pending before the FCC.

III. **The FCC's enhanced ENT procedures already are in effect**
For those stations eligible to use Electronic Newsroom Technique (ENT) procedures for closed captioning of live programs (that is, stations not affiliated with ABC, CBS, Fox, or NBC or stations affiliated with one of these four networks but outside the top 25 markets), new enhanced ENT procedures came into effect June 30, 2014. These ENT requirements, which are codified under Section 79.1(e)(11), are as follows:

In-studio produced news, sports, weather, and entertainment programming will be scripted.

For weather interstitials where there may be multiple segments within a news program, weather information explaining the visual information on the screen and conveying forecast information will be scripted, although the scripts may not precisely track the words used on air.

Preproduced programming will be scripted (to the extent technically feasible).

If live interviews or live on-the scene or breaking news segments are not scripted, stations will supplement them with crawls, textual information, or other means (to the extent technically feasible).

The station will provide training to all news staff on scripting for improving ENT. The station will appoint an "ENT Coordinator" accountable for compliance.

These ENT rules do not relieve a broadcast station of its obligations under Section 79.2 of the FCC's rules regarding the accessibility of programming providing emergency information (that is, the requirement to provide at least some type of visual rendering of critical details provided aurally and to make accessible to individuals who are blind or visually impaired emergency information provided visually).

(*Courtesy Ben Henson*)

1.4.2 Emerging Accessibility Directions

With the advent of more powerful computer systems, entirely new captioning strategies have emerged that are dynamic and rapidly evolving. For example, in 2001 the BBC began experimenting with "voice writing" techniques for "subtitling" (as captioning is typically known outside of the United States).[11]

Speech-to-text is frequently viewed by laypersons as the "holy grail" for good and fast captions, largely due to common use of "Google Talk," "Siri," and other smartphone applications that do a decent basic job of converting speech to text. Such applications do not share the typical broadcast challenge of responding to multiple speakers, background music, and on-location sounds, and emotive content. For years, the common joke among captioning professionals has been that "Speech-to-text for broadcast captioning is only five years into the future—and always will be."

However, significant progress has been made. To date, the primary advantage of voice recognition–based captioning techniques is a significant lowering in hourly captioning costs. However, there are few captioning and transcription houses that use these more cost-effective techniques.

Significantly, both accuracy metrics and latency are close to that achieved by the best captioning stenographers. However, some practitioners report being able to train voice writers, screened in advance through a battery of suitability tests, in just 6 weeks versus the typical years required to achieve proficiency as a stenographic transcriber.

In both stenographically based and voice writer–based systems, the quality of output is dependent on the skill of the operators. It is commonly agreed that there is no such thing as 100% accuracy for live captioning. Even for prerecorded programs, different captioning houses and production companies have different conventions for how they handle nonspeech content, in particular.

One end-to-end transcription and captioning solution has been spearheaded by cognitive psychologist Dr. Ellyn Sheffield, based at Towson University. The system has been named Verb8tm and is marketed by BTS Solutions. This system grew out of demonstrations funded by the Department of Education to explore the potential to caption broadcast radio under a strategic alliance between NPR and Towson University. Although NPR no longer has any ownership interest in the technology, Verb8tm now handles all of NPR's transcription needs. Sheffield indicates the process achieves a 30% reduction in costs, while maintaining all of the "daily news of record" accuracy requirements of NPR.

NPR, through its NPR Labs research and development department, has led several initiatives that may point the way for future broadcast accessibility services:

- Beginning in March 2014, Latino USA became the first NPR program to offer captioned radio broadcasts.
- In the summer of 2014, NPR Labs and a consortium of 25 Gulf States Stations demonstrated the ability to use Radio Data System to transmit CAP compliant messages to the deaf to warn of impending threats to safety and property.

Serving the Deaf-Blind—A Final Frontier?

The deaf-blind are unique among those with sensory disabilities in terms of both the overall small numbers involved and the significance of the challenges in serving them with live media programming. Statistics on the number of deaf-blind in America vary between 42,000 and 700,000, depending on the severity of impairment definition employed.[12]

In 2013, NPR Labs completed a demonstration project in association with the Helen Keller National Center demonstrating how the "Captioned Radio" content could be simultaneously ported to Refreshable Braille Displays for use by the nation's deaf-blind users.

Concerning services for the deaf-blind, the CVAA established a National Deaf Blind Equipment Program at the FCC to support the availability of refreshable braille displays for this communications-underserved population.

---

[11] See http://www.intralinea.org/specials/article/Respeaking_for_the_BBC
[12] See http://www.aadb.org/FAQ/faq_DeafBlindness.html#count

1.5 CONCLUSION

The regulatory accessibility requirements for broadcasters are likely to ramp up as indicated by widespread adoption globally of the CRPD's guidance on information and communications technologies as well as the United States' passage of the CVAA.

As the population and computer technologies continue to mature, new ways of using the latter, to better serve the former, appear inevitable.

CHAPTER 1.5
THE EMERGENCY ALERT SYSTEM

Larry Wilkins
CPBE, Alabama Broadcasters Association, Hoover, Alabama

1.1 INTRODUCTION

Since the early 1950s, the Federal Government has worked to create an effective means of alerting the population in case of a national emergency. The first such program, created by President Harry S. Truman, was labeled Control of Electromagnetic Radiation (CONELRAD). In case of a national alert, all stations (which were mostly AM stations at that time) would go off the air. Selected stations would return at either 640 or 1240 kHz. After a few minutes, these stations would go off the air and other stations would come back on using 640 kHz or 1240 kHz. This design was to "confuse enemy aircraft," which used radio stations as direction finders.

Some of the older radios actually had a "CD" symbol on the dial positions of 640 and 1240 kHz (see Fig. 1.5.1). Thankfully, this system was never used.

There are reports that the Federal Government created a different alerting system (to overcome the deficiencies of CONELRAD) with the name National Emergency Alarm Repeater. This system was to use the national power grid to send alert signals to receivers plugged into electrical sockets. The plan was discarded.

The CONELRAD system was replaced in the early part of 1960 with the Emergency Broadcast System (EBS). This system used a "daisy chain" approach to relay messages around the country, and it also required stations to test their equipment once a week. This was done by a series of transmission "off and on" commands after which a 1000-Hz tone was transmitted, followed with the emergency message.

This transmitter "off and on" procedure proved to be a source of technical problems (as some stations could not get their transmitters to come back on). As a result of this and other problems, EBS version 2 was put into operation around 1976. The alert signal was changed to a dual tone to reduce technical problems and false alerts.

National-level alerts were to be originated by the White House and distributed by various radio and television networks, as well as the new major wire services (Associated Press and United Press International). While the system was developed for national alerts, over the years EBS was used mostly by the National Weather Service to issue severe weather warnings. The EBS test message "slide" for television use is shown in Fig. 1.5.2.

1.32 REGULATORY ISSUES

FIGURE 1.5.1 Artifacts from the CONELRAD program.

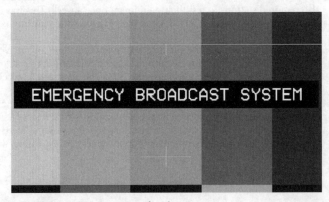

FIGURE 1.5.2 EBS test message for television.

1.2 EMERGENCY ALERT SYSTEM

In 1997, a completely new system was put into place. The Emergency Alert System (EAS) improved the capabilities for national, state, and local alert transmissions (see Fig. 1.5.3). By using a special digitally encoded header, Specific Area Message Encoding, it allowed the origination and distribution of alerts for specific situations and locations. This header code, transmitted three times for redundancy, includes information concerning the Originator identification, the Event code, Location code, and the valid time period of a Message.

FIGURE 1.5.3 EAS program logo.

Once the header is received, the station decoder can either ignore the message or relay it on the air if the message applies to their area. These message rules are determined by each station. If the decoder is set to relay the message, regular programming is interrupted and the alert message is transmitted, preceded by an attention alert of a "two-tone" combination of 853 and 960 Hz sine waves. Once the message is complete, an end-of-message (EOM) data package is transmitted three times, returning the station's normal programming to air.

The backbone for distribution of the national-level Presidential Emergency Alert Notification (EAN) is the network of Primary Entry Point (PEP) stations. This program is administered by Federal Emergency Management Agency (FEMA) under the Integrated Public Alert and Warning System (IPAWS) (see Fig. 1.5.4).

FIGURE 1.5.4 Integrated Public Alert Warning System logo.

The system consists of various FEMA-designated broadcast stations that are tasked with receiving and transmitting "Presidential-Level" messages initiated by FEMA. As the entry point for national-level

EAS messages, these PEP stations are designated "National Primary" stations. PEP stations are equipped with additional and backup communications equipment, and power generators designed to enable them to continue broadcasting information to the public during and after an event.

In 2006, an Executive Order from the President instructed the Department of Homeland Security (DHS) to increase the ability to reach the greatest number of people possible during a national emergency. Because not everyone is watching television or listening to the radio at all times, only delivering an alert message to broadcast and cable operations would not reach the masses. A joint effort was undertaken by DHS and FEMA to develop the means to meet the provisions of the executive order.

An entirely new distribution system was developed that would allow information concerning national- and state-level alerts to be consistently disseminated simultaneously over many warning systems. These would include highway display signs, mobile phones and tablets, and reverse 911 systems, as well as broadcast and cable operations. The system is labeled Common Alerting Protocol (CAP) (see Fig. 1.5.5).

FIGURE 1.5.5 Common Alerting Protocol logo.

The major benefit, according to the FEMA.gov web site, is that a single alert can trigger a wide variety of public warning systems, increasing the likelihood that intended recipients receive the alert by one or more communication pathways.

CAP provides the capability to include rich content, such as photographs, maps, streaming video, and more. In addition, it has the ability to geographically target alerts to a defined warning area, limited only by the capacity of the delivery system used.

Because CAP provides the capability to incorporate both text and equivalent audio, CAP alerts can better serve the needs of hearing or visually impaired persons. Although IPAWS does not provide translation services, CAP does provide the capability to issue alerts in multiple languages.

Under the CAP program, IPAWS created IPAWS-OPEN, which is an application and data center infrastructure that provides alert aggregation, authentication, and dissemination to multiple communications media. Authenticated government officials at all levels nationwide can access the system.

The Federal Communications Commission (FCC) along with FEMA oversees the operation of the EAS/CAP system at the federal levels. The FCC's role includes prescribing rules that establish technical standards for the EAS, procedures for EAS participants to follow in the event the EAS is activated, and EAS testing protocols. Additionally, the FCC ensures that the EAS state and local plans developed for broadcasters and cable companies conform to FCC EAS rules and regulations.

Basically, all broadcast and cable operations must adhere to the following rules concerning EAS/CAP:

1. Participants must install the proper equipment to monitor and relay any alerts or tests required by the FCC rules.

   All EAS Participants must monitor two EAS sources. The monitoring assignments are specified in the EAS State Plans and are determined according to FCC monitoring priorities. Participants are also required to monitor the IPAWS-OPEN system.

2. Participants are required to test their ability to receive and distribute EAS messages and to keep records of all tests.

   The FCC requires all broadcast and cable operations to perform scheduled test procedures to check the readiness of the Emergency Alerting equipment. The following is a list of the required test and logging:

   - Participants must receive and log the reception of the Required Weekly Test (RWT) from both local monitoring sources and, in addition, the reception of RWT from IPAWS-OPEN.
   - Participants must originate and log an RWT test on random dates and times. The tests will consist of transmitting the EAS digital header codes and EOM codes once each week.
   - Participants must receive and relay (within a 60-minute window) the state-issued Required Monthly Test (RMT). The RMT consists of the EAS digital header codes, a two-tone attention

signal, and a brief test script followed by the EOM code. Reception and retransmission should also be logged.
- Participants must receive and immediately relay an EAN. The National-Level alert should be aired continually until termination code is sent by the national originator.

All the above items must be included in the station log and then reviewed and signed by the designated Chief Operator each week. Information concerning any malfunction or missed alerts is also required to be entered into the station log. These station logs are required to be maintained for a period of 2 years.

Other requirements include making available a copy of the National EAS Handbook and the State EAS plan. These handbooks should be located at each operator position and should be immediately available to staff responsible for authenticating and initiating emergency action notifications, termination notices, alerts, and tests.

1.3 FOR MORE INFORMATION

The National EAS Handbook can be obtained from the FCC.gov web site. For information about the State Plan, stations and cable operators should contact their state broadcasters association.

SECTION 2
RF TRANSMISSION

Section Leaders

Douglas Garlinger
CPBE, 8-VSB, Past-President, NASB, Inc., Indianapolis, IN

Gary Sgrignoli
Meintel, Sgrignoli, & Wallace, Warrenton, VA

RF (radio frequency)—the very essence of broadcast engineering—requires the transmission of a broadcast through the ether using radio frequency emissions. In the beginning, it was the spark gap of Marconi, and then the dits and dahs of an unmodulated carrier. Today, it is the sophisticated modulation techniques of 8-VSB, OFDM, QPSK, 16QAM, etc.

All require RF transmitters, transmission lines, filters, and antennas. This section enlightens the twenty-first century RF broadcast engineer with very practical information balanced with sufficient theory. If you are a studio engineer or an IT (information and technology) specialist wanting to learn more about RF, this section will offer you a tremendous amount of material that will assist you in your career.

We provide an overview of radio transmission with an emphasis on the practical considerations of AM and FM transmitters. We include information on the selection, installation, operation, and maintenance of rigid and semiflexible coaxial transmission lines for both FM and TV applications. We offer concise information on the selection criteria for FM channel combiners.

There is a chapter on the design consideration and selection of transmitting antennas for FM, very high frequency (VHF), and ultrahigh frequency (UHF) broadcast. Also, a practical guide to understanding AM antenna systems, written especially for those who operate and maintain these systems.

We offer an overview of the often-ignored international shortwave broadcasting service, the ionosphere, and an introduction into the frequency coordination of these transmissions worldwide.

We have a chapter explaining the methodology to predict coverage and interference for 8-VSB transmissions. We provide a chapter that treats the DTV-RF signal in a manner that explains the basics of 8-VSB in concepts already familiar to the reader. Another chapter is a detailed discussion of the ATSC DTV Transmission System Standard.

We then build on that information with a chapter on the practical design of television transmitters, with both solid-state and tube amplifiers. We provide a chapter on the constant impedance filter and DTV Mask. We offer a chapter that provides information about measuring the 8-VSB DTV RF signal, including detailed measurement of the multiple parameters that determine signal quality.

And finally, we touch on the convergence of microwave and cellular with a discussion of hybrid IP and bonded cellular for electronic news gathering.

Chapter 2.1: AM and FM Transmitters
Chapter 2.2: Coaxial Transmission Lines

Chapter 2.3: FM Channel Combiners
Chapter 2.4: Transmitting Antennas for FM and TV Broadcasting
Chapter 2.5: Practical Aspects of Maintaining Medium-wave Antenna Systems in AM Transmission
Chapter 2.6: International Shortwave Broadcasting
Chapter 2.7: Evaluation of TV Coverage and Interference
Chapter 2.8: DTV RF Considerations
Chapter 2.9: ATSC DTV Transmission System
Chapter 2.10: Television Transmitters
Chapter 2.11: ATSC DTV Mask Filter
Chapter 2.12: DTV Television RF Measurements
Chapter 2.13: Hybrid Microwave IP and Cellular Data for Newsgathering

CHAPTER 2.1
AM AND FM TRANSMITTERS

Scott Marchand (FM) and Alex Morash (AM)
Nautel, Hackett's Cove, NS, Canada

2.1 INTRODUCTION

The objective of this chapter is to provide an overview of amplitude modulation (AM) and frequency modulation (FM) transmitters. AM and FM transmission systems have been studied for many years and there are numerous resources that explain the basic theories of operation behind many of the topics discussed below. Focus will be placed on more practical considerations for those involved with AM or FM transmitters.

Throughout the chapter, any reference to broadcast standards will be based on the United States' Federal Communications Commission (FCC) rules and regulations, unless otherwise stated. The rules and regulations for each region are well documented and will not be covered in any significant detail within this chapter. Broadcasters should be familiar with the rules and regulations that govern the license to broadcast in their part of the world.

2.2 THEORY

2.2.1 Amplitude Modulation

Amplitude modulation refers to the process where the amplitude of a "carrier" waveform is varied proportionally to the amplitude of the signal being transmitted. For the purposes of this text, the discussion is focused on AM transmissions in the medium wave band, meaning the carrier is a sine wave with a frequency in the range 531 to 1700 kHz. This range is restricted to span a smaller set of carrier frequencies depending on what part of the world the transmitter is located; in North America, the acceptable carrier frequency range is 540 to 1700 kHz.

Figure 2.1.1 shows a 1 kHz modulating tone, a 20 kHz carrier signal, and the resulting basic AM signal.

It is seen from Fig. 2.1.1 that the amplitude of the resulting AM signal varies with the amplitude and rate of the modulating signal.

The modulation depth is defined as the ratio of the modulating signal's amplitude (K) to the carrier signal's amplitude (A), and is expressed as a percentage:

$$m = \frac{K}{A} \cdot 100\%$$

2.4 RF TRANSMISSION

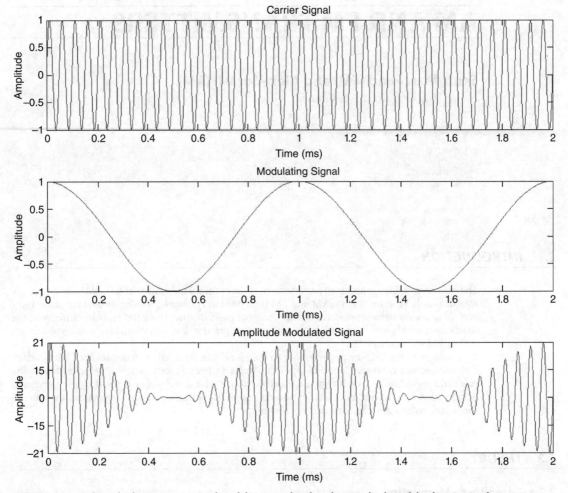

FIGURE 2.1.1 Relationship between carrier signal, modulating signal, and resulting amplitude modulated carrier signal.

Therefore, a modulating signal with a modulation index of 100 percent causes the carrier signal to vary from 0 to twice its amplitude, as shown in Fig. 2.1.1.

Note that when the modulating signal reaches its maximum, this is referred to as "+m" (or +100 percent modulation in the above example), and when the modulating signal reaches its minimum, this is referred to as "−m" (or −100 percent modulation in the above example).

An AM signal is generated by multiplying the carrier signal by the modulating signal plus a DC bias to generate the output AM signal. If the carrier signal is defined as (where A is the amplitude and f_c is the carrier frequency)

$$c(t) = A sin(2\pi f_c t)$$

and the modulating signal is defined as (where K is the amplitude and f_m is the modulating frequency)

$$m(t) = K sin(2\pi f_m t)$$

the resulting AM signal will be

$$y(t) = c(t) \cdot [1 + m(t)] = A\sin(2\pi f_c t) \cdot [1 + K\sin(2\pi f_m t)]$$
$$= A\sin(2\pi f_c t) + \frac{1}{2}AK\sin[2\pi(f_c + f_m)t] + \frac{1}{2}AK\sin[2\pi(f_c - f_m)t]$$

There are two observations that can be made about the final form:

- In the frequency domain, there are three components: (1) the original carrier signal, (2) a "sideband" component at f_m above the carrier signal, and (3) a "sideband" component at f_m below the carrier signal.
- The sideband components have an amplitude that is less than the amplitude of the carrier signal (as long as $m < 200$ percent, which is normal).

These concepts are illustrated in Fig. 2.1.2, shown in the frequency domain, where the carrier frequency is 1000 kHz, the modulating frequency is 1 kHz, and the modulation index is 100 percent. In this case, the sideband components are located at 1000 kHz + 1 kHz = 1001 kHz and 1000 kHz − 1 kHz = 999 kHz. Since the spectrum is shown on a logarithmic scale, the amplitude of each sideband component is

$$20\log\left(\frac{1}{2}m\right) = 20\log\left(\frac{1}{2} \cdot 1\right) = -6 \text{ dBc}$$

or 6 dB below the reference level defined by the carrier.

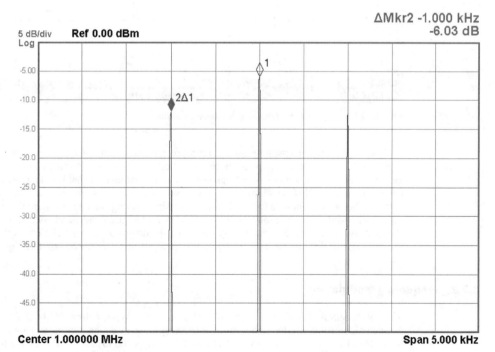

FIGURE 2.1.2 Frequency domain representation of an AM waveform with the parameters given in the text.

2.6 RF TRANSMISSION

In practice, the information being transmitted is audio (e.g., voice or music), hence the modulating signal is an audio signal. The resulting AM signal is therefore not as simple as the examples of tone modulation provided above. The audio signal will have a frequency and amplitude that are constantly varying, causing a "smearing" of the sidebands. An example of an AM signal generated with an audio modulating signal is shown in Fig. 2.1.3 in the frequency domain. In this case, the audio content was limited to 5 kHz, so the signal spans $f_c - 5$ kHz to $f_c + 5$ kHz, and has a total bandwidth of 10 kHz.

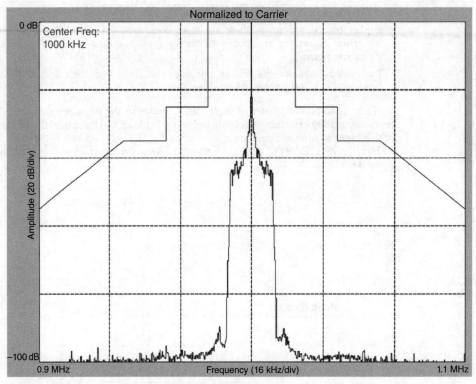

FIGURE 2.1.3 Typical AM waveform with audio modulation.

Historically, many transmitter modulator topologies cannot produce negative modulation troughs of less than −100 percent without generating significant amounts of distortion and spurious emissions. However, many radio stations want to present audio to their listener that is as loud as, or louder than, their competitors. This is normally accomplished by using an audio processor to asymmetrically increase the positive modulation peak amplitude while leaving the negative modulation peak amplitude unchanged. This technique increases the transmitted audio sideband energy, hence increasing the loudness of the sound to the listener. However, the downside of this technique is that because the modulating signal is no longer symmetrical, the audio distortion of the transmitted signal is increased.

2.2.2 Frequency Modulation

The FM broadcast band spans from 87.5 MHz to 108.0 MHz, falling within the VHF (very high frequency) portion of the radio spectrum. Exceptions exist, namely with Japan using 76.0 to 90.0 MHz. In most parts of the world, channel spacing is 200 kHz wide with each adjacent channel falling on the subsequent odd frequency (e.g., 89.1 MHz, 89.3 MHz, …).

With FM, the instantaneous frequency of the carrier varies proportionally to the modulating signal's instantaneous amplitude, and the carrier's rate of change in frequency is a function of the modulating signal's frequency. The relationship between the frequency deviation of the carrier and the modulating frequency is defined as the modulation index:

$$m = \frac{\Delta f}{f_m}$$

where m = modulation index
Δf = frequency deviation ($\pm$ Hz)
f_m = modulating frequency (Hz)

The FCC defines 100 percent modulation as having a peak frequency deviation of ±75 kHz, which is achieved when the modulating signal reaches a specific positive or negative amplitude (based on the audio input type and sensitivity). For example, a 100 MHz carrier modulated by a 1 kHz sine wave at 100 percent modulation would increase to 100.075 MHz and decrease to 99.925 MHz, cycling at a rate of 1000 times per second (1 kHz).

The relationship between the carrier signal, modulating signal, and resulting frequency modulated carrier signal is illustrated in Fig. 2.1.4.

The frequency modulated carrier signal is not a pure sine wave; it is varying in frequency and phase over time, but constant in amplitude. The production of this time domain waveform requires

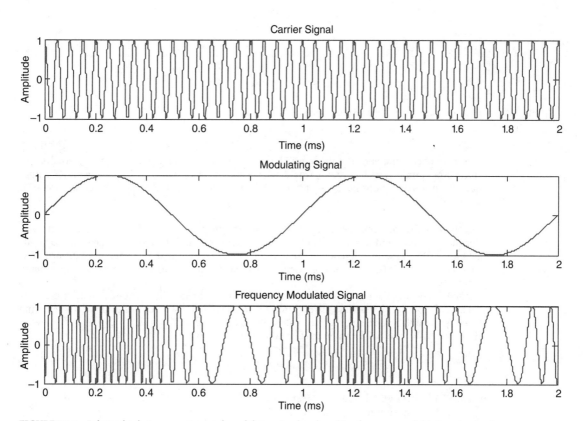

FIGURE 2.1.4 Relationship between carrier signal, modulating signal, and resulting frequency modulated carrier signal.

the addition of many sine waves, of varying amplitude and phase, those waves being the carrier and theoretically an infinite number of sideband components.

The sideband pairs are spaced from the carrier by multiples of the modulating frequency. An FM transmitter's total RF (radio frequency) output power remains constant with modulation because power is distributed between the carrier and the sidebands as a function of the modulation index. A carrier null occurs at very specific modulation indices, where all the radiated power is distributed in the sidebands, and no energy is radiated at the carrier frequency.

The relative amplitudes of the various frequency components are determined by the numeric values of the associated Bessel functions (carrier J_0, and sidebands J_1 through J_n). These values can be found in modulation function charts and are graphically shown versus modulation index in Fig. 2.1.5.

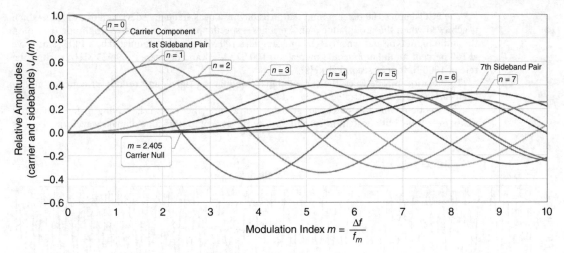

FIGURE 2.1.5 Relative amplitude of carrier and sidebands versus modulation index.

For simplicity, the above example was limited to only one modulating frequency. Analysis of typical programming, such as music, would be very complicated, where the instantaneous RF output spectrum has many more sidebands present at a given time, distributed over the entire occupied bandwidth.

2.2.2.1 Bandwidth Requirements. Any bandwidth limitations in the RF transmission path will alter the amplitude and phase of the higher-order sidebands, and the received signal after demodulation will therefore be different from the original modulating signal. The amount of distortion in any practical FM system depends on the available bandwidth versus the modulation index being transmitted.

The FCC defines the necessary bandwidth (NBW) as the bandwidth required to receive the original modulation signal with less than 1 percent distortion; that distortion is a result of not receiving the entire modulation energy. Approximately 98 percent of the FM signal's energy is concentrated within the NBW. The NBW can be calculated using the formula:

$$\text{NBW} = 2\left(\left|\Delta f_{max}\right| + f_{m\,max}\right)$$

where NBW = necessary bandwidth (Hz)
Δf_{max} = maximum frequency deviation (Hz)
$f_{m\,max}$ = maximum modulating frequency (Hz)

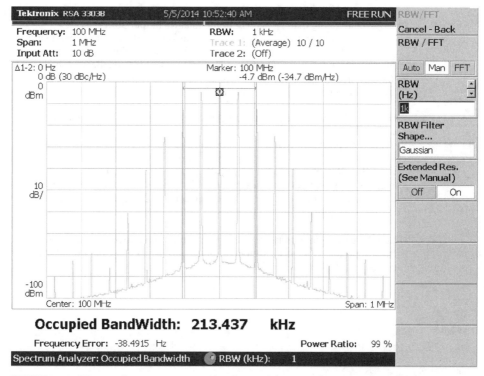

FIGURE 2.1.6 RF output frequency spectrum with 53 kHz modulating signal at 100 percent modulation.

Figure 2.1.6 shows the RF spectrum of an FM transmitter operating with a carrier frequency of 100 MHz modulated to 100 percent by a 53 kHz sine wave (note that the unmodulated carrier reference is 0 dBm). The sidebands are spaced 53 kHz apart starting from the carrier. The modulation index is approximately 1.415 (75 kHz divided by 53 kHz). Using readily available modulation function charts to acquire the Bessel values (see Table 2.1.1), it is seen that the level of the various frequency components relative to the unmodulated carrier agrees with the spectrum analyzer plot

TABLE 2.1.1 Bessel Values and Relative Power Distribution for a Modulation Index of 1.415

Frequency Component	0 (Carrier)	1st	2nd	3rd	4th	5th	6th	7th
Bessel values for $m = 1.415$	0.5587	0.5446	0.2110	0.0520	0.0094	0.0014	0.0002	0.0000
Frequency component's percentage of the total RF output power (percent)	31.2	29.7 (per sideband)	4.5 (per sideband)	0.3 (per sideband)	The power within these sidebands is negligible at less than 0.1 percent of the transmitter's total RF output power.			
Level below the unmodulated carrier (dBc)		−5.06	−5.29	−13.5	−25.7			

shown in Fig. 2.1.6. It is also seen that the carrier (J_0) up to and including the second sideband pair ($\pm J_2$) contain more than 98 percent of the total RF output power.

A common error that engineers make is assuming the occupied bandwidth of an FM signal is simply two times the maximum frequency deviation of the carrier, or −75 kHz to +75 kHz (for 100 percent modulation); therefore, 150 kHz. In actual fact, the Bessel function clearly demonstrates that it is much greater than that, theoretically infinite. In practice, however, a signal of acceptable quality can be transmitted in the limited bandwidth assigned to an FM channel (200 kHz).

To broadcast a monaural signal (left or right audio only), a minimum bandwidth of 180 kHz is required; whereas a much higher-quality stereo multiplexed signal requires a minimum bandwidth of 256 kHz.

The composite baseband signal extends to 99 kHz to support Subsidiary Communications Authorization (SCA) subcarriers, for those who wish to broadcast additional services as part of their signal. The SCA injection level is typically limited to 10 percent, resulting in the modulation index being quite low with few significant sidebands. For this reason, it does not increase the NBW beyond that is required for a stereo multiplexed signal at 100 percent modulation; however, it does increase the risk of SCA distortion.

2.2.2.2 Composite Baseband. The FCC defines the composite baseband multiplex (MPX) signal, which represents the various modulating signals available to the broadcaster and their respective injection levels, as shown in Fig. 2.1.7.

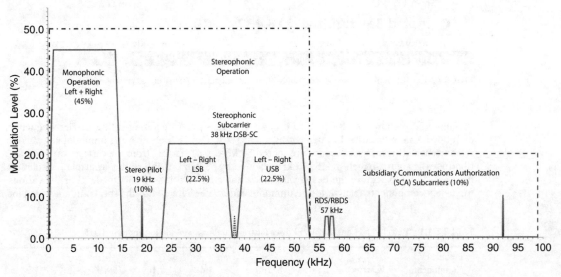

FIGURE 2.1.7 Composite baseband multiplex signal.

2.2.3 HD Radio™

This text will be limited to iBiquity's trademarked HD Radio digital radio technology and provide an overview of transmitter and system requirements. Other digital broadcasting technologies exist, with similar transmitter requirements; however, system requirements are typically quite different.

HD Radio is a digital radio technology that uses an OFDM (Orthogonal Frequency Division Multiplexing) system to transmit a set of digital carriers, either exclusively (all-digital) or in addition to the existing AM or FM analog signal (hybrid or IBOC [in-band on-channel]), and permits simultaneous transmission of audio and data.

Many of today's broadcast transmitters are designed to be compatible with the HD Radio system; however, performance varies between manufacturers.

2.2.3.1 Benefits and Considerations. Some benefits of the HD Radio system are as follows:

- Improved digital "CD-like" sound quality for FM programs and "FM-like" sound quality for AM programs.
- A single FM carrier frequency can simultaneously broadcast up to three additional channels with selection made via the compatible HD Radio receiver (e.g., HD1, HD2, and HD3); one channel, typically the main digital HD1 channel, must simulcast the existing analog signal, allowing for a smooth transition to the analog program in the event the digital signal is lost. This process requires careful synchronization between the analog modulation and the digital data. Due to system limitations, multiple channel broadcasts are currently not possible in the AM IBOC system.
- Additional data services, such as album art, artist information and advertising logos, interactive user functions, as well as weather and traffic updates in real time.
- Content subscriptions are currently free; however, listeners require a compatible receiver.

However, realization of these benefits requires the necessary network infrastructure and HD Radio hardware components, such as an *importer* for adding additional HD channels and advanced digital services. Information on setting up an HD Radio system, along with details about the various hardware components, is extensive and falls outside the scope of this chapter.

2.2.3.2 HD Radio Hybrid Waveform. Currently, HD Radio stations are operating in hybrid mode, where both analog and digital signals are transmitted in order to reach both types of receivers. The transition to full digital mode (dropping the analog signal) has not yet been realized, and is therefore outside the scope of this chapter.

The hybrid waveform adds a block of OFDM digital subcarriers above and below the existing analog signal. Analog receivers filter out the digital subcarriers, and continue to receive the analog broadcast, whereas digital receivers demodulate and realize the benefits of the digital carriers.

For AM transmissions, there are many configurations possible depending on the desired analog and digital transmission qualities. The overall transmission is 29.5 kHz wide, with between 5 and 8 kHz of bandwidth allocated for the analog modulation, and up to 81 subcarriers above and below the analog carrier, in the primary, secondary, and tertiary groups. The primary subcarriers are required, and adding additional subcarriers will improve the digital transmission, but may interfere with the analog transmission. A typical AM hybrid spectral waveform is shown in Fig. 2.1.8.

For FM transmissions, the channel bandwidth allocated for the standard hybrid mode transmission is 400 kHz, comprising of ±100 kHz for the analog signal, a 30 kHz guard band and 70 kHz allocated for 191 digital subcarriers on either side. Additionally, broadcasters can operate in extended hybrid mode which increases the number of digital carriers by up to 76 for additional services; however, the additional carriers are added within the 30 kHz guard band, and may interfere with the analog transmission. Broadcasters who include SCAs in their analog transmission should also consider the risk of extended mode digital carriers interfering with the SCA signal. A representative FM hybrid waveform is shown in Fig. 2.1.9.

For FM systems, some transmitter manufacturers offer the facility to have asymmetrical digital sideband power levels to mitigate the risk of first-adjacent channel interference, where the upper or lower digital power level can be reduced relative to the opposite sideband (e.g., the upper sideband is limited to −20 dBc to prevent interference with an upper adjacent channel, while the lower sideband is permitted to operate at −14 dBc for increased digital coverage). A representative asymmetrical FM hybrid waveform is shown in Fig. 2.1.10.

2.2.3.2.1 Combining Methods. There are four primary techniques for broadcasting the hybrid signal:

Low-level combining: In this technique, both the analog and the IBOC digital carriers are amplified by a single transmitter. This is the only practical option for AM transmitters.

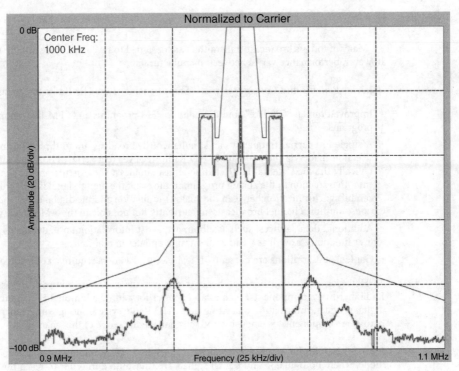

FIGURE 2.1.8 RF spectrum of an amplitude modulation hybrid waveform.

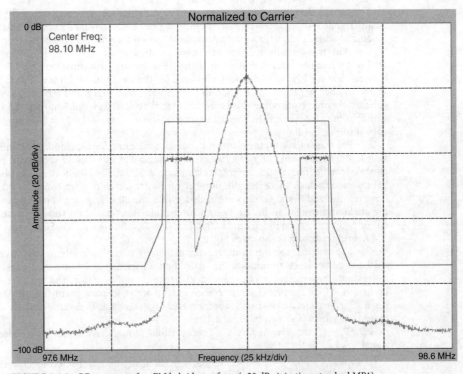

FIGURE 2.1.9 RF spectrum of an FM hybrid waveform (−20 dBc injection, standard MP1).

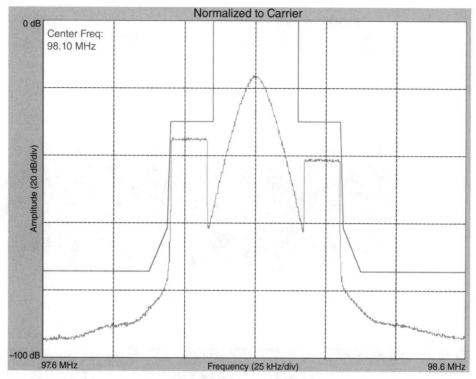

FIGURE 2.1.10 RF spectrum of asymmetrical frequency modulation hybrid waveform (−14 dBc LSB [lower sideband], −20 dBc USB [upper sideband], extended MP3).

High-level combining: In this technique, one transmitter amplifies the analog signal while another transmitter amplifies the IBOC signal. The transmitters are combined before the antenna using a hybrid coupler or injector with specific coupling factor. This, paired with the transmitter power ratio, determines the injection level. This is ideal for customers who already have an existing analog-only transmitter and do not want to purchase a new low-level combined system; however, the infrastructure for this type of system is typically expensive and yields very low efficiency due to the wasted reject power.

Split level combining: This technique is the same as high-level combining, except the digital transmitter transmits a hybrid signal rather than all-digital, providing an increase in analog-radiated power once combined with the analog-only transmitter. It has only been implemented in a small number of systems.

Spatial combining: This technique is similar to high-level combining except each transmitter is connected to its own antenna. The analog and digital antenna patterns must be similar in order to maintain the injection ratio in the broadcast area and at least 35 dB of isolation between the antennas is recommended. Of the three high-level combined systems, spatial combining yields the highest efficiency.

2.2.3.2.2 Peak Power Requirements. The IBOC RF envelope varies with time, where the instantaneous voltage is the vector sum of the analog signal plus the numerous digital carriers: a peak occurs when the vectors align, and a trough occurs when the vector sum nulls (see Fig. 2.1.11).

A broadcast transmitter must be able to pass the peaks of the signal at the desired average output power level. The ratio between these two values determines the capability of the transmitter to pass a digital signal, and is quantified by the peak to average power ratio (PAPR). In practice, the peak power

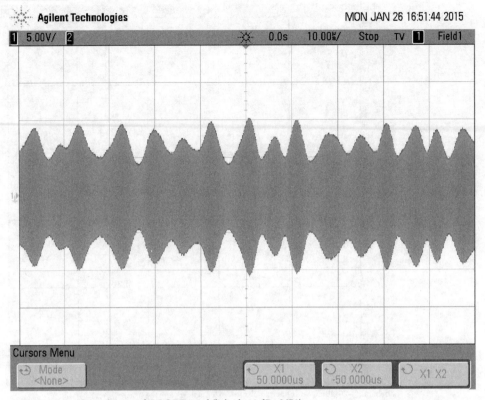

FIGURE 2.1.11 Time domain of IBOC RF signal (hybrid −10 dBc, MP3).

capability of a transmitter limits the average output power capability for a given PAPR, and is a hardware limit, determined by the number of final stage power amplifiers (PAs) and their output saturation point.

2.2.3.2.3 Determining Digital Power Requirements for FM Transmitters. The FCC has approved a voluntary increase in the maximum digital power level an FM broadcaster can radiate from the standard −20 dBc (1 percent digital power) up to −14 dBc (4 percent digital power). An experimental injection level of −10 dBc (10 percent digital power) is also possible for those stations that can guarantee against interference with any other channel.

For low-level combined systems, the FM transmitter manufacturer's product specification clearly indicates the transmitter's power capability versus specific injection levels (e.g., −20, −14, −10 dBc); however, the way in which the power capability of the transmitter is presented can be confusing, and it is important to understand which power measurement is being referenced. Manufacturers may state their power capability as either an "analog total power output" (analog TPO), or a "total average power" versus injection level. Even though both may be describing the exact same transmitter power output, the average power measurement appears larger (especially as the injection level increases) as it is the addition of the digital and the analog power, whereas the analog TPO will appear smaller and decreases with increasing injection levels, due to the higher peak power requirements of the signal.

The following formula determines the total digital power and total average power for a given analog TPO and desired injection level:

$$\text{Total Digital Power} = \text{Analog TPO} \cdot \left(10^{\frac{\text{Injection Level}}{10}} \right)$$

$$\text{Total Average Power} = \text{Analog TPO} + \text{Total Digital Power}$$

For high-level or spatial combined systems, the injection level is defined by the effective radiated power ratio of the digital to analog signal. When determining the power requirement for the IBOC transmitter (or analog transmitter), any attenuation in the transmission path must be accounted for.

2.3 TRANSMITTER OVERVIEW

Even though AM and FM modulated carriers are generated through different processes, the basic functional blocks of each transmitter type are very similar. Figure 2.1.12 shows the basic building blocks of an AM or an FM broadcast transmitter.

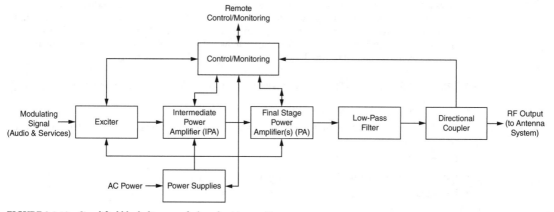

FIGURE 2.1.12 Simplified block diagram of a broadcast transmitter.

2.3.1 Power Supply

There are many variants of power supplies within a transmitter, which are responsible for converting the incoming AC or DC power to a regulated AC or DC output voltage. There are generally two classes of power supplies used in broadcast transmitters:

- High-power output supplies used for powering the RF amplifier stages.
- Low-power output supplies used for powering ancillary circuits and devices, such as control logic and fans.

The low-power output supplies are generally switching or linear power supplies that can be stand-alone modules or implemented as part of a circuit card assembly. The high-power output supplies are critical to transmitter reliability and directly impact the transmitter's efficiency. There are two standard solutions for these high-power output supplies:

- "Linear" implementations that utilize a single high-power transformer, and a single high-power rectifier and filter to obtain the desired DC voltage for powering the RF amplifiers. Until about a decade ago, these were the standard for higher-power transmitters, as they are very efficient (typically 96 percent or higher), robust, and highly reliable.
- Switch mode power supply (SMPS) implementations, which use multiple power supply modules, combined to generate the required power in higher-power transmitters. The SMPS style has been slowly gaining market share over the linear style due to its advantages of universal input ranges, high-power factor and modularity, providing redundancy and easily serviceable hot-pluggable

designs. In addition, the newest generation boasts much higher reliability figures and power supply AC–DC efficiency of 94 percent or higher. The availability of relatively lower cost and higher-efficiency power supplies at +48 V has been the main motivator for manufacturers of FM transmitters to use these power supplies almost exclusively in their designs.

2.3.2 Exciter

The basic function of the exciter is to accept the modulating signals and generate the signal(s) required to drive the subsequent power amplification stages. Current exciters use advanced signal processing devices such as digital signal processors (DSPs) and field programmable gate arrays (FPGAs) to generate high-quality (low-distortion) drive signals. In addition, because of the available processing power, exciters are used to implement additional features that improve transmitter performance and make basic maintenance requirements easier, such as precorrection algorithms, power reduction algorithms, basic audio processing, spectrum analysis, and impedance measurements.

Because of the differences in modulation schemes, AM and FM exciters are different in some ways. Each exciter type is described in more detail in the following sections.

2.3.2.1 AM Exciters. AM exciters generally accept the following basic modulation inputs:

- Balanced analog audio inputs, mono or possibly stereo.
- AES/EBU (Audio Engineering Society/European Broadcasting Union) audio (left, right, left + right, or stereo)
- I/Q over AES
- Dedicated Exgine connection for HD Radio
- Local media playback (e.g., MP3 via USB)
- Streaming audio

The exciter then uses one (or possibly two for HD Radio Hybrid or DRM simulcast operation) of these audio inputs and synthesizes the magnitude and RF drive signals. The magnitude signal is the modulation signal that is to be mixed with the carrier signal at the final stage amplifier. The RF drive signal is the carrier, which is amplified by the final-stage amplifiers, and may or may not be phase modulated, depending on the modulation scheme in use.

2.3.2.2 FM Exciters. FM exciters generally support the following audio input types:

- Analog left/right (monaural or stereo; integrated stereo generation or via an external box)
- MPX
- AES/EBU or AES3 (digital format)
- MPX over IP (Internet Protocol) (requires Internet connectivity with sufficient bandwidth)
- MPX over AES (MPX delivered using AES/EBU digital format; requires a specialized processor)
- RDS/RBDS (radio data system/radio broadcast data system) serial or over IP (internally generated or via an external encoder)
- SCA (allows for broadcasting additional services as part of the main analog signal)
- Dedicated Exgine connector for HD Radio (audio and data delivered via Exporter and Importer if applicable)
- Local media playback (e.g., MP3 via USB)
- Streaming audio over IP (requires Internet connectivity with sufficient bandwidth)

The exciter receives the modulating signal(s) from one or more of the above external source(s) and frequency modulates the internally generated RF carrier signal to develop the RF drive signal that is amplified by the RF amplifiers. For digital modulation schemes, the RF drive signal is also amplitude modulated with the IBOC signal information.

Current FM transmitter manufacturers utilize a direct FM modulation technique for generating the FM waveform. The direct FM modulation is generated in older transmitters via a voltage-controlled oscillator whose frequency changes in proportion to an applied voltage (the modulating signal), and more modern transmitters use direct digital synthesis, which utilizes numerically controlled oscillators for direct-to-channel RF generation with a high-speed RF digital-to-analog converter. The later implementation offers improved noise immunity, as synthesis of the FM signal is handled in the digital domain, where the analog domain is prone to amplitude and phase noise effects.

The transmitter's RF bandwidth limitations and internal power supply regulation are responsible for the overall AM noise specification, synchronous and asynchronous, respectively; however, the exciter is the most critical component in high-quality audio broadcasting.

In addition to the exciter's basic audio-processing capabilities, some transmitter manufacturers also offer internal low-cost audio processors, for those who want some minimum level of audio quality enhancement and security against over-modulation, but cannot afford a high-priced stand-alone audio processor.

2.3.3 Intermediate Power Amplifiers

Because the output signal level from the exciter is often too low to drive the gates of the final-stage amplifiers directly, there are typically multiple intermediate stages of amplification throughout the transmitter's RF power chain; however, the term intermediate power amplifier (IPA) tends to be reserved for a discrete module that amplifies the exciter RF output to sufficiently drive the final-stage PAs. Of course, an increase in the number of intermediate amplification stages negatively affects the system's overall reliability, and price, and represents yet another tuned circuit with limited bandwidth that affects overall amplitude and phase response. As semiconductor technology advances for FM transmitters, increases in the gain and power capability of a single device allow the number of intermediate gain stages to be reduced.

2.3.4 Final-Stage Power Amplifiers

The final-stage PAs are responsible for amplifying either the exciter or IPA RF output signal to the final desired transmitter RF output power level. Previous generation PAs were built with frequency-dependent components that required tuning based on the desired carrier frequency of the transmitter. Modern PAs are broadband and do not require tuning based on the carrier frequency. The implementation of the amplifier stages is significantly different between AM and FM transmitters.

2.3.4.1 AM Power Amplifier. Solid-state AM RF amplifiers are normally class-D amplifiers due to their high efficiency (up to 98 percent efficient amplifiers have been demonstrated in practice). This high amplifier efficiency is critical to obtain overall AC to RF efficiencies of 90 percent and higher in modern transmitters. Although class-D amplifiers are nonlinear, this can be compensated for using good design practices and precorrection algorithms. The RF drive (possibly phase modulated) is applied to the gates of the RF FETs (field-effect transistors), effectively amplifying the carrier signal.

Mainly due to device limitations, and depending on the desired output power of the transmitter, a single RF amplifier may not be capable of generating the full desired output power of the transmitter. In this case, multiple PA outputs are combined (added) together to obtain the full desired output power. There are then two commonly used methods of applying the modulating (magnitude) signal to the carrier:

- Amplifying the magnitude signal and applying it directly to the drain of the FET(s) in the top half of the amplifier.

- Switching amplifiers on and off to change the instantaneous output power of the transmitter.

There are pros and cons to each method, but the exact scheme chosen is secondary to the actual performance of the transmitter.

2.3.4.2 FM Power Amplifier. Typical FM transmitters use multiple class-C amplifiers in a push–pull configuration, as they yield higher efficiency. However, the class-C amplifier is nonlinear by design, which is detrimental to digital modulation schemes, where amplitude modulation is required. Therefore, in digital applications (i.e., HD Radio), the PAs are operated in class-AB, providing a more linear characteristic and minimizing distortion, but negatively impacting efficiency.

The simplified block diagram of a broadcast transmitter shown in Fig. 2.1.12 does not include reference to RF splitters and combiners; components required to drive and combine more than one PA. The number of splitting and combining stages is always symmetrical; however, the types of splitters and combiners used depend upon the transmitter topology, with many manufacturers preferring to use hybrid couplers for their simplicity, high reliability, broadband design, and very high isolation.

Many FM broadcast transmitter manufacturers have made the move from VDMOS (vertically diffused metal-oxide semiconductors) devices to LDMOS (laterally diffused metal-oxide semiconductors) for their higher efficiency, higher gain, higher power density, lower junction-to-case thermal resistance, and an environmentally conscious benefit as beryllium oxide (BeO) is no longer used as the insulating substrate.

FM transmitters just 5 years ago were boasting 60 to 65 percent AC–RF efficiency with VDMOS. Today, they are competing at record levels, putting 70 percent AC–RF efficiency or higher on their specification sheets. This is in big part, thanks to LDMOS devices, yielding 83 percent or higher DC–RF efficiency, approximately 10 percent higher than VDMOS.

2.3.4.3 Low-Pass Filter. Because today's highly efficient PAs are also very nonlinear, they generate a significant amount of energy at harmonics of the carrier frequency. In order to meet the harmonic emissions standards, a low-pass filter (LPF) is required to significantly attenuate these harmonics before the output of the transmitter, without affecting the modulated carrier. The required aggressiveness of the LPF is based on the transmitter's final-stage PA RF output spectrum and their level of harmonics relative to the carrier frequency.

2.3.4.4 Directional Coupler. Directional couplers are realized using different methodologies between AM and FM transmitters; however, their function is similar. The directional coupler provides an isolated sample of the forward and reflected power signals at the RF output of the transmitter, which is used by the control/monitoring system for metering, output power regulation, and protection from high VSWR (voltage standing wave ratio) conditions. RF monitor ports are also fed by directional couplers, and used for RF output spectrum analysis and demodulator samples.

No matter the implementation, the most important element of any directional coupler is that the forward signal couples poorly into (i.e., is isolated from) the reflected port and vice versa. This performance metric is characterized by directivity, which is the ratio of the device's isolation to coupling factor. Ideally, isolation, and hence directivity, is infinite, and there is no coupling between the forward signal and the reflected port; however, in practice there is always some amount of unintentional coupling between the forward and reflected signals.

Because the forward and reflected signals are time varying, the exact magnitude and polarity of the error introduced depend on the phase relationship between the signals. It is important to note that for this reason, the measured reflected power and VSWR will vary with the position of the directional coupler, since the phase relationship between the forward and reflected signals varies throughout an RF system. When the signals are in phase, the maximum positive error will be introduced to the reflected port, and when the signals are 180° out of phase, the maximum negative error will be introduced to the reflected port. For this reason, if there are two directional couplers installed in a transmission system, unless they are perfectly placed where the forward and reflected signals have the same phase relationship, they will never measure the same reflected power. Additionally, because the reflected signal level increases as VSWR increases, the introduced error actually decreases, since the level of forward power coupled to the reflected port does not change.

High-quality couplers typically specify a 30 dB directivity or higher; however, these levels of performance can still introduce significant error. Table 2.1.2 demonstrates the possible error in perceived

TABLE 2.1.2 Error in Perceived VSWR Versus Actual VSWR for a Given Directivity

Actual VSWR	Directivity (dB)	Peak Positive Error (percent)	Highest Perceived VSWR	Peak Negative Error (percent)	Lowest Perceived VSWR
1.1	−30	+180	1.17	−89	1.03
1.1	−40	+49	1.12	−39	1.08
1.5	−30	+36	1.61	−31	1.40
2.0	−30	+22	2.17	−20	1.85
3.0	−30	+15	3.32	−14	2.73
3.0	−40	+6	3.12	−6	2.88

VSWR (which is related to the error in reflected power reading) versus the actual VSWR for a transmitter. The VSWR measured by the directional coupler will be somewhere between the highest and the lowest perceived VSWR, and as previously mentioned, will depend on the phase relationship between the forward and reflected signals.

Note that the forward and reflected power meter accuracy has to be taken into account in addition to the directivity error.

2.3.5 Control/Monitoring

The control system's primary function is to manage the transmitter's internal systems (such as the power supplies or the exciters) so that the transmitter operates as desired by the user. The controller takes user input through means such as discrete wire parallel interfaces, local and remote user interfaces, or SNMP (Simple Network Management Protocol).

However, control/monitoring as an individual block in Fig. 2.1.12 does not accurately represent its function within the transmitter. There is typically a primary control module and a number of subdevices running their own control/monitor functions that send critical operating parameters to the main controller. The control system will monitor every critical parameter, use this information to determine if there is a problem, and if so, determine the required action. The object is to avoid critical failures, such that the transmitter remains on-air for as long as possible (although maybe at reduced power) even during a fault condition.

The transmitter also provides information to the user for monitoring, through the same control interfaces. Some transmitters even include proactive user monitoring systems, where the transmitter will send an email or text message to the site manager indicating there is a problem.

2.4 FACTORS AFFECTING PERFORMANCE

The following sections discuss opportunities for the broadcaster to improve transmitter system performance.

2.4.1 Pre-Emphasis

Noise affects the higher-frequency components in the baseband most significantly. To lessen these effects and improve signal-to-noise ratio, the higher frequencies are boosted using pre-emphasis before transmission and then attenuated at the receiver using de-emphasis. Pre-emphasis and de-emphasis are inverse functions and must be applied concurrently to negate the effects. For FM

2.20 RF TRANSMISSION

transmission systems, the time constant is based on an ideal RC (resistor/capacitor) network, having a 6 dB/octave response above its corner frequency. North America uses 75 μs (2122 Hz), whereas other parts of the world typically use 50 μs. Pre-emphasis is applied to the left and right audio channels, prior to multiplexing if for stereo transmission.

2.4.2 Transmitter Hardware Limitations and Precorrection

If a transmission system were perfectly linear, the transmitted signal would be spectrally pure and the transmitted data would be perfectly replicated at the receiver. However, because of device and process limitations, nonlinear processes are inherent in AM and FM transmitters and cause the following undesirable effects:

- They alter the signal being transmitted, causing the information being reconstructed at the receiver to be inaccurate. In analog transmission systems, this manifests as audio distortion, and in digital transmission systems, this manifests as data errors.
- In digital systems, they cause intermodulation products, which are spectral spurs that were not present in the original modulating signal (see Fig. 2.1.13). In the case of IBOC, the third (IM3) and fifth (IM5) order intermodulation products occur near the carrier and can cause adjacent channel interference.

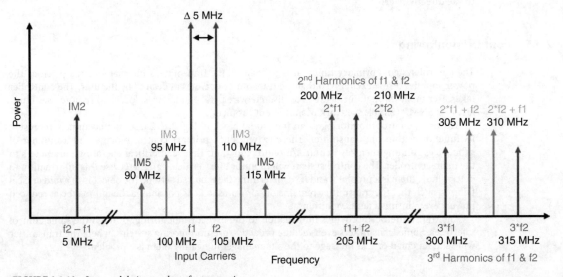

FIGURE 2.1.13 Intermodulation products for two carriers.

In order to standardize the performance of transmission systems, regulatory bodies impose limitations on the spurious emissions and distortion or errors in the transmitted signal. In order to meet these transmission standards, the modern transmitter's exciter will manipulate the modulation signals in an attempt to compensate (or precorrect) for some of the undesirable characteristics of the transmitter. Such precorrection techniques can be fixed values that are set through the transmitter user interface, or they can also be dynamic curves that are determined by monitoring the output of the transmitter and comparing it to the input signal; the latter being termed *adaptive precorrection*. A dynamic curve may be "trained" using a test signal to determine the required compensation characteristic, or it may determine the characteristic continuously using the normally transmitted signal. These signal processing compensation techniques are generally programmed onto powerful

DSPs and/or FPGAs that are part of the transmitter's exciter system. To illustrate this point for both AM and FM transmitters, Figs. 2.1.14 and 2.1.15 show AM and FM transmitter output signal spectra with and without properly configured precorrection.

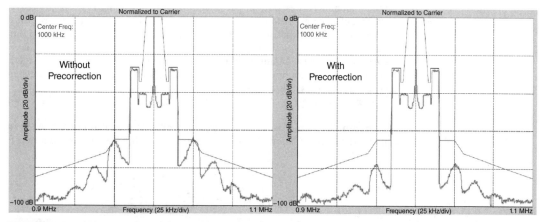

FIGURE 2.1.14   AM transmitter RF output signal with and without precorrection.

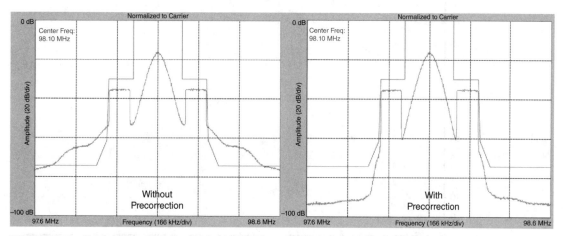

FIGURE 2.1.15   FM transmitter RF output signal with and without precorrection.

Note that the primary benefit of adaptive precorrection is that it continuously attempts to reduce intermodulation products by maintaining system linearity under changing transmitter operating conditions, typically due to internal factors such as amplifier operating temperature (e.g., gain variations) or external factors such as antenna VSWR.

2.4.3 Transmitter Linearity

Linear amplification requires the output signal to be proportional in amplitude and phase to the input signal. However, inherent nonlinear parasitics of the modulation and amplification chain in the transmitter (i.e., nonlinear parasitic capacitance in FETs) cause the signal path to have a nonlinear characteristic, altering the final output signal, and causing the effects described previously. In AM transmitters, amplitude linearity directly affects audio distortion.

Additionally, gain limitations in the amplification chain limit the peak power capability of the transmitter, and if the input signal requires that the output signal exceed this limit, the amplifier will hit its maximum voltage limit and "clip" the signal. This occurs due to slightly different processes in AM and FM transmitters, but the effect is similar.

In FM transmitters, amplifier linearization can be improved by changing the PAs to class-AB operation; however, amplifier compression, and to a greater extent output saturation, still causes nonlinearity at peak powers. The output waveform actually clips as it reaches the saturation limit, thus distorting the signal and increasing the level of intermodulation products. Moving from class-C operation to class-AB operation also significantly reduces the efficiency of the amplifier. Negative peaks can also be problematic for nonlinear effects as the semiconductor devices begin to back out of conduction; increasing bias also tends to lessen these effects.

Figure 2.1.16 demonstrates the typical gain behavior of an FM amplifier versus idle current (bias).

In AM and FM transmitters, amplitude modulation to amplitude modulation (AM–AM) and amplitude modulation to phase modulation (AM–PM) precorrection techniques are used to correct for

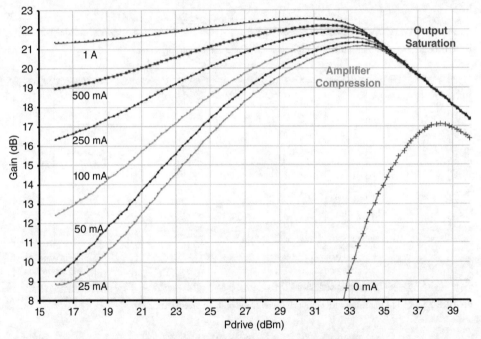

FIGURE 2.1.16 Typical FM power amplifier gain characteristics versus idle current (bias).

any amplitude or phase nonlinearities, respectfully. To illustrate how these compensation techniques work, Fig. 2.1.17 shows an example of an amplitude nonlinearity, the required precorrection curve, and the resulting overall linear system response.

The peak power capability of a transmitter is a hard system limit and cannot be corrected for. In this case, a different approach is taken where instead of allowing the transmitter to hard clip, a "soft-limiter" is applied to the input signal by the signal processing unit. The soft-limiter still clips the signal (the amplitude is lower than the original signal), but instead of having the sharp edges of a hard clip, the resulting signal is shaped such that the resulting distortion and intermodulation products are reduced.

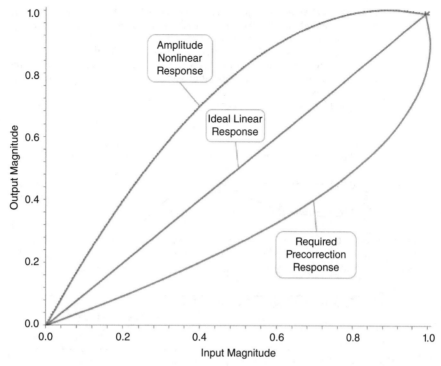

FIGURE 2.1.17 Illustrative behavior of precorrection technique.

Additionally, some transmitter manufacturers implement aggressive PAPR reduction algorithms on the input signal to reduce the PAPR requirement of the digital signal and to maximize the power capability of the transmitter. With these algorithms, as well as the soft-limiter algorithms, data errors are introduced because the signal is being altered by the process, but these algorithms are implemented in such a way that the error correction in the receiver can generally compensate for the introduced errors.

2.4.4 Frequency Response

Due to the nonlinear modulation processes in both AM and FM transmitters, the "bandwidth expansion" resulting from these processes requires a transmission system with infinite bandwidth to ensure a modulating signal is exactly reproducible by the receiver. Of course, these ideal transmission systems do not exist in practice; there is always some frequency response in the signal path limiting the overall bandwidth of the system and/or applying a nonconstant attenuation across the bandwidth of the transmitted signal.

In order to maximize the bandwidth of the magnitude path through the transmitter, a digital filter can be applied to the input signal that compensates for the frequency response of the magnitude path (up to a point). The signal processing path generally has a limited amount of gain that can be applied, and there is therefore limitation on how much roll-off can be compensated for. In AM transmitters, these curves are used to directly compensate for the frequency response through the transmitter, due to elements such as modulator filters, improving audio frequency response.

In FM systems, combining multiple transmitters to transmit using a single antenna is often necessary. This typically requires the use of bandpass or bandstop filters to provide sufficient isolation between each transmitter. Having these filters in the transmission path creates undesired frequency response and variances to group delay. As channel bandwidth requirements increase from 200 kHz for traditional FM to 400 kHz for HD Radio systems, the amplitude and group delay response of the channel combiner become more of a factor in the quality of the broadcast signal.

In this case, some FM transmitter manufacturers offer the ability to equalize the entire system, effectively precorrecting the FM signal within the exciter to account for nonlinearity in the combiner network. This improves stereo separation, synchronous AM noise, and spectral performance for HD Radio systems.

Figures 2.1.18 and 2.1.19 illustrate the effects of equalization on FM channel combiner performance.

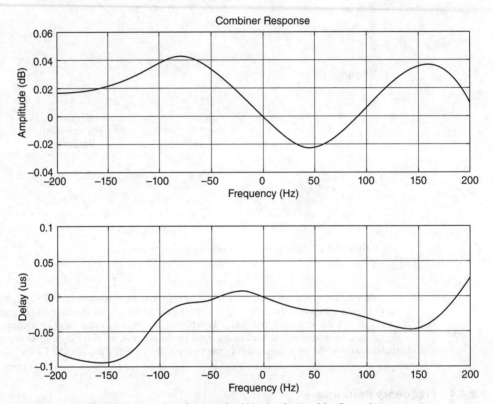

FIGURE 2.1.18 Typical FM channel combiner amplitude (top) and group delay (bottom) response.

2.4.4.1 Delay Matching. In AM transmitters, the complex digital modulating signal is separated into magnitude and phase modulation components; however, the delay through the magnitude and phase modulation paths may not be equal, meaning the signal does not properly re-combine at the RF amplifier. To compensate for this effect, an adjustable delay can be added to one of the paths in the exciter to ensure that the two signals combine properly at the RF amplifier.

2.4.5 AM Antenna Impedance Requirements for Digital Transmission

For simplicity, the above topics on transmitter linearity and precorrection are generally discussed with the transmitter running into an ideal 50-Ω load that has a flat frequency response across the bandwidth of modulation and beyond. However, the AM antenna networks that transmitters are connected to at transmission stations have a limited bandwidth with relatively high Q, and present

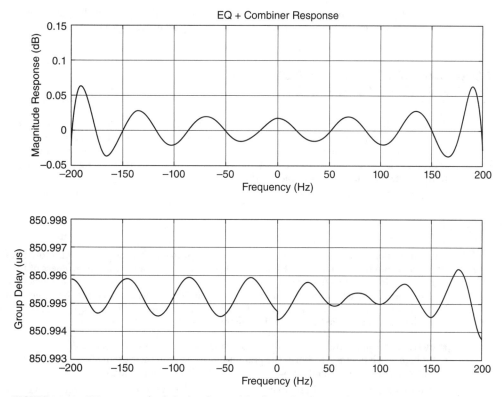

FIGURE 2.1.19 FM system amplitude (top) and group delay (bottom) response with equalization.

nonideal complex impedances to the transmitter. These impedances degrade the performance of the transmitter and alter the signal being transmitted. If these effects are not considered when designing the antenna and matching networks, they alone can easily be enough to cause the system to fail the regulatory performance requirements.

For digital transmission systems, the defining standard will often provide recommendations on the impedance presented to the transmitter. Of course, the lower the VSWR presented to the final amplifier at these frequencies, the better the performance of the system.

Furthermore, experimentation has shown that as a starting point to obtain optimal performance, the impedance characteristic presented to the RF amplifiers should be Hermitian in nature, meaning that the impedance at $f_c + f$ is the complex conjugate of the impedance at $f_c - f$ [1]. The exact rotation for optimal performance may be a few degrees above or below this starting point, and should be discussed with the transmitter manufacturer before designing the necessary impedance matching networks.

It is important to note that each of these recommendations discusses the impedance characteristic presented to the RF amplifiers, *not* the output of the transmitter. It is therefore important to account for all networks between the antenna and RF amplifier, including the transmitter impedance matching and filtering network. To obtain the recommended impedance characteristics, careful design of the antenna impedance matching network is required, in conjunction with a phase rotation network to obtain the correct rotation of the impedance characteristic.

2.4.5.1 AM Antenna Impedance Requirements for Analog Transmission. Load impedance requirements for analog transmissions are typically much less stringent than those for digital transmission. Most modern transmitters are capable of running continuously with very high VSWR at the sidebands (e.g., 2:1). However, as with digital transmission, running into high-Q antenna networks can degrade the quality of the transmitted signal, causing an increase in audio distortion.

2.5 REDUCING THE TOTAL COST OF OWNERSHIP AND EXTENDING TRANSMITTER LIFE

A broadcast transmitter allows the broadcaster to offer a service to advertisers to reach a target audience, with the objective of earning revenue. Every decision that the broadcaster makes with respect to a transmitter purchase has a direct effect on their bottom line. The cost of ownership of a radio transmitter is distributed between energy, maintenance, and parts; the following sections discuss how to minimize the expenditures in each of these areas.

2.5.1 Efficiency

The transmitter is after all just a power converter; its primary purpose is to convert low-frequency AC energy into higher-frequency RF energy for signal transmission. The effectiveness of the transmitter in performing this function is quantified by AC–RF efficiency, and gains in efficiency can lead to significant financial savings. In addition, as transmitter efficiency increases, waste heat into the building will decrease, resulting in reduced cooling requirements and overall operating cost savings. Other benefits of improved efficiency include the following:

- Increased reliability due to lower device operating temperatures. Manufacturers with a reputation for high reliability may be worth the premium price tags on their products, as the long-term support costs of a low-reliability product will far outweigh the initial additional cost.
- Reduced service calls and less maintenance due to increased reliability. Frequent site visits are costly, especially for sites that have limited accessibility. Additionally, reduced power conditions could cause listeners to change the dial, leading to lost ad revenue.

Some FM transmitter manufacturers are developing very intelligent algorithms that manage power consumption while operating in HD Radio mode, which is inherently less efficient. These algorithms are using the existing transmitter control/monitoring system and making adjustments to critical parameters in order to continually optimize transmitter performance, with the objective of reducing waste power when possible. Savings realized using these algorithms can be significant.

For AM transmitters, another opportunity for reducing transmitter power consumption is implementing Dynamic Carrier Control (DCC), also known as Modulation Dependent Carrier Level (MDCL). With AM transmission, the majority of the power output from the transmitter generates the carrier signal, which contains no actual data (all of the transmitted data is contained in the audio modulation which generates the sidebands). DCC is an algorithm that is implemented in the transmitter's exciter, which reduces the level of the carrier (or both the carrier and the sidebands), depending on the level of modulation being applied, without significantly affecting the received audio quality. Depending on the type of modulation being applied to the transmitter, a power savings of 30 percent or higher can be realized.

There are five standard MDCL algorithms:

- Amplitude Modulation Companding
- Enhanced Amplitude Modulation Companding
- Dynamic Amplitude Modulation (DAM)
- DAM Full
- DCC1

The exact algorithm used to provide the maximum power savings depends on the type of audio signal being applied (speech, classical music, etc.), and choosing the best algorithm may require some research and/or experimentation. Certain algorithms may also be used when running IBOC MA-1 (hybrid).

To illustrate the importance of efficiency to the cost of ownership of a transmitter, the following example demonstrates the possible savings between an older generation FM transmitter operating at 40 kW with a typical AC–RF efficiency of 64 percent versus a newer generation at 72 percent. A conservative electricity rate of 15 cents per kilowatt-hour is assumed. The example also shows the reduction in waste BTU/hr where additional cost savings would be realized if the site were using an air conditioner, possibly allowing for a reduction in size of the air conditioner. Actual calculations for electrical cost savings due to the air conditioner will not be shown as air conditioner performance factors vary by application and unit.

$$\text{Waste Power Recovered} = \frac{40 \text{ kW}}{64\%} - \frac{40 \text{ kW}}{72\%} = 62.5 \text{ kW} - 55.6 \text{ kW} \cong 6900 \text{ W}$$

$$\text{Waste} \frac{\text{BTU}}{\text{hr}} \text{Recovered} = 6900 \text{ W} \cdot 3.412 \frac{\frac{\text{BTU}}{\text{hr}}}{\text{W}} \cong 23{,}500 \frac{\text{BTU}}{\text{hr}}$$

$$\text{Waste Tons Recovered} = 23{,}500 \frac{\text{BTU}}{\text{hr}} \cdot \frac{1 \text{ Ton}}{12{,}000 \frac{\text{BTU}}{\text{hr}}} \cong 2 \text{ Tons}$$

$$\text{Yearly Electrical Cost Savings} = 6.9 \text{ kW} \cdot 24 \frac{\text{hr}}{\text{d}} \cdot 365 \frac{\text{d}}{\text{yr}} \cdot \frac{\$0.15}{\text{kW} \cdot \text{hr}}$$

$$\cong \$9100/\text{year (excluding air conditioning savings)}$$

2.5.2 Sophisticated Control/Monitoring and Fault Diagnostics

Transmitters that provide comprehensive remote control/monitoring and highly advanced instrumentation for fault diagnostics allow for quick troubleshooting of faults and minimize the number of reduced power or off-air conditions—and required trips to the site. Simple designs with fewer components and more automated functions can reduce the amount of investment required by skilled technicians to service and maintain the transmitters. Some manufactures have such advanced fault diagnostic and remote monitoring systems that notifications can be sent to the technician via email immediately, and the transmitter will contact the manufacturer to have warranty replacement parts sent directly to the site without any customer involvement.

2.5.3 Staffing Requirements

Highly specialized technical staff can be difficult to acquire and maintain. Turnover in the industry is frequent, and a high degree of site or studio technical support experience and skill is not easy to come by, or afford if found; therefore, systems that are designed with simplicity to install, operate, maintain, and have a low failure rate reduce the need for broadcasters to employ specialized technical staff. Some transmitter manufacturers offer services to monitor and maintain equipment on behalf of the broadcaster. These fees can be quite reasonable compared to employing full-time staff who are only responsible for site maintenance and servicing on an occasional basis.

2.5.4 Buying Redundancy Options and Spares

As discussed previously, every minute a transmitter is off-air represents a loss in revenue to the broadcast company. It is therefore important to consider the options for reducing the time to repair when a failure occurs, and weigh the upfront cost of these options against the revenue lost during off-air time. There are a few options available to minimize the off-air time of a transmitter, including the following:

- Have spares available for immediate replacement to minimize reduced power or off-air time when a fault occurs. Having to wait for replacement parts to arrive, or trying to repair failed units through service calls, even if under warranty, can represent significant lost time. Talk with the manufacturer; they should be able to help determine where to best spend money by identifying high-risk components that represent single-point failures, and those components that tend to fail most often. Request charts that indicate the percentage reduction in transmitter RF output power for a given subcomponent failure, and ask for failure rate data, or *mean-time between failure* (MTBF) results for a given product or component.
- Purchase redundancy options for components within the transmitter. Many transmitters have options to include standby power modules, power supplies, exciters, etc., which often include automatic switch over in the event of a failure. Additionally, transmitter manufacturers often design redundancy into their system such that a single high-risk component failure may result in a reduction in output power, rather than the transmitter being completely off-air.
- Purchase redundancy of the overall system, by choosing main/standby, $N+1$, or combined configurations. This tends to be the most expensive option, but provides the highest redundancy.

2.5.4.1 Consider Recurring Expenses.
When considering different transmitter options, be sure to factor in lifetime recurring costs in addition to the upfront cost of the transmitter. Vacuum tubes, for example, have a finite life expectancy and replacement costs should be factored in to the long-term operating costs. Semiconductor devices can operate for significantly longer periods, as life expectancy relates inversely to junction temperature. Additionally, there are typically numerous semiconductor devices operating in parallel; therefore, one failure does not represent a significant drop in power or a significant cost to replace. Fans are another device that have a finite life expectancy, but may or may not represent a single-point failure.

Understanding the expected recurring costs also allows for the purchase of spares when they are most affordable; spares are often sold at a significantly discounted price at the initial purchase of the transmitter.

2.5.5 Manufacturer's Recommendations and Industry Best Practices

Proper planning can minimize unforeseen changes, which are costly at any time in the project, and can have a ripple effect; discuss requirements with the equipment manufacturers, and consult with professional site planners prior to investing. As well, following installation, operation and maintenance recommendations can significantly reduce total cost of ownership and off-air time. Review and adhere to the minimum site and system requirements provided in the manufacturer's product manuals. Some key factors to consider are discussed in the following sections.

2.5.5.1 Available Space.
When considering purchasing a new transmitter, ensure the available space is sufficient. If the building is pre-existing, it may be possible to make modifications to accommodate installation of the transmitter, but this may depend on whether the space is leased or owned. Not taking the following into consideration could be quite costly:

- Is the transmitter, within its crates, going to fit through all access ways?
- Will internal subcomponents need to be removed to reduce weight for transport and installation?

- Are there height restrictions where the transmitter must lay on its back or side, and if so, is this acceptable?
- Is sufficient floor space available for the transmitter's footprint, including doors that need to swing open?
- Is the floor of the transmitter room rated for the additional weight of the transmitter?
- Have considerations been made for the ancillary equipment?
- Will it be possible to service and maintain the equipment, or is space going to be too restrictive?
- Will the cabinet need to be accessed from the sides as well as the front and rear?

2.5.5.2 Lightning Protection. Lightning strikes can cause catastrophic failures if a site is not properly configured to route lightning energy away from critical components, such as RF FETs, which are susceptible to failure due to the high voltage and current they are subjected to during a lightning strike. There are many resources available for proper site installation to minimize the impact and related revenue lost due to lightning strikes; most manufacturers make this information readily available and offer training if requested. These practices have proven to minimize the damage caused to a transmitter by a lightning strike and should be adhered to as closely as possible to minimize the possible repair costs. Items to consider are as follows:

- Transient suppression devices should be installed on all external cabling connected to the transmitter, including AC distribution, RF output feed line, remote control, and audio cabling. Generally, correctly selected ferrite toroids are applied for this purpose on all cables, and active devices such as MOVs (metal oxide varistors) are also used on the AC feed lines.
- A correctly configured low-impedance grounding system should be connected to the transmitter's station reference ground point.
- Spark gaps in the transmitter and the antenna system should be present and correctly configured.

2.5.5.3 Environment. The environment in which a transmitter exists plays a critical role in determining the life expectancy of components within the transmitter. Some critical transmission site environmental factors to consider are as follows:

- With solid-state semiconductor devices, ensuring the device junction temperatures are as low as possible for as long as possible will have the biggest effect on extending the life of these components, and minimizing the replacement cost over the lifetime of the product. It is first and foremost critical to ensure the normal ambient temperature operating range (typically 0 to 50°C, but derate versus altitude by 2°C/1000 ft or 3°C/500 meters) is not exceeded. However, in order to minimize junction temperatures, hence maximizing life expectancy, the following factors should be considered:
  - When planning for installation, determine if the cooling systems available at the site are sufficient, or if additional cooling—or even a dedicated closed loop system—is necessary. Ensure ducts can be routed to the transmitter as required.
  - Install air-conditioning systems to maintain a low ambient room temperature (considering of course the cost tradeoff between minimizing failures and paying for air-conditioning energy). Once air conditioning is installed, inspect on a regular basis to ensure it is in good working condition.
  - Ensure intake or exhaust ducting is installed by a professional to minimize the static pressure on the intake or exhaust ports of the transmitter, and ensure the exhaust system presents a slightly negative pressure, maximizing the air flow through the transmitter.
  - Ensure air filters are properly installed—oil is applied if required—and are cleaned or changed regularly. Dirty air filters restrict the air flow through the transmitter, increasing device operating temperatures.

- If the transmitter has an external power transformer, ensure its cooling requirements are considered.
- Maintain a clean site, ensuring that outside air is sufficiently filtered, and any dirt generated by work being performed in the site is properly contained. Not only do dirt and dust clog transmitter air filters, but they could cause tracking and flash-over failures in high-voltage circuits.
- Ensure the humidity levels in the site meet transmitter requirements (i.e., typically 0–95 percent, noncondensing). Also, ensure that humidity is not condensing on equipment above a transmitter and causing water to drip down into the transmitter. Common issues include turning on a cold transmitter immediately after it has been installed in a warm humid site, or cold air-conditioned air blowing directly on warm humid exhaust ducts.

In addition to the above, some transmitter manufacturers have intelligent control systems that continuously monitor the operating conditions of the transmitter and make adjustments to extend the life of various components within the transmitter, such as fans, semiconductors, and graphical displays.

2.5.5.4 AC Power. The standard input on most AM and FM transmitters is AC power, and it is important to ensure that the AC power feed is correctly selected and installed to maintain a high level of transmitter performance, and protect critical downstream devices during fault conditions. Some factors to consider when installing and maintaining the transmitter's AC power feed are as follows:

- Ensure there is appropriate means of routing AC cabling to the transmitter's entry point (i.e., does the transmitter have top or bottom entry?).
- Ensure the AC supply, circuit breaker, and wiring are properly sized and comply with the local electrical code; consult with the transmitter manufacturer and a qualified electrician.
- Ensure the transmitter AC input options support the site AC supply configuration.
- Ensure the AC voltage and frequency available at the site are appropriate for the transmitter. If applicable, ensure the transmitter AC power transformer is correctly tapped for the available voltage.
- Ensure the AC supply is well regulated and transient free; if not, surge suppression devices or a voltage regulator may be necessary to maximize transmitter MTBF.
- Where possible, install an uninterruptible power supply (UPS) to maintain power for critical control systems through brief interruptions in the main AC source. This can reduce transmitter recovery time.
- When operating on three-phase power, and the transmitter's main power supply is linear, most transmitter manufacturers recommend avoiding open-delta AC power source transformer configurations. Reasons for this include that open-delta transformers generate higher than expected peak voltages, are more susceptible to transients, and their behavior is unpredictable [2]. In some cases, connecting a transmitter to an open-delta source may void the manufacturer's warranty. Confirmation from the power utility is the best way to verify what source configuration has been provided.

If it is imperative that the transmitter remain on-air during power outages, or at least that the amount of off-air time is minimized, the installation of a generator may be necessary. However, pairing a transmitter with a generator is nontrivial, as there are many possible interactions between the transmitter's power supply and the generator's control electronics that may cause instability.

Depending on the transmitter's power supply configuration, it may be necessary to oversize the generator, or select a specific winding configuration. In addition, selecting the correct generator control system and ensuring it is properly configured are critical factors to maintaining a stable system. For these reasons, generators to be used for operating transmitters should be selected carefully with assistance from the transmitter and generator manufacturers.

2.5.5.5 HD Radio System. Selecting the correct transmitter power level is more complicated for HD Radio systems compared to a purely analog transmitter. It is difficult and expensive to replace a transmitter if the power level was not properly selected. To help with this process, consider the following points:

- Does the transmitter have margin for an increase in digital power versus the current licensed analog TPO in order to improve digital coverage?
- Will the broadcast be standard MP1 or extended MP3 hybrid mode for added services (the latter requiring more peak power capability and ancillary equipment)? Higher modes are being developed for future use; will they be supported with the selected transmitter?
- Are there any geographical issues with the transmitter site location versus the intended audience that may put the system at risk for not reaching listeners?
- Consider any adjacent channels that may be interfered with while operating in HD Radio mode. If any, what asymmetrical sideband configuration will be required to prevent interference and will it still meet coverage requirements? Is the transmitter capable of asymmetrical sideband operation?
- Account for all HD Radio licensing fees.

2.5.5.6 Additional Considerations. In addition to the above, the following should be considered:

- Ensure maximum audio input levels as well as remote input/output circuitry limits are not exceeded in order to prevent equipment circuitry damage.
- Ensure the transmitter is terminated with the correct load impedance to optimize performance (e.g., $50\ \Omega + j_0\ \Omega$); this improves audio performance, system efficiency, and transmission coverage, reduces the likelihood of damage to external feed line and antenna equipment, and minimizes RF amplifier FET junction temperatures. Degradation in load impedance can be caused by antenna icing or grounding issues where the antenna array impedance is degrading. Transient events where there is a rapid change in impedance can be caused by an arc in the RF output feed line or antenna system. Poor mechanical connections for high-current joints can result in extremely high temperatures that can melt colocated components, such as bullets. Although a transmitter will attempt to protect itself from damage due to any of these events, good maintenance will minimize the risk of catastrophic faults and off-air time.
- Consider Internet access to the site and the bandwidth required for enhanced control/monitoring; most modern equipment supports IP connectivity with web and mobile apps which may require significant Ethernet bandwidth.
- Consider bandwidth requirements and network component limitations if IP-based audio is being proposed.
- Rodents and vermin can cause catastrophic damage to equipment. Install traps or fencing to restrict access, and ensure equipment covers are left installed whenever possible.
- Quite often, workers will forget to set up a transmitter for remote operation prior to leaving the site, which can mean an additional trip to the site to resolve the issue. Consider a reminder to help prevent this situation, such as a bright indicating light near the exit, indicating the remote control status.

2.5.5.7 Maintenance and Repair. Follow the manufacturer recommended set of activities and schedule for maintenance. A strategically scheduled amount of off-air time for maintenance will ensure the unscheduled off-air times are minimized. When repair is necessary, follow the manufacturer's guidelines for equipment repair. In addition, to minimize the time and money spent on troubleshooting and repair, consider the following:

- Periodically keep a detailed record of the transmitter's operating state and critical parameters (i.e., meter readings) to track trends in performance.

- Keep a detailed record of any maintenance activity that took place; document who performed the activity, why, and when.
- Determine warranty status prior to servicing equipment; making changes could void the manufacturer's warranty.
- Only use manufacturer recommended parts and equivalent replacements (e.g., air filter MERV [minimum efficiency reporting value] rating or capacitor voltage ratings).
- Review the manuals and understand the expectations of the manufacturer before attempting to service or repair equipment.
- Ensure the transmitter site has a suitable workbench and troubleshooting manuals in the event work must be performed during routine maintenance or to resolve failures.
- Torque electrical connections, specifically high-current connections, on a regular basis to prevent overheating and possible damage/fire (e.g., AC input terminals and RF output feed-line hardware).
- Periodically lubricate moving parts related to the cooling system (e.g., louvers, bearings, etc.).
- Change batteries (if applicable) every second season (e.g., Spring and Fall) to prevent an unexpected loss of stored settings if AC power loss occurs.
- Maintain the site as well as the transmitter, addressing site issues as soon as possible.
- Ensure safe work practices are followed, as there are electrical hazards that may be an inherent part of the transmitter's design. For this reason, consideration should be given to safety features available from the transmitter manufacturer, including electrical interlocks, mechanical interlocks, and antenna grounding switches.

REFERENCES

[1] du Treil, Lundin and Rackley, Inc., Sarasota, FL. 2004. Evaluation and improvement of AM antenna characteristics for optimal digital performance, Sarasota, FL.

[2] Technical Staff. 2005. Susceptibility of the open-delta connection to third harmonic and transient disturbances, Harris Corporation, Mason, OH.

CHAPTER 2.2
COAXIAL TRANSMISSION LINES

Derek Small, Nicholas Paulin, Philip Young, and Bill Harland
Electronics Research, Inc., Chandler, IN

2.1 INTRODUCTION

The purpose of this chapter is to provide information on the selection, installation, operation, and maintenance of rigid and semiflexible coaxial transmission lines with an emphasis on those transmission lines used for terrestrial broadcast service. This chapter includes tables and general product specifications to support system planning, but always refer to the supplier's specifications for system planning and use their instructions and recommendations to install and commission systems using their products.

In general, the selection of transmission lines is based on the following points:

- Operating frequency
- Power rating
- Characteristic impedance
- Efficiency (attenuation)
- Tower loading (size and weight)

Today digital transmission is more prevalent and the higher peak-to-average ratios (PARs) require special consideration, particularly for systems that combine several television or frequency modulation (FM) channels to a single transmission line.

Figure 2.2.1 illustrates the cross section of coaxial transmission line. A selection of common lines is shown in Fig. 2.2.2.

2.2 TRANSMISSION LINE TYPES

Coaxial transmission lines are constructed of two concentric conductors on the same axis (coaxial). The mode of transmission is transverse electromagnetic (TEM), where both the electric (E) and the magnetic (H) fields are entirely transverse (orthogonal) to the direction of propagation. Figure 2.2.3 illustrates the E-field (inner to outer vector) and the H-field (counterclockwise circular vector) for a TEM line with the energy flowing out of the page. The magnitudes of the E- and H-fields are strongest at the center conductor. Waveguide, a high-power hollow transmission line, operates in a TE mode where only the E-field is transverse to the direction of propagation.

2.34 RF TRANSMISSION

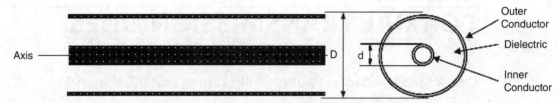

FIGURE 2.2.1 Cross section of coaxial transmission line.

FIGURE 2.2.2 Rigid, semiflexible air dielectric, and semiflexible foam dielectric transmission lines.

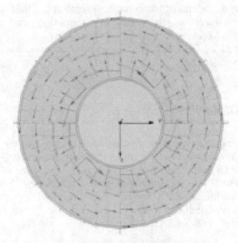

FIGURE 2.2.3 Transverse electromagnetic and magnetic fields.

Most terrestrial broadcast transmission systems use either semiflexible coaxial transmission line or rigid coaxial transmission line. The inner conductor in all coaxial transmission lines is made of high-conductivity copper. Semiflexible lines consist of two soft-tempered copper conductors. The outer conductor is seam welded and corrugated so that it can be coiled onto a reel for transportation from factory to the transmitter site. The center conductor in foam dielectric semiflexible line is supported or held in place by the dielectric material, usually polyethylene (PE). The center conductor in air dielectric semiflexible transmission line is supported by a continuous ribbon made of PE, polypropylene (PP), or PTFE (Polytetrafluoroethylene; Teflon®). Semiflexible transmission lines are usually jacketed with PE that includes carbon black for protection from ultraviolet radiation.

Rigid transmission line is constructed of two smooth walled tubes. Outer conductors are made of either copper with brass flanges or aluminum. EIA (Electronic Industries Alliance) line size refers to the nominal outside diameter of the outer conductor referenced in inches. PTFE disks or pegs support the center conductor in rigid coaxial transmission line.

2.3 ELECTRICAL AND OPERATIONAL PARAMETERS

When a system designer selects transmission lines for a specific application, a number of performance and operational factors need to be considered. Performance considerations might be line size and impedance as it affects efficiency and power handling. Operational considerations might be the number of channels, modulation, tower load, expansion, and various de-rating factors. The sections below outline specific electrical and operational parameters to consider when choosing a coaxial transmission line.

2.3.1 Cutoff Frequency

Coaxial transmission lines propagate energy in the TEM mode as described previously. The TEM mode has no cutoff frequency and can therefore operate down to 0 Hz. The TEM mode has no upper limit either making coaxial transmission line ideal for wideband applications. However, coaxial line can generate undesirable modes of propagation above a certain frequency and therefore limits its upper frequency of operation. The first undesirable, higher-order mode of operation is a TE11 mode, and is called the cutoff frequency (f_c) of coaxial line. While the TEM mode is still present and propagating energy, some of the energy can couple into the higher-order mode and result in signal loss due to differing propagation characteristics. The cutoff frequency is inversely proportional to the inner and outer conductor dimensions (and dielectric constant):

$$f_c = \frac{7514}{\sqrt{\varepsilon}(d+D)} \text{MHz} \tag{1}$$

where ε = dielectric constant
D = inside diameter of the outer conductor (inches)
d = outside diameter of the inner conductor (inches)

Larger-diameter lines have cutoff frequencies below 700 MHz, which can make some line sizes unsuitable for operation at higher-UHF (ultra-high frequency) television channels or in microwave applications. Transmission line manufacturers use different factors of safety in specifying maximum operating frequency and so this should be considered when making a specific product selection. The safety factor takes into account odd line geometries in elbows, insulator design and in some cases transformers that step to lower impedances. Theoretical and

TABLE 2.2.1 Rigid Transmission Line Cutoff Frequencies

Line Size	f_c (MHz) Theoretical	f_c (MHz) Useful	Outer I.D. (inches)	Inner O.D. (inches)
7/8 inches 50 Ω	6673	6000	0.785	0.341
1-5/8 inches 50 Ω	3429	3000	1.527	0.664
3-1/8 inches 50 Ω	1731	1600	3.027	1.315
4-1/16 inches 50 Ω	1331	1262	3.935	1.711
6-1/8 inches 50 Ω	876	806	5.981	2.600
6-1/8 inches 75 Ω	977	830	5.981	1.711
7-3/16 inches 75 Ω	835	752	7.000	2.000
8-3/16 inches 75 Ω	730	704	8.000	2.290
9-3/16 inches 50 Ω	582	552	9.000	3.910
9-3/16 inches 75 Ω	649	615	9.000	2.580

operational cutoff frequencies for common rigid transmission line are provided in Table 2.2.1. Refer to manufacturers specifications for cutoff frequency of line being used.

2.3.2 Characteristic Impedance

The characteristic impedance of a transmission line is determined by the relative diameters of the inner and outer conductors and is expressed in Equation (2):

$$Z_o = \frac{60}{\sqrt{\varepsilon}} \ln\left(\frac{D}{d}\right) \qquad (2)$$

For reference, relative values of attenuation, voltage breakdown, and peak power handling are plotted in Fig. 2.2.4 versus the ratio of coaxial diameters, D/d, in an air dielectric. Minimum attenuation on a transmission line is optimum when $D/d \sim 3.6$. Applying the above formula, one finds mathematically that for minimum attenuation the characteristic impedance is 77 Ω. Note, however, the curve is relatively flat allowing for quite a good range for lowest possible attenuation for line sizes around the optimum.

Maximum peak voltage V_p applied to a transmission line before breakdown is shown to be optimum when $D/d \sim 2.718$, or 60 Ω. The maximum power that can be applied to a line is a function of V_p^2/Z_o and is optimum at $D/d \sim 1.65$, or 30 Ω. As a compromise, and to provide the greatest utility, the characteristic impedance of commercially available coaxial transmission line is either 50 Ω ($D/d \sim 2.3$ in air dielectric) or 75 Ω ($D/d \sim 3.496$ in air dielectric), when minimum attenuation is required.

The vast majority of semiflexible and rigid transmission lines have a characteristic impedance of 50 Ω. In general, 50-Ω transmission line is used in VHF (very high frequency) digital television and FM broadcast services as attenuation is lower at these channels and they provide high average power handling capability relative to their physical size. Rigid transmission lines in larger sizes are usually 75 Ω for UHF digital television service as the attenuation is less than 50-Ω transmission lines of the same size.

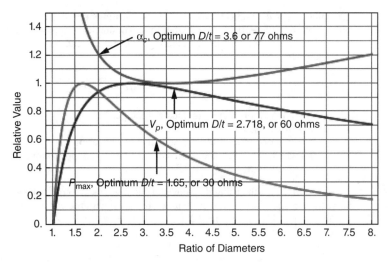

FIGURE 2.2.4 Relative values of attenuation, peak voltage, and peak power.

2.3.3 Attenuation

The attenuation of transmission line is expressed as a power ratio in dB per unit length of line at specified frequencies. Larger transmission lines have lower attenuation than smaller lines. Attenuation increases with the increase in frequency for all coaxial transmission lines. It is interesting to note that attenuation decreases with increasing frequency for waveguide transmission lines. Attenuation results from dielectric and conductor losses. Since air occupies most of the space between the inner and outer conductors in rigid transmission lines, and dielectric loss factors are negligible at UHF and below, the effects of dielectric losses can be ignored. The primary source of attenuation in a terrestrial broadcast transmission line is conductor loss due to material dimensions, skin depth (depth of current on a conductor), and surface finish. At RF (radio frequency) and microwave frequencies, current flow is on the conductor's outermost surfaces, with the greatest density at the surface. The current density drops exponentially to an insignificant amount below what is called the *skin depth*. The skin depth is a function of frequency and resistivity of the material as indicated in Equation (3):

$$\delta = \frac{1}{2\pi} \sqrt{\frac{\lambda \rho}{30\mu}} \text{ inches} \qquad (3)$$

where λ = wavelength (inches)
ρ = resistivity (0.6772×10^{-6} Ω inch for copper)
μ = permeability (1 for copper)

The attenuation constant for rigid and air dielectric transmission line can be calculated as shown below (Equation (4)). The formula allows for conductors manufactured of different materials:

$$\alpha_c = \frac{27.2}{\lambda} \left(\frac{\delta_d \mu_d}{d} + \frac{\delta_D \mu_D}{D} \right) \frac{\sqrt{\varepsilon}}{\ln\left(\frac{D}{d}\right)} \text{ dB/unit length} \qquad (4)$$

Note that most manufacturers de-rate their attenuation specifications to 95% to account for conductor surface conditions and connection losses. Attenuation increases with temperature, the generally accepted practice is to publish specifications based on an ambient temperature of 20°C (68°F) with no differential for the higher inner conductor temperature during operation. The attenuation correction factor for higher operating temperatures can be calculated with the formula shown in Equation (5):

$$M_\alpha = \sqrt{1 + \sigma_o(T_t - T_o)} \qquad (5)$$

where M_α = attenuation correction factor at temperature
T_t = inner conductor temperature (°C)
T_o = inner conductor temperature at standard rating (20°C)
σ_o = coefficient of resistance at standard rating (copper conductors at 20°C. $\sigma_o = 0.00393/°C$)

For example, results using the above formulas are used to calculate attenuation of 3-1/8 inch EIA 50-Ω transmission line at ambient temperature, and with the correction factor for the inner conductor operating at 100°C; see Fig. 2.2.5.

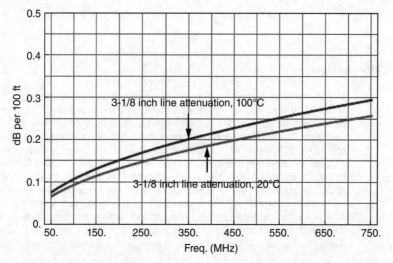

FIGURE 2.2.5 Attenuation versus frequency and inner conductor operating temperature for 3-1/8-inch rigid transmission line.

Once the attenuation constant and any correction factors have been applied, the efficiency of the system can be determined. The total attenuation (α_{total}, dB) is found by multiplying the attenuation constant by the length of transmission line in the system. The total attenuation is converted to efficiency with the formula in Equation (6):

$$\text{Efficiency} = \frac{100}{10^{\alpha_{total}/10}} \% \qquad (6)$$

One final note regarding attenuation is that semiflexible coaxial transmission lines used for broadcast service usually always have corrugated outer conductors and often have corrugated inner conductors. The corrugations actually increase the distance the signal needs to travel, when compared to rigid transmission line, which incorporates smooth walled inner and outer conductors, and has a tendency to increase attenuation.

2.4 POWER HANDLING

The power rating for coaxial transmission line is expressed as two separate operating specifications, average and peak power, and provided to permit safe operation and prevent unnecessary operational failures:

- Average power-handling capability is a function of the amount of heat created by loss. Terrestrial broadcast transmission line average power specifications typically allow the inner conductor to reach 100°C with an ambient temperature of 40°C. This operational parameter is limited by the performance or reliability of the dielectric supporting materials.
- Peak power handling capability is determined by the maximum voltage gradient present prior to breakdown with appropriate safety factors. The maximum voltage gradient allowed is limited by line imperfections, dielectric strength, voltage standing wave ratio (VSWR), and altitude.

2.4.1 Average Power

Transmission line size for single channel operation is primarily determined by the signal's average power since there is plenty of headroom for the peak voltages that may occur. When multiple channels are combined to one transmission line, voltage becomes the limiting factor due to the PARs of 8-VSB (vestigial sideband) and OFDM (orthogonal frequency-division multiplexing) modulations. Average power handling capability is determined by the amount of heat created by loss and the ability to remove that heat. Current density is the greatest on the inner conductor and generates the most heat as a result. line ratings are typically based on an inner conductor temperature of 100°C in a 40°C ambient with power. Table 2.2.2 illustrates typical line ratings for various transmission line sizes. Manufacturer's published average power ratings can vary based on differences in allowable inner conductor temperature.

TABLE 2.2.2 Typical Inner Conductor Operating Temperatures by Line Size

Type	Inner Conductor Operating Temperature °C (°F)
Rigid coaxial transmission line	100 (212)
Air HELIAX* (<2-1/4 inches) semiflexible	100 (212)
Foam HELIAX* semiflexible	100 (212)
Air HELIAX* (>3 inches) semiflexible	121 (250)
High-power air HELIAX* semiflexible	150 (302)
High-temperature air HELIAX* semiflexible	200 (392)

*HELIAX is a registered Trademark of CommScope.

Dielectric materials that support the inner conductor and the ability to transfer heat from the inner to the outer conductor are limiting factors for the line's average power handling capability. Operational performance of the dielectric material must be considered when exposed to elevated temperatures as it impacts service life of the transmission line. Average power handling is de-rated at elevation due to the change in the heat transfer from the inner to the outer conductor resulting in an elevated inner conductor temperature for a given power.

Average power, P_{ave}, can be estimated using formula (7) given below; an example for 3-1/8-inch EIA 50-Ω transmission line is provided with the inner conductor operating at 100°C:

$$P_{ave} < \frac{16380 \, \sigma_{heat} \text{Dia}}{M_\alpha \alpha_c} \text{ Watts} \tag{7}$$

where σ_{heat} = the heat transfer coefficient of the line, W/inch²
Dia = the outside diameter of the line

Average power handling versus frequency for 3-1/8-inch rigid transmission line is shown in Fig. 2.2.6.

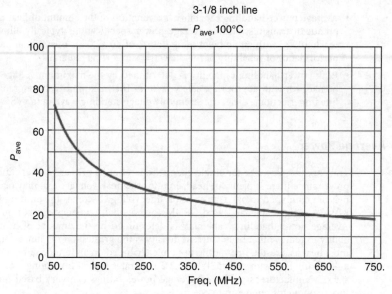

FIGURE 2.2.6 Average power handling versus frequency for 3-1/8-inch rigid transmission line.

Table 2.2.3 provides typical heat transfer coefficients for rigid transmission line of specific sizes.

TABLE 2.2.3 Heat Transfer Coefficients of Specific Line Sizes

Line Size (inches)	Z_o	Outer I.D.	σ_{heat}	Inner Material	Outer Material
7/8	50	0.785	0.128	Copper	Copper
1-5/8	50	1.527	0.120	Copper	Copper
3-1/8	50	3.027	0.107	Copper	Copper
4-1/16	50	3.935	0.104	Copper	Copper
6-1/8	50	5.981	0.097	Copper	Copper
6-1/8	75	5.981	0.077	Copper	Copper
7-3/16	75	7.000	0.076	Copper	Copper
8-3/16	75	8.000	0.074	Copper	Copper
8-3/16	50	8.000	0.092	Copper	Copper
9-3/16	50	9.000	0.090	Copper	Copper
9-3/16	75	9.000	0.069	Copper	Copper

Transmission line components are usually designed to provide a useful life of 20 years or more. If the transmission system is operated at power levels that exceed manufacturer's ratings, useful life will be significantly reduced. The sum of the transmitter outputs is used for power planning for systems with multiple users combined to a single line run since average power rating is based on temperature rise, provided the peak voltage rating of the line is not exceeded.

2.4.2 Peak Power

Peak power handling capability is determined by the maximum voltage gradient present prior to breakdown along with appropriate safety factors. The maximum voltage gradient allowed for a given line size is limited by imperfections, dielectric strength, altitude, temperature, and VSWR. Characteristics of the inner and outer conductor surfaces can have a dramatic effect on breakdown and it is easily demonstrated in a laboratory using a DC Hi-Pot tester. Moisture, contamination, and burrs that are not properly finished during manufacturing can significantly reduce withstanding voltage of a given line despite the published rating. The theoretical dielectric strength of air is 30,000 V/cm and is the limiting factor when compared to the Teflon, PE, or PP supports that are used to support the inner conductor. Variation in the density of air with altitude affects the withstanding voltage. Gas molecules can ionize, become conductive, and initiate an avalanche of current between the inner and outer conductors with excessive voltage across the line. The voltage required to initiate an avalanche increases with increased air density, or likewise, air pressure. Air pressure and, therefore, density decrease with altitude and de-rating must be applied for a given line size.

System VSWR for a single-channel broadcast facility is usually relatively low and not too much of a concern due to overhead in voltage withstanding in the transmission line. Multichannel facilities can become a problem due to the elevation of VSWR as channels are added. Modern DTV (digital television) and FM stations using 8-VSB and COFDM (coded orthogonal frequency-division multiplexing) modulations have very high peak-to-average power ratios, PAR. The probability that the voltages of individual carriers will add at their peaks is very small but must be considered. This peak voltage is exacerbated by elevated transmission system VSWR, and de-rating is required to protect the line. Line manufacturers use different techniques to calculate peak voltage/power probabilities for digital transmission systems and how they are applied for de-rating line.

Line peak power rating is derived from a DC production test voltage, E_p, which includes safety factors for properly manufactured transmission line. Theoretical withstanding voltages cannot be used due to line manufacturing imperfections mentioned above and insulator junctions where the voltage gradients can become substantial near the insulator. The formula used for DC production test voltage is as follows:

$$E_p = 31700 \, d \, \xi \, \text{Log}\left(\frac{d}{D}\right)\left(1 + \frac{.273}{\sqrt{d \, \xi}}\right) \qquad (8)$$

where E_p = production test voltage
ξ = air density factor = 3.92 B/T
B = the absolute pressure, cm of mercury
T = temperature, K
ξ is 1 at sea level at 23°C

Transmission line and components are tested at the factory using a DC Hi-Pot tester. Electrodes are connected to the inner and outer conductors and the DC production test voltage, E_p, is applied to confirm the component will not arc over. Test voltages (rounded) are listed in Table 2.2.4 for common rigid transmission line sizes.

The production test voltage, E_p, has become an accepted standard in broadcast for evaluating the voltage handling capacity of a line. E_p, a DC test voltage, must be converted to an RF RMS (root mean square) voltage so that maximum power handling of a line can be calculated. The formula used to convert E_p to RMS RF voltage is as follows:

$$E_{rf} = \frac{.7 E_p}{SF\sqrt{2}} \qquad (9)$$

TABLE 2.2.4 Typical DC Production Test Voltages by Transmission Line Size

Line Size (inches)	Impedance (Ω)	Test Voltage, E_p (kV)
7/8	50	6
1 5/8	50	11
3 1/8	50	19
4 1/16	50	24
6 1/8	50, 75	36
7 3/16	75	42
8 3/16	75	47

where .7 is a DC-to-RF factor, empirically verified
$1/\sqrt{2}$ converts to an RMS voltage for power calculation
SF is a safety factor
   = 1.4 for semiflexible cables
   = 2 for rigid transmission line

The maximum RF RMS operating voltage is then converted to a maximum peak power using

$$P_{peak} = \frac{E_{RF}^2}{Z_o} \tag{10}$$

Typical maximum power handling capabilities of common size transmission lines are found in Table 2.2.5.

TABLE 2.2.5 Typical Power Capacities of Transmission Lines by Size

Line Size	P_{peak} (kW)
7/8 inches 50 Ω	41
1-5/8 inches 50 Ω	132
3-1/8 inches 50 Ω	440
4-1/16 inches 50 Ω	710
6-1/8 inches 50 Ω	1536
6-1/8 inches 75 Ω	1069
7-3/16 inches 75 Ω	1426
8-3/16 inches 75 Ω	1825
9-3/16 inches 50 Ω	3296
9-3/16 inches 75 Ω	2271

Peak power can be misleading. The P_{peak} values are the highest envelope powers available for a given line size and must be de-rated as a function of modulation. For example, COFDM modulation typically specifies a PAR of 10–12 dB resulting in envelope powers, P_{env}, that are 10–16 times the average power. The resulting peak voltage is then

$$V_p = \sqrt{2 Z_o P_{env}} \tag{11}$$

For lines with combined COFDM signals, a very conservative approach to choosing transmission line would be to sum V_p for each channel and choose a line with appropriate maximum voltage capability, $\sqrt{2}E_{rf}$, after appropriate de-rating factors. Several papers discuss de-rating the voltage sum V_p based on probability, for a less conservative, more practical approach. Refer to transmission line manufacturer for method that best fits their line.

2.4.3 Connector Power Rating

Connectors on semiflexible cable assemblies may have a lower peak or average power rating than the cable. Connector power ratings for popular interfaces are shown in Table 2.2.6.

TABLE 2.2.6 Connector Power Ratings (*Courtesy of CommScope*)

Connector Type	DC Test Voltage (kV)	Average Power (kW)*	Peak Power (kW)
SMA	1	0.1	2.5
BNC	1.5	0.1	5.6
TNC	1.5	0.3	5.6
UHF	2	0.3	10
N	2	0.6	10
HN	4	0.6	40
SC	4.2	1.2	44
7-16 DIN	4	1.3	40
4.1/9.5 DIN	2.5	1.2	16
LC	5	3.5	63
7/8 inches EIA	6	1.7	90
1-5/8 inches EIA	11	4.9	300
3-1/8 inches EIA	19	16	902
4-1/2 inches IEC	21	27	1100
6-1/8 inches EIA	27.5	57	1890

*Average power ratings of the connector interfaces are based on an operating frequency of 900 MHz. The values shown in this table are typical for most applications.

2.4.4 Voltage Standing Wave Ratio

Transmission line inherently has minor deviations in its characteristic impedance, Z_o. Deviation in line impedance is caused by corrugations, production variations, dents, flange reflections, inner conductor supports, or due to an impedance mismatch between line and the antenna. Deviation from the characteristic impedance causes power to be reflected back toward the source. The reflection coefficient, ρ, is a measure of the reflected power as follows:

$$\rho = \frac{Z_l - Z_o}{Z_l + Z_o} \qquad (12)$$

where Z_l and Z_o are load and characteristic impedances and are complex numbers, i.e., they have a magnitude and phase associated with them. Since characteristic and load impedances are complex numbers, ρ is complex and the reflected signal traveling toward the source creates a "standing wave" by creating voltage maximums when the incident and reflected signals add (in phase), and voltage minima when they subtract (opposing phase). The VSWR is a measure of the reflected power and is

$$\text{VSWR} = \frac{1+|\rho|}{1-|\rho|} \quad (13)$$

The power returned to the source in dB is

$$RL = 20 \text{ Log}|\rho| \text{ dB} \quad (14)$$

For a single impedance mismatch, the mismatch loss at the load due to the reflected portion never making it there is

$$L_{mis} = 10 \text{ Log}(1-\rho^2) \text{ dB} \quad (15)$$

The mismatch loss for multiple reflections is based on the phase and magnitude of the sum. A perfect VSWR would be 1:1 and the mismatch loss would be 0 dB; no practical system can achieve this level of performance. System VSWR of 1.1:1 or less is generally considered to be good; however, system performance can vary significantly depending on complexity.

2.4.5 Velocity Factor

The velocity factor in an air cable, with no dielectric supports (not practical), is the same as the velocity of a wave traveling in free space and is c, the speed of light. The addition of dielectric material causes the signal in a transmission line to propagate more slowly. The velocity factor is expressed as a percentage of the speed of light

$$v_f = \frac{1}{\sqrt{\varepsilon_{eff}}} \quad (16)$$

where ε_{eff} is the effective dielectric constant.

Since Teflon supports, pins and discs, are not homogeneous across a given cross section of transmission line, and the fact that they are spaced along the line, the velocity factor for transmission line lengths is determined in the laboratories of transmission line manufacturers. Published velocity factor specifications for transmission lines require consideration when phase matching line runs. For example, a transmission system may use dual line runs to provide additional power handling capability and the velocity factor is used to match the line electrical lengths for proper system operation.

2.5 DIFFERENTIAL EXPANSION

The inner conductor of a transmission line runs substantially hotter than the outer conductor and it will expand to a greater length. For a long useful service life, the transmission line must incorporate some method to accommodate the differential expansion between the two coaxial conductors. Corrugated inner and outer conductors in semiflexible cables allow for the differential expansion. The inner conductor of a 20-foot rigid line section will expand 0.166 inches at rated average power. The inner conductor of a 20-foot rigid line section will expand 0.166 inches at rated average power. This movement causes degradation to the contact surfaces of each line section connector and over time the metal particles can lead to a catastrophic failure of the transmission line system. Most rigid transmission line designs accommodate this movement with either a watchband spring attachment to the inner connector or with a bellows attachment to the inner connector. Figure 2.2.7 shows one example.

FIGURE 2.2.7 Aluminum outer conductor 3-1/8-inch rigid transmission line with bellows expansion compensation.

2.5.1 De-rating Factors

When planning a transmission line system, it is important to understand the line operating specifications and how they are derived. Most manufacturers employ standard conditions for their power and operating specifications. In general, attenuation, average, and peak power ratings are based on a VSWR 1.0 and at atmospheric pressure. As a part of the selection process, de-rating factors must be applied to the published average and peak power specifications.

2.5.1.1 De-rating Peak Power for Modulation and VSWR. The peak power rating of transmission line must be de-rated for VSWR and the modulation of the broadcast service it will carry. The following equations illustrate methods for de-rating peak power for broadcast service:

$$\text{For AM service: } P_{max} = \frac{P_{pk}}{\text{VSWR}(1+M)^2} \quad (17)$$

$$\text{For FM and DTV service: } P_{max} = \frac{P_{pk}}{PA\ \text{VSWR}} \quad (18)$$

$$\text{For analog TV service: } P_{max} = \frac{P_{pk}}{(1+AU+2\sqrt{AU})\text{VSWR}} \quad (19)$$

where P_{max} = the de-rated peak power
P_{pk} = the line peak power rating
M = the modulation index, 1 for 100%
PA = the peak-to-average power ratio, $10^{PAR/10}$:
  PA = 4 for 8-VSB
  PA = 10 for COFDM
AU = the aural-to-visual ratio

2.5.1.2 De-rating Average Power for VSWR. The average power rating of a transmission line should be de-rated for VSWR. It is recommended that FM and DTV transmission line systems assume a VSWR of 1.5:1 to ensure sufficient safety margin in the system. The formula to derive the VSWR de-rating factor is

$$\text{D.F.} = \frac{2\,\text{VSWR}}{(1+F^1)\text{VSWR}^2 + 1 - F^1} \quad (20)$$

where D.F. = the de-rating factor
F^1 = factor based on line size and frequency (from Fig. 2.2.8)

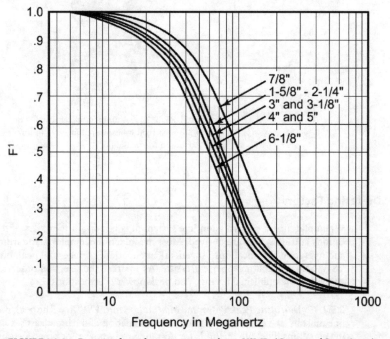

FIGURE 2.2.8 De-rating factor for average power due to VSWR. (*Courtesy of CommScope.*)

2.5.1.3 De-rating Attenuation for Ambient Temperature. The attenuation specifications for transmission lines are derived at an ambient temperature of 20°C. The graph below provides a correction factor that can be applied to the published attenuation value to reflect loss performance in actual conditions. For a given TPO (transmitter power output), and using Equations (4) and (7), a good approximation to the line operating temperature and resulting attenuation can be determined; see Fig. 2.2.9.

2.5.1.4 De-rating Average Power for Ambient Temperature and Solar Radiation. The transmission line average power rating must be adjusted from the 40°C value in published specifications to the actual temperature conditions present at the site. The de-rating factor is shown in the graph in Fig. 2.2.10.

COAXIAL TRANSMISSION LINES **2.47**

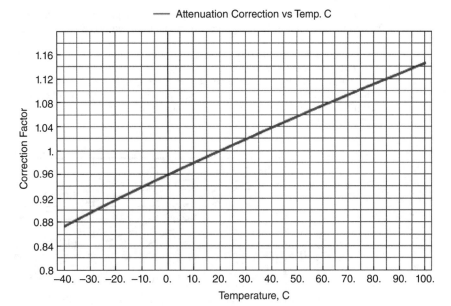

FIGURE 2.2.9 Attenuation correction factor for ambient temperature.

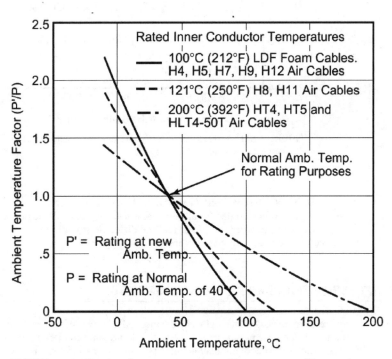

FIGURE 2.2.10 Variation of average power rating ambient temperature. (*Courtesy of CommScope.*)

Direct exposure to solar radiation reduces the average power handling capability of the cable or rigid transmission line. Painting the transmission line white can reduce the effects of solar radiation but it becomes important to maintain the exterior coating applied to ensure continuing protection. The average radiation intensity for moderate climates is 200 W/m^2 or less, while hot, dry climates have radiation intensities as high as 1000 W/m^2 or higher. The values for average direct solar radiation for locations in the United States are shown in Fig. 2.2.11. To determine the de-rating factor, locate the normal solar radiation value on the map in Fig. 2.2.12, multiply that value by 11.6 to convert it to W/m^2, and look up the de-rating factor on the graph in Fig. 2.2.11.

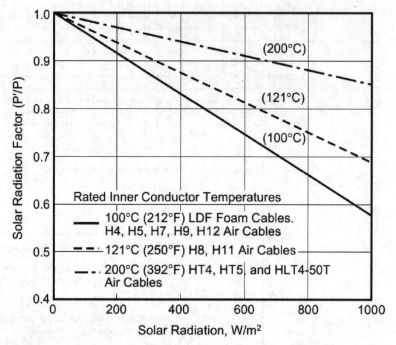

FIGURE 2.2.11 Variation of average power rating with intensity of direct solar radiation. (*Courtesy of CommScope.*)

2.5.1.5 De-rating Average and Peak Power for Altitude. Transmission line average and peak power ratings must be de-rated for altitude because lower atmospheric pressure reduces heat transfer from the inner to the outer conductor and the dielectric strength of the air inside the transmission line. Recommended de-rating factors are plotted in Fig. 2.2.13.

2.6 SEMIFLEXIBLE TRANSMISSION LINE SYSTEMS

Semiflexible transmission line is globally the most popular type of transmission line for terrestrial broadcast transmission systems. The large variety of cable sizes and connector types provide suitable products for many different types of installations and applications. The wide variety of attachment accessories and other hardware allows the cable to be mounted on many different structures. Jacketed semiflexible cables can be installed in conduit and its construction is suitable for direct underground burial.

COAXIAL TRANSMISSION LINES   **2.49**

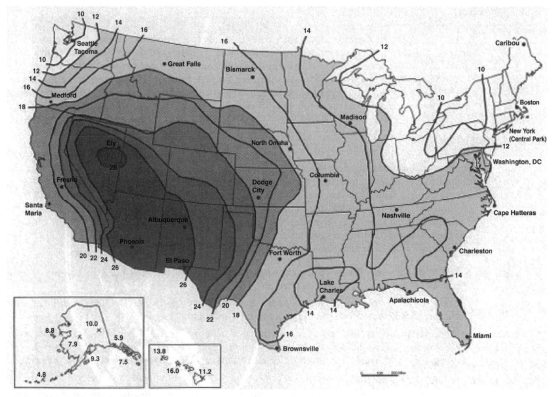

FIGURE 2.2.12 Average daily direct normal solar radiation (MJ/m^2), annual. (*Courtesy of CommScope.*)

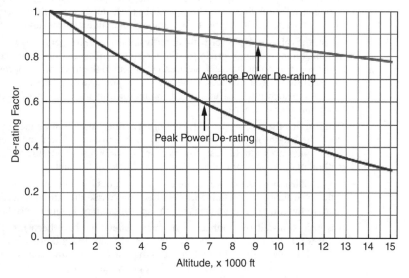

FIGURE 2.2.13 De-rating average and peak power handling capacity versus altitude.

2.6.1 Selection

In general, if the antenna does not require pressurization, foam dielectric cables are preferred if they are able to handle the power level required, as this eliminates the need for pressurization with a dehydrator or bottled nitrogen. Since foam dielectric transmission lines are only available in smaller cable sizes, they often are not selected for systems where the total transmission line length is great because of the higher attenuation of small diameter cables, particularly with UHF digital television systems. If the antenna must be pressurized, then air dielectric cable is preferred.

2.6.2 Installation

For most installations, it is desirable to have the antenna connection factory installed as this is done in a controlled environment, by a trained and experienced technician who is equipped with the correct tools for the job. This reduces the potential for leaks in pressurized air dielectric cables, and the quality control and test procedures provide additional assurance of proper installation. In most installations, the connector at the transmitter end of the transmission line is installed after the cable is hung on the tower and the horizontal run has been attached to the ice bridge and the cable fed through building entry port.

General installation guidelines include the following:

- When the shipment is received, inspect the cable for damage and inventory the accessories to be sure all necessary parts have been provided.
- Check air cables to be sure they are under pressure. The pressure caps should remain intact until they are installed on the tower and about to be attached to the antenna.
- Review the complete set of installation instructions provided with the cable and its installation accessories prior to the start of the installation process. Precisely, follow all manufacturer recommendations and instructions for hoisting and securing the cable to the tower and the ice bridge.
- During installation, support the cable reel on a suitable stand that will allow the reel to rotate.
- Confirm that the hoist and load line have adequate capacity, including the required safety margin, to hoist the weight of the entire vertical run. Support the cable's weight during hoisting by using properly sized hoisting grips, installed at intervals of no more than 200 feet as the cable is hoisted up the tower. It is also recommended that the cable is secured to the load line at 50-foot intervals between the hoisting grips to be certain that the cable is not being stressed with its own weight during installation. Secure the hoisting grips to the tower by a suitably sized chain and turnbuckle to provide permanent support for the transmission line vertical run.
- Secure the cable to the tower with butterfly and snap-in hangers at intervals recommended by the manufacturer, and appropriate for the location wind speed and icing conditions. Figure 2.2.14 shows some common hangers. Secure the hangers to the tower with appropriate attachment accessories. In some circumstances, if the tower does not have members that provide attachment points at recommended intervals, additional conduit or iron angle may need to be installed on the tower to support the vertical run.
- Confirm that the cable is properly supported at intervals recommended by the cable manufacturer along the length of the horizontal run including inside the building after the entry port. Under no circumstances should hose clamps or wraplock be used to secure the cable to the tower as this method of direct attachment will damage and deform the cable.
- Semiflexible transmission line can be installed in conduit but be certain that the pulling force applied to the cable during installation does not exceed the tensile strength specifications of cable, and that the bends in the conduit run are greater than the minimum bend radius specified for the cable selected.
- It is best to install the connectors on cables after they have been installed in the conduit; this requires that the ends of the cable are protected from dirt and other contaminants while it being installed. Apply cable pulling lubricant to the transmission line as it is being fed into the conduit. The conduit should have either drain holes or vents to prevent any moisture from collecting in the conduit. If water should accumulate in the conduit, and it freezes, the cable may be damaged.

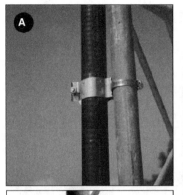 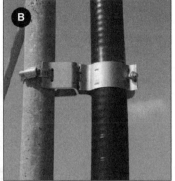 

A. Standard Hanger attached with Round Member Adapter (hose clamp).

B. Standard Hanger bolted to Stand-Off attached with Round Member Adapter (hose clamp).

C. Snap-In-Hanger in Stand-Off attached with Round Member Adapter (hose clamp).

D. Standard Hanger attached with bolt-on feed tab

E. Rigid Line Spring Hanger attached with Round Member Adapter RLA002 for 3-1/8 to 6-1/8 inch rigid line. For member sizes from 1-1/4" to 2" use RLA002-287K; For member sizes from 2-1/4" to 3" use RLA002-387K; For larger member sizes and some non-standard tower sections custom adapters will be required.

FIGURE 2.2.14 Typical transmission line hanger attachment configurations.

- If the transmission line is to be buried, protect the cable entrance and exit into the ground, and install a proper underground electrical warning tape in the trench above all buried cable. Always hand tamp on gravel and on sand. A vibrating plate compactor may be used, but when used on sand, a 6-inch minimum clearance to the transmission line must be maintained.
- Place a 2-inch thick concrete pad over sand when the transmission line is underneath a roadway or parking lot. A 20-foot minimum row is required above the trench to allow for access, and this area should be kept free of trees, large shrubs, and buildings. The frost line at the site may require a greater minimum depth than shown in Fig. 2.2.15. The trench must not be in submerged area, and must have sufficient clearance from other underground structures.

The transmission line run needs to be electrically grounded to the tower near the antenna input, the base of the tower, and outside the building entry port. If the vertical or horizontal run is greater than 200 feet, then additional grounding kits should be installed so that the system is grounded at intervals of no more than 200 feet. If the transmission line passes through the aperture of a high-power FM or TV antenna, it should be grounded at intervals of no more than 18 inches or installed in a grounded metal shield through the antenna aperture. Be sure to ground the system in accordance with local electrical codes and be certain that you do not attach copper ground leads directly to tower steel as there is a potential for a galvanic reaction that could damage the tower member.

Once the cable has been installed, the system should be purged and pressurized, if required, according to the manufacturer's recommendations. Performance should be measured by a qualified technician or engineer to ensure the system has been installed properly and to provide a benchmark for future maintenance inspections and system sweeps.

A typical installation is shown in Fig. 2.2.16.

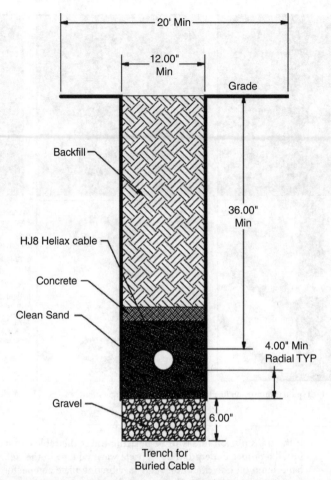

FIGURE 2.2.15 Typical buried installation for semiflexible cables.

2.7 RIGID TRANSMISSION LINE

Rigid transmission line is selected when either the peak and/or average power capacity of semiflexible coaxial transmission line is not adequate or the transmission line length dictates using rigid line for its lower attenuation to minimize line losses.

2.7.1 Selection

Once the size of the transmission line has been determined based on the peak voltage and average power handling requirements for the system, the next step is determining the proper section length. Rigid transmission line is manufactured in flanged sections of a fixed length. At each flange section, all rigid coaxial inner connectors exhibit a minor deviation from the characteristic impedance of the transmission line. This deviation causes a small amount of power to be reflected back to the RF source (VSWR). By using the correct fixed line length, the VSWR buildup occurs outside the system's

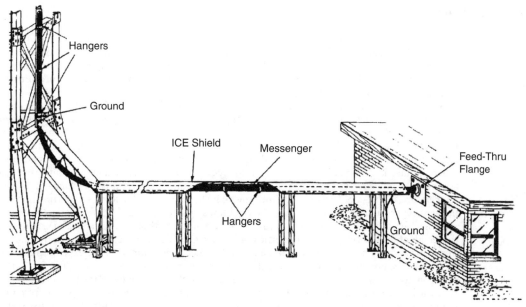

FIGURE 2.2.16 Typical horizontal run installation for semiflexible cables. (*Courtesy of CommScope.*)

designed operating frequency. This needs to be considered for both digital television and FM service. The proper section length for VHF and UHF television channels is shown in Table 2.2.7.

TABLE 2.2.7 Recommended Rigid Transmission Line Section Lengths for US Television and FM Channels

US Television Channels	Section Length
Channels 2, 3, 5, 6, 8, 9, 12, 14, 15, 18, 19, 22, 23, 27, 31, 32, 35, 36, 39, 40, 43, 44, 47, 48, 51	20.00 feet (6.096 m)
Channels 16, 20, 24, 28, 33, 37, 41, 45, 49	19.75 feet (6.020 m)
Channels 4, 10, 13, 17, 21, 25, 26, 29, 30, 34, 38, 42, 46, 50,	19.5 feet (5.944 m)
Channels 7, 11	18.67 feet (5.689 m)
FM Radio Frequencies	Section Length
88.1–95.9 and 100.3–107.9 MHz	20.00 feet (6.096 m)
96.1–98.3 MHz	19.50 feet (5.944 m)
98.5–100.1 MHz	19.00 feet (5.791 m)
88.1–107.9 MHz	17.50 feet (5.342 m)

For multichannel combined FM systems, 17.5-foot rigid line sections are used. In the case of multichannel digital television systems, if all of the channels in the system do not use a common standard rigid line section length, then the system to accommodate two channels can be accommodated with a custom line length used throughout the system. Or, if a true broadband system is required, then an optimized system can be designed that uses multiple section length line sections to minimize the build-up of reflections within the band of operation.

2.7.2 Installation

Rigid transmission line is shipped on individual cartons or on skids with multiple line sections. In addition, the installation will also include elbows to transition from the vertical to the horizontal run and for the elbow complex at the top of the vertical run to connect to the input of the antenna. The elbow complex also allows the transmission line to be disconnected to perform system measurements without the influence of the antenna.

Generally, the vertical run of the transmission line system is installed first. The installation can be started from either the base of the tower, temporarily supporting the transmission line with one or more fixed hangers at the base, or it can start at the top of the run, near the antenna input. The first section at the top of the vertical run is always supported by a rigid fixed hanger. The tower steel and the copper transmission line expand and contract with variations in ambient temperature at different rates. The form of differential expansion requires that the vertical and horizontal runs must accommodate the greater expansion of the transmission line run. Over a 100°F change in ambient temperature, a 1000-foot transmission line grows 4 inches more than the tower supporting it. A rigid hanger near the antenna input prevents the rigid transmission line run from damaging the antenna input and forces all the expansion to occur at the base of the vertical run. There is also expansion of the length of the horizontal run. The vertical spring hangers (E in Fig. 2.2.14) allow the transmission line to expand and contract relative to the tower. In order to accommodate the expansion and contraction of the horizontal run and prevent damage at the base of the vertical run, the first vertical spring hanger (E in Fig. 2.2.14) is installed some distance above the elbow at the base of the vertical run. The lateral installed near the base of the vertical run prevents the expansion section from moving the horizontal or vertical runs laterally. The minimum suggested distance for the first vertical hanger can be found in Table 2.2.8.

TABLE 2.2.8 Minimum Distance to the First Vertical Spring Hanger versus the Length of the Horizontal Run

Horizontal Run Length		3-1/8 inch		4-1/16 inch		6-1/8 inch		7-3/16 inch		8-3/16 inch	
20 feet	(6.1 m)	5 feet	(1.5 m)	6 feet	(1.8 m)	9 feet	(2.7 m)	10 feet	(3.0 m)	12 feet	(3.7 m)
40 feet	(12.2 m)	6 feet	(1.8 m)	7 feet	(2.1 m)	11 feet	(3.4 m)	12 feet	(3.7 m)	15 feet	(4.6 m)
60 feet	(18.3 m)	7 feet	(2.1 m)	8 feet	(2.4 m)	13 feet	(4.0 m)	15 feet	(4.6 m)	17 feet	(5.2 m)
80 feet	(24.4 m)	8 feet	(2.4 m)	9 feet	(2.7 m)	14 feet	(4.3 m)	17 feet	(5.2 m)	20 feet	(6.1 m)
100 feet	(30.5 m)	9 feet	(2.7 m)	10 feet	(3.0 m)	15 feet	(4.6 m)	18 feet	(5.5 m)	22 feet	(6.7 m)

To accommodate the expansion of the vertical run, the horizontal run must be greater than the minimum length shown in Table 2.2.9.

TABLE 2.2.9 Minimum Recommended Horizontal Run Lengths for Various Vertical Run Lengths

Vertical Run Length		3-1/8 inch		4-1/16 inch		6-1/8 inch		7-3/16 inch		8-3/16 inch	
100 feet	(30.5 m)	15 feet	(4.6 m)	15 feet	(4.6 m)	15 feet	(4.6 m)	20 feet	(6.1 m)	20 feet	(6.1 m)
500 feet	(152.4 m)	25 feet	(7.6 m)	30 feet	(9.1 m)	25 feet	(7.6 m)	40 feet	(12.2 m)	40 feet	(12.2 m)
1000 feet	(304.8 m)	35 feet	(10.7 m)	40 feet	(12.2 m)	50 feet	(15.2 m)	60 feet	(18.3 m)	60 feet	(18.3 m)
1500 feet	(457.2 m)	40 feet	(12.2 m)	50 feet	(15.2 m)	60 feet	(18.3 m)	70 feet	(21.3 m)	70 feet	(21.3 m)
2000 feet	(609.6 m)	45 feet	(13.7 m)	60 feet	(18.3 m)	70 feet	(21.3 m)	80 feet	(24.4 m)	80 feet	(24.4 m)

When the installation of the hangers of the vertical run is complete, the spring hangers should be set according to the manufacturer's instructions. The vertical spring hanger tensions are expressed in terms of the total vertical run length and the ambient temperature at the time they are being set.

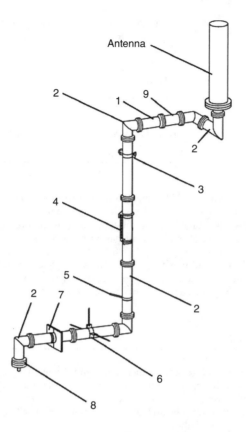

Legend

1 Rigid Line Section
2 90 Degree Elbow
3 Rigid Hanger
4 Vertical Spring Hanger
5 Lateral Brace
6 Horizontal Spring Hanger
7 Wall/Roof Feed Thru
8 Gas Barrier
9 Fine Matching Section

FIGURE 2.2.17 Simplified rigid transmission line system drawing.

To avoid having large variations in spring tensions, the spring hangers should be set within hours of one another.

Figure 2.2.17 shows a simplified rigid transmission line system. The simplified system illustrated in the diagram is provided as a guide for selection of the components required. Each system should be engineered individually.

In the process of assembling rigid transmission line sections be sure to never have more than one section of rigid line supported by a flange joint without support from a hanger. Follow the manufacturer's instructions for hanger spacing. When joining line sections together, align pins with the corresponding flange alignment holes and join mating sections. Be sure to firmly push the line sections together ensuring that the O-ring seal remains in place and the inner connector slides into the inner conductor, and the inner conductor insulator seats properly in the mating flanges. Install and alternately snug mounting hardware at 180° intervals while maintaining a uniform gap between flanges. When performing the final torqueing, sequence in a circular pattern. Do not over-tighten (see Table 2.2.10).

TABLE 2.2.10 Commonly Accepted Hardware Torque Specifications for Transmission Line Systems

Hardware Size	Torque Value
1/4 inches (6 mm)	7 lb-ft (9 N m)
5/16 inches (8 mm)	12 lb-ft (16 N m)
3/8 inches (10 mm)	21 lb-ft (28 N m)
1/2 inches (13 mm)	46 lb-ft (62 N m)
5/8 inches (16 mm)	76 lb-ft (103 N m)

When properly installed, a small uniform gap should be noted around the flange circumference. The use of antiseize compound on all stainless hardware is recommended to prevent galling. If hardware becomes galled during the tightening procedure, remove damaged hardware by sawing or breaking and install replacement hardware to ensure proper electrical contact between mating surfaces.

When the installation of the vertical run is complete—including installation of the rigid hanger at the top of the run—the rigid hangers temporarily used to support the line during installation must be immediately removed to prevent damage to system.

2.8 PRESSURIZATION

Rigid transmission lines and air dielectric semiflexible lines must be pressurized. If these systems are unpressurized, changes in temperature can permit the system to breathe, bringing in moisture laden air. This will cause condensation inside the system. Moisture will accumulate and result in possible failure. The system must be pressurized at the time of installation and remain pressurized at all times to avoid this type of failure.

The transmission line and the antenna should be purged prior to placing the system in service, and at any other time that moist air could have entered the system. Purging may be accomplished by pressurizing the transmission line from the transmitter end to approximately 5–10 psi. Water vapor will rise above the nitrogen in the system, so purge at the highest point in the system first. Bleed the system at the bleeder valve, if one is included in the antenna, or by temporarily loosening the four bolts in the brass plate on the top of the highest antenna line block to allow slow leakage. If the antenna is center fed, the lower half of the array should also be bled by loosening the four bolts in the brass plate on the bottom of the lowest antenna line block.

Allow at least one full tank of dry nitrogen to be used for this purging. If a dry air compressor is used, allow it to purge the system slowly for a minimum of 1 h. In the event that there has definitely been water in the transmission line and/or antenna, the use of several tanks of dry nitrogen is recommended. At minimum, the purging process should allow at least three exchanges of the full volume of the system. After the purging process, tighten all bolts and connections in the system and pressurize the system. Be certain that pressure level is maintained below the lowest rated component in the system.

An automatic membrane dehydrator is shown in Fig. 2.2.18.

Manufacturer transmission line peak and average power ratings are based on a standard condition of atmospheric pressure but it is good practice to confirm this when selecting the components used in a system. Pressurizing the transmission line increases average and peak power ratings. Since air dielectric semiflexible and rigid transmission lines require pressurization, this provides additional power handling capacity. The effect of pressurization on average power rating is shown in Table 2.2.11. If the system is operated at power levels where pressurization is required to maintain safety margins, the system will require control interlocks to prevent damage in the event of a leak or a failure of the pressurization system.

FIGURE 2.2.18 Andrew automatic membrane dehydrator. (*Courtesy of CommScope.*)

TABLE 2.2.11 Average Power Uprating Factors with Pressurization

Pressure (PSIG)	Average Power P^1/P Factor	
	50-Ω Line	75-Ω Line
0	1.00	1.00
5	1.09	1.08
10	1.16	1.15
15	1.21	1.22
20	1.26	1.28
25	1.31	1.33

Peak and average power ratings can also be increased through the use of exotic gases in lieu of dry air or nitrogen but the expense, safety issues, and environmental concerns have made this practice extremely rare.

2.9 MAINTENANCE AND INSPECTION

Immediately following installation, the transmission line system should be inspected to verify that it has been installed according to the manufacturer's instructions. The system should also be measured to establish a baseline VSWR and time domain response. Additional recommendations include the following:

- Visually inspect the system annually, including a climbing examination of each flange connection, grounding kit, and hanger. Replace any missing or damaged hardware and/or hangers. In addition to annual inspections, inspect the system after any severe wind and/or ice storm event, or other extreme loading condition.

- Inspect for hot spots that would indicate the beginning of failure of the inner conductor connections.
- Verify proper operation of the dehydrator or gas supply valve used to pressurize air dielectric transmission lines on an annual basis.
- Measure the VSWR and time-domain response of the system on a regular schedule.

It is recommended that the transmission line and antenna system be protected by a VSWR protection circuit that will remove transmitter power in the event of high reflected power. It is important that if an overload occurs, the cause of the overload is determined before reapplying power. Under no circumstances should these protection devices be overridden.

A final note, the installation, inspection, maintenance, or removal of antenna and transmission line systems requires qualified, experienced personnel. Manufacturer installation instructions and mounting assemblies have been designed for these specially trained individuals. Take steps to verify that personnel providing services that include climbing towers have the proper certifications, equipment, and insurance coverage to protect themselves and your station.

CHAPTER 2.3
FM CHANNEL COMBINERS

Derek J. Small
Dielectric LLC, Raymond, ME

2.1 INTRODUCTION

A multistation frequency modulation (FM) combiner, single feed line, and broadband (BB) antenna provide a solution for the increasing demand of FM spectrum and tower space. The purpose of this chapter is to provide concise information on the selection criteria for FM *channel* combiners. Today, proliferation of the FM spectrum and the amount of information contained within a channel through the use of IBOC (in-band on-channel) places unique requirements on combiner efficiency, amplitude/delay variation, intermodulation (IM) suppression, and isolation. The chapter will start with a brief outline of combiner types and proceed to discuss isolation requirements and resulting amplitude/delay variations for the combining function and IM reduction.

2.2 COMBINER TYPES

High-power FM channel combiners can be separated into two types—junction and directional. Each type has benefits over the other as discussed below.

2.2.1 Junction

The junction combiner is illustrated in Fig. 2.3.1, and can take the form of a "star point" or "manifold"-type junction. When space and initial cost are of concern, junction combiners provide the best solution. Each channel filter in a star point design is connected to the junction using $\frac{n\lambda}{2}$ line lengths and optimized for performance. Manifold combiners utilize $\frac{n\lambda}{2}$ line lengths off all junctions and optimized for performance. Optimization is required to improve input VSWR (voltage standing wave ratio) performance of all filters at the junctions. Each filter and the line to the junction are dependent on the other filters and lines to junctions, and therefore, it limits the ability to easily add channels in the future. When properly matched, junction combiners provide the lowest loss solution for combining multiple stations due to the limited amount of components. One key factor to keep in mind is system VSWR performance when adding a large number of channels to a junction. Manifold junctions provide the best VSWR performance for a large number of stations combined to a line.

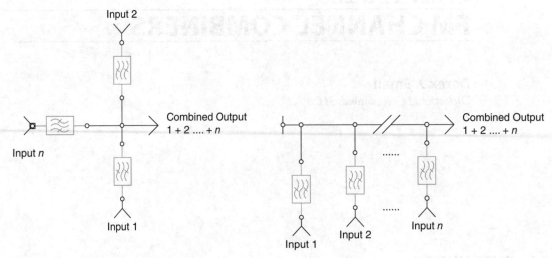

FIGURE 2.3.1 FM starpoint and manifold junction combiner.

2.2.2 Directional

A typical FM directional filter is a four-port device that utilizes two filters, two hybrids, and a load to direct signal flow from one port to another. Figure 2.3.2 illustrates directional filter modules cascaded to combine three channels. The BB input is terminated or can be used to add channels in the future, provided the last hybrid is capable of handling the additional power. Other directional filter types, not used for high-power FM broadcast, include the use of waveguide with *E*- and *H*-coupling instead of hybrids for $\frac{\pi}{2}$ inputs to the filters. Circulators and filters can also provide directionality but are inefficient, and nonlinear mixing within the ferrite circulator introduces spurious signals.

Directional filter modules provide many benefits over junction combiners. Due to the directional nature of signal flow and out-of-band impedance match, modules are easily combined without optimization since VSWR is low and interactions are significantly minimized. This allows future expansion by adding additional modules in place of the BB port termination. Directional filters also reduce IM products generated by transmitters over junction combiners using the same filter order due to the additional isolation provided by the hybrids. Power handling is another benefit since the signal is split into two filters. Directional filters have a size and cost disadvantage when compared to junction combiners, particularly when channel spacing is tight and higher-order filters are used.

Signal flow through a directional filter is well documented; a brief review follows. A signal entering the narrowband port (input 1 in Fig. 2.3.2) is split in the hybrid and exits with equal amplitudes and quadrature in phase, one at 0° and one lagging 90° (indicated by −90° in Fig. 2.3.2). Each signal passes through a bandpass filter where amplitude and phase relations are maintained as it enters the output hybrid. The signal at 0° enters the output hybrid where it travels 90° and recombines in phase with the other signal entering the hybrid at 90°. A signal entering the BB port splits and is reflected off the filters to recombine at the output port.

Isolation from the narrowband input to the BB port is a function of hybrid quality and how well filter transmission and reflection amplitude and phase are matched. Filters that are matched well in both amplitude and phase can provide upward of 40 dB isolation. Isolation from the BB port to narrowband input port (of a single module) is similar with additional isolation provided by the filter rejection.

Figure 2.3.3 shows a high-power four-station combiner with an output splitter.

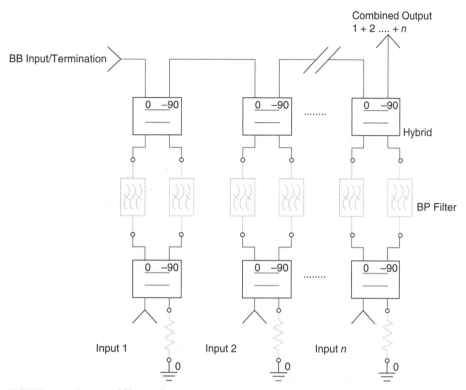

FIGURE 2.3.2   Directional filter combining system.

FIGURE 2.2.3   High-power four-station combiner with an output splitter.

2.3   FREQUENCY RESPONSE

Today, most junction or directional combiners are designed and manufactured using bandpass filters. Combiner design begins with selecting a filter response providing the most efficient combining function with minimal channel degradation. Chebyshev filter functions have been used for years and have provided enough rejection for use in combiner circuits. Generalized Chebyshev filter functions have recently been developed for closely spaced channels due to the ability to place transmission zeros for increased rejection. Channel spacing and power drive filter order, size, and loss. Typical filter responses required for IM suppression and channel spacing are discussed below.

2.3.1 IM Suppression

When channels are combined, it is important to keep channel-to-channel isolation at a sufficient level to reduce IM products that result from mixing within the transmitter. The dominant IM product from a transmitter is generated by the mixing of the interfering transmitter frequency, F_i, with the second harmonic of the transmitters' carrier frequency, F_c, $(2F_c - F_i)$ [1]. The "mixing loss" or "turn-around" loss of the generated IM product can vary from 5 dB to significantly higher and is a function of channel separation and transmitter type (tube or solid state). Other mixing harmonics are generated but have higher turn-around loss and are attenuated further by greater combiner isolation with frequency.

Federal Communications Commission (FCC) 47 CFR § (section) 73.317(d) 0150 FM Transmission System Requirements states that

"Any emission appearing on a frequency removed from the carrier by more than 600 kHz must be attenuated at least $43 + 10\text{Log}_{10}$(Power, in watts) dB below the level of the unmodulated carrier, or 80 dB, whichever is the lesser attenuation."

TPO (transmitter power output) versus required IM attenuation is plotted in Fig. 2.3.4. The break point TPO of whether the formula or the 80 dB value is used for attenuation is solved from the following equation:

$$80 \text{ dB} = 43 + 10\text{Log}_{10}(\text{TPO, in watts}) \text{ dB}$$

TPO levels below approximately 5 kW will use the formula for IM attenuation, while TPOs above 5 kW require 80-dB attenuation of IM products.

The interfering signal in a channel combiner is attenuated by the sum of isolation between channels, turn-around loss, and filter rejection. In the case of a junction combiner, the isolation between channels

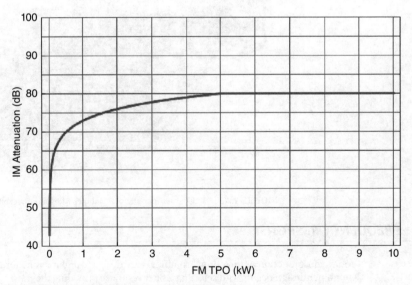

FIGURE 2.3.4 FCC 47 CFR § (section) 73.317(d), plotted.

is equivalent to the filter rejection. As mentioned previously, a directional system will provide more isolation due to the hybrids, typically another 30 dB, and relieves the burden on the number of poles in the filter for a given channel spacing. Table 2.3.1 summarizes the minimum number of poles required to provide 80 dB of IM suppression for junction and directional combiners. *Note*: The table assumes a filter VSWR of 1.05 (see the next section for filter order and VSWR).

TABLE 2.3.1 Number of Poles Required to Meet 80-dB IM Suppression, FCC 47 CFR § (section) 73.317(d)

Number of Poles	Spacing (MHz)	
	Directional Combiner	Junction Combiner
2	8.4	9.0
3	1.6	2.4
4	.8	1.2

2.3.2 Channel Spacing and Rejection

One of the first questions asked by the designer is "what are the channels to be combined, or minimum channel spacing desired?" Spacing between channels sets up the number of poles in the filter, and consequently, channel-to-channel isolation. Most FM combiner applications are fulfilled with 3- or 4-pole filters; however, today's proliferation of the FM band now sometimes requires the use of higher-order filters, or cross-coupled designs to meet the needs of very closely spaced channels.

IBOC and the possibility of eventually going all digital now means that combiner design must include the entire channel band, unlike the past when an FM signal only partially occupied the channel. Therefore, the filter passband must be at least 400-kHz wide and defines one aspect of the combiner. Second and third considerations are tied to one another as tradeoffs—acceptable VSWR and required rejection. If greater VSWR is allowed, more rejection is achieved for a given number of filter poles. These three parameters define the combiner filter response for a given channel spacing.

Typical 2-, 3-, and 4-pole Chebyshev filter responses are illustrated in Fig. 2.3.5. Filter responses are designed around the minimum bandwidth of 400 kHz with a VSWR of 1.05:1.

Rejection can improve by trading VSWR; for example, Fig. 2.3.6 illustrates rejection of 2-, 3-, and 4-pole filters with a VSWR of 1.1:1, and channel bandwidth is still 400 kHz. Directional filters can take advantage of elevated filter VSWR since it is directed to the reject load where it is absorbed.

The only disadvantage is its lower module efficiency, and is characterized by filter mismatch loss given by

$$-10 \, \text{Log}_{10}(1-\rho^2)$$

where ρ is the reflection coefficient of the filter VSWR.

2.3.2.1 Required Rejection. Figures 2.3.5 and 2.3.6 illustrate filter responses for 2-, 3-, and 4 pole filters with 400-kHz wide passband with 1.05 and 1.1 VSWR in the passband. Section 2.3.1 discusses IM suppression requirements in junction and directional filter combiners. Junction combiners will require 40-dB isolation, or rejection, between combined channels. This includes a safety factor, the transmitter "turn-around" loss, to hit 80-dB suppression. Filter rejection in a junction combiner is also necessary for the combining function; in other words, high rejection is required to entirely reflect other channels off the filter to recombine at the junction. Higher rejection also helps improve VSWR variation for the reflected channel; typically 30 dB or more is required for combiners with good input VSWR.

Directional filter design can provide 30–40-dB isolation without considering filter rejection. Filter rejection is provided primarily to direct signal flow. For example, the throughput loss from the BB input to combined output, of a single module, is equivalent to the return loss response of the module

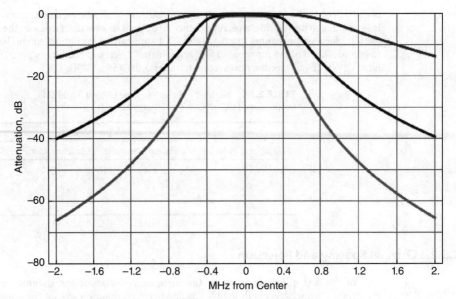

FIGURE 2.3.5 Typical 2-, 3-, and 4-pole Chebyshev filter responses, 400-kHz passband, and 1.05 VSWR.

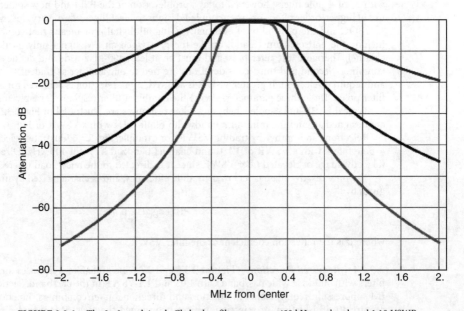

FIGURE 2.3.6 The 2-, 3-, and 4-pole Chebyshev filter response, 400 kHz passband, and 1.10 VSWR.

filters. Figure 2.3.7 depicts the return loss for the 2-, 3-, and 4-pole filter rejection plots in Fig. 2.3.5. From the plots, it is evident that a higher-order filter provides a sharper slope on the return loss resulting in lower BB port insertion loss closer to the filter passband.

Another way of looking at this is to use conservation of energy and plot BB port loss as a function of filter rejection. Figure 2.3.8 shows that 27–30-dB rejection is required for low module BB port throughput loss.

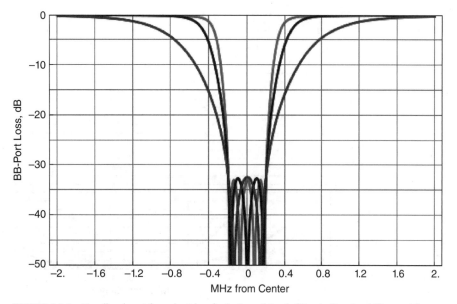

FIGURE 2.3.7 Broadband port throughput loss for 2-, 3-, and 4-pole filters in directional filter module.

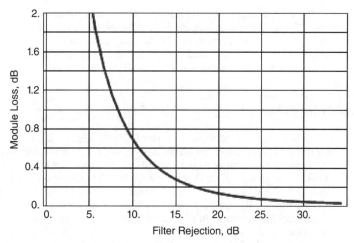

FIGURE 2.3.8 Module BB port loss as a function of filter rejection.

2.3.3 Filter Order and Group Delay

Group delay is the derivative, or rate of change, of phase with frequency and a measure of how long it takes a signal to traverse a transmission system. Filters in FM channel combiners contribute significant group delay to the RF (radio frequency) system due to narrowbandwidths the filters have. The delay itself is not necessarily a problem; variation over the band can be. Delay can vary radically across a passband and is a function of rejection, VSWR, and filter order. Group delay variation for the 2-, 3-, and 4-pole rejection plots in Fig. 2.3.5 is shown in Fig. 2.3.9. The delay plots are referenced to mid-band delay. Mid-band delay increases with filter order and decreasing bandwidth.

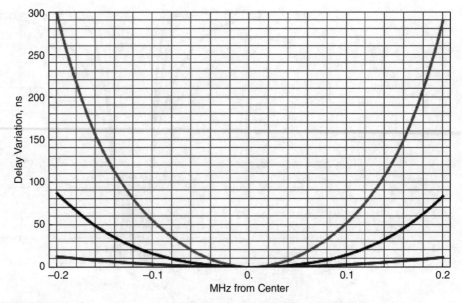

FIGURE 2.3.9 Delay variation for 2-, 3-, and 4-pole filters with 400-kHz bandwidth and a 1.05 VSWR.

Stereo separation and THD (total harmonic distortion) can be effected by excessive delay variation. Transmitter manufacturers have different correction systems to compensate for variations. IBOC sidebands are very tolerant of the delay and resulting amplitude variation due to the robust nature of the COFDM (coded orthogonal frequency division multiplexing) signal. The following section discusses techniques to correct for delay and resulting amplitude variation.

2.3.4 Filter Order and Loss

Combiner loss is due to the finite cavity Q (cavity Quality factor) of the filter and the time it takes the signal to pass through the filter. Cavity Q is a function of cavity size, construction materials, and technique. Larger cavities reduce current densities and have higher Q factors, thereby reducing loss. As discussed above, tighter channel spacing requires higher-order filters that have greater mid-band delay and greater variation of that delay over the passband and result in more passband loss. Insertion loss plots for 2-, 3-, and 4-pole filters (400-kHz bandwidth and a 1.05 VSWR) designed using large FM cavities are illustrated in Fig. 2.3.10. The same filter responses designed in smaller, lower-power cavities are shown in Fig. 2.3.11. Note higher mid-band losses and greater loss variation. The losses result in heat. The FM filter manufacturers typically use blowers to cool cavities and increase power handling. A designer may also provide the widest possible bandwidth for a given application to reduce delay, thereby reducing loss.

2.4 CROSS-COUPLED FILTERS

Proliferation of the FM spectrum has led to greater demand for combining closely spaced FM channels. Most FM combining applications are solved with 3- and 4-pole filters, and occasionally 5-pole filters for tighter spaced channels. It was shown previously that losses can become high with higher-order filters, particularly for lower-power applications. When space, loss, and power become a

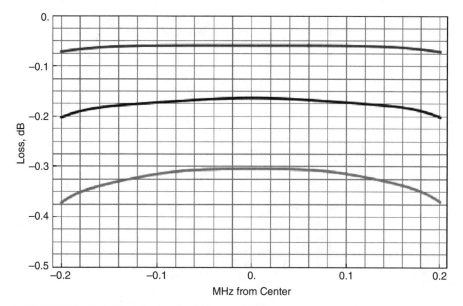

FIGURE 2.3.10 Typical insertion loss for a high-power FM filter.

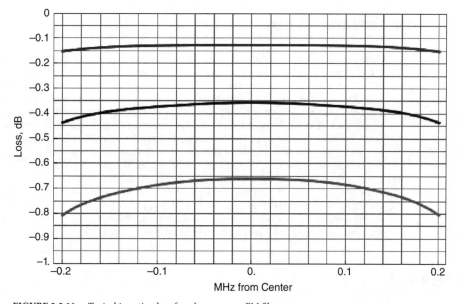

FIGURE 2.3.11 Typical insertion loss for a lower-power FM filter.

concern, cross-coupled filters provide an economical solution. Cross-coupling refers to coupling nonadjacent cavities to provide transmission zeros, or notches in the reject band. Figure 2.3.12 illustrates the placement of transmission zeros using cross-coupling. Both filter responses are 4-pole with a VSWR of 1.05 with a bandwidth of 400 kHz. Note how the cross-coupling provides required rejection for IM isolation for channels that might be spaced 800 kHz on either side in a junction combiner.

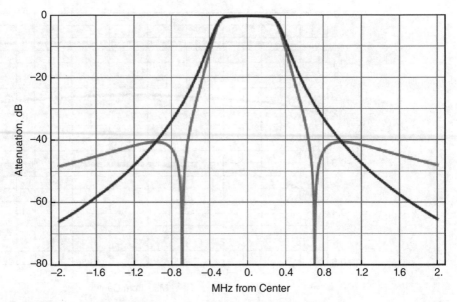

FIGURE 2.3.12   The 4-pole filter comparison, Chebyshev and X-Coupled Chebyshev.

2.5 DELAY/LOSS CORRECTION

Delay and amplitude distortions can degrade stereo separation, THD, and to some extent, put asynchronous amplitude modulation on the FM signal. Modern transmitters can correct for some delay and loss variation. Another correction technique is to utilize delay/loss equalizers at low level. Equalizers provide a mirror image of the group delay, where there is more delay at mid-band frequency and less delay at the band edges. High-level equalization is not a practical solution as it adds significant loss to the RF system. Low-level equalizers are preferred and use a single hybrid with identically tuned cavities at the 0° and 90° ports. The number of cavities off each port is the same and can vary from one to several (rarely) and depends on the amount of equalization required. The low-level signal enters the hybrid input port and is split into the 0° and 90° ports and delivered to the cavities. The cavities are shorted, with a short distance varied with frequency, i.e., the short distance is longest at the center of channel since the signal passes through the cavities to a physical short. At band edges, the signal does not travel the entire distance through the cavities and results in less delay. The exciter output feeds into the equalizer and provides a predistorted signal that is amplified and passed through the combiner where the signal is recovered or properly leveled out.

REFERENCE

[1] Mendenhall, Geoffrey N., "A Study of RF Intermodulation Between Transmitters Sharing Filterplexed or Co-located Antenna Systems," Broadcast Electronics (1983), Quincy, IL.

CHAPTER 2.4
TRANSMITTING ANTENNAS FOR FM AND TV BROADCASTING

Kerry W. Cozad
Continental Electronics, Dallas, TX

2.1 INTRODUCTION

Broadcast, by definition, means to scatter over a wide area. While this definition applies to both farming (spreading seeds) and over-the-air television (TV) and frequency modulation (FM) radio signals, the broadcaster of TV and FM signals is significantly more concerned about where the signals end up than the farmer. And for over-the-air transmission, the broadcaster has one primary tool to work with—the transmitting antenna. Since each market has its own coverage needs and physical characteristics, a single tool design will not satisfy the requirements for every broadcast station. Fortunately, based on over 75 years of design and application developments, the transmitting antennas available today are more like Swiss Army knives, being able to address several needs in one device and also having the choice of several models based on the frequency band of use.

The primary objective of any transmitting antenna is to provide a usable signal to a receiver within a defined coverage area. There are many criteria that can be used for defining a usable signal but they all will typically require that the antenna: (1) creates minimal distortion to the signal that is provided to its input connection, and (2) provides maximum signal levels in the direction(s) of desired coverage.

For minimal distortion, the first concern is the mechanical stability of the antenna structure and its relationship to the RF (radio frequency) electrical connections. For example, loose connections can create signal distortion just from the effect of a connection being good one moment and not being connected in the next moment (intermittent). Improper assembly of the antenna can also produce high reflected signals that then create distortion from interaction with the primary signal. The second concern is the antenna output response as a function of frequency. The output response can be affected by the actual design of the radiation characteristics of the antenna (its *illumination function*) or by the physical environment of its surroundings. The radiation characteristics include the frequency stability of the azimuth and elevation patterns as well as the input VSWR (voltage standing wave ratio). The physical environment includes whether the antenna is top- or side-mounted on its support structure and whether there are other structures nearby, such as other antennas, in the same vertical aperture.

The antenna characteristics that impact providing maximum signal levels are (1) azimuth pattern, (2) elevation pattern, (3) polarization, and (4) efficiency. Consolidated, these characteristics make up the *gain* of the antenna in the direction of interest.

For new installations, the concepts described above are typically key points in discussions among the broadcast station's engineering staff, the station's consulting engineer, and the various antenna manufacturers. At the time of a purchase decision, these concepts will impact the costs of the antenna, which include the costs of shipping the antenna to the transmission site, tower modifications based on the mounting and windload characteristics of the antenna, and the actual installation.

There is a third criterion that affects the primary objective of providing a usable signal and that is the reliability of the antenna once it has been installed and put into service. Mechanical stability was mentioned previously but the reliability of the antenna over the long term is also impacted by environmental effects on the antenna and the maintenance regimen employed by the station. Environmental effects can be the result of weather (lightning, high winds, ice, and snow) and/or corrosives in the air (salt and industrial or agricultural chemicals). In any case, once the initial process of deciding on a model, installing and commissioning the antenna is complete, a regular maintenance regimen that includes annual inspections and inspections immediately after any significant weather occurrence is the key to maintaining the optimum performance of the transmitting antenna and keeping costs down.

2.2 GENERAL ANTENNA CHARACTERISTICS

2.2.1 Radiator Types

The type or model of the antenna is first determined by the frequency band of operation. For FM and TV broadcasting[1] in the United States, the bands are as follows:

- TV Channels 2–3 (low VHF [very-high frequency]): 54–66 MHz
- TV Channels 4–6 (mid VHF): 66–72 MHz and 76–88 MHz
- FM Channels 201–300: 88–108 MHz
- TV Channels 7–13 (high VHF): 174–216 MHz
- TV Channels 14–51 (UHF [ultra-high frequency]): 470–698 MHz

Because the principal performance characteristics are optimized based on the antenna design as a function of wavelength (electrical size) and the physical size is inversely proportional to the frequency (the higher the frequency, the smaller the physical size), typical antenna models are grouped according to the bands described above. Some examples are as follows:

Low and Mid VHF: These use dipoles mounted against a ground plane (panel) or dipoles mounted on a pole mast (batwing or superturnstile). Modifications of these designs are used for transmitting both horizontal and vertical polarizations such as crossed dipoles for panels and transmission dual mode for pole mounted design.

FM: Most FM stations utilize both horizontal and vertical polarizations so the antenna design is typically a slant dipole, a ring style, or a multiarm short helix. For multichannel and higher-power applications, crossed dipole panel/cavity or a crossed dipole pole mounted design is used.

High VHF: This channel band includes frequencies that are at least double those in the previously discussed bands. This reduces the physical size of the radiator. While the use of dipoles for panels and pole mast designs is still widespread, the use of slot radiators and helical elements can also be found. The choice of which radiator design to use is driven by radiation pattern considerations, windloads, polarization, and costs. Maintenance considerations and long-term reliability are also important factors.

UHF: The primary radiator design used for UHF applications is the slot. The antenna supporting mast is an aluminum, brass, or steel tube that contains an aluminum or copper inner conductor. This creates a coaxial transmission line. Then, the outer tube is slotted to allow the signal that

---

[1]Actual available channels are subject to change based on DTV (digital television) transition and spectrum auction results.

is present inside the coaxial line to be radiated into free space. Each antenna manufacturer has its own methods for coupling the signal through the slot. The advantage of this design is that it results in a very low cross-sectional area, keeping the antenna windloads low. It also eliminates the need for individual feed lines and connections to each of the radiators. For transmitting both horizontal and vertical polarizations, a common technique is to add small dipoles to the antenna array that are either fed directly from the main coaxial mast or parasitically from the slots. For multichannel applications, panel antennas can be used for both horizontal only and horizontal/vertical polarization transmissions.

2.2.2 Antenna Gain

As mentioned in the introduction, the gain of the antenna directly affects the signal level at the receiver. Therefore, it is very important to understand what affects the antenna gain. There are numerous references available that describe the derivation and measurement of the gain of an antenna, and the reader is encouraged to review the "Bibliography" for more detailed information. A more general description of antenna gain follows.

The azimuthal pattern of an antenna is sometimes referred to as the *horizontal pattern*, as it is determined by the radiation characteristics of the signal in the plane parallel to the earth. With the use of both horizontal and vertical polarizations for broadcasting, it is preferred to use the terms *azimuth* and *azimuthal pattern* to avoid confusion. If the pattern represents equal signal strength in all directions of the plane, then the pattern is omnidirectional. For actual antennas, equal signal strength does not occur due to the physical size of the mast and the location of the radiators relative to a wavelength. Therefore, an antenna that is designed to radiate equally in all azimuthal directions is more generally described as being *nondirectional*. If the antenna has been designed to concentrate the signal in specific directions with greater strength than others, then it is considered to be *directional*. Using a directional antenna can be significantly more efficient in the application of transmitter power. An example would be if the transmitting location is very near a large body of water, radiating the signal over the water where there are no viewers would be a waste of that energy. An antenna design that restricts the radiation toward the water allows a greater amount to be radiated where the potential receivers are located. Even where geographical barriers may not be present, the knowledge of specific concentrations of potential receivers can result in a directional azimuthal pattern being a more efficient antenna design. Another use of directional antennas is to optimize coverage while still providing protection in a direction where potential interference can be caused and is restricted due to licensing constraints.

The shape of the azimuth pattern is determined by four primary factors. The first is the azimuth pattern of the individual radiator such as a slot or dipole panel. The second factor is the relative positions of the radiators to each other and the center of the array (the *phase center*). The third is the relative amplitude and phase of the signal feeding each radiator. The fourth factor is the effect of the mounting structure.

The elevation pattern is determined by how the signal is radiated in the plane perpendicular to the earth. The elevation pattern is controlled by adjusting the phase and amplitude of the signal feed to each radiating bay in the vertical plane of the antenna (the *illumination*) and the vertical spacing between bays. For most antennas, the elevation pattern is the same in all azimuth directions. Exceptions would be antennas that have an unequal number of radiators in each vertical face or are intentionally fed with nonsymmetrical illuminations. These techniques may be used when special protection is required in specific directions or the coverage distribution near the transmitting site is more concentrated in some directions than others. To provide maximum efficiency, the antenna should be designed so that the signal is directed toward the receiver locations. Since the antenna is typically mounted as high above the coverage area as possible to permit the greatest line-of-site coverage, as much of the radiated signal as possible should be concentrated below the horizontal plane (downward). This is normally accomplished by phasing the antenna elements in the vertical plane such that the signal beam is electrically tilted in all azimuthal directions. The amount of tilt desired is determined by the height of the antenna above the coverage area, the coverage area's maximum radius from the transmitter site, and the width of the elevation main beam; see Fig. 2.4.1.

2.72 RF TRANSMISSION

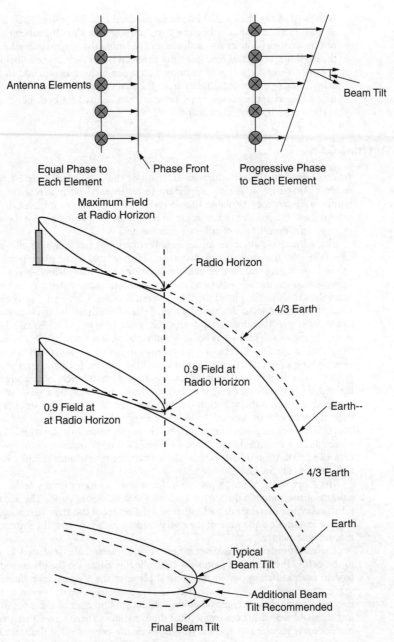

FIGURE 2.4.1 The principles of electrical beam tilt.

Figure 2.4.2 illustrates optimum beam tilt as a function of maximum range. Figure 2.4.3 charts depression angle as a function of distance for various antenna heights.

The shapes of the azimuth and elevation patterns are based on how the input signal is radiated from the antenna. Notwithstanding any losses (I^2R) within the antenna array or polarization splits, only the signal that is input to the antenna can then be radiated. A nondirectional antenna provides

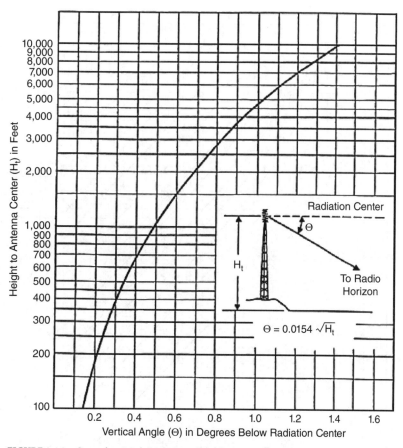

FIGURE 2.4.2 Curve showing the optimum tilt for the maximum range.

equal signal strength in all azimuth directions. If the radiated signal is concentrated in specific directions, then the actual signal strength must be greater in those directions and lower in others relative to a nondirectional antenna. This increase in signal strength is a simplified description of the antenna gain. By concentrating the signal in a specific direction, less transmitter power is required to provide the same signal level that would be provided by an antenna that was not directionalized. As a result, the choice of transmitting antenna gain can have an impact on the amount of transmitter power required (transmitter size).

Factory testing to confirm that the antenna meets the pattern and gain specifications has evolved significantly since the late 1970s. Factory testing at that time typically involved the use of a far field test range where the patterns could be measured directly. The gain could also be measured on the basis of a comparison method using a standard reference gain antenna. Since the gain is a direct result of the azimuth and elevation patterns, it can be determined by calculating the amount of directionalizing that has occurred to those patterns versus a reference standard. For broadcasting, the standard used is a half-wave dipole. So the measured azimuth pattern is compared to an omnidirectional pattern and the elevation pattern is compared to a pattern that approximates a cosine function. By calculating the ratios of the areas, the directivity of the patterns can be determined. Then by multiplying the directivities, the peak directivity for the antenna is determined. To obtain the actual antenna gain, any signal losses that occur within the antenna must be subtracted from the gain. For antennas that radiate both horizontal and vertical polarizations, the amount of polarization split must also be

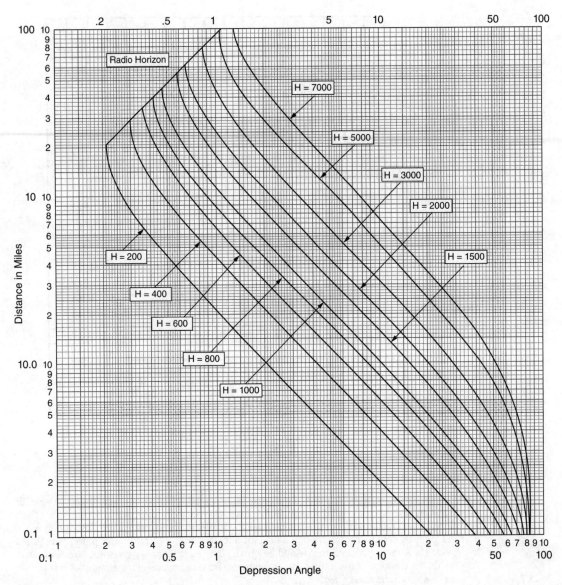

FIGURE 2.4.3 Depression angle as a function of distance for various antenna heights.

factored into the gain calculation. Once the peak gain for the antenna is determined, the gain at any azimuth and elevation angle can be calculated by multiplying the relative pattern level expressed as a power ratio by the peak gain.

Over the last several years, computer programs have been developed that automatically take into account the antenna patterns, gains, and input power levels to calculate signal levels at receive locations. But even with these calculations now being automated, it is important to understand how the parameters are developed so that the results can be checked for reasonableness. It is also helpful if a coverage or reception issue arises to be able to do some simple calculations to determine if some change in the antenna may be a cause for the issue.

2.2.3 Factory Testing

The purchase of a transmitting antenna is a significant investment by the station in terms of both cash outlay and in meeting the needs of the station's customers, its listeners and viewers. It is important to understand what testing can and should be performed at the manufacturer's facility prior to shipment of the antenna to the transmitting site.

Factory testing usually involves two methods of confirming the performance of the antenna: direct measurements of the fully assembled antenna array and calculations based on individual component testing. The process for direct measurements is straightforward. The fully assembled antenna is placed on a calibrated range that includes the ability to rotate the antenna on at least one axis at one end of the range and a calibrated source or receive antenna at the other end of the range. To meet the necessary criteria for accurate measurements of the antenna patterns, these ranges can be anywhere from several hundred to several thousand feet in length, and must be clear of any buildings, hills, or other obstructions that can affect the propagation of the testing signals. Use of this method for broadcast antennas has become very rare due to the cost associated with maintenance of the range as well as time required to actually take the measurements. The second method, which uses direct measurements of the individual components and then uses well established methods of calculating the array characteristics, is the primary method being used by manufacturers today. The exception to this method is the measurement of the input VSWR or impedance of the array, which is normally performed on a fully assembled antenna that has been placed an appropriate height above the ground to minimize ground reflection effects. While the component measurement method could apply, mechanical tolerances in the fabrication and assembly of the antenna can have a significant impact on the RF coupling along the array, resulting in variations in the measured input impedance that cannot be easily accounted for in the calculations. It is much easier to assemble the antenna on the ground and make adjustments to confirm the VSWR performance meets the specifications than have to make these adjustments after the antenna has been installed on top of a tower or building.

To determine the free space azimuth pattern, initial configuration designs are made using antenna array design programs or using a catalog of previously measured antenna patterns. For pole-mounted antennas for VHF TV, there is a long history of direct measurements so the confidence level is extremely high of what to expect for the azimuth pattern without having to take additional direct measurements. For UHF TV, slot antenna azimuth patterns are usually confirmed using a 1–4 bay full-scale model on an outdoor range or anechoic chamber. Most panel style antennas for both TV and FM rely on amplitude and phase measurements at the input to each panel and then this is combined with a measured or calculated element pattern for the panel to calculate the resulting azimuth pattern for the array. Most FM antennas are side mounted on the supporting tower, and because of the interaction between the antenna and the tower it is important to document what the resulting patterns are. These measurements are typically made using accurate one-fourth scale models of both the antenna radiator and the tower (one-fourth physical size and four times the frequency of operation).

Most antenna manufacturers confirm the elevation pattern performance by taking measurements of the amplitude and phase that either is feeding the radiating element (e.g., a panel radiator) or is in the very near field of the radiated signal (e.g., a slot). This information provides the illumination characteristics of the antenna and a direct comparison can be made between the design based on the purchase specifications and the resulting illumination of the as-built array. This method has been proven to provide excellent correlation with direct measurements of full arrays and it is much easier to control environmental conditions that can affect the direct measurements on a range.

Figure 2.4.4 shows an antenna undergoing single-layer impedance testing. Figure 2.4.5 shows an antenna undergoing single-layer pattern tests.

FIGURE 2.4.4 Factory single-layer impedance testing. (*Photo courtesy of Electronics Research, Inc. (ERI).*)

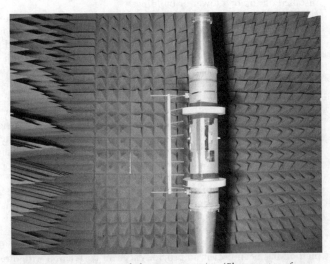

FIGURE 2.4.5 Factory single-layer pattern testing. (*Photo courtesy of Electronics Research, Inc. (ERI).*)

2.3 INSTALLATION

The quality of the installation will have a direct impact on the performance and long-term reliability of the antenna. The station engineer should choose installation crews with experience and a known track record of handling FM and TV transmitting antennas. Installation companies that are primarily focused on working with lighter weight antenna systems, such as cellular, mobile radio, and microwave, will typically not have an adequate understanding of the assembly of high-power broadcast antennas. This inexperience can lead to mistakes that can result in additional work during the antenna commissioning process or catastrophic failures later during high-power operation. If the station does choose to use a less experienced crew, it is recommended that either an independent inspector or a manufacturer's representative be hired to be on site during the installation in order to address questions and perform inspections at key milestones during the process.

Figure 2.4.6 shows an example of manufacturer's installation details for an FM antenna system. Because guy wires can physically interfere with the antenna or cause distortions in radiation patterns if they pass through the aperture of the antenna, guy wire placement needs to be part of the installation plan, as shown in Fig. 2.4.7.

Prior to the actual installation, several reviews should be performed in order to avoid problems on site:

(1) Once the model of antenna has been determined, the mechanical parameters and windloads should be reviewed by the station's tower structural engineer to confirm what, if any, modifications to the tower will be needed. These modifications include the tower structure itself to meet structural codes as well as the specific areas where the antenna will be mounted. For top-mounted antennas, new tower top assemblies and plates may be needed. For side-mounted antennas, it should be confirmed who is responsible for the mounting brackets and that there will be no interference with existing components, such as conduits and lighting, presently in the areas on the tower where the brackets will be attached.

(2) If the tower is painted, confirm that the color(s) of the antenna (radomes, masts, and brackets) is compliant with FAA (Federal Aviation Administration) and FCC (Federal Communications Commission) codes (see Fig. 2.4.8 as an example of matching radome color to the tower).

(3) For top mount antennas, confirm who is responsible for the top beacon, what color (red or white strobe), and who is responsible for supplying the wiring harness and/or cables to connect the beacon to the wiring on the tower.

(4) For top-mounted antennas, confirm that the antenna mounting flange (bolt hole size, number of bolt holes, bolt circle radius, and orientation of the bolt holes relative to the desired antenna pattern) is matched to the tower top plate (see Fig. 2.4.9). Also confirm who is responsible for supplying the mounting hardware. Additionally, the thickness of the tower top plate and the antenna mounting plate should be compared to the length of the proposed mounting bolts to be sure there will be an adequate number of threads available for attaching the washer and nut. While it may seem that this would have a high level of visibility, it is not unusual to arrive on site and see one of these issues occur, which can have a significant impact on the schedule and cost.

(5) For side-mounted antennas, in addition to item 1 above, interference with guy wires should be reviewed. This can occur either where the guy wire just above the antenna may actually touch the top of the antenna or may rub against it in heavy winds.

(6) All antennas should be reviewed for pressurization requirements. Most high-power FM and TV antennas require their feed systems to be pressurized even if the radiators are not.

(7) For antennas that are provided with electrical deicing, confirm the electrical requirements (voltage and amperage) and the appropriate wiring on the tower.

(8) Confirm that adequate space on the site is available for off-loading and assembly of the antenna.

(9) Confirm that road access to the site will accommodate the length of the shipping trailer.

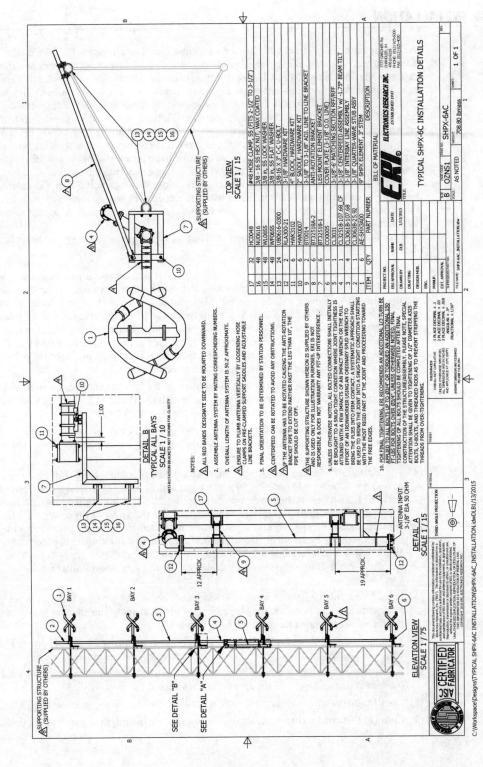

FIGURE 2.4.6 Typical installation details for an FM antenna system. (*Photo courtesy of Electronics Research, Inc. (ERI).*)

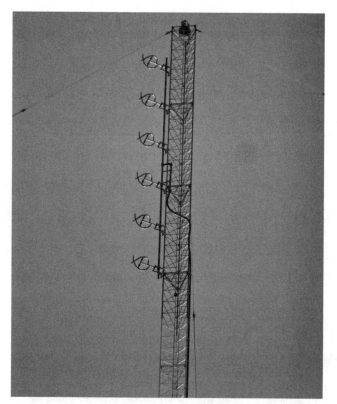

FIGURE 2.4.7 Antenna located to minimize any interference from the guy wires. (*Photo courtesy of Electronics Research, Inc. (ERI).*)

FIGURE 2.4.8 Antenna installation using top-mounted and side-mounted antennas. (*Photo courtesy of Electronics Research, Inc. (ERI).*)

Once the antenna has arrived on the site, the manufacturer's instructions for handling and installation should be followed. Depending on the model of the antenna, some of the key concerns are as follows:

(1) Check the shipment for any signs of damage. This includes wood and cardboard boxes, antenna radomes, antenna radiating elements, and any brackets that have been supplied. In most instances where damage has occurred, it is due to shifting of the load or improper placement of tie downs (straps or chains). Normally, for larger antennas that are shipped on dedicated trucks, the manufacturer will assist in the loading and will give instructions to the driver for proper care of the components and damage occurs only when there is an accident or a tie down fails. In any case, immediately notify the manufacturer if damage is discovered. This will reduce the time it takes to make arrangements for repairing the damaged parts and can also assist the station in making any insurance claims that may be necessary.

(2) If any components are shipped pressurized, check the pressure levels. If there is no pressure, notify the manufacturer. If you need to add pressure, be sure to use nitrogen or dry air.

(3) If the antenna must be stored prior to installation, prepare an area that is level and large enough to accommodate the antenna mast with adequate supports. No parts of the antenna should be in contact with the ground, and it should be kept clean and dry. Remember, the antenna is primarily designed to withstand forces and the environment while in a vertical position. So, while stored horizontally, the weight must be properly supported to avoid excessive stresses on the antenna

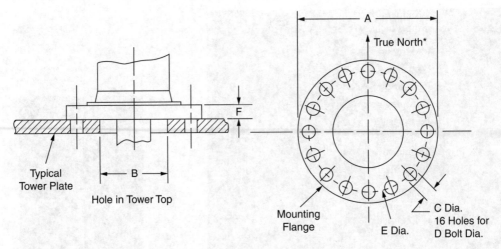

FIGURE 2.4.9 Example antenna base to tower top plate attachment.

components and mast, and should not be sitting in dirt, water, or mud to avoid contamination and deterioration of the antenna's electrical components (radomes, dielectric feed points, etc.).

(4) Any time the antenna is lifted from a horizontal position, extreme care must be taken to avoid overstressing the antenna mast and components. Carefully review the manufacturer's guidelines and use only the lifting devices and locations that are part of the antenna. If there are any questions or concerns over the instructions, review with the manufacturer before proceeding.

(5) When lifting the antenna for installation on the tower, do not add any additional weight to the antenna. The lifting devices are designed based on the actual weight of the antenna so any additional weight will impact the margin for safely lifting the antenna.

2.4 MAINTENANCE

Regardless of the location where the antenna is installed, it must withstand extremely harsh environmental conditions. Wind, snow, ice, hot and cold temperatures, lightning, sunlight, chemicals, sand, and other abrasives in the air, birds, insects, and tower climbers looking for a foothold; all these conditions will eventually produce detrimental effects to the performance of the antenna. Due to these conditions, it is important to perform annual inspections, at the very minimum, in order to identify potential problems early. Damage or material deterioration of any antenna element will only get worse over time and in more extreme environmental conditions, the deterioration will accelerate very quickly. The end result is typically a catastrophic failure that takes the antenna off the air and costs tens of thousands if not hundreds of thousands of dollars to repair. A simple but comprehensive annual review that includes a physical inspection of the antenna and a VSWR measurement from the transmitter facility may only cost from $2000 to $5000. Certainly, significant savings over a catastrophic repair could have been avoided.

Because the antenna manufacturer realizes that these conditions will be present, they take various levels of care to design countermeasures to the effects that can eventually occur. For maintenance purposes, it is important for the station engineer to understand what countermeasures have been used so that these elements can be checked regularly to confirm their continued effectiveness. The station engineer should also compare the antenna countermeasures against their knowledge of the environment to be sure the appropriate design considerations have been addressed. Such countermeasures include the following:

- Radomes or slot covers made of UV-resistant materials to protect the radiators and feed points
- Galvanized or stainless steel to prevent corrosion
- Paint or other chemical coatings
- Lightning protection
- Pressurization of feed systems and full radomes
- Gaskets or caulking (RTV [room temperature vulcanizing]) used to prevent water ingress at overlapping covers and electrical connections
- Mechanical braces/brackets for feedlines

This list is not necessarily exhaustive but does provide key elements for review. Additional items and a sample Inspection Check List is included in Table 2.4.1. If damage to any of these items is found, immediate action should be taken to avoid further damage. Having spare hardware, paint, slot covers, gaskets, etc. on site can save time and costs by having the repairs made during the inspection.

Figure 2.4.10 shows some actual installation problems discovered in the field.

TABLE 2.4.1 Field Engineering Check List

Date		
Call Letters		
Location		
Station Contact Person		
Field Engineer		
Equipment Inventory		
RF Equipment		
Transmitter Manufacturer and Model:		
TPO:		
RF System Manufacturer:		
Combiner (Y/N)		
Output Patch Panel (Y/N)	Output Switch (Y/N)	
Main Transmission Line Manufacturer:	Size:	Impedance:
Main Antenna Manufacturer:	Model:	
Standby Transmission Line Manufacturer:	Size:	Impedance:
Standby Antenna Manufacturer:	Model:	
Pressurization Equipment		
Dehydrator (Y/N)	Manufacturer:	Model:
Nitrogen (Y/N)		
Is a manifold used to distribute pressure to several antenna/line systems? (Y/N)		
Describe:		
Inspection Checklist		

(Continued)

TABLE 2.4.1 Field Engineering Check List (*Continued*)

Pressurization

Pressurization system operational (Y/N)

Pressurization to main transmission line: _____ psig

Confirm pressure is reaching gas inlet at gas barrier (Y/N)

Confirm positive pressure at top of vertical run by opening valve or loosen flange (Y/N)

Is system pressure tight? (Y/N)   If not, estimate leak rate: _____ psig/hour:

Confirm main line and antenna was purged (Y/N) or evacuated (Y/N) prior to leaving site

VSWR Monitoring

Location of couplers for sampling reflected power:

 Between transmitter cabinet and filter (Y/N)

 Between filter and transmission line (Y/N)

Note reflected power/VSWR transmitter alarm level:

Note differences between measured reflected levels from check out and readings from transmitter:

Rigid Transmission Line Installation (coax and waveguide)

Visually inspect horizontal run and note any concerns regarding line and elbow support:

Is there adequate clearance for the line entering the tower to allow for thermal changes? (Y/N)

From base of tower, do vertical spring hangers appear to have been installed correctly? (Y/N)

Confirm fixed rigid hanger has been properly installed at top of vertical run (Y/N)

 Note: no elbows should be installed below the fixed rigid hanger.

Request digital photos of fixed hanger installation and elbow complex to confirm proper installation and support

Photos included in field report (Y/N)

Flexible Transmission Line Installation

Horizontal run properly supported (Y/N)

Drip loop at base of tower properly installed (Y/N)

Cable grounding properly installed at base of tower (Y/N)

Spacing on vertical run hangers appropriate for area (closer spacer for high winds) (Y/N)

Antenna Installation

Top mounted antenna: has plumbness been checked prior to electrical checkout? (Y/N)

Side mounted antenna: is antenna mounted in proper orientation based on latest version of installation drawings? (Y/N)

Modular Antenna Model: confirm modules are mounted in proper order and right side up (Y/N)

Modular Antenna Model: confirm feed cables are feeding appropriate module (Y/N)

Modular Antenna Model: confirm feed cables and power dividers are properly supported (Y/N)

Electrical Deicing checked (Y/N)

Notes and Additional Comments

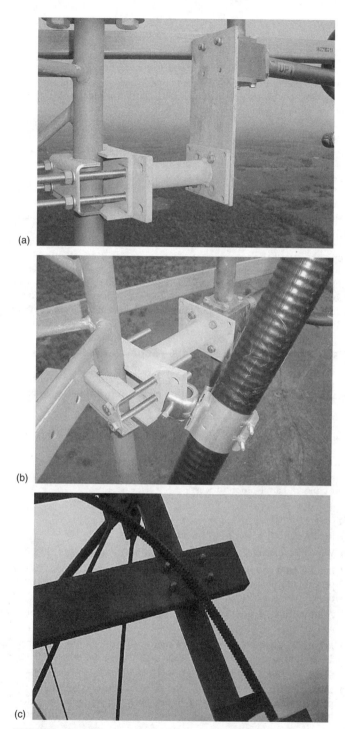

FIGURE 2.4.10 Antenna system and feedline installation errors: (a) missing hardware, (b) transmission line clamp attached to antenna element support bracket adding additional loads, (c) feed cable resting against bolt and electrical tape being used as a hanger.

2.4.1 Radomes and Slot Covers

For most UHF antennas and FM/VHF antennas located in areas subject to snow and ice, the radiators and feed points are protected by fiberglass or plastic (polycarbonate) radomes or slot covers. These components should be inspected for damage such as cracks, bullet holes, loose hardware, and loose fitting seams. For antennas using radomes that are pressurized, the system pressure response over time should be reviewed daily to identify possible leaks. If water is allowed to enter the protected area, arcing due to the failure of a contaminated insulator or pooling of water in a low spot will result in significant damage. Fortunately, quick action when a pressurization leak occurs can avoid these problems.

Antennas that use slot covers or unpressurized radomes can also suffer damage if water is allowed to run freely into the antenna radiator or feed area. These need to be checked for proper fit at the seams and joints where water flowing down the antenna could seep through and into the radiator area. Many of the same issues as described above may occur if large amounts of water are present.

Repairing damage to radomes and slot covers should be reviewed with the manufacturer. Radomes and slot covers are critical components of the radiating system. Their dielectric characteristics directly impact the antenna performance.

If the color of the radome or slot cover fades, it may be possible to paint them to bring the devices back into compliance with FCC and FAA requirements. This should be reviewed with the manufacturer for specific guidelines. For reference, Fig. 2.4.11 lists general guidelines that can be used for discussions regarding the painting of fiberglass-based radomes and slot covers.

2.4.2 Antennas Using Corporate Feed Systems—Superturnstile, Panels, and FM

For corporate feed systems, special attention must be paid to the support and routing of the individual feedlines and the brackets attaching the antenna elements to the tower or support mast. Improper

---

Caution. The information discussed below is a suggested guideline for applying paint to fiberglass radomes for broadcast antennas. It is important that the paints and chemicals used do not contain metallic components that will interfere with the RF performance of the antenna. Contact the antenna manufacturer and local representative of the paint provider to confirm paint types and application process.

Application Notes

(1) Remove all RF power from antenna.
(2) Wipe down radomes with denatured alcohol using a cotton cloth.
(3) Lightly sand the radome surface using 60–80 grit sandpaper to remove oxidation and to prepare the surface for painting. *Caution*: Do not sand aggressively; do not break through the radome gel coating; do not expose raw fiberglass fibers.
(4) After sanding, rewipe the radomes again with denatured alcohol to remove sanding dust and dirt.
(5) *Primer*: Apply one coat only of primer to the radome via spray, brush, or roller and allow 24–36 h to dry. *Note*: When applied correctly this material will be transparent. If the application starts to become opaque you are applying too much primer.
(6) Mixing instructions for paint, catalyst, and reducer: thoroughly mix the catalyst with the paint first and then add the reducer. Before applying paint, refer to the proper mixing instructions listed in the product bulletin and in this work instruction.
(7) Apply top coat paint using spray, brush, or roller. If required, a second coat of finish paint may be applied after the first coat has sufficiently dried.

FIGURE 2.4.11 Example application note for painting a radome.

assembly of cable supports results in loose cables that are then damaged due to vibration or abrasion against other objects. Loose hardware results in damage to the radiating elements due to vibration. When doing a physical inspection of these antenna types, the inspector should have spare hardware and mounting components for the feed cables and the antenna mounting brackets as these replacement repairs should be performed immediately. For additional information on this subject, see the paper by Kerkhoff included in the Bibliography section.

2.4.3 Antenna VSWR and Pressurization Measurements

Regular measurements of the antenna system VSWR and pressurization levels are the best preventive care actions. Both of these measurements can provide early detection of problems that will only get worse over time.

Any consistent loss of pressure over time (taking into account temperature variations) is usually an indication of a loose flange connection, a failing gasket or o-ring, or a crack/hole in the radome or a feedline. In many cases, loss of pressure is the first indication of a developing problem that can be caused by loose hardware, lightning damage, or a pinched o-ring that is beginning to burn from RF heating. If the problem is discovered in time, a quick inspection and repair can avoid significant problems later.

VSWR measurements should include the individual channel as well as wider frequency response sweeps and time-domain responses. Changes that may occur to the RF feedlines, insulators, and radiators can be detected by careful and regular review of these VSWR measurements. Such changes include deterioration within the antenna as well as physical damage to external components, such as the dipoles or pattern shapers.

2.5 SUMMARY

There are many types and configurations of TV and FM transmitting antennas for the broadcast station to choose. Many of the critical characteristics have been described to assist the broadcast engineer during discussions with consultants and antenna manufacturers. For long-term reliability and optimum performance, it is critical that the installation and maintenance of the antenna be carefully planned. With the proper regimen in place for inspections, the station can expect the antenna to provide consistent performance for many years.

BIBLIOGRAPHY

Benson, Blair K. *Television Engineering Handbook*, McGraw-Hill, New York, NY, 1986.

Cozad, Kerry W. "A Technical Discussion of Andrew Broadcast Antenna Measurement Techniques," SP42-03, Andrew Corporation, Orland Park, IL, 1989.

Cozad, Kerry W. "Mounting Your Television Broadcast Antenna for Optimum Reception and Costs," *NAB Engineering Conference Proceedings*, National Association of Broadcasters, Washington, DC, 1992.

Johnson, R. C., and H. Jasik. *Antenna Engineering Handbook*, 2nd ed., Mc-Graw-Hill, New York, NY, 1984.

Kerkhoff, William, A. "Care and Feeding of FM Multichannel Antennas," *57th Annual NAB Broadcast Engineering Conference Proceedings*, National Association of Broadcasters, Washington, DC, 2003.

Kraus, J. D. *Antennas*, McGraw-Hill, New York, NY, 1950.

Walker, Prose A. *National Association of Broadcasters Engineering Handbook*, 5th ed., McGraw-Hill, New York, NY, 1960.

Whitaker, Jerry C. *Standard Handbook of Broadcast Engineering*, McGraw-Hill, New York, NY, 2005.

Williams, Edmund A. *National Association of Broadcasters Engineering Handbook*, 10th ed., Elsevier Focal Press, New York, NY, 2007.

CHAPTER 2.5
PRACTICAL ASPECTS OF MAINTAINING MEDIUM-WAVE ANTENNA SYSTEMS IN AM TRANSMISSION

Phil Alexander
CSRE, AMD, Broadcast Engineering Service and Technology, Indianapolis, IN

2.1 INTRODUCTION

This chapter is a guide to understanding AM antenna systems, written for those who operate and maintain these systems. It is not intended as a comprehensive review of all AM antenna system types nor does it explore many specialized systems. There are several excellent texts suitable for studying medium-wave antenna design and its underlying theories, including books dating back to those published during the resurgence of radio after World War II. However, expanding upon theoretical and design information is not our purpose, practical application is our goal.

In this chapter, we focus on the practical aspects of typical systems and the knowledge needed for repairing failed systems *that have worked in the past*. This brief review excludes some information found in reference works that describe theories and design principles for new systems, but we limit our focus to a summary we hope will help you understand the basics needed for restoring a system to the air without learning all the complex details needed for designing a new medium-wave antenna system "from scratch."

What is an AM antenna? A practical definition must include the complete system from the transmitter output connection to the air. Thus, we refer to *antenna systems* rather than the less inclusive and sometimes confusing term "antenna." It may be useful to find where you have the term "AM antenna" stored in your memory and replace it with a term that has more meaning!

It is important to remember that "AM" is a simple abbreviation for *amplitude modulation* and that except for peak voltages within the transmission system, the method of modulation has little bearing on the technical aspects of transmission within the frequency spectrum we are discussing. Thus, what we know as an "AM antenna system" can be a misleading title, especially at a time when other methods of modulation are under study as a replacement for "AM." Using a more fundamental difference between so-called "AM antenna systems" and all other types is masked by this commonly used but very misleading name would be helpful. In fact, before World War II, the name of what we now call the "AM band" was officially "the Standard Broadcast band." The rise of VHF/FM made that term obsolete at a time when promotion of FM and FM stereo had led to market adoption of "AM" and "FM" as band identifiers for the general public.

Calling the "AM band" the "medium-wave," "medium frequency," or simply "MF" or "MW" (medium wave) band would have been more consistent with other divisions of the spectrum into "bands." Thinking of the so-called "AM band" as the *medium-wave band* is a constant reminder that the spectrum we use in this band is much lower than all other bands used for domestic commercial broadcasting. Appreciation that the "AM band" is the middle part of the medium-wave band with a typical wavelength of 300 meters versus a typical wavelength of 3 meters in the VHF/FM broadcast band helps highlight the very dramatic differences you will find between it and your experience with the higher broadcast frequencies employed in FM and television. This is the reason physical components used for medium-wave transmission are very large in comparison with all others, which helps us understand why a five foot piece of three inch copper strap may work as a perfectly good grounding jumper in MW service.

While we are correcting ideas and terminology about "antennas," I suggest we apply the name "radiators" to those vertical steel towers or masts that many call "AM antennas." The FCC (Federal Communications Commission) generally uses the "radiator" terminology in most of its published rules.

2.2 ELEMENTS OF THE ANTENNA SYSTEM

Radiators are an integral part of a medium-wave antenna system, and regardless of the terminology used, transfer power through the system, its networks, transmission lines, and other components to "on air" with minimal loss to assure the best coverage as authorized by the station license.

Several areas are common to all operations in medium-wave systems while others are customized to meet station needs and authorizations. We will begin with those that are generally applicable to most stations and then work up to the more complex cases that utilize directional antennas employing RF signal phasing and power division within the antenna array to shape the actual radiated pattern.

One area that is extremely critical is grounding. A radiator alone is not an antenna system without an extensive and often complex metallic ground. Unlike higher frequency systems, grounding is a stand-alone component of medium-wave systems, vitally important to efficient system operation but essentially invisible. Grounding also is prey to aging and corrosion, damage, and vandalism, as well as both intentional and unintentional installation errors, thus requiring high standards in both ground system installation and maintenance over time.

2.2.1 Ground System Design

When Dr George H. Brown (later of RCA) investigated antennas and grounding in the early 1930s, he determined that 120 equally spaced copper radials at least 90° in length would serve as a satisfactory ground for a vertical MW radiator. This remains the FCC standard for antenna grounding today. Since the late 1940s inclusion of additional grounding near the radiator using either expanded metal mesh approximately 50 feet^2 or 120 interspersed radials typically 25 to 75 feet in length between the longer radials has become fairly common, especially for radiators more than 90 electrical degrees tall. A typical example system is shown in Fig. 2.5.1.

In a typical MW grounding system, uninsulated ground radials are buried 6 to 12 inches below the soil surface. Special plows have been developed for ground radial burial and are used by those specializing in ground system construction. Burial depth and its effect on grounding system efficiency are a topic of some debate. There is evidence from systems installed in the Great Plains in the 1930s at depths of 4 to 6 feet to permit agricultural operations over the radials that were reported as working well 30 to 40 years later. However, these were reportedly lower frequency regional stations and reports were sketchy. What is known is that deep burial potentially reduces the overall antenna efficiency and that increasing frequency compounds the reduction. Although some computer modeling programs can model ground systems as part of a medium-wave antenna, there is little validation data. In this regard, those who work with computer modeling including the author are more skeptical about the ground functions than other simulations. The best advice is burying ground systems similar to the

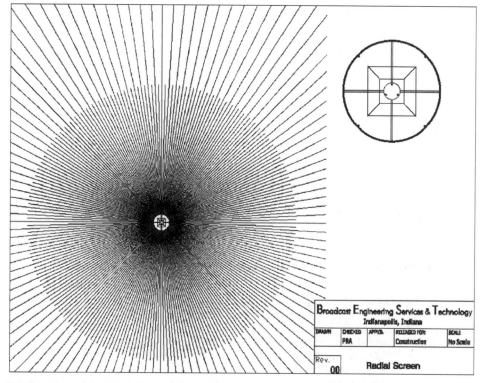

FIGURE 2.5.1 Typical MW radial ground screen.

typical one shown at a shallow depth unless there are unusual considerations, e.g., very high potential for theft or unusual soil conditions.

Ground system bonding is critical and should be done by welding the copper directly or soldering it with a hard alloy such as bronze brazing rod, one of the silver solder alloys or other hard conductive substitute. Never use soft solder intended for copper pipes or electrical/electronic connections.

While radials of #10 uninsulated, soft drawn copper wires are generally specified, this is primarily a mechanical consideration for the wire to permit plowing. Smaller copper diameters are electrically satisfactory. At times when ground system theft has become a significant problem, other materials (copper plated steel, aluminum, etc.) have been tried with no general consensus for a suitable replacement.

Elevated ground systems are another possibility although they are unique engineering designs that are best done by one of the few consulting engineers experienced in the design and installation of these specialized ground systems. Elevated ground systems require application and approval by the FCC prior to installation.

2.2.2 Feeder Lines and Coupling Networks

Another area worthy of special consideration is determining the electrical length of coaxial cables used in feeding medium-wave radiators. Often, even very long cables may be less than a full wavelength, and require special manufacturing processes and precise trimming to assure identical electrical performance in "identical" cables because operating conditions may magnify small performance variations.

Nearly all stations use RF coupling networks for matching transmitters to transmission lines when needed, and for matching those lines to vertical towers or other radiating devices. Understanding RF coupling networks assists us in understanding how they are used in making the "magic" that lets us control exactly the power and phase of RF connected to various radiators of directional arrays. That understanding is also critical in improvising emergency repairs for these "mysterious" arrays when they are struck by wandering off-road vehicles, bolts from above, or any of the other accidents that seem to befall them at the worst possible times.

The key fact to keep in mind is that both nondirectional and directional systems use very similar networks and parts. While the exact bill of materials may be different for each station, all use the same kinds of parts for the same purposes.

Customized networks built for each station are what set medium-wave broadcasting apart. Once you understand how these networks are assembled, what they do and how to repair them, you understand what makes medium-wave facilities different from all other commercial radio installations. In that same way you may also understand how any part of a system that uses RF coupling networks is very similar to another even when one is an antenna coupling unit for a one kilowatt, nondirectional, daytime station while the other is the phasor of a 50 kW six tower directional station.

2.3 TECHNICAL PRINCIPLES

Antenna system discussions are where you usually find the section of the book that looks like it belongs in a math text with a steep learning curve. Since this chapter is not a theoretical design text, we will skip many of the math explanations, but caution you that the formulas we do show are essential. The equations must be solved in the order shown and should be followed carefully.

You may notice that we have abandoned conventional radical symbols where the root sign is widely used. This method has been replaced by the more modern presentation using fractional exponents. For example, showing the square root of -1 using a root sign in the traditional way is difficult for some computers, thus we show it $-1^{0.5}$ which programs easily into a spreadsheet and gives exactly the same answer.

What follows is an exception to our minimal math approach, and that is one that may take you back to a memory your earliest days in electronics. It is Kirchhoff's Law which says: "*The sum of the voltage drops around a circuit will be equal to the voltage drop for the entire circuit. The sum of all currents entering a circuit node will equal the sum of all currents leaving that node.*"

In other words, if the entire circuit is analyzed, current nodes and voltages will sum to zero. *Thus, if voltage is present, current must flow, and Ohm's Law will determine that flow.* In antenna system work, remembering Ohm's Law is very useful, and understanding the implications of Kirchhoff's Law leads to better understanding of the matrix that is the essence of a group of complex networks assembled together for specific related tasks, for example, operating as a directional phasor.

If we have a simple two tower directional array where radiator A receives twice as much power as radiator B, we can accomplish this by building two networks. The output resistance of both would be 50 Ω while the input resistance of B would be twice as large as A. If B is set to 150 Ω and A is set to 75 Ω and the inputs are paralleled, the combined common point will be 50 Ω and the output of A will be twice that of B. Ohm's Law assures that outcome. There are many other power division scenarios, and some are more economical where a large number of radiators must be driven, but an Ohm's law divider is a clean, simple, easily controllable way of dividing power for two to four or five radiators. Cost and space for components are the only real drawbacks of larger Ohm's law power division in more complex systems.

Never forget this fact: While some currents are real and others are mathematically imaginary, remember that shorting or overloading a branch carrying an imaginary current will cause the same damage and make as much smoke as one with real current. For example, maladjustment of some series LC (inductor/capacitor) circuits may cause a current above component design limits just as high voltage may develop in some equally abused parallel circuits.

2.3.1 System Components

The typical components of a network are coils and capacitors that we will call inductive and capacitive reactance. Both of these devices have two descriptions for their values. Their electronic/physical values are typically micro-Henries (μH) for inductors and micro-Farads (μF) for capacitors. Note: These units when used with mega-Hertz (MHz) simplify calculations by cancellation of multipliers.

Reactance values of these devices are described in Ohm's reactance *at the operating frequency* where inductance is a positive value (+j) and capacitance is a negative (−j). Combined with the resistance value in Ohms (Ω), the result is a complex number representing the impedance of any point in an antenna system. In this representation the Ω (resistive Ohms) portion is conventionally stated first followed by the reactive Ohms stated as + j (inductive) or −j (capacitive).*

As an example, a complex impedance of 70 Ω + j15 would mean that at the measurement point the resistance measures 70 Ω with an inductive reactance of 15 Ω. Note that some may omit the Ω and write the impedance as 70 +j15. Rarely, even the j symbol is omitted and a notation on a schematic may say 70 + 15.

Other methods exist for representing complex impedance values such as vector and magnitude, etc. However, vector or polar notation, the notation seen in the Smith chart output of a network analyzer, called $R\theta$ (R theta) notation is less useful for our purposes than rectilinear or Cartesian (or simply j) notation. Thus we will omit discussion of those other methods.

Typical "cold" RF bridges of the passive or GR type and other measuring instruments such as the "hot" or active Operating Impedance Bridge show reactance results directly in the rectilinear form as (±) j. These bridge readings may require correction for the actual measuring frequency while the instrument shows the value at a standard frequency, typically 1.0 MHz. If this is the case, the bridge operating manual will clearly indicate this requirement and the instrument may have a placard on the operating panel or reactance dial face. Do not overlook this important step.

Conversion of rectilinear reactance values from j to physical μH or μF is unnecessary until time for component selection near the end of the process.

An advantage of rectilinear or j (reactance) notation is that components of opposite sign cancel in algebraic addition. As simple math indicates, a fixed capacitor and variable inductor may be connected in series to form the practical effect of a very reliable and less expensive variable capacitor. For example, a variable inductor adjustable from j0 to +j100 in series with a −j50 fixed capacitor functions as a simple series network effectively variable from −j50 to +j50. Thus, the feed point of an antenna with a large inductive reactance, such as a slant wire fed radiator or a skirt-excited radiator, can be brought near j0 with a −j value similar to the magnitude of the +j value of the antenna feed point. A small rotary inductor can then be used for trimming the combined sum to exactly j0. While the antenna may have a feed point impedance of 35 Ω +j550, a rotary inductor variable from j0 to +j50 and a fixed capacitor having a reactance value −j570 will provide a control permitting adjustment of the feed point to exactly 35 Ω j0.

Using a simple RF network, the resistive component can be transformed from 35 to 50 Ω and the load may be adjusted exactly to the transmitter requirement. The same is true for matching a transmitter to a directional array common point, transferring power to and from transmission lines, or any other impedance matching job needed in any antenna system. RF networks are the real work horses of any medium-wave antenna system, and often the least understood of all the critical component groups in a transmitting plant.

2.3.2 RF Networks

RF networks can be divided into simple and complex types. The simple ones are usually called "L" networks considering their schematic layout. Figure 2.5.2 shows one used for matching a modern transmitter with a fixed 50 Ω output impedance to an older plant directional array with a common

---

*Note for the mathematically inclined: (±) j indicates use of the operator, $j = -1^{0.5}$, where j, rather than i, is the convention in antenna systems avoiding confusion with i which indicates current.

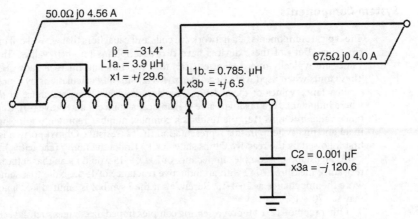

FIGURE 2.5.2 Matching network using the input/output values shown.

point resistance of 67.5 Ω, selected to give an easily read ammeter indication when operating at its authorized power of 1 kW plus a directional array antenna system loss allowance of 8 percent. This common practice in older directional facilities results in a common point input of 1.08 kW with a common point current of exactly 4.00 A. When this practice began, the output impedance range of many tube type transmitters was as much as 30 to 230 Ω for matching anything from a 40 or 50 Ω slant wire antenna feed to a 230 Ω open wire transmission line. Some early solid state transmitters could match load impedances in the 50 to 75 Ω range to accommodate older coaxial lines and elevated common point impedances. As transmitters have become more sophisticated, a 50 Ω fixed output has become the modern standard.

A major transmitter failure in an older unit can force an urgent transmitter replacement. Unlike an older unit, that may mean matching the phasor common point to a 50 Ω transmitter output. Because the administrative and certification costs of a common point impedance change can be significant, some stations may prefer deferring this work until it can be scheduled more economically with other antenna system work under less urgent conditions. In this situation a simple "L" network may be an easy, economic solution.

The "L" network is a very simple one with only two components, one inductor and one capacitor. More complex networks have three basic components, sometimes with more added for special purposes or to simplify adjustment. These are basically "T" and "Pi" networks, so named for their fundamental schematic arrangement. These networks with formulas for calculation and adjustment are shown in Fig. 2.5.3.

Part of a network's definition is also its frequency pass band, i.e., either low pass or high pass. An important fact to remember is that low pass networks delay or retard the phase of the energy passing through them while high pass networks advance phase. These may also be called lagging or leading networks. Knowing this is useful because it is possible to get a rough idea of *phase* effect simply by looking at the physical construction of a network.

The β (beta) function of a network describes the effect it has on phase precisely. In "L" networks β is strictly a function of network input and output resistance; thus it is not useful where phase adjustment is a requirement. "T" and "Pi" networks do allow phase adjustment, typically over a range from about ±70° to around ±130°. This range is a key consideration for networks designed for phase adjustment.

Years ago, many phase adjusting networks were generally designed for 90° phase shift, often with the same fixed input and output resistance. This made the calculations simpler and was very useful in the days of slide rule design; but there is nothing magic about these numbers, especially at a time when computers and scientific calculators are available on nearly every desktop. However, centering a network too far away from 90° phase shift needs careful consideration because input resistance, output resistance and phase shift may affect network efficiency unfavorably.

PRACTICAL ASPECTS OF MAINTAINING MEDIUM-WAVE ANTENNA SYSTEMS IN AM TRANSMISSION

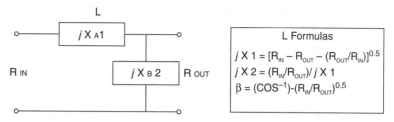

"L" Network Rules

R_{OUT} must be > R_{IN}
$jX1$ must not have the same sign as $jX2$
Phase shift (β) is fixed by R_{IN} and R_{OUT} values
If $+jX1$ network delays phase (lagging/low pass)
If $-jX1$ network advances phase (leading/high pass)

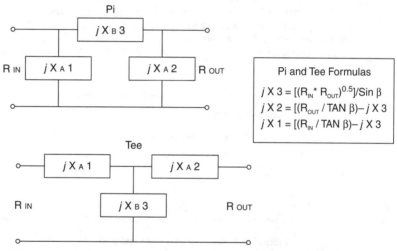

FIGURE 2.5.3 Calculating and adjusting "L," "Pi," and "T" networks.

Generally, the range mentioned above is conservatively safe, but phase shifting networks need careful considerations at the extremes of their operational adjustment limits. For example, in some plants adjustment for changing soil moisture is not uncommon. If that adjustment needs several degrees removed from a lagging network that centers at −75° to bring the array to normal operating parameters, this might cause an efficiency problem that could cause excessive heating in the network itself.

In the network schematics blocks of Fig. 2.5.3, jX_A or jX_B represent the reactances because they may be either $+j$ (inductive) or $-j$ (capacitive). The subscripts "A" and "B" denote two groups of reactive components where one group may be one type reactance while the other must be the opposite.

For example, in the "T" network the reactances in series with the input and output connections are labeled jX_A and the grounded center or shunt branch is labeled jX_B; thus, in any network the two series branches must be the same type while the shunt branch must be the opposite sign component. In this example, if jX_A is $+j$, then the network will have a low pass frequency response, which means it has a lagging phase delay while if all is reversed the network will be high pass and will advance phase.

In practice, capacitive branches usually contain an inductive component for adjusting the branch to the exact $-j$ required; however, the branch remains capacitive. This is simply the most economical method for adjusting the effective value of a fixed capacitor.

For setting network values, a noninductive, 50 Ω standard resistor, a "cold" or passive RF bridge, bridge oscillator, and a bridge detector are essential. A "hot" or active RF bridge may be used in the passive mode by fabricating an output cable terminating in an "unknown" clip lead and driving the bridge input with a bridge oscillator. While accuracy is affected by setup, and is generally inadequate for engineering data submitted to the FCC, it may be useful in urgent circumstances when better gear is not available. The output level of the bridge oscillator should be around 3 to 10 V_{RMS} if radiation from other sources interferes with bridge readings. If a field intensity meter is used as an amplified detector, keep the connector cable as short as is practical and not more than three feet long so that cable capacitance does not significantly affect bridge or meter operation. Observe grounding and other good bridge operating procedures to obtain accurate measurements. An RF network analyzer may also be used; however, this may require conversion of results from radial ($R\theta$) to rectilinear (Cartesian) notation.

A large pair of long nose, bent nose pliers can be very helpful for setting clip taps inside coils. General purpose hand tools, especially a screwdriver and wrench for working with 1/4 to 20 hardware, commonly used with coils and capacitors, will also be very helpful.

For reference, basic equations for capacitive and inductive reactance are given in Fig. 2.5.4.

Useful Information

Basic Reactance

Capacitive Reactance ($-j\,\Omega$) : $X_C = 1/(2\pi f C)$

Inductive Reactance ($+j\,\Omega$) : $X_L = 2\pi f L$

Where: $C = \mu F$
$f\ = MHz$
$L\ = \mu H$

Inductive Reactance of a Physical Coil

Calculating L from physical dimensions
Note: Valid for single layer air wound RF coils only.

$$L\ (\mu H) = (r^2 \times n^2)/(9\,r + 10\,l)$$

Where: r = mean (avg) coil radius in inches
l = winding length in inches
n = number of turns

Note: For edge wound (ribbon) coils, determine the mean diameter by measuring O.D. & I.D. assume $r = 0.5 \times$ mean diameter

FIGURE 2.5.4 Equations for capacitive and inductive reactance.

2.3.3 Preliminary Network Setup and Adjustment ("T" Example)

Note that multiple solutions can have the same approximate result. Unless one branch is held constant, getting too far from a satisfactory solution is a risk. Thus, we suggest careful, exact setting of the input branch in the preliminary adjustment. Leaving this branch undisturbed during final adjustment will assure the network does not stray too far from the theoretical component values.

1. Determine the phase shift desired if the network is part of a DA. If nondirectional, use 90° for a "T" or "Pi." "L" network phase shift is not adjustable. If a specific phase shift is required, a three component network is needed. The phase shift limits of a single network are usually between ±70° to ±130°.
2. Determine input and output resistance—typically 50 Ω_{IN} and measured radiator resistance (R_{OUT}) if used as an antenna coupler.
3. Using the equations shown in Fig. 2.5.3 solve for $jX3$, $jX2$, and $jX1$. At this point, it is useful to know the physical/electrical component values of the network. Thus, convert the jX values to X_L and X_C at the authorized frequency. In marginal sizing cases, increase the value of μH 5 to 15 percent thus increasing available $+j$ and *reduce* the value of μF 5 to 15 percent *increasing* $-j$. It is wise to measure the j reactance of delivered capacitors, noting it in pencil on the upper aluminum cap where it may easily be seen when installed.
4. Invert the sign of the load reactance, which in this case is the j of the *measured* radiator base impedance. In a directional array, the base impedance must be measured with the array active at low power, operating as nearly as possible at authorized array parameters (as noted below in the Complex Conjugate Impedance measurement procedure).
5. Add the inverted load reactance to $jX2$. Do not use the total value in calculations. Use it only for determining the total effective reactance connected to the network output leg. In some cases, it may be simpler to separately correct the radiator reactance to separate that from the network and avoid confusion.
6. Using "J" plugs, if available, disconnect and ground network input and output.
7. Disconnect and "float" center junction leads.
8. Arrange the RF bridge for easy individual connection to each of three "floated" leads.
9. Measure and record reactance of each lead noting "input," "center," and "output."
10. Remove the bridge and reconnect the center tie junction.
11. Unless proceeding immediately with Complex Conjugate Impedance measurement of the completed network, reconnect network output and input normally.

2.3.4 Final Adjustment of Network's Complex Conjugate Impedance

Note that multiple solutions can have the same approximate result. Unless one branch is held constant, getting too far from a satisfactory solution is a risk. Thus, we suggest careful, exact setting of the input branch in the preliminary adjustment and leaving that branch undisturbed during final adjustment will assure the network does not stray too far from the theoretical component values.

1. Using "J" plug, if available, disconnect the network input and ground it through a noninductive precision resistor with value of network's nominal input resistance (e.g., 50 Ω in the case of a newer antenna coupling unit).
2. Using "J" plug, if available, disconnect network output from load and connect it closely to RF bridge "unknown" terminal.

Note: In the case of an antenna coupling network, this work may be more accurately done at the point of antenna/radiator resistance measurement. For example, the location might be the external connection of the coupling unit feed-thru insulator bowl or a connection to the measured line passing through the coil assembly of a toroid ammeter. In these cases, the coupling unit output "J" plug remains connected. If the radiator reactance has been corrected separately then for the purpose of this work only consider it part of the radiator and measure the correction with the radiator set so the radiator appears to be $j0$.

3. With the RF bridge connected to the network output as noted above, adjust (trim) the network components to yield a bridge measurement having resistance of the normal load and the value of load reactance with the *j* sign inverted.

For example, in the case of a single, nondirectional radiator having a measured impedance of 43Ω + *j*60 the network is adjusted for a bridge reading of 43Ω −*j*60. The principal resistance adjustment is *j*X3 while changing *j*X2 affects reactance.

In the case of a directional array radiator, this becomes a reiterative process where the array is connected normally and is operated at low power using the preliminary network adjustment while adjusting the array for operation at normal authorized parameters, or as near them as possible, to permit an active RF bridge measurement of the radiator under low power.

The coupling unit is adjusted as indicated for the nondirectional case above for the actively measured radiator impedance with *j* sign reversed. The array is then brought online at low power and the active bridge measurement repeated while the array is operating at authorized antenna parameters. The coupling unit is adjusted passively to the results of the active radiator measurement. Thus, the procedure becomes active measurement followed by passive network adjustment to the last active measurement with reactance (*j*) sign reversed followed by another active measurement until the array may be operated with authorized antenna parameters without difficulty. The objective is looking into the network output passively and adjusting the coupling unit to the measured impedance of the radiator with reactance (*j*) sign reversed.

2.3.5 Monitoring the Directional Antenna System

The directional properties of a directional array derive from the physical location of each active radiator in the array, the phase angle of the RF energy arriving at each radiator, its relative power level, the electrical height of the radiators, adequate grounding, and the environment of "volunteer" or passive radiators, such as cell towers or structures. Principally, most of these except for phase and power are fixed in the construction of the plant.

The phasor of a directional array adjusts power division and phasing of transmission energy for each radiator within reasonable limits. Generally, larger phase changes may be possible in the radiator coupling unit. If the β of the coupler, also called ACU (antenna coupling unit) or ATU (antenna tuning unit), must be shifted to put the phasor control nearer its mid-range, this should be done last after the directional procedure above has been completed.

Only the power division and phases are variable; thus, they are the parameters that the antenna monitor system measures. The only other monitoring method is via actual field intensity measurement either at monitor points detailed on the station license if the license grant is based on the older method of field measurements, or at typically three check points if the license was granted under the newer method of moments (MoM) computer modeling procedure.

In both cases, the antenna monitor (frequently called a phase monitor) is the major antenna array management tool. In the older, more conventional system there is a heavy dependence on field monitor points and "skeleton" or partial proof of performance.

In a *method of moments* system, total reliance for system monitoring is placed upon the antenna monitor. For this reason, the monitoring system must be inspected and recertified at two-year intervals by a competent authority and the written report placed in the station's public inspection file. In MoM systems, antenna monitor system accuracy is a very serious matter.

Read and understand those sections of the following FCC Rules and Regulations that apply to any directional stations with which you are affiliated:

- Sections 73.61 through 73.69
- Sections 73.150 through 73.158

When chronic problems with array management are discovered it is prudent to determine if siting of the array radiators is exactly correct. Errors in construction of radiators may result from a poor construction survey, especially in older facilities, either from measurement error or incorrect alignment with true north. While very little can correct errors after the fact, knowing the errors that exist may help understand persistent operating challenges and may provide a starting point to determine the scope of possible corrections.

2.4 TROUBLESHOOTING A SINGLE RADIATOR PROBLEM

Problems happen in antenna systems. A lightning strike, collision from an ATV, or intentional damage or vandalism may cause coupling network failure. There also may be longstanding chronic problems that have persisted for years in the antenna system.

2.4.1 Case in Point

There are no visible physical signs of a problem, but radiator base current of a long-established station drifts upward or downward while the transmitter checks out normally, delivering rated power into a dummy load at the transmitter output and from the output end of the coax providing the input to the antenna coupling unit.

A. Informal measurement of the radiator indicates a result consistent with older recorded measurements in the files.

If the radiator measurement is essentially unchanged from file data and the coupler network is a low pass "T" (as most of them are), look hard at the shunt capacitor in the grounded center ($jX3$) leg of the network. Two possibilities are corrosion that may be hidden under the base of the capacitor, especially if there is no copper strap under the capacitor, or a change in capacitor value. Remove the capacitor, clean its caps with a fine wire brush, fine abrasive cloth or paper of at least 200 grit then measure its j value with an RF bridge. Convert the j value to µF. If the µF *value* is consistent with that indicated on the capacitor, and is one that satisfies the capacity required by calculation of the matching network needed for coupling the radiator, the capacitor may be reused.

B. Informal measurement of the radiator indicates a result NOT consistent with older recorded measurements in the files.

Loss of continuity in radiators may not be immediately obvious because it sometimes manifests itself as a slow drift in tower base current. If the radiator measurement has changed significantly from previous measurements, the problem is probably located in the radiator or other attachments to it such as the lighting choke bypass capacitors, a microwave or FM iso-coupler, or other ancillary device. Visually check to confirm good ground connections between the ground radial system and the network.

With these facts the probability of a radiator problem is high; however, it is prudent to confirm correct adjustment of the radiator coupling network by complex conjugate measurement. Break or unplug the connection between the network input and transmission line, and ground the coupler input through a 50 Ω precision resistor. Measure the output of the network with the line connecting the radiator disconnected. The expected measurement is the value of the radiator base impedance with the sign of reactance inverted.

By confirming that the transmitter performs correctly when loaded by a resistive load (dummy load) at the input and output ends of the transmission line, that there is no grounding problem, and that the network is adjusted to the correct impedance match, the radiator is all that remains in the circuit. Contact a tower crew experienced in doing mechanical work and field welding on structures because it is likely that at some level the radiator continuity is poor or nonexistent. The time-honored cure for this problem is welding one leg of the radiator with a short bead about 1 to 2 inch long at each section joint. The length of the weld depends on the joint construction. For example, radiator sections with robust flange joints generally tolerate a weld 1-1/2 to 2 inch in length around their outer periphery while thinner slip joints may need careful evaluation by a structural engineer with specific instructions for the welder. *The objective is unbroken continuity of the radiator from bottom to top without compromise of structural integrity.*

For smaller stations with an operating power less than 2.5 kW, in the event that more confirmation is required or a competent tower crew cannot arrive on site quickly, there is a test which can also serve as a temporary repair. Have one end of heavy galvanized steel wire, e.g., "clothes line wire" fastened to a cleaned spot under a bolt at the top of the radiator, stretched down one leg inside the tower to the base of the structure and secured to a cleaned spot under a bolt.

With the line in place measure the base impedance of the radiator and compare with those in the station's technical file and licensed value. The value should agree closely with the license. A low radiator resistance indicates a shorter height actually indicating poor continuity. If the value was measured and impedance relicensed at some time in the past several years, it is possible strapping a wire to the radiator might increase the impedance substantially above licensed value nearer the original radiator value.

In this event, it is helpful to compare with the oldest impedance value that can be found for the radiator. If the oldest reading and the result of strapping the radiator with a wire are similar, then welding or radiator replacement will be necessary. If the wire is to be used as a temporary operating repair, the wire should be tied to the radiator leg at 10 or 20 feet intervals to prevent damage by high winds. The station must follow the indirect power measurement (FCC 73.51(d)) until a permanent repair is made and the impedance relicensed by a direct power measurement application (FCC 73.51(a)).

2.5 TROUBLESHOOTING DIRECTIONAL ARRAYS

First, consider the first rule of directional antenna systems. Never, ever, touch the controls of the station's phasor or components of an antenna coupling unit at a radiator unless you first record the "before" data. The most convenient method for a coupling unit mounted at a radiator is marking the positions of clip taps on the coils with a colored magic marker. For the phasor it would be the indicator dial readings of the counters on the phasor controls. These readings and antenna monitor indications should be logged. Generally, keeping a separate log for the phasor is more convenient. This log should show the crank setting for each radiator's phase and power control if applicable, and it is wise to include a blank for those which are not applicable. In addition to crank readings, the log should also show the radiator's phase and power indications and there should be a separate blank for date and time, power, pattern if applicable, common point current, common point impedance if applicable, and any other instrumentation that is applicable. It is also helpful to include space for each monitor point or check point reading that may be taken. The value of this log will become more apparent as you become more acquainted with the station and begin making minor adjustments to keep the array in alignment. Keeping the log on a computer or paper spreadsheet will be very useful and time saving when adjustments are needed.

Since the power (division) and phase angle are the only parameters that can be varied, they are the parameters that are monitored by the antenna monitor system. The only other monitoring method is via actual field intensity measurement, either at monitor points detailed on the station license (if the license grant is based on the older method of field measurement) or at typically three check points (if the license was granted under the newer MoM computer modeling procedure).

Usually, a problem in a directional antenna system manifests as either an incorrect antenna monitor indication or an over-limit monitor point reading in an older field intensity proofed system.

In a directional array the phase monitor is the primary monitoring instrument, but when it indicates an error it is important to find if the problem is real or a false indication. There is a simple method for making a basic functional check of the antenna (phase) monitor.

Make a test cable that will accept a single signal from the largest signal sample of the array samples and parallel connect it to all inputs on the antenna monitor. This can be made from coaxial cable tee connectors, couplers and one short jumper cable for each monitor input. Each jumper should be the same length. Something between 18 and 24 inches works well because the cable phase length is not significant and it allows easy connection to all inputs.

The procedure consists of connecting one signal, preferably the strongest, to all inputs and observing the power and phase indications of all channels. All power indications should be the same and all channels should show a phase indication of approximately zero.

If the monitor appears to need re-calibration, see FCC Section 73.69.

If the monitor seems correct and the array is operating within normal parameters, but one of the monitor points is over-limit, do nothing until you have inspected the area within a 3-km radius around the array and a similar radius around the errant MP for new construction, especially cellular site and power line construction. Other construction such as water towers, steel framed buildings, or large structures with metallic siding may also be problematic.

The next step is bringing the array parameters as near as is practical to their exact licensed values and, if the MP remains over-limit, reduce operating power to bring it within limit. If the necessary power level is less than 90 percent of licensed power, the FCC must be notified and other requirements of FCC Section 73.1560 must be observed. Especially, see 73.1560(d).

If the antenna monitor is in calibration and it shows the station is operating at its specified parameters with licensed transmitter power, it is very unusual to have over-limit monitor points unless the pattern is disturbed by external effects. If no cause is obvious, see 73.158 and 73.154. It may be necessary to seek assistance of an experienced professional in the event that a survey must be made pursuant to 73.154. In this event, a full evaluation of conversion of the station to MoM computer modeled operation per 73.151 may be advantageous as this will eliminate the requirement for checking monitor points.

CHAPTER 2.6
INTERNATIONAL SHORTWAVE BROADCASTING

Douglas Garlinger
CPBE, Past-President, NASB, Inc., Indianapolis, IN

2.1 INTRODUCTION

Many of us interested in electronics experience a deep stirring in our souls at the prospect of receiving radio signals from other countries. It is an unforgettable experience the first time you picked up HCJB from Quito, Ecuador, Radio Moscow, the BBC, or the Voice of America (VOA). International broadcasters have used shortwave frequencies to broadcast to the world for over 80 years.

No matter how many high-tech communications satellites or broadband Internet systems are developed for the worldwide exchange of information, none of these advancements will take the magic out of shortwave. For some of us, the allure of shortwave is so strong that we become ham radio operators with shortwave stations of our own.

2.2 FCC REGULATION

The U.S. shortwave privately owned broadcasters are licensed by the Federal Communications Commission (FCC) under Title 47, Part 73, Subpart F, International Broadcast Stations. The VOA is a government-owned U.S. Broadcaster and is not licensed by the FCC.

Any qualified U.S. citizen, company, or group may apply to the FCC for authority to construct and operate a high frequency (HF) (shortwave) international broadcasting station. Application for a construction permit should be filed on FCC Form 309 and application for license, upon completion of construction, on FCC Form 310.

For many years, there was a moratorium in place and new licenses were not issued. There were only four privately owned shortwave broadcast stations in operation prior to 1982: WYFR, KGEI, WINB, and KTWR. In the early 1980s, citing Public Law 80-402, Engineering Consultant George Jacobs led the efforts to persuade the FCC to issue licenses again. WRNO in New Orleans became the first privately owned shortwave station to be licensed in many years. By 1990, the number of licensed stations had grown to 17.

2.2.1 Differences from Domestic AM

There are some immediate differences in shortwave broadcasting that are not characteristic of domestic AM stations. Specifically:

- Stations are not assigned a specific permanent frequency.
- Stations typically use more than one frequency throughout the broadcast day.
- Stations are licensed at 5 kHz increments.
- Stations are limited to 4.5 kHz audio bandwidth.
- Stations may not broadcast solely to a U.S. audience.
- Stations do not have a Public File.
- Stations do not broadcast EAS (Emergency Alert System) alerts, but must sign off air for an Emergency Alert Notification.
- Stations have "target zones" and are not licensed with "coverage areas."
- Stations must operate with a Directional Antenna with at least 10 dB gain in the main lobe.
- Station antennas typically do not have ground radials and rely upon sky wave propagation.
- Stations use primarily 300-Ω open wire transmission line and a few use coax with a balun.
- Stations must operate at a *minimum* TPO (transmitter power output) of 50 kW AM.
- Stations may not exceed 100 percent modulation on positive peaks.
- Stations may operate on single sideband (SSB) with a *minimum* peak envelope power of 50 kW.
- Stations may operate digital modulation (DRM) with a *minimum* average power of 10 kW.

2.3 INTERNATIONAL SHORTWAVE BANDS

There are several shortwave bands distributed throughout the radio spectrum from 3.2 to 26.1 MHz (see Table 2.6.1). They are often referred to by their original wavelength names of the 90-meters to 11-meters band. Frequency bands below 49 meters are primarily used in the tropics and rarely used by U.S. Broadcasters.

TABLE 2.6.1 International Shortwave Bands and Frequencies

Bands (meters)	Operating Frequency Range (MHz)	Notes
11	25.670–26.100	Rarely used
13	21.450–21.850	Used in periods of High Sunspot Activity
15	18.900–19.020	Used in periods of High Sunspot Activity
16	17.480–17.900	
19	15.100–15.800	
22	13.570–13.870	
25	11.600–12.100	
31	9.400–9.900	
41	7.200–7.450	
49	5.900–6.200	
60	4.750–4.995	Primarily used in tropical regions
75	3.900–4.000	Primarily used in tropical regions
90	3.200–3.400	Primarily used in tropical regions

2.3.1 Seasonal Shortwave Schedules

U.S. shortwave broadcasters have a seasonal frequency schedule that is authorized by the FCC and is coordinated internationally with the High Frequency Coordination Conference (HFCC) based in Prague, Czech Republic. The HFCC meets semi-annually at various venues in the world to negotiate frequency usage for the upcoming broadcast season. There are two Seasonal Broadcast schedules each year, designated as "A" or "B." For instance in 2014, there is an A14 schedule from 30 March 2014 to 26 October 2014, and a B14 schedule from 26 October 2014 to 29 March 2015.

2.3.2 International Frequency Coordination

Frequency coordination first began in 1960 with the Eastern Block Iron Curtain nations. In 1962, Western nations began coordinating in a separate group. After the end of the Cold War, the two groups met for the first time in 1990 in Bulgaria. In 1991, a combined group was formed that later became the HFCC. In 1996, in Asia a sister group was formed in Kuala Lumpur within the Asia-Pacific Broadcasting Union (ABU). There is also an Arab States Broadcasting Union (ASBU) founded in 1969. Since 1998, the ASBU coordinates with ABU and the HFCC. Most international shortwave broadcasters in the world are represented within these three groups.

In 1990, several of the U.S. private shortwave broadcasters joined together to create the National Association of Shortwave Broadcasters, Inc. (NASB). Since that time, the NASB has grown in influence. Table 2.6.2 shows that the NASB has hosted or cohosted eight HFCC Conferences around the globe since 2005.

TABLE 2.6.2 HFCC Conferences Since 2005

Conference	Location	Host Organization	Dates: DD–DD MO YR
A05	Mexico City, Mexico	NASB	7–11 February 2005
B05	Valencia, Spain	Radio Nacional de Espana	22–26 August 2005
A06	Sanya, China	RTPRC	13–17 February 2006
B06	Athens, Greece	ERT	28 August–1 September 2006
A07	Abu Dhabi, United Arab Emirates	Emirates Media (EMI)	5–9 February 2007
B07	Birmingham, UK	Christian Vision	27–31 August 2007
A08	Kuala Lumpur, Malaysia	VT Communications	4–8 February 2008
B08	Moscow, Russia	GRFC	25–29 August 2008
A09	Tunis, Tunisia	ASBU	2–6 February 2009
B09	Punta Cana, Dominican Republic	NASB	17–21 August 2009
A10	Kuala Lumpur, Malaysia	ABU	1–5 February 2010
B10	Zurich-Regensdorf, Switzerland	NASB and Thomson Broadcast AG	2–6 August 2010
A11	Prague, Czech Republic	HFCC	14–18 February 2011
B11	Dallas, USA	NASB and Continental Electronics	12–16 September 2011
A12	Kuala Lumpur, Malaysia	ABU	9–13 January 2012
B12	Montrouge/Paris, France	TDF and NASB	27–31 August 2012
A13	Tunis, Tunisia	ASBU	28 January–1 February 2013
B13	Bratislava, Slovakia	RTV/RSI and NASB	26–30 August 2013
A14	Kuala Lumpur, Malaysia	ABU – HFCC	20–24 January 2014
B14	Sofia, Bulgaria	Spaceline Ltd. and NASB	25–29 August 2014
A15	Muscat, Oman	ASBU and Oman Radio TV	1–5 February 2015
B15	Brisbane, Australia	Reach Beyond Australia and NASB	24–28 August 2015

Most of Latin America, Equatorial Africa, and some Central Asian nations do not participate in any of the HF coordination groups.

2.3.3 CIRAF Target Zones

Throughout the course of the 11-year sunspot cycle, each station must assess its desired audience and select target zones from the CIRAF Zone Map. The name CIRAF is derived from Spanish *"Conferencia Internacional de Radiodifusión por Altas Frecuencias."* The map was established at the World Administrative Radio Conference (WARC) in Mexico in 1948. See Fig. 2.6.1.

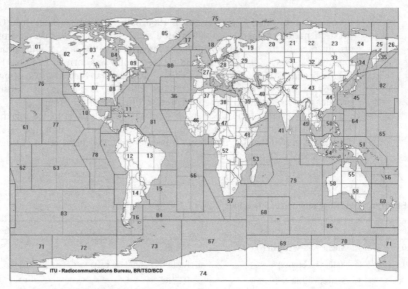

FIGURE 2.6.1 CIRAF zone map (not the same as Ham Radio DX Zones). (*Courtesy: International Telecommunications Union.*)

In years of lower sunspot activity, the propagation is not as good on the upper bands and stations tend to compress their broadcast operations in the bands below 10 MHz. The demand for desirable shortwave frequencies can exceed supply. Large government-owned international broadcasters often have powerful 250 and 500 kW transmitters and the most sophisticated high-gain antennas. Many private U.S. broadcasters also have 250 and 500 kW transmitters and sophisticated antennas. However, some private U.S. broadcasters have only 100 or even 50 kW transmitters with less expensive antennas. On occasion, the FCC authorizes the use of frequencies just outside, but adjacent to the allocated bands, on a noninterference basis. This permits the private broadcaster the advantage of being a few kilohertz away from the main congestion of the allocated band and permits their signal to reach its target area with less interference.

2.4 FREQUENCY MANAGEMENT

Each nation has a Frequency Management Organization (FMO) that coordinates with the HFCC for frequency usage. Some countries have more than one. For the U.S. private shortwave broadcasters, the FMO is the FCC. For VOA, it is the International Broadcasting Bureau (IBB). Many nations of the world do not permit shortwave broadcasting by anyone except the government. Each shortwave broadcaster in the world has a three-letter broadcaster name code identifying it. In some instances the code is the same as the broadcaster FMO code. Table 2.6.3 lists the FMO names. Table 2.6.4 lists the broadcaster name codes.

INTERNATIONAL SHORTWAVE BROADCASTING

TABLE 2.6.3 Frequency Management Organizations (FMO) Names

Code	FMO Name
ABC	Australian Broadcasting Corporation
AIR	All India Radio
ARS	Saudi Arabian Radio & Television
ERU	Egypt Radio and TV Union
FCC	Federal Communications Commission, USA
IBB	International Broadcasting Bureau, Voice of America
IRB	Islamic Republic of Iran Broadcast
JRT	Jordan Radio and Television
KBS	Korean Broadcasting System
NHK	Nippon Hoso Kyokai, Japan
ONT	Tunisian Radio and Television
REE	Radio Exterior de Espana
RTC	R & T of Peoples Republic of China
TDF	Telediffusion de France
TRT	Turkish Radio-Television Corp.
TWR	Trans World Radio
VAT	Vatican Radio
VOR	Voice of Russia
VOV	Vietnam Radio and Television

TABLE 2.6.4 Broadcaster Name Codes

Code	Broadcaster Name
ABC	ABC-Radio Australia
AIR	All India Radio
AWR	Adventist World Radio
BBC	BBC Worldservice
DWL	Deutsche Welle
HBN	Republic of Palau
HCA	HCJB Australia
HRI	LeSea Broadcasting Corporation
IBB	International Broadcasting Bureau
IRB	Islamic Republic of Iran Broadcasting
NHK	Nippon Hoso Kyokai Japan
ORF	Oesterreichischer Rundfunk
RFI	Radio France Internationale
RMI	Radio Miami International
RNZ	Radio New Zealand
SAB	South African Broadcasting Corporation
TWR	Trans World Radio
VAT	Vatican Radio
VOR	Voice of Russia

2.5 FREQUENCY REQUESTS

In the weeks and months leading up to an HFCC meeting, the FMOs will submit frequency hour requests for the upcoming season. Three examples of information in the request are shown below:

Frequency	Start	Stop	CIRAF	Broadcaster	FMO	Language
9560	0000	0100	12–14	China Radio International	RTC	Por
11,775	2300	2400	46–48, 52, 53	LeSea Broadcasting Corp	FCC	Eng
9495	0000	0200	10–13	Radio Miami International	FCC	Spa

Broadcaster	Transmitter	Latitude/Longitude	Beam	Power	Days of Operation
China Radio International	Beijing	39N57/116E27	318	500	1,234,567
LeSea Broadcasting Corp	Furman, SC	32N41/081W08	85	250	123,456
Radio Miami International	Okeechobee, FL	27N28/080W56	142	100	1,234,567

China Radio International. China Radio International wishes to broadcast from a Beijing transmitter located at 39°57′ N 116° 27′ E on an azimuth of 318° with 500 kW TPO all 7 days of the week. They wish to broadcast in Portuguese to zones 12 through 14, which are South America and Brazil in particular. They wish to broadcast on 9560 kHz from 0000 UTC to 0100 UTC.

LeSEA Broadcasting. LeSEA Broadcasting Corporation wishes to broadcast from the Furman, South Carolina transmitter (WHRI) located at 32°41′ N 81°08′ W on an azimuth of 85° with 250 kW TPO on Sunday through Friday. They wish to broadcast in English to zones 46 through 48 and zones 52 and 53. The target is West Africa, Central Africa, and Madagascar. They wish to broadcast on 11,775 kHz from 2300 UTC to 2400 UTC.

Radio Miami International. Radio Miami International wishes to broadcast from the Okeechobee, Florida transmitter (WRMI) located at 27°28′ N 80°56′ W on an azimuth of 142° with 100 kW TPO all 7 days of the week. They wish to broadcast in Spanish to zones 10 through 13 which are Central America, the Caribbean, and the northern half of South America. They wish to broadcast on 9495 kHz from 0000 UTC to 0200 UTC.

2.5.1 Collisions

The HFCC receives thousands of such "frequency hour" requests for each season. Extensive computer analysis is performed in Prague. When an unacceptable interference issue is detected it is referred to as a "collision." Often, many of the collisions are worked out before the conference. Face-to-face meetings at the conferences result in the resolution of thousands of frequency collisions. These informal personal contacts are extremely useful for effective and amicable agreements to resolve collisions. Each day at the conference, a new collision report is generated as each FMO representative negotiates with other FMO representatives to reduce or eliminate the collisions in the week-long conference.

Sometimes, a broadcaster will request frequencies that they do not use. The unused but authorized frequencies are called "wood". This is a practice that is frowned upon. A complete shortwave frequency schedule for all HFCC participants can be found at http://www.hfcc.org/data/guidepost.phtml.

2.6 SHORTWAVE TRANSMITTERS

AM shortwave transmitters are in principle very similar to domestic AM transmitters in design and modulation schemes. Continental Electronics is a well-known U.S. manufacturer. In the past there was Harris, Gates, Collins, and RCA in the United States. HCJB Global designed and built their own transmitters in Elkhart, Indiana. Some of the Internationally known manufacturers, past and present, are Brown-Boveri, Telefunken, NEC, Riz, and Thomson (now Ampegon).

The primary difference is the transmitter must be frequency agile and that requires a modified design with variable inductors, variable capacitors, and servo motors. Typically, a frequency change needs to be accomplished in about 45 s. Final Output tubes are typically tetrodes and can be quite expensive. A 100 kW transmitter tube such as the 4CV100000C or 4CV100000E may cost $18,000. A 500 kW transmitter high power tetrode such as the CQK-650 may cost $145,000. Fortunately, there are a few companies that can rebuild these high power tetrodes.

AM shortwave transmitters are normally at least 50 kW. Typical output power levels are 100, 250, 300, and 500 kW. Most transmitters use 300-Ω balanced transmission line to feed the antenna. On some occasions, coax is used and at the antenna a tapered balun transforms the 50-Ω coax to the 300-Ω antenna feed point.

2.7 ANTENNA TYPES

The propagation analysts at the station and the computer collision analysis in Prague take advantage of an extensive database of antenna types. The number of bays, the gain, the beamwidth, the antenna take-off angle, and many other factors are used in the analysis. Frequently used antennas include dipole arrays, log-periodic, and rhombic.

The rhombic has a low take-off angle of 10° and is generally the most economical to construct. It can be "homebrewed" by a station's engineering staff. However, the rhombic takes a large amount of real estate and can waste as much as one-half of the transmitter power.

The log-perodic or LPA antenna can be designed to cover a wide range of frequencies such as 5.95 to 26.1 MHz. This LPA has a typical take-off angle from 12° to 22°. Sometimes a higher take-off angle is desirable for a shorter transmission path. Radio Miami International employs a log-periodic antenna with a 45° take-off angle and an 86° beamwidth to broadcast from southern Florida to Cuba. It has a lower frequency range of 3.9 to 18.0 MHz that can be more advantageous in tropical regions.

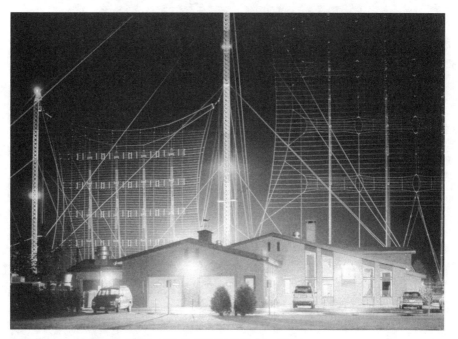

FIGURE 2.6.2 WHRA World Harvest Radio "4 × 4 × 1" dipole curtain arrays.

The dipole array is the most sophisticated antenna and the choice of major broadcasters. The nighttime picture of Fig. 2.6.2 is from WHRA in Greenbush, Maine. This is actually two antennas, each sharing the common 364 feet support tower in the center. The antenna on the right covers one octave from 6 through 11 MHz. The smaller antenna on the left covers one octave from 11 through 21 MHz. This picture is taken from the back of the antennas. There is a reflector screen behind the dipole arrays. Both antennas are 16 dipole arrays, 4 dipoles across, and 4 dipoles stacked vertically. The lowest dipole is one full wavelength off the ground. It is referred to as a "4 × 4 × 1" curtain array and is identified as type 218 in the antenna database. It has a gain of 22 dB, a beamwidth of 24° and it has a very desirable low take-off angle of 7° for long distance transmission.

This antenna is slewable. Note the 300-Ω feed lines going up to the dipoles. The feed lines connect to a 5-position slew switch capable of electronically slewing the antenna direction between −30°, −15°, 0°, +15°, and +30°. These antennas were constructed naturally pointing at 75°. Figure 2.6.3 demonstrates the various main beam azimuths available from this antenna site.

45°	Moscow, Eastern Europe, and Iran
60°	Western Europe, Jerusalem, and Saudi Arabia
75°	Northern Africa
90°	Central and West Africa
105°	West Africa and Southern Africa

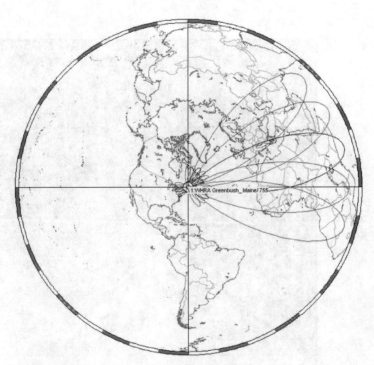

FIGURE 2.6.3 Five-position slew-switch for antenna bearings.

2.8 SINGLE SIDEBAND

The WARCs of 1979, 1984, and 1992 encouraged the development of SSB Transmissions. There was a desire to alleviate frequency congestion and reduce transmitter operational costs. Several major shortwave broadcasters experimented with SSB transmission. Among them were the BBC, Radio Netherlands, and HCJB from Quito, Ecuador. Ultimately, the testing was not fruitful and attention turned to developing a digital shortwave transmission standard.

2.9 DRM® DIGITAL RADIO MONDALE™

DRM is a form of digital broadcasting for shortwave. Transmitters must have an average power of at least 10 kW to be authorized by the FCC. The DRM system uses COFDM (Coded Orthogonal Frequency Division Multiplex). The audio data and associated signals are contained within a large number of closely spaced carriers contained within the permissible broadcast channel. Time interleaving is used to alleviate signal fading. The parameters of the coding can be optimized to operate in different propagation conditions. Parameters can be selected that optimize data capacity, robustness, and transmitter power. A number of U.S. and international shortwave broadcasters are involved in testing of DRM. It has not yet gained wide acceptance due to the lack of widely available economical DRM receivers. You can learn more about DRM at http://www.drm.org/.

2.10 IONOSPHERE

The miracle of shortwave communications is made possible by an invisible layer of electrically charged gases in the upper atmosphere called the *ionosphere*. The earth's atmosphere located 30 miles or more above the surface is primarily made up of oxygen and nitrogen with small amounts of other gases. Ultraviolet radiation from the sun striking these rarefied gases dislodges electrons creating electrically unbalanced atoms called ions. This process is called ionization.

The ionized condition of the ionosphere varies widely and continuously as a result of the amount of ultraviolet radiation it receives from the sun. These variations occur as a result of the daily rotation of the earth, seasonal variations, and geographical variations in the earth or oceans below. As the sun rotates upon its axis every 27 days, it creates a 27-day cycle of solar radiation. In addition, there is an 11-year sunspot cycle. Sunspots are essentially storms on the surface of the sun that emit staggering amounts of ultraviolet radiation. When there are many sunspots on the sun, the propagation of shortwave signals is very good up to 21 MHz (13 meters) and higher. When sunspot activity is very low, shortwave propagation is not reliable above 10 MHz (31 meters).

As ultraviolet radiation from the sun penetrates deeper and deeper into the ionosphere, it causes electrons to be detached from its atoms. The ultraviolet radiation gives up energy during this process and finally reaches a point above the surface of the earth where it exhausts itself. As the earth rotates, during the night, the upper atmosphere receives no sunlight. The dislodged atoms recombine with the atoms of the gases and neutralize their electrical state by a process called *recombination*. The ionization and recombination of the earth's atmosphere normally take place between 30 and 300 miles above the surface of the earth. The ionosphere tends to be strongest and thickest near the equator. The ionosphere is virtually nonexistent over the polar regions of the Arctic and Antarctic. The ionosphere becomes thinner as you leave the equator and approach 60° latitude.

As all radio frequencies enter the ionosphere, they may be reflected, absorbed, or passed through the ionosphere. Typically VHF, FM broadcast, UHF, and microwave frequencies simply pass through the ionosphere into space. AM broadcast and LF signals are dissipated and absorbed when they enter the ionosphere.

There is a special range of radio signals from 3 to 30 MHz called "shortwave." This is a range of frequencies that does not pass through and is not absorbed by the ionosphere, but is reflected back to earth by the ionosphere. This reflection between the earth's surface and the ionosphere is referred to as "skip." For a transmitted shortwave broadcast to reach its intended target audience, the signal may have to skip between the earth and ionosphere several times.

Not all frequencies in the shortwave range are always reflected back to earth. Sometimes they too are absorbed, or they are partially absorbed and partially reflected. At other times they may partially pass through the ionosphere and be partially reflected back to earth.

It becomes a very complex "art" as well as a science to predict just when and which frequencies will skip the best. We need to carefully examine the composition of the ionosphere if we are to learn how to select frequencies which will skip back to the earth. The study of signals that travel this way is called propagation analysis.

The types of gases that make up the upper atmosphere accumulate in several layers. These layers may be separated from each other by several miles, but occasionally they overlap each other. As the sun's ultraviolet radiation passes through these layers, there tends to develop as many as four layers of peak intensity ionization. The height of any given layer varies widely depending on many factors. The four layers are designated D, E, F1, and F2. See Fig. 2.6.4.

D Layer. The D layer is located from 30 to 55 miles above the earth's surface. It appears at local sunrise and reaches maximum intensity at local noon. It weakens or disappears shortly after local sunset. The D layer is not helpful in the propagation of radio signals. A strong D layer will weaken a shortwave signal as it passes through it to the F Layer.

E Layer. The E layer is located from 55 to 75 miles above the earth's surface. The height of the E layer varies due to the season. The E layer is highest in equatorial regions. The E layer reaches maximum ionization at local noon and virtually disappears during the nighttime hours.

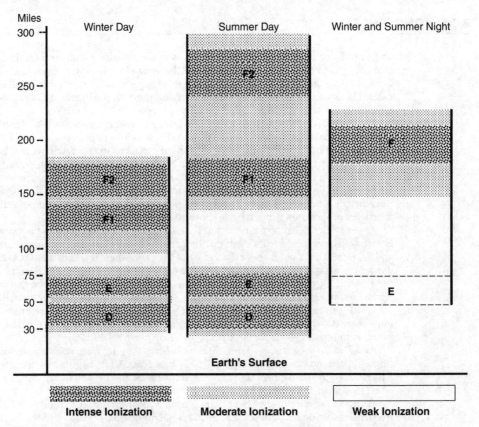

FIGURE 2.6.4 Typical layers of the ionosphere.

F1 Layer. The F1 layer is located 90 to 175 miles above the earth's surface. It is at its greatest altitude during a summer day. In the winter, the F1 layer merges with the F2 layer in the temperate latitudes. Its intensity peaks at local noon and it decays significantly during the night. It contributes to the propagation of shortwave signals through the daylight hours.

F2 Layer. The height of the F2 layer varies seasonally. It is at its lowest of 150 miles during a winter day and reaches 250 to 300 miles during a summer day. Unlike the other layers of the ionosphere, the F2 layer does not disappear at night. At night, the F2 layer is typically 200 miles above the surface. It is the F2 layer that makes shortwave communications possible during the nighttime hours. It is the most highly ionized of the four layers.

2.10.1 Frequency of Optimum Traffic

For long distance communications, you want an antenna that transmits at a very low take-off angle of 7° to 25°. This will allow the signal to travel a farther distance before intercepting the ionosphere, and it will be reflected at a similar angle akin to a banked shot on a pool table. This will permit the signal to arrive at its distant destination with the least number of skips or "hops." Different antenna types have different take-off angles.

For any given amount of ionization, a frequency exists that is the highest frequency that will be reflected back to the earth by the ionosphere. It is called the *maximum usable frequency* or MUF. This is the frequency that will be reflected by the ionosphere with the greatest signal strength. If the transmitted frequency is higher than the MUF, the ionosphere will not be strong enough to return the signal to the earth. The signal will pass through the ionosphere to space. If the transmitted frequency is lower than the MUF, then the ionosphere will begin to absorb some of the energy from the signal. The signal returned to the earth will be weakened.

Another way to look at it is that the MUF is the frequency at which the signal path may begin to fail. In actual practice, this predicted MUF has been found to support shortwave communications for only 50 percent of the time. By lowering the predicted MUF by only 15 percent, reliable shortwave communications can be achieved 90 percent of the time. This frequency is called the *Frequency of Optimum Traffic* or FOT. For instance, if the MUF is determined as 11.75 MHz, then:

```
Frequency of Optimum Traffic (FOT) = 11.75 X 0.85 = 9.97 MHz
```

Eventually, as the frequency is lowered, the absorption of the signal becomes so great that communication cannot be sustained. The lowest frequency that can be successfully used over a particular signal path is called the *Lowest Usable Frequency* or LUF. Absorption is more intense in the equatorial regions and is normally greater during the summer.

We have now defined a range of shortwave frequencies between the MUF and the LUF that can be used for communications. Ham radio operators, the military, aviation, and other nonbroadcast services have the flexibility of deciding each day and hour which frequency can be used for the most reliable communications.

International broadcast stations must decide which frequencies to use many months in advance. Ionospheric propagation analysts or frequency managers are employed to develop frequency schedules that will allow the transmitted programs to reach the desired target areas. Sophisticated computer programs have been developed to aid the analysts.

2.11 SMOOTHED SUNSPOT NUMBER

The sunspot cycles are never identical. Some cycles have been as short as 9 years and others as long as 14 years. Records of sunspot cycles have been kept since 1749. Cycle 24 is generally agreed to have begun in December 2007. Swiss astronomer Rudolph Wolf developed a method, published in 1849, to record the number of observed sunspots as a scientific index called the Smoothed Sunspot Number (SSN). The SSN represents an averaged value of the number of sunspots (see Fig. 2.6.5).

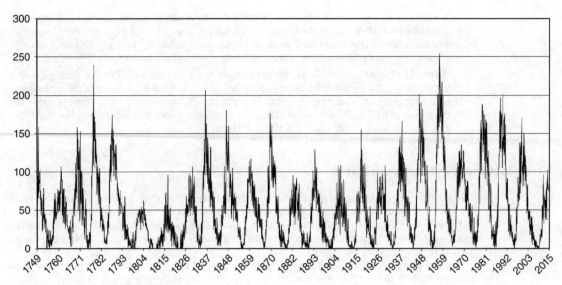

FIGURE 2.6.5 International Smoothed Sunspot Numbers (SSN) since 1749.

The SSN is an important factor in predicting the condition of the ionosphere. The higher the SSN, the greater the level of ionization. The highest peak of sunspots occurred during Cycle 19 in March of 1958. Shortwave communications were better from 1957 through 1959 than they have ever been since the invention of radio.

As you analyze the intended target area from your transmitter location, some signal paths have advantages over others. If you are in the southeastern United States and wish to broadcast to South America, you will be broadcasting to a similar time zone where the condition of the ionosphere is subject to similar daylight or nighttime conditions. You will also be propagating your signal over the stronger tropical ionospheric paths.

If you are broadcasting from Beijing to Brazil, you may be propagating your signal from daylight to darkness or vice versa, and that creates more complex conditions to predict. If you are broadcasting from New England to Scandinavia, you have the additional complication of a weaker ionosphere in the higher latitudes and interference from the aurora borealis when present.

There are computer programs available to assist shortwave broadcast propagation analysts, such as FIELDPLOT, WPlot2000ex, and VOACAP.

Many broadcasters rely upon a version of the Voice of America Coverage Analysis Program (VOACAP) to predict propagation. This program was developed using decades of actual observations of ionospheric conditions to produce a comprehensive computer model of the ionosphere. This program is available from the U.S. Department of Commerce (NTIA/Institute for Telecommunication Sciences; Boulder, Colorado) at http://www.voacap.com/. This program will use the predicted SSN for the desired month, the latitude and longitude of the transmitter and receiver, the take-off angle and directional bearing and type of the transmit antenna, the transmitter power, and the desired receiver signal to noise level. There are several versions of the SSN calculated in different ways by different agencies. It is essential to enter the proper SSN into the VOACAP program. These predicted SSN figures are based on the Lincoln-McNish smoothing function, and can be found at the National Geophysical Database Center in Boulder, Colorado.

2.12 INTERVAL SIGNALS

It is a well-established tradition for shortwave broadcasters to transmit an "interval signal" prior to their broadcasts. This goes back to the days when receivers had slide-rule type dials and it was not possible to precisely tune in a specific frequency. The interval signal was a highly recognizable musical interlude for that station. For Radio Canada, it is the first four notes of their national anthem, "*Oh Canada.*" For the VOA, it is "*Yankee Doodle.*" For the BBC, it is the sound of Big Ben; for Radio Australia, it is "Waltzing Matilda"; and for World Harvest Radio, it is "Onward Christian Soldiers." Station Engineers can use the time that these identification signals are playing immediately after a frequency change, to make final adjustment in the transmitter tuning before the actual programming starts.

2.13 RECEPTION REPORTS

Shortwave stations also rely upon reception reports directly from listeners. There is a hobby called Shortwave Listening (SWL) that is devoted to keeping stations informed of their reception quality. These SWL DXers regularly provide reception reports to the station's engineering department using the SINPO code. They rate the signal in five different categories from 1 to 5, with 5 being the best. SINPO stands for Signal/Interference/Noise/Propagation/Overall Merit (see Table 2.6.5).

TABLE 2.6.5 SINPO Code

Signal	(Man-Made) Interference	(Static) Noise	Propagation	Merit
5 = Very Strong	5 = Clear	5 = None	5 = No Fading	5 = Excellent
4 = Strong	4 = Light	4 = Light	4 = Light Fading	4 = Good
3 = Moderate	3 = Moderate	3 = Moderate	3 = Moderate Fading	3 = Moderate
2 = Weak	2 = Heavy	2 = Strong	2 = Deep Fading	2 = Poor
1 = Unusable	1 = Severe	1 = Severe	1 = Unusable	1 = Unusable

Shortwave stations acknowledge reception reports by sending a verification card or "QSL" card. This confirms the listener's information as to the program details and time of the broadcast. Some SWLs amass a large collection of QSL cards. A sampling of historical cards is shown in Fig. 2.6.6.

Some shortwave stations have regular radio programs for the SWL DXer to encourage them to send in a reception report and collect a QSL card. Stations will often have special promotions commemorating some event and design a special QSL card to induce listeners to send in reports. The DX

FIGURE 2.6.6 Historical QSL cards.

programs often have special information about the station's broadcasts as well as listening conditions for other broadcasters. These program often end their broadcast with a message of "Best Regards" to the listener by using the phrase *"73's and Good DX'ing."*

2.14 DEFINING TERMS

CIRAF Zones—The world is divided into many CIRAF zones. A geographic area is specified by broadcasters to identify the audience they wish to target.

Collision—When an unacceptable interference issue is detected, Frequency Managers negotiate with others to reduce or eliminate the collisions in the HFCC Conference.

Curtain Array—This sophisticated antenna consists of an array adjacent and stacked dipoles that cover one octave of frequencies, such as 9 to 18 MHz. It has the characteristics of high gain, narrow beamwidth, and low take-off angle.

FOT (Frequency of Optimum Traffic)—This is the frequency that reliable shortwave communications can be achieved 90 percent of the time. It is 85 percent of the MUF.

Interval Signal—This is a highly recognizable musical interlude used by shortwave stations prior to a scheduled broadcast that permits listeners to tune in to the station more easily.

Log-perodic (LPA) —This antenna is designed to cover a large range of frequencies such as 5.95 to 26.1 MHz.

MUF (maximum usable frequency)—This is the frequency that will be reflected by the ionosphere with the greatest signal strength.

SSN (Smoothed Sunspot Number)—This is a scientific index that represents an averaged value of the number of observed sunspots.

Take-off Angle—For long distance communications, it is advantageous to have a very low take-off angle of 7° to 25°. This will allow the signal to travel a farther distance before intercepting the ionosphere and reflected at a similar angle back to the earth.

Wood—A frequency request authorized that the broadcaster does not use. This is a practice that is frowned upon.

BIBLIOGRAPHY

Jacobs, George, *The Shortwave Propagation Handbook*, 2nd ed., CQ Publishing, Inc., Hicksville, NY, 1982.
Scholz, Horst, "Shortwave Coordination Developments," HFCC, Prague, Czech Republic, 2012.
World Radio TV Handbook 2001, 55th ed., WRTH Publications Ltd., New York, NY, 2001.

FOR FURTHER INFORMATION

Digtal Radio Mondiale (DRM), www.drm.org
High Frequency Coordination Conference, http://www.hfcc.org/
National Aeronautics and Space Administration, http://solarscience.msfc.nasa.gov/
National Association of Shortwave Broadcasters, Inc., www.shortwave.org
North American Shortwave Association, http://www.naswa.net/
FCC Rules and Regulations, Title 47, Part 73, Subpart F, International Broadcast Stations.
VOACAP Quick Guide, www.voacap.com

CHAPTER 2.7
EVALUATION OF TV COVERAGE AND INTERFERENCE

Bill Meintel
Meintel, Sgrignoli, & Wallace, Warrenton, VA

2.1 INTRODUCTION

In order to insure that television stations are able to meet the goal of providing reliable service to the intended viewers, it is necessary to have a methodology to predict coverage and interference. Likewise, it is also important that all involved parties are using the same methodology. This is important not only for the station operators but for the viewers and the regulatory authority that authorizes the stations' operations, which in the United States is the Federal Communications Commission (FCC).

Prior to the transition to digital television broadcasting, the FCC allotted full-service television channels based on separation distances as shown in Table 2.7.1. The cochannel distances shown in the table assume that the actual carrier frequencies of the stations are offset 10 kHz from one another; otherwise, the required separations are significantly larger.

TABLE 2.7.1 Analog Television Allotment Spacings

	Cochannel	
Zone	Channels 2–13, km (miles)	Channels 14–69, km (miles)
I	272.7 (169.5)	248.6 (154.5)
II	304.9 (189.5)	280.8 (174.5)
III	353.2 (219.5)	329.9 (204.5)
	1st Adjacent Channel	
I, II, III	95.7 (59.5)	87.7 (54.5)

This system was designed to insure that there was little, if any, interference between stations by using large separation distances. In the early days of television there were relatively few stations and spectrum efficiency was not a significant concern. In addition, computers were not available to perform the complex analyses required for a more efficient use of the spectrum.

Prior to the digital transition, coverage and interference predictions generally relied on predicting the location of service contours and any overlapping interfering contours. Service was assumed at all points within the service contour so long as an interfering contour did not extend into the area encompassed by the service contour.

The service contours for analog stations of various classes and frequency are shown in Table 2.7.2.

TABLE 2.7.2 Analog Protected Contours

Band	Full Service Stations (dB µV/m)	Low Power Stations Including Class A (dB µV/m)
Low VHF	47	62
HIGH VHF	56	68
UHF	64	74

The prediction of the extent of these service and interfering contours can be accomplished within a reasonable time frame without the use of computers. The methodology for determining the distance to a contour of a specific field strength value is detailed in the FCC Rules and Regulations Section 73.684, and requires the following information:

- The television channel
- The effective radiated power (ERP) in the direction of interest
- The height of the radiation center above mean sea level (RCAMSL)
- The height above average terrain (HAAT) in the direction of interest
- A set of field strength charts (FCC R&R Section 73.699)

The HAAT is the height that the center of radiation is above the average height of the terrain elevation above mean sea level between 3.2 and 16.1 km (2–10 miles) from the transmitter site. The average is determined by selecting at least 50 elevation points over the distance noted above. In the past, this was done manually by determining elevations from topographic maps; however, today it is done using computers by extracting elevation points from a terrain database. The software currently used by the FCC for processing proposed new or changed television facilities relies on a United States Geological Survey (USGS) database of points spaced at 3-arc second intervals. The HAAT is computed by selecting points from this database at 0.1 km intervals between the required 3.2 and 16.1 km path.

2.2 THE NEED FOR A MORE SOPHISTICATED MODEL

In the early 1990s, the broadcast industry was faced with the possibility of losing more spectrum to the land mobile radio service. It had previously lost TV channels 70 to 83 as well as TV channels 14 to 20 in some of the major U.S. cities. The industry convinced the FCC that it planned to develop a new and improved standard for television that would ultimately free up additional spectrum. However, in that the new system would not be backward compatible additional spectrum would be needed during a transition period to allow simultaneous operation of both the old and new systems. The FCC agreed to postpone any further reallocation of television broadcast spectrum until the standard was developed and the transition completed.

In order to make the needed transition allotments, as well as meeting the promised posttransition spectrum efficiency, it was recognized that greatly improved coverage and interference models would be required. This new model would need to perform a much more sophisticated analysis.

The broadcast industry developed the initial new model and that model formed the bases of the methodology now known as OET-69. The OET-69 name stems from the methodology detailed in the FCC Office of Engineering and Technology Bulletin No. 69. Bulletin No. 69 is only an outline of the methodology to be followed; many of the details were left to the FCC's software developer as the analysis program was created. The FCC has recognized that other software implementations following the OET Bulletin 69 methodology may yield different results due to decisions made in the coding or the computer platform on which the software was implemented. In view of this, an FCC Public notice dated August 10, 1998 and titled "Additional Application Processing Guidelines For Digital Television (DTV)" stated that compliance with the rules will be based on the FCC's implementation of the software.

As previously discussed, the goal was to perform a more sophisticated and accurate coverage and interference analysis that would allow for stations to be more closely spaced while maintaining the existing service to the viewers. One key to reaching that goal was to identify and protect only the actual population being served, rather than the total area within the service contour. This was an entirely new concept since the previous approach was generally to protect service area without regard to population. It is noted that the initial model development, as well as the initial FCC OET-69 model, was only for the evaluation of full-service analog and digital stations. More recently OET-69 has been updated to include low power television stations (LPTVs), including both analog and digital LPTVs, as well as Class A stations. It was also updated to include full-service digital stations that employ a distributed transmission system (DTS) with multiple synchronized transmitters.

2.2.1 Establishing the Service Area to Be Protected

Since the traditional practice had been to define a station's service area as that inside its service contour, it was decided that this would define the evaluation limit for the new approach. In other words, the total population inside the traditionally predicted contour as discussed above would be the protected population.

Digital television, however, requires a lower level of field strength for reception. The factors that went into determining the required field strengths are given in Table 2.7.3a; the actual field strength values are given in Table 2.7.3b. Using the values in Table 2.7.3a the field strength values in Table 2.7.3b can be determined by the following formula:

$$\text{Field strength} = C/N - K_d - K_a - G + L + N_t + N_s$$

TABLE 2.7.3a Planning Factors for DTV Reception

Planning Factor	Symbol	Low VHF	High VHF	UHF
Geometric mean frequency (MHz)	F	69	194	615
Dipole factor (dBm – dBu unit conversion factor)	K_d	−111.8	−120.8	−130.8
Dipole factor	K_a	None	None	See text
Thermal noise (dBm)	N_t	−106.2	−106.2	−106.2
Antenna gain (dBd)	G	4	6	10
Downlead line loss (dB)	L	1	2	4
System noise figure (dB)	N_s	10	10	7
Required carrier to noise ratio (dB)	C/N	15	15	15

TABLE 2.7.3b Digital Protected Contours

Band	Full Service Stations (dB μV/m)	Low Power Stations Including Class A (dB μV/m)
Low VHF	28	43
High VHF	36	48
UHF	41−20 log(615/channel mid-frequency in MHz)	51−20 log(615/channel mid-frequency in MHz)

For example, the service threshold field strength for channel 38, where $K_a = 0$ (see below for an explanation of K_a determination), would be

$$15 - (-130.8) - 0 - 10 + 4 + (-106.2) + 7 = 40.5 \text{ (rounded to 41.0)}$$

The new methodology for digital did, however, include a change in the basic definition of the service contour for ultrahigh frequency (UHF) stations. It was recognized that the level of field strength that defines the service threshold varies by frequency; therefore, an adjustment (dipole factor) is applied to the nominal threshold field strength for channels above and below the geometric mean for the band. At the time the methodology was developed, the UHF band encompassed channels 14 (470–476 MHz) to 69 (800–806 MHz); therefore, the geometric mean frequency is 615 MHz, which falls within channel 38. The adjustments that are applied are determined by the following formula and are reflected in the field strength shown for the UHF band in Table 2.7.4:

$$\text{Dipole factor adjustment} = 20 \log \frac{615}{\text{channel frequency in MHz}}$$

The smallest unit of area for population counting by the Census Bureau is called the *block level*. The blocks are of random size and can be as small as a single apartment building or can cover a fairly large sparsely populated area. It would have been possible to perform coverage and interference analyses for each block; however, it was decided to use a different approach for a couple of reasons. The first reason was computation time as the number of blocks inside the service contour can be quite large, especially in densely populated areas of the country. The second reason is that the computation of field strength to a number of closely spaced points is not expected to vary significantly, thereby negating any advantage to the extra computations.

TABLE 2.7.4 Service Threshold Field Strength (dB μV/m)

Channel	Analog	Digital	Channel	Analog	Digital
2	47.00	28.00	27	63.02	40.02
3	47.00	28.00	28	63.11	40.11
4	47.00	28.00	29	63.20	40.20
5	47.00	28.00	30	63.30	40.30
6	47.00	28.00	31	63.39	40.39
7	56.00	36.00	32	63.48	40.48
8	56.00	36.00	33	63.57	40.57
9	56.00	36.00	34	63.66	40.66
10	56.00	36.00	35	63.74	40.74
11	56.00	36.00	36	63.83	40.83
12	56.00	36.00	37	63.92	40.92
13	56.00	36.00	38	64.00	41.00
14	61.69	38.69	39	64.08	41.08
15	61.80	38.80	40	64.17	41.17
16	61.91	38.91	41	64.25	41.25
17	62.02	39.02	42	64.33	41.33
18	62.12	39.12	43	64.41	41.41
19	62.23	39.23	44	64.49	41.49
20	62.33	39.33	45	64.57	41.57
21	62.43	39.43	46	64.65	41.65
22	62.53	39.53	47	64.73	41.73
23	62.63	39.63	48	64.81	41.81
24	62.73	39.73	49	64.88	41.88
25	62.83	39.83	50	64.96	41.96
26	62.92	39.92	51	65.03	42.03

Instead of performing an evaluation at each Census block, the area inside the service contour is divided into a grid of essentially square cells with the transmitter site at the center of the grid. The population inside the cell is total population of the Census blocks whose reference coordinates fall within the cell boundaries. The geographic coordinates of the evaluation point within the cell is either the centroid of the cell population or the center of the cell if it has zero population. Using the population centroid provides for a more accurate evaluation of the service population in that the computation is being performed at the population center.

Initially the grid is established to encompass an area slightly larger than the service contour. The determination of whether a grid cell is within the service contour is based on the location of the evaluation point; therefore, it is possible that a grid cell with an initial center point outside the service contour will ultimately be included in the analysis because the population centroid point is within the contour. Because of this, it is also possible that some population that is technically outside the service contour may be included depending on the location of the cell evaluation point; likewise, the inverse may also occur. Once it is determined which cells are within the service contour by virtue of their evaluation point location, those cells outside the contour are ignored.

The size of the cells has been limited to a maximum of 2 km on a side. Initially, the FCC allowed the use of any size cell up to the 2 km maximum; however, in a December 2007 Report and Order the FCC set the permissible cell sizes to 2, 1, or 0.5 km. It is further noted that if the station being evaluated is a low power television station (LPTV), including translators and Class A stations, then the maximum allowable cell size is 1 km. Since many of these stations have very small service areas, using the larger (2 km) cell size will not allow for a sufficient number of cells to obtain a reasonable evaluation of the service.

The reason for using a cell size other than the maximum is technically to obtain better accuracy; however, there is also another possible advantage. Changing the cell size also changes the evaluation points as well as the total cell population and this can have an impact on the predicted coverage and interference. In some cases this can mean the difference in whether a particular proposal passes the permitted interference test. The main reason that the FCC has limited the number of permissible cell sizes is eliminate the practice of "shopping" the cell size to find a size that will pass the interference test. As noted previously, changing the cell size changes the evaluation points and the cell population, and it is not unusual for a proposal to fail the interference test at all of the three allowable sizes but pass at some intermediate cell size due to the difference in the evaluation points and the population distribution among the cells. Therefore, by setting the allowable cell sizes at three (two if the station is an LPTV) the number of "shopping opportunities" has been greatly reduced.

2.2.2 Computation of Service and Interfering Field Strengths

Once the grid has been established, the next step is to compute the desired station's predicted field strength at each of the grid evaluation points. These computations are made using the Institute for Telecommunication Sciences (ITS) Irregular Terrain Model, or as it is more commonly known the Longley–Rice model. This model, which will be discussed in greater detail later in this chapter, is designed to compute path loss between the transmitter and a receive point, taking into consideration the intervening terrain. The calculated path loss predicted by Longley–Rice is then subtracted from the field strength predicted by a free space evaluation to determine the final predicted field strength at the individual evaluation points (receive locations).

The path loss returned by the Longley–Rice model is statistical, meaning that the value is for a given percentage of locations, time, and situational variability. For example, predictions of path loss used to determine the service field strength for a digital television station are on the basis of 50 percent of the locations, 90 percent of the time, with a 50 percent situational variability represented as $F(50, 90, 50)$. It is not important that you fully understand this complex statistical relationship.

It is only important that you know which statistical base is used. For our purpose, here there are only three sets that you need to remember.

- Analog service field strength prediction is based on $F(50, 50, 50)$.
- Digital service field strength prediction is based on $F(50, 90, 50)$.
- Interference (analog or digital) field strength prediction is based on $F(50, 10, 50)$.

There are a number of inputs to the Longley–Rice model, some of which vary from study to study and some that are fixed at least for its use in OET-69. The variable inputs are:

1. Frequency of the facility being studied in MHz
2. Height of the transit antenna center of radiation above ground level in meters
3. The terrain elevation profile between the transmit and the receive sites in meters
4. The number of points in the terrain profile
5. The distance increment between the terrain elevation points in km
6. The appropriate statistical values for the desired computation as discussed above

The fixed inputs to the Longley–Rice model as used in OET-69 are:

1. Height of the receive antenna above ground level in meters = 10 meters
2. EPS (relative permittivity of ground) = 15
3. SGM (ground conductivity in Siemens per meter) = 0.005
4. ZSYS (coordinated with setting of EN0 below) = 0
5. EN0 (surface refractivity in N-units (parts per million)) = 301
6. IPO (denotes horizontal antenna polarization) = 0
7. MDVAR (Code 3 sets broadcast mode of variability) = 3
8. KLIM (Climate code; 5 is for continental temperate) = 5

The terrain elevation profile between the transmit and the receive sites is a list of the terrain elevation above mean sea level retrieved from the same USGS 3-arc second database as previously mentioned. The default retrieval increment is 1.0 km although smaller resolutions are permitted by the FCC. The smallest practical increment is approximately 0.1 km since smaller increments would be less than the 3-arc second resolution of data.

In addition to computing the path loss from the transmitter to a specific receive location, it is also necessary to know the power being transmitted over the specific path. The effective radiated power (ERP) depends on the transmitter output power (TPO), the transmission line loss, and the gain of the antenna in the desired direction. For stations that are contained in the FCC database, most of the required information is readily available. The database provides the maximum ERP value, and if the station employs a directional antenna pattern, a tabulation of the relative field values is also provided for the horizontal plane. Unfortunately, until very recently, the FCC has not stored the elevation (vertical plane) antenna pattern relative field values; therefore, that information is only available in a handful of cases.

During the development of the OET-69 and its predecessor models, the lack of elevation pattern data was recognized. The elevation pattern data is a critical part of making accurate predictions for a large number of points over a station's entire service area. In view of that, the FCC Advisory Committee for Advanced Television Service (ACATS) developed a set of typical elevation patterns that can be applied in place of the actual patterns. Typical patterns were provided for each of the three television broadcast bands (low VHF, high VHF, and UHF) with separate patterns for analog and digital for high VHF and UHF (see Table 2.7.5 for full service stations and Table 2.7.5a for Class A and LPTV stations).

TABLE 2.7.5 Standard Antenna Elevation Patterns for Use with OET-69 Analysis Model

Depression Angle	Low VHF		High VHF		UHF	
	Analog	Digital	Analog	Digital	Analog	Digital
0.75	1.000	1.000	1.000	1.000	1.000	1.000
1.50	1.000	1.000	0.950	0.970	0.740	0.880
2.00	0.990	0.990	0.860	0.940	0.520	0.690
2.50	0.980	0.980	0.730	0.890	0.330	0.460
3.00	0.970	0.970	0.600	0.820	0.220	0.260
3.50	0.950	0.950	0.470	0.730	0.170	0.235
4.00	0.930	0.930	0.370	0.650	0.150	0.210
5.00	0.880	0.880	0.370	0.470	0.130	0.200
6.00	0.820	0.820	0.370	0.330	0.110	0.150
7.00	0.740	0.740	0.370	0.280	0.110	0.150
8.00	0.637	0.637	0.310	0.280	0.110	0.150
9.00	0.570	0.570	0.220	0.280	0.110	0.150
10.00	0.480	0.480	0.170	0.250	0.110	0.150

TABLE 2.7.5a Standard Antenna Elevation Patterns for LPTV for Use with OET-69 Analysis Model

Depression Angle	Low VHF		High VHF		UHF	
	Analog	Digital	Analog	Digital	Analog	Digital
0.75	1.000	1.000	1.000	1.000	1.000	1.000
1.50	1.000	1.000	1.000	1.000	1.000	1.000
2.00	1.000	1.000	1.000	1.000	1.000	1.000
2.50	1.000	1.000	1.000	1.000	0.660	0.920
3.00	1.000	1.000	1.000	1.000	0.440	0.520
3.50	1.000	1.000	0.940	1.000	0.340	0.470
4.00	1.000	1.000	0.740	1.000	0.300	0.420
5.00	1.000	1.000	0.740	0.940	0.260	0.400
6.00	1.000	1.000	0.740	0.660	0.220	0.300
7.00	1.000	1.000	0.740	0.560	0.220	0.300
8.00	1.000	1.000	0.620	0.560	0.220	0.300
9.00	1.000	1.000	0.440	0.560	0.220	0.300
10.00	0.960	0.960	0.340	0.500	0.220	0.300

The difference between the analog and digital patterns, as can be seen in Table 2.7.6, is that the digital pattern does not decrease as rapidly as the analog. It is noted that these patterns only have values for depression angles between 0.75° and 10°. For angles less than 0.75°, it is assumed that the relative field value remains constant, and likewise for angles greater than 10°, the relative field value is also assumed to remain constant.

TABLE 2.7.6 Minimum Cochannel D/U Ratios for Analog Interference to DTV

DTV Signal-to-Noise Ratio (S/N) in the Absence of Interference, dB	Desired-to-Undesired Ratio to Protect DTV Reception from Cochannel Analog Transmissions, dB
16.00	21.00
16.35	19.94
17.35	17.69
18.35	16.44
19.35	7.19
20.35	4.69
21.35	3.69
22.35	2.94
23.35	2.44
25.00	2.00

The ERP for a specific path is computed as follows:

$$ERP = ERP_{(max)} \times (\text{Horizontal Relative Field})^2 \times (\text{Vertical Relative Field})^2$$

where $ERP_{(max)}$ is the maximum ERP at any angle, Horizontal Relative Field is the relative field value from the horizontal plane antenna pattern (azimuth pattern) in the direction of the evaluation point, and Vertical Relative Field is the relative field value from the vertical plane antenna pattern (elevation pattern) at the depression angle toward the evaluation point. The depression angle is the angle below (or above) horizontal from the transmitting antenna center of radiation to the evaluation point.

After completing the computations of the service field strengths for each grid point, an evaluation of the results is made to determine which points have a field strength above the threshold for service. The threshold for service varies from band to band and at UHF from channel to channel as given in Table 2.7.3.b. It should be noted that the full service threshold is applied to all stations.

It should be noted that on some paths, due to the particular combination of terrain elevations and antenna heights, the Longley–Rice model will raise a flag that the computation may be unreliable. When this happens the OET-69 methodology assumes that the particular point is above the service threshold but assigns no field strength value to that point. The significance of this is that it will not be possible to determine if this point is receiving interference from other stations because the field strength of the desired station is unknown.

The next step is to assess if any of the grid cells that were determined to have a field strength above the threshold for service are receiving a field strength from other stations that would be of sufficient intensity to disrupt service (cause interference).

Once again the Longley–Rice model is used in the same manner as discussed above to compute the field strength of the undesired (interfering) field strengths. The only difference being that the computations are made on the statistical basis of 50 percent of the locations, 10 percent of the time with a 50 percent situational variability, represented as $F(50, 10, 50)$. It is also possible on these paths that the computation may be flagged as unreliable and in that case the OET-69 policy is to assume no interference is caused.

The actual determination of whether interference is caused at the individual locations is by determining the ratio of the desired field strength to the undesired field strength (D/U ratio). The D/U ratios that define the threshold for interference are given in Table 2.7.7; however, in the case of cochannel interference to digital, the D/U ratio shown in the table is only for the

TABLE 2.7.7 Interference Criteria Permitted Ratio of Desired-to-Undesired Field Strength, D/U Ratios (dB)

Offset of Undesired Channel from Desired Channel	Analog into Analog	DTV into Analog	Analog into DTV	DTV into DTV
Criteria for Co- and 1st Adjacent Channels				
−1	−3	−14	−48	−28
0	28	34	2	15
+1	−13	−17	−49	−26
Criteria for UHF Analog Taboo Channels (NC means Not Considered)				
−8	−32.0	−32.0	NC	NC
−7	−30.0	−35.0	NC	NC
−4	NC	−34.0	NC	NC
−3	−33.0	−30.0	NC	NC
−2	−26.0	−24.0	NC	NC
+2	−29.0	−28.0	NC	NC
+3	−34.0	−34.0	NC	NC
+4	−23.0	−25.0	NC	NC
+7	−33.0	−43.0	NC	NC
+8	−41.0	−43.0	NC	NC
+14	−25.0	−33.0	NC	NC
+15	−9.0	−31.0	NC	NC

TABLE 2.7.7a Digital LPTV First Adjacent Channel—Interference Criteria

	Permitted Ratio of Desired-to-Undesired Field Strength, D/U Ratios (dB)	
	Simple Mask	Stringent Mask
Into Analog	10	0
Into Digital	−7	−12

case where the desired signal has a signal-to-noise ratio (SNR) equal to or greater than 28 dB for digital-into-digital or 25 dB or greater for analog-into-analog. The SNR is provided by the following formula:

$$SNR = FS - DP\_NLC$$

where FS is the computed field strength for a specific evaluation point and DP_NLC is the threshold for service including the dipole factor adjustment.

Below these SNR levels an adjustment is added to the values shown in Table 2.7.7a. The adjustment for digital-into-digital is provided by the following formula:

$$\text{Adjustment to D/U ratio} = 10\log_{10}\frac{1.0}{1.0 - 10^{-SNR/10}}$$

For example, if a station is operating on channel 26 and the computed field strength is 44.5 dB μV/m, then from Table 2.7.8 the channel 26 dipole factor adjusted service threshold = 39.92 (41.0 − 1.08); therefore, SNR = 44.5 − 39.92 = 4.58.

TABLE 2.7.8 Front-to-Back Ratios Assumed for Receiving Antennas

TV Service	Low VHF (dB)	High VHF (dB)	UHF (dB)
Analog	6	6	6
Digital	10	12	14

Inserting this SNR value into the above formula yields an adjustment of 1.86; therefore, in this case the D/U ratio for digital-into-digital = 16.86 (15.0 + 1.86) and not 15.0.

If the undesired station is analog, then the D/U ratio for SNR values less than 25 is determined by a linear interpolation of the values given in Table 5B of FCC Bulletin OET-69 and repeated here in Table 2.7.6.

There is an exception to the values in Table 2.7.7 for digital LPTV stations. The D/U ratio used to protect other station on a first adjacent channel to a digital LPTV is given by the values in Table 2.7.7*a*. The specific value is dependent on the out of band emission mask filter being employed by the interfering station. These stations have three mask filter options designated as simple, stringent, and full-service. If either a simple or stringent mask is used then the D/U ratio is taken from Table 2.7.7*a*; however, if the station has opted to use the same filter as used by full-service stations then the values in Table 2.7.7 are to be used.

2.2.3 Inclusion of Receive Antenna Pattern

The OET-69 methodology assumes that a receive antenna located outside at 10 meters above ground level is being employed at each receive location. Therefore, before computing the D/U ratio, the undesired signal level is adjusted to take into consideration the orientation of a standard receive antenna with respect to the location of the undesired station. It is always assumed that the receive antenna is pointed in the direction of the desired station. The receive antenna pattern was developed by the FCC Advisory Committee for Advanced Television Service (ACATS) during the development of the digital television system. The formula that is used to determine the antenna discrimination is as follows:

$$\text{Relative gain (dB)} = 80\log(\cosine(\text{angle}))$$

where the angle is the difference between the antenna orientation and the direction of the undesired station. Since the cosine of the angle will never be greater than one, the gain will be less than or equal to zero. This formula will produce rather large negative values as the angle increases; therefore, the adjustment is capped at minus the antenna front-to-back ratio given in Table 2.7.8.

For example, if the desired station is digital on a UHF channel and the angle to the undesired station is 60° off axis, the formula would indicate an adjustment of approximately –24 dB; however, using the front-to-back ratio from Table 2.7.8, the adjustment will only be –14 dB.

Once the receive antenna adjustment to the undesired station's field strength has been made, the ratio of the desired station's field strength to the adjusted undesired station's field strength is determined and compared to the permitted level given in Table 2.7.8 to determine if interference is caused.

All interference assessments are based on a single interfering station. There is no aggregation of the field strengths from multiple interferers, with one exception. The exception is if the undesired station is utilizing a distributed transmission system (DTS) employing multiple synchronized transmitters. In that case, the field strengths contributed by the individual transmitters are combined using the root sum square (RSS) method.

The overall predicted station coverage and interference is determined by summing the results for the individual computation points that were determined to be within the desired station's service contour. An example of the results provided by the standard OET-69 software is shown in Fig. 2.7.1. Line 1 of this example indicates that the desired station has a population of 2,848,395 within its service contour that encompasses an area of 34,010.9 km^2. Line 2 shows the results of the Longley–Rice service analysis where due to the effects of the terrain the predicted service population has been reduced to 2,288,704 and the area reduced to 28,141.4 km^2. This means that the analysis has

Sample OET-69 Study Results

	POPULATION	AREA (sq km)
within Noise Limited Contour	2848395	34010.9
not affected by terrain losses	2288704	28141.4
lost to NTSC IX	0	0.0
lost to additional IX by ATV	74147	1193.9
lost to ATV IX only	74147	1193.9
lost to all IX	74147	1193.9

FIGURE 2.7.1 Sample results.

determined that cells covering an area of 5869.5 (34,010.9 − 28,141.4) km² containing a population of 559,691 had a predicted field strength below the service threshold. The following lines contain information concerning any additional service losses due to interference (IX) from other stations. Line 3 shows that no interference was being caused by analog (NTSC) stations. Line 4, however, indicates that interference from one or more other digital (ATV) stations is causing a loss of service to an area of 1193.9 km² containing a population of 74,147. Line 6 provides the total service loss combining the losses caused by both analog and digital stations. Therefore, this station is predicted to be providing service to an area of 26,947.5 km² containing a population of 2,214,557.

Line 5 shows the predicted service loss that is caused by digital stations ignoring any overlapping interference from analog stations. In this example, it is the same as line 4 since no interference is received from analog stations.

This software was developed at a time when both full-service analog and digital stations were still in operation and line 5 was provided to show what interference would be present when analog operation was terminated. Now that there are no longer any full service analog stations in operation and many of the low power stations have also transitioned to digital, the information on line 5 has diminishing value.

2.2.3.1 Performing a Study. Up to this point, the discussion has been about the details concerning the computation of desired and undesired field strengths in order to determine if service is to be expected at individual points within a station's service contour. With that completed, it is possible to determine the interference-free predicted service, in terms of both population and area. The total interference-free predicted service is just the sum of either the population or the area within the cells that have a field strength above the threshold for service and are not subject to interference.

This chapter is about evaluation of TV coverage and interference and up to this point we have discussed how service and interference are predicted. When performing these studies there needs to be a desired station and a potential interfering station or stations. Due to the large number of authorized stations, a typical study will include a number of potential interfering stations. Because of the geographic separation distances between stations, as well as the channel separations, it is not necessary to include every station as a potential interferer; therefore, a criterion exists that will narrow the list of potential conflicts and significantly reduce the computation time.

The first step is to assess which channels are to be included, and this depends on the type of desired station—analog or digital. Table 2.7.8 will let us determine which channels to include. Only the channel relationships that have a D/U ratio specified in this table need be included. For example, if the desired station is digital then only stations on the same channel or on a first adjacent channel need to be included. On the other hand, if the desired station is analog then a much greater number of channel relationships will have to be considered.

After reducing the universe of potential interfering stations based on their channel relationship to the desired station, the next step is to determine the separation distance between the desired station and the remaining group of potential interfering stations. Based on studies of the potential for interference, a table of maximum separation distances has been developed. Stations separated by more than the distances in this table are highly unlikely to impact one another and therefore are not required to be included as potential interfering stations. The maximum separation distances are shown in Tables 2.7.9 and 2.7.9a.

TABLE 2.7.9 Maximum Search Distance for Potential Interfering Stations

Band	Channel Relationship	Desired Station Type	Distance (km)
Low VHF	Cochannel	Digital	429
Low VHF	Cochannel	Analog	429
Low VHF	1st-Adjacent channel	Digital	229
Low VHF	1st-Adjacent channel	Analog	229
High VHF	Cochannel	Digital	429
High VHF	Cochannel	Analog	420
High VHF	1st-Adjacent channel	Digital	229
High VHF	1st-Adjacent channel	Analog	220
UHF	Cochannel	Digital	429
UHF	Cochannel	Analog	407
UHF	1st-Adjacent channel	Digital	229
UHF	1st-Adjacent channel	Analog	207
UHF	Taboo channels*	Analog	142

*Taboo channels = $N \pm 2, 3, 4, 7, 8$ and $N + 14, N + 15$

TABLE 2.7.9a Maximum Search Distance for Potential Interfering Stations

Analog LPTV to Analog LPTV (Including Class A)			
	Distance (km)		
Channel Relationship	Low VHF	High VHF	UHF
Cochannel	225.0	188.0	247.0
First Adjacent Channel	45.0	39.0	48.0
UHF TABOO Channels			59.0
Taboo channels = $N \pm 2, 3, 4, 7, 8$ and $N + 14, N + 15$			
Analog LPTV to Digital Full Service			
	Distance (km)		
Channel Relationship	Low VHF	High VHF	UHF
Cochannel	394.0	394.0	394.0
First Adjacent Channel	134.0	134.0	134.0
UHF TABOO Channels			NC

The above culling defines the list of potential interfering stations; however, when the actual analysis is performed there is a further culling involving the distance between the undesired station and the desired station cell evaluation points. Table 2.7.10 shows the maximum distance to the cell where evaluation is to be performed. If the distance from the undesired station to the cell is beyond this distance, then no computation is performed and it is assumed that no interference would be caused.

TABLE 2.7.10 Maximum Site to Cell Interference Evaluation Distance (km)

Offset Relative to Desired Channel N	Undesired Channel	Analog into Analog	Digital into Analog	Analog into Digital	Digital into Digital
−8	$N-8$	35	35	NC	NC
−7	$N-7$	100	35	NC	NC
−4	$N-4$	NC	35	NC	NC
−3	$N-3$	35	35	NC	NC
−2	$N-2$	35	35	NC	NC
−1	$N-1$	100	100	100	100
0	N	300	300	300	300
+1	$N+1$	100	100	100	100
+2	$N+2$	35	35	NC	NC
+3	$N+3$	35	35	NC	NC
+4	$N+4$	35	35	NC	NC
+7	$N+7$	100	35	NC	NC
+8	$N+8$	35	35	NC	NC
+14	$N+14$	100	35	NC	NC
+15	$N+15$	125	35	NC	NC

NC = not considered.

Although it is useful to know the predicted service of an individual station, there is also the need to asses if a new proposal or the modification of an existing station will fit into the existing environment. In other words, does the proposal cause an unacceptable level of new interference to other station(s) and/or will it be able to provide the desired level of interference-free service?

Typically, the evaluation of a new proposal will entail an interference study to a number of other stations which are determined in a manner similar to that discussed above. In other words, the separation distances in Tables 2.7.9 or 2.7.9a are applied to find the potentially affected stations. Once the list is determined then each station in the list is evaluated to determine its current level of service, and then a second evaluation is performed to assess the predicted service with the new proposal considered. If the desired station's predicted service population is reduced when the proposal is considered, a percentage difference is computed to determine if the increase in interference is within the permitted range.

The FCC has defined certain levels of new interference as de minimis, meaning that it is small enough to cause only a very minor impact and is therefore acceptable. The de minimis level depends on the class of stations involved. Table 2.7.11 shows the permitted levels of new interference.

One additional point to be aware of is the fact that many times a particular station may have more than one current record in the FCC database for the same station. It is possible for a station to have a licensed facility with a set of operating parameters as well as a construction permit for a new set of parameters. In some cases, there may even be an application for yet another set of parameters. This will typically be for modification of an outstanding construction permit. All of these facilities will need to be included in the study of a proposed new station.

The station license must be protected since it is the current operating facility; the construction permit facility is also required to be protected since the station has been given the authority to build the new facility. The application will also need to be protected until it is acted upon by the FCC, provided that it was filed before the proposal being studied.

These multiple records make the analysis of a new proposal very complicated since there are typically a number of such situations. This means that not only do we need to study the proposal toward

TABLE 2.7.11 Allowable New Interference Percentages

Class of Station		New Interference Permitted,
Desired Station	Interfering Station	Less Than (percent)
Full Service	Full Service	0.5
Full Service	Analog Class A	0.5
Full Service	Digital Class A	0.5
Full Service	Analog LPTV	0.5
Full Service	Digital LPTV	0.5
Analog Class A	Full Service	0.5
Analog Class A	Analog Class A	0.5
Analog Class A	Digital Class A	0.5
Analog Class A	Analog LPTV	0.5
Analog Class A	Digital LPTV	0.5
Digital Class A	Full Service	0.5
Digital Class A	Analog Class A	0.5
Digital Class A	Digital Class A	0.5
Digital Class A	Analog LPTV	0.5
Digital Class A	Digital LPTV	0.5
Analog LPTV	Full Service	No Protection Required
Analog LPTV	Analog Class A	2.0
Analog LPTV	Digital Class A	2.0
Analog LPTV	Analog LPTV	2.0
Analog LPTV	Digital LPTV	2.0
Digital LPTV	Full Service	No Protection Required
Digital LPTV	Analog Class A	2.0
Digital LPTV	Digital Class A	2.0
Digital LPTV	Analog LPTV	2.0
Digital LPTV	Digital LPTV	2.0

each of these records, but the existing service of the stations the proposal potentially affects will be different depending on what stations are included in its existing interference. This requires that the analysis evaluate all of the possible combinations of authorizations.

In the time period when the initial digital television applications were being filed and analog stations were still operating and required protection, the number of scenarios in a typical study sometimes ran into the tens of thousands and could take an entire day or more to study one application. Today, the number of such situations has diminished greatly because the transition from analog to digital has been completed and only a small number of changes are being considered at any one time; however, that situation could change if the spectrum is repacked. The one thing to remember is that making a change in one facility has the potential to impact multiple other stations.

In addition to protecting other television broadcast stations, there are some additional protections that must be taken into consideration. There is a requirement to protect land mobile operations on UHF channels 14–20. This protection is based on a required distance separation between the broadcast station transmitter site and the reference coordinates assigned to the area where the television channel is being used for land mobile. The required separation distance is 250 km for cochannel operation and 176 km for first-adjacent-channel operation. The reference coordinates and assigned channels are listed in the FCC Rules and Regulation Section 73.623(e).

Protection and/or notifications are required when operating stations in close proximity to various radio astronomy, research, and receiving installations. These are defined in FCC Rules and Regulations Section 73.1030, which also details the specific requirements when operating near these locations.

Requests for new allotments on channel 6 must submit an engineering study demonstrating that no interference would be caused to existing FM radio stations on FM channels 200–220 (see FCC R&R 73.623(f)).

Finally, construction of a new tower near an existing AM broadcast facility must be carefully planned so as not to disturb the AM station's antenna radiation pattern.

2.3 THE LONGLEY–RICE MODEL

As stated previously in this chapter, the initial improvements to coverage and interference analysis software were developed for the broadcast industry. Development of a completely new propagation model for broadcast television would have been an enormous undertaking. It would have been well beyond the industry's financial resources, much less the time constraints dictated by the urgency to move forward the planning of a new standard for television delivery. Therefore, a decision had to be made concerning the selection of an existing propagation model. The major contenders that were considered were the Longley–Rice model and the Terrain Integrated Rough Earth Model (TIREM) that had been developed for the U.S. Department of Defense. It is noted that, at the time there was more than one version of TIREM and the version considered was the one that was freely available to the public.

Both models were tested in a nationwide study of coverage and interference and the results were compared. Although there were individual differences in the results, the overall results for the nationwide study were very similar. Therefore, the Longley–Rice model was chosen mainly because it was widely used and there was a single stable and well-documented version publicly available.

The ITS Boulder, Colorado, is part of the U.S. Department of Commerce and was responsible for the development of the ITS Irregular Terrain Model, or as it is more commonly known the Longley–Rice model. The model gets its name from the main developers—A.G. Longley and P.L. Rice. Longley and Rice are the authors of a 1968 work entitled "Prediction of Tropospheric Radio Transmission Loss Over Irregular Terrain: A Computer Method," which forms the basis of the propagation model used in OET-69.

The model is best described in the following taken from NTIA REPORT 82-100.*

"The ITS model of radio propagation for frequencies between 20 MHz and 20 GHz (the Longley–Rice model) is a general purpose model that can be applied to a large variety of engineering problems. The model, which is based on electromagnetic theory and on statistical analyses of both terrain features and radio measurements, predicts the median attenuation of a radio signal as a function of distance and the variability of the signal in time and in space."

A large part of the Longley–Rice model is based on basic propagation theory. In view of this, it is best to start with a review of the underling theory. There are four types of propagation.

1. Free Space—straight line path in a vacuum
2. Refraction—bending caused by the transmission medium
3. Reflection—path altered by an obstacle
4. Diffraction—bending caused by an object in the path

---

*NTIA Report 82-100, "A Guide to the Use of the ITS Irregular Terrain Model in the Area Prediction Mode," George A. Hufford, Anita G. Longley, and William A. Kissick.

A brief discussion of each of these four will provide a clearer understanding of the underlying theory.

Free space propagation requires that the path be clear of absorbing or reflecting objects and is typically expressed as

$$E(\text{V/m}) = \sqrt{\frac{30gP}{d}}$$

where g = gain of antenna, P = power delivered to the antenna in Watts, and d = distance in meters. In broadcast work, the values are typically converted to dB above 1 μV/m.

The next type of propagation is refraction by the medium (e.g., air in the troposphere). Bending is caused by changes in density of the medium that is a function of height. This type of refraction typically causes radio waves to be bent back toward the earth as arcs of circles with bending extending to the radio horizon, as shown in Fig. 2.7.2. Changing the earth's radius (typically 4/3 radius) allows these paths to be considered as straight lines.

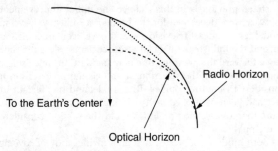

FIGURE 2.7.2 The optical and radio horizon.

The third type of propagation is reflections from an object such as the earth, as illustrated in Fig. 2.7.3. Reflections can be either *constructive* or *destructive*. The effect depends on the difference in path length and is a function of path clearance when objects are encountered. The required clearance is called the Fresnel–Zone radius. This radius is determined by the following formula where the terms are taken from Fig. 2.7.4:

$$R = \sqrt{\pi \frac{d_1 d_2}{d}}$$

The final type of propagation is diffraction where a radio wave is bent by an obstacle, which allows for transmission around an object, although with some loss. There are two types of diffraction of radio waves and the effects are based on optic theory. The first is diffraction caused by a smooth earth and the second is caused by hills or mountains. The second is modeled as a knife edge diffraction.

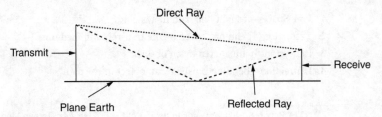

FIGURE 2.7.3 Illustration of reflection of a radio wave.

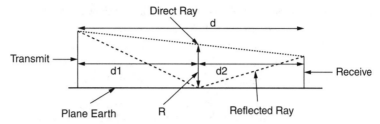

FIGURE 2.7.4 Fresnel–Zone radius.

The various modes of propagation are generally a function of the distance from the transmitter. The radio horizon distance is typically assumed as the line of site region where the expected impairment is multipath. In the region on either side of the radio horizon, distance diffraction over earth's curvature and terrain features become a factor. Beyond the radio horizon the main impairment becomes tropospheric scatter.

The Longley–Rice model was developed in the 1960s and is semiempirical with the theoretical treatment of reflection, refraction, diffraction, and scatter adjusted to fit the measured data. It is applicable to a wide variety of "normal" conditions; however, it is not intended for predictions in urban conditions or dense forest. The latter two conditions require additional obstacle information that was not taken into account in the modeling. It is also not intended to predict short-term variations in that its basis is the yearly median.

There are actually two prediction models contained in the Longley–Rice software. These are an Area Model, where the results are more generalized, and a Point-to-Point Model, which is the model used in the OET-69 methodology.

As noted above, the model is semiempirical and relies on measurement data gathered from VHF Land Mobile type service in Colorado and northern Ohio, and on the VHF and UHF television measurements gathered in studies by the Television Allocations Study Organization (TASO). It also includes climatic data from ITU CCIR work.

The actual determination of the loss factor that is to be applied to the free-space computation in order to arrive at the predicted field strength is a three-step process:

1. Set up the problem
2. Compute the reference attenuation in dB
3. Compute adjustments to reference attenuation in dB

The set up portion involves providing the various data parameters related to the particular computation to be performed, as discussed earlier in this chapter and will not be repeated here. The next step is to compute the reference attenuation (the median attenuation relative to free space). The computation methodology for this step depends on the propagation mode (Line-of-Sight Region, Diffraction Region, or Scatter Region) and the specific mode depends on the distance, antenna heights, and the intervening terrain.

This step deviates from basic calculations in that antenna heights are adjusted to an effective height based on terrain, and the reflecting plane is based on a straight line fit of the terrain profile. In addition, the step includes some additional modification and weighting factors of some basics based on fits to the measurement data.

The final step in the three-step process is the adjustment to the reference attenuation determined in step two to account for:

1. The climate in the area of concern is selected from the options below (OET-69 uses Continental Temperate for all computations):

   Equatorial (derived from CCIR developed curves)

   Continental Subtropical (derived from CCIR developed curves)

Maritime Subtropical (derived from CCIR developed curves)
Desert (derived from CCIR developed curves)
Continental Temperate (based on analysis of actual data)
Maritime Temperate, over land (based on analysis of actual data)
Maritime Temperate, over sea (based on analysis of actual data)

It has been noted that there is very little difference between Continental and Maritime Temperate in the first 100 km. It has also been noted that longer paths in Maritime Temperate are subject to super refraction and ducting 10 percent of the time.

2. The requested statistical computation (percent of time, locations, and confidence). For example, DTV service predictions are for 50 percent of the locations, 90 percent of the time. For broadcast work the Confidence (situation variability) is always set to 50 percent, since broadcast allocations are based on the median situation.

Statistical adjustments are based on the empirical data for climate, distance, and antenna height. A large number of actual measurements were evaluated and the model matches the measurement statistics and situation to parameters of the current evaluation to determine the appropriate adjustment.

In summary, the prediction of the field strength using the Longley–Rice model requires the following:

- Define the problem
- Compute the reference attenuation
- Compute adjustments based on climate and the desired statistic
- Subtract the result from free space field strength

2.4 FINAL THOUGHTS

Propagation prediction is very complex, making modeling extremely difficult. In view of this complexity, all propagation models are a compromise and should only be viewed as one tool in prediction of coverage and interference. One also needs to remember that model predictions are statistical and based on medians over a long time frame, whereas actual measurements are a "snapshot" in time and therefore rarely agree with one another.

CHAPTER 2.8
DTV RF CONSIDERATIONS

Douglas Garlinger
CPBE, 8-VSB, Indianapolis, IN

2.1 INTRODUCTION

The Advanced Television Systems Committee (ATSC) developed the 8-VSB standard to permit the transmission of HDTV (high-definition television) pictures or multiple SDTV (standard-definition television) channels within the same 6 MHz channel spectrum that had been occupied by an NTSC analog channel. The digital signal is transported to the 8-VSB transmitter exciter as an MPEG-2 bitstream with a data payload rate of 19.39 Mbit/s. Television exciters accept this input signal as a SMPTE-310M signal or an ASI (asynchronous serial interface) signal.

This handbook contains in-depth treatments of the ATSC DTV transmission system and DTV television RF measurements in other chapters. This chapter will treat the 8-VSB signal in a different manner to aid some readers in understanding the basics of 8-VSB in concepts already familiar to the reader. As a result of this approach, this chapter will cover some of the material found in those other two chapters.

2.2 8-VSB SIGNAL

The 8-VSB transmitter is an AM transmitter. The transmitted 8-VSB signal is a modulated analog carrier whose radio frequency (RF) envelope varies with the data modulation. The "8" in 8-VSB refers to eight digital levels that correspond to very specific analog amplitude levels of the 8-VSB modulated RF envelope. The polarity and eight specific levels of I-component amplitude during each clock period of the I-signal are used to symbolize three bits of digital data. There are four positive levels (+7, +5, +3, and +1) and four negative levels (−7, −5, −3, and −1). The clock rate for the 8-VSB signal is 10.762 MHz, or 10,762,238 symbols each second. This is often referred to as a *symbol rate* of 10.76 Megasymbols/s or a symbol period of 92.9 ns.

At 3 bits per symbol, the symbol rate of 10.76 MHz would appear to permit 8-VSB data transmission at the rate of 32.28 Mbit/s. More than one-third of this data is used for data synchronizing signals and *forward error correction* (FEC). If you are an IT person, it may help you to think of the 8-VSB transmitter and receiver as simply a one-way modem. The receiver has no way to inform the transmitter that it did not properly receive data. A forward error correction of 2/3 is employed to aid the receiver's ability to correct data errors. The actual payload of the 8-VSB systems is 19.39 Mbit/s.

The 8-VSB receiver can produce a perfect picture in the presence of a high noise floor. The *carrier-to-noise* (C/N) threshold of data errors for the 8-VSB RF signal is specified at 15 dB. The received 8-VSB picture will either be "studio quality," or the 8-VSB receiver will sense an excessive "bit error rate" in the received data signal. If the bit error rate is excessive, the received picture will begin to display pixelization or a "freeze frame." This sudden loss of the received signal is called the *cliff effect*.

2.2.1 Fundamentals of SSB

Before examining the 8-VSB signal more closely, let's take a look at the basic principles of the well-known single-sideband (SSB) transmission. An SSB signal used on the ham bands can be thought of as an audio signal translated to an RF signal. You will remember much of this from your basic electronics theory. If a single audio tone of 1000 Hz is applied to a balanced modulator and an RF CW (continuous wave) carrier of 455 kHz is applied to the other input of the balanced modulator, sum and difference RF frequencies would be created as products of the mixing process. The sum frequency of 456 kHz (455 + 1), the difference frequency of 454 kHz (455 − 1), and the original 455 and 1000 Hz signals would all be present within the modulator. These two additional RF frequencies, appearing in the upper and lower sidebands as shown in Fig. 2.8.1, are separated from the carrier by the audio modulating frequency. The balanced modulator is adjusted, or "balanced," so that the carrier signal cancels itself out. The audio signal unbalances the modulator at an audio rate, producing a double sideband (DSB) output with the RF carrier suppressed. The sidebands are shown in Fig. 2.8.1, but the important phase relationship of the sidebands is not shown.

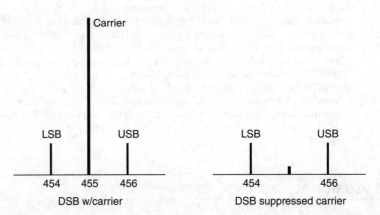

FIGURE 2.8.1 AM sideband characteristics.

An SSB signal can be generated by a technique known as the *phasing method*. Consider the original 1000 Hz audio signal as the in-phase (I) signal. A quadrature (Q) signal delayed 90° is generated from the I-signal. A 90° delayed signal is also created for the incoming RF carrier. When one RF and one AF signal are applied to two separate balanced modulators, the result is a DSB output from each modulator (see Fig. 2.8.2). When the two DSB signals are combined in an amplifier, their phase relationship cancels one sideband and strengthens the other.

The result is a *single-sideband suppressed carrier* (SSSC) signal. Further suppression of the carrier and bandwidth limiting can be obtained by passing the signal through a sharp bandpass filter. The filtered SSB signal can then be fed directly to a linear RF amplifier for transmission.

In a perfect SSB system, a single audio tone will produce an output from the bandpass filter that is a simple CW signal. The *peak-envelope power* (PEP) of the transmitted signal will be the same as the average power. When additional tones are added, the interaction of the various frequencies beating with each other will produce multiple sidebands. Two tones in an SSB system produce a PEP equal to twice the average power, while the PEP produced by four tones is four times the average power.

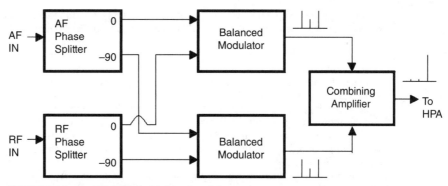

FIGURE 2.8.2 Simplified SSB dual-balanced modulator.

Since a square wave is composed of an infinite number of odd-harmonics, application of a square wave to an SSB modulator would theoretically produce an infinite PEP with infinite bandwidth. Figure 2.8.3 emphasizes the point that the shape of the SSB RF envelope may bear no resemblance to the original modulating signal.

A voice signal is a complex assortment of audio frequencies producing many small sidebands when applied to the input of an SSB modulator. Some of these small sidebands are "in-band" and will

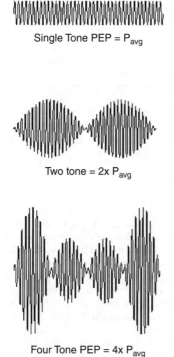
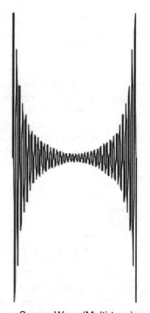

FIGURE 2.8.3 SSB waveforms.

pass through the SSB bandwidth filter unattenuated. "Out-of-band" frequencies, occurring outside the SSB channel's assigned bandwidth, will be attenuated by the SSB bandpass filter.

Average speech has an instantaneous-to-average power range of 15 dB. The linear RF amplifiers following the SSB bandpass filter must be able of handle significantly higher power than that demanded by average speech. With proper speech processing, a power ratio of two to one, or 3 dB, has been found adequate for amateur radio SSB application. Therefore, a transmitter capable of 2000 W PEP can handle an SSB signal whose average power is 1000 W.

If a voice peak occurs that is beyond the capability of the RF amplifier to provide linear amplification, the signal will be clipped by the linear amplifier. This produces a small distortion, or nonlinearity, in the transmitted signal. Additional and undesirable in-band and out-of-band sidebands are produced within the linear amplifier. The out-of-band sidebands may cause "splatter" to adjacent SSB channels.

This potential problem can be largely eliminated through the use of proper speech processing. Modifying the dynamic range of the audio signal before it is applied to the SSB balanced modulator will leave the received SSB signal perfectly acceptable to the ear of the listener.

8-VSB transmissions pose many of the same challenges encountered in a simple SSB transmission. Instead of using a complex speech waveform, 8-VSB uses a complex I-signal waveform. The 8-VSB signal may require a peak power output 6–8 dB above the average power. If the RF amplifier is saturated, clipping the signal, undesirable in-band and out-of-band products will be generated. The 8-VSB receiver—which is really just one-half of a modem—is not as forgiving as the ear of the SSB listener. The 8-VSB receiver will suffer a decrease in signal quality. Uncorrectable data errors may occur and disturb the receiver's video and audio reproduction.

2.2.2 8-VSB Channel Occupancy

The 8-VSB RF signal is essentially an SSSC transmission. Its major advantage over conventional AM transmission is its greater efficiency. This is very similar in principle to ham radio or military SSB transmissions.

Suppression of the carrier and lower sideband allows virtually all the available transmitter power to be used to convey the information contained in the upper sideband. Transmission of only the upper sideband uses half the spectrum required for conventional AM. The suppressed carrier is referred to as the "pilot" and is transmitted approximately 310 kHz above the lower band edge of the channel. A portion of the lower sideband is also transmitted.

8-VSB uses a complex I-signal waveform created from the digital audio and digital video MPEG-2 transport stream. A digital technique known as a *Hilbert transform* is used to provide a uniform 90° phase shift for all of the signal components required for 8-VSB modulation. It is this 90° phase shift that allows for the cancellation of the lower sideband.

2.2.3 I-signal

The actual 8-VSB RF signal demodulated by the 8-VSB receiver is a signal vector that is constantly changing in amplitude and phase. Its instantaneous value can be represented as the vector sum of an I-component signal and a Q-component signal. The I-signal is in phase with the original RF carrier and contains all the digital data information. The Q-signal is a quadrature-phase signal that contains no data information and is related to the I-signal by a fixed transfer function.

The demodulated 8-VSB RF signal shown in Fig. 2.8.4 illustrates 40 symbol periods of the I-component signal. The markers (dots) on the I-signal waveform indicate the precise mid-point of each symbol's duration. The 8-VSB receiver samples the I-component signal at this precise symbol time and measures only the desired symbol amplitude corresponding to one of the eight specific data levels. The digital transmitter must be able to transmit the 8-VSB signal without distortion to these eight amplitude levels. Any nonlinearity in the transmitter can contribute to data errors.

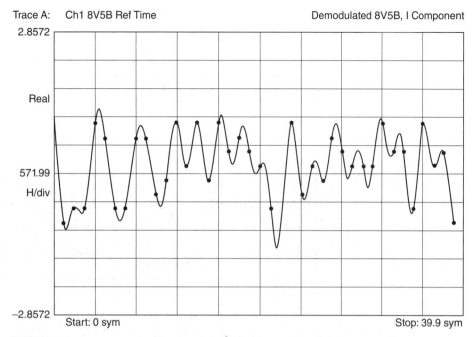

FIGURE 2.8.4 I-component signal for 40 symbol periods. (*Courtesy: Comark Communications.*)

2.2.4 Pilot Insertion

The 8-VSB RF signal makes use of a small in-phase pilot carrier that is actually the suppressed carrier used during the 8-VSB modulation process. This pilot carrier is specified as 11.6 dB down from the average 8-VSB power, and it adds only 0.3 dB to the average data power. The pilot signal is useful for assisting with the receiver's carrier lock and for pointing the receive antenna. It also eliminates a potential ambiguity in the receiver; were the I-signal to become inverted, confusing data could be presented to the receiver (e.g., a data level of −7 could be mistaken for +7).

A DC offset signal of +1.25 is added to the I-data signal. This shifts the axis of the I-signal, offsetting the data to more positive levels; for example, the −1 data level becomes a +0.25 level. This DC offset signal prevents the carrier from being completely canceled out in the modulation process.

2.2.5 VSB Filter

The digital output of the exciter data processing circuitry is converted to an analog signal. The I-signal is driven positive or negative for the entire duration of the 92.9 ns symbol period. There are four positive levels (+7, +5, +3, and +1) and four negative levels (−7, −5, −3, and −1).

If the data symbols were to occur in consecutive order, the I-signal levels would bear some resemblance to a linearity stair-step at the output of the D/A converter (see Fig. 2.8.5). Such a successive pattern, however, would not occur in an actual signal owing to the randomized nature of the data. If two of the same data levels did occur consecutively, the digital I-signal would remain at the same level for the duration of two symbols.

Transmission of the digitally generated I-signal with its sharp rise times would require a great deal of bandwidth. The I-signal is specially filtered into a pulse shape that limits the bandwidth while carefully preserving the I-signal's analog value at the precise mid-point of the symbol period. One type of filter which can perform this function is the raised cosine filter, also known as a *Nyquist filter*.

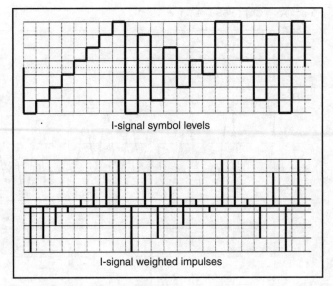

FIGURE 2.8.5 I-signal characteristics.

This is a filter with a flat amplitude response over most of its bandwidth, and a raised cosine frequency response at the bandpass edges.

Before the I-signal is applied to the Nyquist filter, it is processed into a series of weighted impulses occurring precisely at the symbol times (see Fig. 2.8.5). The amplitude of the impulses are "weighted" by the data value of the symbol. Passing a single impulse through the filter produces a raised cosine impulse response (Fig. 2.8.6). Considerable "ringing" is associated with this pulse shape, with one data symbol ringing into another. The nature of the raised cosine pulse is that the ringing frequency of 5.38 MHz contributes zero value to the I-signal at the precise mid-symbol point of other symbol periods.

A close examination of this pulse response (Fig. 2.8.7) reveals that it travels through zero at equally spaced intervals. If a symbol rate is selected so that the data symbol spacing matches the filter response zero crossings, symbols will not interfere with each other, resulting in zero *inter-symbol interference* (ISI). The transmitter filter and the receiver IF filter actually employ root-raised-cosine filters which combine to produce the desired raised cosine filter response for the overall 8-VSB system.

FIGURE 2.8.6 Raised cosine filter impulse response. (*Courtesy: Zenith Electronics.*)

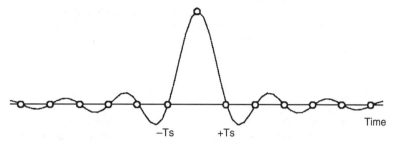

FIGURE 2.8.7 Expanded raised cosine filter impulse response. (*Courtesy: Zenith Electronics.*)

The data weighted impulses are applied to the transmitter root-raised cosine filter. The filter determines the roll-off characteristics of the upper and lower edges of the 8-VSB channel spectrum. The filter can be implemented as a *finite impulse response* (FIR) digital filter. The most attractive advantage of an FIR filter is that it can be designed with exactly linear phase. Ideally, it has zero group delay and provides flat response.

This filtered I-signal carefully maintains proper analog amplitude levels at the precise mid-point of the symbol duration. The demodulated I-component signal shown in Fig. 2.8.8 illustrates a series of 800 symbols. The markers on the I-component signal waveform clearly show the data levels at the symbol times.

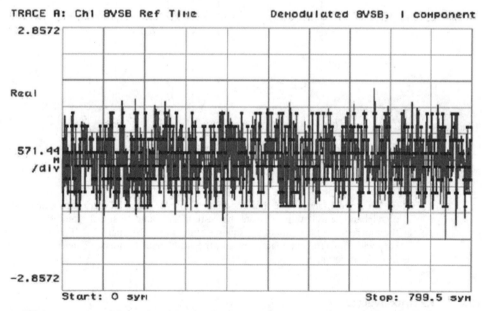

FIGURE 2.8.8 I-component signal (without pilot) showing 800 symbol values. (*Courtesy: Comark Communications.*)

The 8-VSB receiver samples the I-signal at the precise symbol time and measures only the desired symbol amplitude corresponding to one of the eight specific 8-VSB data levels. This is known as *orthogonal* when only one symbol value contributes to the RF envelope at the precise instant of signal sampling.

The combined ringing of successive symbol transitions between the eight possible I-signal data levels will occasionally produce RF envelope transients 6–8 dB above the average value of the 8-VSB RF signal. The use of the raised cosine filter prevents these transients from occurring at the receiver sample points.

2.2.6 Q-Signal

The VSB modulator generates a Q-signal that is orthogonal to the I-signal, with all frequencies shifted 90°. The relationship between the I and Q-signals in an 8-VSB system resembles a Hilbert Transform (see Fig. 2.8.9). The Q-signal, which has zero average value, is constructed to help form the 8-VSB spectrum emerging from the modulator into the desired shape to cancel out the undesired RF sideband. Unlike the I-channel impulse response, the Q-channel impulse response does not travel through zero at every symbol time.

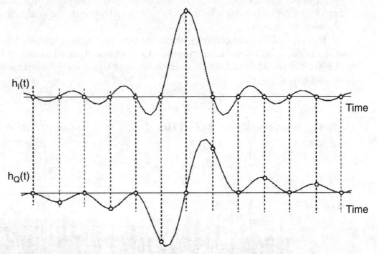

FIGURE 2.8.9 Approximate Hilbert Transform between I and Q-signals. (*Courtesy: Zenith Electronics.*)

The I and Q-signals are then modulated on quadrature IF carriers by using the phasing method of sideband cancellation to create the vestigial sideband IF signal. The resulting 8-VSB envelope is the square root of the sum of the squares of the I and Q-signals. The frequency of the IF carrier may vary with different manufacturers and the number of frequency conversions used in their 8-VSB modulator design.

2.2.7 RF Upconverter

The IF signal will be upconverted to the on-channel frequency, resulting in an upper sideband 8-VSB signal. Some manufacturers may employ a *surface-acoustical wave* (SAW) filter in the IF signal path to assist in adjacent channel suppression of out-of-band emissions.

The 8-VSB signal is then applied to the RF power amplifier stages. It may first go to an intermediate power amplifier (IPA). The final power amplifier (PA) may be comprised of solid-state devices or tubes such as an inductive output tube (IOT) or multi-stage depressed collector IOT.

From the PA, the signal passes through a *constant impedance filter* (CIF) also referred to as the "mask filter" and then to the transmission line and up to the antenna. The biggest source of *linear errors* in the transmission system is the mask filter and the biggest source of *nonlinear errors* is the transmitter power amplifiers.

A sample of the RF output is normally taken from a directional coupler at the output of the transmitter power amplifier and the output of the mask filter and fed back to the exciter. A "precorrection" signal is created and used to modify the RF signal to improve both nonlinear and linear distortion in the transmitter output.

The nonlinear feedback sample is taken from the transmitter output before the mask filter. This sample is compared with the modulator input signal. Opposite distortion is applied at the exciter output. This "predistortion" cancels out the nonlinear distortions caused by the IPA and HPA devices.

The linear feedback sample is taken from the mask filter output. The channel response is analyzed and an inverse filter response is applied to the exciter output to cancel out the linear distortions of the mask filter.

2.3 FCC SPECTRAL MASK

The FCC's rigid spectral mask is shown in Fig. 2.8.10. All transmitter emissions must fall within the skirts of this spectral mask. The power level of emissions on frequencies outside the authorized channel of operation must be attenuated at least the specified amounts below the average transmitted power within the authorized channel.

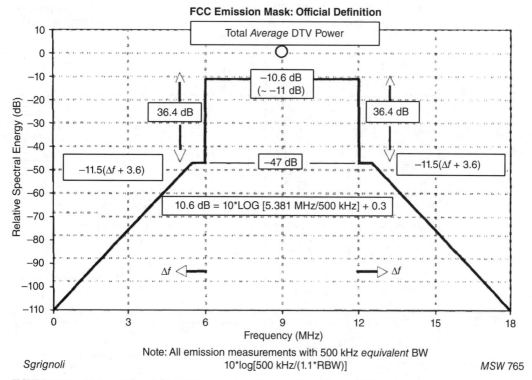

FIGURE 2.8.10 FCC spectral mask for 8-VSB transmitters. (*Courtesy: Meintel Sgrignoli Wallace.*)

2.3.1 8-VSB Spectral Display

In the first 500 kHz from the channel edge, the emissions must be attenuated at least 47 dB. For more than 6 MHz from the channel edge, emissions must be attenuated at least 110 dB. At any frequency between 0.5 and 6 MHz from the channel edge, emissions must be attenuated at least the value determined by the formula

$$\text{Attenuation in dB} = -11.5(\Delta f + 3.6) \qquad (1)$$

where Δf = frequency difference in MHz from the edge of the channel.

This attenuation is based on a measurement bandwidth of 500 kHz. Other measurement bandwidths may be used as long as appropriate correction factors are applied. Measurements need not be made any closer to the band edge than one half of the resolution bandwidth of the measuring instrument. Emissions include sidebands, spurious emissions, and radio frequency harmonics. Attenuation is to be measured at the output terminals of the transmitter (including any filters that may be employed). In the event of interference caused to any service, greater attenuation may be required.

2.3.1 8-VSB Spectral Display

The 8-VSB signal is a combination of a periodic signal and a random signal. An 8-VSB spectral display is shown with pilot carrier in Fig. 2.8.11. The pilot is a continuous wave (CW) signal, while the rest of the channel spectrum is uniformly filled with noise-like information. The pilot signal contributes 0.3 dB to the total signal power. It is common practice for the user of a spectrum analyzer

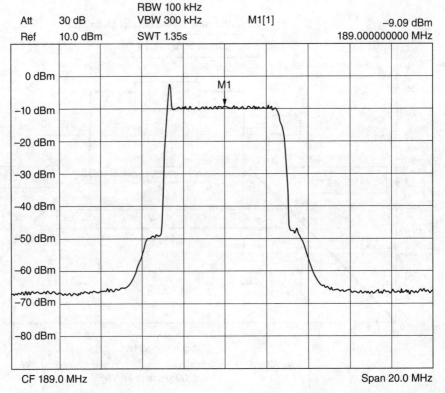

FIGURE 2.8.11 8-VSB spectral display.

to adjust the resolution bandwidth (RBW) to attenuate the noise while preserving the level of the desired signal under test. However, in 8-VSB, the signal under test appears as noise. For this reason, the apparent level of the CW pilot signal with respect to the rest of the noise-like data signal is a function of the RBW filter setting of the spectrum analyzer.

Obtaining a proper spectral display from an actual 8-VSB transmitter requires very good linearity. Intermodulation distortion (IMD) in the RF amplifier of the 8-VSB transmitter will produce emissions in the adjacent channels. The CIF is employed at the output of the transmitter to aid in filtering. The CIF provides a constant impedance response that provides the transmitter with good VSWR on either side of the channel. The physical size of the CIF is a function of channel frequency and transmitter power output.

2.4 8-VSB TRANSMISSION MONITORING

A *vector signal analyzer* (VSA) can be used to monitor the demodulated 8-VSB signal and produce several useful displays. In a perfect 8-VSB RF system, at the precise mid-point of the symbol period, the I-signal value will be one of the eight data amplitude levels.

Any linear or nonlinear distortion in the transmitter modulation process causes the I-signal amplitude to be incorrect at the time it is sampled. Transmitter distortion such as group delay, frequency tilt, and ripple will also lead to sampled amplitude to be incorrect and can result in data errors at the receiver.

Ideally, the instantaneous amplitude of the 8-VSB RF envelope contains the power of just one symbol at the time of sampling. However, between symbol sample times the 8-VSB RF envelope can rise to very high values as a result of adjacent symbols and the combined ringing amplitudes of many nearby symbols. This is why it is very important that the 8-VSB power amplifier have considerable headroom above the average power of the 8-VSB signal. The Q signal lacks these discrete levels and contains no data information for the receiver. Both the I and Q signals are processed by the 8-VSB receiver to determine phase errors in the received signal.

I and Q signals can be displayed in the form of a graphical vector map called an I and Q *constellation*. A VSA was used to generate the displays in this section.[*] As shown in Fig. 2.8.12, the VSA can simultaneously show several 8-VSB displays. The I and Q constellation display is in the upper left corner and the I-signal eye pattern is displayed in the upper right corner.

In a typical 8-VSB constellation display, the I-component of the 8-VSB signal in-phase with the pilot carrier is displayed on the horizontal (X) axis and the quadrature phase Q-signal is displayed along the vertical (Y) axis.

The I and Q constellation display is generated by plotting the 8-VSB in-phase component (I) against the quadrature component (Q). The eight vertical lines along the horizontal (I) axis correspond to the eight VSB data levels. If samples are taken precisely at the symbol times, then each data point should be precisely aligned along the vertical lines. An I and Q constellation display demonstrating some misalignment owing to an imperfect 8-VSB signal is shown in Fig. 2.8.12 (upper left corner).

Constellation sample points do not always fall precisely on the vertical lines. The extent to which the actual I-component sample point departs from the ideal sample point is measured in terms of *error vector magnitude* (EVM). Every effort should be made at the transmitter to keep the EVM as low as possible.

The EVM for 800 data points is shown in Fig. 2.8.12 (lower left corner). The average error vector is 3.6393 percent. A VSA also provides a tabular display of the monitored signal. Pilot level, average EVM, maximum EVM, signal-to-noise level, and other signal parameters can be displayed and printed.

Signal-to-noise ratio (SNR) is defined as the power ratio between the ideal 8-VSB signal and the average "noise" power determined by calculating the average of the squared symbol errors. The "noise" can result from a combination of linear and nonlinear distortion, as well as "white noise" and carrier phase noise.

---

[*]Hewlett-Packard HP 89441A.

2.144 RF TRANSMISSION

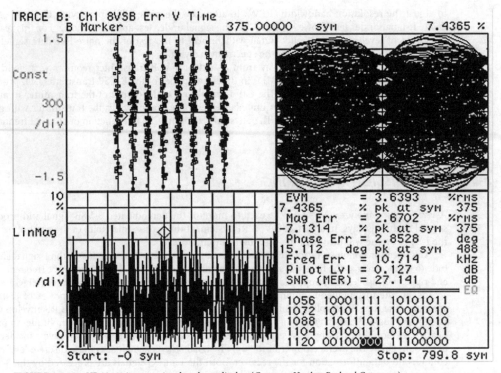

FIGURE 2.8.12 HP 89441A vector signal analyzer display. (*Courtesy: Hewlett-Packard Company.*)

The SNR indicated in Fig. 2.8.12 (lower right corner) is 27.141 dB, which is greater than the 27 dB minimum recommended by the ATSC compliance document.[†] This SNR value produces "open eyes" (upper right corner).

Some instruments display *modulation error ratio* (MER) (see Fig. 2.8.13). This is a parameter expressed in dB similar to SNR. It differs somewhat in that the MER is calculated using the errors in both the I and Q data while the SNR is calculated with only the I data errors (MER in dB ≤ SNR in dB).

2.4.1 Error Vector Magnitude

An EVM below 3 percent is excellent, EVMs of between 3 and 5 percent are acceptable, and an EVM in excess of 5 percent is cause for concern. A graphical illustration of EVM is given in Fig. 2.8.14.

2.4.2 The Importance of Linearity

The received signal of an 8-VSB signal is directly affected by the linearity of the digital transmitter. Any factor that contributes to the distortion of the 8-VSB signal decreases signal quality at the receiver and ultimately results in premature loss of the picture. Although nonlinearities may not

---

[†]ATSC: "Recommended Practice—Transmission Measurement and Compliance for Digital Television," Doc. A/64B, Advanced Television Systems Committee, Washington, D.C., 26 May 2008.

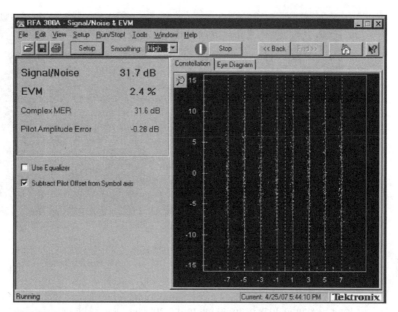

FIGURE 2.8.13 Example of measurements using a Tektronix Model RFA300A. (*Courtesy: Tektronix.*)

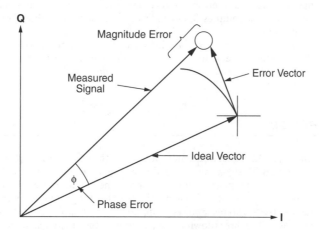

FIGURE 2.8.14 The error vector. (*Courtesy: Tektronix.*)

compromise signal quality in the first thirty or forty miles of the coverage area, near the fringe, even small nonlinearities will move the "cliff effect" closer to the transmitter.

In Fig. 2.8.15a of the constellation display, note the two phasors from the center of the display to the upper left (−7 level) and upper right (+7 level). These phasors would appear to represent the instantaneous sample where the 8-VSB signal has the largest RF carrier envelope at a symbol sample time.

The phasors appear to be of equal amplitude for both the sampled −7 level and the +7 level. You may think this is true because you are accustomed to seeing a constellation display that does not take the 1.25 offset for the pilot into account. A +7 is actually a +8.25, a −7 is −5.75, and a −1 is a +0.25. The true zero point is just to the left of the −1 symbol band. When you look at Fig. 2.8.15b, it is now

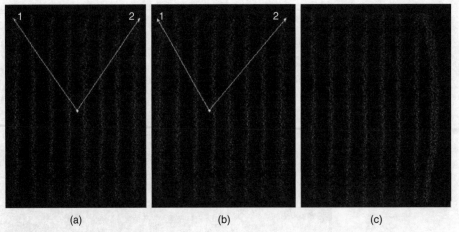

FIGURE 2.8.15 Constellation display examples.

quite clear that phasor 2 is of greater amplitude. A power amplifier that can easily handle the phasor for a −7 symbol may begin to fail in the presence of a +7 phasor.

This is a useful piece of information when you observe a constellation display showing a "bowing" of the +7 level such as on the right side of Fig. 2.8.15c. It is likely that this power amplifier device has reached saturation and is beginning to clip the 8-VSB signal. It is somewhat analogous to observing sync compression in an analog television transmitter. You have simply run out of amplifier head-room.

It is worth noting that the smallest vector would occur with a −1 symbol phasor that was along the x-axis in-phase with the pilot. It would actually be very close to zero carrier at +0.25 and would be susceptible to *incidental carrier phase modulation* (ICPM).

2.5 8-VSB POWER

The 8-VSB RF signal peaks must not exceed the peak envelope power (PEP) rating of the transmitter. The 8-VSB RF signal peaks are identified and equated with a CW sine wave whose voltage peaks equal the voltage peaks of this CW signal. The average power of this CW signal is then said to be the peak power of the 8-VSB signal. The 8-VSB RF signal peaks occur randomly and are statistical in nature. The 8-VSB peak power is important for nonlinear considerations in the transmitter amplifier and for voltage breakdown considerations in the antenna and transmission line. Over time, 8-VSB average power is constant and is defined as

$$V_{\text{RMS}}^2/R \tag{2}$$

Digital transmitter power must be rated in average power because this is the only consistent characteristic of the pseudo-random 8-VSB signal. The 8-VSB average power is indicated in Fig. 2.8.16 in which many peak excursions appear above the average power level.

The PEP of the 8-VSB signal is the result of the combined ringing of the various combinations of symbol values. Data level transitions occurring over a period of plus and minus 35 symbol periods contribute to the PEP. Peak excursions as high as 9 dB (8×) above average have been reported. The duration of these peaks could be measured in nanoseconds, and occur very rarely. The peaks appear to have no capacity to sustain an arc, and 99.9 percent of the 8-VSB signal peaks are typically within 6.3 dB (4.25×) of the 8-VSB average power (see Fig. 2.8.17).

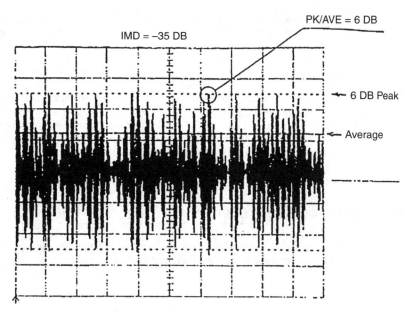

FIGURE 2.8.16 8-VSB RF envelope, 6 dB peaks, −35 dB IMD. (*Courtesy: GatesAir.*)

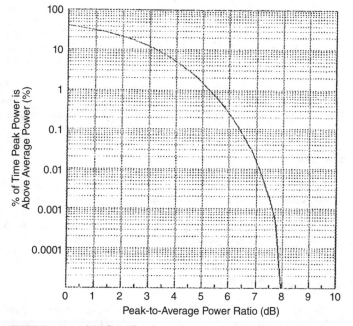

FIGURE 2.8.17 Cumulative distribution function of 8 VSB peak-to-average power ratio. (*Courtesy: ATSC.*)

An 8-VSB transmitter should have a power handling capacity at least 6.3 dB above average power (4.25×). This is the analogous to the SSB ham transmitter power requirement of 2 kW of PEP to accommodate a 1 kW average SSB signal. "Precorrection" techniques used in the 8-VSB modulation process can lower the required headroom of the RF high-power amplifier device used in the transmitter.

Two methods are commonly used to measure the power output in an 8-VSB transmitter. In a liquid cooled transmitter, the calorimeter method can be used. The coolant system will be designed to use thermometers or temperature probes to precisely measure the temperature of the coolant as it enters and exits the transmitter test load. The heat transfer characteristics of the particular coolant and water mixture must be known. The temperature differential and coolant flow rate is then used to calculate the average transmitter power.

In a liquid or air-cooled transmitter, a calibrated power sensor/meter can be used. A precision directional coupler is designed into the RF system near the output of the constant impedance mask filter. The precise attenuation value of the coupler tap is manually entered into the power meter as an "offset" value. The 8-VSB average power can be read directly from the face of the instrument. The transmitter output power should be maintained within the limits of 80 to 110 percent of the licensed power level.

2.6   HIGH POWER AMPLIFIER DEVICES

Several basic technologies are available for high power amplifying devices in television service. The most common approaches are discussed in the following sections.

2.6.1   Klystrons

Klystrons were the high power amplifier of choice for UHF NTSC operation but they are not practical for 8-VSB. A steady electron beam leaves the cathode and is accelerated by a high direct current (DC) potential. The 8-VSB power envelope is a pseudo-random signal. The Class A nature of klystrons would require that the beam current be biased for full power all of the time. The average power of the 8-VSB signal would have to be set well below the 1 dB compression point of the klystron transfer characteristic curve. A 60 kW klystron might produce only 5 kW of 8-VSB power with a DC to RF efficiency around 5 to 10 percent.

The efficiency of the klystron is also reduced by the broad bandwidth tuning required to properly pass the 8-VSB signal. The 1 dB bandwidth of a klystron is typically around 5 MHz. A klystron could be used for 8-VSB service but it is impractical from an economic standpoint.

2.6.2   Tetrode

The tetrode had been the workhorse of VHF television transmitters since the beginning of U.S. broadcasting. Tetrode refers to a broad category of tube construction using four elements: plate or anode, screen grid, control grid, and a cathode heated by a filament. In recent years, the VHF tetrode transmitter has been replaced by the solid-state VHF transmitter. Tetrodes were also used extensively in the low and medium power UHF NTSC transmitters.

2.6.3   Inductive Output Tube (IOT)

The IOT was designed with the goal of combining the smaller size and efficiency of the tetrode with the long life, high power output and reliability of the klystron. The concept of the IOT was proposed in the 1930s, but the actual development did not begin until the 1980s.

The IOT uses a cathode and filament structure similar to a conventional klystron. The RF drive required is 400 to 800 W. The RF input is applied between the cathode and grid by means of an RF input cavity. A density-modulated electron beam is accelerated toward the anode by a 30 to 36 kV DC potential and passes through the anode without interception. The electron beam is accelerated directly into the RF output ceramic region of the tube toward the collector. The RF energy is extracted in the output cavity as the beam is decelerated before it reaches the collector. The electron beam in an IOT is focused by a magnetic field.

The IOT uses a double-tuned output cavity to achieve the necessary bandwidth. Figure 2.8.18 shows an IOT mounted in a frame. The primary output cavity surrounds the RF output ceramic of the tube. The RF signal is coupled from the primary output cavity to the secondary output cavity and then to the 3-1/8 inch coaxial output coupler. Forced air or liquid keeps the cavities cool. One tube covers the entire UHF band. The collector is liquid cooled in the higher power versions. The collector can be air-cooled or liquid-cooled in the lower power versions. The liquid cooling the IOT is a water/glycol mix to prevent the liquid from freezing in cold climates. The IOT has undergone several generations of development which have reduced the overall size and weight.

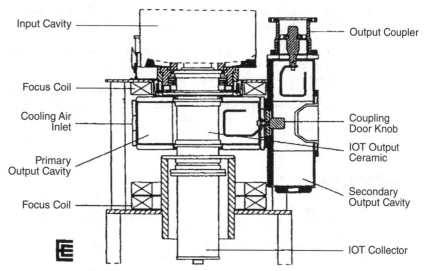

FIGURE 2.8.18 IOT in frame. (*Courtesy: e2v technologies.*)

A characteristic of the conventional IOT is the "crowbar." A shunt device shorts out the high voltage supply to protect the IOT from the occasional internal arc. This is a very violent event and several thousand amperes of current can flow during the crowbar period that can last as long as 135 ms. Two common forms of crowbar protection are the triggered spark gap and the Thyratron. The spark gap deteriorates in its ability to protect the tube with each necessary use or testing of the spark gap, The Thyratron is another method of protecting the IOT. When the likelihood of an internal IOT arc is detected, the Thyratron "fires" and shunts the high voltage supply to create the crowbar. The Thyratron is a vacuum tube and it too deteriorates in effectiveness over time. Some IOT transmitter manufacturers use a silicon-controlled-rectifier (SCR) to control the high voltage beam supply. When a crowbar occurs, the high current flow can be extinguished within 8 ms at the next zero crossing of the incoming alternate current (AC) voltage. This is a much less violent event.

IOT's on a UPS System. It is worth noting that due to the violent crowbar events experienced by a conventional IOT, a station contemplating the use of a UPS system should be very careful in its selection. The UPS manufacturer must completely understand the characteristics of a crowbar and guarantee that the UPS system can safely bypass itself to the direct power company line without causing all transmitter cabinets on-the-air to drop RF output power, and possibly damage the UPS. A common configuration of an IOT transmitter is to use two cabinets combined to produce 100 percent of the total RF output power. When a crowbar occurs in one cabinet, the remaining cabinet normally remains on the air and the result is that 25 percent RF power continues to be broadcast and 25 percent RF power goes into the reject load. If a crowbar occurs in one of the cabinets fed by the UPS system, the violent current surge may cause the UPS to momentarily take all cabinets off the air resulting in 0 percent RF power being broadcast. This defeats one of the primary reasons for installing the UPS.

2.6.4 Multi-Stage Depressed Collector IOT (MSDC IOT)

The conventional IOT may approach 35 to 40 percent efficiency at full power and that efficiency drops significantly when operated at lower power levels. A multi-stage depressed collector inductive output tube (MSDC IOT) has been developed that can improve efficiency. The MSDC IOT was developed by modifying the collector of the conventional IOT. It requires five different high voltage levels. The efficiency of an MSDC IOT may achieve 60 percent at full power and drops only slightly to 50 percent efficiency at half power.

A conventional IOT uses a water/glycol mixture for cooling the collector. The MSDC IOT can use a synthetic oil for cooling that also acts as a dielectric. All the collector segments at different high voltage potentials can share a common dielectric oil bath.

A modern MSDC IOT transmitter can be designed using solid-state switching on the high-voltage beam supply. The solid-state switches are very fast-acting and can extinguish potential crowbars quickly. This prevents tube damage and the dramatic release of energy that can contribute to other seemingly random failures in a conventional IOT transmitter.

2.6.5 Solid-State Devices

Given the typically low transmitter power levels required for VHF 8-VSB service, the solid-state power amplifier is the device of choice for 8-VSB VHF transmitters. Bipolar and *metal oxide silicon field effect transistor* (MOSFET) are the basic semiconductor technologies available. Each class of devices has its own advantages and disadvantages. All are normally constructed in a fashion that permits the entire RF module to be easily replaced. The transmitter may be capable of operating at full power with one or two of the modules removed. The redundancy and ease of service is an attractive advantage to the solid-state transmitter.

The actual power line AC-to-RF efficiency of a solid-state transmitter may not offer any advantage over a tube transmitter. A considerable amount of RF efficiency is lost in the solid-state combining process. A large number of transistors must be combined together to achieve the desired transmitter power output. In many ways, the design of a tube amplifier is simpler and requires fewer parts than a solid-state transmitter. Solid-state transmitters are usually designed to contain several RF modules per cabinet and per power supply. Most RF modules are "hot-swappable" permitting the modules to be replaced while the transmitter remains on the air.

The bipolar device is a mature technology with a long track record in NTSC, PAL, and SECAM solid-state service worldwide.

LDMOS. The *laterally diffused metal oxide silicon* (LDMOS) transistor is of the MOSFET family of semiconductors. It was developed by Motorola in the early 1990s. LDMOS offers linear gain up to the saturation region and provides low intermodulation distortion. Saturation is smooth with LDMOS, which is a desirable characteristic when the ratio of peak to average power is high as with the 8-VSB signal. These devices can generate higher output power and their efficiency has increased at UHF frequencies since their initial development. This transistor can deliver 125 W of linear power and 600 W peak output power. LDMOS transistors are widely used today.

In 1936, William H. Doherty developed an amplifier design that combined an average power Class AB amplifier with a Class C peaking amplifier. This "Doherty" architecture has been developed since 2011 to provide exceptional efficiency in UHF amplifier service. LDMOS transistors used in a Doherty configuration can achieve an amplifier efficiency greater than 40 percent with an overall transmitter efficiency greater than 34 percent. Several manufactures offer LDMOS Doherty transmitter designs of 60 kW, rivaling the efficiency of tubes.

2.7 8-VSB SPECIALIST CERTIFICATION

The Society of Broadcast Engineers offers an *8-VSB Specialist* Certification. If you hold the Certified Broadcast Television Engineer (CBTE) or higher level certification, you are eligible to take the Examination. Much of the material on the *8-VSB Specialist* exam is covered in this chapter.

The SBE also offers an 8-VSB online course at the SBE University. By taking that course and passing the final test, you can obtain a Certificate of Completion indicating you have passed that course. No previous level of SBE Certification is required to take the online course. Much of the material in the 8-VSB SBE University online course is covered in this chapter. The completion of a course through SBE University qualifies for 1 credit, identified under Category I of the Recertification Schedule for SBE Certifications.

For further information you can contact the SBE at 317-846-9000 or check out their website at www.sbe.org.

2.8 DEFINING TERMS

Error vector magnitude (EVM)—The extent to which the actual I-component sample point departs from the ideal sample point is measured in angular terms.

Forward error correction (FEC)—Data error correction achieved by adding extra bits of data to the data stream to overcome noise, interference, echoes, distortion, and other 8-VSB signal impairments.

Hilbert transform—A digital technique used to provide a uniform 90° phase shift for all of the signal components required for 8-VSB modulation.

I-signal—A signal in phase with the original RF carrier that contains all the digital data information.

Mask filter—A constant impedance filter (CIF) that all transmitter emissions must fall within. The FCC specifies the skirts of this spectral mask.

Nyquist filter—A filter that determines the roll-off characteristics of both upper and lower edges of the 8-VSB channel spectrum.

Pilot carrier—A small in-phase pilot signal transmitted approximately 310 kHz above the lower band edge that is actually the remnant of the suppressed carrier used in the modulation process.

Raised cosine filter—The combined response of the root-raised cosine filter used in both the 8-VSB transmitter and the receiver.

Q-signal—A quadrature-phase signal that contains no data information and is related to the I-signal by a fixed transfer function.

REFERENCES

Garlinger, Douglas W.: *Introduction to DTV RF*, Society of Broadcast Engineers, Indianapolis, IN, 1998.

LDMOS Transistor Delivers Highest Performance in UHF Broadcast Transmitters, Freescale Semiconductor, Austin, TX, 2011.

Stefanik, Fred M., and W. Gordon Gummelt: *A New Generation, Ultra-Efficient, High- Power DTV Transmitter*, Comark Communications, Southwick, MA.

Understanding 8VSB, Comark Communications, Southwick, MA.

CHAPTER 2.9
ATSC DTV TRANSMISSION SYSTEM

Gary Sgrignoli
Meintel, Sgrignoli, & Wallace, Warrenton, VA

2.1 INTRODUCTION

The National Television System Committee (NTSC) television standard, described in the FCC (Federal Communications Commission) rules that are contained in Title 47, Part 73 (full power stations) and Part 74 (low power stations, Class A stations, and translators) of the Code of Federal Regulations (CFR),* successfully served as the U.S. analog television standard from 1941 (for black and white) and 1953 (for color) to June 2009 when the analog full-power television stations ceased broadcasting. The advantage of analog NTSC was its rugged synchronization in severe reception environments (white noise, impulse noise, multipath, and interference). This was due to the use of large and repetitive supplementary horizontal and vertical synchronization pulse signals that allows the standard-definition picture to synchronize at impairment levels well below the loss of "acceptable" picture quality. In many cases, significant impairments can be present while "watchable" analog pictures are still possible. The use of narrowband FM modulation for the audio (including stereo subcarriers after 1984) often allowed the sound quality to remain reasonable in severe propagation environments.

However, the cost of this robust analog NTSC synchronization is high in terms of both power and bandwidth efficiency, especially in comparison to today's digital standards. Another disadvantage of NTSC is that this desirable rugged synchronization does not prevent impairment and distortion of the video but rather just allows a stable impaired picture to be displayed.

In the 1990s, technology advanced to the point where highly compressed high-definition (HD) video and 5.1 channel surround sound audio could be digitally transmitted error-free in a 6 MHz bandwidth with relatively low transmission power to fixed location receivers, i.e., *not* mobile or handheld devices. After much discussion and planning by the FCC-created Advisory Committee on Advanced Television Services (ACATS) in conjunction with the Grand Alliance (GA), a new digital television (DTV) system was designed, analyzed, prototyped, laboratory-tested, and field-tested. The new DTV system was subsequently standardized by the Advanced Television System Committee (ATSC), and is described in an ATSC document called A/53.† The FCC adopted this standard (except for the specific video formats) into their rules on December 24, 1996.‡ This ATSC standard described all aspects of the

---

*FCC Rules, Title 47 CFR Part 73, Subpart E (73.601–73.699) for full-power stations, and Title 47 CFR Part 74 Subpart G (74.701–74.798) for low-power stations, Class A stations, and translators, www.fcc.gov.

†ATSC A/53—Part 1: 2007, ATSC Digital Television Standard, Part 1—Digital Television System, January 3, 2007; see main body for system overview, and Annex A, Section 2 "Historical Background" for the history of the standard's development, www.atsc.org.

‡FCC Rules, Title 47 CFR Part 73, Subpart E (73.601–73.699) for full-power stations, and Title 47 CFR Part 74 Subpart G (74.701–74.798) for low-power stations, Class A stations, and translators, www.fcc.gov.

television system, including video and audio formats and their compression, MPEG-2 transport stream packetization, ancillary data (including program metadata), and radio frequency (RF) transmission.

The new ATSC terrestrial broadcast digital television system includes a wide variety of subsystems required for originating, encoding, transporting, transmitting, and receiving HD and standard definition (SD) video as well as compact disc (CD) quality or better audio. The ATSC system has a layered architecture, with each layer having its own distinct context. Well-defined interfaces exist between the various subsystem layers. The single-carrier DTV transmission system layer, called 8-VSB (8-level vestigial sideband), was developed and standardized for this purpose.§

Highly compressed digital video and audio signals are especially susceptible to transmission errors. Data error rates must often be 2 to 3 orders of magnitude better (i.e., lower) than uncompressed data to provide acceptable viewing and listening experiences. An advantage of the new DTV system is the capability of over-the-air transmission and reception of error-free 20 Mbits per second (Mbps) picture and sound signals in the presence of severe impairments (e.g., noise, multipath) and interference (e.g., DTV-into-DTV). This is possible due to: (1) powerful forward error correction (FEC) made up of inner and outer concatenated codes separated by data interleaving, and (2) robust data synchronization codes. These two transmission features allow error-free reception at low signal levels (i.e., low SNR values).

The goals in creating the new all-digital television system were to:

(1) Replicate in the same 6 MHz channelization *fixed* NTSC broadcast coverage and service primarily with *outdoor* receive antennas located 30 feet above ground level (AGL).
(2) Provide greater video resolution (e.g., 720p and 1080i high definition) and audio fidelity (e.g., CD-quality stereo sound or 5.1 channel surround sound) in affordable consumer DTV receivers.
(3) Allow interoperability with other media systems.
(4) Exhibit system extensibility (i.e., flexibility) so that the system could be easily expanded in the future to allow new backwards-compatible enhancements.

Spectrum compatibility was also required, which means the same very high frequency (VHF) and ultrahigh frequency (UHF) television bands that were used for analog television could simultaneously be employed for DTV signals as well as nontelevision signals (e.g., white space devices and cellular smart phone signals). A reduction in the number of allocated RF television channels (above channel 51) was planned and implemented after the analog television turnoff (June 2009) so that other services would have additional spectrum at their disposal (e.g., broadband cellular data). This meant a very crowded spectrum for television signals where interference avoidance and tolerance was paramount.

Therefore, the need for both spectrum and power efficiency in the new DTV transmission system was vital while simultaneously meeting the need for increased television data information traffic (e.g., HDTV, multicasting SDTV, surround sound, datacasting, etc.). A relatively high data rate ($\approx$20 Mbps) was needed for these new services. In order to meet these requirements in a reduced spectrum environment, the DTV system needed to use much lower transmitted power to reduce interference to analog stations (during the transition) and other digital television stations (during and after the transition), while still providing robust DTV reception in the presence of significant impairments and interference.

2.2 SYSTEM DESCRIPTION

The A/53 terrestrial digital television transmission standard describes the processing required to transmit high data rate video, audio, and ancillary information from broadcast stations to viewer homes. The specific purpose of the transmission system is to provide a means of accurately and

---

§ATSC Digital Television Standard—Part 2: RF/Transmission System Characteristics, Doc. A/53 Part 2:2011, December 15, 2011; downloads of the standard can be obtained at www.atsc.org.

robustly conveying DTV data over a 6 MHz RF television channel from the high-power DTV transmitter at the broadcast station to the DTV receiver in a viewer's home. The transmission system, which uses the single-carrier 8-VSB modulation scheme, serially transmits data symbols that fill up the complete RF channel.

To accomplish over-the-air transmission, a number of system design parameters were considered such as occupied channel bandwidth, transmitted and received signal power, receiver input signal-to-noise ratio (SNR) at threshold of visible (TOV) and audio errors, and acceptable probability of data error. Unlike analog transmission where the figure of merit is signal fidelity at the receiver, digital transmission seeks to provide the lowest probability of data reception errors for a given set of transmission parameters, even in severe propagation conditions. To optimize the final ATSC system design, various parameter tradeoffs were made. A summary of the final ATSC transmission system *parameters* can be found in Table 2.9.1.

Digital communication has the advantage of potential regeneration of the original transmitted data signal if reception is performed above some threshold of impairment (e.g., RF filter distortions or propagation multipath) or interference (e.g., additive white Gaussian noise (AWGN) or adjacent channel DTV-into-DTV). For propagation-related errors to occur, imperfections must cause the received digital signals (representing ones and zeroes) to cross the data level threshold comparison point and be misinterpreted. When the errors do occur, additional data processing can be employed to detect and even correct many of them, thus diminishing their effect on the video and audio signal quality. To further diminish the effects of data errors, receiver synchronization (pilot carrier, symbol clock, and data fields) must reliably perform well below the data error threshold so that recovery is quick from momentary excursions below this reception threshold.

Robust digital data transmission can be accomplished by channel coding that is made up of structured data sequencing (FEC and interleaving that optimize error detection and correction in one-way links by providing coding gain) and waveform coding (critical raised-cosine filtering for minimal intersymbol interference (ISI) and matched filtering that optimizes analog waveforms for symbol detection in noisy environments). Of course, this channel coding and its benefits come with a "cost" because design tradeoffs must be made in bandwidth, data rate, and system complexity.

In the ATSC system, a standard fixed 188-byte MPEG-2 transport data packet consists of a 4-byte header plus 184-byte payload. These packets are created in the transport multiplexer which combines outputs from the video and audio compression encoders as well as ancillary data sources. All of these sources are frequency-locked to a common 27 MHz clock. This combined transport data stream from the multiplexer conveys compressed video and audio signals (from one or more programs), program and system information, and other ancillary data to the DTV transmitter for broadcast over the air to a large number of DTV receivers. Upon reception in a DTV receiver, the inverse process is performed to provide a digital television experience to viewers that can include high definition pictures with accompanying 5.1 channel surround sound.

In addition to the A/53 Part 2 standard, other tutorial documents are available that further describe the DTV RF transmission system in detail.[¶,**,††,‡‡]

2.2.1 DTV Transmitter

Figure 2.9.1 illustrates the 8-VSB transmitter block diagram.

The main function of the DTV transmitter is to encode and modulate digital baseband data, and convert it to an RF signal on a desired 6 MHz television channel. The digital input signal is a 188-byte

---

¶ Sgrignoli, G., W. Bretl, and R. Citta: "VSB Modulation Used for Terrestrial and Cable Broadcasts," *IEEE Transactions on Consumer Electronics*, vol. 41, no. 3, August 1995.

** Citta, R., and G. Sgrignoli: "ATSC Transmission System: VSB Tutorial," *Montreux Symposium Proceedings*, June 1997.

†† Whitaker, J.: "The ATSC Digital Television System," Section 1.11, *National Association of Broadcaster Engineering Handbook*, 10th Edition, NAB, Washington, D.C., 2007.

‡‡ Collins, G.: *Fundamentals of Digital Television Transmission*, John Wiley & Sons, New York, NY, 2001.

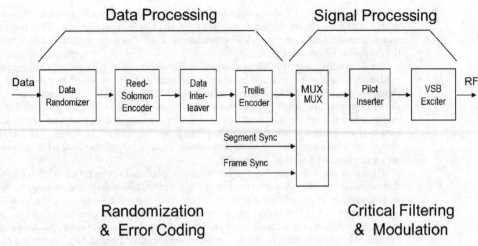

FIGURE 2.9.1 ATSC transmission system transmitter block diagram.

MPEG transport data packet stream with a constant 19.393 Mbps rate. This data packet stream is often conveyed in one of two formats:

(1) A *synchronous* serial interface (SSI): 75-Ω, single coaxial cable with BNC connectors, 0.8 V_{pp}, 38.786 MHz self-clocking interface rate, 8-bit link character, bi-phase mark binary coding, uni-directional SMPTE-310M standard,[§§] or

(2) An *asynchronous* serial interface (ASI): 75-Ω, single coaxial cable with BNC connectors, 0.8 V_{pp}, 270 Mbps self-clocking link rate, 10-bit link character, 27 Mbps data rate with null stuffing, unidirectional DVB-ASI standard.[¶¶]

The transmitter output signal is a high-power modulated single-carrier RF signal on a desired 6 MHz television channel that can be transmitted over the air to a large population of receivers.

The digital transmitter consists of two unique processing sections that improve reception under imperfect propagation conditions: data processing and signal processing.

The *data processing* section deals with data information and creates a stream of randomized data symbols with structured redundancy that provides the basis for "on the fly" FEC in the receiver. Concatenated coding is employed that utilizes two cascaded error-correction codes separated by an interleaver that provides powerful data error correction for error-free operation at low signal levels (i.e., at low SNR values). This concatenation approach has the effect of providing the same error-correcting performance of a longer single code without the loss of spectral or data rate efficiency.

The *signal processing* section deals with critically filtered and optimally modulated signals with minimal ISI that represent the various data symbols distinctively, while simultaneously fitting the signal into the required 6 MHz RF channel bandwidth. This provides additional margin for error-free operation to be achieved.

The ATSC standard calls for the transmitter data bit, data byte, and symbol clocks to be locked to the incoming MPEG-2 transport stream clock so that no data packets are required to be added or dropped.[***]

[§§] SMPTE 310M, "Synchronous Serial Interface for MPEG-2 Digital Transport Streams," Society of Motion Picture and Television Engineers, White Plains, NY, www.SMPTE.org, 1998.

[¶¶] DVB-ASI, "Guidelines for the Implementation and Usage of the DVB Asynchronous Serial Interface (ASI)," ETSI TR 11 891, Digital Video Broadcasting, 2001–2002.

[***] ATSC Digital Television Standard—Part 2: RF/Transmission System Characteristics, Doc. A/53 Part 2:2011, December 15, 2011; downloads of the standard can be obtained at www.atsc.org.

Data Randomizer. The data randomizer, which is the beginning of the *data processing* section, randomizes all of the incoming data (except MPEG 47_{hex} sync bytes) on a byte-by-byte basis according to a polynomial function that is based on a 16-bit maximum length pseudo-random binary sequence (PRBS). This guarantees eight equiprobable symbol levels for over-the-air transmission regardless of the input data stream characteristics (e.g., long strings of fixed data bits), which facilitates signal recovery in the receiver. The randomizing process, which essentially "scrambles" only the payload data, is synchronized to the robust binary data field sync at the beginning of the data field, thus meaning that the randomizing code repeats every data field. No randomization is applied to the binary data segment syncs (which replaces the 47_{hex} MPEG sync byte at the start of every ATSC data packet) or the binary data field syncs since they are inserted into the VSB baseband signal at a later point in the exciter. These data syncs are required to remain fixed repetitive signals for use in the receiver.

The predefined and known randomization code is generated and used for each data field. It is easily reproduced in each receiver for proper de-randomization (after synchronization to the data segment and data field syncs). This process provides reliable receiver performance.

Advantages of data randomization include the guarantee of a transmitted flat (i.e., uniform), noise-like spectrum across the entire 6 MHz channel for any MPEG input data transport stream. This facilitates optimal performance of recovery and correction loops in the receiver (e.g., consistent pilot carrier amplitude and carrier recovery, segment and field sync correlation and detection, and accurate phase noise tracking). Data randomization also minimizes the effect of cochannel interference into analog NTSC signals (required during the digital transition years), and also allows the DTV transmitter to always operate at a constant average power.

Reed–Solomon Encoder. The Reed–Solomon (RS) linear block encoder takes the randomized 188-byte MPEG-2 data packets and encodes the entire packet (*except* for the MPEG sync byte) using 8-bit bytes rather than individual bits. It is part of the *outer* code of a concatenated error-correction system. The result at the RS coder output is the addition of 20 redundant parity bytes at the end of each MPEG data packet (additional 10.6 percent length) for use in forward error detection and correction in the DTV receiver in order to provide coding gain. The result of this FEC coding requires an increase of the 19.392 Mbps payload data clock rate by a factor of 208/188 (i.e., 21.524 Mbps).

Each data packet is coded independently from other packets. Specifically, the RS coder uses only 187 of the payload bytes (not 188) to encode the packet, ignoring the MPEG sync byte which is ultimately replaced later in the processing chain by the data segment sync. Repetitive binary data segment and data field syncs are not included in RS processing either, since they can be easily and robustly extracted from a noisy, randomized data sequence by powerful correlation techniques. This Reed–Solomon code is described as a (207, 187, t = 10) linear code, which adds 20 additional parity bytes at the end of the 187 payload data bytes for error correction of up to a total of 10 byte errors within a 207 byte (i.e., 187 + 20) data packet (note that byte errors are corrected, regardless of how many bit errors occur within a corrupted byte).

Although Reed–Solomon block codes provide error correction in the presence of white noise, they are especially good at correcting contiguous data *burst* errors in the received signal. Data burst errors can be caused by electrical interference (impulse noise), dynamic multipath, or even interference from other signals.

Data Interleaver. The intersegment convolutional byte interleaver provides time diversity by spreading adjacent payload data and RS FEC bytes over many data segments so that one MPEG data packet has its bytes placed on 52 different transmission data segments. Segment syncs, which represent the MPEG sync bytes, are *not* interleaved. The convolutional code is described as a (B = 52, M = 4, N = 208) code. This means that a group of four bytes is moved at a time to a new data segment depending on its original position in the transport packet. The interleaving process is synchronized to the robust binary field sync.

Since the RS block length is 207 bytes and the interleaver segment length is 208 bytes, the positions of successive data segments *precess* with respect to the interleaver process. This interleaving process disperses the payload and RS data bytes over 52 data segments; i.e., over a 4 ms interval ($\approx$1/6 of a data field). The length of the interleaver is such that the interleaving pattern will repeat exactly on successive data fields without requirement of further synchronization to field syncs. However, even after initial MPEG packet phasing is accomplished, resynchronization in the receiver can still be redundantly performed every field to add reliability under severe propagation conditions.

A *convolutional* byte interleaver is employed, as opposed to a simple block interleaver, which minimizes the amount of memory needed in the transmitter and the receiver. Even though it is synchronized to the robust binary data field sync for reliability, convolutional interleaving operates continuously on data from one field to the next (i.e., overlapping data fields).

The advantage of dispersing the data over many data segments is to protect against burst errors in the receiver caused by transmission channel characteristics (e.g., impulse noise or multipath) that often affect contiguous symbols in the data segments. The combination of interleaving and deinterleaving allows de-correlating of the burst errors such that they are spread out in time and thus more likely to be corrected by RS packet decoding. Locking the interleaver to the robust binary data field sync adds reliability in the DTV receiver. The tradeoff to the benefits of error dispersion interleaving is that it increases data decoding latency (delay).

Trellis Encoder. Trellis-coded modulation (TCM) is an error-correction technique invented by Gottfried Ungerboeck in 1984, and combines convolutional trellis coding with modulation symbol selection. It is part of the *inner* code of a concatenated error-correction system. TCM adds redundancy in the form of extra data symbol levels without bandwidth expansion.

The ATSC transmission system uses a 2/3-rate, 4-state, linear code for TCM. This means that two data bits (payload and RS parity) are transformed into 3 bits using a short 4-state feedback coder (i.e., two memory flip flops that have four different possible states, and whose feedback structure linearly combines output and input bits with exclusive OR gates to determine data coding). In other words, rather than dividing the data into large independent blocks like RS coding does, the TCM coder output data *bits* are continuously affected by both present and past data as structured redundancy is inserted into the data stream for use in error correction. A short 4-state trellis code is employed to minimize the lengthening effects of impulse noise or burst errors in the trellis decoder caused by the reacquisition time of the receiver's Viterbi decoder.

Specifically, one input bit is coded into two output bits using a ½-rate convolutional code while the other input bit is independently precoded (i.e., differentially encoded) in a feedback manner. Together, this combined processing provides the three output bits for every two input bits. All data bits are coded except for the binary segment syncs (which replace the 47_{hex} MPEG transport stream sync) and binary field syncs (which are subsequently added to the stream).

In addition to the trellis coding, the TCM technique performs the data symbol mapping function, which transforms the three binary data bits into the various 8-level modulation symbols (±1, ±3, ±5, and ±7 relative constellation units). Two data bits can be represented in a symbol by four levels, but the additional TCM parity bit doubles the number of data symbol levels to 8.

The trellis encoder represents the end of the *data processing* section. It should be noted that TCM is applied only to the 207 data bytes in each data segment, not to the binary data segment syncs and field syncs which must remain strictly binary signals with no TCM processing.

The TCM processing is actually carried out by the equivalent of 12 TCM encoders, each encoder processing 1/12 of the data. TCM processing has the effect of providing intrasegment interleaving protection against very short burst errors in the data signal. The 12-symbol code interleaving process feeds each of the 8 bits within a given input byte (carried by four symbols) through the same TCM encoder for optimum FEC. The addition of the extra TCM bit (from 2 to 3) at the output of the TCM coder for use in FEC in the DTV receiver requires an increase of the 21.524 Mbps RS-coded data (plus sync) clock rate to 32.287 Mbps. However, no additional transmission bandwidth is required since the extra bit is added by increasing the number of symbol levels.

The use of 12 parallel TCM coders allows for the *optional* use of a 12-symbol NTSC rejection subtractive comb filter in the receiver that has notches in its frequency response near the NTSC visual, chroma, and aural carriers in order to remove a majority of any cochannel NTSC interference that might have been present at the receiver input (during the digital transition). With the cessation of terrestrial analog NTSC, comb filters are not automatically activated in receivers, but the same TCM processing in the transmitter must be performed for backward compatibility in existing DTV sets.

The addition of the 1 parity bit to the 2 data bits (i.e., 2/3-rate coding) for every symbol provides excellent error correction under white noise conditions such as occurs at the noise-limited contours of broadcast service areas. The tradeoff of the extra complexity by doubling the number of unique symbol levels through the addition of the extra FEC bit (which effectively doubles the required receiver signal power for the same error threshold) is offset by gaining about 1.5 dB net lower effective error threshold in the DTV receiver. This is accomplished without a required increase of signal bandwidth or signal power. However, TCM is not as robust against burst errors in the receiver, and tends to lengthen them due to decoding lockup time. Therefore, the presence of interleaving and Reed–Solomon coding to the system design minimizes any TCM-induced effect of receiver burst errors.

Synchronization Code Insertion. At the beginning of the *signal processing* section, the 8-VSB signal has special binary synchronization codes inserted into the data stream by a multiplexer in order to provide robust DTV receiver synchronization in severe reception conditions (e.g., weak signals, multipath impairment, signal interference, etc.). Short (4-symbol) binary data segment sync codes and long (832-symbol) binary pseudo-random sequence (PRS) data field sync codes are all synchronized to the symbol clock and inserted at periodic intervals.

Binary data segment sync bytes *replace* the 47_{hex} MPEG-2 sync in every data packet by inserting a special 4-symbol code every 832 symbols, which is every 208 bytes (188 data bytes plus 20 parity bytes). Each data symbol carries 2 data bits (plus 1 TCM parity bit). Therefore, the presence of the data segment sync does not reduce the payload data rate of the MPEG transport packet since the 47_{hex} sync is automatically reinserted at the receiver.

Binary field syncs have a duration of 1 entire data segment (832 symbols), and are inserted once every 313 data segments (only a 0.32 percent data loss) to assure that an integer number of data bytes (64,584) appear in each data field. They are made up of various PRBS codes, and are required for synchronization of the data randomization, RS coding, convolutional data byte interleaving, and byte-to-symbol conversion processes. They also help DTV receivers properly find and identify data segment and data field boundaries in the presence of severe impairments and interference, allowing correct receiver operation.

Both of these types of robust binary sync signals have amplitudes equal to constellation levels (±5) that are well within the maximum range of the 8-level data symbols (±7), thus conserving peak RF envelope power during transmission. Both types of syncs have multiple purposes. One purpose is to allow receivers to synchronize to the incoming data field and data frame. A second purpose is to allow the receiver's equalizer to use the segment and field syncs as a known training signal to remove static or slowly moving dynamic linear distortion from the signal (e.g., due to the fixed transmitter emission mask or receiver IF filter as well as propagation multipath). Data segment and field syncs are both locked to the symbol clock, which is locked to the incoming MPEG transport stream clock.

Pilot Inserter. The pilot inserter adds a small, in-phase pilot carrier to the signal, often accomplished by DC shifting the equiprobable 8-level (±1, ±3, ±5, ±7) baseband data and sync symbols by +1.25 constellation units just prior to the modulation process. The small pilot carrier appears about 309 kHz above the lower edge of the RF channel, and is about 11.6 dB below the *total* average power of the entire signal (i.e., below pilot plus data power). While the pilot carrier may appear to be a large signal when viewed on a spectrum analyzer utilizing a relatively low resolution bandwidth (RBW), it actually adds only 0.3 dB to the total average *data* signal power level.

The pilot carrier provides the DTV receiver with a means to easily and quickly synchronize the RF carrier recovery circuitry to the incoming signal carrier. This type of simple narrowband recovery and tracking circuit allows for removal of most types of interference as well as the undesired effects of the DTV signal's own data, which acts as noise-like interference to the receiver's carrier recovery circuitry.

VSB Exciter. The VSB exciter performs various functions, including the critical root-raised cosine (RRC) Nyquist filtering, vestigial sideband modulation with pilot carrier insertion, and upconversion from IF to the final television RF channel. While linear and nonlinear precorrection is *optional* in the ATSC transmission standard, all modern exciters will offer some form of precorrection for signal optimization.

Often, the VSB exciter will use an internal intermediate frequency (IF) such as the traditional 44 MHz (center) value, but it can also be implemented with a low-IF value. The transmitted RF signal will fit into a 6 MHz bandwidth and reside on the desired RF channel (either VHF or UHF) with its *lower* sideband essentially removed, leaving only a small vestige of about 309 kHz. With the cessation of analog NTSC transmission in the television band, no precision RF carrier frequency offset is required.

Typically, the exciter feeds an intermediate-power amplifier (IPA) and a high-power amplifier (HPA) tandem that has good linearity and flat frequency response as well as a sharp-tuned emission mask filter that limits the amount of adjacent channel intermodulation splatter. Harmonics of the RF signal are often reduced by a separate harmonic (low-pass) filter at the output of the HPA. From here, the signal is conveyed by either coaxial or waveguide transmission line to the antenna for radiation over the air.

VSB modulation (as compared to double-sideband modulation with redundant sidebands) and steep RRC filtering (with a linear phase response) together offer excellent spectral efficiency for use in a crowded RF spectrum. Unlike the analog NTSC system where the Nyquist filter is completely in the receiver, an RRC filter is used both in the transmitter and in the receiver to provide the overall raised-cosine (RC) system response that minimizes inter-symbol interference (ISI) in the transmission signal as well as the matched filtering in the receiver that optimizes data reception in noisy conditions.

Good linearity in the final high-power output stages is essential in order to provide a clean DTV output signal and minimize first adjacent channel splatter interference, thus allowing better spectrum packing. However, nonlinear distortion such as AM-to-AM and AM-to-PM (which varies with signal amplitude) does exist, and is primarily generated in the high power amplifier.

Likewise, a linear phase filter response in the RF circuitry is essential in order to provide a clean DTV output signal. However, linear distortion such as nonflat signal group delay across the entire 6 MHz channel (which does not vary with signal amplitude) does exist, and is primarily generated in the very sharp emission mask band-pass filter.

Since both linear and nonlinear distortions exist in real-world hardware, linear and nonlinear automatic *precorrection* is often used in high-power DTV transmitters to allow radiation of a very clean DTV signal from the antenna. With transmitter precorrection, equivalent SNR values (as affected by ISI) are typically better than 30 dB at the output of the emission mask filter.

Low RF carrier phase noise from exciter local oscillators is also essential for good reception performance in DTV receivers. Transmitter phase noise is generally much lower than receiver phase noise so that low-cost consumer DTV receiver tuners are typically the limiting factor for phase noise error threshold. However, VSB modulation inherently has carrier phase modulation that is dependent on the noise-like data, which is due to the quadrature component created by the absence of the lower sideband. Additionally, this quadrature component can be thought of as a signal that cancels the lower RF sideband while enhancing the upper RF sideband.

2.2.2 DTV Receiver

Figure 2.9.2 illustrates the 8-VSB receiver block diagram.

Like the transmitter, the receiver also consists of two unique processing sections: signal processing and data processing. The receiver must perform the inverse operation from that performed in the transmitter. The *signal processing* section deals with RF channel tuning, critical (matched) filtering, synchronization, and equalization of signals that have likely experienced multipath propagation distortion or signal interference or both. The *data processing* section deals with data information to be decoded from a coded sequence of randomized and interleaved data symbols using "on the fly" FEC.

While the receiver must perform the complementary functions of the transmitter in order to provide a proper digital output signal, it often must do so in severe reception environments.

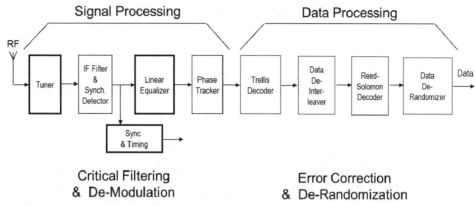

FIGURE 2.9.2 ATSC transmission system receiver block diagram.

Tuner. The DTV receiver tuner converts the desired 6 MHz RF channel to an IF signal, traditionally 44 MHz although some recent receivers use a much lower IF frequency. Either single conversion or double conversion tuning can be used. The same tuning bands are currently used for DTV as they were for NTSC, except the upper end of the UHF band was reduced from CH 69 to CH 51 in June 2009 when full-power analog NTSC television station signals were turned off. The tuner determines both reception sensitivity (noise figure and phase noise) and overload capability (cross-modulation and intermodulation).

The required tuner performance parameters for DTV are the same as for NTSC, except some of the performance values are tighter. For instance, phase noise must be much better (e.g., ≤ -90 dBc/Hz @ 20 kHz offset) in digital ATSC tuners than in analog NTSC tuners due to DTV's more efficient use of the spectrum. In the FCC planning factors for DTV allocations, tuner noise figure was assumed to be ≤ 10 dB for the two VHF bands but ≤ 7 dB for the UHF band.

With further spectrum repacking scheduled and more DTV signals plus nontelevision signals (e.g., white space devices as well as cellular phone LTE signals), robust adjacent channel interference performance is necessary, requiring better intermodulation specs. This is often achieved with better IF filtering to reduce adjacent channel leakage, improved 3rd order intermodulation (IM_3) to minimize an increased noise floor, input tracking filters for mixer overload protection as well as better image rejection (in single-conversion designs), and noncoherent wideband automatic gain control (AGC) to reduce adjacent channel interferer levels before reaching the mixer for reduced IM_3 generation. The wideband AGC typically uses energy from the desired on-channel signal as well as any undesired adjacent channel signals that pass through the tracking filter and mixer (but before any narrowband IF filtering) to control the front-end RF attenuator. This protects the mixer from overload and therefore minimizes intermodulation distortion.

Typically there is a direct tradeoff between noise figure and overload protection that must be optimized. Since DTV service is primarily interference-limited rather than noise-limited, especially after further FCC reductions of the number of terrestrial DTV channels and subsequent spectrum repacking are enacted, overload protection is likely to be a higher priority for DTV tuners than sensitivity.

IF Filter and Synchronous Detector. The IF band-pass filter provides signal selectivity when tuning to a desired RF channel while the VSB synchronous detector performs synchronous carrier recovery, desired signal (narrowband) IF AGC, and matched RRC filtering for optimum coherent decoding.

The IF filter eliminates the sum frequencies generated in the heterodyne mixing process as well as any adjacent channel interference signals, and passes only the difference frequencies that contain the desired signal. Significant IF signal attenuation of adjacent channel interferers, particularly for the first adjacent channel, is desired.

Synchronous carrier recovery and signal demodulation can be achieved by frequency- and phase-locking to the small, in-phase pilot carrier independently of the randomized data although some designs may *optionally* allow the 8-level data to be used in carrier recovery (with some performance degradation) if the pilot is canceled due to a large negative echo. Frequency- and phase-locking of the pilot carrier can occur independently before any other loops are synchronized and can operate below the receiver's data decoding threshold.

This robust carrier recovery and synchronous detection is possible due to relatively narrowband pilot carrier-locking techniques that minimize the effects of significant noise or interference present at the tuner input as well as the large noise-like "interference" (in a 6 MHz bandwidth) that the desired randomized data presents to the carrier-locking circuitry. One possibility for carrier recovery is a frequency- and phase-locked loop (FPLL). With an FPLL, relatively wideband carrier acquisition (e.g., ≈100 kHz) insures frequency lockup, and narrowband carrier phase-locking (e.g., ≈2 kHz) rejects the wideband data symbol phase modulation inherently present in a VSB signal. Additionally, tracking any low-frequency phase noise from the tuner's local oscillator is also achieved. Importantly, carrier recovery can be achieved under conditions of severe impairments (white noise or multipath) or interference (cochannel or adjacent channel).

Narrowband noncoherent AGC, based on average power of the *desired* randomized DTV signal at the output of the IF filter, is performed in parallel with synchronous detection, but independently of data symbol decoding. This AGC must be quick enough to handle not only channel changes, but also fast signal fading conditions.

The RRC matched filtering can take place in the IF band-pass filter (e.g., surface acoustic wave (SAW) filter with large signal overload capability) or in a complex baseband filter (e.g., digital filter with accuracy and repeatability) or a combination of both, and may be aided by the linear data signal equalizer for final "tweaking" of the data pulse shaping. Together with the RRC filtering in the transmitter, the receiver's RRC filtering creates a cascaded raised-cosine (RC) system response that optimizes data detection for zero (minimal) ISI, and provides matched filtering for optimum sensitivity in the presence of noise.

Data Synchronization and Clock Timing. Symbol clock and data field synchronization is necessary for proper data decoding. Symbol clock recovery is often accomplished with band-edge methods. This entails using a narrow band-pass filter centered at the upper Nyquist frequency (i.e., 5.381 MHz) in conjunction with zero-crossing detectors to achieve proper sampling of the data signal for minimal ISI. This captures energy at half the symbol rate in order to frequency- and phase-lock to the symbol clock at twice this band-edge frequency (i.e., $2 \times 5.381 = 10.762$ MHz). The narrowband nature of this process adds robustness.

Data field synchronization can be accomplished using powerful correlation techniques that look for repetitions of the various binary pseudo-random-sequence (PRS) codes that repeat every 313 data segments. The 512-bit PRS code and three 63-bit PRS codes can be used together for correlation in this synchronization process.

Both of these tasks can be accomplished independently of data, and even below data decoding threshold. Synchronization can be achieved in conditions of severe impairments (white noise or multipath) or interference (cochannel or adjacent channel).

Linear Equalizer. Data equalization is required to remove any *linear* distortion that has impaired the DTV signal either due to imperfect transmitter or receiver signal processing (e.g., impedance mismatches, nonflat amplitude or group delay filter characteristics) or due to propagation distortions (e.g., static or dynamic multipath). The severest impairment is due to multipath propagation effects such as complex multiple static or dynamic echoes caused by signal reflections from objects (fixed or moving) in the signal path between transmitter and receiver.

Data equalizers are often a combination of digital feedforward and feedback transversal filters that remove the linear distortion and restore proper sampling of the data signal for minimal ISI.

Transversal filters are tapped delay lines with variable gain amplifiers at each tapped output that are then all summed together at the output. The gain of each tap amplifier is under control of the equalization algorithm. While many DTV receiver equalizers operate in the time-domain, some are implemented in the frequency domain.

Since synchronization is very robust, even in significant multipath conditions, an equalization algorithm can use the known repetitive and spectrally flat binary PRS field sync as a training signal to iteratively direct these equalizer tap weights (amplifier gain values) to converge on a solution that reduces the linear distortion effects. However, since the data field syncs only repeat every 313 data segments (i.e., a 41-Hz repetition rate), only static or quasi-static (e.g., less than 10 or 20 Hz Doppler effect) can typically be tracked successfully using the field sync as a training signal. The typical alternative is to use data-directed equalizers that slice the always-present and plentiful 8-VSB randomized data. This allows an estimate of the correct (i.e., "ideal") symbol levels to be obtained in order to determine equalizer tap weights to cancel fast-moving echoes. Sometimes both methods are employed, either simultaneously or sequentially. Regardless of the specific methods used, a spectrally flat reference signal (known PRS sync or randomized sliced data) allows equalization of all frequency components over the entire 6 MHz channel. Typically, equalizers initially enter an acquisition mode to achieve convergence, and then begin a tracking mode to follow any time-varying multipath distortion.

The equalizers can be real only (in-phase signal processing only) or they can be complex (both in-phase and quadrature-phase signal processing). The time delays of the taps can be symbol spaced (i.e., 91.9 ns apart) or fractionally spaced (e.g., 45.95 ns apart). A variety of equalization algorithms are typically used, including the traditional least mean squares (LMS) version as well as very advanced algorithmic techniques. The DC value at the equalizer input, due to the demodulation of the pilot (plus any effects of multipath), must be subtracted before passing through the equalizer in order for the multipath to be properly reduced.

Modern DTV data equalizers typically must handle echo delays in the range of ±40 μs or longer, often removing echoes with amplitudes in the 90 percent range or greater. When removing significant amounts of linear distortion, equalizer tap gains can be large and subsequently enhance random noise that exists in the receiver's tuner. This effectively degrades (increases) the noise-free 15 dB SNR error threshold. Likewise, in addition to working in severe multipath environments, equalizers must also operate in noisy and interference-laden conditions.

During the digital transition, analog cochannel interference was commonly experienced in the field. A separate NTSC rejection filter is *optionally* available for use in DTV receivers, which is comprised of a 12-symbol delay single-tap subtractive comb filter. Its frequency response has nulls near each of the three large NTSC carriers (visual, chroma, and aural), and is very effective in mitigating NTSC cochannel interference. However, the comb filter has a 3-dB white noise penalty due to the subtraction of the two filter paths, and is therefore switched out of operation when strong NTSC cochannel interference is not detected in order to maintain the nominal 15 dB SNR error threshold. Another method for removing cochannel interference is to let the equalizer do its best to simultaneously remove the cochannel interference while it is equalizing the desired DTV signal. Following the completion of the digital transition, this NTSC cochannel mitigation process has become a moot point.

Phase Tracker. Receiver phase tracking is an important process that removes carrier phase noise that can creep into the system from any frequency heterodyning process that occurs during modulation, demodulation, upconversion, or frequency translation. Perturbations from the expected (i.e., ideal) in-phase and quadrature-phase 8-VSB data levels due to carrier phase rotation direct the phase tracker to remove the phase noise by de-rotating the I-data and Q-data. Since commercial transmitter equipment is not under a severe space and cost restriction like consumer receiver hardware, it is typically the DTV receiver's tuner that generates most of the system phase noise.

Tracking phase noise is accomplished by directly using the output of the equalizer where linear distortion errors have already been removed from the signal and only the effects of phase noise, white noise, or nonlinear intermodulation remain. For receivers with only in-phase detectors, an *estimate* of the received quadrature signal can be created by simple pseudo-Hilbert transform filtering of the in-phase signal. Error estimates of both the in-phase and quadrature signals are obtained at every symbol time, followed by applying the phase tracking algorithm that directs a complex multiplier (I and Q components) to appropriately de-rotate the data signal.

Noise enhancement can occur at low SNR values due to incorrect slicing estimates. Under these conditions, correction loop gain can be reduced or turned off, if desired. With the use of improved (i.e., lower) phase noise oscillators in receivers, noise enhancement becomes less of an issue. Assuming white noise and intermodulation noise levels are reasonably small (i.e., >15 dB below the average desired signal power), the generated error estimates will be primarily due to phase noise and therefore allow good phase tracking performance.

Phase noise is more destructive when the Q-channel signal is large since any I and Q constellation rotation due to phase noise will cause larger deviations in the I-channel component due to quadrature cross-talk. While the 8 discrete data levels in the I-channel signal are essentially equiprobable during symbol sampling times due to transmitter randomization, the Q-channel signal level has a Gaussian probability characteristic. This means that the Q-channel signal spends more time during symbol sampling points at lower values rather than higher ones, helping to minimize the degrading effects of phase noise cross-talk. After the phase noise is reduced or eliminated, the best signal possible enters the FEC decoding circuitry.

The phase tracker, along with the pilot carrier recovery loop, both work together (in tandem) to reduce phase noise. The carrier recovery process removes any frequency errors and low-frequency phase noise (e.g., 2 kHz or less) and the phase tracker removes any high-frequency phase noise (e.g., up to 60 kHz or higher). This concatenated loop system allows the phase tracker to use the known binary field sync as a reference signal to lock quickly and remove any phase noise during this time, and then begin slicing 8-level data immediately following the field sync for high-speed tracking capability. The removal of phase noise allows for low-cost consumer-grade tuners to be used in consumer DTV receivers.

Trellis Decoder. The convolutional trellis decoder uses both hard and soft decoding as well as the TCM structured redundancy to provide the best estimate of the transmitted data symbols. This is accomplished using past state transition sequences (sequential decoding) and amplitude error measurements taken from the eight possible symbol levels (squared Euclidean distance calculations). Soft-decision slicing of the data symbols is performed to create analog-like error signals. Then the total summed squared-errors are compared at each symbol for the various possible limited ("finite") past transition paths ("path metrics") over four or five previous symbols ("trace-back length"), with the most likely symbol being selected (Viterbi maximum likelihood detection) in order to decode the two data bits per symbol. The accuracy of this process depends on how many past symbol transitions are used in the trace-back process, with a tradeoff existing between accuracy and hardware complexity.

The receiver uses 12 parallel TCM Viterbi decoders to decode the data. Twelve different symbol streams are sequentially created by appropriately parsing 3-bit (2 data plus one TCM parity) symbols to different TCM decoders. Each TCM decoder processes a different symbol stream, and consists of data taken every 12 symbols so that all the data bits from one byte are processed by the same TCM decoder. This process provides a short ($\approx$1.1 μs) intrasegment interleaver anti-burst effect. Since there are an integer number of data bytes in every data field (258,336), the robust field synchronization assures that proper TCM byte processing is achieved in the receiver.

The TCM decoder works best in an AWGN environment where the errors are essentially random and evenly distributed, but does not work well under burst noise conditions where long consecutive losses of data occur. When the TCM decoder is ultimately overrun by errors and fails in a burst mode condition (e.g., impulse noise or dynamic multipath), it takes extra time for it to resynchronize and begin outputting error-free symbols again, thus lengthening the effect of the burst error. To minimize this lengthened burst error effect, the employed trellis code design was limited to a short 4-state code for faster recovery.

The reason that TCM coding, interleaving, and RS coding are concatenated is to take advantage of the synergy of the different types of error-correction techniques, using their individual strengths to overcome their weaknesses. This has the effect of increasing the overall system coding gain as well as effectively combating burst errors. The TCM coding provides around 1.5 dB net improvement beyond that of RS coding alone.

Data Deinterleaver. The intersegment convolutional deinterleaver, which is synchronized to the robust data field sync, reassembles the data bytes dispersed in the transmitter. By putting the data bytes back into their original order, any contiguous transmission burst errors in the receiver are spread out over many data segments and therefore de-correlated. This byte error dispersion creates a more uniform distribution of errors (i.e., more similar to random errors) and increases the chance that they will be error-corrected by the Reed–Solomon block decoder (if there are ≤10 byte errors in a given data packet). In other words, the power of the RS decoder is extended by the error dispersion. A convolutional interleaver requires less memory than a simple block interleaver.

While no data overhead (i.e., data rate reduction) is needed in this interleaver process, extra memory is required which adds latency to the data since the entire interleaved block must be received before the packets can be decoded. Continual resynchronization of the convolutional deinterleaver in the receiver is unnecessary after initial synchronization since the interleaver process is locked to the robust binary data field sync in the transmitter. However, deinterleaver resynchronization is often performed (redundantly) in the receiver to add reliability in severe reception environments, thus improving performance.

Burst errors can be caused by external sources (e.g., impulse noise) or internal sources (trellis decoder overrun). Due to the length of the convolutional deinterleaver, errors are spread out over a 4 ms interval, thus allowing the RS decoder to mitigate moderate length burst errors (up to 193 μs) that occur during impulse noise or dynamic multipath conditions.

Reed–Solomon Decoder. The Reed–Solomon decoder performs block FEC on each MPEG-transport stream packet (187 data bytes plus 20 Reed–Solomon parity bytes). While Reed–Solomon decoding, which uses a very efficient decoding algorithm, works well in random noise conditions with many single bit errors, it is very powerful in burst error conditions. The deinterleaver spreads out any occurring burst errors over many data packets, making error decoding much more robust by increasing the chance of MPEG transport stream packets having less than 10 *byte* errors, which can be corrected. RS byte error correction has the advantage that when byte errors are detected and corrected, it doesn't matter how many bits are in error in each of those byte errors. In other words, error correction occurs whether 1 bit is in error or all 8 bits are in error within a corrupted byte.

The combination of a *short* convolutional TCM code (4-state) and a *long* RS block code ($t = 10$) provides a good combination for burst error mitigation. The RS decoder also has the advantage of providing a packet error signal for the MPEG decoder to flag *uncorrectable* packet errors, and thus avoid use of the specific packet in error for decoding video, audio, or ancillary data. This can aid the receiver's video error concealment process to produce better images on the viewer's screen. This packet error flag can also be used with some processing to indicate the current packet error rate (PER) in a DTV receiver.

Data Derandomizer. The data derandomizer has the straightforward task of putting the randomized data back into its original form using the known PRBS randomization code employed in the transmitter, thus "descrambling" data for use by subsequent circuitry. Syncs are not processed since they have already been removed from the baseband data stream. Additionally, the MPEG-2 transport stream sync (47_{hex}) is reinserted into the stream before being sent to the subsequent data processing circuits.

The performance of the derandomizer is further enhanced in severe environments by being locked to the robust binary data field sync.

2.3 VSB BASEBAND DESCRIPTION

The ATSC 8-VSB transmission system transmits data in a power- and frequency-efficient manner using 8-VSB amplitude modulation. Data symbols are created at a 10.762 MHz rate (i.e., every 92.9 ns) with a recommended frequency tolerance of ±2.79 ppm.[†††] The transmitted DTV symbol rate must be fractionally locked to the incoming 19.392659 Mbps MPEG transport data stream rate[‡‡‡] so that no data packets are added or deleted in the 8-VSB modulator. The frequency of all clocks (video encoder, audio encoder, transport, and transmission) is derived from the master studio clock, which is nominally 27.000 MHz.

The 8-VSB transmission system uses 8 unique amplitude levels to represent three bits of information per each transmitted symbol. Two of the three bits are actual payload data bits while the third bit is a parity bit due to TCM that adds one layer of FEC to the signal. Therefore, the raw data rate of the 8-VSB baseband signal is 32.287 Mbps (i.e., 3 bits/symbol × 10.762 Msymbol/s).

The amplitude levels, as shown in Fig. 2.9.3, are described by constellation units (CU), a dimensionless number that represents the combination of 3 bits, or 8 levels (i.e., $2^3 = 8$). Traditionally, these eight levels are represented as ±1, ±3, ±5, and ±7, with a long-term average value of 0, assuming that all of the data is equiprobable due to data randomization. Since these symbol levels average to zero, the baseband 8-VSB signal would produce no discrete RF carrier signal when ultimately modulated; i.e., suppressed carrier modulation would be achieved. However, just prior to modulation, all eight unique amplitude levels of the baseband 8-VSB signal are offset in a positive direction (i.e., they are increased) by a value of +1.25 constellation units. This "DC shift" creates a small, in-phase pilot carrier during modulation that contributes about 0.3 dB to the total average of the 8-VSB signal. Transmitting only a very small pilot carrier (compared to the total data signal power) contributes to signal power efficiency.

The transmitted data is packetized into 188-byte MPEG-2 packets, with each MPEG packet consisting of 184 payload data bytes plus a 4-byte header. The first byte of the header is the MPEG-2 synchronization byte, which is represented by the value 47_{hex}. Before TCM occurs, each MPEG-2 packet has 20 Reed–Solomon parity bytes added to the end of the packet (called a systematic code) for another layer of FEC protection. Each 208-byte MPEG packet is transmitted in a single 8-VSB data segment, as illustrated in Fig. 2.9.3.

However, before modulation takes place the MPEG-2 transport stream sync byte (47_{hex} representing 8 bits) is removed and replaced with a fixed, 4-symbol (4 symbols × 2 data bits/symbol) 8-VSB binary segment sync. The binary data segment sync is buried within the outer limits of the 8-VSB signal; i.e., the ±5 syncs do not extend beyond the ±7 data signal region like the NTSC syncs extend beyond the video region. Since the syncs are less than the outer data levels (±7), they do *not* increase the peak transmitted RF power of the modulated signal the way horizontal syncs do in the analog NTSC system. Therefore, use of these "buried" 8-VSB segment syncs contribute to the overall signal power efficiency.

The very short 4-symbol data segment sync, which repeats every 77.3 µs (12,935 segments per second), takes the place of the MPEG-2 transport stream sync. It only lasts for 0.37 µs (compared to the 10.9 µs horizontal blanking interval in NTSC), and therefore contributes to spectral and time efficiency. Unlike payload data, the data segment syncs are *not* included in any FEC schemes, with their repetitiveness providing the desired robustness.

In addition to data segments, the baseband data is further broken down into data fields and data frames (which has *nothing* to do with video fields and video frames). A data frame is made up of two consecutive data fields: an even field and an odd field, with each beginning with its own unique binary field sync (middle PRS sequence inversion). The binary field sync, as shown in Fig. 2.9.4, consists of multiple codes that reside in an entire data segment, and therefore it has a duration of 77.3 µs like every other data segment. Except for the last 12 symbols that use eight levels (a repeat from the end of the previous data segment), the entire data field sync uses two levels that fall within the amplitude range of the entire data signal. That is, ±5 constellation units are used for the two binary

---

[†††] "ATSC Recommend Practice: Transmission Measurement and Compliance for Digital Television", Document A/64B, May 26, 2008, Section 5.1.3, www.atsc.org website.
[‡‡‡] "ATSC Digital Television Standard—Part 1: RF/Transmission System Characteristics, A/53, December 15, 2011, www.atsc.org.

ATSC DTV TRANSMISSION SYSTEM   **2.167**

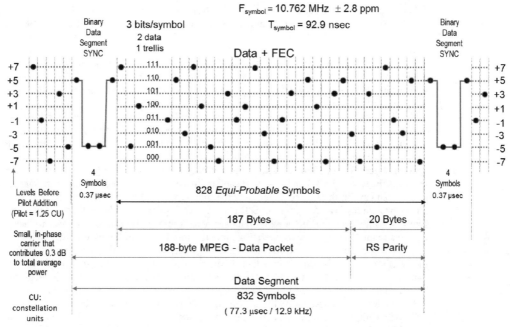

FIGURE 2.9.3 8-VSB data segment format.

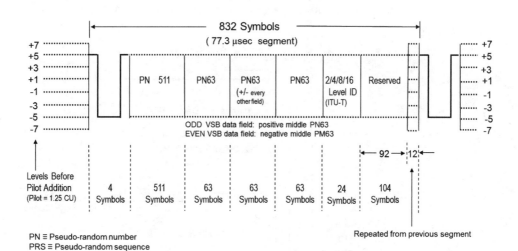

FIGURE 2.9.4 8-VSB data field sync format.

data field sync levels just as they were used for the data segment sync levels. Since they are once again less than the ±7 outer data levels, they do not increase the peak transmitted RF power the way vertical syncs do in the analog NTSC system. Therefore, use of these "buried" 8-VSB data field syncs contribute to the overall signal power efficiency.

The binary data field sync, even though employing an entire data segment for its sync codes, is only transmitted every 313 data segments (24.2 ms) at a 41.3 Hz rate, and therefore contributes to spectral and time efficiency. Unlike payload data, the data field syncs are *not* included in any FEC schemes, with their repetitiveness providing the desired robustness.

Multiple PRS, sometimes referred to as pseudo-random numbers (PN), reside in the field sync. The first is a long 511-bit PRS code, which is followed by three short 63-bit PRS codes. The middle code of the three PRS codes alternates in polarity every other field to identify even and odd data fields, and also allows certain signaling in extended parts of the ATSC transmission standard (e.g., the mobile/handheld system). After these one long and three short PRS codes, a signal identification code is inserted which identifies the type of VSB signal (2, 4, 8, 16, and trellis-coded 8) that is being transmitted according to the ITU-T standard. In the United States, only the trellis-coded 8-VSB system is employed. The remaining binary bits are reserved for future use. The last 12 symbols are 8-levels, and just a repeat of the last 12 symbols from the end of the previous data segment. These are used in the receiver to keep the trellis-coding and interleaving systems operating properly since they must "skip over" the binary field syncs.

Data organization is illustrated in Fig. 2.9.5, which describes the data frame that is employed in the ATSC transmission system. In this diagram, which shows two data fields that comprise a data frame, note the 4-symbol (1 byte) data segment syncs that are placed at the beginning of *each* 77.3 μs data segment (12.94 kHz repetition rate). They are followed by an additional 828 symbols (187 data bytes plus 20 RS parity bytes), which is 1656 bits (excluding trellis-coding bits), for a total of 207 bytes. Unique even and odd data field syncs (each 24.2 ms) are inserted every 313 data segments (41.33 Hz rate), with even and odd fields having an embedded unique field sync identification code (alternating middle PRS polarity). A large number of the binary PRS symbols in these field syncs is known by receivers, and is used to equalize the data signal for removal of linear distortion encountered in transmission, such as complex multipath echoes.

A data frame is made up of two consecutive data fields, and has a time-duration of 48.4 ms. An important aspect of the RF transmission data frame is that every VSB segment carries data byte from

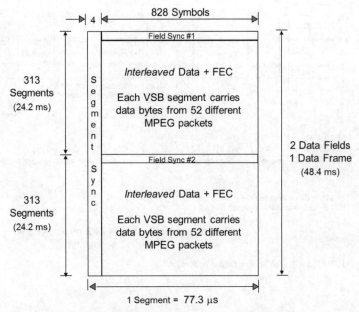

FIGURE 2.9.5 8-VSB data frame made up of an odd and even data field.

52 *different* MPEG packets due to the convolutional interleaver that spreads the data over one-sixth of the field in order to combat burst errors caused during propagation. Despite the DTV data field and frame structure looking similar to the analog NTSC field and frame structure, there is no direct correspondence between DTV data fields and frames and NTSC video fields and frames. One data field carries many different data packets: some are video packets (from one or more programs), some are audio packets (from one or more programs), and some are ancillary data packets (that may or may not be related to the programming). In this sense, the 8-VSB DTV transmission system is "data agnostic" in that it "does not know nor does it care" about the digital information that it conveys over-the-air. The primary goal of the DTV transmission system is to convey as much digital information in an error-free manner as possible from the transmitter to each receiver, regardless of the source or type of the data.

The overall data payload efficiency of the ATSC transmission system can be seen to be about 90 percent (i.e., [188/208] × [312/313] × 100) based on the addition of 20 RS parity bytes every data segment, and a field sync inserted every 313 data segments. If the effect of the 2/3-rate TCM processing is also included, then the efficiency would be rated at about 60 percent. Note that there is no loss in efficiency for using the 4-symbol data segment syncs since these four symbols, which represent 8 data bits (no trellis coding is applied to the segment syncs), replace the MPEG-2 transport stream byte (47_{hex}), which is removed in the transmitter before transmission and reinserted in the receiver after reception.

2.4 VSB BASEBAND SPECTRAL DESCRIPTION

According to FCC rules, the ATSC 8-VSB RF spectrum must fit within a single 6 MHz RF channel that resides in the traditional VHF and UHF television bands. RF signal characteristics are affected by baseband waveform shaping prior to amplitude modulation. Therefore, a detailed understanding of the spectral characteristics of the DTV baseband signal is required.

The baseband signal, with it mathematical complex characteristics (i.e., real and imaginary components), can be thought of as a *baseband representation of an RF phenomenon*. In other words, the real component of the baseband signal represents the in-phase component of an RF modulated signal while the imaginary part represents the quadrature-phase component of an RF modulated signal. Therefore, this section will focus on baseband spectral shaping that occurs in the ATSC digital transmission system.

Simultaneous conditions of band-limited spectrum with no ISI require particular system filtering characteristics that properly shape the data pulses which convey digital information over the transmission system. An individual digital symbol can be thought of as a single impulse (e.g., at time $t = 0$) with a given amplitude (e.g., one of eight discrete levels), infinitesimal time width, and infinite bandwidth. A data stream is made up of a series of these weighted, time-delayed impulses. However, it is unacceptable to use an infinite-bandwidth impulse to solely represent digital symbols in a band-limited channel such as the required FCC 6 MHz television channels employed in the United States.

Therefore, the data impulse must be frequency limited. If an impulse is put through a theoretical but unrealizable Nyquist "brick-wall" low-pass filter to limit its bandwidth to F_N, the equivalent time-domain impulse response is a sinc pulse (i.e., a [sine X]/X shape) with significant signal time "stretching" and amplitude "ringing." The Nyquist bandwidth ($BW_N = F_N$) is defined as the *ideal* minimum required spectral bandwidth that passes a signaling rate of F_S (where $F_S = 1/T_S = 2BW_N$). T_S is the data symbol sampling time and F_N is the Nyquist frequency. In other words, the Nyquist bandwidth is the *ideal* system bandwidth that passes a properly sampled band-limited data signal with zero ISI at a maximum signaling rate (a zero ISI requirement is explained below).

For the ATSC DTV system, the Nyquist frequency (F_N) is 5.381119 MHz and passes a symbol rate (F_S), of 10.762238 Mbps (i.e., $T_S = 1/F_S$ = 92.9 ns). This brick-wall filter's unrealizable nature is due to the part of the time-domain impulse response that occurs *before* time $t = 0$ when the impulse is applied to the filter, thus making it noncausal, i.e., requiring a predictive or anticipatory response.

However, a better (i.e., easier) approximation and thus "more" realizable system filter response would be one where the infinitely steep frequency band transitions are replaced with ones that "roll off smoothly" yet still are relatively steep. A commonly used mathematical representation of such a frequency-domain transition region is the RC response, which is sometimes referred to as a Nyquist slope. This filter has a frequency response (Fig. 2.9.6a) and a time-domain impulse response (Fig. 2.9.6b) that are related by the Fourier transform and it's inverse.

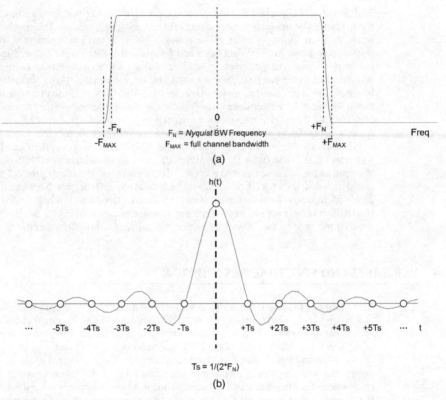

FIGURE 2.9.6 Raised-cosine baseband system low-pass system filter: (*a*) frequency-domain amplitude response, (*b*) time-domain impulse response.

While this system filter is also noncausal, it is easier to *approximate* than the brick-wall filter due to the raised-cosine filter time-domain impulse response falling off much quicker than the brick-wall filter impulse response (i.e., requiring a much shorter filter impulse response length to reasonably approximate it). Note that the required baseband bandwidth needed is no longer the minimal value of F_N, but rather a slightly larger value F_{MAX}, which demonstrates the tradeoff that is required in order to use reasonable filters in real applications. More excess bandwidth means a shorter and "more realizable" filter impulse response, but it also means that the trade-off is a slightly less efficient system in terms of data rate (for a fixed channel bandwidth). One figure of merit for system spectral efficiency is the amount of excess bandwidth that is used beyond the ideal Nyquist bandwidth, which for the ATSC system is about 11.5 percent (i.e., $100 \times [F_{MAX} - F_N]/F_N = 100 \times [6.000 - 5.381]/6.000 = 100 \times 0.115$). This excess system bandwidth of 11.5 percent was once thought to be very aggressive (in the 1990s when analog filters were primarily used to shape the data pulses), but in today's digital communication systems, this value of excess bandwidth is now considered typical and easily achievable.

The baseband time-domain impulse response can be seen to have significant ringing due to the steep low-pass filter transition regions, with zero crossings occurring at regular intervals $1/(2F_N)$. The ringing is seen to stretch out the single discrete impulse (which represents one data symbol) at $t = 0$ over a lengthy amount of time before and after the data impulse (i.e., over many data symbol intervals). This has the *potential* for one data symbol to interfere with many previous and subsequent data symbols, causing ISI. In order to eliminate ISI, the system's symbol clock frequency is specifically selected to be exactly $2F_N$, such that the symbol sampling times fall directly on all of the zero crossings of the impulse response plus the one nonzero data symbol (at $t = 0$). This critical impulse response shaping can be considered "controlled ringing," and has the effect of eliminating this type of ISI when a stream of data is transmitted using very specific sampling times. Note that ISI is only absent at each individual *symbol* sampling time, and not between symbols where the ISI can build up depending on the specific pattern of data stream symbols. However, the sensitivity to precision symbol clock timing can be seen in this example (Fig. 2.9.6b) since any static symbol clock phase error or dynamic symbol clock jitter can cause ISI due to nonoptimum symbol sampling.

The results of the critical ringing can be seen in the five-symbol binary data stream example in Fig. 2.9.7a showing the individual impulse responses (for illustrative purposes) and in Fig. 2.9.7b illustrating the *composite* signal resulting from the five symbols.

In this example (Fig. 2.9.7a), the ±1 constellation unit levels represent the two binary impulse symbol levels initially created before RC filtering is performed. This filtering process expands each individual symbol in a ringing fashion. When sampled at the proper symbol times in a receiver, it can be seen that the original binary symbols can be retrieved precisely, with no ISI. Therefore, this raised-cosine system filtering method allows efficient transmission of high-speed data with minimal ISI through a band-limited system. This principle works with multiple data levels as well (e.g., four levels, eight levels, etc.).

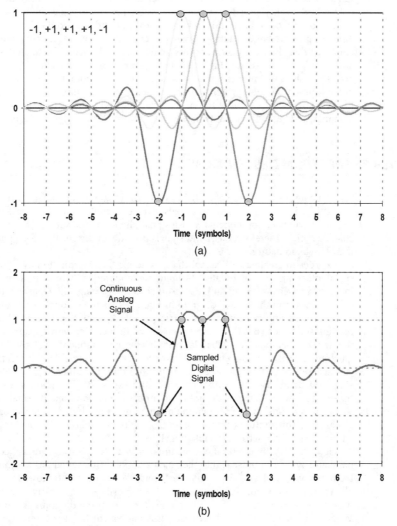

FIGURE 2.9.7 Stream of data symbols: (*a*) shown individually, with no ISI; (*b*) shown combined, with no ISI.

Note that the resulting composite digital baseband signal (Fig. 2.9.7b) is a continuous analog waveform, with levels extending above and below the ±1 binary constellation unit levels *between* the symbols, but yet representing a digital signal at the exact symbol times (solid dots) after being sampled in a digital receiver. When many more symbols are considered in this sequence, the peaks will rise significantly. Since the digital data will be essentially random and therefore noise-like, the signal (and peaks) will vary randomly between symbol times, and can be expressed statistically by a cumulative distribution function (CDF). For 8-VSB, 99.9 percent of the time the RF signal envelope is within 6.3 dB of the average power. However, while the DTV signal *peaks* vary statistically, the *average* power of the digital signal will remain constant regardless of the video, audio, or ancillary information being conveyed by the data symbols. This is unlike analog NTSC where the *peak* envelope signal power (represented by horizontal and vertical syncs) remains constant with video information while the *average* power varies.

If this composite digital signal is displayed in a repeated manner every symbol time, and many of these filtered symbols are allowed to overlay each other (e.g., as would occur on a scope trace), a binary "data eye" diagram is achieved, where the distinct data levels can be seen as well as the optimum symbol sampling point (see Fig. 2.9.8a). If eight different symbol levels are transmitted instead of just two symbol levels, then an 8-level data eye diagram is obtained (see Fig. 2.9.8b). Between symbol times, the data signal looks random due to ISI, and signal peaks extend beyond the maximum (outer) symbol levels. This can produce large peak-to-average power ratios. When more transmitted data symbol levels are employed that allow a higher transmission data rate to be achieved, smaller data "eyes" (i.e., openings) are created, and thus require higher SNR values for error-free detection in receivers. These data eye diagrams also show that proper symbol time sampling has an important effect on data reception, requiring not only proper symbol clock frequency and phase but also minimum symbol clock jitter. Any sampling time perturbations will cause the signal to be sampled in areas that do not have maximum data eye openings, and thus allow data errors to occur in the presence of less noise, impairments, or interference.

2.5 VSB RF SPECTRUM DESCRIPTION

The ultimate spectral shape of an RF-modulated signal is often dictated primarily by baseband signal processing. Frequently, baseband signals involve "complex" mathematical representations (i.e., with real and imaginary components) of RF signal phenomenon, such as in-phase and quadrature-phase characteristics. Therefore, baseband signals provide an insight into RF signal characteristics despite their baseband nature.

In order to create an RF signal that is suitable for transmission over-the-air, the baseband data signal must first be modulated. Amplitude modulation, which is a multiplicative process that involves a baseband signal and a sinusoidal RF carrier signal, creates sum and difference frequencies. As shown in the phasor diagram in Fig. 2.9.9a, simple double-sideband amplitude modulation of a baseband sinusoid results in a carrier signal (A) as well as two equal-amplitude but oppositely rotating sidebands (B). This creates a modulated RF carrier with no carrier phase modulation since the two equal-amplitude sidebands create in-phase components that add (horizontal axis) while the quadrature-phase components cancel (vertical axis). Figure 2.9.9b illustrates the vestigial sideband case where one of the sidebands is smaller (C) than the other (B), while Fig. 2.9.9c illustrates the single sideband case where only one sideband exists (B). In these last two scenarios where equal upper and lower sidebands do not exist, a quadrature component remains in the ideal modulated signal. This means that any amplitude modulation other than double-sideband always creates an RF signal with a quadrature component, indicating the presence of RF carrier phase modulation that is a function of the baseband modulation signal.

Figure 2.9.10a illustrates the spectrum of a "real" (i.e., no "imaginary" component) band-limited baseband system impulse signal (i.e., data symbol). This signal has a *symmetrical* amplitude spectrum (shown) and an *anti-symmetrical* phase response (not shown) around zero frequency as all real signals do. For data systems, both the positive and negative frequency transition regions have raised-cosine shapes for minimal ISI (shown in the diagram as a linear line for simplicity). When modulated on to an RF carrier, this equivalent baseband representative signal becomes a double-sideband signal, with equal and redundant upper and lower sidebands. However, this is

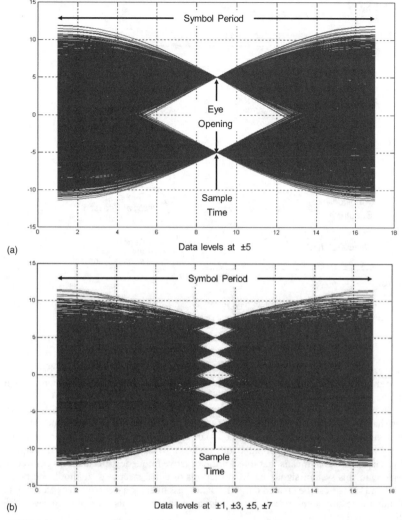

FIGURE 2.9.8 Data eye diagram: (*a*) binary data eyes, (*b*) 8-level data eyes.

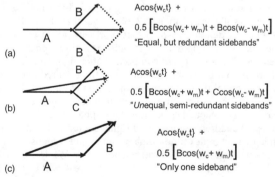

FIGURE 2.9.9 Phasor diagram representing: (*a*) double-sideband modulation, (*b*) vestigial-sideband modulation, (*c*) single-sideband modulation.

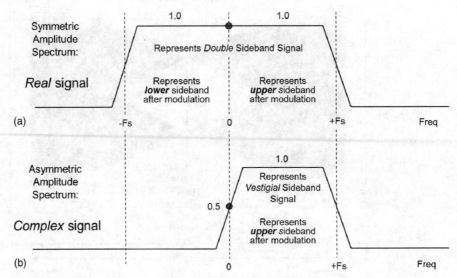

FIGURE 2.9.10 Baseband system representation of: (*a*) double-sideband, (*b*) vestigial-sideband data spectrum signals.

not very spectrally efficient. In theory, only one complete sideband (with no carrier) needs to be preserved in order to distinctively convey the information carried by the data while the other sideband can be totally suppressed.

In practice, however, complete removal of one sideband is typically not feasible, which is the reason that vestigial sideband modulation is often used instead of single sideband. Therefore, in the VSB transmission system, a small vestige (i.e., "part of" or "remnant") of the lower sideband is transmitted along with the complete upper sideband, as shown in Fig. 2.9.10*b*, which in this equivalent baseband representative signal means that most of the negative frequency spectrum is removed. Both transition regions in this system signal are RC functions of equal steepness. It should be noted that vestigial sideband signals can be created either by band-pass filtering a double sideband RF signal or by equivalently processing a complex baseband signal as shown in this example. However, the latter is often the preferred method.

When the inverse Fourier transform is mathematically performed on this *asymmetrical* spectrum amplitude signal, the lack of the negative frequency amplitude response will result in a complex time-domain impulse response, i.e., the signal will have both a real component and an imaginary component, as illustrated in Fig. 2.9.11. Note that the real time-domain impulse response signal (Fig. 2.9.11*a*) is denoted by $h_I(t)$ that represents an in-phase component after signal modulation and the imaginary time-domain impulse response signal (Fig. 2.9.11*b*) is denoted by $h_Q(t)$ that represents a quadrature-phase component after signal modulation. It can be seen that the in-phase time-domain impulse system response has the required periodic zero crossings at all previous and subsequent symbol times that meets the criterion for zero ISI (i.e., open data eyes result), while the quadrature-phase system time-domain impulse response does not (i.e., closed data eyes result). However, this is acceptable since it is the in-phase VSB signal component that carries all of the data while it is the quadrature-phase VSB signal component that essentially cancels one of the sidebands.

While these complex time-domain signals and responses may initially be disconcerting to some, it actually has roots in real-world phenomenon. While a complex number is denoted by "a + jb", it should be noted that both "*a*" and "*b*" are real, while "*j*" is the imaginary mathematical operator ($\sqrt{-1}$). A pair of complex baseband signals, $h_I(t) + j \times h_Q(t)$, can be heterodyned by a pair of orthogonal (quadrature-phase) sinusoids to create an RF signal, i.e., the real baseband component (e.g., $h_I(t) =$ "*a*") can be thought of as simply being multiplied by a cosine wave to create a real in-phase RF signal while the imaginary baseband component (e.g., $h_Q(t) =$ "*b*") can be thought of

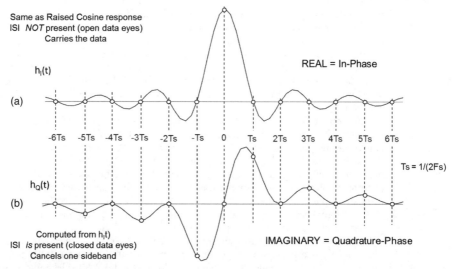

FIGURE 2.9.11 Baseband impulse response of a VSB signal: (*a*) in-phase (real) component, and (*b*) quadrature-phase (imaginary) component.

as being multiplied by a sine wave to create a real quadrature RF signal. This type of straightforward heterodyne operation is referred to as quadrature modulation. All operations described above use real signals. The mathematical "*j*" operator represents the Hilbert transform process, which phase shifts all frequency components by 90°.

However, before the modulation result is described, it is useful to first individually analyze the frequency-domain magnitude and phase characteristics of these two independent real baseband impulse responses (i.e., Fourier transform of I and Q channels), as are shown in Fig. 2.9.12. The in-phase

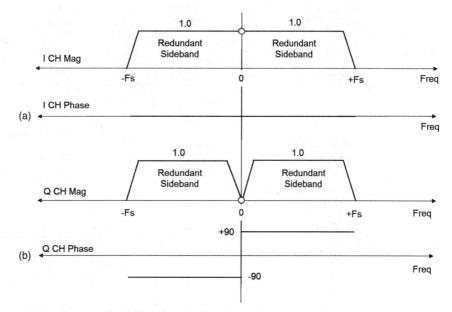

FIGURE 2.9.12 Baseband: (*a*) in-phase (I), (*b*) quadrature-phase (Q) spectral responses of a VSB signal.

magnitude response passes a DC signal (which includes DC required to create the small in-phase pilot) while the quadrature-phase magnitude response passes no DC component (i.e., response is zero at zero frequency). Also, the quadrature-phase frequency response shifts all frequency components by 90° (i.e., pseudo-Hilbert transform) while the in-phase frequency response does not phase shift any of the frequency components (i.e., it has a constant 0° phase response). The reason that the quadrature signal is considered to be a *pseudo*-Hilbert transform is because the ideal (i.e., theoretical) Hilbert transform has an amplitude response that is unity for all frequencies (i.e., there is no attenuation zone around DC).

After direct modulation of the $h_I(t)$ and $h_Q(t)$ impulse responses by a cosine wave and sine wave, respectively, the result is two independent (real) double-sideband RF signals with their respective amplitude and phase characteristics, as shown in Fig. 2.9.13 with their respective spectra centered at the carrier frequency F_C. The amplitude of all of the modulated sidebands is 0.5 due to half the energy being in the upper sideband and the other half of the energy being in the lower sideband. Since the quadrature-phase impulse response was multiplied with a sine wave instead of a cosine wave like the in-phase impulse response, an additional −90° of phase shift is added for both the upper and lower sideband frequencies, resulting in a quadrature-phase impulse response with 180° for the entire lower sideband and 0° for the entire upper sideband.

If these two (real) RF signals are simply added together (i.e., a linear sum), it can be seen that the lower sidebands cancel due to opposite phase angles (180° difference) while the upper sidebands add due to identical phase angles (0° difference). This method of VSB sideband creation is similar to the phasing method for analog single-sideband amplitude modulation.

Complete elimination (i.e., suppression) of the RF carrier hinders carrier lock and demodulation performance in a receiver. Therefore, to facilitate carrier regeneration and signal recovery, a small in-phase pilot carrier is also transmitted with the data. As described earlier, this is easily achieved by offsetting only the in-phase baseband signal by a positive DC offset, selected to be equal to +1.25 constellation units in the 8-VSB system.

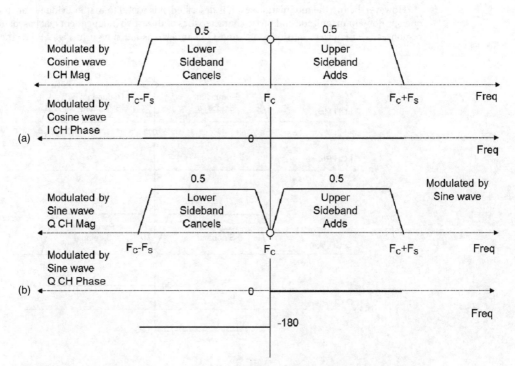

FIGURE 2.9.13 Complex modulated RF: (*a*) in-phase (I), (*b*) quadrature-phase (Q) spectral responses.

This signal representation described above, which illustrates the cascaded RC net system response, is one method of VSB RF signal generation, and it provides insight into its two RF components. That is, the in-phase channel carries the 8-VSB data signal with no ISI as well as the small pilot carrier, and the quadrature-phase channel carries the signal that cancels most of the lower sideband.

The ideal 8-VSB RF signal spectrum that is *transmitted* over the air is defined in Fig. 2.9.14. The 6 MHz RF signal is expected to have a flat, noise-like spectrum, with a small in-phase pilot carrier 309.441 kHz above the lower band edge and steep 618.881 kHz transition regions at each end of the band. The transition regions of the transmitted signal are RRC (i.e., the *square root* of the raised cosine magnitude function) since the overall raised-cosine *system* function required for zero ISI is a combination of cascaded RRC filters shared between the DTV transmitter and the DTV receiver. The RRC's 3-dB points define the 5.381119 MHz Nyquist bandwidth (for zero ISI), which is also equal to the transmitted signal's equivalent noise bandwidth (NBW). The excess bandwidth can be seen to be 11.5 percent (i.e., 6.000/5.381 = 1.11501).

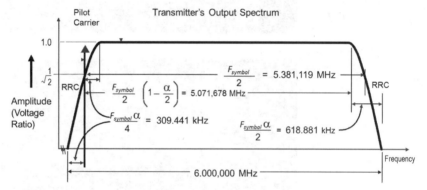

FIGURE 2.9.14 ATSC transmitted signal spectral definition, with RRC transition regions, small in-phase pilot carrier, and 5.381 MHz Nyquist (and noise) bandwidth.

The transmitted RF spectrum of an actual 8-VSB signal is shown in Fig. 2.9.15, where the noise-like flat spectrum region is due to the randomized data signal and the small in-phase pilot carrier signal 309 kHz above the lower band edge of the RF channel is due to the DC offset of the baseband data signal. Both spectral transition regions in the *transmitted* signal are described by a steep RRC function, with a 5.381 MHz Nyquist bandwidth. The entire signal fits within a 6 MHz television channel and the out of band "splatter" energy is at a minimum. The small CW pilot power is 11.6 dB below the total (pilot plus data) 6 MHz signal power. The pilot carrier appears to be much larger than the data power in Fig. 2.9.15 only because of the small (10 kHz) RBW selected for the spectrum analyzer display of the RF signal.

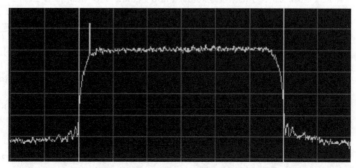

FIGURE 2.9.15 Actual RF spectrum of an 8-VSB signal. (The vertical axis is 10 dB/div and the horizontal axis is 1 MHz/div.)

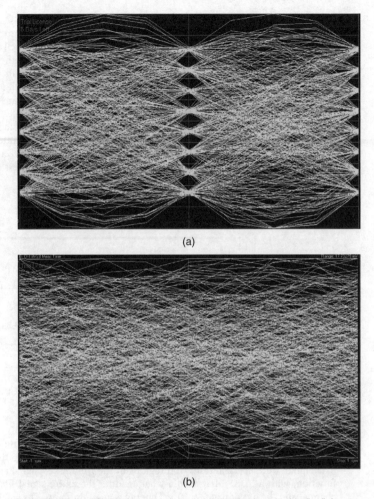

FIGURE 2.9.16 Characteristic waveforms: (*a*) 8-VSB in-phase (I) data "eye" diagram ("open eyes"), (*b*) 8-VSB quadrature-phase (Q) data "eye" diagram ("closed eyes").

Figure 2.9.16 illustrates actual I and Q data "eye" diagrams created in the laboratory from a relatively pristine 8-VSB RF signal, as viewed on a VSB reference analyzer. Note that the in-phase signal has the desired open data eyes (minimal ISI) due to the symbol sampling of in-phase impulse response at zero crossings (see Fig. 2.9.16*a*), which illustrate 8 data levels and 7 data eyes. On the other hand, the quadrature-phase signal has no open data eyes due to symbol sampling of the quadrature impulse response on nonzero crossings (see Fig. 2.9.16*b*). Therefore, all of the 8-VSB data is transmitted on the I-channel while the Q-channel is the pseudo-Hilbert transfer function of the I-channel data, and results in the cancelation of the lower RF sideband.

The signal peaks, which occur between the sampling times of the randomized data symbols, appear random in nature. The average power of the transmitted DTV signal remains constant but the existing peaks can be statistically described by a complementary cumulative distribution function (CCDF). The RF envelope of the 8-VSB signal *exceeds* 6.3 dB above the average power of the signal a total of 0.1 percent of time (i.e., 99.9 percent of the time, the 8-VSB envelope is below 6.3 dB above the average power).

One final baseband diagram can be used to describe the characteristics of the RF-modulated waveform. Figure 2.9.17 contains an 8-VSB constellation diagram made up of the in-phase (I) and

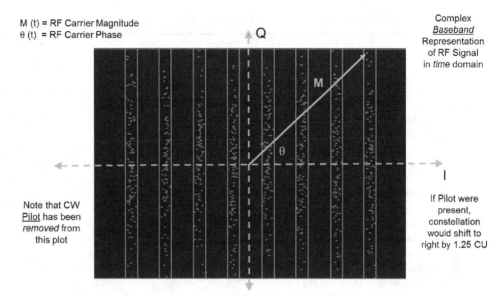

FIGURE 2.9.17   8-VSB constellation diagram with pilot removed.

quadrature-phase (Q) RF components plotted on a Cartesian coordinate graph, from which the RF signal magnitude and phase (relative to the unmodulated in-phase pilot carrier) can be discerned for different symbol times. Each dot on the graph represents a sampling of the demodulated symbols at a given instance of symbol time. The vector from the origin to any dot represents the magnitude M (i.e., $\sqrt{[I^2 + Q^2]}$) and relative phase θ (i.e., $\tan^{-1}[Q/I]$) of the RF signal compared to the unmodulated carrier at a given sampling time. The constellation diagram is one of the most useful diagnostic tools for determining the 8-VSB transmitter signal quality.

Note that the 8 amplitude data symbol levels on the in-phase component, represented by the vertical lines at specific constellation units (±1, ±3, ±5, ±7), are symmetrically situated around the in-phase zero carrier point. This is due to the fact that the small DC offset, which represents the pilot carrier, has been optionally subtracted from all of the sample points on this piece of test equipment by user selection. If the pilot were not subtracted, the 8 vertical lines would be shifted to the right by 1.25 constellation units. Conversely, note that the quadrature component is continuous (i.e., not discrete), with no discrete data levels present since no data is carried in this component. Once again, the quadrature channel can be thought of as the canceling mechanism for one of the sidebands.

2.6   ATSC DTV TRANSMISSION SYSTEM PARAMETERS

Table 2.9.1 contains the pertinent ATSC digital television transmission system characteristics and basic performance parameters.

Note the spectral efficiency (3.232 bits/Hz) of the high data rate (19.4 Mbps) 6 MHz channelized signal with small (11.5 percent) excess bandwidth. The combination of synchronization pulses (segment and field) and FEC (RS and TCM) together reduce the payload rate to about 60 percent of the net raw data rate, but provide stable and reliable performance in severe propagation conditions. According to the ATSC standard, the 10.762 MHz symbol clock should be fractionally locked ([1/2] × [208/188] × [313/312]) to the incoming 19.392 Mbps MPEG transport stream clock, with both having a recommended tolerance of ±2.79 ppm.§§§

---

§§§A/64, Rev. A, "Transmission Measurement and Compliance for Digital Television," May 2000, www.atsc.org.

RF TRANSMISSION

TABLE 2.9.1 8-VSB Transmission System Characteristics

Parameters	Values	Units	Comments
RF Channel BW (BW_{CH})	6.000	MHz	FCC channelization requirement
RF Nyquist/Noise Bandwidth (F_N)	5.381119	MHz	$684 F_H = 684\,(4.5\,\text{MHz}/286) = F_{SYM}/2$
RF Excess BW (BW_X)	11.501	%	6/5.381119; (RF CH BW)/(Nyquist BW)
NTSC Horizontal Rate (F_H)	15.7343	kHz	4.5 MHz/286
Data Symbol Rate (F_{SYM})	10.7622378	MHz	$684 F_H = 684\,(4.5\,\text{MHz}/286) = 2*F_N$
Data Symbol Period (T_{SYM})	92.917479	ns	$1/F_{SYM}$
MPEG-2 Transport Stream Packet (L_{TSP})	188	bytes	4-byte header + 184-byte payload
MPEG-2 Sync Byte (D_{SYNC})	47	hex	$(1000\,0111)_b$; MPEG standard
Data Segment Payload Data	188	bytes	Same as MPEG-2 packet length
Data Segment Symbol Length (L_{SEG})	832	symbols	$(188 + 20) \times (8/2)$
Data Segment Time Duration (T_{SEG})	77.307342	µs	$832/F_{SYM}$
Data Segment Repetition Rate (F_{SEG})	12.935382	kHz	$F_{SYM}/832$; MPEG transport packet rate
Data Segment Sync Symbol Length (L_{SS})	4	symbols	±5 levels
Data Segment Sync Time Duration	0.37167	µs	$4 T_{SYM}$
Data Field Sync Symbol Length (L_{FS})	832	symbols	820 are ±5 levels; last 12 are 8-level
Data Field Sync Time Duration (T_{FS})	77.307342	µs	$832 T_{SYM}$
Data Field Sync Repetition Segments (R_{FS})	1/313	–	Every 313 data segments
Data Field Sync Repetition Rate (F_{FS})	41.3270988	Hz	$F_{SEG}/313$
Data Field Duration (T_{FIELD})	24.197198	ms	$313/F_{SEG}$
Raw Pre-Trellis Data Rate	32.286713	Mbps	$3 F_{SYM}$
Raw BW Efficiency	3	bits/symbol	Design parameter
Trellis-Coding Rate (net)	2/3	–	FEC design parameter
Convolutional Interleaver Length (L_{CON})	52	segments	FEC design parameter
Convolutional Interleaver Duration	4.020	ms	$L_{CON} \times T_{SEG} = 52 T_{SEG}$
Pre-Reed–Solomon BW Efficiency	2	bits/symbol	Design parameter
Pre-Reed–Solomon Data Rate	21.5244755	Mbps	$2 F_{SYM}$
Reed–Solomon FEC	$t = 10$ (207,187)	–	FEC design parameter
Reed–Solomon Parity Bytes (L_{RS})	20	bytes	$2t = (2 \times 10)$
Data Segment Net Payload (inc. segment sync)	208	bytes	MPEG Payload (188) + RS Parity (20)
FEC and Sync Efficiency (RS only)	90.096	%	$[188/208] \times [312/313]$
FEC and Sync Efficiency (TCM, RS)	60.064	%	$(2/3 \times [188/208] \times [312/313])$
Payload Data Rate	19.3926585	Mbps	$([188/208] \times [312/313]) \times 2 \times F_{SYM}$
Overall Data Spectral Efficiency	3.232	bps/Hz	19.3926585 Mbps/6.0 MHz
Total Power Increase due to Pilot	0.31	dB	$10\log[(\{1.25\}^2 + 21)/21]$
Pilot Power Relative to Data Power	≈ -11.3	dB	$10\log[\{1.25\}^2/21]$
Pilot Power Relative to Total Power	≈ -11.6	dBc	$10\log[\{1.25\}^2/(\{1.25\}^2 + 21)]$
Pilot Frequency Location	309.440559	kHz	Above lower RF channel band edge
Peak-to-Average Power Ratio	≈ 6.3	dB	@ 99.9% probability
Data Threshold Packet Errors/s	≈ 2.5	PER	@ TOV, lab measurement
Data Threshold Packet Error Rate	1.93268×10^{-4}	–	@ TOV; $2.5/F_{SEG}$, lab measurement
SNR @ Data Error Threshold	≈ 15	dB	AWGN; 6 MHz measurement channel
Phase Noise Data Error Threshold	≈ -78	dBc/Hz	At 20 kHz offset
Burst Noise Threshold	≈ 193	µs	White noise burst @ 10 Hz repetition rate

Figure 2.9.18 illustrates the 8-VSB probability of the data packet error curve versus SNR, with both signal and noise power referenced to a 6 MHz RF channel bandwidth. As the DTV signal level decreases either due to farther distance from the transmitter or large objects between the transmitter and receiver attenuating the signal, it approaches the DTV receiver's white noise floor (which is dependent on its effective system noise figure) and the SNR decreases dB-for-dB. This figure shows a comparison of the gradual degradation for analog NTSC signals versus the steep cliff effect for digital ATSC signals. The steepness of the ATSC curve is due to the inherent nature of digital communication along with the significant amount of FEC incorporated.

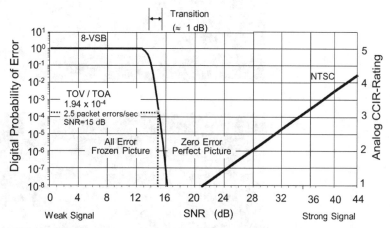

FIGURE 2.9.18 Analog NTSC gradual degradation comparison to digital 8-VSB "cliff effect".

The TOV errors and threshold of audio (TOA) errors for the ATSC system both occur essentially at the same approximate SNR value of 15 dB (data eyes essentially closed, and fully dependent on FEC). The threshold value is defined (via laboratory measurement without any video error masking in the receiver) as about 2.5 transport packet errors per second (1.93×10^{-4} packet error rate), where the effects of errors are just discernible (but where programming is still watchable and hearable). This type of noise performance is possible since receiver synchronization (pilot carrier, symbol clock, segment syncs, and field syncs) is robust well below this 15 dB SNR data error threshold. However, it is clear that the probability of error curve for 8-VSB is very steep (hence the name "cliff effect") in that the digital signal goes from a zero-error perfect picture and sound to an all-error frozen picture and muted sound with only about a 1 dB drop in RF signal level. It is important in practice for SNR values at receive sites to typically be well above the 15 dB value to allow for signal level variation over time caused by a number of factors, including signal fading, dynamic multipath, and varying interference levels.

As a comparison, an analog NTSC picture gradually gets noisy as the signal level decreases. CCIR picture ratings (now referred to as ITU ratings) describe the quality of analog television pictures on an *impairment* scale of 1 (very annoying), 2 (annoying), 3 (slightly annoying), 4 (perceptible but not annoying), and 5 (imperceptible). A "slightly annoying" picture (CCIR rating 3) and an "annoying" picture (CCIR rating 2) occur at approximately SNR values of 34 dB and 28 dB, respectively, both well above the 15 dB 8-VSB SNR error threshold. The analog NTSC audio system, which uses narrowband FM modulation, remains relatively usable until SNR conditions are much lower than that of acceptable pictures. Note that the much lower SNR value where loss of digital picture and sound occurs compared to that of analog NTSC is often offset somewhat in practice by the lower transmitted DTV power allowed for broadcast transmissions by the FCC (at least 7 dB lower and often more). Additionally, while the FCC assumed that the ATSC digital system would replicate CCIR-rating 3 analog NTSC service ("slightly annoying"), it should be noted that, due to the gradual degradation effect, much worse analog pictures can be viewed stably (even down to a CCIR-rating 1.5), although the impaired video is very annoying. This is contrasted with digital ATSC pictures which are, for all intents and purposes, either viewable or not viewable due to the cliff effect.

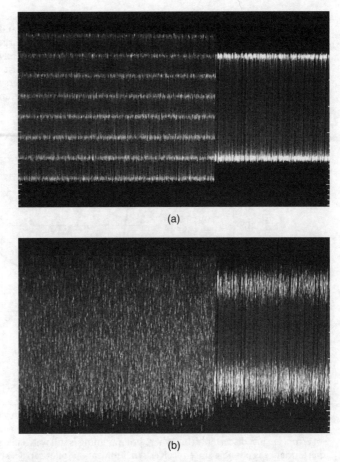

FIGURE 2.9.19 Trace (*a*) shows a pristine signal at a DTV receiver's equalizer output, well above TOV, provides perfect digital picture and sound. Trace (*b*) shows a very noisy signal at a DTV receiver's equalizer output, barely above TOV, also provides perfect digital picture and sound.

Finally, Fig. 2.9.19 illustrates the equalizer data output signal in a DTV receiver under pristine conditions and under heavy white noise conditions. In Fig. 2.9.19*a*, the eight distinct equiprobable symbol levels are clearly identifiable in this pristine condition for the data segment that is transmitted just prior to the embedded binary field sync segment. This is because the data eyes are essentially completely open. However, Fig. 2.9.19*b* illustrates the same digital signal, but under a very noisy condition where the signal is slightly *above* the 15 dB white noise threshold. While the binary field sync signal is again still clearly identifiable, although obviously noisy, the eight distinct levels are no longer discernable since the data eyes are *essentially* closed. Note that *both* of these DTV signals represent reception conditions where perfect picture and sound are achieved since both signals are above the 15 dB VSB system white noise data error threshold, even if in the latter case by only a few tenths of a dB. In the latter case, while the data eyes are mostly closed and raw symbol (and thus raw data) decoding is far from perfect with many slicing errors, a vast majority of the data errors are being identified and corrected by the FEC in the receiver. This illustrates the power of the digital cliff effect as a result of strong FEC in a digital communication system.

CHAPTER 2.10
TELEVISION TRANSMITTERS

John Tremblay
UBS/LARCAN, Toronto

2.1 INTRODUCTION

Previous chapters have touched on the design of the digital television (DTV) modulator, and the modulation basics as well as the specific devices used in the broadcast transmitter design. The basic broadcast transmitter system shown in Fig. 2.10.1 is a simplified version of what is found at a typical broadcast transmission site. The basic building blocks function the same regardless of whether the transmitter is TV or radio, and more importantly, regardless of whether it is analog or digital.

The transmitter must be considered as a complete system—that is to say that from the modulator through the amplifier and mask filter, and in some cases the complete radio frequency (RF) system, must be considered as an integral system. Amplifiers can be optimized for output power and highest efficiency, but only if the modulator can provide the appropriate amount of predistortion to compensate for the nonlinearities of the amplifier, and only if the filter can attenuate the resultant out-of-channel emission to prevent interference.

Typically, a complete, transmission facility starts with a feed from the studio, usually a microwave link, or other method of feeding the signal to the site. The feed from the studio/transmitter link gives us what we generally refer to as baseband information. In a digital broadcast facility, this information is encoded into a transport stream (described in previous chapters). We apply this transport stream to a modulator and *upconverter* (we will refer to this as an "exciter"). Once the RF signal is modulated, it is amplified, filtered, and applied to the antenna. In an ideal world, each of these building blocks performs perfectly. However, in reality, there are always limitations and co-dependencies.

2.1.1 Transmitter System Considerations

We can think of the transmitter—at least the signal path of the transmitter—in terms of three distinct sections: the exciter, power amplifier, and output filter. These subsystems provide very distinct functionality within the transmitter system; however, they are inextricably linked. And of course, there is all the support circuitry for the transmitter, which may or may not prove to be significant in the overall reliability and performance of the transmitter.

While the modulator must perform the actual digital signal processing, converting, and encoding the transport stream such that the RF signal can propagate with minimal information loss across the airwaves to the receivers, it also typically performs a signal conditioning function we call predistortion (also referred to as precorrection). Predistortion enables the RF-modulated digital signal to pass through the amplification process to yield an output signal that is reasonably undistorted. We will first cover the limitations of the rest of the system, then return to the subject of predistortion.

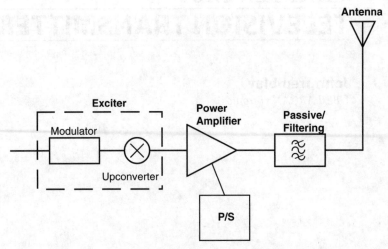

FIGURE 2.10.1 Basic broadcast transmitter diagram.

The amplifier is designed to be as transparent to the signal as possible, given the constraints of technology, cost, and efficiency. With high power transmitter systems, the limitations of any single amplification device necessitate the use of multiple stages of amplification. Virtually every transmitter amplifier section will use intermediate power amplifiers (IPAs), preamplifiers, drivers, or whatever the manufacturer has decided to designate the drive stages to the final power amplification stage of the transmitter. While much attention is paid to the technology used in the final stage of amplification of a TV transmitter power amplifier (PA)—whether it is inductive output tube (IOT) technology, MSDC (multi-stage depressed collector), LDMOS (Laterally Diffused MOSFET), Doherty, Envelope Modulation and so on—the design of the drive stages in the transmitter is also an important consideration.

The output mask filter is present to remove the distortion byproducts of conversion to RF and amplification. The very nature of digital transmission allows a certain amount of signal degradation without impeding the receiver's ability to "pick up the pieces" of the signal so-to-speak and reconstruct a perfect picture/sound. Some of these degradations are propagation-specific but some are transmitter-specific, allowing the designers to sacrifice linearity for power in the amplification chain. This inevitably causes distortions that also fall outside of the channel, and potentially interfere with neighboring spectrum users. Enter the mask filter. Within the physical limitations of filter design, mask filters contribute to the overall system performance.

The entire system, comprising modulator, amplifiers, and filter, are designed to work in concert with one another to provide a high power, high quality signal to the antenna. From the broadcaster's point of view, the system must transmit an acceptable quality signal, be reliable, cost-effective to operate, and simple to maintain. Each television transmitter manufacturer has designed their system to operate and perform well using its own equipment. Because advances in digital signal processing, better understanding of how distortion manifests itself within the amplification devices, and new advances in materials and filter design are constantly evolving, the balance between these factors is always changing.

In the following sections, we will look at the various subsystems of a transmitter, including the support systems, which also affect both the performance and reliability of the overall transmitter.

2.2 THE AMPLIFIER

It is beyond the scope of this handbook to provide a truly in-depth discussion of amplifier design, or discuss the physics of semiconductor design; however, it is hoped that the following outline will give the reader a reasonable understanding of the "how's and the why's" of the amplifiers used in the TV transmitter application.

2.2.1 The Amplification Chain

Power amplifiers can range quite widely in power levels and in technology. The design of an RF power amplifier is quite specialized. Over the past few decades, the shift in technology has been from using vacuum tube-based technology to solid-state technology. There are still frequencies and power levels for which it makes more sense economically to use tube-based amplifiers, and new developments in solid-state devices are creating new opportunities to implement concepts both new and old (such as Doherty techniques and envelope modulation) that have not been possible or too complex to achieve easily with previous technologies.

Figure 2.10.2 shows a block diagram of a typical amplification section of a solid-state digital transmitter.

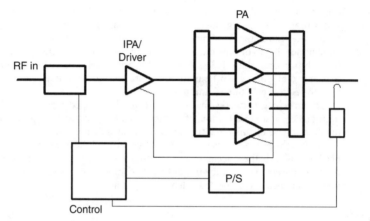

FIGURE 2.10.2 Amplifier block diagram.

The RF level at the exciter output requires that the signal be amplified significantly. The exciter output is typically less than a watt. Thus, the gain required to amplify the signal to the tens of kilowatts level required to achieve the required coverage, and given that an amplifier stage gain is typically 20 dB (100×) or less, multiple stages are usually required. In the design of the RF amplification sections of a broadcast transmitter, due consideration must be given to each of these stages and their contribution to the performance of the overall system. Many of the stages are interdependent.

2.2.2 Technologies for RF Amplification

A number of technologies are available for modern transmitter design. Some of the more common technologies are examined in the following sections.

Solid-State Devices. Final amplification stages constructed from solid-state amplifiers have been in use since the mid-1980s. The output level of a single transistor stage is low relative to the power output required for transmitting television signals, and this necessitates having multiple transistors operating in parallel to deliver the required transmitter power output (TPO). This need for multiple devices operating in parallel inherently adds to the reliability of a solid-state final PA, since a failure of a single device in a parallel combination has less impact than the failure of the *only* device (as in the case of a vacuum tube final). Both bipolar devices and field-effect transistors (FETs) have been used in solid-state PAs, though in almost all current designs, MOSFETs are now used because of their ruggedness and higher gain.

The current technology is predominantly the LDMOS device. LDMOS technology features higher breakdown voltages, and thus higher operating voltages and higher outputs per device. The design of the LDMOS structure has inherently lower capacitances and thus better linearity performance.

The design of an amplifier requires matching the impedance of the device itself to the required impedance of standard transmission lines (50 Ω). Typical devices have impedances that are very low in comparison. Matching circuitry in some cases can enhance the gain of the circuit by providing resonance at the desired frequency. Modern solid-state amplifiers, however, tend to be somewhat wideband in nature, invariably covering at least a multitude of channels in a single configuration. Some designs cover the entire spectrum of their design (i.e., ultrahigh frequency (UHF), channels 14 through 69).

There are advantages to both broadband designs and narrowband designs. Traditional amplifier design involves matching the relatively lower impedance of the device (consider the FET as an example), both on the gate and on the drain to the 50 W feed and load. This maximizes power transfer in the circuit. In early designs, the impedances were such that narrowband matching circuits were required to match the devices properly. This gave maximum gain and performance to the design.

As higher gain devices and particularly FETs evolved, broadband designs also became more capable. The amplifiers could be made to cover a wider band while still retaining the performance and without sacrificing gain significantly. There are a number of advantages of a broadband design; a broadcaster can change channels if needed (or the transmitter can be repurposed to another channel) without needing to be concerned about retuning all the amplifiers in the final stage. When an amplifier is broadband, there is less concern for group-delay variation across the channel of operation. Broadband amplifiers can be made more stable, and of course, the cost of manufacturing is lower than the equivalent "tuned" amplifier.

The matching networks that transform the low impedance of the devices to the required 50-Ω impedance are a combination of inductive and capacitive tuning components. At the frequencies involved, much of this circuitry can be printed into the printed circuit board layout. As shown in Fig. 2.10.3, the printed transmission line traces around the devices are an integral part of the matching network. As such, the placement and orientation of components on these circuits is critical. Because of the high powers involved, the use of high-Q components is sometimes required; thus, replacement of these components with the equivalents is important.

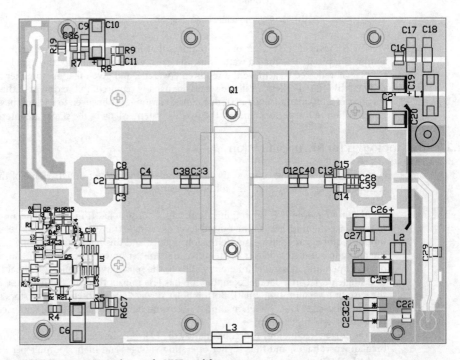

FIGURE 2.10.3 Circuit layout of a UHF amplifier.

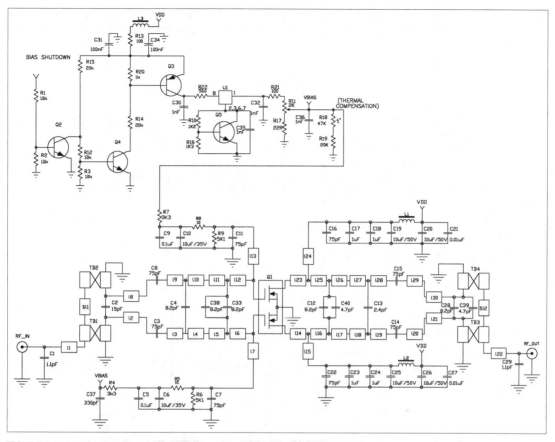

FIGURE 2.10.4 Schematic representation of the amplifier layout shown in Fig. 2.10.3.

Figure 2.10.4 shows the equivalent schematic of the amplifier portion of the layout in Fig. 2.10.3. Note the circuit traces are designated as inductances. The amplifier is a push–pull configuration, designed to be operated in Class B mode (more on this later).

The direct current (DC) voltage is applied to the drain (in the case of the FET amplifier) via decoupling circuits, as is the bias voltage. In order to compensate for any variations in bias, or gain with temperature variations, a thermal compensation circuit is typically designed into the bias network. FET amplifiers generally tend to have a very high gain when they are cold, and this gain drops as the device warms up. Changing the bias voltage on the gate can alleviate this to some extent.

VHF. VHF solid-state amplifiers were the first to evolve as a cost-effective solution for power amplification. This was predominantly due to the evolution of the state of the art for RF solid-state amplifier devices. Power levels and gain were generally higher with VHF power devices. Combined with this, the transmitting power levels required to cover broadcast areas using VHF frequencies were significantly lower than those at UHF. With VHF amplifiers, because of the frequencies, the printed traces tend to be longer and take up more printed circuit real estate. As such, some matching networks may use component parts such as wire-wound cores, or ferrite balun transformers rather than print this circuitry. As mentioned previously, the orientation, and placement of the components, and in many cases even the spacing between the windings of an air-core inductor become part of the tuning of the circuit.

UHF. UHF amplifiers were a number of years behind in the evolution of solid-state devices for RF. The gain was originally relatively low in comparison with VHF solid-state devices. With the advent of LDMOS FET devices, the gain of UHF transistors improved dramatically. With UHF amplifiers, the layout and the components become smaller, and printed traces can be structured to form much of the tuning circuitry. Comparing the circuit layout in Fig. 2.10.3 to the schematic representation in Fig. 2.10.4, this is quite evident.

Inductive Output Tube. IOTs have traditionally been a cost-effective solution to high power UHF amplification. The IOT is a specialized vacuum tube that works in combination with a specialized system of tuned cavities and magnet assemblies to achieve extremely high output power levels. The IOT is cooled by a combination of air and liquid, and utilizes beam voltage levels approaching 35 kV.

IOTs can produce much larger output levels with a single electron tube than other technologies. At current levels, an IOT can be a cost-effective solution in many cases.

The IOT is a sealed vacuum device. A simplified schematic is shown in Fig. 2.10.5.

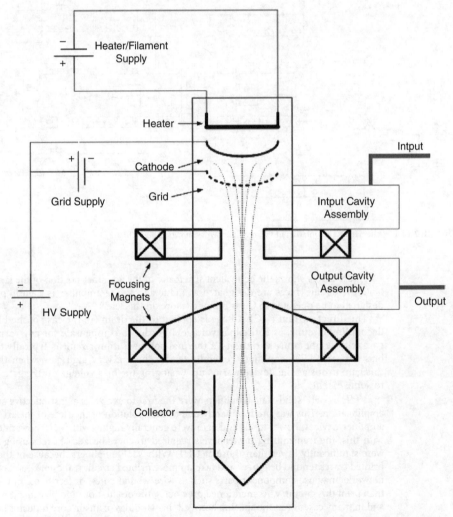

FIGURE 2.10.5 IOT schematic diagram.

A beam of electrons are emitted from a cathode and controlled by a grid spaced close to the cathode. The cathode is maintained at –35 kV and the grid is at approximately –100 V with respect to the cathode. This assembly is held in place and isolated by a ceramic insulator (part of the tube envelope). The input signal passes through this ceramic and "excites" the grid, allowing electron flow. As the grid is made less negative with respect to the cathode, more electrons (current) flow from the cathode to the anode.

Part of the assembly that supports the IOT itself, along with the input and output tuning mechanisms, contains electromagnet assemblies that are used to help focus the electron beam inside the tube. The beam is directed through the tube by the magnetic field generated by the electromagnet assemblies.

Input to the IOT is via a tuned input cavity (a large canister on the top of the assembly). The cavity is tuned in one of multiple ways depending on the input channel. The input cable is at DC ground potential, both on the shield and the inner conductor (via a grounded coupling loop inside the cavity). The cavity is designed to protect against oscillation, since its possible tuning range covers the entire UHF band.

A double-tuned output system is used to couple the signal and transform it to the required 50 Ω. The output cavity assembly is clamped around the output section of the tube. It is extremely important that no electron beam current be drawn without the output cavity assembly being properly installed because it also serves as a shield to RF and X-ray radiation.

Due to the high voltages used, there is a potential for internal arcing between the tube elements, particularly with new IOTs and when the IOT ages, or in the event of overdrive. This arcing can be extremely destructive to the inner workings of the tubes if left unchecked. To protect the IOT against self-destruction in the event of an internal arc-over, there is a circuit built into the transmitter control system that removes the beam voltage from the tube when an arc is detected. This protection generally takes the form of a crowbar circuit that shorts the supply voltage to ground. Crowbar circuits are traditionally either Thyratron-based or vacuum gap-based. The function of the circuit is to remove the voltage, and remove the residual energy or divert the energy of the power supply before the inner workings of the tube are destroyed. A key test for these circuits is known as the "40-gauge wire test," in which a specified length of 40 SWG (British) wire is protected during a direct short circuit across the –35 kV beam supply.

Safety note: A high voltage arc or a crowbar event can create a massive electromagnetic pulse which may cause problems with electronics—in particular, electronic pacemakers.

The Multi-Stage Depressed Collector IOT. Another more efficient version of the IOT was introduced, which borrows from technology and research into Klystron efficiency. This is the Multi-stage Depressed Collector (also known as the Energy Saving Collector IOT, or ESCIOT, from E2V/Marconi). This is essentially an IOT equipped with a collector that utilizes different levels of collector, at different voltages in order to gather those electrons that diverge as the tube is driven harder. These collectors divert these stray electrons back to the power supply, and this increases the tube's efficiency. The typical depressed collector IOT uses three or four of these additional collectors.

The MSDC IOT can show significant improvements in efficiency over a traditional IOT design (on the order of 30 to 40 percent better than an IOT); however, this comes at the cost of adding more power supply stages/components. Note also that crowbars to protect against internal arcing are still required as in the IOT. The cooling system typically requires de-ionized water, or at least highly purified water, or in some cases dielectric fluids (oils). This results in a more complicated cooling system arrangement, particularly in climates where below freezing temperatures are common.

2.2.3 Power Amplifiers

The final power amplification stage has the most significant effect on the signal. The design of the amplifier involves a tradeoff between output power, power efficiency (including cooling considerations), and linearity/performance. Designers strive for the best performance with the highest efficiency. Combined with the precorrection circuitry of the modulator and the output mask filter, the power amplifier stage is optimized to provide the best performance at the output of the system.

Linearity. Because of the nature of the digital television signal, the transmission chain needs to pass the signal through with minimal degradation. As stated before, this is often a tradeoff between the power amplifier design, the predistortion circuitry used, and the filtering after the amplifier. Typically, the design starts with a reasonably linear amplifier 0151 or more specifically, one for which the linearity characteristics are reasonably manageable with predistortion techniques. The combination of predistortion and power amplification produces an output that will inevitably require some filtering in order to meet FCC spectral emission requirements.

The most linear amplifier design is a Class A biased amplifier. In this configuration, the amplifying device (the transistor or the tube) is biased at roughly the halfway point of the expected voltage swing. The input voltage applied to the device and the output voltage swing of the device takes place along the linear portion of the transfer curve (see Fig. 2.10.6). Unfortunately, a Class A biased amplifier is the least efficient design, since even at zero output, the device is biased to draw a significant quiescent current. As such, Class A amplifiers are not used in the output stages of transmitters due to their power consumption and cooling considerations.

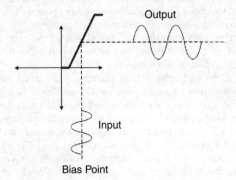

FIGURE 2.10.6 Class A amplifiers operate along the linear portion of the transfer curve of the device.

A more common amplifier design is that of push-pull class B. A pure Class B design has two devices operating on each half of the cycle. The signal is split into two 180° out of phase signals. Each half of the waveform is amplified and then recombined as shown in Fig. 2.10.7. The bias point is zero, resulting in a more efficient amplifier. Note that because the transistors are off, there is a resulting "crossover" distortion. To overcome this distortion somewhat, transistors are given a small bias current; such amplifiers are classed as AB.

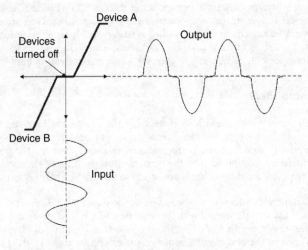

FIGURE 2.10.7 Class B amplifier waveforms.

To achieve the maximum economy in a design, amplifiers are the most cost efficient when they are designed for maximum power output. In modern designs, the power efficiency of a traditional Class AB amplifier is generally higher when the amplifier is operated at or near its maximum.

We can see, however, that as signal approaches the transistors' saturation point, the transfer gain drops. This is often described as "saturation power," or P_{1dB} or P_{3dB}, depending on the application and the design intent.

Figure 2.10.7 shows a linear transfer between input and output, with sharp transitions in the cutoff and the saturation regions. In fact, transistors do not exactly behave in this manner. The transitions are not as sudden. In fact, the transition into the saturation region can be much more subtle. Amplifiers create many different types of distortions, some subtle and some not-so-subtle.

The typical distortions experienced at the power amplification level can take many forms. The frequency response of the amplifier and accompanying group delay effect are factors that need to be considered when dealing with tube-type designs, which are inherently nonbroadband. A tube-based amplifier typically consists of tuned cavities to match the impedances of the tube to the input and output circuits and is tuned (in order to achieve reasonable efficiencies) to the operating channel. Mistuning of these circuits or the matching and loading of these cavity circuits affects the frequency response as well as the linearity of the circuit. Modern solid-state amplification circuits are broadband in nature and thus frequency response is not typically an issue at the amplifier level. These frequency response-related effects are referred to as linear distortions.

RF power amplifiers introduce AM–AM and AM–PM distortion. These are typically classed as nonlinear distortions. These distortions can be straightforward gain compression and phase distortion, and they can also be very dynamic types of distortions, caused by the internal capacitances and inductances and thermal dynamics of the devices themselves. These latter distortions are often referred to as "memory effect" nonlinearities. Without entering into lengthy dissertation about this, the simplest explanation is that memory effect is the device's (or ultimately, the amplifier's) ability to react to the modulated signal transitions between the various digital levels. In short, memory effect is a very dynamic version of both the AM–AM and the AM–PM distortions.

AM–AM Conversion.   As the signal through an amplifier is increased beyond a certain limit, the gain of the amplifier drops. Typical performance measurements include the "1 dB compression point". This is the point at which the gain of the amplifier has dropped by 1 dB relative to the input drive increase. This compression point is commonly referred to as *saturation* and results in what is known as AM–AM distortion. And, as discussed previously, for amplifiers biased toward Class B, there is distortion introduced at lower levels—commonly referred to as *crossover distortion*—also a form of AM–AM distortion. This is illustrated in Fig. 2.10.8. In the analog transmission world, this distortion would have been somewhat quantified by differential gain or LF linearity measurements.

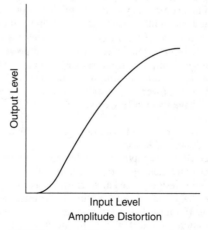

FIGURE 2.10.8   AM–AM conversion.

AM–PM Conversion. Another form of nonlinearity encountered in RF power amplifiers is a change in the phase of the signal with amplitude. This introduces what is referred to as AM–PM distortion. This type of distortion ultimately results in unwanted phase modulation on the signal (see Fig. 2.10.9). In the analog world, it can be somewhat quantified in differential phase and incidental carrier phase modulation measurements. AM–PM conversion can occur at any amplitude of the signal but is more pronounced at crossover and at saturation.

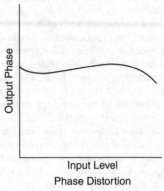

FIGURE 2.10.9 AM–PM conversion.

Design Considerations. Ultimately, we would like to have amplifier design that is as linear as possible. Economics of both design/manufacturing and operation/reliability dictate that compromises must be made.

Careful amplifier design involves the selection of devices with the highest power capabilities and reasonable gain. The choice of bias point for the amplifiers has a significant effect on both linearity and operating efficiency. The actual operating point (the output power level backoff) has a very large impact on both the nonlinearity of the amplifier and the efficiency. Many amplifier designs are at their best efficiency when operating at their designed output power (more about this later). One of the most critical aspects of design is the matching of the amplifying device impedances to the output networks. This is critical for power transfer and efficiency; it affects the dynamic linearity of the amplifiers significantly. Placement of components on the PC board (as shown in Fig. 2.10.3) can be critical to the matching of the circuit.

Each of the techniques discussed here has associated costs or tradeoffs. Ultimately the amplifier is integrated into a complete system with predistortion circuits that compensate (somewhat) for nonlinearities, and with a filter system that can help in eliminating out-of-band distortions.

The Special Case of the Intermediate Power Amplifier (IPA). Because the gain required to increase the input signal level to the desired power output is typically beyond the capabilities of a single stage of amplification, transmitters use multiple stages. These are called preamplifiers, IPA, drivers, and so on. In some solid-state transmitter topologies, the drive stages are integral to the individual power amplifier modules. In this case, each of the parallel power amplifier modules will have its own IPA/driver. Figure 2.10.10 shows a simplified version of two different architectures.

There are advantages and disadvantages to these different topologies. Perhaps the largest advantage is the redundancy factor.

In Fig. 2.10.10b, the amplifier modules have inherently higher gain and, depending on the amount of gain, special care must be taken to minimize the chance of oscillation from feedback in these modules (both in the design phase and later, when servicing).

IPA drive stages have been traditionally as linear as practical in order to minimize their effect on the signal. Traditional approaches have been to use Class A designs for this purpose. One of the key considerations is that they not only need to be linear enough to pass the signal with minimal

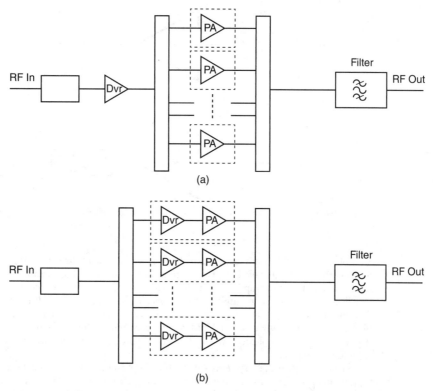

FIGURE 2.10.10 Different methodologies for drive stages.

distortion, but that they need to pass a predistorted signal to the power amplifier stage without affecting it. Often times, this predistorted signal will have peak levels that are artificially higher than normal to compensate for the compression inherent in the final stages. If the IPA cannot pass these peaks, then the effect of the predistortion circuitry can be severely limited.

High Efficiency Techniques Commonly Used. Transmitter efficiency is a prime concern, and with amplifiers that can deliver tens of thousands of watts, efficiency is an important design consideration. Assuming that the lowest loss passive components have been used and the highest gain antennas are deployed while still achieving the desired coverage, then we focus our attention on the RF amplifier chain. Power efficiency not only saves on the electricity costs for operation of the transmitter but also the cooling costs and ultimately the life of the equipment itself. There are many ways of optimizing the efficiency of an amplifier, including operating closer to the device/amplifier's saturation point (with appropriate predistortion techniques).

Doherty. Doherty amplifier topologies have been around since the 1930s. The basic concept is to use two different amplifiers for different levels of the signal. A high efficiency Class C stage is used to amplify the peaks of the signal, while a more linear Class AB stage is used for the lower levels of the signal. This is an extremely effective design for digital signals, since peak levels comprise only a minimal percentage of the signal. Rather than have an amplifier suitable for the entire signal, we can design one that is more efficient operating at the lower levels and one that handles the peaks, which is biased off most of the time. Figure 2.10.11 shows a basic block diagram of a Doherty amplifier.

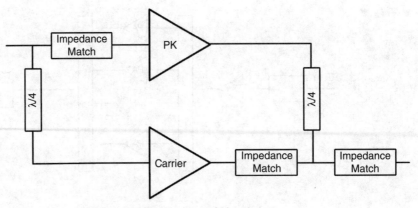

FIGURE 2.10.11 Block diagram of the basic Doherty amplifier.

Doherty amplifiers tend to be relatively narrowband due to the combining circuitry, which needs to properly match and isolate the two sections of the amplifier. For manufacturers, this means more time spent tuning the amplifiers, possibly with different components to be kept on hand; for the broadcaster, this means an amplifier that is specific to the channel and more complex than a more conventional design. That said, new designs are emerging in the UHF range that boast broadband performance with a single design. Utilizing the consistency of printed circuit components and matching elements, the complexity—while still there—tends to be a nonissue in terms of service and/or troubleshooting.

The efficiency for these systems are reliably in the 50 percent range for the amplifiers, whereas the typical efficiency for conventional design can range from 30 to 40 percent. Doherty amplifiers generally produce less power 'per package' than conventional amplifiers designed with the same devices, and thus the costs tend to be slightly higher.

Envelope Tracking. Envelope tracking, also known as *drain modulation* when used in the context of an FET-based amplifier, is the technique of modulating the power supply voltage as the modulation envelope of the signal changes. This has the effect of dynamically keeping the amplifier closer to saturation levels. Figure 2.10.12 shows the tracking supply voltage overlaid on a modulated waveform.

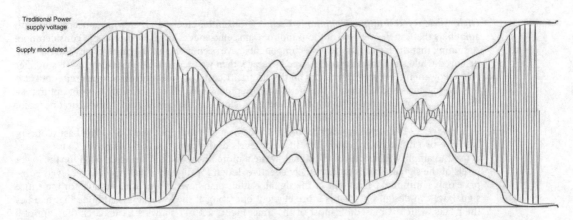

FIGURE 2.10.12 Modulating the power supply voltage to match the modulated signal.

There are many considerations in the design of such an amplifier. The power supply must be able to change levels at frequencies approaching the bandwidth of the modulation signal. The power supply is fed a modulating signal that approximates the envelope of the modulating waveform. The dynamic response of the power supply must be such that it can track and stay in phase with the envelope of the signal passing through the amplifier. High power, high frequency switching devices are not common in today's technology and this necessitates the modulation to be done at lower power levels. Typically we find multiple supply modulation circuits (as many as one per amplifier device), all fed the same modulating signal. This can tend to be very complex with high power transmitters. Careful consideration must be given to the efficiency of the supply modulator circuitry and the phase of the modulation envelope signal to each amplifier.

Amplifier efficiencies can be significantly improved using envelope tracking, but at the cost of complexity. As higher power devices and technologies improve, this technique—perhaps combined with other techniques such as Doherty—will no doubt become more commonplace.

Of course, both the Doherty technique and envelope modulation cause distortions to the signal, and therefore require a certain amount of predistortion to be applied to the signal.

2.2.4 Predistortion/Precorrection

The goal in the design of a transmitting system as stated previously in this chapter is to broadcast an "acceptable" signal over the air such that modern receivers within the coverage area can reliably demodulate and interpret the digital information.

The ATSC signal itself has a certain amount of error tolerance built into it. Forward error correction and interleaving in the signal ensure that if some information is "lost in the translation" so-to-speak, there is sufficient redundancy in the encoded signal that allow this information to be recovered. In addition to the distortion of the signal contributed by the transmission chain, we must also consider the contribution of the signal path to the degradation of the signal. Fading and man-made noise all have a negative effect on the signal. The goal is to keep the transmitted signal as clean as is practical.

Amplifiers cause certain distortions as described in the preceding sections. Bandpass filters (mask filters) can be used to eliminate some of the interference caused by these distortions, which fall outside the channel, but in turn can also contribute distortions of their own. Distortions are generally classed as nonlinear distortions (those caused by the amplifier dynamics) and linear distortions (those caused by frequency-limiting devices, whether it be tuning circuits or mask filters).

Nonlinear Effects. In order to take full advantage of the capabilities (and efficiencies) of the amplifier design, the system should be run at as high an output as is feasible. As we approach saturation, the gain drops and the phase relationship changes (as we saw in Figs. 2.10.8 and 2.10.9). Predistortion (also referred to as precorrection) is just what it says. The signal is passed through a network that (in theory) has characteristics that are the inverse of what the signal will experience in the PA stages. The idea is that the predistortion effect will cancel distortions caused by the amplifier, and the output signal will be as intended (i.e., distortion-free). This method has its limits, however. As we have seen in Figs. 2.10.6–2.10.9, the characteristic curve of an amplifier is such that the harder the amplifier is driven, the more the gain "compresses." That is, there is a limit to how much power the amplifier can produce, no matter how large the input signal. So, increasing the peak levels of the signal to compensate for a gain drop at higher levels will only work up to a certain point, beyond which even greater distortion effects occur (or ultimately, damage to the amplifier device).

Nonlinear distortion can be a dynamic characteristic of the amplifier/device. We can think of this in terms of the characteristics of the signal being amplified. At higher average powers, heating effects within the device create distortions that differ depending on the signal dynamics. A peak signal encountered following a high level is amplified differently than the same peak which follows a lower level. For digital, because of the random nature of the signal, this is much less predictable than for an analog signal. The predistortion for this type of characteristic requires complex modeling and understanding of the amplifier.

Linear Effects. Compensation for the linear distortions in the system involves the compensation for filter (or tuning) roll-off, frequency response slope, and accompanying group delay effect. Mask filters are employed to attenuate the adjacent-channel and out-of-band products. The tuning of the filters sometimes is such that the edge of the channel is attenuated slightly, but more critically, the group delay is sloped at the channel edge.

On the surface, this sounds more straightforward than the predistortion required for the amplifiers. If an in-depth understanding of the filter characteristics is known, then an appropriate compensation network can be configured. This type of predistortion seems relatively straightforward until we consider that filters may drift with ambient temperature variations. Careful attention to filter design can minimize drift with temperature. Depending on the tuning of the filter across the channel, any drift of this nature may or may not have significant effect on the signal quality.

Implementation. In the days of analog television signals, predistortion circuits were set up manually, with the visual feedback of being able to observe a relatively stable waveform. In fact, test waveforms were developed that allowed observation directly of the effects of the transmitter distortions on the signal. With a more-or-less random digital signal and the peak-to-average powers encountered, this is not so easy to do. Manual adjustment, while feasible to some extent if one is intimately familiar with amplifiers and the various effects of their distortions on the observable part of the ATSC signal, is a very lengthy and involved procedure.

In the digital world, we now have the power of digital signal processing and the computational abilities of the processors being used. The RF can be sampled from the output of the amplifier and from the output after the mask filter, and these can be compared with the desired output. Digital processors can capture the signal perform the measurements, analyze the signal, determine the correction factors to apply, and adjust the necessary parameters in the signal processing of the modulator; in effect, automating the feedback and the correction process (see Fig. 2.10.13).

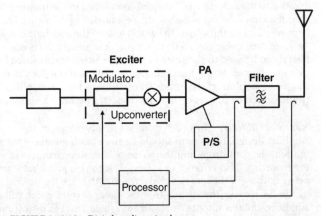

FIGURE 2.10.13 Digital predistortion loop.

Because the predistortion setup can be done automatically with a digital feedback loop, this leads to the concept of creating a continuous feedback loop. There are many names for this type of correction system (and just as many trademarks). Adaptive predistortion can automatically compensate for minor changes in system performance (and attempt to correct for changes in system performance). Algorithms for predistortion are closely guarded secrets, yet inherently amplifiers generally exhibit similar characteristics (in fact, many manufacturers use the same devices).

There are a number of ways to configure a digital predistortion network. Each has its advantages and disadvantages.

A continuous adaptive process can be set up that constantly monitors and adjusts the predistortion settings of the modulator to compensate for changes in performance, whether related to environmental factors (e.g., daily temperature fluctuations) or equipment changes (e.g., tube aging).

Continuously recorrecting the performance has the advantage of keeping the transmission system at peak performance. The disadvantage of this type of a system is that now there are system components not directly in the signal path that, if they fail, may cause signal degradation. This may necessitate the addition of more monitoring equipment that can signal alarms if any parameters fall outside acceptable values.

With modern solid-state amplifier designs and filter technologies, and with sound engineering practice for building HVAC and cooling design, the environmental effect on the transmission path can be minimized. If we can be assured that even under the most extreme conditions, the signal quality never falls below an "acceptable" level, then the digital predistortion can be a set-and-forget value. With somewhat regular maintenance visits to the transmitter, these predistortion settings can be readjusted from time to time if needed. The idea is that when a correction routine is run, the user is present to verify that the performance is indeed acceptable. The disadvantage to this method is that the performance parameters of the transmitter are allowed to drift. The equipment must have, to begin with, very good and reliable performance.

A hybrid of the set-and-forget and the fully adaptive configurations is a continuous monitoring algorithm that, should performance drop below a certain preset value, a predistortion process is activated that corrects the performance. This has the advantage of acting like a watchdog over the transmitter performance. Again, however, the disadvantage is that any failure in the monitoring equipment may set the predistortion algorithm on a course for disastrous performance.

Some manufacturers offer all of these options and leave it to the end user to decide which works best for their circumstances. In every case, an automatic correction routine is much easier and less time-consuming than manually adjusting the performance of a high power transmitter.

2.3 POWER SUPPLIES

One major part of the transmitter is the DC power supply that feeds the RF amplifiers. These power supplies need to be very stable, rugged, and reliable. Variation in power supply voltage can often show as distortion on the output signal, particularly if this variation is load-related (in the days of analog transmitters, this was much more of an issue). Because of the random nature of ATSC modulation, the signal through the amplifiers—and thus the load on the power supplies—tends to be relatively constant.

The typical transmitter site may be miles from the service facility or studio, often times at the end of long runs of utility power lines. The transmitter is thus subject to abuse from lightning and spikes on the power lines caused by inductive motors and other machinery.

Most modern transmitters have multiple power supplies, often one per power amplifier module. Multiple power supplies are typically used to increase redundancy. Switching power supplies are used mainly because of their small size and efficiencies, which are better than the equivalent linear supplies. The tradeoff is that switching power supplies are inherently more complex and are typically a replace rather than repair item. Linear supplies, while less common, are relatively straightforward designs and can often be higher in power. Higher current linear supplies tend to be less efficient than switching power supplies. The advantage with this type of supply is that the components used are generally very rugged.

2.4 CONTROL AND METERING

Transmitter systems require control circuitry to switch them on and off, particularly where turning a transmitter on or off requires a sequence of related circuits to be controlled (as example, cooling pumps, filament voltages, beam and bias supplies, etc.). An IOT requires the application of voltage to the filament and the anode in a proper sequence and with appropriate warm-up times. The control circuitry performs all of this in the correct order.

The on–off functions will be available on a remote control interface. Transmitter safety interlocks should always be connected where safety is a concern. For example, RF patch panels or coax switches, which allow the switching between a main and standby transmitter, will require RF to be disconnected during the patching or switching process. Temperature monitors, either internal to the transmitter or externally installed, can be connected to an interlock system to turn off power if the ambient temperature is too high. With liquid-cooled transmitters, there may be a coolant level interlock, flow interlock, and possibly a leak detector interlock.

Metering circuits will be provided for all the relevant levels within the transmitter. A sample port on the output of the transmitter will feed metering detectors with forward and reflected power readings. There may be readings of power at various other points in the amplification chain such as the drive stages (and possibly at each power amplifier module). Voltage and current readings for power supplies are typically provided. The purpose of all metering levels is ultimately to give a reading of the health status of the transmitter. As such, it is good practice to have a "baseline" of readings recorded somewhere (e.g., the station log) for comparison should problems arise later. These metering levels, the on/off status, and the status of interlocks should be available on remote control connections as well.

With high powers, high currents, and high voltages present in the transmitter circuitry, both personnel safety and circuit protection are important. Crowbar protection circuitry as described in the section on IOTs will be part of the transmitter control and protection network. A recycle of the "on" sequence will be part of the routine for this circuit.

Power supply circuitry will have overcurrent protection and the amplifier chain will have, at minimum, protection against excessive reflected power. Too much power reflected back in the output system of the transmitter can degrade performance and cause damage to the output devices. Samples from the output (typically the same ones that drive the metering circuitry for output forward and reflected power) are monitored for high VSWR conditions and reduce the drive—and/or shut down the amplifiers—before damage can occur. Front panel indications of overload conditions aid in troubleshooting any off-air conditions.

2.4.1 Troubleshooting

When an off-air condition happens, the transmitter's control indications should be the first step in determining the problem. Red lights will indicate any overloads that may have happened causing a latch-out condition. Each step in the interlock chain is indicated, and if there is a problem with an interlock, this should be easily traced. Having a block diagram or network diagram of the interlock chain readily available can be valuable in troubleshooting.

Metering levels are the next step in troubleshooting a transmitter problem. Comparing them to the normal baseline will quickly pinpoint the troublesome module.

Restoring a transmitter back to an on-air condition should only be done after the cause of the problem has been found/eliminated. If there are failures in the amplifier chain, it may be possible (particularly in the case of a solid-state transmitter employing multiple parallel PA stages) to return the transmitter to air at a reduced power. Care should be taken not to overload the remaining stages and cause more damage. As with all troubleshooting of an installed system, the key is to keep in mind that the system was, at one time, working. Having a standby solution in this situation reduces the pressure of getting everything back to working condition immediately, and reduces that chance of more serious damage due to hasty workmanship.

2.5 MASK FILTERS

As mentioned earlier, the RF power amplifier will ultimately contribute some amount of distortion and as a result, not only does this affect the signal waveform but it also has the effect of causing out-of-channel artifacts. Even minor amounts of distortion, which would not affect the receivability of the signal, will cause emissions in the spectrum outside the channel (out-of-band-emissions).

The output of a transmitter inevitably has some amount of out-of-band emission, including harmonics and intermodulation distortion. These must be filtered out in order to reduce interference with the neighboring spectrum users. Filters are discussed more in-depth in other sections of this chapter but it should be reiterated that the mask filter is an integrated part of the transmission system and if specified separately, it must be matched with the characteristics of the transmitter—ultimately so that the entire system performs well and the output spectrum conforms to FCC requirements.

In some cases where a number of stations are sharing a common antenna, the output mask filter is integrated into a channel combiner. In this case, the mask filter becomes part of the channel combining process. Obtaining sampling signals discussed previously for linear predistortion feedback is more complicated in this case. The signal after the filtering element will contain multiple channels and these signals may create problems for the processing algorithm of the predistortion network. This may be a situation where a continuously adaptive process will simply not work, and one-time setup (with all other signals turned off) may be required. Channel combiners, in general (even those not incorporating the mask filter), may contribute linear distortions in the signal path.

2.6 COOLING

The cooling of electronics is likely the most important aspect to ensure long life and reliability. The transmitter installation must be planned carefully to make certain that the airflow or liquid cooling system is adequately sized and installed properly. When planning and installing a transmitter, it is recommended that the services of an experienced air conditioning contractor/engineer be engaged for the design and installation of the cooling and ventilation systems within the transmitter building. The transmitter manufacturer will have recommendations for amount of airflow, the air conditioning requirements, and in most cases will be also supplying the liquid cooling system (if the transmitter is liquid-cooled).

There are significant differences between the approaches to cooling with air versus cooling with liquid. While this is not meant to be a dissertation on one versus the other, a short comparison of the benefits and drawbacks of each is in order.

2.6.1 Air

Forced air cooling of electronics has been in use since the beginning of broadcasting. The principle is simple—blow air across the hot parts and take away the heat. It can rely on one single air moving device or multiple devices. The advantage of air-cooling is essentially the disadvantages that can be listed for liquid-cooling; there are fewer moving parts and minor air leaks are significantly less of a problem than minor fluid leaks. The disadvantages of air cooling are that in some hot-climate situations, extensive air conditioning may be required. Forced air cooling can be noisy in situations where the sites are staffed.

The electronics will require a certain amount of air flow across heatsinks and components in order to keep device temperatures at an acceptable level. This cooling air will pick up heat from the electronics as it flows. The internal air flow dynamics within the cabinet of the transmitter is designed such that if the proper amount and temperature of air is introduced at the inlet (and taken away at the outlet), then proper cooling of all the subassemblies will take place.

The transmitter itself will present a certain amount of airflow resistance (static pressure) that the supply or exhaust fan must work with. These cooling fans may be included inside the transmitter itself, or they may be external and integrated into the air system of the building installation. In addition, "helper" fans may be used in some situations where long ducting runs are required. The balance of the fan/blower and the resistance of the *entire* air system is critical to supplying the correct amount of airflow.

2.6.2 Liquid

In the case of a liquid-cooled transmitter, the majority of the heat produced by the electronics is removed via the cooling fluid. This is dissipated in a heat exchanger, typically outside the transmitter building. The cooling fluid is circulated through the transmitter by a pump system.

Liquid-cooling of electronics has certain disadvantages; however, if properly designed, installed, and maintained, these can be minimized. Amplifiers can be cooled significantly better through the use of coldplates that route the fluid directly past the hot spots. Heat transfer is much better than relying on the spreading of heat to fins, and then transferring this heat to air. If amplifiers can be cooled better, then powers can be increased, efficiencies increased, and so on. The disadvantage of liquid, beyond the proximity of liquid and electronics in the same space, is that we now have pumps moving fluid to a heat exchanger where a fan moves air. Additionally, there may also need to be some extra air movement in order to remove heat from those components which are not coldplate-mounted. There can be many more moving parts in the system, and this needs to be factored in to reliability considerations.

The system coolant pumps are usually arranged for ($n + 1$) redundancy, so that part of the system (e.g., one exchanger fan and/or one pump) can be shut down for maintenance with no effect on the others.

The liquid coolant is a mixture of distilled water and pure ethylene glycol; usually 50 percent of each.*

Always follow the manufacturer's recommendation for coolant. While it may seem similar to the cooling system in your car, automotive antifreeze mixtures are not generally recommended as they contain additives that may be acidic (disposal of used coolant should be done by EPA-approved firms that recycle automotive fluids).

2.6.3 HVAC

Thorough consideration and analysis must be given to the entire subject of transmitter building ventilation. Due to the complexity of the entire discipline of "heating, ventilation and air conditioning" (HVAC), it is recommended for best results that the services of an experienced air conditioning contractor/engineer be engaged for the design and implementation of your ventilation system. Whether your transmitter is forced-air cooled, or liquid cooled, some attention is needed to the heating and cooling of the building itself.

Should you choose to duct the heat from the cabinet exhaust fan, be aware there is a certain amount of heat classified as unavoidable heat loss to the transmitter room, due largely to radiation from the cabinets, pipes, and so on. This causes a heat rise in the room regardless of the transmitter cooling system configuration.

When the exhaust is ducted outside, particularly in the case of forced-air cooled transmitters, ensure that at least enough filtered replacement air is available to the transmitter room to replenish the exhausted air. If the transmitter room is kept at a slightly higher pressure than the outside, it can minimize the ingress of dust into the room.

REFERENCE

Aitken, Steve, and Greg Morton: "Depressed Collector IOTs: Proven Digital Performers—But What Of Analog," e2v technologies.

---

*Note: for climates where outside temperatures never drop below freezing, the system can use distilled water.

CHAPTER 2.11
DTV MASK FILTERS

Daniel S. Fallon
Dielectric, Raymond, ME

2.1 INTRODUCTION

High power television transmitters, in addition to the licensed in-channel signal, generate substantial wideband noise on channels not licensed by the broadcaster operating the transmitter. The out of channel noise is generated by intermodulation (IM) products in the output high power amplifier of the transmitter. The most significant IM products are 3rd and 5th order. The purpose of a mask filter (also known as an IM filter) is to attenuate the IM products from the transmitters so they do not interfere with neighboring channels.

2.2 FILTERS AND EMISSION LIMITS

The emission limits for full power and Class A television stations are defined in the FCC (Federal Communications Commission) rules document 47 CFR Part 73. In Part 74, the emission limits are defined for LPTV (low power television) stations. Figure 2.11.1 shows the output spectrum of a typical DTV transmitter (solid line) as well as the FCC full service emission limit—also known as a "mask" (dashed line). The level of noise emission outside the desired channel is known as the *shoulder level* and is typically anywhere from −36 to −45 dB from the mid-band or *head* power level. A shoulder level of −36 dB is shown in the figure. This shoulder level provides FCC compliance for the first 500 kHz from the channel edge but not farther from the channel.

Figure 2.11.2 shows the transmitter output attenuated by a six-section mask filter. Clearly, the addition of the mask filter provides FCC compliance for the transmitter.

Figure 2.11.3 shows a comparison of the FCC full mask with the low power TV simple and stringent masks. A six-section filter is required to meet both the full service and the stringent mask; the simple mask can generally be met with a four section.

In some cases, the attenuation provided by a six pole filter is not sufficient and an eight pole filter is used instead. The first case is if the output amplifiers are pushed into a slightly nonlinear range and the shoulders do not meet the −36 dB band edge requirement on their own. The second application of eight pole filter is to combine adjacent channels onto the same transmission line/antenna. An advantage of using the eight pole filters is to substantially reduce the noise broadcast into the adjacent channel. This reduces interference into the adjacent channel, improving the signal to noise ratio for comarket or adjacent market broadcasters operating on these channels (e.g., LPTV). Figure 2.11.4 shows a transmitter spectrum with an eight pole at the output.

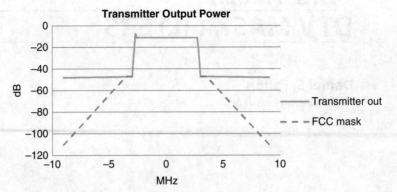

FIGURE 2.11.1 Transmitter output power and the FCC full service mask.

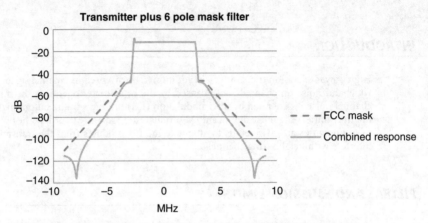

FIGURE 2.11.2 Transmitter with six-section mask filter and the FCC full service mask.

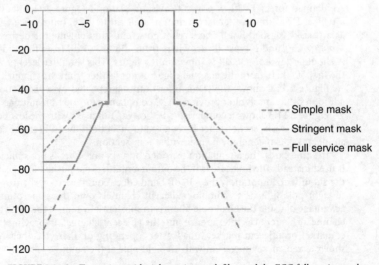

FIGURE 2.11.3 Transmitter with eight section mask filter and the FCC full service mask.

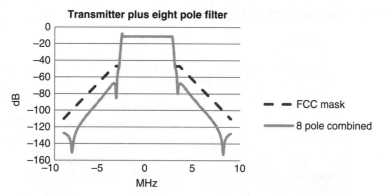

FIGURE 2.11.4 Transmitter with eight-section mask filter and the FCC full service mask.

2.3 TYPES OF FILTERS

Figure 2.11.5 shows the frequency response (attenuation vs. frequency) for several different filter types. For the purpose of this comparison they are all selected to have a 6 MHz passband width. It is clear that as the number of sections in a filter increases, the out of band attenuation also increases. Note the difference between the "pole placed" six- and eight-section filters and their Chebychev counterparts. The pole placed filters have attenuation nulls (transmission zeros) close to the passband. This increased close-in attenuation is exchanged for less attenuation farther from the passband. Filters are composed of resonators and couplings between the resonators. For standard Chebychev responses, the resonators are coupled in series from input to output. For responses that include transmission zeros (i.e., a "pole placed" filter), nonadjacent resonators are also coupled (also known as *cross-coupling*). Except for the LPTV "simple" mask, the most prevalent television mask filters are either six-section pole placed or eight-section pole placed. The disadvantage of higher order filters is they have slightly more passband loss and group-delay variation.

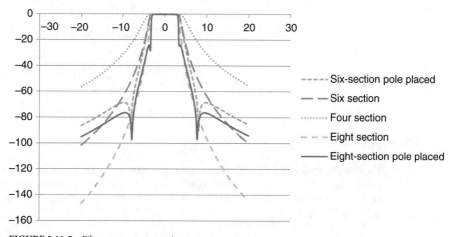

FIGURE 2.11.5 Filter response comparison.

2.3.1 Reflective and Constant Impedance Filters

A filter by itself is a two port device and is also known as a *reflective filter*. This is because any power not in the passband of the filter is reflected back to the transmitter. Reflective filters generally work fine for solid-state transmitters. They do not work well with tube transmitters.

A configuration in which there are two filters, two hybrids, and two loads, as shown in Fig. 2.11.6, is a *constant impedance filter* (CIF). They are also known as *balanced filters*. The moniker "constant impedance" comes from the fact that the transmitter sees a good match at all frequencies. These filters can be used with tube transmitters and have the advantage of doubling the power rating of a reflective filter because each side of the CIF sees only half of the input power. The disadvantage with respect to reflective filters is the cost associated with the extra components and the size of the CIF in the transmitter room. The CIF can also be used as a channel combiner by replacing the ballast load with a second transmitter. Eight-section pole placed filters in a CIF configuration is the standard way of combining adjacent channels.

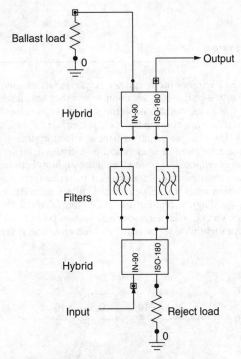

FIGURE 2.11.6 CIF schematic.

2.3.2 Filter types

There are several different filter geometries that can be applied to television mask filters. These geometries include comb-line, interdigital, coaxial cavity, rectangular waveguide, and dual mode cylindrical waveguide. Each of these has their specific merits that make them the choice for specific applications. Figures 2.11.7–2.11.11 show the internal geometry of each of these types of filters.

The interdigital, comb-line and coaxial resonator filters have a lot in common. They all consist of round rod resonators shorted to the case at one end and open circuited at the other. The length of the rod determines the resonant frequency and filter channel. The unloaded Q of the resonator (and filter loss) is determined by the size of the resonator and cavity. The differences can be seen in the figures.

The interdigital filter (Fig. 2.11.7) has adjacent resonators shorted at opposite ends of the cavity. This configuration allows increased coupling between resonators and therefore wider filters.

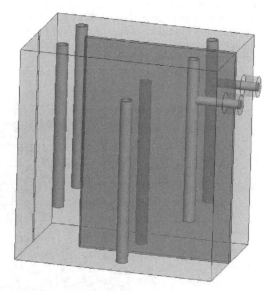

FIGURE 2.11.7 Interdigital filter geometry.

The comb-line filter (Fig. 2.11.8) consists of a series of rods shorted at the same end of the case. These filters are good for narrow to moderate bandwidths. The prime advantage of a comb-line filter is its simplicity.

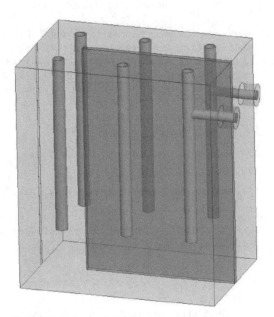

FIGURE 2.11.8 Comb-line filter geometry.

The coaxial cavity filter (Fig. 2.11.9) takes the comb-line filter and introduces a metallic wall between the resonators. The center conductor is round but typically the outer is square; the resonator is actually a shorted square coaxial line. In the coaxial cavity filter, the coupling between resonators is accomplished either with an iris or coupling loop. The advantage of a coaxial cavity filter over a comb-line is a wider spurious-free range. Spurious responses are resonances above the filter passband that reduce the filter attenuation (sometimes below 1 dB) in the desired stopband. These resonances are caused by either waveguide modes or one of the filter tuning structures appearing as a resonator at the higher frequency. The wide spurious range allows either second harmonic suppression or a wide tuning range for a tunable filter.

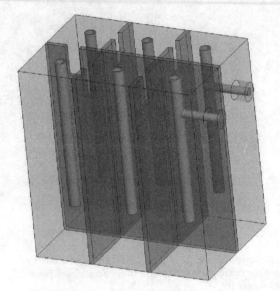

FIGURE 2.11.9 Coaxial cavity filter geometry.

The rectangular waveguide filter is shown in Fig. 2.11.10. The filter consists of a series of rectangular cavities coupled by either irises, posts, or obstacles. Figure 2.11.10 depicts the post type coupling. There is a tuning screw in the center of one broad-wall of each resonant cavity; it serves to fine-tune the resonance. The resonant frequency is set primarily by the size of the cavity; hence these filters are sized to tune to a single channel. A filter sized for channel 21 cannot be properly retuned to

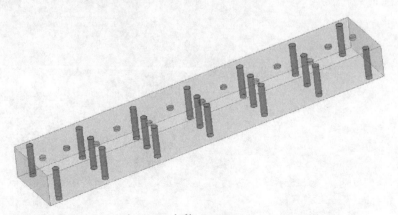

FIGURE 2.11.10 Rectangular waveguide filter geometry.

channel 22. These filters are larger than coaxial cavity filters and have larger unloaded Q and therefore lower loss. Due to their size, they are only practical for ultrahigh frequency (UHF) applications. Due to the difficulty of cross-coupling, a rectangular waveguide filter they are used more commonly for channel combiners than mask filters.

The dual mode cylindrical waveguide filter is shown in Fig. 2.11.11. A photograph of a dual mode CIF with waveguide hybrids is shown in Fig. 2.11.12. This is also a UHF-only filter, and each filter is sized to a single channel. It is also larger than any of the other filters discussed. It is, however, the most commonly used mask filter for high power UHF transmitters. The reason for this is the extremely high Q_0 of the cylindrical waveguide cavity that results in very low insertion loss; thus, these filters exhibit the lowest loss and highest efficiency of any practical filter layout. They are called dual mode because each cylindrical cavity supports two physically orthogonal waveguide modes that can be configured as two independently tuned filter sections. A six-section filter has three physical waveguide cavities; an eight-section has four. The cavities are coupled by irises cut into plates that separate the cavities. By proper selection of iris and tuning screw location, these filters can be easily cross-coupled. Most dual mode responses include transmission zeros.

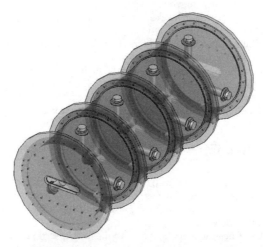

FIGURE 2.11.11 Eight-section dual mode cylindrical waveguide filter geometry.

FIGURE 2.11.12 Eight-section dual mode cylindrical waveguide CIF.

2.3.3 Band Considerations

The frequency band in which the filter needs to operate narrows down the choices considerably.

Low-band very high frequency (VHF) (channels 2–6) is characterized by very long wavelengths and very large bandwidths. Band width in percent is defined as the channel band width divided by the center frequency. For low-band VHF, this varies from 10.5 percent at channel 2 to 7.0 percent at channel 6. The insertion loss of a filter is inversely proportional to bandwidth. This means that for low band (high bandwidth), the loss is much less dependent on resonator unloaded Q (Q_0) than it is for UHF filters. The natural filter geometry for low-band VHF is interdigital since it can easily realize the large inter-resonator couplings required by the large band widths. The large wavelengths at low band require physically large filters. The low loss and relatively low power levels mean that heating due to losses in a low band filter is minimal.

High-band VHF (channels 7–13) is characterized by shorter wave length and narrower bandwidth than low band. The wavelength is still too long for practical waveguide structures. The bandwidth range is 3.4 to 2.8 percent over the channel range. Comb-line structures are typically chosen as the most practical for this channel range. The advantage of comb-line resonators is they can easily be made band tunable so one filter can be tuned to any channel in the 7–13 range. Interdigital filters can also be made band tunable for channels 7–13. The loss of the filter is dependent on the Q_0 of the resonators, so for larger power levels larger filters are required. Typical power ranges for high-band VHF are in the 5–7 kW range, although there are installations and associated filters for up to 30 kW.

UHF has much narrower percent bandwidth per channel, the largest range of channels, and typically higher power levels than VHF. These factors explain the larger variety of filters used for UHF filters. Lower power options (up to about 10 kW) for UHF are typically band tunable coaxial cavity filters. Typically, these filters can be tuned to any channel in UHF from 470 to 860 MHz. They can also be tuned to any required bandwidth—6 MHz for North/South American applications and 8 MHz for Europe/Africa/Asia. For the high power North American market, the dual mode cylindrical waveguide filter is most common. The very low losses of these filters realize efficiencies of 97 percent or better. This efficiency cannot be matched with any other filter technology. Rectangular waveguide and comb-line filters have been used for UHF mask filters but are less common.

2.4 THERMAL STABILITY

Narrowband filters are sensitive to changes in temperature, due to changes in their physical dimensions. The temperature change in television mask filters can be driven either by changes in room ambient or by "self-heating" (i.e., heating of the surface and components of a filter due to ohmic losses of RF power).

The self-heating is proportional to the filter insertion loss. Lower power coaxial cavity filters will have the highest loss, and their power limit is generally set by the maximum acceptable temperature of the filer enclosure. This temperature is somewhat subjective but most manufacturers have selected a temperature of around 60°C as the maximum hot-spot case temperature at maximum rated power. The filters are fabricated from materials that can withstand temperatures well in excess of 60°C—the selection is more from a personnel safety consideration. It is not desirable to have very hot surfaces exposed to people working in a transmitter room.

Filters are sensitive to changes to temperature because they are made from metallic cavities. Metal generally expands as it is heated and this will expand the size of the cavity. The resonant frequency of the filter is inversely proportional to cavity size, so as the cavity expands the resonant frequency drifts down. Equation (1) gives the change in frequency (Δf) as a function of the center frequency (f_0), the change in temperature (ΔT), and the coefficient of thermal expansion of the metal (α):

$$\Delta f = f_0 \left(\frac{\alpha \, \Delta T}{1 - \alpha \, \Delta T} \right) \qquad (1)$$

Aluminum is a common material for filter construction. The coefficient of thermal expansion (CTE) for aluminum is 22 parts per million (ppm) per °C. Using this number, a center frequency of 599 MHz (channel 35) and an average change in surface temperature of 30°C, the change in resonant frequency is 396 kHz. This is a significant amount of drift and could cause a response tuned at 20°C to deviate from the specified performance in both the passband and the stop band. Note that the drift is proportional to center frequency so drifting filters are less of a concern at VHF.

There are four commonly used approaches to combat frequency drift. The first is somewhat primitive but allows the use of all aluminum construction. The next two involve the use of Invar which is a low CTE material, and the last is a complicated implementation.

The first way to account for the passband drift is to build "guard band" into both the passband and stop band requirements. This means if the filter drifts due to heat, it will still meet the performance requirements. This is accomplished by using an eight-pole filter to meet a specification typically met by a six-pole cross-coupled filter. The eight-pole tuning is shifted high in frequency at the tuning temperature to account for the drift down in frequency at the operating temperature. This approach has been proven adequate in several transmitter installations using dual mode cylindrical waveguide filters. The caveat is that the transmitter room needs good HVAC control, and if the HVAC malfunctions- the filter would likely drift out of specification.

The second and simplest solution is to build the filter out of Invar. Invar is a nickel steel alloy with a very low coefficient of expansion (1.6 ppm/°C). The drift, according to Equation (1), for a 30°C temperature rise at channel 35 is 34 kHz. This is small enough that it will not impact filter specification compliance. Most dual mode cylindrical waveguide eight-pole and six-pole cross-coupled filters are made of Invar cavities. The disadvantages of Invar are the weight and cost of the material. The advantage is the simplicity and reliability, even with the loss of HVAC.

The third way to correct for frequency drift is used in most coaxial cavity, comb-line and interdigital filters. It is to use a "temperature-compensated" resonator assembly. Figure 2.11.13 shows a cross-section view of a temperature-compensated probe. A small diameter Invar rod is run down the center of the resonator. This rod ties the top of the resonator to a probe tip that is able to slide up and down the inside of the resonator tube (e.g., on spring fingers). As the resonator heats up it will expand but the probe tip is held in place stabilizing the frequency of the resonator.

The fourth way to correct passband drift with heat is to use any of a number of mechanical configurations to retune the expanded cavity so it stays on the required frequency. Various solutions have involved bi-metallic springs, cantilever mechanisms, and flexible end plate assemblies. These techniques have been applied to coaxial cavity filters, rectangular waveguide filters, and dual mode cylindrical waveguide filters. They may or may not use Invar rods for part of the assembly. The issue with most of these approaches is that they are complicated and have labor-intensive setups.

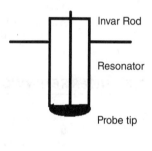

FIGURE 2.11.13 Temperature compensated resonator.

Even though Invar is a more expensive material, an Invar filter is simple and reliable.

2.5 INSTALLATION CONSIDERATIONS

Most mask filters are designed for either floor or ceiling mount. The large waveguide CIFs are typically put in the ceiling of a room, allowing transmitters or walkways below them. Smaller coaxial cavity filters can be installed alone in the ceiling or placed in a frame with couplers, patch panels, test loads, etc. The latter is known as a "unitized" RF system. These are typically floor mount. The layout of the transmitter room and the equipment in it will dictate how a mask filter is mounted. The mask filter should be relatively close to the transmitter output but there can be other coaxial components between them. A key thing to remember is to avoid mounting a filter directly in the outlet of a heater or air conditioner; an even temperature around a filter is preferred. If fan cooling of a filter is required, it will come with the fans already in place. The filter should be mounted to allowing access to the tuning probes if maintenance is required. It is wise to avoid running conduit or water piping directly underneath a ceiling installation just in case any of the filter components need to be dropped to work on them.

The heat dissipated by a filter system should be accounted for in sizing the HVAC system of a transmitter room. For low power filters (<10 kW), the dissipated heat will be less than a few hundred watts. For high power filters, the dissipated power could be 2–4 kW. The manufacturers' catalog will have insertion loss numbers that can be used to calculate the heat dissipated with the transmitter power output (TPO).

If a filter is cooled with water or with a fan, and its operation relies on it, a thermal interlock should be included. This interlock will trip if the temperature of the filter exceeds a preset maximum. The interlock should be wired into the transmitter interlock system and used to reduce the transmitter power ("fold-back") should it trip. The interlock is intended to signal a failure of the fans or water cooling.

The transmitter proof of performance is the test where a filter's performance is verified. There are a few different ways of doing a transmitter proof, but one of the steps in the proof is to certify that the transmitter meets the emissions mask. This is generally done at installation of the transmitter by a skilled field engineer.

2.6 MAINTENANCE

Most filters are very reliable and should outlast the transmitter they are paired with. The two items that will require a minimal amount of maintenance are oil loads on CIFs and fan-cooled filters. The oil load included with high power filters comes with a recommended maintenance schedule—checking the DC resistance and inspecting the oil. Fan-cooled filters should be checked periodically to assure the fans are operating properly. The fan blades should also be checked for dust build up and cleaned periodically.

2.7 FILTER RETUNING

There is a subset of filters that are band-tunable. These generally are high band VHF comb-line filters and coaxial cavity UHF filters. Tuning a filter to a new channel is not for the in-experienced; it is a time-consuming process and requires specialized equipment (an RF network analyzer). For station engineers or station groups that do not have the equipment or in-house expertise and need to change the channel of their transmitter, there are a couple of options. One is to hire an experienced field engineer to come and retune the filter. The second is to send the filter back to the manufacturer to have it retuned. The advantage of the first is time—a field engineer should be able to retune a mask filter in a few hours. The advantage of the second is cost—a factory technician can retune a filter in less than an hour so the labor cost is minimal; the prime cost is shipping the filter to and back from the factory. The choice of how to approach retuning a band tunable filter will be driven by the resources of the station, the schedule, the availability of a back-up transmitter/filter, and the availability of a skilled field engineer.

CHAPTER 2.12
DTV TELEVISION RF MEASUREMENTS

Linley Gumm
Consultant, Beaverton, OR

2.1 INTRODUCTION

This chapter provides information about measuring the 8-VSB DTV RF (8-level vestigial sideband digital television radio frequency) signal. Very useful measurements may be made with general purpose instruments such as power meters and spectrum analyzers; however, specialized test equipment such as a TV analyzer is required to make detailed measurement of the multiple parameters that determine signal quality.

2.2 MEASUREMENTS USING GENERAL PURPOSE TEST EQUIPMENT

2.2.1 Power Meter Measurements

The power meter is the instrument of choice for measuring 8-VSB signal power. Its measurements are easy and accurate with only a modest amount of care. Start by making sure that the meter selected is designed to accurately measure true average power. The most accurate meters use a thermo-electric sensor. Most diode-based units, while having the ability to read peak power, give less accurate average power readings. Typical accuracies of average-reading thermo-electric meters are somewhat larger than ±1 percent (i.e., ≈ ±0.5 dB) while diode-based units are often four times that value.

The power meter is relatively fool-proof. But, like all RF instruments, some care must be taken. First and foremost, the power meter reads the power of the *entire* applied signal; desired and undesired alike. Therefore, caution must be taken that only the desired signal is the only one present at the power meter's input.

Second, the power meter (like all RF test equipment) can accommodate a limited amplitude range. A typical thermocouple sensor can make accurate measurements for amplitudes above −35 dBm but can handle a maximum power of only +20 dBm (100 mW) for a 55 dB dynamic range. Sensors are available that incorporate an input attenuator that increases the maximum allowable input power, but the 55 dB measurement window remains the same.

Power meters are often used when it is desired to validate a transmitter's built-in power output monitoring reading.* Some form of attenuator or directional coupler must be used to adapt the power

---

*Instrumented liquid cooled dummy loads are sometimes used to make calorimeter-based transmitter power measurements. This is well known and will not be dealt with here.

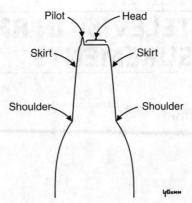

FIGURE 2.12.1 Names of parts of the 8-VSB signal. (© Linley Gumm, used by permission.)

meter's limited input power measuring capability to the transmitter's large output power. This can take the form of a well calibrated power attenuator or a calibrated directional coupler. The FCC requires that the total transmitter power measurement uncertainty be less than 5 Percent.[†] Therefore, the attenuation or coupling uncertainty between the transmitter's antenna terminals and the power meter must be 5 percent minus the meter's uncertainty, or about 3.5 percent for typical thermal-electric power meters.

Some vendors provide a calibrated power meter permanently connected to a dedicated transmission line coupler. When operated within the device's frequency and power ratings, the 5 percent specification is guaranteed.

2.2.2 Spectrum Analyzer Measurements

Spectrum analyzers are the most useful RF measurement instrument. Analyzers allow the visualization and measurement of the RF spectrum by way of a display of signal amplitude vs. frequency. Conventional spectrum analyzers are essentially electronically tuned receivers that are swept over a selected frequency range; a range that is determined when the user selects the frequency at the center of the display (i.e., the *center frequency*) and width over which the analyzer is swept (i.e., the *span*). The analyzer then electronically sweeps a *Resolution Bandwidth* filter of known width over the selected range. At each point along the sweep, the analyzer places a dot in the display representing the amplitude of the signal falling within the Resolution Bandwidth filter.

The *reference level* refers to both the (normally) top line of the graticule and the analyzer's amplitude control. The control is thus used to select what amplitude is represented by that line. The vertical display is linear in dB. With a display linear in dB, and with a known dBm amplitude represented by the display's reference level, the amplitude of any smaller sine wave signal may be directly read off of the display. Markers are normally available that present the user with a numerical amplitude readout of selected points in the display.

Signal Terminology. When seen on the spectrum analyzer, the 8-VSB signal has several characteristic parts. Figure 2.12.1 is provided to define the names used in this chapter.

The flat portion of the signal is (often) called its *head*. The 8-VSB signal is unique in that it has a *pilot* carrier at its lower edge. The *skirt* is the area where the signal is falling in amplitude near each channel edge. Just beyond the channel's edge, virtually every transmitter has *shoulders* where intermodulation products created by the transmitter fall into the adjacent channel.

[†]FCC 47CFR§73.664(b)(2). An exception is when calorimeter measurement is used. In that case the FCC requires an accuracy of 4 percent.

Dynamic Range and Intermodulation. A spectrum analyzer's most important specifications are its frequency range and its dynamic range. Frequency range is fairly obvious. Dynamic range or amplitude range over which measurements may be made is more complicated.

Like every receiving device, the analyzer's ability to measure low amplitude signals is limited by its own noise floor. Likewise, multiple high amplitude signals can cause intermodulation (IM) products created in the analyzer's front end to be displayed as if they are real signals. This limits the analyzer's ability to simultaneously measure large and small signals. The analyzer's dynamic range is thus defined as the range of amplitudes from its internal noise floor to the amplitudes where its IM products start to emerge above that noise floor.

IM is created by small linearity errors in the analyzer's circuitry. A very useful model is to describe the device's output as being proportional to the instantaneous amplitude of the input signal plus a small signal proportional to the square of the instantaneous input amplitude plus another small signal proportional to the cube of the instantaneous input amplitude, etc.

The signal proportional to the square of the input signal is said to have second-order curvature. If two (or more) signals are present, second-order curvature will create new signals called intermodulation beats (or products) at their sum and difference frequencies plus beats at the signals' second harmonics. For example, if signals at 100 and 101 MHz are applied, these signals will be present in the output. But, in addition, second-order curvature will create sum and difference signals at 201 and 1 MHz plus second harmonic signals at 200 and 202 MHz.

Likewise, the signal proportional to the cube of the input signal is said to have third-order curvature. If two (or more) signals are present, this curvature also creates new beats at $2F_1 \pm F_2$, at $2F_2 \pm F_1$, and harmonics at $3F$. That is, if 100 and 101 MHz signals are applied, besides the two input signals themselves, the output will contain intermodulation beats at 99 MHz, 102 MHz, 301 MHz, and 302 MHz plus third harmonics at 300 MHz and 303 MHz.

The amplitude of a spectrum analyzer's second and third-order IM beats vary in a very orderly way.[‡] When all the fundamental signals are increased in amplitude by 1 dB, all second-order beats increase by 2 dB in *absolute* amplitude while third-order beats increase by 3 dB. Note that the amplitude *difference* between the fundamental signals and their second-order beats gets 1 dB smaller for every 1 dB increase in fundamental amplitude; or 2 dB for third-order beats.[§] If the decrease in amplitude *difference* between the fundamentals and their beats is extrapolated, there is an input amplitude where the beats will (in theory) increase to the same amplitude as the fundamentals. This is called the intercept power or IP. Many spectrum analyzers have a second-order intercept (IP2 or SOI) and third-order intercept (IP3 or TOI) power specification. These are very useful indexes to the analyzer's dynamic range.

Digital Signal Overload. Using the spectrum analyzer to measure the amplitude of sine wave signals is easy. Tune to it, adjust the reference level so its amplitude is on screen, then read its amplitude directly from the screen or use a marker. As long as the entire signal is narrow enough to fit entirely within the analyzer's Resolution Bandwidth, the measurement will be accurate.

When measuring DTV, to see detail, a Resolution Bandwidth much smaller than DTV's 6 MHz width must be used. As it is swept across the DTV signal, the analyzer correctly gives the amplitude of the portion of the DTV signal within its resolution bandwidth. But, since the resolution bandwidth is much smaller than the signal's 6 MHz bandwidth, the analyzer shows it to be much lower in amplitude than it really is. Because the 8-VSB is very noise-like, the amplitude difference is proportional to 10log of the ratio of bandwidths. For example, if a 10 kHz resolution bandwidth is used, the head of the 8-VSB signal will appear to be $10\log\left(\dfrac{5.38 \text{ MHz}}{10 \text{ kHz}}\right) \approx 27$ dB below the signal's actual amplitude.[¶]

The signal's apparent amplitude varies with Resolution Bandwidth, changing by 10 dB with each factor of 10 change in bandwidth and 5 dB for each factor of 3. Because the IM created by the analyzer's internal

---

[‡]Simons, Keneth A.: "The Decibel Relationships Between Amplifier Distortion Products," *Proceedings of the IEEE*, Vol. 58, No. 7, pp. 1071–1086, IEEE, New York. July 1970.
[§]Intermodulation products increase much faster when the amplitudes start to overload the system.
[¶]"IEEE Standard 1631-2008: "Recommended Practice for Measurement of 8-VSB Digital Television Transmission Mask Compliance for the USA," IEEE, New York, 10016-5997.

distortions are proportional to the DTV signal's *total* power while the displayed amplitude is proportional to only the signal within the resolution bandwidth, it is easy to inadvertently overload the analyzer.

Most spectrum analyzers have an attenuator between their input and the input mixer. When the analyzer is operated in the default or Automatic mode, the RF attenuator's value is automatically selected by the reference level control. The RF attenuator will be set to zero (or a small value) until the reference level is adjusted upward to some specific value; a value that is selected by the manufacturer to minimize IM distortion. Above that value the IF gain is kept constant and the RF attenuation is increased in step with reference level increases.

To show how easy it is to overload the spectrum analyzer, assume a 0 dBm 8-VSB signal is connected to a typical medium performance analyzer. Further, assume that this analyzer is being operated as shown in Fig. 2.12.2 with a Span of 20 MHz (i.e., 2 MHz per division) and a resolution bandwidth of 10 kHz.** The RF attenuator is set to its *auto* mode.

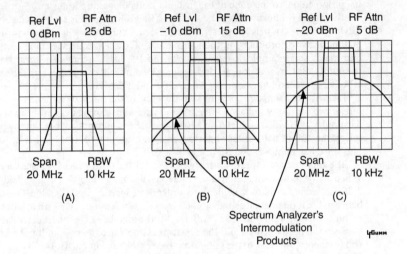

FIGURE 2.12.2 Digital signals can overload spectrum analyzers well before the signal reaches the reference level. A 0 dBm 8-VSB signal is displayed in each case. (A), (B), and (C) show the results with different reference levels. See text for details. (© Linley Gumm, used by permission.)

At (A), the reference level is set to 0 dBm. At this setting, this mythical analyzer sets the RF attenuator to 25 dB. That reduces the amplitude of the 8-VSB signal at the input mixer to −25 dBm. In a 10 kHz bandwidth, the head of the 8-VSB signal is correctly displayed at −27 dBm. This mythical instrument's displayed average noise level (DANL) in a 10 kHz bandwidth is −105 dBm, which is 5 dB below the bottom of the screen. The third-order intermodulation caused by the −25 dBm mixer signal is just at the bottom of the screen. Therefore, the shoulders observed on the signal at 37 dB below the signal's head are the transmitter's IM products as shaped by the transmitter's channel filter. Therefore, everything displayed is real.

But, having the head of the 8-VSB signal almost 30 dB down screen *looks odd* to many users. Novice users may change the reference level control to −10 dBm in order to move the signal up screen as in (B). The analyzer makes this change by removing 10 dB of attenuation (which raises the signal at the mixer to −15 dBm) and rescaling the display. The signal's head is still correctly displayed at −27 dBm but that amplitude is now a bit less than two divisions from the reference level.

However, with a −15 dBm mixer signal, the analyzer's IM products now are only 50 dB below the signal's head at mid-channel. They have a shape similar to the broad curve of the simple emissions mask shown in Fig. 2.12.7. The transmitter's own IM shoulders are larger than the analyzer's but

---

**This is an invented example. However, it shows the performance expected of an analyzer with a TOI or IP3 intercept of about +13 dBm.

rapidly fall away because of its channel filter. At about 3 MHz from the channel's edge the analyzer's IM products become larger than the transmitter's, indicated by the change of slope of the adjacent channel emissions.

At (C), the reference level control has been again lowered by 10 dB. Even though the 8-VSB signal's head is still below the reference level, the mixer is now being driven with a −5 dBm signal which creates a huge amount of intermodulation in the analyzer, overwhelming everything.

When looking at a DTV signal for the first time, avoid overload problems by adjusting the analyzer to its maximum reference level before connecting the signal to the analyzer. Then, lower the reference level in steps. Stopping as the signal's head moves up the screen to the appropriate level; a level dependent on the analyzer's intermodulation performance. As a rule of thumb, if a 100 kHz resolution bandwidth is being used, the signal's head should be 10 to 20 dB below the reference level; if 30 kHz, 15 to 25 dB; if 10 kHz, 20 to 30 dB. Above all, watch for the shape changes along the signal's skirts and shoulders as the reference level is reduced, indicating overload.

Desirable Spectrum Analyzer Specifications. Almost any spectrum analyzer will make useful 8-VSB measurements. However, if a unit is to be purchased, consideration should be given to obtaining one that can be used to make FCC emissions mask measurements. The following features/specifications are recommended for that purpose:††

- Frequency coverage of the TV bands; i.e., <50 to >700 MHz.
- Has a 10 kHz resolution bandwidth (typically one RBW among many).
- Has the ability to automatically measure the average power between two specified frequencies (i.e., with a Band Power or Channel Power mode, etc.).
- The numerical difference in amplitude between its DANL in a 10 kHz resolution bandwidth and the specified third-order intercept power (i.e., IP3 or TOI) must be ≥110 dB.
- A real plus but not required is an internal RF attenuator with a ≤5 dB step size.
- Not on IEEE 1631-2008's list but highly recommended by the author:
  - Internal tracking generator. Highly useful for verifying/adjusting the frequency response of almost everything RF in a transmitter facility including filters, test cables, couplers, etc., and initial tuning of IOT amplifiers. When spectrum mask measurements are made, frequency response tests of the filter used in that process must be made just before the test is performed. A tracking generator is ideal for that task.
  - Marker frequency count. A mode, when enabled, where the spectrum analyzer actually counts the frequency of the signal designated by the marker.

2.2.3 Spectrum Analyzer Measurements

DTV Signal Amplitude. Most spectrum analyzers have built-in channel power measurement functions to measure noise-like broadband (digital) signals. This is especially useful when measuring the amplitude of off-air DTV signals where the signal's shape is often irregular. Settings:

- If Span selection in the channel power mode is allowed, select a 10 MHz span.
- Turn on the channel power or band power function and adjust it for a 6 MHz integration bandwidth.
- Use the detector and video filter settings recommended by the manufacturer for maximum accuracy when measuring noise-like signals. This typically is either no or a wide video filter setting plus an RMS or sampling mode detector setting.
- Adjust the center frequency to center the signal in the analyzer's channel power measurement window.
- If available, enable signal averaging to achieve a more stable and accurate result.

---

††IEEE Standard 1631-2008, op.cit., p. 14.

- When measuring signal amplitude, the resolution bandwidth selection is not too important. A value of 100 to 10 kHz is recommended. Figure 2.12.3 shows a typical total signal power measurement in a 6 MHz channel bandwidth.

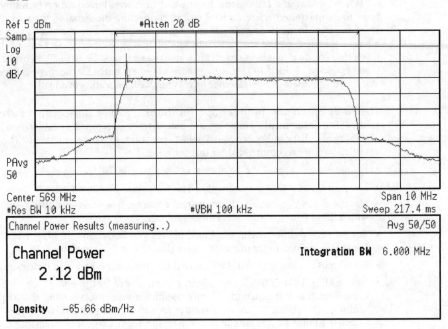

FIGURE 2.12.3 Typical use of measurement mode to measure total signal power.

If the analyzer does not feature a band power (or similar) mode, the amplitude of the 8-VSB signal may be *approximated* if the signal's head is flat. Measure the pilot signal's amplitude using a ≤10 kHz resolution bandwidth. Calculate the DTV signal's total amplitude by adding 11.6 dB to the pilot signal's amplitude.[‡‡] It should be noted that this result is an *estimate* and, *at best*, has an accuracy of about ±1 dB.

Pilot Frequency. A very convenient way to determine the frequency of an 8-VSB signal is to measure its pilot frequency using an analyzer's marker frequency count mode.

In a nominal 8-VSB channel, the pilot signal is 309.441 kHz above the lower channel edge.[§§] Over a 0 to 40°C range and power line variations of ±15 percent, transmitters for digital low power TV, TV translator, and TV booster stations licensed under CFR47§74 must maintain a frequency accuracy of ±10 kHz.[¶¶] The frequency accuracy required for DTV transmitters licensed under CFR47§73 is unclear. There are no specific frequency requirements for these transmitters except for the situation where there is a (physically) nearby analog transmitter on the lower adjacent channel. In that case, the DTV signal's pilot frequency must be maintained at 5.08218 MHz ± 3 Hz above the analog TV station's picture carrier.[***] Even with most analog stations now gone, this requirement is still operative under certain conditions when analog low power and analog translator stations are present.

[‡‡]IEEE Standard 1631-2008, op.cit., p. 27.
[§§]ATSC: "Recommended Practice: Guide to the Use of the ATSC Digital Television Standard, Including Corrigendum No. 1," Doc. A/54A, Section 8.5.1, p. 76, Advanced Television Systems Committee, Washington, D.C., 4 December 2003.
[¶¶]47CFR§74.795(b)(4).
[***]47CFR§73.622(g).

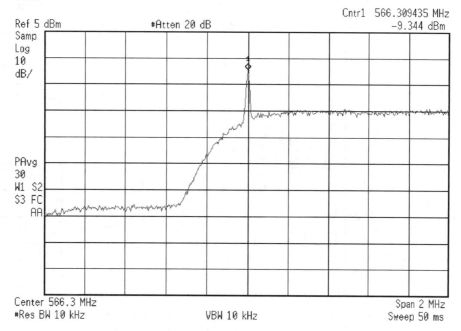

FIGURE 2.12.4 Measurement of pilot frequency.

Opinions vary for all other CFR47§73 licensed transmitters. Some choose to apply the FCC's analog TV frequency accuracy specification (±1 kHz) to DTV.[†††] Others assume that lacking a specific DTV rule, the only requirement is to meet the FCC's emission mask specification.[‡‡‡] The ATSC recommends particular offsets for reducing mutual interference between cochannel DTV stations which are not recognized by the FCC.[§§§] The ATSC does not make specific frequency accuracy recommendations for nominal DTV stations. As an aside, IEEE Standard 1631-2008 notes that relatively small offsets from the signal's nominal frequency makes it more difficult to meet the emission mask specification.[¶¶¶] In this uncertain situation, the selection of an appropriate pilot frequency accuracy specification is left to the discretion of the reader.

As shown in Fig. 2.12.4, to actually make the measurement, the settings should be:

- Adjust the center frequency to bring the pilot signal to the middle of the spectrum analyzer's display.
- The span should be 2 MHz or less.
- Adjust the reference level to be roughly equal to the signal's total power (see Section "S/N as a Function of Flatness").
- Adjust the resolution bandwidth to 10 kHz or less.
- Turn on a marker and position it at the peak of the pilot signal. Enable the marker frequency count mode.
- The counter's resolution should be adjusted to a value between 100 and 1 Hz.
- If available, enable measurement averaging to enhance accuracy.

[†††]47CFR§73.687(c)(1).
[‡‡‡]47CFR§73.622(h).
[§§§]ATSC A/54, op.cit., Section 8.5.5, p. 79.
[¶¶¶]IEEE Standard 1631-2008, op.cit., "Annex B, Effects of 8VSB Pilot Offsets on Meeting FCC Emissions Mask Requirements," pp. 57–63.

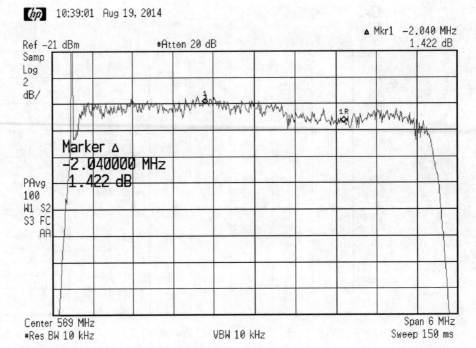

FIGURE 2.12.5 Flatness variation measurement.

Careful attention must be given to the analyzer's inherent frequency accuracy which is normally specified in terms of fractional error (e.g., 1×10^{-6}). That is, with that accuracy a measurement of a channel 7 pilot signal (174.3094 MHz) may be in error by as much as $\pm 174.3094 \times 10^6) \times (1 \times 10^{-6}) = \pm 174.3$ Hz. Similarly the analyzer's error could be as much as ± 692.3 Hz when measuring a channel 51 pilot signal at 692.3094 MHz. Most spectrum analyzers will accept an external frequency reference signal (e.g., 10 MHz). The use of a trusted frequency reference such as a GPS-disciplined oscillator or a precision source will drastically improve the accuracy.

The spectrum analyzer's uncertainty must be subtracted from the desired setting tolerance. For example, using a (1×10^{-6}) spectrum analyzer, a channel 51 transmitter must be adjusted to be within $\pm (1000 - 692) = \pm 308$ Hz of the desired frequency to ensure the transmitter is actually within a ± 1000 Hz window.

Signal Flatness. The spectrum analyzer is useful for measuring the 8-VSB signal's flatness. The DTV signal is specified to be flat over the central 4.762 MHz of the TV channel.**** That is, the spectrum is flat from 618.9 kHz above the lower channel edge to the same distance below its upper edge.

The measurement is relatively simple, but there is a complication because most analyzers use too little attenuation in the automatic RF attenuator mode when measuring DTV signals with small vertical scale factors. As shown in Fig. 2.12.5:

- First adjust the reference level to be roughly equal to the signal's total power (see Section "S/N as a Function of Flatness"). Now switch the RF attenuator to its manual mode.
- Adjust the Vertical Scale Factor to 2 or 1 dB/div.
- Select a span of either 6 MHz or 10 MHz.
- The resolution bandwidth should be 30 or 10 kHz.

****IEEE Std 1631-2008, op.cit., p. 6.

- Adjust the center frequency and reference level controls to center the 8-VSB signal in the display.
- If available, heavily average the signal (e.g., 50 times). Otherwise, make sure the analyzer's sweep is in an auto mode and select the narrowest video filter available.
- Turn on a marker. Even with extreme averaging the trace will be *rough* with small variations caused by the noise-like signal. Adjust the marker's frequency to the highest or lowest point in the middle portion of the signal's head, carefully selecting a point where the marker is in the middle of the trace's fine grain roughness.
- Now select the Δ marker function. Move the Δ marker to the other extreme along the middle portion of the signal's head. Again, position the marker to be roughly in the middle of the trace's fine grain roughness. The magnitude (ignore the sign) of the delta marker readout now shows the peak signal's to peak unflatness.

If a Δ marker function is not available, place a first marker on the lowest amplitude spot in the above frequency range and another on the highest. Note the difference in amplitude between the markers.

See Section "S/N as a Function of Flatness" of this chapter to interpret the results of this measurement. When finished, do not forget to return the RF attenuator mode to automatic.

Shoulder Amplitude Ratio. One of the key indicators of an 8-VSB transmitter's performance is its head to shoulder amplitude ratio (i.e., the difference in amplitude when measured in dB). As will be discussed in the next section, it's important for the transmitter to create a clean signal to meet the FCC's emissions mask. Typically, to make this measurement, a ≥−25 dBm sample of the transmitter's antenna signal is required.

It's critical when making this measurement to avoid spectrum analyzer overload. Settings:

- Ensure that the RF attenuator mode is set to auto.
- Use the techniques given in Section "S/N as a Function of Flatness" above to determine the signal's power. Once known, adjust the reference level to that value.
- Adjust the span to 5 MHz.
- Use the analyzer's center frequency control to roughly center the pilot signal horizontally on screen.
- Select a 10 kHz resolution bandwidth. Use signal averaging if it's available (>50 times); otherwise use a very narrow (10 Hz) video filter.
- Select the detector mode recommended for power measurements.

As shown in Fig. 2.12.6, position a marker exactly at the center frequency of the channel. Then, using a Δ marker mode, adjust the Δ marker to a frequency of 3 MHz plus twice the resolution bandwidth setting.†††† The Δ marker then reads out the head-shoulder amplitude difference. Repeat, measuring the upper shoulder.

If a Δ marker mode is not available, place one marker in the middle of the signal's head and another on the signal's two RBWs outside the channel edge. Compute the amplitude difference between the two markers.

Figure 2.12.7 shows that the minimum head-shoulder ratio to meet the FCC's mask is 36.4 dB for the full service and stringent masks and 35.4 dB for the simple mask. Realistically, transmitters need to operate with a modest amount of margin, making a measured difference of ≥38 dB highly desirable.

Emissions Mask Compliance. DTV transmitters operated under the FCC's authority must operate within one of the three emissions masks shown in Fig. 2.12.7; which mask applies to a given transmitter must be determined from the station's licensing information.

---

††††This is greater than the FCC's specified value of one-half of the Resolution Bandwidth but results in a more reliable value as the skirt of the analyzer's Resolution Bandwidth filter picks up less energy from within the TV channel itself.

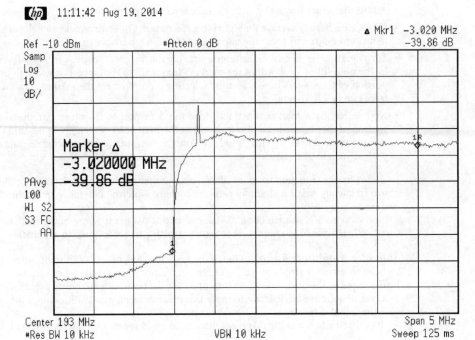

FIGURE 2.12.6 Shoulder amplitude measurement.

Measurement of a DTV transmitter's emissions is a difficult and exacting process, well beyond the scope of this chapter. The interested reader should refer to the IEEE's Recommended Practice which provides detailed instructions on how to perform and interpret the measurements as well as comprehensive theory and background information.‡‡‡‡

2.2.4 Adjacent Channel Power

An estimate of the transmitter's mask compliance can be made if the spectrum analyzer has the ability to automatically measure adjacent channel power ratio (ACPR), which is the ratio of the transmitter's power within its assigned channel vs. the *total* amount of power it is splattering into each adjacent channel. Because the transmitter's large adjacent channel emissions near the channel's edge dominate the numeric power integration, for an estimate it is unimportant that the analyzer cannot measure the very low power emissions further away.

To make the measurement, a sample of the transmitter's output signal with an amplitude of at least −25 dBm must be available. This sample must be free from any signals in the adjacent channels. The setting should be:

- Use the analyzer's center frequency control to center the signal within the window designated by the analyzer.
- Use the techniques given in Section "S/N as a Function of Flatness" above to determine the signal's power.
- Once known, adjust the reference level to this value.
- Select a 10 kHz resolution bandwidth. Use trace averaging if it is available (>20 values); otherwise use a very narrow (10 Hz) video filter bandwidth.

---

‡‡‡‡IEEE Standard 1631-2008, op.cit., Entire Document.

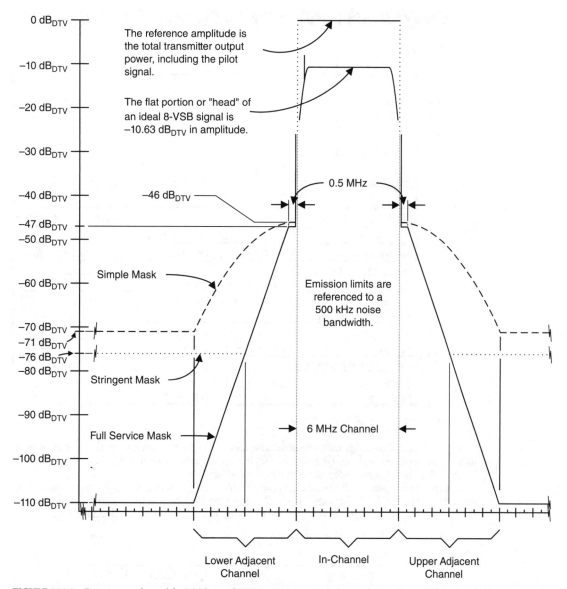

FIGURE 2.12.7 Emissions masks used for FCC licensed DTV transmitters. (© Taylor and Francis Group LLC Books, used by permission.)

- Select the detector mode recommended for power measurements.
- Enable the spectrum analyzer's ACPR measurement mode. Adjust this mode for
  - 6 MHz main channel integration bandwidth
  - 6 MHz channel to channel frequency spacing
  - 5.98 MHz adjacent channel integration bandwidth
- Enable averaging if available.

2.222 RF TRANSMISSION

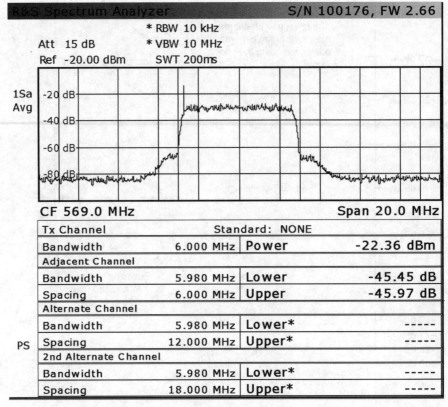

FIGURE 2.12.8 Adjacent ACPR measurement.

Figure 2.12.8 shows this measurement. The *unusual* (5.98 MHz) adjacent channel integration bandwidth specified is designed to avoid making measurements within one-half of a resolution bandwidth of the channel's edge as per the FCC's regulation.§§§§

The spectrum analyzer will measure the power in the two adjacent channels as well as the main channel and calculate the ACPR for each. Table 2.12.1 shows the amount of adjacent channel power (relative to the desired signal) allowed when each of the FCC's masks are integrated over the adjacent channel.

TABLE 2.12.1 Maximum ACPR Values as Determined by Applicable FCC Mask

FCC Mask	Maximum Adjacent Channel Power Ratio
Full Service	−44.6 dBc
Stringent	−44.6 dBc
Simple	−39.5 dBc

It must be emphasized that the values in Table 2.12.1 are for emissions exactly conforming to the mask; which, in the case of the sharp corners in the full service and stringent masks, is impossible. Realistically, a transmitter using a full service or stringent mask should have a measured ACPR less than these values.

---

§§§§47CFR§73.622(h).

2.3 DETERMINING SIGNAL QUALITY

ATSC 8-VSB, like all digital communications systems, sends a sequence of symbols between the transmitter and the receiver using carefully designed filtering that maximizes the data rate while minimizing inter-symbol interference and limiting emissions to a specific bandwidth. Measuring the transmitter's ability to create the correct signal requires special test equipment. The name of this type of equipment varies. Typical names are Measurement Test Set, Vector Analyzer, TV Analyzer, etc. These instruments decode the signal by using idealized digital filtering and then sampling the resulting data signal at the instant in time when it is intended to be correct.

At each symbol instant, 8-VSB transmitters send one of eight different steps in amplitude. When the channel is ideal and noiseless, the amplitude value measured at the receiver will be *exactly* one of those ideal values. When the received signal contains noise in addition to the signal, the recovered symbol values become uncertain. However, all sorts of other transmission errors can also cause similar uncertainty; uncertainty that, to the receiver, looks and acts just like noise. Thus, *noise* in DTV is *anything* that causes the amplitude of the recovered symbols to vary from their ideal value.

2.3.1 Signal Quality Measures

The concept that *noise* is anything that causes the signal to be nonideal allows the creation of very useful all-in-one digital signal quality measures or indexes. Three such measures are in common usage; all measure essentially the same thing and all tell essentially the same story. They are signal-to-noise (S/N or SNR), modulation error ratio (MER), and error vector magnitude (EVM).

2.3.2 Signal to Noise (S/N or SNR)

S/N or SNR was the first 8-VSB quality measure and is still used in ATSC documentation. Mathematically it is defined as

$$\text{SNR} = 10 \log \left(\frac{\sum_n I_n^2}{\sum_n \delta I_n^2} \right) \text{dB} \qquad (1)$$

where I_n = the amplitude of the *n*th *ideal* real-axis symbol value of a very long record of ideal real-axis values (with the pilot offset removed)

δI_n = the amount of real-axis *error* between the *n*th actual and ideal symbol value

The Σ sign indicates a summation of the values of I_n^2 and δI_n^2 over the entire record length. Since all the values are proportional to the signal's *voltage*, their square is thus proportional to their *power*. Thus, S/N is the ratio, expressed in dB, of the summation of the ideal *in-phase* signal power divided by the summation of the signal's *in-phase* noise power.

Since ATSC demodulation is performed only on the signal's real axis, S/N has the advantage of measuring exactly what the receiver's demodulator sees. Its disadvantage is that it ignores quadrature axis errors.

2.3.3 Modulation Error Ratio (MER)

MER is more often used. It is defined as

$$\text{MER} = 10 \log \left(\frac{\sum_n \left(I_n^2 + Q_n^2 \right)}{\sum_n \left(\delta I_n^2 + \delta Q_n^2 \right)} \right) \text{dB} \qquad (2)$$

where I_n = the nth ideal real-axis value of a very long record of ideal real-axis values with pilot offset removed
Q_n = the nth ideal imaginary-axis value of a very long record of ideal imaginary-axis values
δI_n = the amount of real-axis error of the nth value
δQ_n = the amount of imaginary-axis error of the nth value

As in S/N, all values are voltages that are squared to obtain values of power. MER is very similar to S/N except that the signals and errors from the quadrature axis are included in the calculations. Normally, S/N and MER are nearly identical.

2.3.4 Error Vector Magnitude (EVM)

EVM is slightly different. It is defined as

$$\mathrm{EVM} = \sqrt{\frac{\frac{1}{N}\sum_{n=1}^{N}\left(\delta I_n^2 + \delta Q_n^2\right)}{S_{\max}^2}} \times 100\% \tag{3}$$

where δI_n = the amount of error that the nth real-axis value exhibited
δQ_n = the amount of error that the nth imaginary-axis value exhibited
$S_{\max}$ = the maximum value or state along the real axis (i.e., 7 since the pilot signal offset is removed before the EVM calculation is made)

EVM is thus similar in concept to S/N and MER but uses a different value to scale the error power and its results are typically expressed in terms of percent. EVM is sometimes expressed in terms of dB:

$$\mathrm{EVM}_{\mathrm{dB}} = 10 \times \log\left[\frac{\frac{1}{N}\sum_{n=1}^{N}\left(\delta I_n^2 + \delta Q_n^2\right)}{S_{\max}^2}\right] \mathrm{dB} \tag{4}$$

Note that EVM expressed in dB is always a negative value.
MER (and S/N) can be approximately equated to $\mathrm{EVM}_{\mathrm{dB}}$:

$$\mathrm{EVM}_{\mathrm{dB}} \approx -(\mathrm{MER} + 3.68)\mathrm{dB, and}$$
$$\mathrm{MER} \approx -(3.68 + \mathrm{EVM}_{\mathrm{dB}})\mathrm{dB}$$

(remembering that EVM in dB is a negative value).

2.3.5 Relationship of Response Errors to S/N

There are many parameters that, when incorrect, will cause DTV *noise*. Of these there are perhaps four which are within the control of the transmitter's operator.

- Deviation of the transmitter's frequency response from the ideal raised root-cosine shape.[¶¶¶¶]
- Deviation of the transmitter's phase response or envelope delay from the ideal flat response.

[¶¶¶¶]ATSC: "ATSC Digital Television Standard, Parts 1–6, 2007", Part 2, Figure 6.4, p. 12, Advanced Television Systems Committee, Washington, D.C.

- Transmitter amplitude and phase errors as a function of the transmitter's instantaneous output amplitude, commonly called transmitter IM which is due to transmitter nonlinearities.
- Reflected signal noise caused by reflections within the transmitter's antenna system.

The relationship between each of these parameters and S/N is only approximately known. However, even fuzzy data is often useful when trying to determine the causes of poor performance. The data given here was obtained by using laboratory equipment to create and measure a series of 8-VSB signals, each with only one type of error.***** The measurements were made in units of S/N but MER values should be very similar. The channel errors measured were mathematical constructs that may not exactly represent reality but do give guidance as to the sensitivity of each type of error in lowering S/N.

S/N as a Function of Flatness. The approximate relationship between flatness and S/N is given in Table 2.12.2. Δ dB, p–p refers to the maximum peak to peak variation in flatness across the central 5.38 MHz of the 6 MHz TV channel. Two shapes of flatness errors were simulated. The first was a linear tilt across the channel; the second was a crowned response, peaked at mid-channel. Both gave essentially the same result.

TABLE 2.12.2 Peak to Peak Flatness Error vs. S/N

ΔdB, p–p	0.4 dB	0.6 dB	0.8 dB	1.0 dB	1.2 dB	1.4 dB
S/N	36 dB	33 dB	31 dB	29 dB	28 dB	27 dB

S/N as a Function of Envelope Delay. The shape of the envelope delay curve strongly affects the value of the resulting S/N. This can be seen in Table 2.12.3. Two rows of data are shown. The simulated error in the Δt Ramp row is the peak to peak amount of a linear ramp of envelope delay across the central 5.38 MHz of the 6 MHz width of the channel. The simulated error in the Δt parabola row is the peak to peak amount of a parabola of envelope delay with a maxim at the channel edges and a minimum at mid-channel.

TABLE 2.12.3 Envelope Delay vs. S/N

Δt Ramp p–p	4 ns	7 ns	10 ns	15 ns	22 ns
Δt Parabola p–p	10 ns	16 ns	20 ns	27 ns	40 ns
S/N dB	35 dB	32 dB	30 dB	28 dB	25 dB

Transmitter IM. Transmitter amplitude and phase errors cause intermodulation distortion (IM). Because of the broadband nature of the 8-VSB signal, this IM distortion creates adjacent channel splatter. Typically this splatter is said to be caused by the transmitter's third-order distortion. However, in a transmitter's class B power amplifier, *any* nonlinearity will create adjacent channel splatter.

Figure 2.12.9 shows how to make an estimate of the transmitter's S/N performance caused by IM based on its head to shoulder ratio. The solid line shows the transmitter's actual signal including adjacent channel splatter as shaped by the transmitter's channel filter. The dotted line shows its unfiltered IM signal at the power amplifier's output.

The spectral shape of the unfiltered adjacent channel splatter or IM is well known. What is not well known is that the IM signal also extends *under* the transmitted signal (i.e., within the channel) and is thus a source of noise to the desired signal. Lab measurements and simulations indicate that the S/N caused by the transmitter IM can be estimated by subtracting 1 or 2 dB from the transmitter's head to

---

*****Sgrignoli, Gary and Gumm, Linley: *NAB Engineering Handbook*, 10th Edition, chapter 8.4, "DTV Transmitter Measurements", Focal Press, © 2007, pp. 1913–1921. Also: Linley Gumm, *Signal to Noise Relationships in 8-VSB*, Tektronix Technical Brief, http://www.tek.com/document/technical-brief/signal-noise-relationships-8-vsb

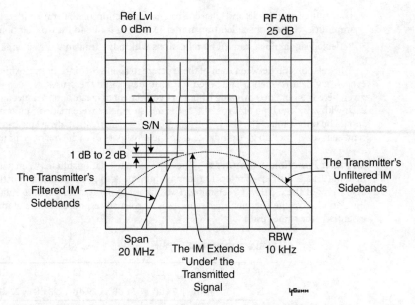

FIGURE 2.12.9 Estimating S/N caused by transmitter nonlinearities. (© Linley Gumm, used by permission.)

shoulder ratio. That is, if the head to shoulder ratio is 38 dB, the S/N caused by transmitter IM will be about 36 dB.

S/N as a Function of Antenna VSWR. Transmitters are connected to their antenna via a transmission line. Thus, the transmitter creates a signal which arrives at the antenna after it's sent. A small portion of that signal is reflected from the antenna and travels back down the transmission line, arriving back at the transmitter. Because the transmitter does not provide a particularly good reverse match to the transmission line, that small reflected signal is mostly reflected once again and travels back toward the antenna along with the rest of the transmitter's current output signal.

A new 8-VSB symbol is sent every 93 ns. Because 8-VSB is very noise-like, if the round trip delay to the antenna is longer than 93 ns (i.e., a transmission line longer than 45 feet), the rereflected signal will be completely random with respect to the current output signal and is thus a source of *noise* to it.

The effect is just large enough to require some caution. To estimate it, the signal first experiences a loss of perhaps 1 dB in the filtering system and transmission line on its way to the antenna. If the antenna has a VSWR of 1.10:1, the reflected signal will be about 26 dB below the arriving signal. On its way back to the transmitter, the reflected signal experiences another 1 dB loss. And, lastly, the transmitter's return loss is probably between 3 and 10 dB. Thus, the rereflected signal is 1 + 26 + 1 dB + (3 to 10 dB) = 31 to 38 dB below the outgoing signal, which is the S/N from this effect. All is well if the antenna VSWR is low. Many transmitters incorporate built-in correction for this problem.

2.3.6 Transmitter S/N Budget

The ATSC recommends that DTV transmitters be operated with at least a 27 dB S/N. At this S/N, the noise contributed by the transmitter itself will cause the receivers to be about 0.25 dB less sensitive.[†††††]

[†††††]ATSC: "Recommended Practice: Transmission Measurement and Compliance for Digital Television," Doc. A/64B, Section 5.1.2, p. 10, Advanced Television Systems Committee, Washington, D.C., 26 May 2008.

Since DTV gives ideal pictures any time the signal is above threshold, viewers tend to use antennas that provide signals just nicely above threshold. Therefore, this 0.25 dB loss of threshold may affect users over the entire viewing area, not just in the fringe areas.

Experience has shown that to a first order, the effect of each of the S/N parameters is independent of the others. That is, the *noise* power contributed by each effect more or less acts independently of the others and can thus be added together to find their cumulative effect.

The effective *noise* power of a given S/N cause can be determined by taking the antilog of the negative of the S/N value. That is, the noise power caused by a given cause is

$$\text{Noise Power} = 10^{\frac{-S/N}{10}} \quad (5)$$

with respect to the total signal power. To determine the final S/N resulting from a number of S/N causes, find the noise power for each source, add them together and then take the log of the result:

$$\text{Output S/N} = 10\log\left(10^{\frac{-S/N_1}{10}} + 10^{\frac{-S/N_2}{10}} + 10^{\frac{-S/N_3}{10}} + \cdots\right) \quad (6)$$

where $S/N_1, S/N_2, S/N_3$, etc. = the various S/N values caused by the effects

Output S/N = the value measured at the transmitter's output

As an example, Table 2.12.4 assumes that a transmitter has a variety of S/N sources.

TABLE 2.12.4 Example of Contributions to an S/N Budget for a Typical Transmitter

Item	Performance	S/N
Flatness	0.6 dB p–p	33 dB
Envelope delay	16 ns p–p parabolic	32 dB
Amplitude of shoulder at channel edge	38 dB	36 dB
Antenna VSWR	1.10:1	35 dB
Assumed everything else (phase noise, DSP noise, etc.)		35 dB
Output S/N		**27 dB**

Since the output S/N is what is important and since it is made up of a number of line items, the exact value of any given item is not too important. When a transmitter is operating with a large output S/N, the operator should note the performance of each S/N contributor to allow easier troubleshooting when S/N falls.

2.4 SIGNAL QUALITY MEASUREMENTS

Many transmitters have built-in measurements for many vector-based measurements such as MER, using them to automatically correct their output signal. When performing initial acceptance tests or when troubleshooting when a transmitter performs below its capabilities, it may sometimes be necessary to make independent measurements.

The items to be measured include:

- The various parameters that determine S/N (and/or MER and/or EVM)
- An accurate determination of the resulting output S/N, etc.

```
R&S ETL Digital Overview                      S/N 100176, FW 2.66
Ch: ---  RF 183.000000 MHz  ATSC/ATSC Mobile DTV (RF Layer)

  * Att   0 dB
    ExpLvl -32.50 dBm
```

Level   -31.3 dBm

	ATSC Parameters			ATSC Main and M/H Data detected		
	Pass	Limit <	Results	< Limit	Unit
	Level	-60.0	-31.3	10.0	dBm
	Constellation		8VSB / Normal		
	MER (rms)	24.0	36.8	-----	dB
	MER (peak)	10.0	15.1	-----	dB
	EVM (rms)	-----	0.95	4.40	%
	EVM (peak)	-----	11.50	22.00	%
OLim	BER before RS		0.0e-9(27%/1e10)	2.0e-4	
	BER after RS		0.0e-8(16%/1e7)	1.0e-10	
	Packet Error Ratio		0.0e-6(16%/1e7)	1.0e-8	
	Packet Errors		0	1	/s
PS	Carrier Freq Offset	-30000.0	-20.1	30000.0	Hz
	Symbol Rate Offset	-10000.0	-1.1	10000.0	Symb/s
	MPEG Ts Bitrate		19.392656		MBit/s

Lvl -31.3dBm | BER 0.0e-9 | MER 36.8dB   DEMOD   MPEG

FIGURE 2.12.10 Typical table of results from a TV analyzer.

- Symbol rate frequency
- Pilot amplitude

The measurements of flatness and head to shoulder ratio outlined in Section 2.2 of this chapter are part of the necessary tests. Their results can be utilized with the information in Section 2.3.5 to determine their effect on S/N (MER).

Specialized test equipment such as a vector analyzer, a TV analyzer, or a measurement test set is necessary to measure the resulting S/N (and/or MER and/or EVM). While expensive, this type of equipment provides an accurate and independent measurement of the transmitter's performance. Furthermore, this type of equipment may also provide direct measurement of other key operating parameters such as bit error rate, symbol clock frequency, and pilot amplitude.

Measurement Tip: Vector measurement systems often have the ability to make measurements with or without a data signal equalizer inserted in the signal path. When an equalizer is used, it corrects any linear distortion such as amplitude flatness and/or envelope delay problems that may be present. However, it does not correct for S/N loss caused by the transmitter's nonlinearities.

The fact that the equalizer corrects one type of distortions and not the other is a powerful troubleshooting tool. If a transmitter has a low S/N (MER) value with the equalizer disabled but exhibits a distinct improvement when it is enabled, it is a sign that flatness and/or envelope delay issues are the major cause of the low S/N (MER) value. If the S/N (MER) is roughly the same with and without the equalizer being enabled, it is a sign that the major cause of the signal's low S/N (MER) value is caused by transmitter nonlinearities.

2.4.1 Measurements with a Vector Analyzer or TV Analyzer

Because of large variations between products, it is impossible to give procedures to use these specialized instruments. To show the possibilities presented by them, typical data is presented here.

Figure 2.12.10 is a table of results presented by a typical TV analyzer, showing many of the measures necessary for a healthy DTV transmitter. The first line shows the measured signal power. Below that the reader will find an MER measurement taken over a long time period (rms) as well as the peak value of MER experienced during the measurement followed by similar EVM data. Reading down, there are values for the bit error rate before and after the Reed–Solomon error correction used 8-VSB demodulation as well as information about MPEG data packet errors. Near the bottom of the table, the transmitter's carrier frequency error is reported as well as its symbol rate error. The final value is the calculated value of the MPEG bit rate.

Another table presented by this unit is given in Fig. 2.12.11. New data as well as a repetition of some of the previous results is presented. At the top, the signal's amplitude is repeated. The pilot value in the second row gives the measured value of the data field offset that creates the pilot signal (nominally

R&S ETL Modulation Errors   S/N 100176, FW 2.66
Ch: --- RF 183.000000 MHz ATSC/ATSC Mobile DTV (RF Layer)

* Att 0 dB
ExpLvl -32.50 dBm

	Pass	Limit <	Results	< Limit	Unit
	Level	-60.0	-31.3	10.0	dBm
	Pilot value	1.20	1.29	1.30	
	Data Signal/Pilot	11.0	11.0	11.6	dB
OLim	Pilot Amplitude Error	-0.3	0.3	0.3	dB
	MER (rms)	24.0	37.0	-----	dB
	MER (peak)	10.0	16.4	-----	dB
	EVM (rms)	-----	0.93	4.40	%
PS	EVM (peak)	-----	9.93	22.00	%
	Signal/Noise Ratio (Low Q)	24.0	37.8		dB

Lvl -31.3dBm | BER 0.0e-8 | MER 37.0dB   DEMOD   MPEG

FIGURE 2.12.11 Typical table of channel errors from a TV analyzer.

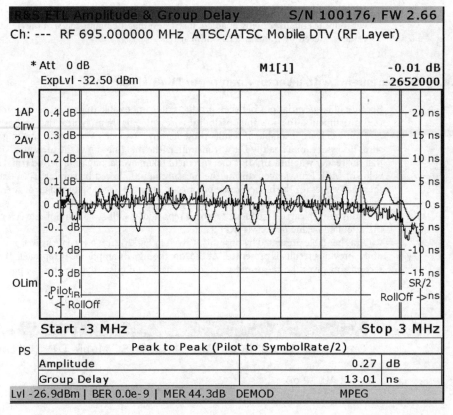

FIGURE 2.12.12 Detailed frequency response and flatness from a TV analyzer.

1.25 units).[‡‡‡‡‡] The next value is the ratio of the power of the DTV data signal power to the pilot signal power (nominally 11.3 dB as per the ATSC).[§§§§§] The last pilot value is its amplitude error. The rest of the data in this table is a repeat of the previous table except the bottom line which gives the S/N value for this signal, which (as expected) agrees quite closely with the MER value given in Fig. 2.12.10.

One of the many things that a TV analyzer type product can do that cannot be performed by general purpose instruments are a detailed measurement (based on digital signal processing) of the transmitters performance. An example is the frequency response and envelope delay measurement shown in Fig. 2.12.12. Lacking color, the fuzzy line across the middle of the display is this signal's flatness *error*. The line swinging back and forth across the center two divisions is the envelope delay. Because envelope delay is the *difference* in phase with frequency, its measurement tends to be noisy. In general, one should ignore the rapid variations in the envelope delay curve and concentrate on the broad shape of the curve.

REFERENCE

[1] IEEE Std 1631-2008: *Recommended Practice for the Measurement of 8-VSB Digital Television Transmitter Mask Compliance for the USA*, published by IEEE, 9 Park Ave., New York, NY 10016-5997.

[‡‡‡‡‡]ATSC A/53 Part 2, op.cit., Section 6.9.1, p. 40.
[§§§§§]Ibid.

CHAPTER 2.13
HYBRID MICROWAVE IP AND CELLULAR DATA FOR NEWSGATHERING

Nuraj Lal Pradhan and John Wood
VISLINK, Inc., North Billerica, MA

2.1 INTRODUCTION

The evolution of microwave, satellite, and cellular transmission technologies has enabled the convergence of microwave and cellular data connection technology. This hybrid IP newsgathering technique is capable of simultaneously combining microwave bandwidth with multiple cellular data connections for *electronic newsgathering* (ENG). Each transmission technology has its own unique advantages for newsgathering. Combining these technologies provides a versatile alternative for broadcasters. Live venues previously inaccessible to news crews without timely preplanning and setup can now be easily reached. This hybrid newsgathering technique meets the ever-present desire for greater resiliency and flexibility.

In the past 20 years, there have been a number of key technologies that have shaped the evolution for high definition (HD) contribution newsgathering, which has impacted today's products and technology migration to hybrid microwave IP solutions; notably:

- Migration from analog frequency modulation (FM) to digital Coded Orthogonal Frequency Division Multiplexing (COFDM) modulation
- Migration of 4G Long-Term Evolution (LTE) networks and the capability to aggregate multiple 4G cellular channels
- Migration from Ku band satellite applications to Ka band satellite applications

These technology trends have combined to change the traditional ENG service as illustrated in Fig. 2.13.1.

2.2 MIGRATION FROM ANALOG FM MODULATION TO DIGITAL MODULATION

For many years, FM modulation was the main transmission methodology for microwave newsgathering within the 2 GHz Broadcast Auxiliary Services (BAS) spectrum. FM microwave links were used to support standard definition (SD) video contribution, where the composite video 1 V_{pp} signal was

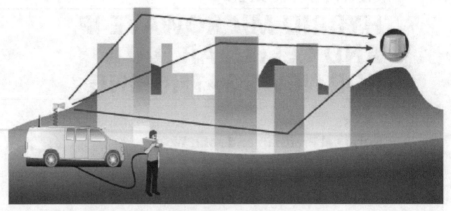

FIGURE 2.13.1 Traditional ENG application drawing.

directly modulated onto a 2 GHz phase-locked oscillator and amplified for microwave transmission. Carson's bandwidth rule is expressed by the relation

$$\text{CBR} = 2(\Delta f + f_m), \qquad (1)$$

where CBR = the bandwidth requirement
Δf = the peak frequency deviation
f_m = the highest frequency in the modulating signal

Carson's bandwidth was always at the tip of the tongue for every broadcast maintenance RF engineer at the time. For analog FM microwave, a 4 MHz peak deviation was used with a 4.2 MHz highest modulated frequency, which represented the video bandwidth [1].

This was the original driving factor to support 17 MHz channels within the BAS spectrum (Fig. 2.13.2). While very efficient, it had its drawbacks when the demand and transition for the

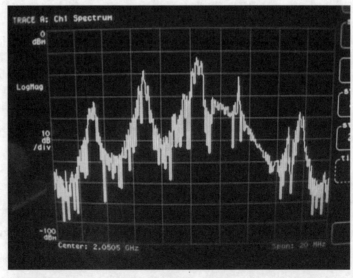

FIGURE 2.13.2 17 MHz FM spectrum plot.

carriage of HD content was required and spectrum reduction was demanded. The main drawback was its inability to cope with multipath and overall spectrum efficiency, which led to the utilization of digital modulation schemes.

2.2.1 The Benefits of COFDM

In the late 1990s, the need to utilize a digital modulation system for ENG was recognized. It was necessary to overcome many of the deficiencies of multipath and improve the overall *quality of service* (QoS) for broadcasters. The use of COFDM for mobile and ENG grew out of the Digital Video Broadcasting-Terrestrial (DVB-T) standards, which was adopted internationally for over-the-air digital broadcast. At the time, the two applications were quite different, but the common thread was that COFDM techniques were used to overcome multipath, non-line-of-sight, and signal quality limitations in both situations. COFDM systems were tested worldwide and in a collaboration of several manufacturers of Moving Picture Experts Group (MPEG) II and COFDM equipment providers. The results were generally very good with regard to link performance, but the operational system had its limitations. In each city that was visited, the testing showed that in comparing legacy analogue FM versus COFDM DVB-T microwave shots, COFDM shots won out 99 percent of the time. There were a number of key benefits in using COFDM for ENG applications that the testing showed:

- The requirement of not having to maintain line-of-sight (LOS)
- Ability to overcome severe multipath dropouts
- Improved video and audio quality utilizing video compression
- Noise free video over the usable fade margin
- Higher available user system gain
- Lower occupied bandwidth (8 MHz COFDM) versus FM
- Flexibility of trading MPEG throughput versus microwave path requirements

A spectrum analyzer display of an 8 MHz COFDM emission is shown in Fig. 2.13.3.

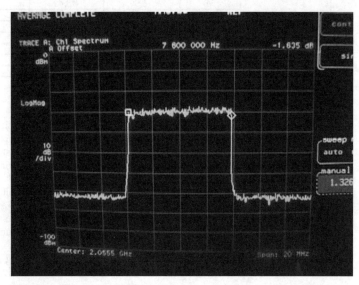

FIGURE 2.13.3   8 MHz COFDM waveform.

At the time, the operational transmitter equipment required the interfacing of a number of third party boxes that were bulky and quite expensive. The lack of one common control interface between the radio, the MPEG encoder, and the COFDM modulator made the user interface cumbersome and hard to manage. The OEM MPEG and OFDM modulators developed for fixed terrestrial applications were quite different from those needed to meet ENG requirements. The results called for a new, compact, lightweight, rugged, and reliable product with a consolidated design approach that integrated the MPEG encoder and COFDM modulator based on the DVB-T standards combined with the radio electronics, under one common control user interface. This single, integrated product is the standard in ENG today.

2.3 COFDM OVERVIEW

The advantages of COFDM technology are apparent in its ability to offer error-free transmission under severe multipath conditions, as well as its ability to occupy less overall bandwidth than its previous analog FM counterpart. Today, COFDM technology utilization for ENG is a mature technology used in the majority of the news shots. A typical COFDM waveform occupies 7.61 MHz of bandwidth at its 1 dB bandwidth points. Its digital implementation is based on the European Telecommunication Standards Institute (ETSI) standard EN300-744 for Digital Video Broadcasting (DVB) framing, channel coding, and modulation architecture. The uniqueness of the standard allows the user to customize data throughput as a function of three main parameters: forward error correction, guard interval, or delay spread and modulation type. Table 17 of the ETSI standard [2] shows how a user can increase or decrease the bit rate as a function of his microwave link requirements (see Fig. 2.13.4).

Modulation	Code rate	Guard interval			
		1/4	1/8	1/16	1/32
QPSK	1/2	4,98	5,53	5,58	6,03
	2/3	6,64	7,37	7,81	8,04
	3/4	7,46	8,29	8,78	9,05
	5/6	8,29	9,22	9,76	10,05
	7/8	8,71	9,68	10,25	10,56
16 QAM	1/2	9,95	11,06	11,71	12,06
	2/3	13,27	14,75	15,61	16,09
	3/4	14,93	16,59	17,56	18,10
	5/6	16,59	18,43	19,52	20,11
	7/8	17,42	19,35	20,49	21,11
64 QAM	1/2	14,93	16,59	17,56	18,10
	2/3	19,91	22,12	23,42	24,13
	3/4	22,39	24,88	26,35	27,14
	5/6	24,88	27,65	29,27	30,16
	7/8	26,13	29,03	30,74	31,67

FIGURE 2.13.4 8 MHz COFDM bit rate table.

Like any other communications application, as the modulation density or bits per hertz efficiency increases, so does the *carrier-to-noise* (C/N) requirement for its receiver. Meeting the C/N requirement is a major challenge toward achieving a reliable transmission system. Figure 2.13.5 compares three different modulation techniques: Quadrature Phase Shift Keying (QPSK), 16 Quadrature Amplitude Modulation (QAM), and 64 QAM—used within a COFDM domain. As expected, QPSK affords the minimum C/N requirement while also offering the lowest bit rate with a C/N of approximately 6 dB and a bit rate of 5.5 Mbits/s.

Also, as with many other digital communications systems, there are ways to improve the overall system C/N requirement for a given microwave path. The main technique that is employed is antenna diversity. Currently, there are a number of diversity receive techniques on the market. They include packet-based switching implementations and Maximal Ratio Combining (MAX RC) techniques, each of which has their pros and cons. A packet-based implementation, while offering errorless switching,

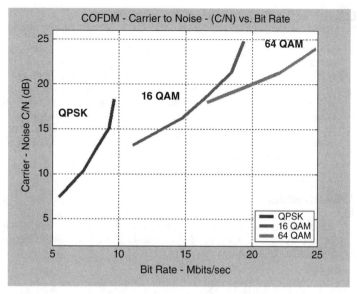

FIGURE 2.13.5 C/N versus COFDM bit rate.

yields minimal improvement in C/N. MAX RC techniques have the potential to offer a user up to 5 dB improvement in C/N requirements with two antennas and a subsequent 8 dB improvement with four antennas. The improvement in C/N offers users the ability to increase overall data throughput for the application. The increase in data throughput allows for a higher quality signal transmission over the link, or the potential for a higher speed downstream IP data transfer. Figure 2.13.6 shows the comparative increase in throughput for a COFDM application with and without two-antenna diversity applied to the microwave communications link.

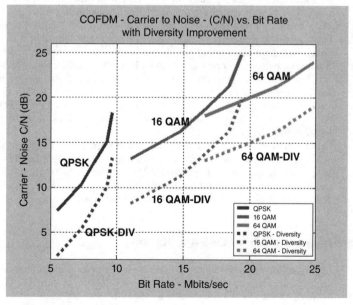

FIGURE 2.13.6 C/N diversity improvement.

2.4 CELLULAR NEWS GATHERING (CNG)

U.S. broadcaster expectations for QoS pertaining to content news gathering has broadly changed over the past several decades as different methods for collecting and transmitting fast breaking news events have evolved. CNG is a great new tool to supplement ENG and *satellite news gathering* (SNG) operations. However, many broadcasters have learned the hard way that the cellular solutions do not afford the same reliable performance that they have come to expect from licensed ENG and SNG systems. CNG transmissions are vulnerable to the unpredictability of network availability, signal fluctuations, overcrowding (especially at large event), and limited network access due to network carrier domain blackout. To ensure consistent, reliable coverage regardless of public infrastructure bandwidth, a hybrid technology has been developed.

2.4.1 4G LTE Technology

Since Verizon launched its major network services in December 2010 in the United States, 4G cellular LTE technology has expanded the capabilities for newsgathering operations. A summary of U.S. commercial LTE bands is shown in Table 2.13.1.

TABLE 2.13.1 U.S. Commercial LTE Bands

Carrier	Launch Date	LTE Band	Duplex Mode	Uplink (UL)	Downlink (DL)
Verizon	December 2010	13	FDD	777–787 MHz	746–756 MHz
AT&T	September 2011	17	FDD	704–716 MHz	734–746 MHz
T-Mobile	May 2013	4	FDD	1710–1755 MHz	2110–2155 MHz

Although most of the daily usage of cellular technologies is downlink-centric, CNG newsgathering operations are entirely uplink-centric, transmitting from the field to the studio. The LTE 3GPP standard promotes a total 50 Mbps downlink (DL) capability versus a 25 Mbps uplink (UL) capability for *user equipment* (UE) category 2, and 100 Mbps DL with a 50 Mbps UL for category 3, which includes *multiple input multiple output* (MIMO) antenna support. The UL/DL data throughput numbers are dependent on many factors, including bandwidth utilization, cell tower congestion, UE equipment, network coverage, and mobile network backhaul, and are more specific to the total network availability of a cell tower. Results using speedtest.net [3] show the variability of a typical LTE network, where the UL and DL throughput numbers can vary significantly (see Fig. 2.13.7). In this particular test, UE category 2 was used with a 5 MHz LTE bandwidth; there is a significant difference between the UL and DL capabilities.

Although a common LTE UL data rate from a UE can vary anywhere from 1 Mbps to up to 7 Mbps per modem dependent on geographic location, the data rate is not persistent enough to support continuous real-time video applications for mobility. Thus, it is necessary to utilize multiple modems simultaneously and aggregate them together to produce a higher continuous persistent data throughput stream for real-time HD video distribution.

2.4.2 User Equipment (UE) for CNG Applications

Typical cellular newsgathering user equipment comes in multiple styles of packages, including portable handheld, mast-mount, and camera back devices, as shown in Fig. 2.13.8.

Each of these UE configurations utilizes up to six independent 4G modems that are used to improve the persistence of the cellular network connectivity. As the user moves between cell towers,

FIGURE 2.13.7 Typical LTE speed test.

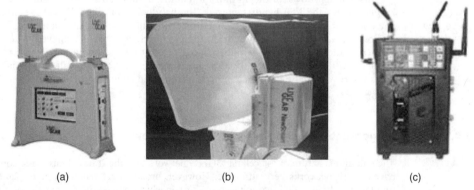

FIGURE 2.13.8 CNG equipment: (*a*) portable/handheld cellular transmitter, (*b*) cellular transmitter for mobile ENG van, (*c*) transmitter camera back. (*Courtesy: VISLINK.*)

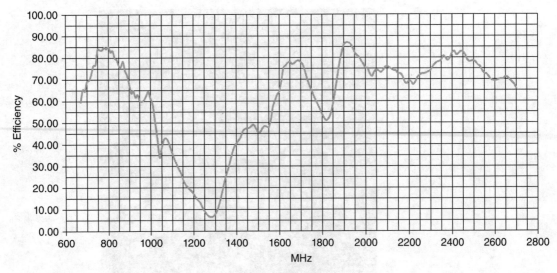

FIGURE 2.13.9 Antenna efficiency (in percent) calculated from spherically integrated gain.

each modem connects simultaneously to each of the networks. The UE equipment provides the HD interface from the camera and utilizes H.264 adaptive bit rate (ABR) encoding for compression and distribution over the six independent modems. A properly designed UE device with multiple modems takes into consideration the placement of the RF antennas for proper antenna de-correlation and isolation between modems. Band 13 and band 17, 4G LTE modems operating within the 700 MHz band require a quarter wavelength spacing of approximately 4 inches:

$$\tau = \left(\frac{c}{f}\right)/4. \qquad (2)$$

This spacing allows for the optimization of the simultaneous antenna efficiency of greater than 70 percent for the LTE bands required. The handheld portable UE package utilizes specialized antennas optimized in clusters to maximize the antenna efficiency and cellular performance. Figure 2.13.9 illustrates the practical antenna efficiency that is achieved. The higher antenna efficiency allows for overall stronger signal reception versus commercial cell phones and USB modems, which achieve roughly 35 percent antenna efficiency, half the signal strength. This helps improve signal persistence in highly congested cellular environments and in the network fringe areas.

2.4.3 Cellular Aggregation Technology and CNG Workflow

The ubiquitous nature of the cellular 3G/4G network and the Internet has made contribution video over LTE IP networks very attractive. However, broadcasters require a high-quality, smooth and uninterrupted video delivery solution. As previously discussed, CNG transmission has drawbacks, including the excessive delay and burst nature of packet losses within the cellular network and bandwidth availability, both of which can have a major impact on the performance of real-time video streaming application. Latency and loss rate can change in a millisecond, even in a stationary location, which may result in video breaks and jitter.

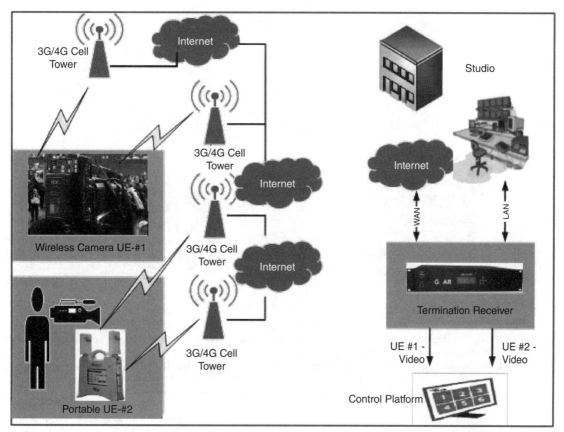

FIGURE 2.13.10 IP CNG workflow.

The solution is to aggregate multiple cellular links, as shown in Fig. 2.13.10, by effectively distributing video streams between the heterogeneous links to deliver sustained, smooth high-quality video streams. The cellular aggregation technology and 4G network capabilities enables IP workflow from the field, where multiple UE newsgathering units can provide simultaneous live streaming video directly into the studio through one common receiver platform can support a full IP workflow with a *wide area network* (WAN) IP input and a *local area network* (LAN) output into a studios ENG ingest automation system.

A proprietary aggregation algorithm effectively balances loads between the available six cell connections, as shown in Fig. 2.13.11. From an IP workflow standpoint, it provides a single virtual IP connection to achieve a common reliable network connection. The algorithm is designed to find a real-time, optimal equilibrium distribution between the multiple connections based on the appropriate link quality metrics, while also providing a single virtual connection with improved bandwidth, persistence and reliability, minimizing the risk associated with network interruptions and unpredictable delay over individual connections.

Another key part of a CNG system is the ability to constantly adjust the video bitrate to adapt to the varying network conditions. As part of the H.264 compression utilization of the video signal, ABR is used as a constant feedback mechanism to supply smooth video performance that is necessary at the studio for rendering accurate video. Controlling the rate of the video stream based on the available network capacity helps to avoid congestion and loss of packets in the network.

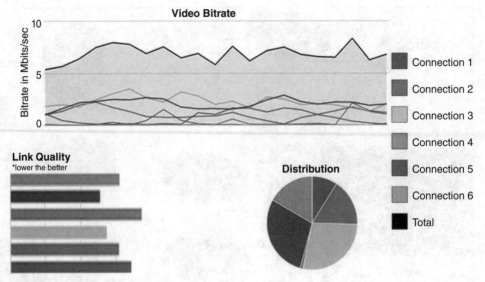

FIGURE 2.13.11 Load balancing of cell modems.

Furthermore, error control is usually needed to cope with packet loss and delay in these dynamic wireless networks. Error control strategies include *forward error correction* (FEC) and retransmission schemes. FEC adds additional information on the original video stream so that it can be constructed correctly at the receiver in the presence of packet loss. Channel-coded FEC applies a block code to a set of packets, *say k*, to generate an *n*-packet block (where $n > k$). The receiver only needs to recover k packet out of the n-packet block to correctly play out the video stream. Retransmission is one of the main methods of recovering lost packets. However, it is not always suitable in real-time video streaming as the retransmitted packet misses its play-out time. So, retransmission scheme can only be used for error control when the one-way trip time is shorter than the maximum delay.

Different network protocols have been adopted to provide a complete streaming service between the end users. The widely used TCP protocol might not be efficient, and its key feature, the retransmission of erroneous or lost packet, might not be applicable. On the other hand, UDP protocol, usually preferred for real-time application, is not reliable and lacks flow control capabilities.

The key issues faced with broadcasting video stream over the cellular wireless network and possible solutions toward achieving low latency broadcast with a certain degree of QoS are listed in [4].

2.5 MIGRATION FROM KU TO KA BAND IP SATELLITE SYSTEMS

Ka band utilization for IP-over-satellite systems is gaining momentum within broadcast for HD contribution applications. Originally introduced to provide high speed Internet connectivity, the technology is finding a place for mobile news van SNG applications for broadcasters. There are a number of key points that have led to the utilization of Ka band satellite as a hybrid technology for newsgathering, including:

1. Infrastructure build out and coverage, similar to the expansion of 4G LTE networks, which provides overall higher bandwidth capabilities. Via Sat-1, launched in 2011, offers a capacity of 140 GB/s over the North American continent.

2. Higher capacity drives down the cost to the users, thus allowing for more bandwidth to support streaming video applications.
3. IP modems with adaptive modulation technologies make IP workflows possible. These modems adapt to their link budget requirements and can dynamically allocate their IP Ethernet traffic as required for layer 2/3 network applications.
4. The higher frequency Ka band provides less congestion and interference, allowing for focused spot beams to be used. Ku band satellites are spaced two degrees apart, while Ka band satellites are spaced further apart, which allows for more frequency reuse capabilities.
5. Higher frequency allows for an overall smaller terminal footprint which equates to portability and capability of man-portable or vehicle-mounted terminals.
6. Higher power spot beams support the migration to small terminal dishes with lower power amplifiers, providing the capability to airline to check these carry cases.
7. IP network distribution format allows news reporters to connect directly into the studio's IT network workflow.

All of these points support the migration to an IP workflow for hybrid applications.

For a typical SNG application, a 0.75 to 1.0 meter mobile dish is used with a 3 to 4 W BUC that enables an uplink speed of up to 10 Mbps and a downlink speed of 20 Mbps, just enough to support HD contribution capabilities from a mobile news van unit to the studio.

2.6 MIGRATION TO HYBRID MICROWAVE SOLUTIONS

The migration toward a hybrid microwave IP solutions workflow for newsgathering allows broadcasters to deliver news more efficiently. Hybrid solutions utilize each technology's strength and combine them to offer the ultimate newsgathering tool for ubiquitous and reliable coverage. Hybrid solutions:

- Provide the capability to use licensed band microwave, thus leveraging the BAS frequency band, which provides the exclusivity, reliability, and performance characteristics that broadcasters have grown accustomed to.
- Utilize 4G LTE networks to provide ubiquitous coverage and IP workflow for integration into broadcaster's IT management structures.
- Utilize Ka band satellites to provide flexibility, network connectivity, IP workflow, and resiliency when other networks are not available.

2.6.1 Hybrid Solution and Technology Tradeoffs

As shown in Tables 2.13.2 and 2.13.3, each solution uses different technologies and offers unique advantages for each of its implementations. For example, licensed band microwave uses dedicated channels that are generally interference-free with highly reliable connections; however, it also requires a high degree of planning and infrastructure implementation requirements. Ka band satellite and 4G cellular solutions both require network infrastructure with layer 2/3 networking capabilities for direct integration into an IP workflow, but they use vastly different core transmission technologies (DVB-S2 and LTE, respectively), each of which has benefits and weaknesses.

2.7 HYBRID AGGREGATION

Utilizing channel aggregation techniques and combining all three IP networks—microwave, cellular, and Ka band satellite—allows broadcasters to take advantage of the strengths of these network types. This approach gives broadcasters the greatest QoS and ubiquitous coverage capabilities. Utilizing

TABLE 2.13.2 Hybrid Solution Tradeoffs

Hybrid Solution	Strengths	Weaknesses	Optimum Bit Rates
Licensed band microwave	(1) Dedicated channels and spectrum utilization (2) Highly reliable connections (3) Interference-free (4) Migration towards IP workflow	(1) Central receive and backhaul infrastructure requirements—CAPEX dependent (2) Highly planned coverage requirements	6 Mbps nominal, up to 31 Mbps capable for an 8 MHz bandwidth
4G bonded cellular	(1) IP workflow integration (2) Ubiquitous coverage (3) Ease of use (4) Lower output power utilization per modem	(1) Reoccurring data costs (2) Dependency on third party network (3) Performance-dependent on network coverage (4) Original versions not optimized for uplink capabilities	Nominal uplink—1 Mbps up to 10 Mbps, downlink—up to 20 Mbps
Ka band satellite	(1) Available bandwidth and capacity in North America (2) IP workflow integration (3) Pay as required bandwidth capabilities	(1) Dependency on third party networks (2) Portability versus other technologies (3) Prone to adverse weather effects	Uplink up to 10 Mbps, downlink up to 20 Mbps

TABLE 2.13.3 Hybrid Transmission Tradeoffs

Transmission Technology	Strengths	Weaknesses
Licensed microwave—COFDM—DVB-T	(1) Long symbol durations up to 224 μs (2) Variable delay spread support (3) Supports long distance transmission capabilities (4) Flexible bit rate support (5) Low C/N capabilities	(1) High peak to average power ratios with 2K FFT operation (2) Prone to Doppler shifts for high mobility applications requires antenna compensation
Cellular—OFDM-LTE	(1) Consumer network standard (2) Layer 2/3 networking support (3) Link connectivity at high mobility (4) Adaptive data rate control	(1) Downlink (DL) data centric (2) PHY designed for short range communications—low symbol duration and delay spreads—66 μs (3) Constrained data rates (4) Requires FDD operation with certain modes
Ka band satellite—DVB-S2	(1) Adaptive modulation support (2) Layer 2/3 networking support (3) Adaptive rate control and power control (4) Low peak to average power ratio	(1) Downlink data (DL) centric (2) Separate UL/DL bandwidths (3) Link margin effects in extreme weather conditions (4) Not designed for NLOS conditions

proprietary cellular channel aggregation techniques, as outlined previously, each of the unique transmission physical layers provides a single, virtual IP connection to achieve a common reliable network connection. The algorithm is designed to find a real-time, optimal equilibrium distribution between the multiple heterogeneous links based on the appropriate link quality metrics, while also providing a single virtual connection with improved bandwidth, persistence, and reliability, minimizing the risk associated with network interruptions and unpredictable delay over individual connections (see Fig. 2.13.12).

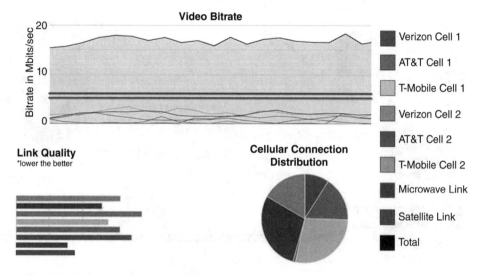

FIGURE 2.13.12 Hybrid aggregation distribution.

By aggregating cellular and microwave channels, the user can move in and out of the microwave range and still broadcast an uninterrupted video stream. The bidirectional aggregated channel can provide a high capacity backhaul link to the studio for features such as video file transfers, mobile hotspot UE connections for multiple videos, and return video from the studio to the mobile ENG van. It also makes it possible to control the transmitter from the receiving studio. This hybrid aggregated system is illustrated in Fig. 2.13.13.

2.8 SUMMARY

The evolution of microwave, cellular, and satellite transmission technologies has made hybrid aggregated microwave solutions a reality for ENG. Broadcasters can utilize all three of the transmission solutions simultaneously within an IP workflow to automatically allow load balancing and aggregation techniques to seamlessly manage the data connectivity (see Fig. 2.13.14).

Hybrid product suites can include a mobile van transmitter, which combines ENG, SNG, and CNG capabilities in a single rack unit; an ENG/CNG camera back; and the cellular newsgathering pack with hybrid technology.

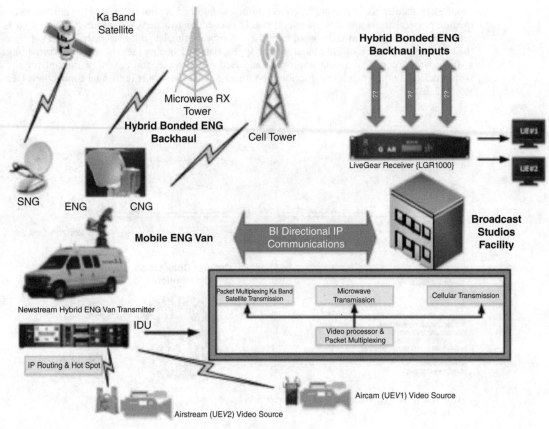

FIGURE 2.13.13 Hybrid aggregation for mobile ENG. (*Courtesy: VISLINK.*)

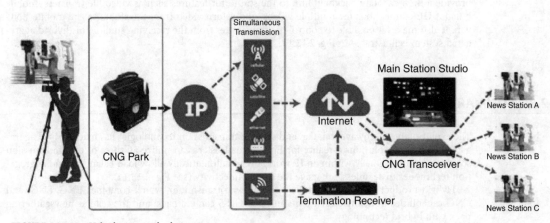

FIGURE 2.13.14 Hybrid aggregated solution.

LIST OF ACRONYMS

ABR	adaptive bit rate
BAS	Broadcast Auxiliary Services
BUC	block upconverter
C/N	carrier-to-noise ratio
CNG	cellular news gathering
COFDM	coded orthogonal frequency division multiplexing
DL	downlink
DVB	Digital Video Broadcasting
DVB-T	Digital Video Broadcasting—Terrestrial
ENG	electronic news gathering
ETSI	European Telecommunication Standards Institute
FDD	frequency division duplex
FEC	forward error correction
FM	frequency modulation
HD	high definition
LAN	local area network
LGR	live gear receiver
LOS	line of sight
LTE	Long Term Evolution
MAX RC	maximum ratio combining
MIMO	multiple input multiple output
MPEG	Moving Picture Experts Group
MRC	microwave radio communications
NLOS	non–line of sight
OEM	original equipment manufacturer
PHY	physical layer
QAM	quadrature amplitude modulation
QoS	quality of service
QPSK	quadrature phase shift keying
SD	standard definition
SNG	satellite news gathering
TCP	Transmission Control Protocol
UDP	User Datagram Protocol
UE	user equipment
UL	uplink
WAN	wide area network

REFERENCES

[1] Valkenburg, M.: *Reference Data for Engineers*, Eighth Edition, Radio, Electronics, Computer and Communications, 1993.

[2] Digital Video Broadcasting (DVB): "Framing Structure, Channel Coding and Modulation for Digital Terrestrial Television," ETSI EN 300744 V1.5.1 (2004–2011).

[3] www.speedtest.net

[4] Pradhan, N., and Wood, J.: "Broadcasting Video Over the Cellular Network and the Internet," 2012 Society of Motion Picture and Television Engineers (SMPTE) Annual Technical Conference and Exhibition, Hollywood, California, October 2012.

SECTION 3
DTV TRANSPORT

Section Leader—Dr. Richard Chernock
Triveni Digital, Princeton Junction, NJ

When television was analog, a single 6 MHz radio frequency (RF) transmission carried one television program (both video and audio). Viewed simplistically, voltages were "wiggled" the right way at the right times to convey the analog information that caused video to appear on the television screen and audio to come out of the speakers. With the transition to digital television (DTV), the situation became much more complicated. For DTV, the same 6 MHz RF transmission could (and usually did) carry many television programs simultaneously—with each composed of video, possibly many audio streams, data, and information about the components transmitted (metadata).

Instead of "wiggling" voltages, a DTV transmission consists of a serial stream of "bits" (ones and zeros). This stream of bits carries a series of "numbers" that allows the receiver to calculate the pictures to be shown and audio frames to be played, as well as the equivalent of a relational database that allows the receiver to understand what it is being sent and the relationships between the various components that make up the DTV emission signal.

A stream of bits needs an imposed structure to be useful—some means for a receiver that is accessing the stream at a random time to begin to understand what it is receiving and how to recreate the information from the components that are being serialized across the transmission. As is common with digital communication systems, DTV is a layered system, ranging from the basic packets that are handed to the RF system for transmission up to the larger structures riding upon them that carry the components and other information. Generally, these packets are typically composed of a header (which carries information about what is in the packets themselves) plus payload (the information itself).

In addition to the layered packetized structure, information is needed to let the receiver know what types of information are in the digital stream, how to identify and associate individual elements, how to compose larger structures (e.g., which audio and video streams make up a television "program"), and characteristics of the different components. Some of this information is used by the receiver in its internal workings, while other information is aimed at the viewer (often promotional "program guide" information that lets the viewer know what will be broadcast at a later time). This type of information is known as metadata (data about data).

The chapters that follow explain DTV transport and metadata in depth.

Chapter 3.1: MPEG-2 Transport
Chapter 3.2: Program and System Information Protocol: PSIP
Chapter 3.3: IP Transport for Mobile DTV
Chapter 3.4: Mobile Emergency Alert System
Chapter 3.5: ATSC Mobile DTV System

CHAPTER 3.1
MPEG-2 TRANSPORT

John R. Mick Jr
Mick Ventures, Inc., Denver, CO

3.1 INTRODUCTION

The MPEG-2 Standard, ISO/IEC 13818, a collection of 10 individual specifications, defines the mechanisms, syntax, and semantics used to encode video, audio, and data, and to combine them in preparation for transmission and/or storage.

The MPEG-2 Systems specification [1], which will be our focus in this chapter, describes the lowest level packet formats and the applicable syntaxes for carrying private data and elementary coded data. This specification defines the mechanism for identifying content carried in a stream, the synchronization and timing model, and the buffer models used to control streams so a receiving decoder does not overflow or underflow.

The MPEG-2 Video specification [2] defines the syntax and semantics necessary for efficient coding of interlaced and noninterlaced pictures having different spatial resolutions [3]. The MPEG-2 Audio specification [4] provides support for multichannel audio encoding compatible with the MPEG-1 syntax and algorithms [3].

The MPEG-2 Video and MPEG-2 Audio specifications have been supplemented by improved specifications capable of higher quality video and audio reproduction in a fewer number of transmitted bits. Examples are the MPEG-4 Part 10 "Advanced Video Coding" standard [5] and the A/52 "Digital Audio Compression (AC/3)" standard [6]. These newer advanced codec specifications still rely heavily on the core MPEG-2 Systems specification for transport. In the remainder of this chapter, we will examine the MPEG-2 Systems Standard as it remains a key digital television foundation.

3.2 MPEG-2 SYSTEMS SPECIFICATION INTRODUCTION

The MPEG-2 Systems specification defines two low-level packet formats: Program Streams and Transport Streams. Each format has been designed and optimized for a particular application and environment with both formats carrying one or more Program Elements.

MPEG-2 Program Streams are designed for low-loss, low-error, high-reliability transmission environments where their large, variable-sized payloads provide optimum transmission performance. Today, we find the MPEG-2 Program Stream's main application is the Digital Video Disk (DVD) also known as Digital Versatile Disk.

MPEG-2 Transport Streams were defined for the carriage of multiple MPEG-2 Programs and applications in potentially lossy environments where the transmission quality varies and the transmission climate may be prone to errors. These environments are similar to those faced by broadcast engineers in terrestrial transmission systems, satellite systems, cable systems, and long distance networks where errors can be both bit value errors and packet loss transgressions [1]. We will focus only on MPEG-2 Transport Streams in this chapter.

MPEG-2 Transport Streams additionally provide the capability to simultaneously carry multiple MPEG-2 Programs within a single stream with each MPEG-2 Program having its own unique time base [1].

MPEG-2 Systems also defines two other fundamental syntactic structures to transport private data and coded bit streams: the MPEG-2 `private_section` and the MPEG-2 Packetized Elementary Stream (PES) packet respectively. We find both structures offer a number of options to accommodate as many stream types as possible.

MPEG-2 `private_sections` are primarily used to asynchronously deliver data using a minimum amount of syntactic overhead through the addition of a header and a trailer in order to encapsulate the data. The MPEG-2 `private_section` is the basis for a majority of the syntactic structures used by the digital television specifications worldwide.

Extensions built on top of the MPEG-2 `private_section` include the MPEG-2 Systems' Program Specific Information (PSI) specification. PSI enables a decoder to identify an MPEG-2 Program and to demultiplex, or separate, the MPEG-2 Program into its individual elementary components (i.e., video, audio, and data streams). Many of the ATSC A/65 PSIP [7] data structures inherit their syntax from the MPEG-2 `private_section` syntax.

Once the data is encapsulated into the MPEG-2 `private_section`, MPEG-2 Systems packetizes, or segments, the `private_section` into one or more MPEG-2 Transport Stream packets ready for transmission. We will discuss MPEG-2 Transport Stream packets, `private_sections`, and PSI in further detail below.

MPEG-2 PES packets are used for synchronizing data delivery with a common time base. Content encoded according to the MPEG-2 Video [2], the MPEG-2 Audio [4], or other newer specifications such as MPEG-4 Part 10 [5] are typically encapsulated in MPEG-2 PES packets. The MPEG-2 PES packet syntax carries coded elementary stream data along with decoding and Presentation Time Stamps (PTSs). The ATSC A/53 Digital (Terrestrial) Television Transmission [8] specification uses the MPEG-2 PES packet format to encapsulate and transmit audio and video.

Elementary streams, which are coded bit streams produced by the compression process, are segmented and encapsulated into MPEG-2 PES packets and then packetized into MPEG-2 Transport Stream packets. The MPEG-2 PES packet prepends only a header to the data payload, and like `private_sections`, segments the data into MPEG-2 Transport Stream packets. We will discuss MPEG-2 PES packets in more detail below as well.

The MPEG-2 Systems specification uses two forms of a Cyclic Redundancy Check (CRC) for error detection. The first is a 32-bit CRC utilized by all long form sections (`private_section`) and the second is a 16-bit CRC employed by the MPEG-2 PES packet syntax.

When we find there is not enough data available to fill an MPEG-2 Transport Stream packet payload field, padding is required. This padding is referred to as stuffing. There are two primary methods for stuffing (padding) an MPEG-2 Transport Stream. The first method is specific to MPEG-2 `private_sections` where the unused MPEG-2 Transport Stream packet payload bytes are set to the value 0xFF. The second stuffing method uses the MPEG-2 Transport Stream `adaptation_field` to fill the remainder of the MPEG-2 Transport Stream packet. This method must be used with MPEG-2 PES packets and it may be applied to any other `stream_type`.

3.2.1 MPEG-2 Segmentation, Encapsulation, and Packetization

We find the MPEG-2 Standard transports data in a similar fashion to those of many existing networks. In these network protocols, data from one layer is fully encapsulated as the payload of the next layer by the next layer protocol's prepending its own header (and possibly trailer information) to the data it receives from the layer above. Additionally, each of these layers may segment the data into smaller units as required for presentation to the next layer. The data is passed from layer to layer until it is finally transmitted. At the receiver, the reverse process takes place.

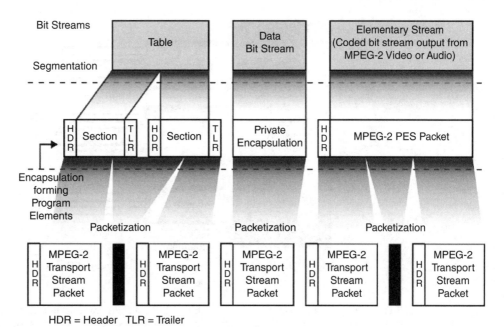

FIGURE 3.1.1 MPEG-2 Standard segmentation, encapsulation, and packetization.

The MPEG-2 Standard uses this same approach, and Fig. 3.1.1 illustrates this process using the MPEG-2 Systems Standard.

3.2.2 The MPEG-2 Network Composition

An MPEG-2 Network consists of one or more MPEG-2 Transport Streams each comprised of one or more MPEG-2 Programs. Each MPEG-2 Transport Stream is a linear sequence of MPEG-2 Transport Stream packets and, if transmitted, is received at a constant bitrate.

The MPEG-2 Program is a collection of Program Elements where each Program Element refers to a single encapsulated bit stream. An MPEG-2 Program is identified by a `program_number`. The scope of a `program_number` is local to its associated MPEG-2 Transport Stream, meaning the `program_number` has significance only within the MPEG-2 Transport Stream where it appears.

A Program Element consists of a basic bit stream encapsulated in one of the many defined encapsulation syntactic constructs. Each Program Element has an associated `stream_type` assisting in identifying the encapsulation method.

A typical MPEG-2 Program might be modeled as a single compressed video stream and one or more encoded audio streams. Each audio stream might carry a separate language audio track such as English, French, and German. Each coded bit stream is a Program Element and the MPEG-2 Program is the aggregation of the Program Elements. Figure 3.1.2 illustrates a collection of MPEG-2 Programs and their Program Elements.

An MPEG-2 Program should not be confused with the meaning of the traditional broadcaster's program. A broadcaster's program is equivalent to a television show such as "Gilligan's Island"® or "Who Wants to Be a Millionaire"®. To a broadcaster, a program is not only a collection of Program Elements; it also has characteristics such as a start time and duration. (The ATSC refers to this concept as an Event.) Looked at a different way, the MPEG-2 Program approximates a channel of programming such as HBO®, or ESPN® where each channel carries a sequence of shows and a show is equivalent to a broadcaster's interpretation of a program.

3.6 DTV TRANSPORT

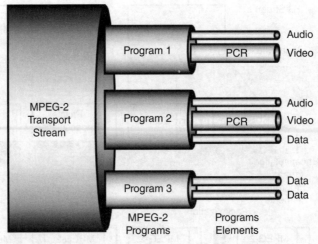

FIGURE 3.1.2 An MPEG-2 Program in an MPEG-2 Transport Stream.

Each Program Element that is part of the MPEG-2 Program is not required to have a synchronization relationship to another Program Element; however, if the Program Element does require a synchronization reference, then the MPEG-2 Systems Standard requires all the program's Program Elements reference a common time base. The common time base is unique to the MPEG-2 Program and is required to enable the decoder to synchronize the presentation of the time-oriented Program Elements (such as video and audio) [1].

In MPEG-2 Systems, the elementary stream is defined as a generic term referring to a coded data stream where the stream must be encapsulated using the MPEG-2 PES packet syntactic structure. An elementary stream refers to a lower-level data coding, placed into the payload field of the MPEG-2 PES packet. An elementary stream is not a Program Element; it is contained as the MPEG-2 PES packet payload. The MPEG-2 PES packet stream is the Program Element. One would think all bit streams, coded or not, could be called elementary streams. However, the MPEG-2 committee defined the term elementary stream to specifically mean a bit stream carried in MPEG-2 PES packets.

3.3 THE MPEG-2 TRANSPORT STREAM

An MPEG-2 Transport Stream is a continuous series of bits—forming bytes and framed together to construct a sequence of MPEG-2 Transport Stream packets. An MPEG-2 Transport Stream packet is 188 bytes in length[1] and always begins with the synchronization byte whose value is 0x47. The constant 188-byte length makes error recovery and stream synchronization easier. Figure 3.1.3 illustrates an MPEG-2 Transport Stream and the MPEG-2 Transport Stream packet structure.

3.3.1 The MPEG-2 Transport Stream Packet

In Fig. 3.1.3, the first four bytes of the MPEG-2 Transport Stream packet, the `sync_byte` through `continuity_counter` fields, constitute the MPEG-2 Transport Stream packet header. This means the MPEG-2 Transport Stream packet's payload area is up to 184 bytes in length and the actual content

---

[1]The MPEG-2 Transport Stream packet structure was designed to be carried by ATM. At the time when the 188-byte packet size was selected, the ATM payload was defined to be 47 bytes. Thus, 47 * 4 = 188.

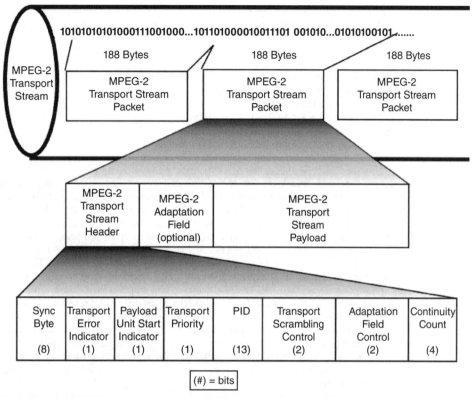

FIGURE 3.1.3 An MPEG-2 Transport Stream and MPEG-2 Transport Stream packet.

definition of an MPEG-2 Transport Stream packet payload varies based on the stream_type and the payload's data encapsulation method.

The MPEG-2 Standards use the term "reserved" to indicate the standard is preserving the field or value for its future use. Syntactic elements defined as reserved almost always set the bits associated with the field to 1b's (i.e., set the required number of bits to 1b). Receivers encountering reserved fields should typically ignore the contents of these reserved fields.

Each MPEG-2 Transport Stream packet may contain the optional adaptation_field syntax and if present, immediately follows the last byte of the MPEG-2 Transport Stream packet header. The adaptation_field is not considered part of the MPEG-2 Transport Stream packet header, nor is it considered part of the MPEG-2 Transport Stream packet payload. When the adaptation_field is present, the MPEG-2 Transport Stream packet payload's size is 184 bytes minus the adaptation_field's total byte length.

The MPEG-2 Transport Stream Packet Identifier (PID) is a 13-bit value used to identify multiplexed packets within the MPEG-2 Transport Stream. The PID allows packets from up to 8192 (2^{13}) separate bit streams to be simultaneously carried within the MPEG-2 Transport Stream and provides a unique stream association for each MPEG-2 Transport Stream packet. Each data stream carried within the MPEG-2 Transport Stream is assigned to a particular PID and a PID carries data from one and only one data stream. Consequently, a PID may carry only one method of encapsulation, for example: private_sections or MPEG-2 PES packets.

A conceptual way you can think about PIDs is to view them as each being a wire carrying a single Program Element. An MPEG-2 Transport Stream would be viewed as a cable with 8192 individual wires bundled together.

The MPEG-2 Systems specification defines a basic PID allocation map identifying PIDs having special meaning or purpose. MPEG-2 Systems reserves PIDs 0x0000 thru 0x000F and PID 0x1FFF for its use and the ATSC and other standards organizations reserve further PID ranges.

The MPEG-2 Systems specification defines the payload_unit_start_indicator field's behavior for usage with MPEG-2 private_sections, such as PSI and data structures inheriting from the private_section syntax, as well as MPEG-2 PES packets. We will discuss the field's interpretation for each of these formats further below.

The transport_scrambling_control indicates if this MPEG-2 Transport Stream packet's payload has been scrambled (i.e., encrypted). The MPEG-2 Transport Stream packet header, the optional adaptation_field, and the payload of the NULL MPEG-2 Transport Stream packet (PID 0x1FFF and discussed below) are never scrambled. A value of 00b indicates the MPEG-2 Transport Stream packet's payload is not scrambled.

The inclusion of the optional adaptation_field is signaled via the two-bit adaptation_field_control field located in the MPEG-2 Transport Stream packet header. The adaptation_field_control value of 10b signals the MPEG-2 Transport Stream packet may be entirely comprised of the adaptation_field and there is no MPEG-2 Transport Stream packet payload. The value 11b indicates the MPEG-2 Transport Stream packet contains both an adaptation_field and payload bytes. If a receiver/decoder obtains an MPEG-2 Transport Stream packet having the adaptation_field_control bits set to 00b, the decoder should discard the MPEG-2 Transport Stream packet as this value is reserved. The adaptation_field is optional in all MPEG-2 Transport Stream packets except the NULL MPEG-2 Transport Stream packet where the adaptation_field is not allowed.

The continuity_counter is a 4-bit rolling counter associated with MPEG-2 Transport Stream packets of the same PID. The counter is incremented by one for each consecutive MPEG-2 Transport Stream packet for a given PID except when adaptation_field_control is set to 10b (adaptation_field only, no payload) or 00b (reserved). When the adaptation_field_control is set to these values, the continuity_counter value must be repeated.

If an MPEG-2 Transport Stream packet is to be repeated (i.e., duplicated), which can only occur when the adaptation_field_control is set to either 01b (no adaptation_field, payload only) or 11b (adaptation_field and payload), the continuity_counter is not incremented. You might consider repeating an MPEG-2 Transport Stream packet if the contents of the packet are vital. Only one repeat occurrence is allowed for any given MPEG-2 Transport Stream packet containing payload. When duplicating the MPEG-2 Transport Stream packet, all fields must be duplicated except the adaptation_field's Program Clock Reference (PCR) which, if included, must have a valid value.

If a receiver encounters a situation where the continuity_counter increments by more than one between successive packets on a given PID, it should typically assume some number of MPEG-2 Transport Stream packets have been lost.

The payload field carries the data content. The data content can be one of many types: a derivative of the private_section such as PSI, other section types, or an MPEG-2 PES packet which itself may contain an elementary stream.

3.3.2 adaptation_field (not illustrated)

The adaptation_field's inclusion in an MPEG-2 Transport Stream packet is signaled when the first bit of the 2-bit MPEG-2 Transport Stream packet header's adaptation_field_control field is set. If the adaptation_field is included, this field immediately follows the MPEG-2 Transport Stream packet header's continuity_counter field. An adaptation_field may never span multiple MPEG-2 Transport Stream packets.

The adaptation_field contains both generic and video specific fields and setting an associated indicator flag bit includes a field. Three of the most important adaptation_field elements are the discontinuity_indicator, the fields comprising the PCR value, and the stuffing byte loop.

The adaptation_field's discontinuity_indicator signals either system time base discontinuities for the Program Elements or MPEG-2 Transport Stream packet header continuity_counter field discontinuities associated with the transport PID. The discontinuity_indicator may also indicate both discontinuity forms are occurring concurrently. The discontinuity_indicator essentially notifies the decoder to expect an intentional upcoming disruption of the Program timeline and `continuity_counter` field, for example as the consequence of switching a video, audio, or data elementary stream associated with splicing.

The `adaptation_field`'s PCR fields, `program_clock_reference_base` and `program_clock_reference_extension`, carry a 42-bit value representing a sampling of the System Time Clock (STC) at a precise instance in time. Decoders use the PCR to reconstruct a local STC that accurately matches the encoder's clock and we will discuss the STC and other related time stamps further below.

The `adaptation_field`'s `stuffing_byte` loop present at the end of the `adaptation_field` is used to pad, or stuff, the `adaptation_field`, thereby forcing the start of the MPEG-2 Transport Stream packet payload to be offset to a desired location.

3.3.3 The NULL MPEG-2 Transport Stream Packet

The NULL MPEG-2 Transport Stream packet is a special MPEG-2 Transport Stream packet designed to pad an MPEG-2 Transport Stream. While individual MPEG-2 Programs within a multiplex can have variable bitrate characteristics, the overall MPEG-2 Transport Stream must have a constant bitrate. A NULL MPEG-2 Transport Stream packet is employed when an MPEG-2 Transport Stream does not have any data packets ready to be transmitted (in order to maintain a constant bitrate output). NULL MPEG-2 Transport Stream packets may be added and/or removed by any remultiplexing process within the data path, meaning a downstream remultiplexing process or decoder cannot assume the presence of these packets [1].

The NULL MPEG-2 Transport Stream packet is always carried on PID 0x1FFF and the MPEG-2 Transport Stream packet's payload may contain any data values. (Note: The term NULL MPEG-2 Transport Stream packet does not translate to mean an MPEG-2 Transport Stream packet with payload bytes set to 0x00.) In practice, the MPEG-2 Transport Stream packet payload is often set to random data that are statistically bit neutral.

The `continuity_counter` of a NULL MPEG-2 Transport Stream packet is undefined and may contain any legal value. Consequently, the `continuity_counter` is always ignored. The payload of the NULL MPEG-2 Transport Stream packet may not be scrambled (i.e., the `scrambling_control` field must be set to 00b). MPEG-2 Systems additionally constrains many of the fields of the NULL MPEG-2 Transport Stream packet's header, such as the optional `adaptation_field` cannot be included.

3.3.4 Multiplex Concepts

The MPEG-2 Transport Stream includes PSI Program Elements that assist in locating the individual MPEG-2 Programs and their associated Program Elements (i.e., the video and audio streams). PSI along with the individual Program Element MPEG-2 Transport Stream packets are interleaved forming the MPEG-2 Transport Stream. Figure 3.1.4 illustrates how two MPEG-2 Programs each consisting of a video and audio stream might be multiplexed into an MPEG-2 Transport Stream.

In Fig. 3.1.4, program P1's video stream consists of three MPEG-2 Transport Stream packets carried on PID 1023. Each MPEG-2 Transport Stream packet has a `continuity_counter` associated with the specific PID. For program P1's video stream, the `continuity_counter` values begin at three and end with five. The individual MPEG-2 Transport Stream packets of this PID are linearly recombined to form the original data buffer, in this case an MPEG-2 PES packet carrying a video elementary stream.

Program P1 also has an associated audio stream carried on PID 1024. Program P1's audio stream consists of two MPEG-2 Transport Stream packets with the `continuity_counter` values of 2 and 3, respectively. Notice the NULL MPEG-2 Transport Stream packets that were included. These MPEG-2 Transport Stream packets are carried on PID 0x1FFF and can appear anywhere in the stream as needed. Similarly, in Fig. 3.1.4, Program P2's packet composition is illustrated.

3.10 DTV TRANSPORT

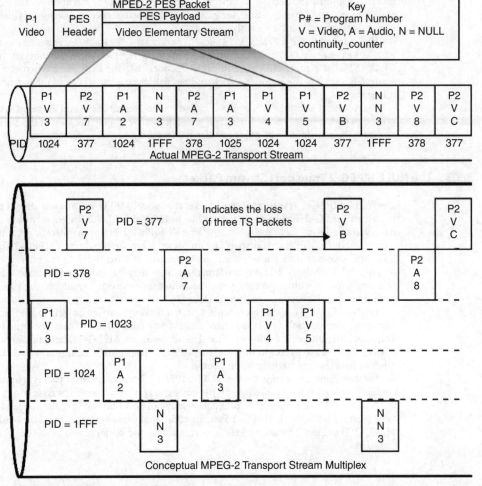

FIGURE 3.1.4 MPEG-2 Transport Stream program multiplex.

In Program P2's video stream on PID 377, the second to last MPEG-2 Transport Stream packet's `continuity_counter` is 0xB rather than the expected value of 0x9. This condition indicates an error and the loss of possibly three MPEG-2 Transport Stream packets on this PID. The next expected and received `continuity_counter` value is 0xC.

MPEG-2 Transport Stream packets cannot simply be rearranged in the MPEG-2 Transport Stream. This limitation exists because the mechanism for recreating the original STC in the decoder depends upon the actual arrival time of the packets carrying the individual PCRs as compared to the value carried in the PCR field. Shifting the packets carrying the PCR induces jitter into the data stream, which may cause the decoder's STC to vary and could eventually lead to a failure in synchronization at the decoder output. For example, variances in the STC can lead to the presentation of audio either too early or too late as compared to the video picture leading to "lip-sync" problems. Additionally, the temporal location of the individual MPEG-2 Transport Stream packet payload delivery conforms to the buffer model associated with the encapsulation type. Shifting or rearranging the MPEG-2 Transport Stream packets potentially causes buffer model violations by either overflowing or underflowing the buffer.

3.4 TABLES, SECTIONS, AND DESCRIPTORS

In MPEG-2 Systems, a *table* refers to a collection of information that is not video or audio and a *section* is a segment of a table. Sections are further divided as necessary in order to be carried via an MPEG-2 Transport Stream packet. All sections inherit their syntactic construct from the MPEG-2 Systems' `private_section` syntax and in practice, the MPEG-2 Systems' `private_section` syntax is never directly utilized. Thus, the term "section" is a generic term referring to any data structure whose syntax is based on the MPEG-2 `private_section`. The MPEG-2 `private_section` defines a data encapsulation method used to place private data into an MPEG-2 Transport Stream packet with a minimum amount of structure [1].

The terms "private_section" and "section" are typically used interchangeably. However, this can be confusing as sections are generic and MPEG-2 `private_sections` always refer to the MPEG-2 Systems syntactic construct. Additionally, it is an unfortunate common practice that the terms, "section" and "table," are often incorrectly interchanged. To be strictly correct, a section is the basic syntactic unit and a table is a logical collection of related sections.

Sections are not required to start in any particular location within the payload of an MPEG-2 Transport Stream packet. When a new section begins in an MPEG-2 Transport Stream packet, a one byte `pointer_field` is included indicating the starting offset of the new section and the `payload_unit_start_indicator` signals inclusion of the `pointer_field`. The first byte of the MPEG-2 Transport Stream packet payload carries the `pointer_field` and the `pointer_field` contains the number of bytes immediately following the field itself to the start of a new section. Thus, the `pointer_field` indicates the offset (in bytes) to the start of the new section and this field may contain a value in the range of 0x00 to 0xB6 (182) bytes. If the `payload_unit_start_indicator` is set to 0b, the MPEG-2 Transport Stream packet's payload does not contain a `pointer_field` and instead, the payload contains the continuation of a previously started `private_section` along with any necessary stuffing bytes.

3.4.1 The MPEG-2 Systems private_section Syntax

Figure 3.1.5 provides an overview of the MPEG-2 Systems `private_section` syntax.

A section, or more specifically the MPEG-2 `private_section`, always begins with an 8-bit `table_id`. The `table_id` uniquely identifies the section's type.

The `section_syntax_indicator` controls whether the "short" section or "long" section `private_section` syntax is selected. The short section includes the `private_data_bytes` immediately following the `private_section_length` field. The long section incorporates the `table_id_extension` field through the `last_section_number` fields prior to the `private_data_bytes` inclusion. In practice, the long version of the `private_section` syntax is generally selected over the short `private_section` syntax as the long section provides a standard mechanism to segment a large table into multiple sections. The short `private_section` is limited to 4093 bytes of total data while the collection of long `private_sections` can accommodate 255 * 4084 payload data bytes or a maximum table size of 1,041,420 bytes. The remainder of this text applies only to long sections.

Each long section is related minimally by its `table_id` and `version_number` fields. In some cases, additional fields are required in order to determine the relationship. All sections having the same `table_id` and `version_number` are collected and reassembled to form the table assuming no other fields are needed for distinction. Sections having the same `table_id` but different `version_number` field values signal the carriage of more than one instance of the named table. In this case, each table version is considered unique and the currently applicable version is signaled by the `current_next_indicator` being set to 1b. The `current_next_indicator` flag set to 0b allows a table to be transmitted ahead of its valid time facilitating preprocessing by the receiver.

The section header refers to the fields from the `table_id` through the end of the `last_section_number` field and the section trailer embodies only the `CRC_32` field.

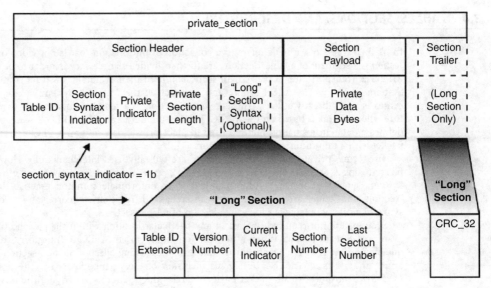

FIGURE 3.1.5 MPEG-2 Systems `private_section` syntax overview.

The `table_id_extension` field may be used as an extension for specifying a sub-table type relative to the primary table identified by the `table_id`. This field was incorporated in the syntax in order to informally increase the number of tables that could be carried in the MPEG-2 Transport Stream. The extension functions by enabling a 16-bit value to be further associated with the `table_id`. For example, the `program_association_section` (discussed later) uses the field to carry the `transport_stream_id`. The extension field is typically renamed in the application of the `private_section` syntax.

The `version_number` identifies the section revision. Any time the section's `private_data_bytes` are modified, the `version_number` is incremented by 1 modulo 32 and the CRC_32 is recalculated for the modified section. (The CRC_32 must be recalculated any time any of the section data changes including the section header.) `version_number` field values are typically scoped to a section, meaning they are applicable to the section in which they appear and tables segmented into multiple sections may simultaneously have `version_number` fields containing different values. This scope limits the amount of recalculation required when only a single section of a multisection table is modified. However, some tables define their section's `version_number` to be applicable to all the sections comprising the table, thereby changing the `version_number` scope from section local to table global. When this is the case, the `version_number` must be the same in all sections forming the table and any table modification requires the version number be updated in all sections and the entire collection of sections is regenerated (or in the case of a receiver, reacquired). The scope of the `version_number` therefore must always be defined in the section's syntactic description.

The `current_next_indicator` is applied in conjunction with the `version_number` field. When the `current_next_indicator` is set to 1b, the section is the currently applicable section and the value located in the `version_number` field indicates the `version_number` currently relevant. The `current_next_indicator` value of 0b indicates the section is not currently active and the `version_number` field contains the version value becoming active next.

The `section_number` is employed to reassemble the individual sections forming the table. The `section_number` identifies this section's reassembly position in view of the entire table. The `last_section_number` specifies the total number of sections comprising the table. Thus, the receiver knows it has all the component sections when it has received the total number of sections indicated by this field. The `last_section_number` value is always the section count minus one. Note these

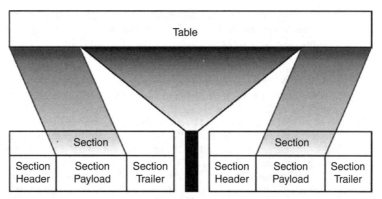

FIGURE 3.1.6 Table segmentation and encapsulation into MPEG-2 Systems `private_sections`.

two fields begin numbering using the value zero and a table comprised of a single section would have 0x00 in both fields. A table composed of three sections would have `section_numbers` of 0x00, 0x01, and 0x02, and the `last_section_number` field would contain the value 0x02 for all three sections.

The `CRC_32`, which constitutes the section trailer, is included only in the long version of the section. The `CRC_32` is calculated over all the bytes comprising a section beginning with the `table_id` through the last byte of `CRC_32` field itself. A `CRC_32` result of zero indicates the section is valid and it was received without error.

3.4.2 Section Segmentation and Encapsulation

Large tables are segmented and placed into the `private_data_bytes` field of a section. Using the long `private_section` syntax, Fig. 3.1.6 illustrates how a large table is segmented into multiple MPEG-2 `private_sections` and encapsulated by the section syntax.

Each table segment is placed into the `private_section`'s `private_data_bytes` field. The MPEG-2 `private_section`'s header and trailer are prepended and appended respectively, thereby encapsulating the data in the MPEG-2 `private_section`. Sections are further packetized (divided) as necessary when they are placed into the MPEG-2 Transport Stream packet payload.

3.4.3 Section Packetization into MPEG-2 Transport Stream Packets

One or more sections may be placed into an MPEG-2 Transport Stream packet depending on the section's size. If the sections are small, there can be multiple sections contained within the single MPEG-2 Transport Stream packet. Sections larger than a single MPEG-2 Transport Stream packet are divided and serialized across multiple MPEG-2 Transport Stream packets. Each MPEG-2 Transport Stream packet containing a portion of the divided or spanning section must contain the same PID in the MPEG-2 Transport Stream packet header. Once a section's packetization commences on a PID, a new section may not start on the same PID until the previous section completes.

A new section may begin immediately following the end of a preceding section in an MPEG-2 Transport Stream packet or the new section's start may be delayed until the beginning of the next MPEG-2 Transport Stream packet. When a new section begins immediately after the end of another section, we say the sections are "section packed" or transmitted as "back-to-back" sections. When sectioning packing is used, the new section must immediately follow the final byte of the previous section. Section packing is the most efficient method for transporting sections in terms of bandwidth utilization.

3.14 DTV TRANSPORT

When a section does not completely fill an MPEG-2 Transport Stream packet's payload area and there is no new section ready for inclusion, the remaining bytes of the MPEG-2 Transport Stream packet are stuffed, or filled, with the value 0xFF. This value represents a reserved (forbidden) table_id value in the MPEG-2 Systems Standard, so this value indicates no new section is contained within this MPEG-2 Transport Stream packet.

The MPEG-2 Transport Stream packet header's payload_unit_start_indicator is set to 1b whenever one or more sections begin within the MPEG-2 Transport Stream packet. Setting this bit signals the inclusion of the pointer_field, which we discussed previously.

If an MPEG-2 Transport Stream packet header's adaptation_field_control bits are set to 01b and a new section begins in the MPEG-2 Transport Stream packet (signaled by the payload_unit_start_indicator set to 1b) then the pointer_field immediately follows the continuity_counter. If the adaptation_field_control bits are set to 11b and a new section starts in the packet, the pointer_field immediately follows the last byte of the adaptation_field.

If the pointer_field's value is zero, the section begins immediately following the pointer_field as illustrated in Fig. 3.1.7. (All illustrations assume there is no adaptation_field signaled in the MPEG-2 Transport Stream packet.) The new section in Fig. 3.1.7 begins at the sixth byte in the MPEG-2 Transport Stream packet. (Four bytes for the header plus one byte for the pointer_field precede the start of the section.) Following the end of the new section, the remainder of the MPEG-2 Transport Stream packet is stuffed with the value 0xFF as the section size in this example was smaller than the MPEG-2 Transport Stream packet's payload.

If the pointer_field is non-zero, the section begins $n + 1$ bytes after the pointer_field where n is the value contained in the pointer_field. In Fig. 3.1.8, the pointer_field contains the value ten. This value indicates the new section begins at the sixteenth byte of the MPEG-2 Transport Stream packet. The sixteenth byte is determined to be the starting point because there are four bytes for the MPEG-2 Transport Stream packet header plus 1 byte for the pointer_field plus ten bytes of data immediately following the pointer_field. Consequently, there are fifteen bytes of data prior to the start of the section.

Figure 3.1.9 illustrates two sections contained in two MPEG-2 Transport Stream packets. The first section begins at the start of the first MPEG-2 Transport Stream packet in a manner similar to Fig. 3.1.7. The second section begins immediately following the first section concludes and the second section continues into the second MPEG-2 Transport Stream packet. In the second MPEG-2 Transport Stream packet, the payload_unit_start_indicator is set to 0b indicating no section begins in the packet. At the end of the section, the remainder of the MPEG-2 Transport Stream packet is stuffed with 0xFF. Note in this example, the payload_unit_start_indicator being set to 0b indicates there is no pointer_field in the second MPEG-2 Transport Stream packet.

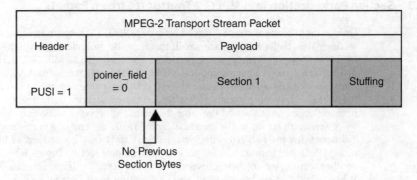

PUSI = payload_unit_start_indicator

FIGURE 3.1.7 A single section starting immediately in an MPEG-2 Transport Stream packet.

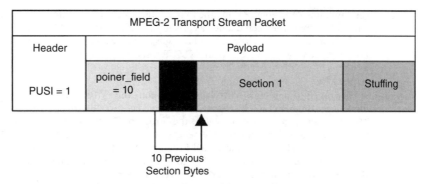

PUSI = payload_unit_start_indicator

FIGURE 3.1.8 A single section offset starting in an MPEG-2 Transport Stream packet.

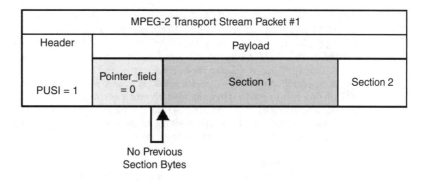

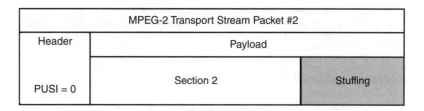

PUSI = payload_unit_start_indicator

FIGURE 3.1.9 Two sections carried in MPEG-2 Transport Stream packets.

3.5 MPEG-2 PROGRAM SPECIFIC INFORMATION (PSI)

We find the MPEG-2 PSI provides metadata necessary to identify an MPEG-2 Program and to demultiplex, or separate, the Program and its Program Elements from the MPEG-2 Transport Stream multiplex. MPEG-2 PSI is comprised of four tables, three of which are defined in MPEG-2 Systems. The three defined tables are: the Program Association Table (PAT), the Conditional Access Table (CAT), and the Program Map Table (PMT). The fourth table, the Network Information Table (NIT), is optional and its contents have been defined as private. The NIT is not used in an ATSC transmission system.

Each of the four tables listed above has a specific named section syntactic instance forming their table. For the three tables applicable to an ATSC Transport Stream, their section syntaxes are the `program_association_section`, the `conditional_access_section`, and the `TS_program_map_section` respectively.

The PAT provides a correlation between the MPEG-2 `program_number` and a PID identifying the MPEG-2 Transport Stream packets carrying the `TS_program_map_section`. The `TS_program_map_section` includes the "Program Definition" [1] along with a list of the PIDs that identify the individual Program Elements comprising the MPEG-2 Program. The PMT is defined as the complete collection of individual Program Definitions where there is one `TS_program_map_section` per MPEG-2 Program. The PMT is unique in that its contents can be spread over multiple PIDs. In comparison, the PAT, CAT, and NIT are each required to be located on individual unique PIDs of 0x0000, 0x0001, and 0x0010 respectively.

3.5.1 The Program Association Table (PAT)

The PAT identifies all the MPEG-2 Programs carried in the MPEG-2 Transport Stream multiplex. Each PAT entry consists of an MPEG-2 Transport Stream unique `program_number` and a PID identifying the location of the MPEG-2 Program's `TS_program_map_section`. Every MPEG-2 Program carried in the MPEG-2 Transport Stream is required to have a `TS_program_map_section` and the collection of `TS_program_map_sections` forms the PMT. Each `TS_program_map_section` may be carried on a unique PID or multiple `TS_program_map_sections` may be carried on the same PID in order to improve system bandwidth efficiency. Consequently, the same `program_map_PID` may be referenced by more than one MPEG-2 Program's `program_number` in the PAT.

In Fig. 3.1.10, the PAT associated with `transport_stream_id` 0xABCD contains five MPEG-2 Program entries where an MPEG-2 Program entry is a `program_number` and PID pair. The first Program entry, `program_number` zero, identifies the `network_PID` carrying the NIT. In Fig. 3.1.10, the `network_PID` is identified as PID W. (In the ATSC, the NIT is not used, and therefore, the `program_number` value zero should not appear in the PAT.)

The remaining entries define four MPEG-2 Programs: P1, P2, P3, and P4 with `program_numbers` 10, 25, 77, and 2003 respectively. Program P1, whose `program_number` is 10, has its `TS_program_map_section` carried on PID X. Program's P2 and P3 each have their `TS_program_map_section` carried on PID Y. PID Z carries Program P4's `TS_program_map_section`.

3.5.2 Program Map Table (PMT)

Each MPEG-2 Program contained in an MPEG-2 Transport Stream is identified by its `program_number` and a "Program Definition" is a mapping between a `program_number` and the Program Elements comprising it along with other Program-related information. The `TS_program_map_section` syntax transports the Program Definition. The PMT is formed by the logical collection of all Program definitions and a PMT is specific to one MPEG-2 Transport Stream.

The Program Definition syntax (i.e., the `TS_program_map_section`) provides for the transport of the following information:

- Program specific data in the form of descriptors
- Identification of the Program Elements comprising the MPEG-2 Program
- The MPEG-2 Transport Stream location for each Program Element
- Program Element specific data in the form of descriptors

Program Elements that are part of an MPEG-2 Program may include video elementary streams, audio elementary streams, and data streams. The Program Definition's Program Element locations

MPEG-2 TRANSPORT **3.17**

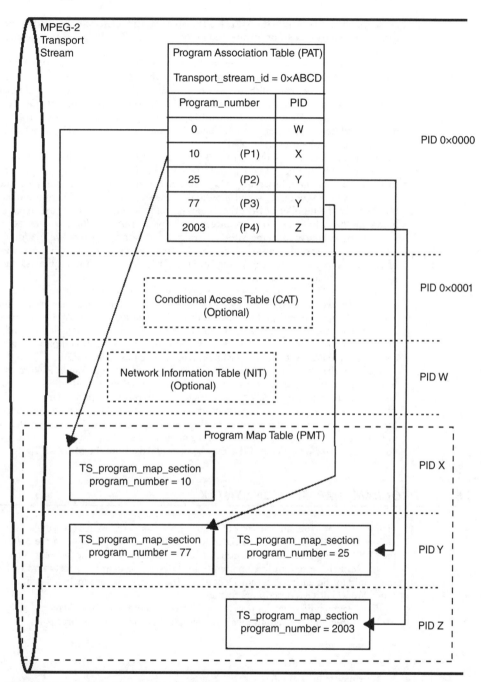

FIGURE 3.1.10 MPEG-2 Program Association Table usage.

are specified as PIDs within the MPEG-2 Transport Stream. Descriptors provide a simple mechanism to include context specific or private data for both the MPEG-2 Program and the individual Program Elements.

The `TS_program_map_section` syntax inherits its structure from the MPEG-2 `private_section` and the PMT is the aggregation of all the `TS_program_map_sections` found in an MPEG-2 Transport Stream. If the `TS_program_map_section` is carried in more than one MPEG-2 Transport Stream packet, the associated MPEG-2 Transport Stream packets must contain the same PID. Each Program Definition is limited to a single `TS_program_map_section` as the `section_number` and `last_section_number` fields are constrained to the value 0x00 and `TS_program_map_sections` are never scrambled.

As an example, if only one MPEG-2 program exists in the MPEG-2 Transport Stream, there can be only one `TS_program_map_section` and the PMT consists of a single section. If we find there are ten MPEG-2 Programs in the MPEG-2 Transport Stream multiplex, there will be ten `TS_program_map_sections` and the PMT consists of 10 sections. There is a one to one correlation between the number of MPEG-2 Programs and the number of `TS_program_map_sections` found in the MPEG-2 Transport Stream multiplex. The collection of the `TS_program_map_sections` combine to form the MPEG-2 Transport Stream multiplex's PMT. Additionally, this same one to one correspondence holds true for the count of PAT `program_numbers` and the number of `TS_program_map_section` in the MPEG-2 Transport Stream multiplex (excluding `program_number` zero indicating the NIT).

While the PMT is similar to the PAT and the CAT in inheriting from the MPEG-2 `private_section` syntax and in the collection of sections forming the table, there are two important differences. First, each Program Definition is limited to a single `TS_program_map_section`. Second, the individual `TS_program_map_sections` may be carried on any available PID and more than one `TS_program_map_section` may be carried on the same PID. In comparison, all of the sections comprising the PAT or the CAT must be carried on their assigned PID.

Each of the `TS_program_map_sections` carried on the same PID must be completed before the next `TS_program_map_section` may begin on the identical PID else there would be no way to discover which `TS_program_map_section` a particular packet belongs to. If a `TS_program_map_section` is packetized across more than one MPEG-2 Transport Stream packet, the next `TS_program_map_section` being carried on the same PID cannot start until the entire previous `TS_program_map_section` has completed. `TS_program_map_sections` that are packetized and carried on different PIDs may be interleaved within the MPEG-2 Transport Stream.

3.6 TS_PROGRAM_MAP_SECTION SYNTAX

Figure 3.1.11 provides an overview of the `TS_program_map_section` syntax.

The PCR_PID identifies the PID containing the PCR if applicable to the MPEG-2 Program. If the MPEG-2 Program does not require or have a reference clock, the field contains the PID 0x1FFF.

The MPEG-2 Program Information (outer) descriptor loop transports zero or more descriptors that apply to the entire Program Definition. The descriptors are included sequentially and in any order. We will discuss descriptors further below.

The Program Element loop associates zero or more Program Elements' definitions with the MPEG-2 Program. A Program Element's definition is comprised of three parts:

- stream_type
- PID
- Descriptor loop

The `stream_type` identifies the bit stream contents and encapsulation method. The PID identifies the location where in the MPEG-2 Transport Stream the information is carried, and the descriptor loop contains zero or more descriptors (the Stream (inner) descriptor loop) applying only to the specific Program Element.

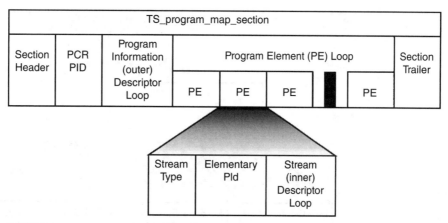

FIGURE 3.1.11 TS_program_map_section syntax overview.

3.6.1 Descriptors

The descriptor is a generic structure used to carry additional information within MPEG-2 constructs. Descriptors may carry either normative or private data and a descriptor's inclusion is often optional. A descriptor cannot stand alone in the MPEG-2 Transport Stream; rather, it must be contained within a larger syntactic structure. Descriptors are typically carried within a descriptor loop. The basic descriptor format is a tag byte, followed by a length byte followed by data {tag, length, data} [1]. The tag byte uniquely identifies the descriptor, the length byte specifies the number of data bytes immediately following the length field and the form of the data varies for each specific descriptor.

The placement of descriptors in a syntactic structure is typically indicated by the following looping construct contained within a syntactic structure.

```
for (i=0; i<N; i++) {
    descriptor()
}
```

Zero or more descriptors may be placed within the descriptor loop. In some cases, a descriptor may be used more than once in the loop, dependent upon the particular descriptor. To iterate through the descriptor loop, a receiver would examine the tag field and if the descriptor is supported, interpret the number of data bytes following the length field (which is signaled by the value in the length field). A receiver not supporting or choosing to ignore the descriptor identified by the tag byte would skip the number of data bytes following the length field (which is signaled by the value in the length field) in order to get to the next descriptor (assuming there is another descriptor in the loop). Remember, the total descriptor length is the descriptor length field value plus two (one byte each for the tag and length fields).

The descriptor location rules are also specifically defined for each descriptor; therefore, you must consult the specific descriptor documentation/standard for its usage rules.

3.7 MPEG-2 PACKETIZED ELEMENTARY STREAM (PES) PACKETS

We find MPEG-2 PES packets transport elementary streams such as video, audio, private data, reserved, and padding streams, along with stream synchronization information. MPEG-2 PES packets enable the elementary stream to be carried transparently in an MPEG-2 Transport Stream. Elementary streams are each independently carried in separate MPEG-2 PES packets; thus, an

3.20 DTV TRANSPORT

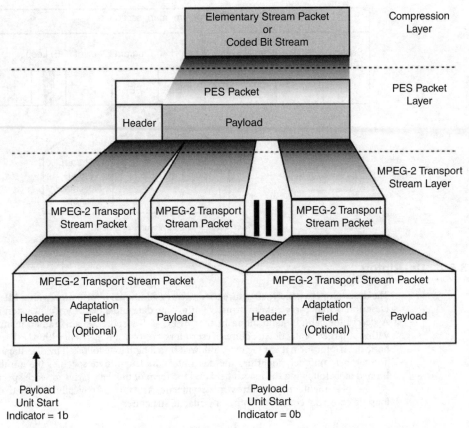

FIGURE 3.1.12 MPEG-2 elementary stream, MPEG-2 PES packet, and MPEG-2 Transport Stream relationship.

MPEG-2 PES packet contains data from one and only one elementary stream. Generally, each MPEG-2 PES packet stream is carried in its own MPEG-2 Transport Stream PID.

The MPEG-2 PES packet consists of a PES packet header followed by the PES packet payload. Each MPEG-2 PES packet can be of variable length, can be considerably longer than an MPEG-2 Transport Stream packet, and is subdivided, packetized, and serialized when placed into the MPEG-2 Transport Stream packet payload field. An MPEG-2 PES packet must always commence as the first byte of the MPEG-2 Transport Stream packet payload and only a single MPEG-2 PES packet may begin in an MPEG-2 Transport Stream packet. Thus, two MPEG-2 PES packets (or portions thereof) are not permissible in a single MPEG-2 Transport Stream packet. Figure 3.1.12 illustrates the elementary stream, MPEG-2 PES packet, and MPEG-2 Transport Stream packet relationship.

MPEG-2 PES packets carry stream synchronization information in the MPEG-2 PES packet header using PTSs and Decoding Time Stamps (DTSs) fields. The time stamps enable decoding the access units and presenting the access units respectively. The PTS and the DTS are each 33-bit long with units in 90 kHz clock periods and we will discuss their usage below.

3.7.1 The MPEG-2 PES Packet Syntax

Figure 3.1.13 provides an overview of the MPEG-2 PES packet syntax.

For MPEG-2 Transport Stream packets carrying MPEG-2 PES packets, the `payload_unit_start_indicator` is set to 1b when an MPEG-2 Transport Stream packet's payload field begins

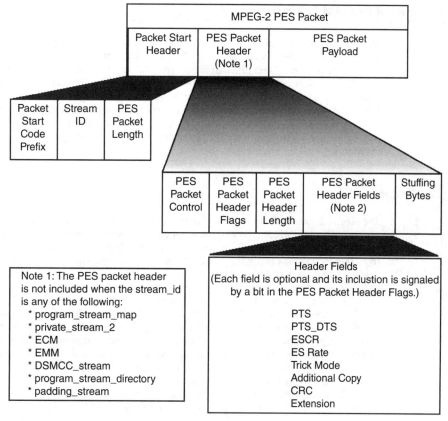

FIGURE 3.1.13 MPEG-2 PES packet syntax overview.

with the first byte of an MPEG-2 PES packet. The field contains a 0b to indicate no MPEG-2 PES packet begins in the MPEG-2 Transport Stream packet. Instead, the packet contains the continuation of a previously started MPEG-2 PES packet.

An MPEG-2 PES packet always begins with a 32-bit start code defined as the combination of the `packet_start_code_prefix` and the `stream_id`. The `packet_start_code_prefix` is always 0x000001 followed by the 8-bit `stream_id` and the `PES_packet_length` immediately follows the PES packet start code.

The MPEG-2 PES packet header follows the `PES_packet_length` only if the `stream_id` is one of the "enabled" types, meaning the PES packet header's inclusion is allowed by the syntax. Otherwise, the `PES_packet_data_byte` field immediately follows the `PES_packet_length`. The MPEG-2 PES packet header carries stream synchronization data along with other information. There are several optional fields in the MPEG-2 PES packet header and each optional field may include a series of fields. A field is included by setting its associated indicator flag bit where the indicator flags are located in the MPEG-2 PES packet header's flags byte.

The MPEG-2 PES packet start code, `PES_packet_length` field, and the PES packet header including all PES packet header's optional fields must never be scrambled. The `PES_scrambling_control` bits located in the PES packet header indicate if the `PES_packet_data_bytes` are scrambled. If the PES packet header is not present, the `PES_packet_data_bytes` are not scrambled.

3.8 MPEG-2 SYSTEM TIMING

One of the basic and important concepts of the MPEG-2 Systems Standard revolves around the system timing model. Video and audio Program Elements are delivered as separate streams, with differing delivery rates and different sized Presentation Units. For video, the Presentation Unit is a picture (a frame or field of video). For audio, the Presentation Unit is a block of audio samples (also known as an audio frame). The timing model was developed to control the presentation rate and to enable the synchronization of video and audio presented to the viewer. As we will discuss, elements enabling the synchronization are clock references allowing the decoder to create a clock exactly matching the encoder and time stamps temporally coordinate the presentation of video and audio Presentation Units.

3.8.1 Timing Model

The MPEG-2 Standard's timing model requires the clock at the encoding side be exactly regenerated at the decoder or more specifically, the input encoder sampling rate be regenerated identically at the decoder output. Video and audio consist of discrete Presentation Units, which must be delivered from the decoder at exactly the same rate as they enter the encoder. Thus, the output rate at the decoder must match the input rate at the encoder.

In developing the timing model, MPEG-2 adopted two basic concepts: a constant end-end delay model and the decoding process is instantaneous, as illustrated in Fig. 3.1.14. Both of these notions are theoretical but they provided a firm basis to build upon. MPEG-2 does not specify how the encoders operate; rather, the specification specifies the contents and delivery of the bit stream. With these concepts applied to the bit stream, it is possible to develop implementations of both encoders and decoders taking real-world constraints into consideration and will interoperate. In real systems, the delay through the encoding and decoding buffers is variable [1] and the decoding process takes a finite (and possibly variable) amount of time.

The data rates and sampling sizes for different components of an MPEG-2 Program can be quite different. As a comparison, for video with a frame rate of 29.97 Hz, the sampling size is one picture

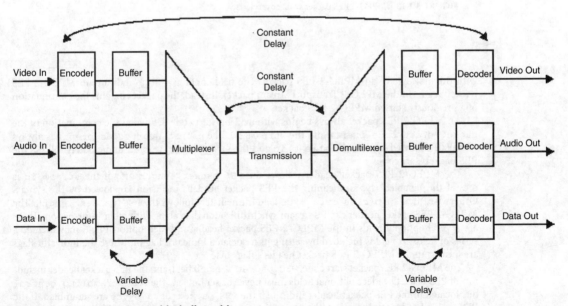

FIGURE 3.1.14 MPEG-2 constant delay buffer model.

and the sample duration is 33.37 ms per sample (picture). A common audio sampling frequency is 44,100 Hz and typically there are 1152 samples per frame, creating an audio frame having a duration of 26.12 ms. Thus, the two sampling rates are significantly different and their decoder output requirements will reflect this characteristic.

The MPEG-2 Systems' timing model solves the issues of synchronization of individual elements by use of a common time reference shared by all individual Program Elements of an MPEG-2 Program. This common time clock is referred to as the STC.

3.8.2 System Time Clock (STC)

The STC is the master clock reference for all encoding and decoding processes. Each encoding source samples the STC as needed to create time stamps associated with the data's Presentation Units. Time stamps associated with the Presentation Unit are referred to as the PTS and we will discuss these below. Time stamps associated with the decoding start time, known as DTSs, may also appear in the bit stream and we will discuss these below as well.

The STC is not a normative element in the MPEG-2 System specification; however, it is required for synchronized services (including video and audio), meaning that all practical implementations require its use. The STC is represented by a 42-bit counter with units of 27 MHz clock periods (27 MHz equals approximately 37 ns per clock period (tick)).

The STC must be recreated in the decoder such that it is identical to the STC at the encoder for both buffer management and synchronization reasons. In order for a decoder to reconstruct this clock exactly, the STC is periodically sampled and transmitted in the MPEG-2 Transport Stream packet `adaptation_field` as clock references known as PCRs. Figure 3.1.15 illustrates a general circuit at the decoder used to recreate the STC. At the time this model was developed, components capable of providing a clock with the required precision were very expensive. This type of circuit provides an elegant way to provide the required accuracy, utilizing inexpensive components.

Each MPEG-2 Program may have its own STC or multiple MPEG-2 Programs may share a common STC (by referring to the same Program Element carrying the PCR values). The STC increases linearly in value in the absence of discontinuities. Since the STC value is contained within a finite size field, it wraps back to zero when the maximum bit count is achieved, approximately every 26.5 h.

Figure 3.1.16 illustrates the STC and the time stamp relationship for the video encoding process using a 29.97 Hz frame rate. As an example, for MPEG-2 Video [2], the encoding process typically uses three types of pictures: Intracoded pictures (I-pictures), Predictive-coded pictures (P-pictures), and Bidirectionally predictive pictures (B-pictures) in compressing and encoding a video stream. The display order and the encoding/transmission order are different and occur as illustrated in Fig. 3.1.16.

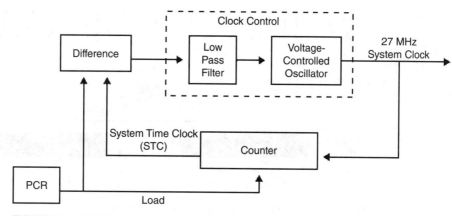

FIGURE 3.1.15 MPEG-2 System time clock.

3.24 DTV TRANSPORT

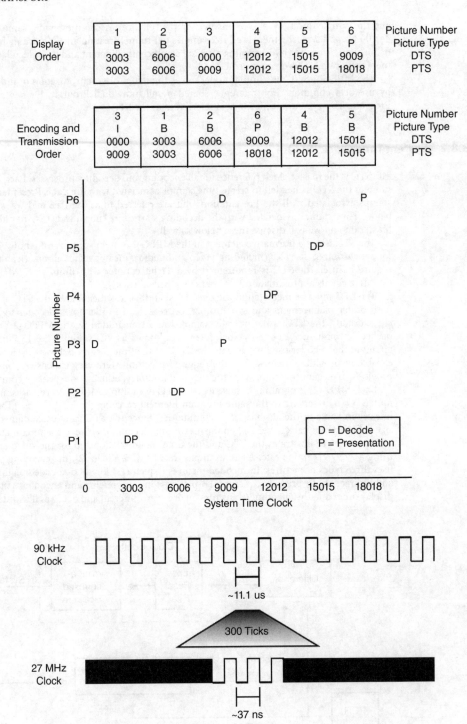

FIGURE 3.1.16 MPEG-2 Video Time stamp to STC relationship.

As described below, the DTS and PTS time stamps control respectively when decoding and presentation occur in the decoder.

In Fig. 3.1.16, the STC increments linearly along the bottom timeline. Picture 3, the I-picture is decoded instantaneously at time zero per the MPEG T-STD decoding model and the DTS value of zero. However, the picture is not displayed until the STC time of 9009. The I-picture must be decoded prior to the B-picture, since the B-pictures reference information contained in the I-picture. The B-pictures, picture 2 and picture 3, are instantaneously decoded and presented at times 3003 and 6006. The P-picture is decoded at time 9009 when the I-picture is displayed but it is not displayed until the time 18018.

The STC is derived from the `system_clock_frequency` specified as 27,000,000 Hz ± 810 Hz. The STC period is 1/27 MHz or approximately 37 ns per clock period.

3.8.3 Program Clock Reference (PCR)

The PCR is a 42-bit value used to lock the decoder's 27 MHz clock to the encoder's 27 MHz clock, thereby exactly matching the decoder's STC to the encoder's STC. The PCR is carried in the MPEG-2 Transport Stream packet's `adaptation_field` using the `program_clock_reference_base` and `program_clock_reference_extension` fields. The MPEG-2 Systems specification mandates the PCR be sent at least every 100 ms or 10 times a second. The PCR may be sent more frequently if desired or required by a complimentary specification. In addition, the specification recommends the amount of PCR jitter be no more than 500 ns. The decoder uses the arrival time of the MPEG-2 Transport Stream packet carrying a PCR value and the PCR value itself compared to the current STC value to adjust the clock control component in Fig. 3.1.15.

The `program_clock_reference_base` is constructed by dividing the value of the 27 MHz clock reference count by 300. This operation creates a 33-bit value in units of 90 kHz clock periods. The `program_clock_reference_extension` contains the remainder of the previous division (i.e., the 27 MHz clock divided by 1 modulo 300).

Each MPEG-2 Program is limited to a single STC; therefore, only a single PID is required for transporting the PCR. The PCR may be carried on the same PID as a video, audio, or data Program Element as the PCR field is independent of the encapsulated data payload. Different MPEG-2 Programs may share the same STC by referring to the same `PCR_PID`. MPEG-2 Programs not requiring an STC (or a common timeline) set the `PCR_PID` field to the value 0x1FFF indicating there is no Program Element carrying a PCR.

3.8.4 Presentation Time Stamp (PTS)

The PTS is a 33-bit quantity measured in units of 90 kHz clock periods (approximately 11.1 μs ticks) carried in the MPEG-2 PES packet header's PTS or `PTS_DTS` fields. The PTS, when compared against the STC, indicates when the associated Presentation Unit should be output. In the case of video, a picture is displayed and in the case of audio the next audio frame is sent to the audio decoder. MPEG-2 Video [2] and Audio [4] require the PTS must be contained in the MPEG-2 Transport Stream at intervals no longer than 700 ms and the ATSC requires the PTS be inserted at the beginning of every access unit (i.e., coded picture or audio frame).

In Fig. 3.1.16, the Presentation Unit (in this case a picture) occurs every 3003 clock ticks and consequently, every 3003 clock tick interval on the STC a new picture is presented.

The PTS, when included, is divided into two 15-bit quantities and a 3-bit quantity spread across 36 bits. There are also three "marker bits," always set to 1b, interspersed amongst the three groups. This division into three parts along with the `marker_bits` inclusion avoids MPEG-2 PES packet header `start_code` emulation. Avoiding `start_code` emulation prevents decoders from incorrectly identifying the beginning of an elementary stream.

3.8.5 Decoding Time Stamp (DTS)

The DTS is a 33-bit quantity, measured in units of 90 kHz clock periods (approximately 11.1 μs), that may be carried in the MPEG-2 PES packet header's `PTS_DTS` field. The MPEG-2 Standard only defines a normative meaning for the DTS field for video. Generally speaking, a video stream is the only `stream_type` to include the DTS, since the bidirectional picture encoding (B-picture) requires the access units (i.e., the coded picture) to be decoded in a different order than they are transmitted. The DTS value is compared to the STC indicating when the access unit should be removed from the buffer and decoded.

The DTS must always be accompanied by a PTS. If the DTS contains the same value as the PTS, the DTS is omitted and the decoder assumes the DTS is equal to the PTS. The DTS, if present, must be contained in the MPEG-2 Transport Stream at intervals no longer than 700 ms. The ATSC mandates the DTS must be inserted at the beginning of every access unit (i.e., coded picture or audio frame), except when the DTS value matches the PTS value.

The DTS is encoded in the same manner as the PTS—splitting the 33-bit quantity into three portions and incorporating the `marker_bits`. (Refer to the PTS section for an explanation regarding why this division occurs.)

In Fig. 3.1.16, the DTS value is associated with Presentation Unit three (i.e., picture 3); however, it must be decoded at time zero so picture 1 may reference the information.

3.8.6 Discontinuities (Program Transitions)

MPEG-2 Program and MPEG-2 Transport Stream discontinuities are a reality in digital television. Planned discontinuities, where the interruption is not the result of an error, can occur in any number of situations. One example is the splicing of a commercial into the video and audio stream. Other planned discontinuity scenarios include switching between content sources or a new MPEG-2 Program commencing. In each of these cases, the STC is interrupted from its normal linear increment pattern to some new random value from which the normal count then continues.

In all of the above instances, the decoder should be notified of the upcoming interruption by the MPEG-2 Transport Stream packet `adaptation_field`'s `discontinuity_indicator`. The `discontinuity_indicator` indicates an STC discontinuity or a disruption in the `continuity_counter`. The signaling of `continuity_counter` disruptions via the `discontinuity_indicator` is limited in its practical usefulness. The multiplexing process can use the `discontinuity_indicator` to indicate a known and expected discontinuity in the program timeline.

The STC interruption results when a decoder receives a new PCR value associated with the MPEG-2 Program. The new STC is provided in the next MPEG-2 Transport Stream packet carrying a PCR as indicated by the `PCR_PID` in the PMT. Decoders receiving new PCR must adjust themselves accordingly. This update may include discarding all pending MPEG-2 PES packets as the PTS and DTS are likely no longer valid. Additionally, the internal STC may be reset and the PCR phase lock loop reset.

Besides the STC reference changing, another discontinuity that may be encountered as part of the stream changeover or Program interruption involves the MPEG-2 Transport Stream packet header's `continuity_counter` value. The `continuity_counter` may skip to a new value when the newly encoded stream is inserted. Thus, the decoder upon seeing the `discontinuity_indicator` is made aware of an upcoming `continuity_counter` change and this change should not be treated as an error or indicative of lost packets.

REFERENCES

[1] ISO/IEC 13818-1, "Information Technology—Generic Coding of Moving Pictures and Associated Audio—Part 1: Systems."

[2] ISO/IEC 13818-2, "Information Technology—Generic Coding of Moving Pictures and Associated Audio—Part 2: Video."

[3] ISO/IEC JTC1/SC29/WG11 Coding of Moving Picture and Audio, Leonardo Chiariglione, "MPEG: Achievements and Current Work," October 2000, www.cslet.it/mpeg_general.htm.

[4] ISO/IEC 13818-3, "Information Technology—Generic Coding of Moving Pictures and Associated Audio—Part 3: Audio."

[5] ISO/IEC 14496-10, "Coding of Audio-Visual Objects—Part 10: Advanced Video Coding."

[6] ATSC: "Digital Audio Compression (AC-3, E-AC-3) Standard," doc. A/52:2012, Advanced Television Systems Committee, Washington, D.C., 17 December 2012.

[7] ATSC: "Program and System Information Protocol (PSIP) for Terrestrial Broadcast and Cable," doc. A/65:2013, Advanced Television Systems Committee, Washington, D.C., 7 August 2013.

[8] ATSC: "ATSC Digital Television Standard," doc. A/53, Advanced Television Systems Committee, Washington, D.C.

CHAPTER 3.2
PROGRAM AND SYSTEM INFORMATION PROTOCOL: PSIP

Dr. Richard Chernock
Triveni Digital, Princeton Junction, NJ

3.1 INTRODUCTION

PSIP (Program and System Information Protocol) is the glue that holds the digital television (DTV) signal together. When broadcast TV was analog, the situation was simple—each radio frequency (RF) channel carried one television program (with the video and audio tightly bound within the signal), so selecting what to watch simply meant selecting the appropriate channel. The transition to digital brought the capability to carry multiple programs within a single RF channel, each with separate audio and video streams. Selecting what to watch now meant tuning to the appropriate RF channel and decoding the correct audio and video streams to make up the desired TV program.

PSIP is metadata inserted into the broadcast stream that allows navigation and access of the channels within the DTV Transport Stream as well as providing information to facilitate efficient browsing and event selection.

Support for an Electronic Program Guide (EPG) is another important function enabled by the PSIP standard. The concept is to provide a way for viewers to find out "what's on" directly from their television sets, similar to the programming guides that are typically available for cable and satellite broadcast services. In a terrestrial broadcast environment, each broadcaster needs to include this type of information within the broadcast stream. Viewers will have the ability not only to choose what channel to watch but also be able to select from multiple options within a broadcast. Examples include selecting from a set of alternative audio tracks in different languages, or choosing one of several programs shown at the same time on different virtual channels.

One strong desire during the transition to digital was to enable a system that would not require retraining of the viewers. In order to achieve this, the following functions were necessary to preserve from the user's viewpoint, which became part of the core design philosophy for PSIP:

- "Where am I?"
  Provide a channel number, channel name
- "Where am I going?"
  Organize channels organized by major/minor groups
  Enable an EPG in the receiver/STB
- What's on now?
  What programs do I want to plan to watch?
- "How can I get where I want to go?"
  Direct entry of channel number
  Navigation on the EPG grid

Additionally, the digital transition involved a change in the broadcast RF channel. During the analog days, television stations became associated with their broadcast channel number and considerable expense was put into branding this number. There was a strong desire to retain this branding, which led to an indirection between the channel number used for tuning and the actual RF channel; the virtual channel concept is explained below.

3.2 VIRTUAL CHANNELS

PSIP introduced the concept of virtual channels, which break the link between RF channel number and user's notion of channel number. For analog broadcast, the "channel number" was the same as the RF carrier designation. With the channel repacking involved in the transition to digital broadcast and the desire to retain existing channel branding, the "channel number" is no longer tied directly to the RF frequency, instead it is indirectly defined by the Virtual Channel Table (VCT). With this indirection, no matter where the broadcast RF frequency shifts in the future, as far as the viewer is concerned, the broadcaster's "channel" will remain the same.

In addition, with digital TV, a single RF channel can include multiple "channels" of programming—thus the notion of utilizing two-part channel numbers. The first part (the "major channel" number) refers to the actual transport (the broadcaster's branding), while the second part (the "minor channel" number) corresponds to the particular stream of programming.

The DTV receiver is expected to use the information in the channel's VCT together with the information contained in other channels' VCTs to build a navigation aid for the viewer so that both analog and DTV programs can be selected.

During setup, the DTV receiver is expected to scan the broadcast band and store the location of each active major channel per its physical channel number (i.e., the RF channel number assigned for that specific 6 MHz channel, or some internal key representing that frequency band), as the major channel number and the Transport Stream ID (a unique number assigned to each station by the FCC). Unlike analog television, the DTV receiver does not use the physical channel number to identify the source of digital programming. Instead, the receiver looks at the major channel assignment in the channel's VCT and utilizes that number as the major channel number for navigation via the virtual channel number. This way, the viewer uses a channel number, labeled by the station itself and stored in the receiver's memory along with the station frequency information, to identify and tune the RF emission from the station. The receiver also looks at and stores the minor channel information carried in the VCT and uses it in conjunction with major channel number for navigation. The stored information for each virtual channel allows the receiver to select the desired video and audio information from among all the other video and audio streams that may be in the DTV station's emission stream.

3.3 PSIP TABLES

PSIP is a collection of tables, each of which describes elements of digital television services. Figure 3.2.1 shows the primary PSIP tables and the notation used to describe them. You can think of PSIP as being akin to a relational database, with specified linkages between tables (such as the `source_id` connecting each Event Information Table (EIT) to the VCT loop it applies to).

Four of the PSIP tables are considered to be "base tables" and the packets of these base tables are all labeled with a base *packet identifier* (PID) (`0x1FFB`). The base tables are as follows:

- System Time Table (STT)
- Master Guide Table (MGT)
- Virtual Channel Table (VCT)
- Rating Region Table (RRT)

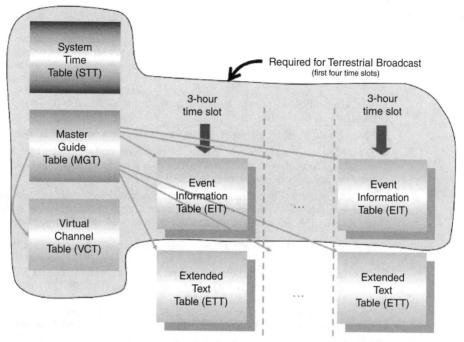

FIGURE 3.2.1 Overall structure of the PSIP tables.

EITs are a second set of tables, whose packet identifiers are defined in the MGT. The Extended Text Tables (ETTs) are a third set of tables, and similarly, their PIDs are defined in the MGT.

3.3.1 System Time Table (SST)

The STT is a small data structure that fits in one Transport Stream packet and serves as a reference for time-of-day functions. It also indicates whether or not Daylight Saving Time is in effect, and signals the day and hour for transitions into or out of Daylight Saving Time. Time in the STT is represented as the count of Global Positioning System (GPS) seconds that have occurred since 00:00:00 January 6, 1980. To convert this count into time of day, one must account for the number of leap seconds that have occurred since the start of GPS time and the present. The STT provides that number in a parameter called "GPS-to-UTC offset." Receivers can use this table to manage various operations and scheduled events, as well as display time of day.

The STT bases its reference for time of day on GPS time, which is measured in terms of seconds since 12:00 a.m. January 6, 1980. This count increments monotonically and hence can be used as a reliable and predictable time base for specification of future times of action.

A receiver needs one other piece of information to derive Coordinated Universal Time (UTC): the current count of the number of leap seconds that have occurred since the beginning of GPS time. The STT delivers this information as well. Leap seconds account for the difference between time based on atomic clocks (as is GPS time) and time based on astronomical events such as the earth's rotation.

A receiver needs two pieces of additional information before it can use the STT data to track local time of day: (1) the offset in hours from UTC (the time zone), and (2) whether or not Daylight Saving Time is observed locally. For a digital television, this information may be entered directly by the consumer via a unit setup function.

3.3.2 Master Guide Table (MGT)

The MGT provides indexing information for the other tables that comprise the PSIP standard. The purpose of the MGT is to provide a convenient reference for receiving devices to the various instances of PSIP tables present in the Transport Stream. The MGT offers the following information for the benefit of receivers:

- A list of all PSIP tables present in the Transport Stream. Tables other than those defined in the A/65 standard may be listed as well.
- For each type of table listed, the PID value for the Transport Stream packets that are used to carry it.
- The version number of each type of table. In some cases, one MGT table type represents more than one table section or even more than one instance. For that case, all tables and table sections must share the common version number indicated in the MGT.
- The total number of bytes transmitted in all transmitted sections of tables of each type. The size parameter is helpful to the receiver so that it can determine beforehand whether or not necessary storage resources are available, and if available, how much storage must be allocated.

Figure 3.2.2 shows a typical MGT indicating, in this case, the existence in the Transport Stream of a VCT, the RRT, four EITs, one ETT for channels, and two ETTs for events.

The first entry of the MGT describes the version number and size of the VCT. The second entry corresponds to an instance of the RRT. Notice that the `base_PID` (0x1FFB) must be used for the VCT and the RRT instances as specified in PSIP.

The next entries shown in the MGT example above (no ordering of elements is implied) correspond to the first four EITs that are required to be supplied in the Transport Stream. The broadcaster is free to choose the PID values (subject to the constraints in A/53 Part 3, Section 6.9) as long as they are unique in the Transport Stream. After the EITs, the MGT example indicates the existence

MGT

table_type	PID	version_num.	table size
VCT	0x1FFB (base_PID)	4	391 bytes
RRT – USA	0x1FFB (base_PID)	1	959 bytes
EIT-0	0x1FD0	6	970 bytes
EIT-1	0x1FD1	4	970 bytes
EIT-2	0x1DD1	2	970 bytes
EIT-3	0x1DB3	7	970 bytes
ETT for VCT	0x1AA0	21	312 bytes
ETT-0	0x1BA0	10	2543 bytes
ETT-1	0x1BA1	2	2543 bytes

FIGURE 3.2.2 Example content of the Master Guide Table.

of a channel ETT using PID 0x1AA0. Similarly, the last two entries in the MGT signal the existence of two ETTs, one for EIT-0 and the other for EIT1.

Descriptors can be added for each entry as well as for the entire MGT. By using descriptors, future improvements can be incorporated without modifying the basic structure of the MGT. The MGT is like a flag table that continuously informs the decoder about the status of all the other tables (except the STT, which has an independent function). The MGT is continuously monitored at the receiver to prepare and anticipate changes in the channel/event structure. A clever "version bubbling" mechanism allows the receiver to discover whether any information in any of the PSIP tables has changed by simply looking for a change in the version of the MGT table itself. When tables are changed at the broadcast side, their version numbers are incremented and the new version numbers are listed in the MGT. Based on the version updates and on the memory requirements, the decoder can reload the newly defined tables for proper operation.

3.3.3 Virtual Channel Table (VCT)

The VCT, also referred to as the Terrestrial VCT (TVCT), contains a list of all the channels that are or will be online, plus their attributes. Among the attributes given are the channel name and channel number. This information is useful for construction of an EPG in a receiver and even for receivers that do not support an EPG, enabling the receiver to identify digital services in a consistent and user-friendly manner.

The VCT is essentially a list containing information about each service that a broadcaster creates or has announced that it will create within the DTV Transport Stream. The VCT consists of one or more virtual channel definitions. The major elements of these definitions are as follows:

- The two-part (major/minor) channel number the user will use to access the service.
- Its short name (up to seven characters).*
- The service `channel_TSID`.†
- Its MPEG-2 `program_number`.
- The type of service (digital TV, audio only, data).
- Its "source ID."‡
- Descriptors indicating what PIDs are being used to identify packets transporting parts of the service and descriptors for extended channel name information.

Other data specific to each terrestrial virtual channel includes a flag that tells whether the service requires one of several special handling conditions, and an indication as to whether "extended text" is available to provide a textual description of the service.

3.3.4 Rating Region Table (RRT)

The RRT has been designed to transmit the rating system in use for each country using the ratings. In the United States, this is incorrectly but frequently referred to as the "V-chip" system; the proper title is Television Parental Guidelines (TVPG). Provisions have been made in the ATSC standard for multicountry systems.

In some cases, such as the United States, a certain region's RRT is unchangeable and defined outside the ATSC standard. In such cases, that region's RRT need not be sent. Receiving devices can be

---

*PSIP provides a descriptor mechanism to define longer channel names as needed.
†This is required to be the same as the TSID of the stream where the DTV service is being transmitted.
‡This is the key link to the announcements in the EITs.

built using the standard definition available for that part of the world. The RRT for `rating_region` 0x01 (US plus possessions) is fully defined in EIA/CEA-766-A.[§]

Note that the EIA/CEA-766-A standard also defines the Canadian RRT (`rating_region` 0x02), but unlike the one for the United States, the Canadian table is specified as changeable. That means at some time in the future, an updated Canadian RRT can be sent. The new table would be identifiable as an update because the `version_number` field would have been incremented.

The function of the RRT is to define a rating system for a given region, where the rating system is characterized by a number of *rating dimensions*, each of which is composed of two or more *rating levels*. An example of a typical rating dimension used on cable is the Motion Picture Association of America (MPAA) system. The levels within the MPAA dimension include "G," "PG," "PG-13," and so on.

Once a receiver learns the dimensions and levels of a rating system, it can do two things:

- Provide a user interface to allow the user to set limits on program content
- Interpret content advisory data on individual program events

Based on a user's preference for certain program content, the receiver can block programming that exceeds a desired threshold.

PSIP does not define the actual dimensions and levels of any rating region; rather, it provides the transport mechanism to deliver the table. The table structure in PSIP allows one or more instances of the RRT to be sent, as needed, where each instance defines one region. For terrestrial broadcast, for many parts of the United States, only the U.S. Rating Region will be applicable. For areas close to national borders, however, a Canadian[¶] or a Mexican rating table may be sent in addition.

Note that inclusion of the RRT corresponding to `rating_region` value 0x01 (U.S. plus possessions) is not required in transmissions in the United States.

3.3.5 RRT Structure

The RRT is a fixed data structure in the sense that its content remains mostly unchanged. It defines the rating standard that is applicable for each region and/or country. Several instances of the RRT can be constructed and carried in the Transport Stream simultaneously. Each instance is identified by a different `table_id_extension` value (which becomes the `rating_region` in the RRT syntax) and corresponds to one and only one particular region. Each instance has a different version number that is also carried in the MGT. This feature allows updating each instance separately.

Figure 3.2.3 shows an example of one instance of an RRT for a region called "Tumbolia," assigned by the ATSC as `rating_region` 20. Each event listed in any of the EITs may carry a Content Advisory Descriptor. This descriptor is an index or pointer to one or more instances of the RRT.

3.3.6 Event Information Table (EIT)

EITs provide program schedule information for any digital channel listed in the VCT. A program schedule consists of the start and end times of each event, its name or title, linkage to a textual description of the event, and an optional list of descriptors that can give further information pertinent to the event, such as its content advisory and the caption services available with the program.

[§]ANSI/CEA: "U.S. and Canadian Rating Region Tables (RRT) and Content Advisory Descriptors for Transport of Content Advisory Information Using ATSC Program and System Information Protocol (PSIP)," Doc. ANSI/CEA-766-C, American National Standards Institute and Consumer Electronics Association, Arlington, VA, April 2008.

[¶]See ANSI/CEA-766, footnote #4.

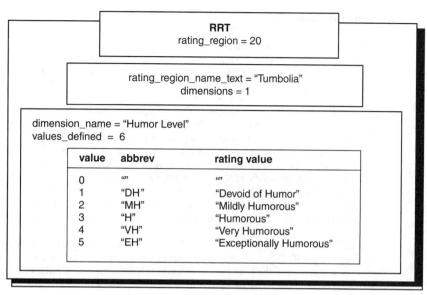

FIGURE 3.2.3 An instance of a Rating Region Table (RRT).

There are up to 128 EITs, EIT-0 through EIT-127, each of which describes the events or television programs for a time interval of three hours. Because the maximum number of EITs is 128, up to 16 days of programming may be advertised in advance. At minimum, the first four EITs are required to always be present in every Transport Stream, and 24 are recommended. Each EIT-k may have multiple instances, one for each virtual channel in the VCT.

Each EIT covers a period of three hours. The PSIP generator should automatically convert from local time to the universal time used inside the system. (The receiver converts back to that receiver's local time.) EIT-0 represents the "current" three hours of programming. The first four EITs (EIT-0, EIT-1, EIT-2, and EIT-3) are required to be present by A/65. The maximum number of EITs is 128, permitting up to 16 days of program information to be delivered to receivers. It is strongly recommended that at least 24 EITs be sent at all times (three days). Following this recommendation ensures that schedule information for the broadcaster will be present in the EPG if the station is watched within the previous 3 days.

EIT-0 plays an additional role beyond announcement (promotion of upcoming events), as it provides signaling information pertaining to the current program and therefore is particularly important with respect to the carriage of closed caption information, ratings information, language information, and other essential data. The connection from broadcast station equipment to the PSIP generator should enable direct updates of current program parameters in EIT-0. By contrast, the EITs for the future are primarily promotional and less critical to system performance as long as the station virtual channel lineup is not changed.

Each EIT has space for event titles. The receiver recommendation is to display the first 30 characters of the title, and it is recommended that the first 30 characters be chosen carefully to maximize the chance of meaningful display by receivers. If it is desired to send additional information about the entire event, this is sent in another structure—the ETT. Such information would optionally be presented to consumers, usually after an action. Receivers may have limited support for descriptive text so there may be a tradeoff between covering more events and more data about each event. Also, the rate this information is sent can be adjusted by setting the time interval between ETTs to make more efficient use of bandwidth.

3.3.7 Extended Text Table (ETT)

PSIP defines a method to include multilingual text in the digital Transport Stream. The bulk of this text is event descriptions, giving a few words, a sentence, or a few sentences of information about the contents of a given program. Depending on the type of program, the text can include the title of an episode, a synopsis of the story line, the names of actors, the year of production, or anything the provider of the guide data wishes to include.

The ETT is an optional component used to provide detailed descriptions of virtual channels or events. While optional, ETTs should be sent, as they are a powerful promotional tool for the broadcaster and many consumer receivers display ETT data when it is available.

These descriptions are called Extended Text Messages (ETMs). The format of the 32-bit ETM identification element tells the receiver whether the ETM describes a channel or an event within the EIT. This format allows the receiver to search for a single description quickly without having to parse the payload of a large table.

Each instance of an ETT carries one text block. Fields in the EIT and VCT link a program event or virtual channel to an ETT instance. As with all text delivered with PSIP, multiple languages are supported.

ETTs are linked to events announced in EITs by the Source ID (to identify the virtual channel) and by the event ID (to reference a specific event on that channel). ETTs may also be linked to data events announced in the Data Event Tables. In this case, the data ID plays the same role as the Source ID. Yet another way ETTs are used is to deliver textual information regarding a virtual channel; this kind of ETT is called a channel ETT.

As illustrated in Fig. 3.2.4, there may be multiple ETTs, one or more channel ETT sections describing the virtual channels in the VCT, and an ETT-k for each EIT-k, describing the events in the EIT-k. These are all listed in the MGT. An ETT-k contains a table instance for each event in the associated EIT-k. As the name implies, the purpose of the ETT is to carry text messages. For example, for channels in the VCT, the messages can describe channel information, cost, coming attractions, and other related data. Similarly, for an event such as a movie listed in the EIT, the typical message would be a short paragraph that describes the movie itself. ETTs are optional in the ATSC system.

3.4 PSIP DESCRIPTORS

Digital television receivers are required to decode closed captioning and to support blocking of programming based on content advisory information. They may also offer the option to decode alternate audio services (such as alternate languages or Descriptive Video services). In order to inform the

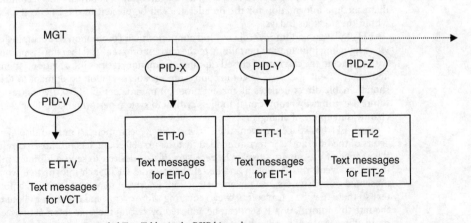

FIGURE 3.2.4 Extended Text Tables in the PSIP hierarchy.

viewer that closed captioning and alternate audio are available from the broadcast station and to provide easy access to these services, receivers depend on data that can be sent by the broadcaster via PSIP. Content advisory information, also sent in PSIP, is directly acted on by the receiver to block programs.

This information is carried in PSIP data structures called *descriptors*. Descriptors are a general-purpose, extensible tool provided by the MPEG-2 Systems standard (ISO/IEC 13818-1). All descriptors have a format that begins with an 8-bit tag value followed by an 8-bit descriptor length and data fields. The tag value tells the receiver what type of descriptor it is. The length field permits devices that do not understand the descriptor (i.e., do not recognize the descriptor tag) to easily bypass it. Delivery of the following descriptors is mandatory when the element described is part of a program:

- The Service Location Descriptor (SLD)
- Content Advisory Descriptor (EIT)
- AC-3 Audio Descriptor (EIT and PMT)
- Caption Service Descriptor (EIT)

3.4.1 Service Location Descriptor

The Service Location Descriptor (SLD) is required to always be present in the TVCT (one SLD per virtual channel). The SLD acts as a directory of components for the virtual channel (somewhat equivalent to the PMT in MPEG-2 PSI)—providing a list of the program elements, stream types, and PIDs.

3.4.2 Content Advisory Descriptor

Parental advisory information is carried in the Content Advisory Descriptor. The receiver uses this information directly to block programs that exceed the ratings selected in a user setup procedure, and may be used by the receiver to provide on screen information about a program's rating for objectionable material.

3.4.3 AC-3 Audio Descriptor

The receiver looks for and uses the AC-3 Audio Descriptor to create viewer information about the audio services that are available. In addition to describing possible alternate audio services that a broadcaster might send, this descriptor provides the receiver with audio setup information such as whether the program is in stereo or surround sound.

Digital television can carry audio streams in different languages. Analog television allowed a clear distinction between primary and secondary audio. A flag has been introduced into the AC-3 Audio Descriptor structure to replicate this function, allowing a broadcaster to indicate which audio stream is to be considered primary. Inclusion of ISO639 language code bytes in the AC3 descriptor provides signaling and announcement of the language used for the audio service.

3.4.4 Caption Service Descriptor

Captioning text itself, which is carried in an area of the video data set aside for captioning, does not carry a description of the captioning language (such as English or Spanish), so receivers are expected to rely on the Captioning Service Descriptor to provide the data needed to create on-screen captioning information.

The intention of the caption service announcement in PSIP data is to indicate to the viewers whether the event (described by the program name and description in EIT/ETT) is scheduled to be captioned, and if so, in what language(s). During an advertising insert within the program, caption data is not required to match the description in the Caption Service Descriptor.

More importantly, the receiver is expected to rely on the Caption Service Descriptor to tell it that the program is captioned in the first place; the presence of captioning text is not used by most receivers to indicate to the viewer that the program is captioned. In addition to viewer information, the Caption Service Descriptor contains important control information needed by the receiver for proper display of captioning.

3.5 THE BIG PICTURE

As mentioned previously, PSIP can be considered to be a set of interlinked tables, as illustrated in Fig. 3.2.5. The MGT serves as a directory of all other PSIP tables, with pointers and links to each of the PSIP tables in the emission. The VCT, which serves as the main "tuning" (or service discovery) element, is linked to the EITs that carry further information about the current events. If there are descriptions of the virtual channels, these are carried in ETTs linked to the virtual channels in the VCT. Other ETTs carry descriptions of the scheduled events and are linked to the appropriate portion of the EITs.

3.5.1 Putting It All Together for the Viewer

Figure 3.2.6 shows a representative EPG identifying the PSIP tables that provide the information making up the elements of the EPG display. Specifically,

- The **STT** provides the current time.
- The **VCT** provides the two part channel number and short name of the currently viewed channel, as well as channel numbers and short names for other available services.

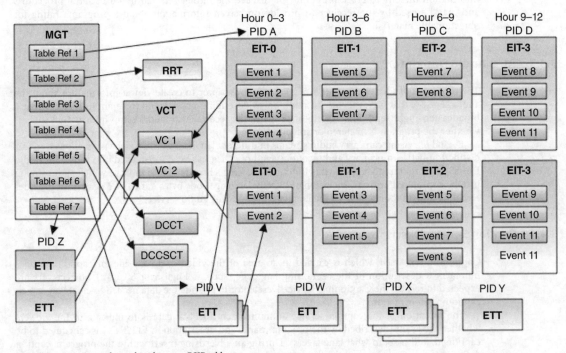

FIGURE 3.2.5 Interrelationships between PSIP tables.

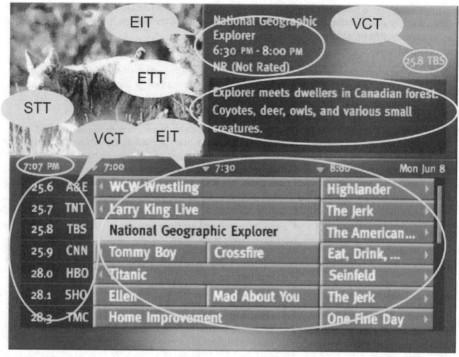

FIGURE 3.2.6   EPG with PSIP elements identified.

- The **EIT** provides information about the currently viewed event (title, start time, duration, and rating) as well as information to fill out the program grid.
- The **ETT** provides a longer description of the currently viewed event.

3.5.2 PSIP Information Sources

In order to generate PSIP, a number of types of information are needed, some relatively static, some quite dynamic—as shown in the list below:

- Branding
  Station identity
  Channel information
- Signaling
  Information necessary for tuning
  Information necessary to make pictures and sound
  "Characteristics" of the current event: such as ratings, captioning, audio language
- Announcement
  EPG/schedule information
  What events (programs) are coming up and when
  Descriptive information for events
  "Characteristics" of events: ratings, captioning, audio language, and so on
- Linkage amongst information

While there are significant differences in broadcast television station workflows, the information listed above typically originates from a small number of possible systems, as demonstrated in the following list:

- Listing services (such as Tribune Media Services and Rovi) collect information about a TV station's upcoming schedule and augment this with further information from other sources (such as the content creators). This augmented information is then made available back to the TV station and often used in PSIP. Information from this source typically includes: channel name/number, program name, time, descriptions, CC, ratings, audio types.
- Traffic systems are commonly at the core of a TV station's scheduling and business workflow. Scheduling is typically initially done within the traffic system and advertisements associated with programs. This schedule may be refined over time as part of the workflow. The traffic system usually contains information about the program name and time, and optionally may include: channel name/number, program descriptions, CC, ratings, audio types.
- The automation system is ultimately responsible for controlling the play-out of all programs and interstitials that go to air. Information contained within the automation system includes the program ID and accurate time, and optionally may include: channel name/number, program name, descriptions, CC, ratings, audio type.
- The multiplexer/encoder/control system is aware of configuration (signaling) information that needs to be encoded in PSIP, such as the PIDs used for elementary streams (as needed in the TVCT SLD) and optionally the channel name and number.
- Human operators are (or may be) aware of all information needed for PSIP; however, with the continual downsizing of station operations, the personnel at a station are already busy with many other things and so interaction with a PSIP system should be avoided in nonintervention operations.

As can be seen above, there are a number of systems involved in the station workflow that contain information needed for PSIP. However, due to the way station operations have evolved, the amount of information known by these systems and the accuracy over time varies. For example, a listing system typically "knows" a lot of information about the television events and their descriptions, but does not usually become informed about last-minute schedule changes. At the other end of the spectrum, the automation system "knows" exactly when each event will air (even in the presence of last minute changes, primarily since the automation system is responsible for the actual play-out), but the automation system often does not have the full information and description of the events.

For these reasons, it is recommended that PSIP generator systems draw information from the sources available to the station and merge the information in a prioritized fashion to take advantage of the depth available in some and the accuracy available in others—especially in the presence of last-minute changes.

3.5.3 Table Intervals

One of the significant components of channel change time for a digital TV receiver is how quickly the receiver can gather the metadata necessary to understand the configuration that has been set up for that virtual channel (for example, which PIDs carry the video and audio components that comprise the specific television program). There is a sequence that the receiver follows to assemble this information.

- The MGT must first be obtained and parsed (the MGT provides the linkage and version information for all other PSIP tables).
- The VCT is then obtained and parsed to get the information about the virtual channel and the PID assignments (via the SLD).
- EIT-0 is obtained and parsed to gather further information about the particular event that is being watched (such things as rating info, language information, and so on).

TABLE 3.2.1 Suggested Repetition Intervals for PSIP Tables

PSIP Table	Suggested Repetition Interval	Required Repetition Interval
MGT	150 ms	≥150 ms
TVCT	400 ms	≥400 ms
EIT-0	Once every 0.5 s	n/a
EIT-1	Once every 3 s	n/a
EIT-2 and EIT-3	Once every minute	n/a
STT	Once every second	1000 ms
RRT (not required in some areas)	Once every minute	≥1 min

The frequency at which these tables are sent (cycle time) will determine how quickly the receiver can set up the configuration to successfully decode the current event. It might seem that sending them as frequently as possible would reduce the channel change time to a minimum. However, in a sense, these tables represent overhead in the system—bits that could be used to carry actual content. Consideration of these factors led to the selection of recommended repetition intervals as shown in Table 3.2.1. The recommended table transmission times listed result in a minimal demand on overall system bandwidth. Considering the importance of the information that these PSIP tables provide to the receiver, the bandwidth penalty is trivial.

CHAPTER 3.3
IP TRANSPORT FOR MOBILE DTV

Gomer Thomas
Gomer Thomas Consulting, LLC, Arlington, WA

3.1 INTRODUCTION

The broadcast multiplex and signaling for the ATSC Mobile DTV system are specified in ATSC document A/153 Part 3 [1]; announcements (ESG data) are specified in A/153 Part 4 [2].

The ATSC Mobile DTV system is a native IPv4 system. Only IPv4 packets come out of the physical layer (although there are features to support other possible types of network packets in the future). All of the service components and signaling are carried over IP (typically IP multicast).

3.2 CONTENT DELIVERY FRAMEWORK

The ATSC Mobile DTV system delivers data in the form of "parades." The number of parades and the Forward Error Correction (FEC) parameters of each parade are determined by the physical layer configuration. Each parade can carry one or two "ensembles" (similar to "data pipes") depending on the configuration. Each ensemble consists of a sequence of Reed–Solomon (RS) Frames. The size of each RS Frame depends on the configuration of the parade in which it appears. Due to the memory and processing requirements of decoding the RS Frames, most receivers will be able to receive only one, or at most two, ensembles at a time. The minimum time between successive RS Frames in an ensemble is slightly under a second.

The payload of an RS Frame is a rectangular array of bytes, containing 187 bytes in each column and a variable number of bytes each row (from 79 to 1636), depending on the configuration. Each row contains a 2-byte header, with the rest of the row available for network packets (e.g., IP packets), as shown in Fig. 3.3.1.

The only two values defined in the initial version of A/153 Part 3 for the Network Protocol field are 0x0 for IPv4 and 0x7 for a "framed packet type" which accommodates any network protocol packets (albeit with higher overhead). The framed packet type is not expected to be used, currently. Future versions of the standard could define values for other network protocols, such as IPv6.

The Error Indicator bit indicates whether the frame decoder has detected an error in the row.

If the Stuffing Indicator field indicates that there are any stuffing bytes, then the first one or two bytes of the stuffing bytes field tell how many stuffing bytes there are (with some clever coding to handle the degenerate cases when there are exactly one or exactly two stuffing bytes).

If a new network packet starts in the row, the Pointer Field gives the starting byte position of the first such packet; otherwise it contains 0x7FF.

| Network Protocol (3-bit) | Error Indicator (1-bit) | Stuffing Indicator (1-bit) | Pointer Field (11-bit) | Stuffing Bytes (k bytes) | Payload (N-2-k bytes) |

FIGURE 3.3.1 Row structure of Reed-Solomon Frames.

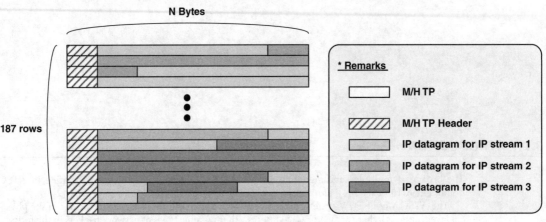

FIGURE 3.3.2 Rows in an RS Frame payload.

Any datagram can be split across multiple RS Frame rows. Figure 3.3.2 shows a typical RS Frame with datagrams spanning rows.

Any datagram other than an NTP*/UDP/IP datagram can be split across RS Frames.

The Maximum Transmission Unit (MTU) is 1500 bytes, i.e., any IP packet larger than 1500 bytes in size must be fragmented. However, fragmentation cannot be applied to RTCP[†]/UDP/IP or NTP/UDP/IP datagrams.

Different ensembles are treated as separate IP subnets with separate network interfaces, i.e., the scope of an IP address in an IP packet header is the ensemble in which the IP packet arrives. (This allows different ensembles to be used by different service providers without the need for IP address coordination.)

3.3 SERVICES

ATSC Mobile services consist of components. Streaming components, such as audio and video, are carried via RTP.[‡] Closed captions are carried in video frame headers. Discrete components, such as files, are carried via FLUTE.[§]

Each service has a 16-bit service identifier (service ID). The service ID can be presented to viewers as a two-part channel number (high-order byte, followed by a delimiter, followed by low-order byte). High-order bytes in the range 0–1 are ATSC reserved; those in the range 2–69 are allocated to local

---

*Network Time Protocol, specified in IETF RFC 1305, March 1992, and RFC 5905, June 2010.
[†]Real-Time Transport Control Protocol, specified in IETF RFC 3550, July 2003.
[‡]Real-Time Transport Protocol, specified in IETF RFC 3550, July 2003.
[§]File Delivery over Unidirectional Transport, specified in IETF RFC 6726, November 2012.

broadcasters according to the rules for assigning major channel numbers in ATSC A/65 (PSIP); those in the range 70–255 are reserved for "regional" services. A service is considered "regional" if it is broadcast in essentially the same form in more than one broadcast area (with possible differences in interstitial materials), and the service provider wants the multiple broadcasts to be identified as the same service. A registration authority will likely need to be set up for such usage.

Each service also has a service name, category (type), status (active, hidden, etc.), some other indicators, and a list of components.

Services are normally contained within a single ensemble, but it is possible for a service to have components in more than one. In such a situation the service is signaled (with the same service ID) in all ensembles where it has any components. The signaling in each ensemble lists all the components of the service that are in that ensemble, and there are indicators in each ensemble showing that the service has components in multiple ensembles. (This accommodates such things as a scalable video encoding with a base encoding in a more robust ensemble and an enhancement encoding in a less robust ensemble.)

3.4 SIGNALING

There are two low-level signaling mechanisms (sometimes called L1 and L2 signaling).

The Transport Parameter Channel (TPC), defined in A/153 Part 2 [3], gives a list of all the parades in the broadcast stream, and it gives the FEC parameters of each one.

The Fast Information Channel (FIC), defined in A/153 Part 3 [1], gives a list of all the ensembles in the broadcast stream. It gives a list of all the services in each ensemble, with the service ID and a few flags for each service. It also indicates the ensemble(s) in which the Service Labeling Table (SLT), Guide Access Table (GAT), and Emergency Alert Table (EAT) appear. (These tables are described below.)

Both of these low level information channels are delivered outside the usual RS-Frame structure, in a way that makes them quickly available when a device tunes to the RF (radio frequency) band, without waiting for a full RS Frame to be delivered.

For upper level signaling, each ensemble contains a "service signaling channel," consisting of IP packets with a dedicated IP address and UDP port (assigned by IANA, the Internet Assigned Numbers Authority). The following six signaling tables are carried in these channels:

- Service Map Table (SMT)—appears in every ensemble; lists all services in the broadcast stream that have at least one component in the ensemble; and describes the components and other properties of each service.

- Guide Access Table (GAT)—appears only in the ensemble(s) indicated in the FIC; lists sources of service guide data and describes each one, including network type (broadcast or broadband) and access information.

- Cell Information Table (CIT)—optional; provides access information for services in adjacent broadcast areas that are the same as, or very similar to, services in the currently watched broadcast, to allow continuity of viewing for a viewer traveling from the coverage area of one transmitter to the coverage area of others.

- Service Labeling Table (SLT)—appears only in the ensemble(s) indicated in the FIC; lists all services in the broadcast stream and provides the service ID, name, and category (type) for each one.

- Rating Region Table (RRT)—defined in ATSC A/65; if transmitted, must be transmitted in the ensemble indicated for the GAT in the FIC and can be transmitted in other ensembles; and describes content advisory rating systems in use in the broadcast stream.

- Emergency Alert Table (EAT)—defined in A/153 Part 10 [4]; appears only in the ensemble(s) indicated in the FIC; and carries emergency alert messages as needed.

The format of all these tables is based on the MPEG-2 private section format.

The combination of the FIC, SLT, and SMT enables three levels of channel scans. A very rapid channel scan can look at only the FIC for each broadcast stream to compile a service list which includes only the service IDs. A fairly rapid channel scan can look at the SLT for each broadcast stream to compile a service list which includes the service IDs, names, and categories (types). A full-channel scan can parse the SMT for each ensemble in each broadcast stream to get a service list which includes full information on each service.

The SMT provides for each component of each service the source IP address, destination IP address(es), and destination IP port(s) of the component, along with a "component descriptor" that contains additional information specific to the type of the component. There are component descriptors for video (AVC/SVC¶), audio (HE AAC v2**), file delivery (FLUTE), service protection key delivery (STKM/LTKM††), NTP time base, and one "wild card" descriptor that can be used to describe new types of components (albeit less efficiently than for known components).

There are also a number of other descriptors that can be put into the SMT to provide additional information about services and current programs (events), such as:

- Content Labeling Descriptor—unique identifier for current content
- Caption Service Descriptor—information about caption services
- Content Advisory Descriptor—parental guidance rating of current content
- Genre Descriptor—genre of current content
- Rights Issuer Service Descriptor—provides access information for rights issuers (for protected services)
- Protection Descriptor—indicates which option must be used for filtering encrypted and clear packets for protected services

3.5 TIMING AND BUFFER MODEL

The broadcast stream carries NTP (Network Time Protocol) timestamps that can be used by receivers to maintain a reference clock. Components are carried in RTP packet streams, and each such stream has an associated RTCP packet stream that is used for synchronization. The RTP packet headers contain "RTP timestamps" that give presentation times of access units in the RTP packets, relative to an RTP clock unique to each component. The associated RTCP packets each contain an RTP timestamp (RTCP_RTP_TS) relative to the RTP time line and a corresponding NTP timestamp (RTCP_NTP_TS) relative to the NTP time line. Then, the presentation time PTS of an access unit relative to NTP time is

```
PTS = RTCP_NTP_TS + (RTP_TIMESTAMP - RTCP_RTP_TS)/(RTP_CLOCK_RATE)
```

where RTCP_RTP_TS and RTCP_NTP_TS are from a recent RTCP packet.

As long as the timestamps are set correctly by the encoders, and the access units are presented at the proper NTP times, the multiple media streams of a service will be synchronized.

The buffer models for streaming content are adapted from the buffer models in the MPEG-2 Systems standard [5]. Incoming IP packets go through a sequence of specified buffers for smoothing, filtering, assembling RTP packets, etc., ending up with a sequence of frames. Then the decoder takes the frames from the final buffer at the proper time for decoding. None of the buffers are allowed to overflow, and the final buffer is not allowed to underflow, i.e., receivers must provide appropriate space for buffering, and encoders and multiplexers must manage the data insertion into the broadcast so that the data is not delivered too early or too late relative to presentation times.

¶Advanced Video Coding/Scalable Video Coding, specified in ISO/IEC 14496-10:2005.
**High Efficiency Advanced Audio Coding version 2, specified in ISO/IEC 14496-3:2005/Amendment 2:2006.
††Short-Term Key Management/Long-Term Key Management.

3.6 ANNOUNCEMENTS

Each mobile broadcast stream can contain service guide data for the services in that broadcast stream, and it can contain service guide data for other services as well.

The service guide data specified in A/153 Part 4 [2] is based on the OMA BCAST Service Guide (SG) [6], with certain restrictions. The Service, Schedule, Content, Access, Session Description, Purchase Item, Purchase Data, Purchase Channel, and Preview fragments are included, with certain elements and attributes excluded.

A classification scheme based on the genre table in A/65 is used for the OMA SG Genre element.

A mapping is defined from the values in the Content Advisory Descriptor of the ATSC PSIP standard [7] to values in the OMA SG Parental Rating element.

Both broadcast and broadband delivery of OMA SG fragments is supported, based on the OMA BCAST SG delivery specifications. Broadcast delivery is via services that are signaled in the SMTs for the ensembles in which they appear, as well as being identified in the GAT. The URLs for broadband delivery appear in the GAT.

3.7 TERMS

ESG—Electronic Service Guide

FEC—Forward Error Correction

FLUTE—File Delivery over Unidirectional Transport

NTP—Network Time Protocol

RTCP—Real-time Transport Control Protocol

RTP—Real-time Transport Protocol

REFERENCES

[1] ATSC: "Mobile DTV Standard Part 3: Service Multiplex and Transport Subsystem Characteristics," Doc. A/153 Part 3:2013, Advanced Television Systems Committee, Washington, D.C., 29 October 2013.

[2] ATSC: "Mobile DTV Standard Part 4: Announcement," Doc. A/153 Part 4:2009, Advanced Television Systems Committee, Washington, D.C., 15 October 2009.

[3] ATSC: "Mobile DTV Standard Part 2: RF Transmission System Characteristics," Doc. A/153 Part 2:2011, Advanced Television Systems Committee, Washington, D.C., 7 October 2011.

[4] ATSC: "Mobile DTV Standard Part 10: Mobile Emergency Alert System," Doc. A/153 Part 10:2013, Advanced Television Systems Committee, Washington, D.C., 11 March 2013.

[5] ISO/IEC 13818-1:2007, "Information Technology—Generic Coding of Moving Pictures and Associated Audio Information: Systems," October 2007.

[6] Open Mobile Alliance OMA-TS-BCAST_Service_Guide-V1_0, "Service Guide for Mobile Broadcast Services," Version 1.0, February 2009.

[7] ATSC: "Program and System Information Protocol for Terrestrial Broadcast and Cable," Doc. A/65:2013, Advanced Television Systems Committee, Washington, D.C., 7 August 2013.

CHAPTER 3.4
MOBILE EMERGENCY ALERT SYSTEM

Wayne C. Luplow
Zenith R&D Labs, Lincolnshire, IL

Wayne Bretl
Zenith R&D Labs, Lincolnshire, IL

Jay C. Adrick
GatesAir, Mason, Ohio

3.1 INTRODUCTION

The M-EAS (Mobile Emergency Alert System) is a means of transmitting emergency alert information to Advanced Television Systems Committee (ATSC) mobile DTV (Digital Television) receiving devices that are enabled with a software application. It is implemented as a set of high-level capabilities to transmit CAP (Common Alerting Protocol) message content and optional additional associated rich media information, in a maximally robust and useful form. CAP is an XML-based data format used to exchange alerting messages originating from various U.S. agencies, and is the form used in the familiar EAS (Emergency Alert System) from which broadcasters receive messages typically scrolled over their normal video programming. M-EAS capabilities include sending the alert message as metadata and generating an overlay in the receiver instead of at the studio/transmitter; restricting alerts to the appropriate affected geographic area; waking up the receiver from standby mode; directing the user to associated information; and general management such as alert message storage and expiration.

It is important to note that M-EAS is an add-on to Mobile DTV and not a replacement for EAS, which is a service provided by traditional fixed DTV broadcast. M-EAS does not impact existing EAS messaging services. Traditional EAS on fixed DTV includes a text crawl that is added to the video before broadcast. M-EAS includes a text string that is transmitted as metadata. The M-EAS receiver generates a visible banner according to its capabilities, formatted in an optimum way for its display.

3.2 M-EAS AS PART OF ATSC MOBILE DTV

M-EAS is added to ATSC mobile DTV by inclusion of certain data tables in the mobile signal. Rich media content for emergency information is carried as ATSC non-real-time (NRT) content, and is referenced by the tables. NRT is a feature developed by ATSC that provides the capability to transmit content (audio, video, etc.) as a file and store it in the receiver for use on demand.

The key addition to ATSC mobile DTV for M-EAS is an Emergency Alert Table (EAT), which carries the basic CAP text and links to any related rich media files. All of the M-EAS content is carried with the added robustness of the mobile portion of the ATSC signal.

3.3 M-EAS RELATIONSHIP TO NATIONAL ALERTING INFRASTRUCTURE

Figure 3.4.1 shows the various parts of the emergency alerting infrastructure in the United States, and the position of M-EAS in the overall structure. M-EAS is designed to use the nearly universally available digital broadcast TV signal to reach widely available personal receivers.

3.4 M-EAS INPUT SOURCES

M-EAS receives input from the IPAWS (Integrated Public Alert and Warning System) aggregator and/or from local sources as indicated in Fig. 3.4.2. Local sources create messages in the IPAWS-compliant CAP format. In addition, M-EAS can add rich media content such as maps and videos for such purposes as giving weather reports, instructing the public on safety procedures, shelter locations, weather radar, amber-alert details, evacuation routes, etc.

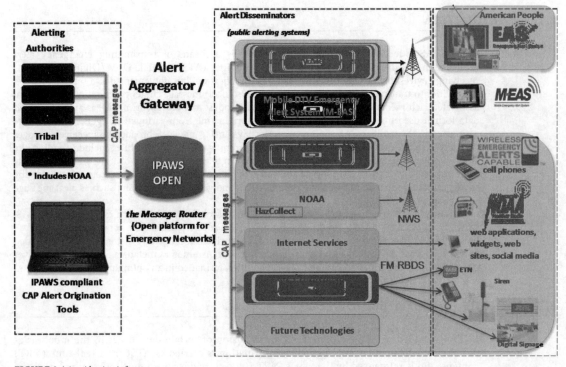

FIGURE 3.4.1 Alerting infrastructure.

MOBILE EMERGENCY ALERT SYSTEM   **3.51**

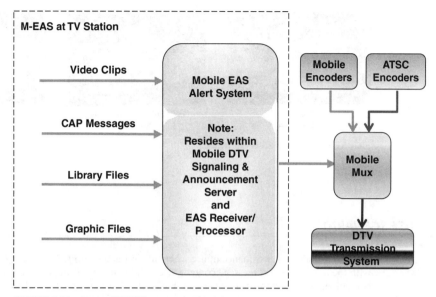

FIGURE 3.4.2   Types of M-EAS messages and content.

Sources of content include the IPAWS system for external input, plus all local TV station resources—SNG, ENG, news copter, traffic cameras, tower camera, live studio, weather radar, graphic systems, electronic still camera pictures, and hyperlinks to internet content.

The general signal flow at the station is shown in Fig. 3.4.3.

For stations that are already transmitting ATSC Mobile DTV (A/153), the addition of M-EAS requires the following system additions:

- Software key enabling a CAP IP output on the station's current EAS receiver/processor (receives/generates alerts)
- Mobile content manager software added to the mobile signaling server (creates M-EAS signaling and M-EAS content menu, associates rich media content with alert messages, manages alerts to station generated rules)
- Possible software updates to the station mobile DTV multiplexer and the mobile DTV exciter

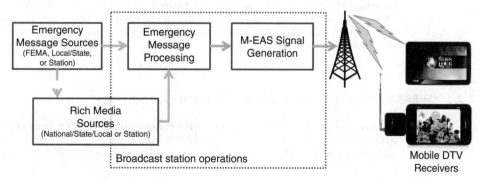

FIGURE 3.4.3   M-EAS signal flow.

3.52 DTV TRANSPORT

         Pop-up Alert   Available Associated   View Alert
         Message   Media   Media

FIGURE 3.4.4 Example user experience.

3.5 USE SCENARIO

Figure 3.4.4 shows the sequence of use when an alert is transmitted. The user first sees a banner text with the basic CAP text message content. Second, the banner is linked to any associated rich media content via the Electronic Service Guide (ESG). Third, the rich media content is selected by the user, downloaded and stored on the receiving device, and displayed.

3.6 M-EAS ADVANTAGES

The major advantages of M-EAS in relation to other alert delivery systems result from its use of the one-to-many transmission of the digital television broadcast system. Broadcast transmitters are not subject to over-capacity outages due to the large number of users in a disaster. They are generally provisioned to withstand extreme natural events without physical damage, and to provide extended operation on substitute power when utility service is lost. M-EAS receivers operate on battery power, which can be recharged either from the mains or from automobile power jacks in an emergency.

3.7 IMPLEMENTATION

Deploying M-EAS requires the addition of software or a server with software to the mobile broadcast system and the interfacing of the M-EAS system to the station's existing EAS receiver/processor. The overall cost of adding M-EAS is small, estimated to be between 5 percent and 10 percent of the cost of implementing mobile DTV.

3.7.1 M-EAS Content Manager and M-EAS Transmission Software

The M-EAS Content Manager is realized in software running on a shared hardware platform. Additional software needs to be added to the station's existing mobile DTV signaling and announcement server as shown in Fig. 3.4.5. The software addition has two levels of functionality: M-EAS content management and M-EAS transmission.

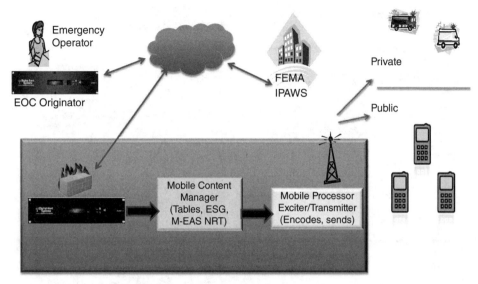

FIGURE 3.4.5 Equipment overview.

3.7.2 Interface to Station's Current EAS/CAP Equipment

Current FCC regulations require all stations to have an EAS/CAP receiver/processor capable of decoding CAP formatted messages from FEMA's IPAWS server under authorized access. Since the EAS/CAP receiver/processor connection is already documented with FEMA, the station needs no additional authorization to receive these messages.

All stations have an EAS/CAP receiver/processor device, which is usually interfaced to a character generator as well as the master control audio. For M-EAS, an output interface on the EAS/CAP receiver/processor is required for CAP-xml file output via IP to drive the M-EAS Content Manager. The EAS/CAP receiver/processor serves to filter alerts, translate non-CAP alerts to CAP, and originate local station-based alerts.

Figure 3.4.6 shows a more detailed view of data flow, from the incoming alert data to the broadcast stream. Figure 3.4.7 shows details of the Content Manager functional blocks and interfaces.

Figure 3.4.8 shows a typical sequence of processes. There are specific processes for retransmitting alerts (for the benefit of receivers that tune in after the initial alert), and for declaring alerts to be expired, which are not illustrated here.

Figure 3.4.9 is a simplified diagram of the M-EAS-related data tables in the broadcast stream. The EAT contains a particular CAP message and references the Service Map Table (SMT), which in turn references related streaming media, the service guide, and related NRT rich media files.

3.8 EMERGENCY ALERTING AS PART OF ATSC 3.0

Advanced emergency alerting is a required integral function of the ATSC 3.0 standard, the next-generation DTV system under development as this book went to press. ATSC 3.0 will carry fixed, mobile and emergency alert services simultaneously in the same RF channel, each with its respective required level of robustness.

3.54 DTV TRANSPORT

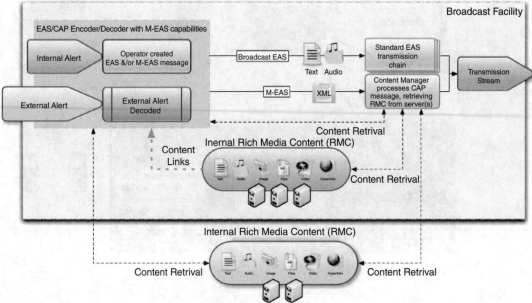

FIGURE 3.4.6 Detailed signal flow.

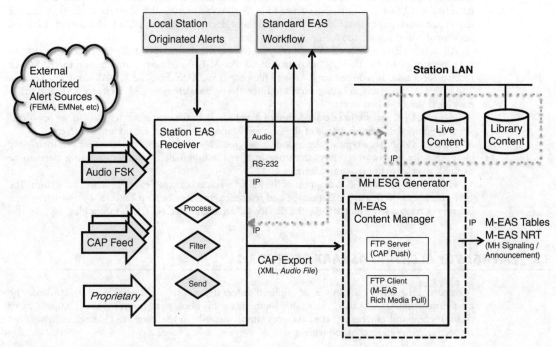

FIGURE 3.4.7 Content Manager interfaces.

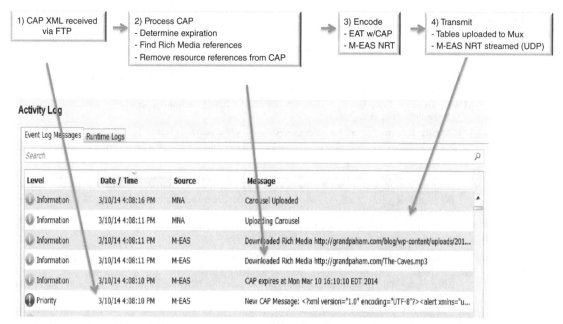

FIGURE 3.4.8 Content Manager CAP process and log.

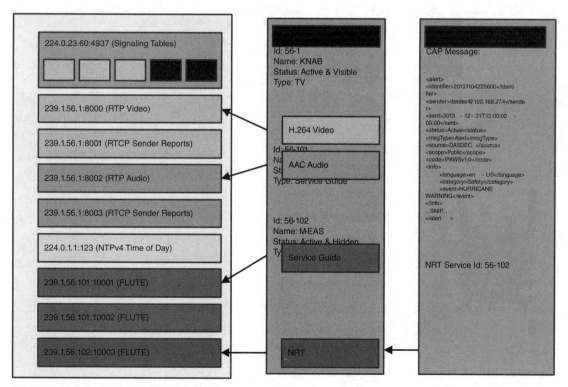

FIGURE 3.4.9 Broadcast stream M-EAS content.

BIBLIOGRAPHY

ATSC: "ATSC Digital Television Standard," Parts 1–6, Advanced Television Systems Committee, Washington, D.C., 2009–2011.

ATSC: "ATSC-Mobile DTV Standard," Parts 1–10, Advanced Television Systems Committee, Washington, D.C., various dates, which included at the time of writing A/153 Part 10:2013, "Mobile Emergency Alert System."

ATSC: "Non-Real-Time Content Delivery," Doc. A/103:2012, Advanced Television Systems Committee, Washington, D.C., 9 May 2012.

ATSC standards are freely available from the ATSC web site at http://www.atsc.org.

CEA: "Mobile/Handheld DTV Implementation Guidelines," Doc. CEA-CEB26-A (or latest version), Consumer Electronics Association, Arlington, VA.

This document can be found at http://www.ce.org by following links for Standards and the R4 Video Systems committee to document CEA-CEB26. At the time of writing, the link to purchase the current version of the document (CEA-CEB26-A) was at: https://www.ce.org/Standards/Standard-Listings/R4-Video-Systems-Committee/CEA-CEB26.aspx.

"Common Alerting Protocol, V. 1.2, USA Integrated Public Alert and Warning System Profile Version 1.0," OASIS Standard, 13 October 2009.

"Common Alerting Protocol, V. 1.2," OASIS Standard, 1 July 2010.

"Joint ATIS/TIA CMAS Federal Alert Gateway to CMSP Gateway Interface Specification," Doc. J-STD-101, Alliance for Telecommunication Industry Solutions and Telecommunication Industry Association, Washington, DC, October 2009.

"Joint ATIS/TIA CMAS Federal Alert Gateway to CMSP Gateway Interface Test Specification," Doc. J-STD 102, Alliance for Telecommunication Industry Solutions and Telecommunication Industry Association, Washington, DC, January 2011.

Bretl, W.; Kim, J.; Kutzner, J.; Laud, T.; Luplow, W.: "Mobile Emergency Alert System–A New Means of Alerting the Public," IEEE Broadcast Technology Symposium, 18 October 2012.

Kutzner, J.; Luplow, W.; Adrick, J.: "Mobile Emergency Alerting via ATSC Mobile DTV," SMPTE Annual Technical Conference and Exhibition, 23 October 2013.

Luplow, W.; Kutzner, J.; Adrick, J.: "Broadcaster Implementation of the ATSC Mobile Emergency Alert Standard," IEEE Broadcast Technology Symposium, 10 October 2013.

Mobile Emergency Alert internet web site: http://mobileeas.org/

CHAPTER 3.5
ATSC MOBILE DTV SYSTEM

Jerry C. Whitaker
ATSC, Washington, DC

3.1 INTRODUCTION

Following the successful transition from analog NTSC to the all-digital ATSC digital television (DTV) system in 2009, the ATSC developed and published a comprehensive standard for mobile and handheld services—known as *ATSC Mobile DTV* (also known as Mobile/Handheld or "M/H"). The services supported by this technology included free (advertiser-supported) television and interactive services delivered in real-time, subscription-based TV and file-based content download for playback at a later time. The standard also supported transmission of new data broadcasting services.

Following completion of the ATSC Mobile DTV Standard, document set A/153 [1], several field trials were held and a number of stations began broadcasting ATSC Mobile signals. Although A/153 has not been implemented on a large-scale basis, the technologies included in the standard led to further development within ATSC that would ultimately impact the next generation digital television service developed by the organization, ATSC 3.0.

Notable among the technologies and concepts developed for ATSC Mobile that have found their way into ATSC 3.0 (or are expected to, at this writing) are the following:

- **Internet Protocol transport.** The legacy (A/53 [2]) system utilized MPEG Transport. In a significant break with the past—and in recognition of the enormous influence of IP in modern communications—the decision was made to base the transport layer of ATSC Mobile on IP. (See Chapter 3.3.) IP transport has also been chosen for ATSC 3.0.

- **Over-the-air + over-the-top.** The traditional television broadcast model has all elements coming from a single source—the broadcast station. The ATSC Mobile DTV system stepped out of that mindset to integrate information (video, statistics, etc.) from the Internet into the broadcast programming to improve the user experience. Such hybrid services are a key element of the ATSC 3.0 experience.

- **Advanced Emergency Alerting.** The creation of a mobile platform opened a range of possibilities to expand the Emergency Alert System (EAS). Extensive work was done to develop a standard that applied new technologies to the task of informing the public in the event of an emergency situation. (See Chapter 3.4.) Advanced Emergency Alerting (AEA) will be an element of ATSC 3.0.

- **Versioning.** The legacy ATSC DTV system contained all the signaling necessary to receive and decode the transmitted information. The concept of signaling a system different from the original one was never a system requirement. The rapid evolution of mobile communications has clearly demonstrated the need to be able to signal to receivers that new services are available.

The capability to signal new technology versions at various layers of the system was incorporated into ATSC Mobile DTV. It is a fundamental element of ATSC 3.0.

- **Multiple operating points.** The legacy ATSC DTV system included one transmission operating point: 19.4 Mbps and a received C/N requirement of ~15 dB. ATSC Mobile expanded the range greatly by providing broadcasters the ability to trade off robustness for throughput. This capability is another fundamental design criterion for ATSC 3.0.

It is not unusual for one technology advance to lead to another. In the case of ATSC Mobile DTV, some of the basic technologies developed and the lessons learned have been applied to work on ATSC 3.0.

3.2 ATSC MOBILE DTV, A/153

ATSC Mobile DTV is built around a highly robust transmission system based on vestigial sideband (VSB) modulation coupled with a flexible and extensible Internet Protocol (IP)-based transport system, efficient MPEG AVC (ISO/IEC 14496-10 or ITU H.264) video, and HE AAC v2 audio (ISO/IEC 14496-3) coding. The ATSC Mobile DTV Standard describes the methodology for new services to be carried in digital broadcast channels along with current DTV services without any adverse impact on legacy receiving equipment.

In addition to live television, the ATSC Mobile DTV system provides a flexible Application Framework to enable new receiver capabilities. Receivers that make use of an optional Internet connection enable new interactive television services, ranging from simple audience voting to the integration of Internet-based applications and transactions with television content.

3.2.1 Documentation

In a tip of the hat to the core ATSC DTV Standard—document A/53—the ATSC Mobile DTV Standard is known as A/153. Like A/53, A/153 is modular in concept, with the specifications for each of the modules contained separate *Parts*. The individual Parts of A/153 are as follows:

Part 1—"Mobile/Handheld Digital Television System"
Part 1 describes the overall ATSC Mobile DTV system and explains the organization of the standard. It also describes the explicit signaling requirements that are implemented by data structures throughout the other Parts.

Part 2—"RF/Transmission System Characteristics"
Part 2 describes how the data is processed and placed into the VSB frame. Major elements include the Reed–Solomon (RS) Frame, a Transmission Parameter Channel (TPC), and a Fast Information Channel (FIC).

Part 3—"Service Multiplex and Transport Subsystem Characteristics"
Part 3 covers the service multiplex and transport subsystem, which comprises several layers in the stack. Major elements include Internet Protocol (v4), UniDirectional Protocol (UDP), Signaling Channel Service, FLUTE* over Asynchronous Layered Coding (ALC)/Layered Coding Transport (LCT), Network Time Protocol (NTP) time service, and Real-Time Protocol (RTP)/Real-Time Transport Control Protocol (RTCP).

Part 4—"Announcement"
Part 4 covers announcement, where services can optionally be announced using a Service Guide. The guide specified in Part 4 is based on an Open Mobile Alliance (OMA) broadcast (BCAST) Service Guide, with constraints and extensions.

---

*File Delivery over Unidirectional Transport.

Part 5—"Application Framework"
Part 5 defines the Application Framework, which enables the broadcaster of the audio–visual service to author and insert supplemental content to define and control various additional elements of the Rich Media Environment (RME).

Part 6—"Service Protection"
Part 6 covers Service Protection, which refers to the protection of content, either files or streams, during delivery to a receiver. Major elements include the Rights Issuer Object (RIO) and Short-Term Key Message (STKM).

Part 7—"Video System Characteristics"
Part 7 defines the AVC (Advance Video Coding) and SVC (Scalable Video Coding) system for the ATSC Mobile DTV video layer. Additional elements covered in this Part include closed captioning (CEA 708) and Active Format Description (AFD).

Part 8—"Audio System Characteristics"
Part 8 defines the HE-AAC v2 Audio System in the ATSC Mobile DTV system.

Part 9—"Scalable Full Channel Mobile Mode"
Part 9 is an extension of the ATSC Mobile DTV system that enables use of the full channel bandwidth for mobile services.

Part 10—"Mobile Emergency Alert System"
Part 10 describes the Mobile DTV EAS, including emergency alert tables, signaling for wake-up, and automatic tuning.

3.2.2 System Overview

The ATSC Mobile DTV service shares the same RF channel as the standard ATSC broadcast service described in ATSC A/53 ("ATSC Digital Television Standard, Parts 1–6"). The mobile system is enabled by using a portion of the total available 19.4 Mbps bandwidth and utilizing delivery over IP transport. The overall system is illustrated in Fig. 3.5.1.

In very simple terms, the system achieves the robustness needed for mobile reception by adding extra training sequences and forward error correction. The total bandwidth needed for the ATSC Mobile DTV service depends on several factors, including the number and type of program services, the quality level, and level of robustness desired—typically ranging from less than one megabit per second to many megabits per second. The ATSC Mobile DTV system converts the current 8-VSB emission into a dual-stream system without altering the emitted spectral characteristics. It does this by selecting some of the MPEG-2 segments (corresponding to MPEG-2 Transport packets in the current system) and allocating the payloads in those segments to carry the M/H data in a manner that existing legacy receivers will ignore.

A block diagram representation of the broadcast chain is shown in Fig. 3.5.2.

ATSC Mobile DTV data is partitioned into *Ensembles*, each of which contains one or more *Services*. Each Ensemble uses an independent RS Frame (an FEC structure), and furthermore, each Ensemble may be coded to a different level of error protection depending on the application. Encoding includes FEC at both the packet and trellis levels, plus the insertion of long and regularly spaced training sequences into the data stream. Robust and reliable control data is also inserted for use by receivers. The system provides bursted transmission of the data, which allows the receiver to cycle power in the tuner and demodulator for energy saving. A simplified block diagram of the ATSC Mobile DTV transmission system is illustrated in Fig. 3.5.3.

In the ATSC Mobile DTV physical layer, the data is transferred by a time-slicing mechanism to improve the receiver's power management capacity. Each Frame time interval is divided into five subintervals of equal length, called *Subframes*. Each Subframe is in turn divided into four subdivisions of length 48.4 ms, the time it takes to transmit one VSB frame. These VSB frame time intervals are in turn divided into four *Slots* each (for a total of 16 Slots in each Subframe).

The data to be transmitted is packaged into a set of consecutive RS Frames, where this set of RS Frames logically forms an Ensemble. The data from each RS Frame to be transmitted during a single

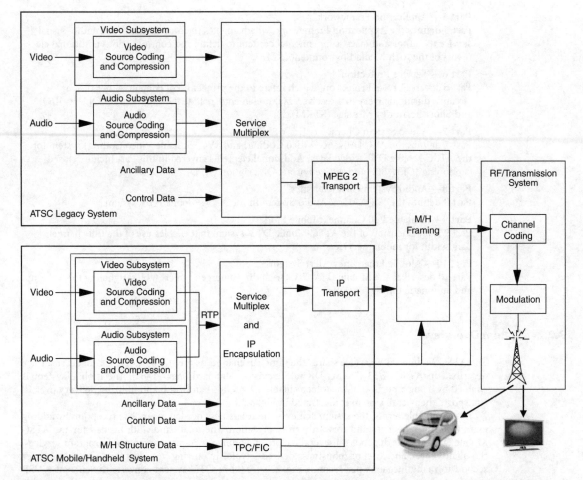

FIGURE 3.5.1 The ATSC Mobile DTV system.

Frame is split up into chunks called *Groups*, and the Groups are organized into *Parades*, where a Parade carries the Groups from up to two RS Frames but not less than one. The number of Groups belonging to a Parade is always a multiple of five, and the Groups in the Parade go into Slots that are equally divided among the Subframes of the Frame.

The RS Frame is the basic data delivery unit, into which the IP datagrams are encapsulated. While a Parade always carries a Primary RS Frame, it may carry an additional Secondary RS Frame as output of the baseband process. The number of RS Frames and the size of each RS Frame are determined by the transmission mode of the physical layer subsystem. Typically, the size of the Primary RS Frame is bigger than the size of Secondary RS Frame, when they are carried in one Parade.

The FIC is a separate data channel from the data channel delivered through RS Frames. The main purpose of the FIC is to efficiently deliver essential information for rapid Service acquisition. This data primarily includes binding information between Services and the Ensembles carrying them, plus version information for the Service Signaling Channel of each Ensemble.

An "ATSC Mobile DTV Service" is similar in general concept to a virtual channel as defined in ATSC A/65 ("Program and System Information Protocol") [3]. A Service is a package of IP streams transmitted through a multiplex that forms a sequence of programs under the control of a

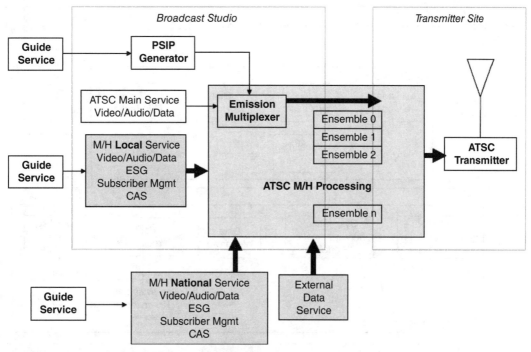

FIGURE 3.5.2 ATSC Mobile DTV broadcast system block diagram.

broadcaster, which can be sent as part of a schedule. Typical examples of ATSC Mobile DTV Services include TV services and audio services. Collections of Services are structured into Ensembles, each of which consists of a set of consecutive RS Frames.

In general, there are two types of files that might be delivered using the methods described in the ATSC Mobile DTV system. The first of these is content files, such as music or video files. The second type is metadata for security, signaling, and announcement. This includes long- and short-term keys for service protection, logos, and Session Description Protocol (SDP) files. In either case, the delivery mechanisms are the same and it is up to the terminal to resolve the purpose of the files.

A simplified block diagram of the organization of the major elements for delivery over the physical transport subsystem is illustrated in Fig. 3.5.4.

Signaling in the ATSC Mobile DTV system provides metadata to the receiver relating to tuning, including whether content should/can be rendered. Key design goals included:

- Keep it compact (low bit rate)
- Make it flexible and extensible
- Support rapid service acquisition
- Support basic functionality even when the receiver does not have up-to-date Service Guide information
- Support the unique requirements of roaming, such as hand-off from one transmitter to another for regional and national services when crossing broadcast area boundaries

In the ATSC Mobile DTV system, the Services available on that system (or another system) can be announced via the Announcement subsystem, called a Service Guide. A Service Guide is a special Service that is declared in the Service Signaling subsystem. A receiver determines available Service

3.62 DTV TRANSPORT

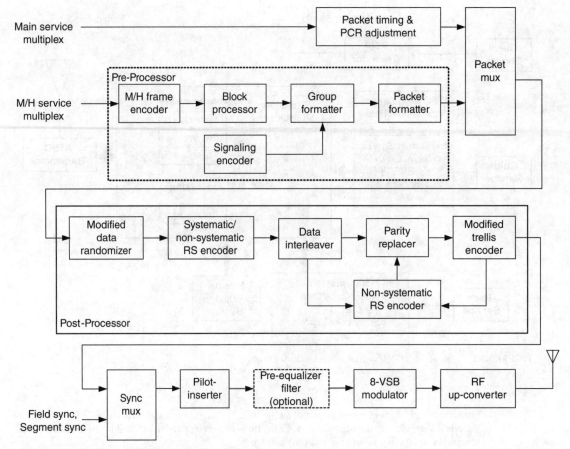

FIGURE 3.5.3 Block diagram of an ATSC Mobile DTV transmission system.

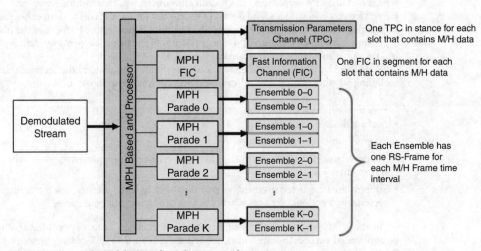

FIGURE 3.5.4 ATSC Mobile DTV physical transport subsystem.

Guides by reading the Guide Access Table. This table lists the Service Guides present in the broadcast, gives information about the service provider for each guide, and gives access information for each guide.

The ATSC Mobile DTV Service Guide is based on the OMA BCAST Service Guide, with certain constraints and extensions. A Service Guide is delivered using one or more IP streams. The main stream delivers the Announcement Channel, and zero or more streams are used to deliver the guide data. If separate streams are not provided, guide data is carried in the Announcement Channel stream. The Service Guide is designed so that it may also be delivered over a separate connection if a device has two-way connectivity.

The Application Framework is a toolkit that allows for the creation of graphical and syntactical elements to be added in conjunction with the delivery of audio and video. It differs from a middleware specification in that the Application Framework allows for an overlay on top of the audio/video plane rather than a true software control layer controlling all upper layers of the client. The Application Framework data is transmitted in-band alongside the audio and video.

This subsystem enables the broadcaster of the audio–visual service to author and insert supplemental content to define and control various additional elements to be used in conjunction with the audio–visual service. It enables the definition of auxiliary (graphical) components, layout for the service, transitions between layouts, and composition of audio–visual components with auxiliary data components. Furthermore, it enables the broadcaster to send remote events to modify the presentation and to control the presentation timeline. The Application Framework enables coherent rendering of the service and its layout over a variety of device classes and platforms, rendering of action buttons and input fields, and event handling and scripting associated with such buttons and fields. Example applications are shown in Fig. 3.5.5.

The Application Framework is important because it allows the ATSC Mobile DTV system to expand beyond the mobile playback of video and audio, and provides a new set of tools for Internet-style personalization and interaction. This entertainment experience can capitalize on whatever data return channel(s) may be available.

The Application Framework is built around the OMA Rich Media Environment (OMA-RME). The OMA-RME, designed around a similar requirement set, is an umbrella standard encompassing elements in application creation, delivery, and control.

Service Protection refers to the protection of content, be that files or streams, during its delivery to a receiver. Service Protection assumes no responsibility for content after it has been delivered to the receiver. It is intended for subscription management. It is an access control mechanism, only.

The ATSC Mobile DTV Service Protection system is based on the OMA BCAST Digital Rights Management (DRM) Profile. It consists of the following components:

- Key provisioning
- Layer 1 registration
- Long-Term Key Message (LTKM), including the use of Broadcast Rights Objects (BCROs) to deliver LTKMs
- Short-Term Key Messages (STKM)
- Traffic encryption

The system relies on the following encryption standards:

- Advanced Encryption Standard (AES)
- Secure Internet Protocol (IPsec)
- Traffic Encryption Key (TEK)

In the OMA BCAST DRM Profile, there are two modes for Service Protection—interactive and broadcast-only mode. In interactive mode, the receiver supports an interaction channel to communicate with a service provider to receive Service and/or Content Protection rights. In broadcast-only mode, the receiver does not use an interaction channel to communicate with a service provider. Requests are made by the user through some out-of-band mechanism to the service provider, such as calling a service provider phone number or accessing the service provider web site.

FIGURE 3.5.5 Examples of user experiences created with an Application Framework in conjunction with audio and video. (*Photos courtesy of MobiTV.*)

The ATSC Mobile DTV system uses MPEG-4 AVC and optionally SVC video coding as described in ISO/IEC 14496 Part 10, with certain constraints. A base format of 240 lines × 416 pixels, 16:9 aspect ratio, progressive scan, is specified, with the ability to increase the resolution or quality through use of the SVC option.

AVC is the international video coding standard ISO/IEC 14496-10 (MPEG-4 Part 10) "Advanced Video Coding," finalized in 2003. It supports a wide range of applications, from low-bit-rate mobile video streaming to high-bit-rate HDTV broadcast and DVD storage.

Audio coding for the ATSC Mobile DTV system utilizes HE AAC v2, specified in ISO/IEC 14496-3—MPEG-4 Audio. Constraints include the following:

- HE AAC v2 Profile, Level 2, maximum of two audio channels up to 48 kHz sampling rate.
- Bit rate and buffer requirements comply with the MPEG-4 HE AAC audio buffer model.
- Supported sampling frequencies are 32, 44.1, and 48 kHz.

HE AAC v2 is the combination of three audio coding tools: MPEG-4 AAC, Spectral Band Replication (SBR), and Parametric Stereo (PS).

3.2.3 Scalable Full-Channel Option

Considering services beyond traditional television broadcasting, ATSC developed an extension of the ATSC Mobile DTV system that enabled use of the full channel bandwidth for mobile services. This standard, A/153 Part 9, adds increased capacity in a scalable manner up to complete channel bandwidth usage. The basic A/153 (Core Mobile Mode, or CMM) requires a minimum of 4.7 Mbps to be transmitted as conventional 8-VSB. Scalable Full Channel Mobile Mode (SFCMM), on the other hand, can scale capacity up to the total available from the channel.

3.2.4 Evolving the ATSC Mobile DTV System

During development of the Mobile DTV system, it was recognized that technology will continue to move forward. As such, a mechanism to evolve the ATSC Mobile DTV system over time is important. This capability was built into the System Configuration Signaling architecture. The goal of this system is to support a configuration change of the protocol stack used by the broadcaster. Key considerations include:

- The system must provide information about each piece of content and how it is transmitted
- A receiving device must be able to determine if it can support such a content before the content is exposed to the user

As designed, the signaling is multilayer and supports two types of changes:

Major Version Change: a non-backward-compatible level of change.

Minor Version Change: a backward-compatible level of change, provided the major version level remains the same. Decoders/receivers can assume a minor change does not prevent them from rendering content.

The following signaling requirements were established for the ATSC Mobile DTV system:

- Capable of signaling the addition of a new elementary subsystem. For example, a Digital Rights Management capability may be added.
- Capable of signaling the removal of an elementary subsystem. For example, service protection is removed and replaced with functionality that resides outside of the ATSC system, i.e., an out-of-band method.
- Capable of signaling the replacement of an elementary subsystem. For example, one encryption is replaced with another encryption—the black box operation is equivalent.
- Capable of signaling service compatibility in an expedient manner, where the receiver is able to determine if it can support a service within one complete frame time.

- Capable of signaling all functionality needed to support a service correctly (i.e., transport, file management, SVC sync, and so on).
- Capability to support the Electronic Service Guide not displaying an event that cannot be decoded by the receiver.
- Signaling of the elementary subsystem functions must be sufficiently complete that the receiver can determine whether it can successfully process the content.
- Capable of signaling multiple generations of service carried concurrently, in the same ATSC Mobile DTV emission.
- Capable of signaling legacy services with optional extensions, such that the legacy receiver ignores the optional functionality signaling, and supports the legacy portion of the service.
- Capability to change the System Configuration Signaling system protocol without adversely affecting products built to the original signaling protocol.
- Capability to support a receiver determining a channel is out of service.
- Capable of signaling that the protocol version of a single elementary subsystem has changed.
- Capable of communicating a code for the version of each elementary subsystem required to decode and correctly display the services offered.

The signaling approach is hierarchal, with the physical of RF layer being considered the bottom of the stack. Much of the signaling is defined as integral parts of the data structures. At the bottommost layer, a simple (one-bit) signaling means was established. A major change of the entire physical layer can be signaled by use of another such bit. Other signaling for the RF layer is implemented with a simple version field in key data structures, each of which enables signaling of changes in the data structure above.

At higher layers, more signaling capability is established, reflecting the increasing likelihood of change in those layers as time progresses.

3.3 SUPPORTING RECOMMENDED PRACTICE

Shortly after the A/153 standard was published, ATSC developed a Recommended Practice (RP) on the ATSC Mobile DTV system. Because of the complexity of mobile DTV, it was recognized that guidelines for implementers would be beneficial. RP A/154 [4] was developed to address this need.

The document provides an overview of the system and detailed guidance for key Parts of the standard, including:

- ATSC Mobile DTV system overview
- Data transport
- Signaling data delivery and usage
- Announcement data delivery and usage
- Streaming data delivery
- File delivery
- Application framework

3.4 TRANSMISSION INFRASTRUCTURE

The RF environment for a mobile system is quite demanding. In order to reach certain portions of a given station's coverage area, more than one transmitter may be needed. The typical arrangement is a conventional high power/high tower transmission plant and one or more additional low power/low tower transmitters strategically located in areas of difficult terrain. ATSC document A/110 [5], "ATSC

Standard for Transmitter Synchronization," defines a method to synchronize multiple transmitters emitting trellis-coded 8-VSB signals in accordance with ATSC A/53 Part 2 (the ATSC DTV Standard) and of both single and multiple transmitters emitting Mobile DTV signals in accordance with ATSC A/153 Part 2. The emitted signals from transmitters operated according to the standard comply fully with the requirements of both ATSC A/53 and A/153.

A/110 specifies the mechanisms necessary to transmit synchronization signals to one or several transmitters using a dedicated PID value, including the formatting of packets associated with that PID value, and without altering the signal format emitted from the transmitters. It also provides for adjustment of transmitter timing and other characteristics through additional information carried in the specified packet structure. Techniques are described for cascading transmitters in networks of synchronous translators.

REFERENCES

[1] ATSC: "ATSC Mobile DTV Standard," Parts 1–10, Doc. A/153, Advanced Television Systems Committee, Washington, D.C., various dates.

[2] ATSC: "ATSC Digital Television Standard," doc. A/53, Advanced Television Systems Committee, Washington, D.C., various dates.

[3] ATSC: "Program and System Information Protocol For Terrestrial Broadcast and Cable," Doc. A/65, Advanced Television Systems Committee, Washington, D.C., 7 August 2013.

[4] ATSC: "ATSC Mobile DTV Recommended Practice," doc. A/154, Advanced Television Systems Committee, Washington, D.C., 30 January 2013.

[5] ATSC: "ATSC Standard for Transmitter Synchronization," doc. A/110, Advanced Television Systems Committee, Washington, D.C., 8 April 2011.

SECTION 4
INFORMATION TECHNOLOGY SYSTEMS

Section Leader—Wayne M. Pecena
CPBE, CBNE, Texas A&M University, KAMU, College Station, TX

Information technology (IT) has had a profound impact upon the modern broadcast plant, whether radio or TV. IT is a field as diverse as broadcast engineering and has become another essential technology that the successful broadcast engineer must master. An understanding of IT systems often begins with an understanding of Internet Protocol (IP) networking technology, as IP networking often is found to be the foundation technology of all IT aspects.

This section begins with an IP networking tutorial for the broadcast engineer authored by Wayne Pecena for those needing a quick start technology briefing. An emphasis is placed on blending the Internet Engineering Task Force (IETF) and the Institute of Electrical and Electronic Engineers (IEEE) standards-based IP networking technology essentials applied in the real-world infrastructure environment. A focus is placed upon gaining an understanding of the Three Letter Acronyms (TLAs) and Four Letter Acronyms (FLAs) found in the industry. TLAs and FLAs often dominate any technology field. IP networking is no different and an understanding of the terminology within the context is an essential element of overall understanding. For those experienced in IP networking technology and implementation, this tutorial should also serve as a quick technology review or refresher offering best practice approaches to network design and implementation.

Once the basic IP networking concepts are established, advanced topics are explained by several seasoned industry veterans. An overview of network systems as found in the broadcast plant is provided by Gary Olsen, where he focuses on the changes in the core broadcast infrastructure as migration to IP occurs. The system demands required by transport of broadcast real-time content are explored. File-based workflow concepts are applied to the IP environment and challenges examined, such as management architectures for file progress monitoring and orchestration through an IP-based plant, as well as Quality Control in a file or IP transport environment.

Synchronization has been a critical aspect of broadcast plants beginning with discrete horizontal and vertical sync signals in the early days of TV engineering. Each technology progression has brought new synchronization methods from composite sync to trilevel sync found in many digital plants today. As a migration to IP occurs, suitable synchronization methods must be provided as Nikolaus Kero examines the IETF IP and IEEE Ethernet-based clock synchronization techniques as applied to pixel frequency synchronization with an SDI (serial digital interface) data stream. IETF Network Time Protocol (NTP) and Precision Time Protocol (PTP) processes are detailed in comparison with IEEE-1588 Transparent Clock (TC) and SMPTE ST2059-2 PTP for broadcast standards. In the end, NTP and PTP are likely to be the foundation component to meet the IP clock synchronization of the broadcast IP plant.

4.2 INFORMATION TECHNOLOGY SYSTEMS

The section wraps up with John Mailhot contributing an overview of video transport beginning with a historical view of T-carrier and SONET carrier techniques commonly used by telecommunications carrier networks. The available bandwidth versus the content bandwidth requirements brought us to compression methods such as MPEG-2 that have become the base standard within the industry. As the broadcast plant evolves into an IP infrastructure, the use of IETF Real Time Protocol and similar SMPTE specifications become the carrier transport standards of today. The appropriate use of forward error correction is examined, as well as the challenges presented by transport of uncompressed video over IP (VoIP) in accordance with the SMPTE ST2022-6 standard. The chapter ends with an excellent quick reference or bibliography of VoIP standards.

Chapter 4.1: Information Technology and the Broadcast Plant

Chapter 4.2: Network Systems Overview

Chapter 4.3: Time and Frequency Transfer over Ethernet Using NTP and PTP

Chapter 4.4: History, Development, and Current Standards for Video Transport over Internet Protocol Networks

CHAPTER 4.1
INFORMATION TECHNOLOGY AND THE BROADCAST PLANT

Wayne M. Pecena
CPBE, CBNE, Texas A&M University, KAMU, College Station, Texas

4.1 INTRODUCTION

It comes as no surprise that the broadcast technical plant of today is vastly different from the broadcast plant infrastructure of the past. Technology has always had a major impact on broadcast plant design and capabilities over the history of the industry. The impact of information technology (IT) has brought a new rate of change and innovation introduction never seen in the past. Moore's law is often quoted as a factor in the rapid change of technology innovation where the capability of an existing technology doubles every 18 months (see Fig. 4.1.1).

Today's modern broadcast technical plant is often based solely or in part on an Ethernet-based Internet Protocol (IP) network as a core technology, or in some cases, a hybrid approach with legacy technology still in place. As legacy technology is depreciated, it is given that IP-based infrastructure will be the replacement choice.

The fundamental utilization of IT in the broadcast plant is often based on the use of an Ethernet physical infrastructure with interconnected IP-based host devices. Regardless of the specific IT area, the IP network becomes the foundation technology for system integration or "internetworking." Numerous reasons can be provided for the widespread use of IP networks in the broadcast plant. The justification can often be summarized in three major areas. These areas include the following:

- System flexibility
- Leveraging the IT industry economies of scale
- System simplification

System flexibility exists in the ability to dynamically change system content flows without a major system rewire to accommodate a work flow change. The core of a broadcast plant is often the traditional X–Y routing matrix. Individual devices in the plant are often hard-wired to a specific matrix source input or destination output. The matrix is a specialized and customized design specific to the type of signals involved whether it is composite video, serial digital interface (SDI) video, analog audio, digital audio, and so forth. Enter the IP world and the expensive specialized routing matrix becomes a "common-of-the-shelf" Ethernet switch available for numerous manufacturers. Because of the size of the IP networking market within the IT marketplace, Moore's law (attributed to an observation of Gordon Moore of Intel in 1965) continues to be valid today with cost reduction and increased performance. The cost of an IP interface within a device of the broadcast plant is often a

4.4 INFORMATION TECHNOLOGY SYSTEMS

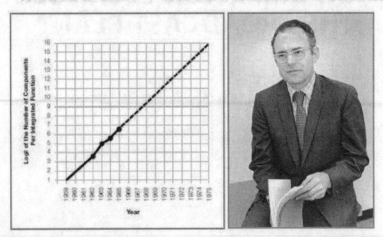

FIGURE 4.1.1 Moore's law (http://www.networkworld.com/article/2166095/computers/intel--keeping-up-with-moore--39-s-law-becoming-a-challenge.html).

fraction the cost of its equivalent baseband example. In 2015, the cost of a 1 GigE interface is approximately one-tenth (1/10) that of a single high-definition serial digital interface (HD-SDI).

The broadcast industry is able to take advantage of the incredible economies of scale that exist in an industry the size of the IT marketplace. Within the broadcast technical plant, the architecture of the physical interconnection becomes simpler and flexible as media formats change. In lieu of traditional copper cabling point–point baseband interconnection techniques, the system interconnection is based on a "stared" approach with an Ethernet switch at the core. The overall plant is often based on numerous switches and interconnected by copper cabling or often fiber optic technology to accommodate high-capacity communication paths. The common IP network can accommodate a wide variety of payload media formats ranging from SDI, to HD-SDI, to UHD (ultra-high-definition) formats. The broadcast plant begins to exhibit many attributes of a conventional "data center" in terms of core devices; internetworking techniques; power redundancy; and heating, ventilation, and air conditioning (HVAC) demands. Again, the IT industry is providing the economies of scale to lower the cost of technology and provide a rapid advancement of future technology. The broadcast plant of the future will be an all-encompassing IP-based facility. The use of common AES (Audio Engineering Society) audio and/or SDI video interconnection will likely be found only in specialized interface applications.

4.2 THE IP NETWORK—A TECHNOLOGY REVIEW

The IT field is diverse and widespread. The impact can be seen in many areas of the broadcast industry, but the network has become the foundation technology of the broadcast plant. Over the past years, many networking schemes and protocols have come and gone. Today, the Ethernet-based IP network has become the de facto standard of the IT industry and in turn the broadcast plant. An understanding of IP networking fundamentals is an essential knowledge area for the broadcast engineer today.

Five components are required to build a network. These components include the following:

- A *send* host device
- A *receive* host device
- A message or data to be sent between the host devices
- A medium to provide connectivity between the host devices
- A set of rules or "protocol" to govern the behavior of the host devices

Communications occur between hosts in one of three forms: unicast, broadcast, or multicast. Unicast is the basic one-to-one host communication scheme. Broadcast communications represent a one-to-many approach whose scope includes all host devices that are the members of a specific network. Multicast is similar to broadcast, except the scope that includes selected host devices or devices that desire to receive the multicast information. Figure 4.1.2 illustrates a simplistic conceptual "unicast" network model. Such a simplistic network model is often discounted as the real world is comprised of networks of tens, hundreds, and even thousands of network devices. The basic network model is often useful to understand and troubleshoot the network with thousands of host devices and should always be kept in the back of your mind as more complex aspects are addressed. Broadcast and multicast communication models will be presented later in this chapter.

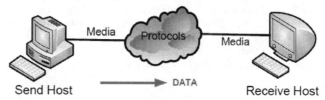

FIGURE 4.1.2 The five major elements of a network.

Icons representing a host device are often used in network illustrations. It is important to note that a host device, whether the send host or the receive host, can in reality be a wide variety of physical devices. In the radio broadcast plant, the host devices might be an audio console, an audio processor, or even a transmitter. In the television (TV) broadcast plant, the host devices might be a camera, a content storage system, or a transmission CODEC (coder/decoder). Regardless of the physical attributes or the functional purpose, a host device is any device that can be connected to the network infrastructure and can share a common addressing scheme with other devices on the network.

4.3 NETWORKING STANDARDS

The networking industry, like many technology-based industries, is comprised of "standards" that govern interoperability and practices that allow systems to be implemented that are comprised of products from many different manufacturers. It should be noted that proprietary standards do exist from many manufacturers and may bring some unique and essential capabilities to the system design, but the use of standard-based devices allows for a flexible system today and tomorrow.

Several international organizations come into play when networking standards are discussed. Some of the standards or practices fall into the "de jure" category and others into the "de facto" category. Three significant organizations that provide the bulk of the industry networking standards include the following:

- The Internet Engineering Task Force (IETF)
- The Institute of Electrical and Electronic Engineers (IEEE)
- The International Organization for Standardization (ISO)

The Internet Engineering Task Force or "IETF" is considered the foundation standard group for IP-based networking. The "Request-for-Comments" series (RFC-xxx) of documents form the bible of IP networking with all aspects of protocol functionality defined. Many RFCs are *required* to be implemented in a host device for that device to be an "IETF compatible IP host," where other devices might have capabilities that are *recommended* and not have features that are *elective*. Some unique devices might contain *limited use* features and technology that has changed over time; thus, many RFCs become *depreciated*.

The IEEE is the governing body for the Ethernet standard. Ethernet was developed over 40 years ago and is the "standard" physical networking medium in use today. The original Ethernet developed in the early 1970s by Bob Metcalf and David Boggs at the Xerox Palo Alto Research Center does not bear much resemblance of the Ethernet varieties of today, other than the underlying action of the protocol across the network. It should be noted that the original contention-based structure, Carrier Sense Multiple Access with Collision Detection (CSMA/CD) adopted from the University of Hawaii "Aloha" wireless network, remains within the Ethernet standards today. This algorithm is invoked when multiple host devices share a common network segment or "collision domain." Whereas once a "bus"-based scheme referred to as "Thick-Net" (comprised of RG-8 50 ohm coaxial cable) was commonly used, today's Ethernet physical medium implementation is found as twisted-pair copper, fiber optic, or wireless media commonly connected in a "star" configuration.

Within the overall standard work of the IEEE, Ethernet standards will be described as "Project 802" standards and are organized under an 802.xx nomenclature system. Table 4.1.1 illustrates a sampling of the twisted-pair copper and fiber optic-based physical Ethernet media in use today. The original 10 Mbps Ethernet of 40 years ago seems pale in comparison to the 10 Gbps and even 100 Gbps Ethernet that are now common. In 2015, 400 Gbps Ethernet had been demonstrated in the development of laboratory environment.

TABLE 4.1.1 Sampling of Ethernet 802.3xx Standards

IEEE Standard	Designation	Cable Type[1]	Speed	Maximum Length (m)
802.3i	10-Base-T	UTP CAT 3	10 Mbps	100
802.3u	100-Base-T	UTP CAT 5	100 Mbps	100
802.3ab	1000-Base T	UTP CAT 5e	1 Gbps	100
802.3z	1000-Base-SX	MM fiber	1 Gbps	500
802.3z	1000-Base-LX	MM fiber	1 Gbps	500
802.3z	1000-Base-LX	SM fiber	1 Gbps	(Several kilometers)
802.3an	10G-Base-T	UTP CAT 6	10 Gbps	55
802.3ae	10G-Base-SR	MM fiber	10 Gbps	300
802.3ae	10G-Base-LR	SM fiber	10 Gbps	(Several kilometers)

Notes: MM: multimode; SM: single mode; UTP: unshielded twisted pair

4.4 THE OSI MODEL

The International Organization for Standardization or "ISO" brought a fundamental model concept to the networking industry that has become the language of this industry today, even 35+ years after its introduction. The Open System Interconnection or "OSI" model is a layered abstract model that describes the process of a host application communicating with the network. The model describes how data traverse each of the seven layers of the model from the Application layer (7) to reach the Network layer (1). Figure 4.1.3 illustrates the OSI model.

INFORMATION TECHNOLOGY AND THE BROADCAST PLANT   **4.7**

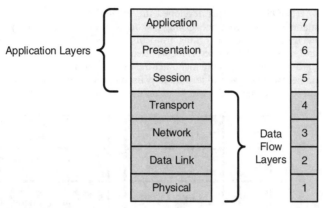

FIGURE 4.1.3 The OSI model.

Each layer of the model describes a functional aspect of a host application communicating with the network. The first four layers (1–4) of the seven-layer model are considered the "Data-Flow" layers, and these layers are of primary interest to the understanding of networking and the terminology used within the industry. The upper three layers (5–7) are considered Application layers. The upper layers are not considered to be an active component of networking and will not be described in detail.

A key aspect of the OSI model governs how a layer can only communicate with an adjacent layer. Layer 1 can only communicate with Layer 2; Layer 2 can only communicate with both Layers 1 and 3, and so forth. Jumping over a layer or layers is not permitted. Thus, Layer 1 could never communicate directly with Layer 7 as an example. Bypassing or jumping around a layer is not allowed. The communication process is illustrated in Fig. 4.1.4.

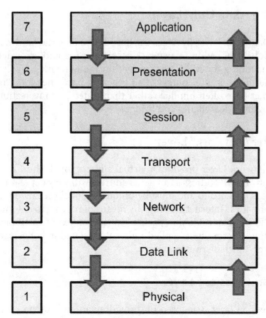

FIGURE 4.1.4 The OSI model interlayer communications.
Note: Jumping around layers is not allowed.

4.8 INFORMATION TECHNOLOGY SYSTEMS

The OSI model was developed in the late 1970s when the industry was filled with proprietary manufacturer-driven networking protocols and media interconnection schemes. Many readers have likely experienced networking protocols such as DecNet, SNA, Banyan/Vines, Novell IPX, AppleTalk, and so on. Each of these protocols often included a proprietary network physical medium requiring hardware from a specific manufacture. The importance of the OSI model remains today as the fundamental language, and terminology used in the industry today is derived from the OSI model. Our study of the OSI model will focus on the Data-Flow Layers as illustrated in Fig. 4.1.5.

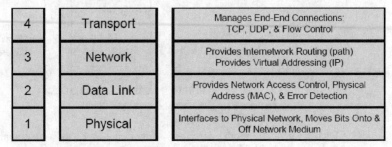

FIGURE 4.1.5 The Data-Flow Layers of the OSI model.

4.4.1 Layer 1—The "Physical" Layer

The Physical layer is often the simplest layer to understand since it deals with "bits" or placing a "1" or "0" onto or taking off of the network medium. The terminology *"placing bits onto or taking off the wire"* is often used to describe this process. Each of the Data-Flow layers has an associated protocol data unit (PDU) that defines how the information is packaged at the respective level. The PDU at Layer 1 is the "bit." Common network devices found at Layer 1 include hubs and repeaters.

4.4.2 Layer 2—The "Data Link" Layer

The Data-Link Layer provides the host device physical hardware address by means of a Media Access Control (MAC) address, network access control, and error detection. The hardware address is fixed by embedding the MAC address within the network adapter firmware. The PDU at Layer 2 is the "frame." Thus, in the context of an Ethernet-based network, the Ethernet frame is the PDU with the Ethernet term often implied rather than stated in conversation. Common network devices found at Layer 2 include switches and bridges.

4.4.3 Layer 3—The "Network" Layer

The Network Layer provides a virtual address for the host device in the form of an IPv4 or an IPv6 address. The IP address as a virtual address will vary depending on the network address that the host device is attached to and must be unique if global routing on the Public Internet is intended. Layer 3 also provides "Internetworking" by means of routing to determine the best path to a distant network. The PDU at Layer 3 is the "packet." In the context of an IP-based network, the IP Packet is the PDU with the IP term often implied rather than stated in conversation. Common network devices found at Layer 3 include routers.

4.4.4 Layer 4—The "Session" Layer

The session layer is focused on providing an end-to-end control of communications flow between host devices. Protocols such as TCP (Transmission Control Protocol), UDP (Unidirectional Protocol), and RTP (Realtime Transport Protocol) are typically utilized in the flow control process with the network application and associated information content determining the best approach. Real-time content will often utilize *nonguaranteed* UDP flow control in order to minimize the latency occurring in *guaranteed* TCP flow control handshaking. The PDU at Layer 4 is the "segment." Figure 4.1.6 summarizes the PDU for each of the Data-Flow layers.

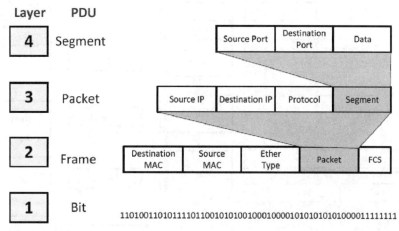

FIGURE 4.1.6 The Protocol Data Unit.

4.5 ENCAPSULATION AND DE-ENCAPSULATION

At the application layer, information or data is generated by a send host to be sent to a companion receive host device on the network. As the data are passed from Layer 7 to Layer 6 to Layer 5 and so forth, header information is added at each layer or the data received by the layer are encapsulated with the layer header information and passed to the next layer. The original application layer generated data grow as they pass through the various layers and finally become bits at Layer 1 or the physical layer.

At the receive host, the bits received at Layer 1 are de-encapsulated and passed in reverse order through the layers and the original send host application generated data presented to the send host application layer. Figure 4.1.7 illustrates this process.

The general philosophy of the layer model approach utilized by the OSI model allows layers to be swapped out as the network environment changes. The changing environment was much more prevalent in past times, and the industry was supporting multiple network media (often proprietary) and network protocols. In practical terms, a host application is neither concerned about the networking medium nor networking protocol utilized. Each layer handles an aspect of the overall process with independence between the groups.

The OSI model Data-Flow layer functions can be summarized as the transport layer ensuring the data are received by the receive host device (assuming TCP is utilized). The network layer is focused on the determination of the best way to reach the receive host network by means of a logical address (IP address). The Data-Link layer package data are based on the network medium type and determine the physical address (MAC address) of a host device on a network. The physical layer simply transmits the data as bits.

The OSI model is often questioned as being valid today in the environment of the Ethernet-based IP network as the de facto industry standard. As a result, additional models have been developed to better describe the processes for a host application to communicate over a network. The TCP/IP Model and its close companion, the Department of Defense (DoD) model, are later "descriptive" model developments that retain the layered approach and are focused on the communication link

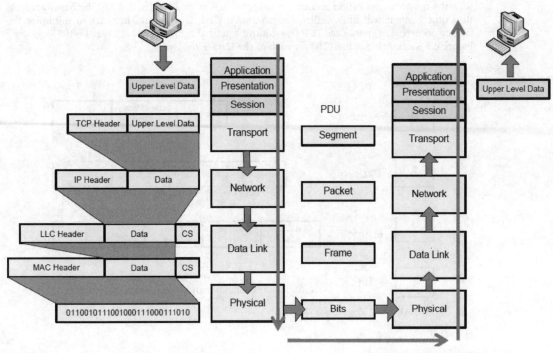

FIGURE 4.1.7 The encapsulation process.

established between a transmit host and a receive host. The four-layer architecture of these models was developed to align closely with the IP protocols as defined by an IETF RFC. Figure 4.1.8 compares the composition of the three models.

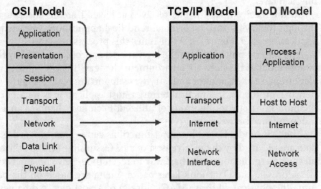

FIGURE 4.1.8 The OSI, TCP/IP, and DoD models compared.

Even when you consider the OSI model as a legacy model, the importance still remains in understanding the functionality and terminology whether utilized in network design or network troubleshooting processes. The numerical layer designation only applies to the ISO model. The TCP/IP or the DoD model does not use numerical layer designations.

It is a commonplace to describe the process of data flowing through the TCP/IP model as flowing through the TCP/IP "stack."

4.6 THE DATA-FLOW LAYERS

In the following sections, the Data-Flow layers will be examined in some detail.

4.6.1 The Physical Layer "PHY"—Layer 1

The physical layer or "PHY," as it is commonly denoted, is unique in that this is the only layer of the OSI model that communicates with the network medium. The physical layer details the specifics of the electrical or optical interface to the physical network medium. The architecture of the network may also be determined by the physical layer interface implementation. An example would be the "bus" structure dictated by early Ethernet implementations versus the star configuration required of current industry implementations using Ethernet switches. This layer also sets the standard of encoding of the network information whether the network medium is copper, optical, or wireless. In reality, the Physical layer and the Data-Link layer comprise the implementation of a specific network medium.

Ethernet is the industry standard Physical or "PHY" implementation. Ethernet can be found using coaxial cable, twisted-pair cable, fiber optic, and wireless media. All are defined by an IEEE 802.xx standard. Today, there are well over 30 different IEEE 802.xx standards and specifications. These standards range from the early experimental Ethernet providing <3 Mbps over a bus-structured coaxial cable to the more common 100-Mbps, 1-Gbps, and 10-Gbps Ethernet versions found today. In 2015, 100-Gbps Ethernet was the latest standard in widespread implementation with 400-Gbps Ethernet in the development stages.

In addition to the IEEE 802.xx specification, a shorthand notation is more commonly used in the industry. The term "Ethernet" can have different meanings depending on the context of its use. The term is often used in a generic state, describing all forms of Ethernet as suggested by the context of use. By the standards, the term Ethernet refers to the legacy 10-Mbps versions. The term "Fast Ethernet" is used to describe 100-Mbps Ethernet, and GigBit Ethernet or "GigE" is used to describe Ethernet operating at or over 1,000 Mbps.

The "Base" designator notation is a shorthand notation often found when Ethernet is described or components are specified. The shorthand notation specifies the data transmission speed, the type of signaling utilized, and the physical medium of the interface. The common "100-Base-TX" notation describes a transmission rate of 100 Mbps, utilizing a baseband signaling scheme, and implemented over twisted-pair cable. The notation is often further shortened to simply "100-Base-T." Likewise, the "100-Base-FX" notation describes a transmission rate of 100 Mbps, utilizing a baseband signaling scheme implemented over fiber optic cable.

Although the use of twisted-pair-based Ethernet over CAT 5, 5E, and 6 is certainly widely deployed, fiber optics is rapidly becoming the PHY medium of choice due to the higher bandwidths and distance demands. Whereas 1- and 10-Gbps connectivity can be provided over Cat 6 twisted-pair cabling, the distance is limited and careful installation must be deployed with regard to cable bend radius and bundling. In many broadcast plants, fiber offers more flexible PHY medium and prepares the plant for ever increasing bandwidth capacity demands, especially in the TV and video broadcast plants. However, twisted-pair cabling is likely to remain in our plants, whether for lower bandwidth demands such as audio over IP or control/monitoring applications. Two twisted-pair cable termination schemes are often found. Both are endorsed as the ANSI/ITA/EIA (American National Standards Institute; International Technology Alliance; Electronics Industries Association) standard groups and are commonly referred to as the EIA-568A or the EIA-568B premise wiring standards. In general, EIA-568B is most prevalent in commercial environments where AT&T popularized in many in-house PBX (private branch exchange) phone system installations. The EIA-568A standard may find more popular use in the home environment as backward compatibility with the Universal

Service Ordering Code or "USOC" two-pair consumer telephone wiring standard is maintained. Either approach can be used in a facility and often is at the discretion of network installation staff or installation contractors without installation specifications in place. However, it is important to recognize that regardless of what scheme is chosen, the same wiring standard must be used within the facility across all physical wiring cabling and connections. In general, EIA-568B is often considered the "preferred" U.S. wiring method.

An exception to use of the same uniform wiring scheme can arise when a "cross-over" cable is required to interface network equipment devices. Typically, a "straight-through" cable (same wiring scheme on each end) is used to connect a data terminal device such as a host device to a data communication device such as an Ethernet switch port. When two data communication devices need to be connected together, a cross-over cable is required when Automatic Media Dependent Interface or "Auto MDI-X" feature is not supported by the network devices. Since this is an optional feature of the standards, it may not be found available in all devices. In such instances, a physical cross-over cable, connecting two Ethernet switches together or connecting two host devices together, is required.

A cross-over cable can be constructed by using the EIA-568A wiring scheme on one end of the cable and the EIA-568B scheme on the other end of the cable. Table 4.1.2 provides a comparison of EIA-568A and EIA-568B wiring.

TABLE 4.1.2 Comparison of EIA-568 Wiring Schemes

PIN #	EIA-568A	EIA-568B
1	White/Orange	White/Green
2	Orange	Green
3	White/Green	White/Orange
4	Blue	Blue
5	White/Blue	White/Blue
6	Green	Orange
7	White/Brown	White/Brown
8	Brown	Brown

Figure 4.1.9 indicates how to identify pin #1 of an EIA-568 A or a B RJ-45 connector. The "Registered Jack" type 45 designation comes from the telephone industry as defined by the Universal Service Ordering Code going back to the early Bell Systems' days requiring compliant equipment to

FIGURE 4.1.9 RJ-45 connector pin #1 identification.

be used for customer-connected equipment. Today, the RJ-45 connector is ubiquitous with networking as adopted by the IEEE for twisted-pair Ethernet. You may also find the terminology "8P8C" used to describe the Ethernet RJ-45 connector, which indicates that eight pins and conductors are utilized.

Table 4.1.1 provided a sampling of IEEE 802.xx standards and their common industry "Base" notations. A complete listing can be found at http://www.ieee802.org/3/.

As gigabit plus speeds become more dominant, fiber is likely to become the typical physical medium deployed. Twisted-pair cabling is likely to only be seen in applications where short cable distances are involved, especially in the 10 Gbps and above installations. Many types of Ethernet available allow deployment flexibility with compatible PHY interfaces for Multimode (MM) fiber or Single Mode (SM) fiber, as well as various optical launch powers to address different physical link distances and the resultant optical budget requirements. As illustrated in Table 4.1.2, Ethernet is available in a wide variety of offerings, each designed for a specific network medium.

The variety of Ethernet offerings can be a challenge for equipment manufacturers, especially when Gigabit Ethernet is utilized. The Gigabit Interface Converter of "GBIC" was developed to allow a manufacturer to produce a network product with a standardized hardware interface slot and allow the end-user to select an appropriate GBIC transceiver module to match their specific needs with regard to fiber type and/or optical launch power required. The GBIC is a "hot swappable" self-contained transceiver module that contains the necessary optical and electrical interface circuitry to provide optical transmit and receive functionality to a standardized network equipment electrical interface. Standard Connector (SC) style fiber optic connectors are utilized with MM or SM fibers to connect the GBIC transmit or receive optical ports to the fiber distribution plant. Figure 4.1.10 shows a typical GBIX optical transceiver module.

FIGURE 4.1.10 A typical GBIC optical transceiver module.

In most network equipment manufactured today, the GBIC has been replaced by the "mini-GBIC" or Small Form Pluggable "SFP" optical transceiver module. The SFP module is considerably smaller than the GBIC and allows a higher interface density to be provided on equipment. The SFP is identical in concept to the GBIC, offered in many varieties to accommodate the network PHY interface needs, but it is simply smaller in physical size. Furthermore, versions of the SFP include SFP+ and XFP (10G Gigabit Small Form Factor Pluggable) varieties to accommodate 10-Gbps-based Ethernet interfaces.

Figure 4.1.11 shows typical SFP modules. Note that the "blue" extractor lever on the leftmost SFP indicates an SFP designed to interface to SM fiber whereas an MM SFP can be found with "black," "beige," or even a silver extraction bar. The rightmost SFP is designed for twisted-pair cabling.

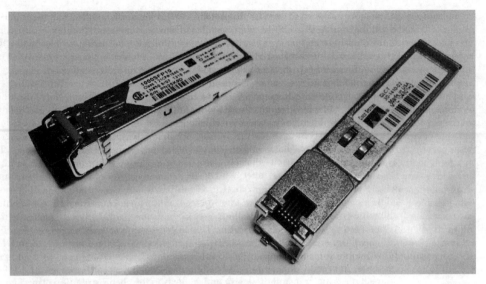

FIGURE 4.1.11 Typical SFP optical transceiver module.

SFP modules are not "standardized" by the IEEE as other Ethernet interfaces are. Rather, their standardization is based on a Multi-Source Agreement or "MSA" between equipment manufacturers. The MSA specification for an SFP contains a small amount of EEPROM (electrically erasable programmable read-only) memory map within the device. The SFP is meant to be a universal optical interface that should be able to be freely exchanged between different network equipment manufacturers. In reality, this is seldom true as most equipment manufactures add specific firmware information to the SFP memory to personalize the module to a specific manufacturer in an attempt to lock products from the manufacturer. Thus, a Cisco-supplied SFP must be used in Cisco network equipment or an HP-supplied SFP must be used in HP network equipment. In some cases, the "foreign" SFP will be found to function, but the network equipment will generate an error message that a noncompatible optical module is in use. In 2015, the Internet market place was full of SFP programming devices and SFP programming services to allow OEM-supplied SFP modules to be utilized in a wide variety of network equipment products. Any of these alternate approaches should be utilized with caution; cost saving is typically the motivation to consider use of alternate supplied products. OEM SFP modules are often one-tenth the cost of a manufacturer's supplied SFP modules.

Some SFP optical transceivers also incorporate Digital Optical Monitoring that allows optical operating parameters to be accessed and monitored such as optical transmit power, optical receive power, Bit Error Rate metrics, and optical emitter temperature. These real-time operating parameters can be useful for troubleshooting optical-link problems and are often an excellent proactive network monitoring parameter for performance benchmarking over time.

4.6.2 The Data-Link Layer—Layer 2

The Data-Link layer is a unique layer of the ISO model in that it is comprised of two sublayers:

- The Logical Link Control SubLayer (LLC)
- The MAC Sublayer

The sublayer approach was the methodology utilized to provide host node–node connectivity in an environment of different noncompatible network protocols (see Fig. 4.1.12).

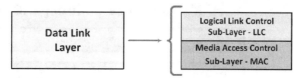

FIGURE 4.1.12 The Data Layer and Sublayers.

The Logical Link Control Sublayer or "LLC" layer is responsible for providing network protocol multiplexing, framing control, and detection of dropped frames and the resultant retransmission action. Protocol multiplexing functionality is often considered a latency capability of the LLC layer from the days of multiple protocols exiting in the network such as IP, IPX, ATM, PPP, and X.25. Today, IP is clearly the dominate protocol in use, although the Ethernet standard continues to support a multiple protocol environment. Today, the use of alternate protocols is usually found in wide area network (WAN) applications.

The LLC layer format data received from the Network Layer into a specific frame format determined by the network protocol and hands the frame to the Physical layer at the send host. At the receive host, the process is reversed. Bits handed to the LLC layer are formatted into a frame for handoff to the Network layer. Frames are detected to contain errors by means of a 32-bit cyclic redundancy check (CRC) routine and are dropped upon detection, whereupon a recreation and retransmission routine can occur.

The MAC sublayer or "MAC" layer is responsible for providing a physical or "logical" address for the host device. For Ethernet, the MAC address is a 48-bit address expressed in hexadecimal notation. Colons are generally utilized to separate the bytes of a MAC address such as A4:67:06:A8:41:D5.

Alternate MAC address expression formats are as follows:

A4-67-06-A8-41-D5

A467-06A8-41D5

A4.67.06.A8.41.D5

Regardless of the format used to express the MAC address, the 48-bit MAC address is split into two 24-bit functional components. The first 24 bits of the 48-bit address are allocated to specify the Organizational Unique Identifier or "OUI." The OUI indicates the registered manufacturer of the host network interface. The actual equipment device manufacturer may be represented or is often the manufacturer of the chipset found in a host device (rather than the actual host manufacturer). The IEEE maintains the OUI registry and details can be found at https://standards.ieee.org/develop/regauth/oui/public.html.

Several web resources provide handy OUI look-up functions such as https://www.wireshark.org/tools/oui-lookup.html.

The remaining 24 bits of the 48-bit MAC address are assigned by the manufacturer to serialize each network interface. Multiple interface Layer 2 devices such as a 24-port Ethernet switch will have a unique MAC address for each interface port.

The overall MAC address scheme is illustrated in Fig. 4.1.13.

It is important to note that the MAC address must be unique and is only valid on the local network segment, or only valid within the "Broadcast Domain" of a specific network. Several special or reserved MAC addresses exist. The most popular is the "Broadcast" MAC address represented by the destination address with all 6 bytes containing "1's" and represented by the hexadecimal character sequence "FF" or "FF:FF:FF:FF:FF:FF." A broadcast frame is delivered to all host devices on the network or "Broadcast Domain" that the send host is connected to.

Ethernet frames can be found in several formats that have evolved over the life of Ethernet. All are similar in concept, but slight differences exist; and the type field is used to specify the specific format of the frame. Ethernet frame formats you might find include the following:

- Ethernet II (DIX)
- IEEE 802.2 LLC

4.16 INFORMATION TECHNOLOGY SYSTEMS

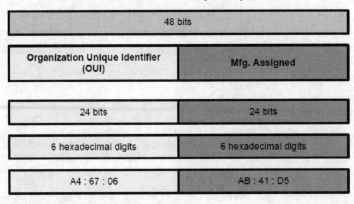

FIGURE 4.1.13   The Data Layer "MAC" address format.

- IEEE 802.3 SNAP (Subnetwork Access Protocol)
- Novell 802.3 (Novell raw)

Today, the Ethernet II frame is often referenced as the "DIX" frame format. The Ethernet II or DIX Frame is widely in use today in IP-based networks. The "DIX" acronym refers to the industry developers—Digital Equipment Corporation, Intel, and Xerox. Multiple Ethernet frame types can exist on a network, but it is important to recognize that a host device can only transmit or receive a single frame type. Thus, all host devices must be configured for the same frame format for communications to occur. The Ethernet II frame format is shown in Fig. 4.1.14.

The Ethernet frame begins with a preamble or a Start Frame Delimiter (SFD) that signifies the start of an Ethernet frame. The SFD is simply a 7-byte sequence of alternating "1's" and "0's" (10101010) with an eighth byte ending in "11" (10101011). It should be noted that the preamble is not used in the calculation of an overall Ethernet frame length.

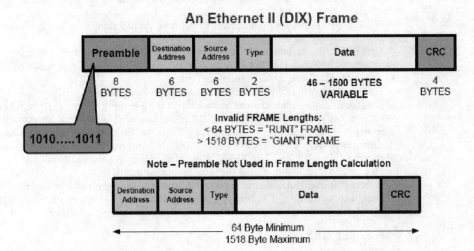

FIGURE 4.1.14   The Ethernet II or DIX frame format.

The Ethernet frame header consists of a fixed length source and destination address fields, a type field, and a variable length field containing the payload data. The 6-byte Source Address field represents the MAC address of the send host device and the 6-byte Destination Address field that of the receive host device. The type field, often referred to as the "EtherType" field, is used to indicate both the payload size and the protocol of the payload data contained in the frame.

The payload length can vary within minimum to maximum limits within the Ethernet frame. If a frame is less than 64 bytes (46 payload bytes + 18 header bytes), the frame is referred to as a "runt" frame. Likewise, a frame greater than 1,518 bytes is considered a "giant" frame. Runt or giant frames occur due to some type of error condition, which is often a collision between send host devices. It should be noted that a "giant" frame should not be confused with a "Jumbo" Ethernet frame, which is a valid frame often used in advanced networking applications where all network hardware offers support for such frame formats.

The end of an Ethernet frame is indicated by the presence of an Inter-Frame Gap or a 12-byte idle period between frames. The idle period is signaled by a loss of carrier in legacy Ethernet types such as the original "Thicknet" Ethernet or by the use of a special symbol set encoding scheme found in later versions of Ethernet. Remember, the Preamble/SFD and the Inter-Frame gap are not considered when the overall length of the Ethernet frame is determined.

An important aspect of network design allows Virtual Local Area Networks (VLANs) to be created. The "VLAN" is a useful feature to be used in network design and architecture implementation that provides architecture flexibility, security enhancement, and performance enhancement by limiting the scope of the broadcast domain.

A VLAN is comprised of a group of switch ports that may be spread across one of more Ethernet net switches that act as a single switch. This capability allows a group of host devices to be grouped together on a dedicated network regardless of physical location. Grouping can occur for several reasons, but often performance and security are the dominate reasons. Performance can be enhanced by limiting the reach of broadcast traffic on the network. Security can be enhanced by isolating certain hosts within a network and controlling across those devices. Each VLAN is an isolated network or "broadcast domain" and no connectivity exists between VLANs within an Ethernet switch. If connectivity is required, external Layer 3 routing is required as each VLAN has its own Layer 3 addressing scheme. Figure 4.1.15 illustrates the assignment of Ethernet switch ports to individual VLANs.

By default, all Ethernet switch ports are members of a single VLAN, often VLAN 1. Thus, a new switch can often be used without any configuration to perform basic switch functions. An individual Ethernet switch port connected to host devices can only belong to a single VLAN and is considered an "access" or "untagged" port. The ports that are utilized to interconnect Ethernet switches and transport individual VLANs between switches are considered "trunk" or "tagged" ports. The differing terminology found is based on an individual switch manufacturer. Figure 4.1.16 provides a virtual view of the VLANs that illustrate in the previous physical interconnection diagram.

VLAN design techniques will be presented in more detail later in this chapter, and device configuration information can be found in the Appendix section. VLAN membership can be created manually or automatically. Manual or "static" VLAN assignment is the most common configuration

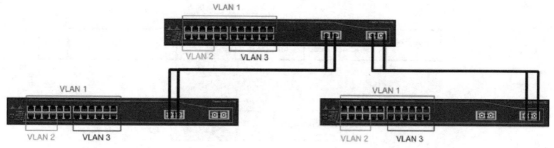

FIGURE 4.1.15 VLAN port assignment.

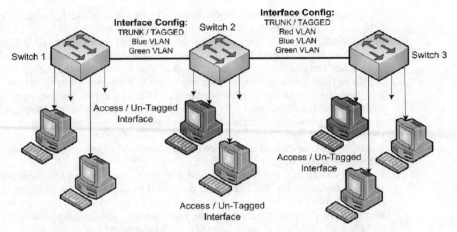

FIGURE 4.1.16   VLAN virtual view.

approach. Each Ethernet switch port is configured for membership in a specific VLAN. Automatic VALN configuration is accomplished by the creation of a database that assigns a host device to a specific VLAN based on the host MAC address. VLANs that traverse network connections between Ethernet switches are created by "tagging" packets across shared physical network links. The "tagged" Ethernet frame signifies what VLAN an Ethernet frame belongs to. Proprietary manufacturer tagging schemes exist from some manufacturers. The IEEE has specified the 802.1Q tagging standard for interoperability between different manufacturers' Layer 2 devices.

Figure 4.1.17 illustrates an 802.1Q "Tagged" Ethernet frame. A 4-byte "tag" is added to the frame header between the source MAC Address field and the type field. Several fields are contained within the 802.1Q tag that provide VLAN membership information and can provide for a priority assignment when quality-of-service features are utilized. With 12 bits allocated for VLAN identification, up to 4,096 individual VLANs can be identified. In reality, some VLAN numbers are reserved and not all Layer 2 network devices support all possible VLANs. However, most Layer 2 devices support at least 1,024 possible VLANs.

Since the Ethernet frame size has now been increased by 4 bytes, a new CRC value must be calculated and added to the frame so that it will be considered a valid Ethernet frame by the receive host device.

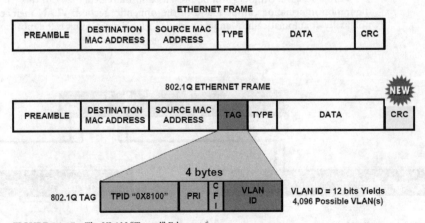

FIGURE 4.1.17   The VLAN "Tagged" Ethernet frame.

4.6.3 The Layer 2 Ethernet Switch

The Ethernet 2 switch is the basic building block of a modern network. The Ethernet switch is based on an early network device: the "bridge." Ethernet bridging or Ethernet switching terminology is often used interchangeably. Ethernet switches can be divided into two general categories: Managed and Un-Managed.

The Managed switch is typically used in a professional network environment such as the broadcast technical plant. A Managed switch allows for custom configuration of the switch operating parameters, configuration and monitoring of individual switch ports, and the establishment of VLANs. The Un-Managed switch is characterized as a "plug-and-play" device using autoconfiguration techniques for integration into the network infrastructure, commonly found in home networks. The Un-Managed switch is also characterized by its lower cost. Regardless of a Managed switch or an Un-Managed switch, the Ethernet switch provides four basic functions. These functions include the following:

- Learning MAC addresses
- Filtering Ethernet frames
- Forwarding Ethernet frames
- Flooding of Frames throughout the network

Additional functions and capabilities often exist in a "Managed" Ethernet switch such as the Spanning Tree Protocol (STP), VLAN creation, port security, and Power-Over-Ethernet (POE). STP is used to prevent "feedback" loops where redundant paths are available between Layer 2 devices. VLANs become a critical network design capability to isolate networks over a common physical infrastructure. Switch-port-based security provisions often become the starting point for ensuring network availability and control. POE can be a useful feature in a facility that has dispersed low-power host devices within the network. It is assumed the Managed Ethernet switches will be utilized in the professional broadcast networks.

Learning MAC addresses is a major function of an Ethernet switch. The switch listens to traffic occurring on each port and captures the source MAC address of any incoming frames on each port. The frame and port information is maintained in an internal table within the switch. This internal MAC address table is automatically created by the normal traffic flow process in the network.

When a frame is received, the switch also looks at the destination MAC address. The switch processing queries the internal MAC address table for the destination MAC addresses. If the MAC address is found in the internal table, the port number is returned and the switch processing forwards the frame to the appropriate port and on to the desired host. If the destination MAC address is not found in the internal table, the switch will flood the frame to all switch ports (except the port where the frame originated). The desired host will in turn respond to the received frame. Upon responded the source MAC address will be captured, and the frame MAC address and port captured and added to the internal table. Thus, the process of filtering frames by forwarding only to a specific switch port. When multiple devices are connected to a switch port, such might occur via a hub, multiple MAC addresses will appear in the internal MAC address table. In the case of a switch port receiving a broadcast frame (destination address = FF:FF:FF:FF:FF:FF), such as an Address Resolution Protocol (ARP) request, the switch will send the frame to all switch ports or "flood" the frame to all ports except the incoming port. An example MAC address table is shown in Fig. 4.1.18.

1	0000.aa67.64c5	DYNAMIC	Fa0/14
1	0000.aa70.d9b9	DYNAMIC	Fa0/7
1	0001.e641.96cd	DYNAMIC	Fa0/2
1	0004.00d5.285d	DYNAMIC	Fa0/18
1	0007.50c4.3440	DYNAMIC	Fa0/2
1	0008.74a5.9ee0	DYNAMIC	Fa0/2
1	0009.0f0a.6974	DYNAMIC	Fa0/8
1	000b.db12.a3f9	DYNAMIC	Fa0/12

FIGURE 4.1.18 An example Ethernet switch MAC address table.

The entries in the switch internal MAC address table are maintained for a specific time frame referred to as the *aging time*. When the aging timer is expired, the MAC address is purged from the table and must then be relearned. A common default value is 5 minutes. In a Managed switch, the default can be changed to reflect the network environment. Devices in the content signal path of a broadcast plant are likely to be considered to be very stable. Thus, the default aging time might be significantly increased. It is common to manually purge the MAC address table during installation and troubleshooting rather than awaiting the table to automatically build after system changes occur.

It is often typical to provide multiple paths between critical Ethernet switches in the network. However, traffic loops or feedback paths can occur when multiple paths exist between switches. Protocols such as STP are used to prevent the feedback paths or loops from occurring, but automatically switch to alternate communication paths if a link failure or port failure should occur. STP is an IEEE standardized protocol (IEEE 802.1d) that is enabled by default in most Ethernet switch products. Switch STP-related configuration parameters simply allow the default operation to be modified to suit a specific network need. Switch ports can be in one of the several modes:

- Disabled
- Listening
- Learning
- Blocking
- Forwarding

The STP is based on "blocking" or disabling a port when multiple paths exist between switch devices. The protocol is based on establishment of a "root" switch. The STP automatically elects a "root switch" often by selecting the lowest MAC address of switches in the network. The root can also be specified by manual configuration. The root is used as a reference point for the remaining Ethernet switches in the network. Nonroot switches then establish a single root port which represents the shortest path (or lower cost path) to the root switch. Switches exchange configuration information by means of Bridge Protocol Data Unit communications between the networks Ethernet switches. All root ports forward frames as normal and any nonroot port block any frame forwarding or activity. If a link failure occurs, the link with the shortest path is then automatically selected by enabling frame forwarding on a previously designated nonroot switch port. STP convergence is on the order of 30–50 seconds depending on the overall network scope, and a later STP implementation (IEEE 802.1w) referred to as Rapid Spanning Tree Protocol provides convergence with a second. Figure 4.1.19

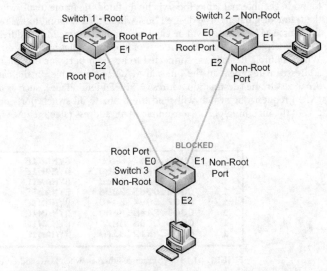

FIGURE 4.1.19 STP port blocking.

illustrates a network with multiple paths between Ethernet switches and the port election status that occurs by the STP action. Ethernet switch #1 is elected (or configured) as the root switch and all ports become root ports. Each remaining switch then becomes a nonroot switch which elects a single root-port that has the shortest path to the root switch. In this example, port E0 is elected as the root port on each nonroot switch. All remaining ports on the nonroot switches are set as nonroot ports.

Power-Over-Ethernet can be useful feature commonly offered by an Ethernet switch in many technical plants to relieve the burden of some network devices requiring local AC power (often utilizing an common "wall wart" power adapter) where CAT-5 twisted-pair cabling is utilized. Similar to phantom power found in many broadcast audio applications, POE allows a network device to be powered with DC voltage supplied through an Ethernet switch port. The IEEE has standardized POE offerings through IEEE802.3af standard developed in 2003 and the later IEEE802.3at standard known as "POE+" or "POE Plus." POE was developed to power VoIP desksets and has expanded to include all types of network devices such as wireless access points, IP cameras, access card readers, building or facility environmental control devices, and the list goes on. The list of POE powered devices will only expand as the "Internet of Things or IoT" evolution moves into the broadcast environment to include audio and video encoders/decoders, router control panels, NTP (Network Time Protocol)-based time clock displays, and devices not yet imagined. The POE standards define how a Power Sourcing Device or "PSD" such as an Ethernet switch supplies DC power to a Powered Device or "PD" such as a VoIP deskset or other network device. DC voltage is carried from the PSD to the PD by either utilization of unused pairs in the 4-pair CAT-5 cable (such as 100-Base-T) or by superimposing the DC voltage on data signal pairs when all four conductors are utilized (such as in 1000-Base-T). The nominal POE voltage is 48 VDC with the available current determined by the IEEE POE standard employed. Essentially, the current delivered is the differentiating factor between the two POE standards in use in 2015. The maximum current is limited by the carrying capacity of the 24 AWG (American Wire Gauge) wire utilized in CAT-5 or 23 AWG found in Cat-6 based cable. Power sourcing equipment (PSE) equipment is designed to sense a PD is connected and only then supply DC power to the device. In situations where non-POE Ethernet switches are installed, an in-line power insertion device can be utilized between the Ethernet switch and the network device. The inserter is a passive DC insertion device and as such Ethernet cabling limitations apply. Such units are often referred to as "mid-span" power inserters and are available in single port and multiport versions. An Ethernet switch supplying host device DC power is referred to as an "end-span" device. Table 4.1.3 highlights the key specifications of the IEEE802.3af and IEEE802.3at standards.

TABLE 4.1.3 POE Standards Comparison

	IEEE802-3af	IEEE802.3at or POE+
PSE voltage	44–57 VDC	50–57 VDC
PD voltage	37–57 VDC	42–57 VDC
PSE Power delivered	15.4 W	30 W
PD Power available	12.95 W	25.5 W
Maximum current	350 mA	600 mA

The use of Ethernet switch port security is considered one of the "best practices" in network operation. In addition to providing security enhancement regarding what can connect to the network, switch port security can often enhance network availability, reliability, and operation. Network availability is enhanced by limiting host devices that can connect to the network. Common threats that originate from a legitimate but compromised host flood a switch port with different source MAC addresses until the internal MAC address table overflows and disrupt network availability. This is a common approach of a "MAC flooding attack" and is basics of a Layer 2 "denial of service" or DoS attack. The limit of one MAC address per switch port also restricts the use of network hubs that can impact performance by creating a bandwidth contention situation. The elimination of hub use in the network provides an added layer of security with regard to traffic monitoring and snooping by unauthorized devices.

Network reliability is enhanced by ensuring that full port bandwidth is available between a host and a switch port in a full-duplex mode of operation. With a single host connected to a switch port there is collision domain without any bandwidth contention, and the result is reliability of operation. Network operation is also enhanced by ensuring that network resources are not overtaxed by too many unattended host devices attempting to access resources.

Another justification for implementation of switch port security is that a single MAC address per switch port will allow easy implementation of IEEE 802.1x port authentication now or in the future. Port authentication requires that a host verify the host identity via a radius server. Port authentication requires that only one host can be present of a switch port. Even if authentication is not desired or used today, you have provided some amount of "future proofing" as security concern mitigations are only likely to increase rather than decrease in the future.

Ethernet switch port security features simply allow limiting the host devices that can be connected to a switch port or limiting a specific host to access the switch port. Controlling access to the Ethernet switch port in turn controls the access to network resources.

Switch port security features also determine what actions are to be taken if a security violation should occur. Options allow for a variety of actions, including simply shutting down the switch port and notifying network administrator.

In most Ethernet switch products, host MAC addresses are not limited when default configurations are used. The only limit that exists is based on the internal MAC address table or Content Addressable Memory (CAM) table memory size. This limit is placed on all switch ports and in many switch products; the limit is over 4,000 MAC addresses per Ethernet switch.

Once port security is enabled, the default limit becomes one (1) MAC address per switch port on most Ethernet switch products. Configuration commands or configuration GUI (graphical user interfaces) dialog screens allow this default to be increased if multiple host devices must be supported by an individual switch post. This practice should be avoided by maintaining the policy of one device per Ethernet switch port.

If more than one host device must be connected to a single switch port, a "best-practice" configuration policy would be to set the Ethernet switch port maximum MAC address configuration parameter to just enough addresses for desired hosts. Additional or growth MAC address space should not be provisioned.

Once port security is enabled and the maximum number of allowable hosts established on a switch port, a couple of options typically exist. A specific MAC address can be "statically" entered into the switch configuration for each port. Or, a MAC address can be "dynamically" learned, thus saving the network administration configuration time and the possibly of data entry errors. "Fat finger" configuration errors are too often commonplace in network equipment configuration. Appendix C provides examples of switch port security configuration using a Cisco Ethernet switch as an example.

Once port security is enabled, the next step is to specify the violation action that is to occur when a switch port violation happens. Common violation action options include the following:

1. Switch port shutdown
2. Switch port restrictions
3. Switch port traffic protection

As the name implies, the switch port shutdown action simply disables the Ethernet switch port by placing in a disable mode. No frames will be forwarded across the interface. In most products this is the default violation action when port security is enabled.

The switch port restriction mode allows the interface to remain enabled and the legitimate host traffic will be forwarded. The unauthorized frames from the host that created the violation will be dropped by the Ethernet switch. Violation notifications are created and available to the network administrator via a device log "syslog" file or a System Network Management Protocol trap generation.

The switch port traffic protection mode impact is similar to the port restriction mode in that when a violation occurs, authorized frames will allow to be forwarded and unauthorized frames are dropped with no logging or notification.

Many products offer an "auto-recovery" feature that saves network administrator intervention to reset the switch port every time a violation occurs. Once a port violation occurs, and causes the switch port to be disabled, an auto-recovery timer begins a countdown. Once the configured time is reached, the port is automatically enabled. If the violation is not cleared by the expiration of the recovery timer, the port will be enabled and then instantly placed in the disable mode.

A final parameter to consider is the MAC table aging. MAC addresses are maintained in the internal switch table memory or CAM for a time period based on an absolute time value or when a MAC address is inactive for a specified period of time. It may be desirable to modify the switch default aging times.

It should be noted that the network administrator may manually clear the internal MAC address table when configuration work has been performed within an Ethernet switch to establish a clear MAC address table build point. A loss of power or a manual switch reboot also clears the table.

The aging parameter may be one of the most important configuration parameters to consider in the broadcast IP network plant. Default configuration parameters of network infrastructure equipment are often determined on the basis of typical use in a business or office network environment. Because of the nature of a broadcast IP network that is transporting program content information as IP streams or time-critical control information in an automation environment, many of these default parameters may need a second look. The office or business network within the broadcast facility will likely find the default parameters adequate.

The broadcast IP network plant is likely very stable, as host devices are not joining and leaving the network over time. In addition, a host is unlikely to be inactive due to the continuous real-time nature of a broadcast content IP transport stream. Thus, it may be preferred to increase the default aging time to the maximum available or disable it entirely by configuring the Ethernet switch port to a static mode. This approach will eliminate any frame traffic disruption as the internal MAC table is refreshed upon a switch port aging timeout value.

As we wrap up study of Layer 2 or the Data Link layer, keep in mind the basic network model presented in the introduction. Figure 4.1.20 represents the basic frame flow through the conceptual network infrastructure. As Layer 2 network devices (Ethernet switches in this example) are added to the end-end network, note that encapsulation and de-encapsulation occur only up to Layer 2 in the "stack" of the network device. Layers beyond Layer 2 are not touched and essentially ignored by the Layer 2 device. Also keep in mind that the MAC address is only valid on a local network segment.

Figure 4.1.21 outlines the communications between a send host and a receive host. Note that the Layer 2 source and destination MAC addresses change as different segments of the overall end–end network are traversed, and that the Layer 3 source and destination IP addresses do not change. Implementation of Network Address Translation or "NAT" is the only situation that will change the IP source and destination addresses as the network is traversed. NAT fundamentals will be presented later in this chapter.

4.6.4 The Network Layer—Layer 3

The Network Layer or Layer 3 is focused on "Internetworking" or determining the best path to a receive host network by means of a Layer 3 "virtual" IP address. The IP families of protocols are also found implemented at Layer 3 and defined by the IETF RFC-791. At Layer 3, a correlation between a host MAC address and an IP address is required. The ARP (IETF RFC-826) maintains a table that maps a Layer 2 MAC address to a Layer 3 IP address so that the header information can be properly formulated. All IP host devices are required to maintain an ARP table. The table is populated by a host broadcasting an ARP Request for a specific IP address; i.e., "Who Has 128.194.247.54?"

4.6.5 Layer 3 IP Addressing

At Layer 2, addressing was focused on a specific host device by means of a physical MAC address. Layer 3 utilizes a virtual address to identify a different network. The virtual aspect comes into consideration as the IP address is dependent on the specific network that the host is connected to. If a host should be moved from a network to another, the IP address must change. At Layer 3 the PDU is the "packet" and in the context of an IP network, an "IP Packet." The Layer 3 packet can only be received from Layer 4 or passed to Layer 2 as defined by the OSI model. The transmit host information received from Layer 4 is appended with Layer 3 or IP header information before being passed to Layer 2. Figure 4.1.22 and Table 4.1.4 illustrate the IPv4 packet header composition. Detailed understanding of all header information can be a complex study, but some basic concepts should be understood. A basic understanding of the IPv4 header makeup is useful when configuring network

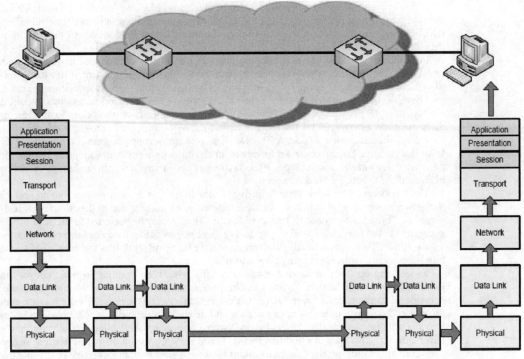

FIGURE 4.1.20 Encapsulation and de-encapsulation within a Layer 2 conceptual network.

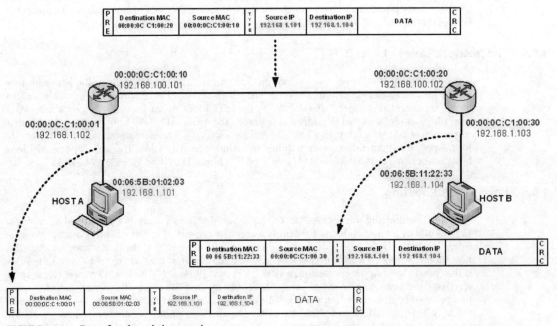

FIGURE 4.1.21 Frame flow through the network.

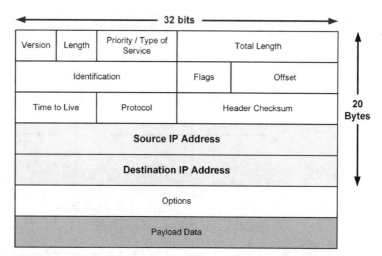

FIGURE 4.1.22 The IPv4 packet header.

TABLE 4.1.4 The IPv4 Packet Header Fields

	Field	Function	Size (bits)
1	Protocol Version	Indicates the version of the packet: IPv4 or IPv6	4
2	Header Length	Overall Header Length: 60 byte maximum	4
3	Type of Service	Precedence and Type of Service Indication	8
4	Total Length	Overall Length: 65,535 byte maximum	16
5	Identification	Unique Identification of the Packet	16
6	Flags	Fragment Control Bits	3
7	Offset	Indicates IP datagram start when fragmentation occurs	13
8	Time-to-Live "TTL"	Indicates number of network segments the packet will pass	8
9	Protocol	Indicates the Layer 4 (Transport) protocol used	8
10	Header Checksum	Packet Integrity Check	16
11	Source IP Address	32-bit Host Source Address	32
12	Destination IP Address	32-bit Host Destination Address	32
13	Options	Indicates "options" in use	Variable

device Access Control Lists or performing network protocol analysis with tools such as Wireshark. The source IP address, destination IP address, and the payload description information are the foundation of the IP header with other fields providing support functions for successful delivery of the IP packet to the desired destination network. An option field is provided which can provide for implementation of special services such as security features.

The IP address is a key component of the Layer 3 functionality, whether IPv4 or the newer IPv6 format. Several key points should be kept in mind with regard to IP addressing, and these points should be considered as rules to observe:

- Each network must have a unique network ID.
- Each host must have a unique host ID.
- Every IP address must have a Subnet Mask.
- An IP address must be unique if global routing is in use.

The IPv4 Address is a 32-bit binary address developed in the late 1970s and provides 4,294,967,296 or 2^{32} possible addresses. The 32-bit binary address is difficult to comprehend and manipulate, so a

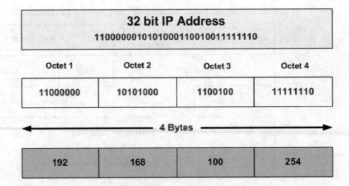

FIGURE 4.1.23 The IPv4 address.

shortcut notion is commonly used. Figure 4.1.23 illustrates the IPv4 address in binary form and the development of its expression in the common doted-decimal format. The 32-bit binary address is divided into four 8-bit sections often referred to as "octets." Each octet represents a byte of data. The octet is converted from binary to a decimal format and each octet separated by a decimal point "." to form the doted-decimal expression format.

The IPv4 Address is actually comprised of two parts that specify the Network Address and the Host Address components. Recall that each IP address must have an Address Mask. The IP address "mask" determines where the separation point between the network address and the host address occurs. The mask for an IPv4 address is also a 32-bit address also expressed in doted-decimal format. Figure 4.1.24 illustrates the network mask determining the Network and Host separation point from the IP address of "192.168.100.254" with an Address Mask of "255.255.255.0."

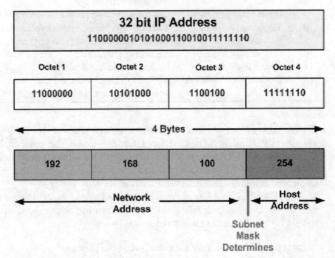

FIGURE 4.1.24 The two-part IP address; in this example, IP address "192.168.100.254" with Address Mask of "255.255.255.0."

The original implementations of IPv4 addressing utilized a "Class" system to organize an entire range of possible IPv4 Addresses. The classification scheme established five classes of addresses beginning with "Class A" through "Class E" with each address class boundary defined by a range of addresses. Address Classes A–C are intended to be used for "internetworking," and each has a default mask address predetermined and referred to as the "Default Mask." Class D address space is reserved for Multicasting and Class E address space is reserved for development or experimentation. Class D and E address space does not utilize a default mask value. Class A address space provides a smaller number of Network addresses but a large number of Host addresses are available. Class C address space provides a large number of Network addresses with a small number of available Host addresses. The subnet mask must be the same for all host devices connected to a given network or subnet.

Figure 4.1.25 illustrates the address classes and the default address masks for Classes A–C. The use of a default mask for each address class was soon realized to present limitations and challenges to network deployment due to the fixed nature of network and host availability within each class.

As network deployment grew, the ability to size a network for the required number of hosts became challenging as globally unique IPv4 address became scarce. The use of "Classful" addressing

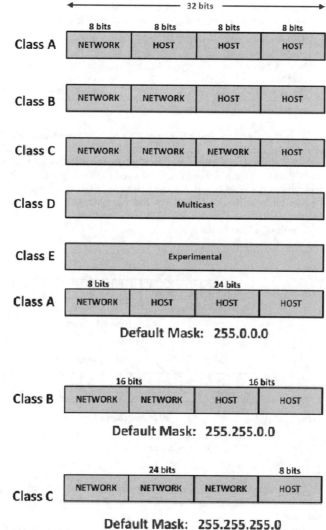

FIGURE 4.1.25 IPv4 address classes and "default" address masks.

has been abandoned today in favor of "Classless" addressing, where globally unique IP addresses are utilized to provide routing across the public Internet. Whereas Classful addressing is not often used in globally routed networks, it is important to keep in mind the default A–C masks, as the terminology is widely used and the concept can be applied to subnetting and supernetting applications. It should be noted that Classful addressing and default masks are commonly utilized in a private network environment where efficient use of address space is not a concern. Table 4.1.5 illustrates the number of networks and the number of hosts available in each class of address space. It should be apparent that the three available class choices allowed limited choices when provisioning real-world networks in an efficient manner.

TABLE 4.1.5 IPv4 Address Class Based Network and Host Possibilities

First Octet Range	1–126	128–191	192–223
Network range	1.0.0.1–126.255.255.255	128.0.0.1–191.255.255.255	192.0.0.1–223.255.255.255
Available networks	126	16,384	2,097,152
Available hosts/network	16,777,214	65,534	254
Network bits	8	16	24
Host bits	24	16	8
Default mask	255.0.0.0	255.255.0.0	255.255.255.0

With the abandonment of "Classful" addressing and the default address mask based on the address class, the address mask could no longer be implied and had to be explicatively stated with each IP address. The ability to specifically state the mask for each IP address allowed flexibility in the "right-sizing" of the number of networks and the hosts per network to better fit the specific network deployment. Variable Length Subnet Masks or "VLSMs" allowed the subnet mask to be specified on a bit-by-bit basis for the 32-bit IPv4 address mask rather than occurring at a fixed boundary as in Classful addressing. Figure 4.1.26 illustrates the fixed Classful address mask and the variable VLSM or "Classless" address mask.

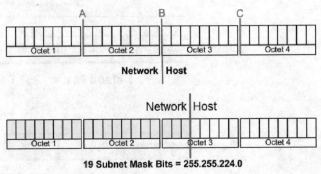

FIGURE 4.1.26 IPv4 addressing: (*a*) fixed "Classful" address mask, (*b*) variable "VLSM" or "Classless" address mask.

Upon establishment of the boundary between the Network address and the Host address, the number of possible networks or the number of available hosts can be calculated based on the available bits.

The number of possible networks is equal to 2 number of network bits.
The number of available hosts per network is equal to 2 number of host bits – 2.

It is important to recognize that the first and last IP addresses of a network are not assignable to a host device. Thus, these addresses are considered unuseable with regard to host assignment. The first IP address represents the network address or commonly referred to as the "wire address." The last IP address is the Broadcast IP address for that network.

A "gateway" IP address is assigned for each network when external connectivity to that network is desired. The gateway host IP address is the IP address assigned to the external router interface providing external connectivity to the network. It is common practice to utilize the first assignable or useable IP address within the network IP address range, although any IP address within the useable network IP range can be utilized as the gateway IP address.

The use of Classless IPv4 addressing also required the various Layer 3 routing protocols to recognize that the address mask could vary on a bit-by-bit basis. All early routing protocols were developed on the basis of the default mask scheme and thus became outdated as Classless addressing schemes were deployed. Classless Inter Domain Routing or "CIDR" was developed and routing protocols quickly updated to reflect the growing and widespread use of Classless IP addressing (IETF RFC-1518). In addition to the updating of the routing protocols, CIDR provided a shorthand ability to express the address mask. A slash notation "/" scheme was developed as a shorthand approach to fulfill expressing the mask in doted decimal notation. The CIDR notation technique allowed an IP address such as "192.168.100.254" with an address mask of "255.255.224.0" to be expressed as "192.168.100.254 /19." The slash notation and the end of the IP address indicated the number of "1" contained in the doted decimal address mask. Converting each octet of the doted-decimal address mask of "255.255.224.0" to binary yields a total of nineteen (19) "1's" in the mask as follows:

Decimal: 255.255.255.0

Binary: 11111111.11111111.11100000.00000000

In summary, the IPv4 address mask may be found expressed in a number of formats as summarized in Fig. 4.1.27.

Classful Addressing:
192.168.100.254
(Implied Mask 255.255.255.0)

VLSM Addressing:
192.168.100.254 255.255.224.0
(Explicit Mask 255.255.224.0

CIDR Notation:
192.168.100.254 /19 ←——— Number of Mask Bits

FIGURE 4.1.27 Expressing the IPv4 address mask.

In practice, it is often commonplace to convert between the doted-decimal address mask format and the CIDR notation format, or vice versa. It is often useful to remember a table to make this process easier when calculator tools are not available. You really do not need to memorize the table, just remember "1" and you can make the table by starting with a "1" in the left-most position or least-significant digit position. Double for the next position to the left and continue the process until all eight positions are filled; see Fig. 4.1.28.

Remember the Powers of 2:							
128	64	32	16	8	4	2	1

FIGURE 4.1.28 Powers of 2.

Subnetting and supernetting are essential aspects of network address planning and design. Subnetting allows smaller networks to be created by borrowing host bits to create a separate network. However, subnetting comes at a cost of two IP addresses for each new network created. Recall that the first and last addresses of a network are not assignable to a host, thus unuseable. The first address is the Network ID and the last address is the Broadcast address for that network.

Within the overall IPv4 address space, the use of "Private" addresses is made available by each class of address space. IETF RFC 1918 details the provisions of private address space often referred to as "RFC 1918" space in the industry. The use of private address space is commonplace in the industry ranging from Class C home or small-office networks to large corporate networks employing Class A or B use. Private address space cannot be used for internetworking or "routing" in the public Internet network environment. Internet Service Provider (ISP) routers are aware of RFC 1918 address space and do not forward any private address space to the public Internet unless a form of Network Address Translation (NAT) is used. A common misconception is that private address space cannot be routed. Private address space can be routed but only within the confines of a private network. Private or RFC-1918 address space is provided in Class A–C space and allocated as follows:

Class A: 10.0.0.0 to 10.255.255.255

Class B: 172.16.0.0 to 172.31.255.255

Class C: 192.168.0.0 to 192.168.255.255

When private address space is used, it is often desirable or necessary to provide translation between the public and the private environment. A form of "NAT" is utilized to provide the translation. NAT is often implemented within a Layer 3 network device such as a router, a firewall appliance, or a dedicated NAT appliance device at the border of the network between the private and public device interfaces. Figure 4.1.29 illustrates a typical NAT device placement.

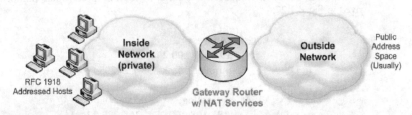

FIGURE 4.1.29 Network address translation device placement.

NAT can be implemented in several methods to serve different purposes or applications. Static NAT or "One-to-One NAT" is the basic implementation and provides a fixed two-way mapping between a Private address and a Public address.

Private Address ← → Public Address

192.168.100.254 → 165.95.240.132

192.168.100.254 ← 165.95.240.132

In this application, an equal number of private IP addresses and public IP addresses are required. This approach is often referred to as "security-by-obscurity" as it essentially hides the real IP address of the private host from the outside public address world. Whereas the host IP address is hidden, the security value to today's network environment can be questioned.

A more popular implementation is Dynamic NAT, where a two-way mapping is provided between a number of private host devices and a range or "pool" of public IP addresses for a given time period or "lease" interval. Often the number of private host devices exceeds the number of public IP addresses available. Common use can be found in an environment where the user host devices come and go such as a wireless network or other Bring-Your-Own-Device environment. IP addresses assigned in such a manner often have a short "lease" time or will expire within a short period of

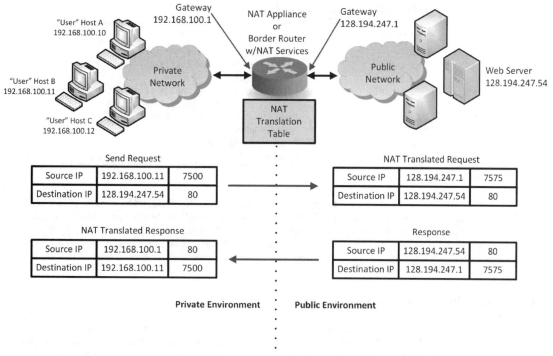

FIGURE 4.1.30 Port address translation or "PAT" operation.

inactivity. A private host device will not have connectivity to the public network once the pool of public addresses is assigned or depleted.

Port Address Translation or "PAT," also referred to as "NAT Overload," is a widely used technique where a single public IP address is available to serve a number of private addressed host devices. PAT is commonly found in the home network environment where the ISP is a cable or telco provider. When PAT is used, the NAT device adds a unique port number to the public IP address that represents a private host device. The PAT device maintains an internal table to properly map the outbound and inbound IP packets to the appropriate host device. Figure 4.1.30 illustrates the PAT process and provides an example of how a PAT device mapping table is developed. The NAT device providing the PAT service assigns a unique port number to the outbound packet and maintains a translation table entry for the response which is in turn translated to the proper internal or private host IP address from the incoming public IP.

The NAT process uses several address types and understanding the terminology used is often the key to understanding the NAT process. The "Inside Local" IP address represents a private IP address assigned to a host that resides on a network behind a NAT translation device such as a border router. An "Inside Global" IP address is used to identify an outside public IP address to a private host. The "Outside Global" IP address is a public IP address assigned to a host on an outside or public network. The "Outside Local" IP address is the IP address used by the outside public host to identify an inside private addressed host. It should be noted that multiple layers of NAT can occur in a network. This technique is often found in large carrier networks that may be mitigating the implications of IPv4 address space shortages. A common term utilized to describe multiple layer NAT implementations is "Carrier Grade Nat" or "CGN" or "Large Scale NAT" or "LSN." The terminology Carrier Grade NAT may suggest premium performance, in reality any form of NAT suffers in terms of scalability, performance, and security. And, most importantly any form of NAT breaks the IP networking end–end model.

TABLE 4.1.6 Reserved and Special IPv4 Addresses

Address	Use
0.0.0.0/8	Network address "This Network or Wire Address"
10.0.0.0/8	Private IP address space (RFC 1918)
127.0.0.0/8	Loopback address
169.254.0.0/16	IETF Zero Configuration Address Space (RFC 3927)
172.16.0.0/16	Private IP address space (RFC 1918)
192.168.0.0/16	Private IP address space (RFC 1918)
224.0.0.0/4	Multicast address space
240.0.0.0/4	Experimental address space
255.255.255.255/32	Broadcast address

A handy IPv4 address to remember is the "loop-back" address. This address is commonly 127.0.0.1, although the entire Class A block of addresses was allocated for this function (i.e., 127.0.0.0 /8). In reality, IPv4 addressing provides slightly over 3.7 billion useable public addresses (3,706,452,992) once the over 580 million (588,514,304) reserved, special use, and private address space is removed. A summary of "Reserved" (IETF RFC 5735) and special use IPv4 addresses is listed in Table 4.1.6.

4.6.6 Layer 3 Internetworking or Routing

Internetworking is a major function of Layer 3 and is implemented as an intermediary network device named a "router." Today, the router can be assumed to be an IP-based device, although early routers were often multiple protocol devices incorporating a number of proprietary protocols such as IPX, AppleTalk, DecNet, and so on. Internetworking is the forwarding of packets between different networks by means of a routing protocol. The routing protocol determines the "best" path to reach a destination network based on a structure that includes both a destination host address and a destination network address. In some cases, there may be only one path to a destination network. In other cases, there may be multiple paths of different bandwidths or performance factors to a destination network. In reality, a combination of the two is often found in the end-to-end network.

The ability of a network protocol such as IP to route packets to a different network is often assumed today; however, not all network protocols are routable. Many networking protocols that we developed in the early years of networking were not routable as the protocol structure did not allow for a network address. Microsoft NetBios is an example of a widely used networking protocol that is not routable and thus its use is limited to host communications within the same network or subnet as the protocol is only concerned with a host address and not a network address.

Routing can be performed in a static or dynamic manner. Static routing is often used in a small network environment where communication links between networks are fixed and stable. Static networks do not add any router traffic overload to a network and can be implemented within inexpensive hardware platforms. The manual configuration and ongoing updates by the network administrator are the major drawback of a static routing environment. Static routing is often found in a "stub" network such as a home or Small Office Home Office (SOHO) network with a single ISP connection. In this case, only one path exists to networks outside the local environment.

Dynamic routing is used in most professional networking applications. Dynamic routing is used where the network environment often changes, where multiple connectivity paths exist in the network, and where growth is expected. Dynamic routing allows the best path determination process to adapt automatically without network administrator intervention. Dynamic routing requires a more robust routing hardware platform and is less prone to human error once the initial configurations are implemented.

Dynamic routing can be implemented by one of two routing protocol categories. A "distance vector" based routing protocol uses network distance or "network hop" count as a metric. The routing protocol metric is the network protocol method of determining the best path to a destination network. A better path is represented as a lower metric. A distance vector routing protocol obtains network information from a neighboring network router and is often referred to as "routing by rumor." A "Link State" based routing protocol uses a network "cost" as a metric. Like the distance vector metric, a lower cost represents a better path to a destination network. The cost metric may include one or more factors that characterize the network. The cost metric factors considered often include hop count, bandwidth of the network paths, the traffic load with the network, and latency by a specific networking protocol.

Today, numerous routing protocols exist in both legacy (often deprecated) and current technology forms. These protocols can be categorized by their use in terms of where they are utilized within an overall network infrastructure. Interior Gateway Protocol (IGP) routing protocols are focused on the use in a local or a corporate network structure defined as an "autonomous" system or the scope of a network system that is under management control. Exterior Gateway Protocol (EGP) routing protocols are used where autonomous systems are connected together such as ISPs in the public Internet network environment. Figure 4.1.31 illustrates the appropriate application of IGP and EGP network protocols for use in the respective network environment.

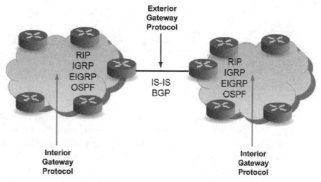

FIGURE 4.1.31 Interior and Exterior Gateway Protocol applications.

The routing protocol performs several fundamental functions including the following:

- Learn the route to each subnet or network in the Internetwork Scope.
- Determine the best route to each subnet or network.
- Remove routes that are no longer needed.
- Update the routing table as new network resources are discovered.
- Prevent routing loops.

Regardless of the routing protocol utilized, each router builds an internal table of information to enable the determination of the "best" route. The "routing table" may be built by the network administrator in the case of static routing or the table is built automatically by a dynamic routing protocol. The routing table will contain the following information as a minimum:

- The destination network address
- The metric or cost information
- The gateway or "Next-Hop" network address
- The route type (direct connected or remote routes)

Figure 4.1.32 illustrates the routing table(s) for two of the routers in a small network. Keep in mind that if static routing is utilized, the network administrator must manually provide all the routing table information. Any updates require manual intervention. If dynamic routing is utilized, the routing table information is automatically populated by the network routing protocol. Any updates are automatically implemented.

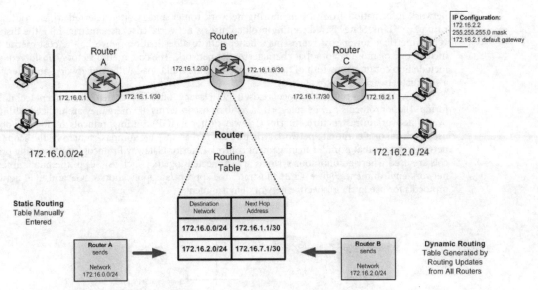

FIGURE 4.1.32 Example of the routing table.

Selection of the routing protocol to use in a specific network is often challenging due to the number of possible choices. There is not usually a single correct routing protocol to use as many variables often come into consideration; the best choice is the choice that fits a specific network environment. The many choices available can be organized as shown in Table 4.1.7. The Classful routing protocols can usually be quickly discounted due to their lack of VLSM support. For most individual or corporate network choices, the selection can be narrowed to RIPv2, EIGRP, or OSPFv2. If the network scope involves connectivity between autonomous systems such as an ISP might require, BGPv4 would be the only choice. And if IPv6 was required, each protocol offers a suitable IPv6-enabled protocol. In many cases, it would not be uncommon to find IPv4 and IPv6 routing protocols utilized in a network requiring IPv4 and IPv6 support.

TABLE 4.1.7 Practical Routing Protocol Choices

	Interior Distance Vector		Interior Link State		Exterior Path Vector
Classful	RIP	IGRP			EGP
Classless	RIP v2	EIGRP	OSPF v2	IS-IS	BGP v4
IPv6	RIPng	EIGRP v6	OSPF v3	IS-IS v6	BGP v4

With a narrowed list of choices, the selection of the correct protocol can be made on the basis of the best fit for the specific network scope and environment. The first consideration to be determined is often that of multivendor interoperability. If the network environment involves Layer 3 routing hardware from multiple vendors, a standard-based routing protocol is required such as RIP or OSPF. The IETF RFC reference denotes that these are standard-based protocols. Note that the version nomenclature is often omitted as current protocol versions are assumed to be utilized. Proprietary protocols are often an appropriate choice where all Layer 3 devices are capable of handling a manufacturer-specific protocol such as all Cisco© devices when EIGRP is chosen as the routing protocol. Proprietary protocols often offer additional features such as a multiple factor cost metric as offered by EIGRP.

The next consideration is typically the overall size of the network in terms of overall Layer 3 devices incorporated. A protocol such as RIP and EIGRP both include hope count limits. The RIP hop count of 15 is often a limiting factor even in a moderate-size network, whereas the EIGRP hop count of 224 is generally not found to be a practical limiting factor in most cases. If the network is of significant size or likely to grow in terms of Layer 3 routing devices, OSPF provides an unlimited size.

The remaining factors may be important to consider or they may not be of concern. RIP is slow to converge across all devices in the network. In a small stable network, this factor is often not a practical concern. The RIP protocol sends routing updates every 30 seconds. The use of network bandwidth for these continuous updates is often an issue when low bandwidths are used for network segments such as Telco carrier provided ISDN and T1 circuits. When Ethernet is used, the bandwidth of the "router housekeeping" data is usually not a concern even if legacy 10-Mbps Ethernet segments are still in service. Router traffic is minimized when EIGRP and OSPF are used and routing updates are only sent when a change has occurred. It should be noted that OSPF sends a "paranoia update" every 30 minutes even when network changes have not occurred. Table 4.1.8 offers a summary comparison of the more popular IGP routing protocol choices.

TABLE 4.1.8 Interior Gateway Protocol Comparison

	RIP v2	EIGRP (Cisco)	OSPF v2
Type	Distance vector	Hybrid	Link-state
Metric	Hop count	Bandwidth/delay	Cost
Administrative Distance	120	90	110
Hop Count Limit	15	224	None
Convergence	Slow	Fast	Fast
Updates	Full table every 30 sec.	Send only changes when change occurs	Send only when change occurs, but refreshed every 30 min.
RFC Reference	RFC 1388	N/A	RFC 2328

A "Layer 3 Switch" is often found in a network. From the name, a conflict of terminology should be apparent as a network switch such as an Ethernet switch is a Layer 2 device and a Layer 3 device is a router in terms of network hardware. The "Layer 3 Switch" is a common marketing term utilized to describe a network device that incorporates Layer 2 network switching and Layer 3 routing is a single box. Layer 3 switches are commonplace in network workgroup applications. These devices provide the same capabilities of a managed Layer 2 switch in terms of port configuration, port monitoring, and VLAN inclusion, and offer additional Layer 3 internetworking capability. Each switch port can be a separate network interface. The Layer 3 routing capability is often a software option to many Layer 2 switches, and the presence of Layer 3 routing capabilities is only indicated by specific manufacturer operating system software versions rather than physical device appearance. Limitations of Layer 3 switches are found in only having Ethernet interfaces available and usually only IETF RFC-based routing protocol use such as RIP or OSPF. No WAN interfaces are provided such as those often found for Telco carrier provided circuits.

Figure 4.1.33 represents the basic packet flow through the conceptual network infrastructure. As a Layer 3 network device (router in this example) is added to the end-to-end network, note that encapsulation and de-encapsulation now occur up to Layer 3 in the "stack" of the network device. Layers beyond Layer 3 are not touched and essentially ignored by the Layer 3 device.

4.6.7 Layer 3 Protocols

In addition to the IP, several other protocols that are considered part of the IP family are implemented at Layer 3 and considered as core protocols by many. These core protocols include IPSec, Internet Group Management Protocol (IGMP), and Internet Control Message Protocol (ICMP). The Internet Protocol Security or "IPSec" protocol implements secure communications by encryption of the Layer 3 IP packet in the communication session between a send host and a receive host across the network. Network gateway devices can also be utilized to provide all hosts within the network secure communications to all other hosts within another network via a secure network-to-network communication path. The IGMP is a protocol used in an IP multicasting network environment to control multicast group memberships in a one-to-many networking configuration.

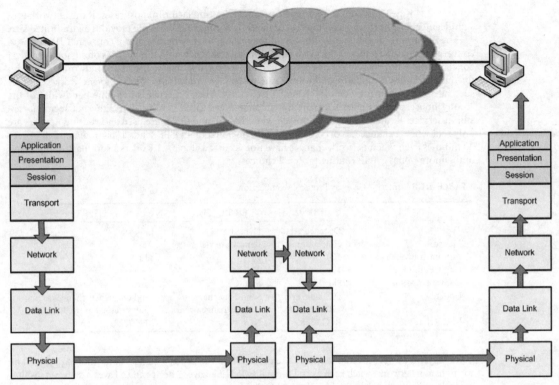

FIGURE 4.1.33 Encapsulation and de-encapsulation within a Layer 3 conceptual network.

The ICMP (IETF RFC-792) is one of the more interesting core protocols in the IP family. This is often called the "tattle tale protocol" as it provides messages between network devices that are used for diagnostic and control purposes. The popular "ping" utility used to test for connectivity or reachability of a receive host is based on the use of the ICMP protocol in its implementation. Ping sends an "Echo Request" control message to the desired receive hosts and relies on a "Echo Response" message from the receive host to verify its presence. But, what occurs when the desired revive host is no longer operational on the network?

Recall our basic conceptual network consisting of a single send host and a single receive host. The send host transmits a packet to the receive host device, but the receive host is no longer present on the network, which the send host device is not aware of. The expected TCP/IP response when a reply is not received within an expected time frame is to retransmit the packet. In this example, the retransmission could go on indefinitely. However, ICMP steps in the last available Layer 3 device (router) closest to the distance receive host network transmits a message to the send host device that the host is unreachable. In the case of the ping utility, the familiar "Destination Host Unreachable" error message is displayed. Other common network diagnostic utilities such as "traceroute" also rely on ICMP for their basic implementation.

4.6.8 A Brief Introduction to Multicast

Multicast is a unique form of IP communications across a network that strives to maximize resources and minimize duplicate network traffic by minimizing unicast traffic flow across the network infrastructure where the same payload content is delivered to multiple receive hosts. Multicast is often found in media and broadcast networks such as Audio over IP, Video over IP

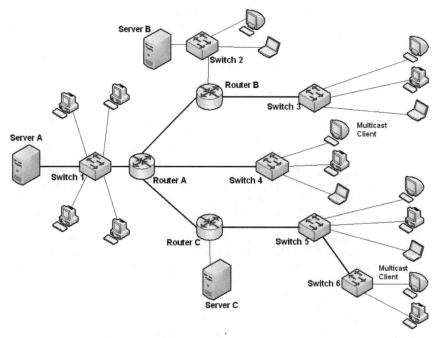

FIGURE 4.1.34 Multicast example system.

or "VoIP," and IPTV environments where UDP or "best effort" transport is utilized. TCP is not possible in a multicast environment. (TCP and UDP are covered in greater detail later in this chapter.) Multicast enables payload content to be sent from a single send host to a specific group of receive hosts that desire or choose the payload content. It is important to distinguish that a broadcast packet is received by all host devices in the network, whereas the multicast packet is received by a select group of hosts or those hosts that desired to receive the multicast packet payload. A receive host selects the desired payload by joining a specific multicast group to receive multicast packets. Recall that multicast utilizes a special section of reserved IPv4 IP address space or Class D address space. Figure 4.1.34 illustrates multicast activity where traffic is minimized across network segments and traffic is only sent to network devices and host devices or "clients" that desire the multicast payload from a send device or "multicast server". Implementation of multicast requires Layer 2 and Layer 3 devices that are multicast capable in terms of feature set, memory, and processor power. Multicast is often promoted as a method to relieve server resources of multiple unicast content stream delivery, where, in reality, increased resource need is seen by the network devices delivering the multicast content.

The IPv4 address space 224.0.0.0 to 239.255.255.255 is further subdivided and allocated for specific multicast applications as listed in Table 4.1.9.

TABLE 4.1.9 IPv4 Multicast Applications

Address Space	Application
224.0.0.0 to 244.0.0.255	Reserved—Routing Protocol Intercommunications
224.0.1.0 to 238.255.255.255	Public Use
239.0.0.0 to 239.255.255.255	Private Use (nonroutable onto the Internet)

Within the "reserved" multicast address space, individual addresses are reserved for specific routing protocol use such as 224.0.0.5 for OSPF enabled routers or 224.0.0.9 for RIP-enabled routers.

Public use multicast IPv4 address can be routed on the "public" Internet, but in reality is not possible. The Internet is comprised of multiple independent networks without any overall coordination or cooperation, thus making multicast impossible to implement in an end–end manner. Multicast is commonly found implemented in private network environments or pseudoInternet networks such as "Internet 2" focused on the higher education and research communities.

By default, a Layer 2 network device forwards or "floods" multicast frames to all device ports whereas a Layer 3 network device blocks any multicast packets. The Layer 2 Ethernet switch must incorporate support for multicast with discovery capabilities in order to designate switch ports requiring multicast frame delivery based on the connected host desire to receive specific multicast payload. This "trimming" process eliminates multicast traffic where it is not needed or desired. A Layer 3 IP router by normal operation builds a "routing" table of destination addresses and the IP routing protocol actions forward packets from a send host to a destination receive host. Special routing tables created by a multicast-enabled routing protocol are required in the multicast environment that provide information regarding the multicast group (or multicast source), interfaces that point toward the multicast sources, and interfaces that point toward multicast hosts. Each multicast group within a network utilizes a unique multicast address to identify a specific multicast send host. In an IPTV multicast enabled network, the multicast group might be thought of as the channel number, thus allowing a host to select the desired multicast payload content.

The IGMP is the key multicast protocol that allows a multicast receive host to send a request to join a multicast group and thus receive the desired multicast payload. "IGMP Snooping" allows a Layer 2 device to listen to multicast activity and learn the multicast group activity. As an example a Multicast Source host such as a streaming content server provides content associated with the multicast address or group "239.1.2.3." The server sends packets utilizing UDP transport to the multicast address 129.1.2.3. A host desiring to receive the multicast content associated with the group identified by the address 239.1.2.3 informs the network devices that the content is desired and joins the group to receive the desired content. Other content available on the network is identified by a different group or multicast IP address. The server or multicast source does not need to know about all the receivers that desire to receive content with the network infrastructure responsible to deliver the desired multicast content to the desired multicast receive host. In comparison, a content server would need to know about all desired receive hosts in a unicast network environment and construct an individual content stream to those hosts.

IGMP is only concerned with group management and a multicast routing protocol is required to actually deliver the desired multicast payload to a specific multicast receive host or client. The Protocol Independent Multicast or "PIM" protocol is a standardized protocol to deliver payload content packets to the desired receive hosts. PIM is based on the creation of a distribution tree that is in turn "trimmed" to eliminate network segments not requiring multicast traffic or "grafting" network segments that need to be added to deliver the desired payload. Figure 4.1.35 illustrates a multicast distribution tree.

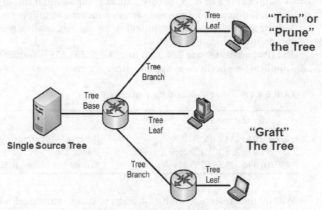

FIGURE 4.1.35 Multicast distribution tree example.

PIM can be found in one of two standardized versions:

- PIM "Dense Mode" can often be found in larger multicast environments where quick tree creation is desired. All segments of the network distribution tree are flooded and then trimmed based on no multicast client need. Dense Mode can typically be found in a network environment where the majority of the client devices desire multicast content such as an IPTV network.
- PIM "Sparse Mode" is often found in smaller networks where there is a concern for bandwidth utilization. A "Rendezvous" point router is designated and multicast servers (send hosts) and multicast clients (receive hosts) register with the Rendezvous point. This method is better suited for environments that incorporate a smaller number of multicast clients and the longer tree trimming or grafting time is not a concern.

It should be noted that proprietary PIM versions exist such as the Cisco "PIM-SM-DM" protocol.

Multicast forwarding or multicast routing is unique and is often referred to as "reverse routing" as the action occurring is opposite from the dominate unicast process. Reverse Path Forwarding or "RPF" organizes routing tables to include paths from the multicast receiver back to the multicast source. The RPF table information may be incorporated into the routers unicast routing table or a separate RPF table may be maintained as implemented by a specific equipment manufacturer.

4.6.8.1 A Brief Introduction to IPv6. The IPv4 addressing scheme developed in the mid-1970s provided over 4.2 billion possible addresses with slightly over 3.7 billion considered to be useable in the context of assignable to hosts participating in Internet global routing. Within 10 years after implementation of IPv4 (early 1980s) it became clear that the available address space provided by a 32-bit address was not going to be adequate to support the rapid growth of IP networks and the "Internet" worldwide. Development of the next IP addressing scheme or "IPv6" (IETF RFC-1883) began in earnest by the mid-1990s and actual implementation began in the late 1990s at somewhat slow pace. Today, IPv6 implementation has gained momentum as essentially all IPv4 address space available for allocated has been depleted. IPv6 provided a vast amount of address space through use of a 128-bit address. In addition to increased address space, IPv6 also was an opportunity to re-engineer the IP. Enhancements were incorporated that addressed secure communications and authentication, traffic or flow-control provisions, improved support for multicasting applications, autoconfiguration, and enhanced support for mobile applications. IPv6 should not be considered any more secure than IPv4, where IPSec can be implemented as an option, but IPSec is an integral component of IPv6 rather than an optional feature. Multicast is an integral feature of IPv6. Mobility application enhancement is often a major feature for use of IPv6, in addition to the need for globally routable address space—especially in cellular networks, which experience rapid growth worldwide. IPv6 returns IP communications to the original end–end communications model without intermediary devices to distort the original send or receive host identity. Routing performance has also been increased by establishing a fixed length for the IPv6 packet header as opposed to a variable length IPv4 packet header. The IPv6 header is fixed at 40 bytes by elimination of the "option" filed found in an IPv4 header. The same functionality is provided by an "Extension Header" when required. The header checksum has been eliminated in the IPv6 header and no packet fragmentation is permitted at intermediate network devices. All of these factors result in faster packet processing and faster IP forwarding. Figure 4.1.36 illustrates the differences in the IPv6 and Ipv4 packet headers. In essence, the IPv6 header was simplified and fixed in size to achieve faster packet processing within the network.

With enhancements, expanded IP address space and scalability are the most significant capabilities provided by IPv6. As previously stated, IPv6 utilizes a 128-bit address providing 2^{128} or 340 undecillion possible addresses. The overall available IPv6 address space size is often impossible to comprehend in terms of scope, although in 1970 the IPv4 32-bit address was thought to provide more IP addresses than could be utilized.

The IPv6 128-Bit Address: 2128 = 340,282,366,920,938,463,463,374,607,431,768,211,456 (a total of 340 UNDECILLION IPv6 addresses) 3.4 × 1038.

Scalability exists in that based on current assignment policy, around 18% of the possible IPv6 address space is currently allocated for assignment by the Internet Assigned Numbers Authority

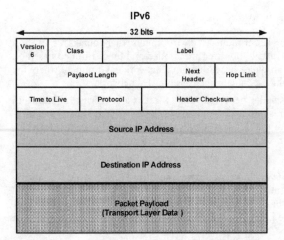
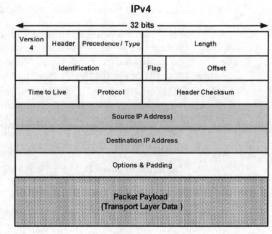

FIGURE 4.1.36 The IPv6 and IPv4 packet headers compared.

worldwide. In North America, the American Registry for Internet Numbers further allocates address space to individual network providers and organizations. Enormous, although not infinite, amount of IPv6 address space is potentially available.

The 32-bit binary IP address was a challenge to comprehend and was represented in a "doted-decimal" format. The IPv6 address presents a bigger challenge to represent a 128-bit binary address. A similar simplified concept is used to represent the IPv6 address.

The 128-bit binary IPv6 address is first divided into eight (8) 16-bits binary groups or "quads" or "chucks". Each 16-bit binary group is then converted to a hexadecimal word to represent the 16-bit group. Each hexadecimal word is then separate by a colon (:) to comprise the complete IPv6 address. Figure 4.1.37 illustrates the representation of a 128-bit binary IPv6 address.

128-Bit Address Binary Format:
00100110000001111011100000000000011110101010000000000011001000011001010110011000100011110111000100100001010001110001

Subdivide Into Eight (8) 16-bit Groups:
0010011000000111 1011100000000000 0000111110101010 0000000000000011
0010000110010101 1001100010000111 1011110001001000 0010100011110001

Convert Each 16-bit Group to Hexadecimal:
(separate with a colon)
2607:b800:0faa:0003:2195:9887:bc48:28f1

FIGURE 4.1.37 The IPv6 address format.

Simplification of the IPv6 may be undertaken in some cases to shorten the address by elimination of redundant information. Figure 4.1.38 provides an example of address summarization. A group or chuck consisting of all zeros ("0's") can be represented as a single zero ("0"). Successive groups of zeros can be represented as a double-colon ("::") regardless of the number of groups. However, all summarized groups must be consecutive and this summarization can only be performed once in an IPv6 address.

2001:0000:0000:0000:0DB8:8000:200C:417A
or
2001:0:0:0:DB8:8000:200C:417A
or
2001::DB8:8000:200C:417A

FIGURE 4.1.38 IPv6 address summarization example.

As in the IPv4 world, each IPv6 address must have a subnet mask. In the case of IPv6 the subnet mask is 128 bits. CIDR or slat notation ("/") is used exclusively to represent the IPv6 subnet mask. The mask indicates the prefix or network component of the address from the host identification component of the address. In the following example, the IPv6 address is divided into two components based on a 64-bit subnet mask value expressed in "/" notation:

2001:0:0:0:DB8:8000:200C:417A /64 Prefix Component: **2001:0:0:0:** Host Component: DB8:8000:200C:417A

With IPv4 addressing, a single IP address is assigned to each host interface. With IPv6, multiple IP addresses are assigned to each host interface with each address representing a specific network function. Addresses are classified as follows:

- Global Unicast addresses
- Link-Local addresses
- Unique Local addresses
- Multicast addresses
- Special addresses

The Global Unicast addresses are often described as being equivalent to the IPv4 public IP address. The Global Unicast address is routable across the public Internet. Link-Local addresses are the equivalent of a private address and they can only be assigned to hosts within the same network or subnet. Unique Local addresses are similar to the Link-Local address in terms of a private nonroutable address to the public Internet, but the scope can extend to all networks within an organization's administrative control. An IPv6 Multicast address is equivalent to an IPv4 multicast address. Special use addresses or "reserved" addresses are designated for a specific functional purpose such as the IPv6 "loop-back" address as "0:0:0:0:0:0:0:1" or summarized as "::1." The IPv6 address "::" (equivalent to 0:0:0:0:0:0:0:0) is considered an "unknown" address representing all hosts on a network similar to the IPv4 address of "0.0.0.0."

Different IPv6 address types are indicated by examination of the first few bits of the IPv6 address and can be identified from Table 4.1.10. It should be noted that there is no Broadcast address in the IPv6 environment. Any IPv6 address is considered a Unicast address except for the designated Multicast address.

TABLE 4.1.10 IPv6 Address Types

Address Type	First Field of Address	Address Use
Global Unicast Address	2xxx: or 3xxx:	Routable Public
Link-Local Address	FE8x:	Private—same network
Unique or Site Local Address	FECx:	Private—same organization
Multicast Address	FFxx:	Multicast Group Participation
Special or Reserved Address	00xx:	Reserved or Special Applications

Autoconfiguration is an additional feature of IPv6. Configuration of an IPv6 host can be accomplished manually, in a stateful manner, or in a stateless manner. Manual host configuration requires all host IP configuration information to be manually entered. This option is commonplace in server host configuration. Stateless autoconfiguration is likely the most popular configuration approach in an IPv6 network. Stateless autoconfiguration allows a host to automatically configure IPv6 parameters based on listening for a Router Solicitation or "RA" announcement or making a RA request. The RA provides the network identification and the host being autoconfigured generates a host ID based on the host MAC address in the IEEE Extended Unique Identifier—64 or "EUI-64" format. The 64-bit EUI address is generated by bit manipulation of the 48-bit MAC address. Combining the network address and the host address creates a Global IPv6 address:

Global Host Address = Network Link Address + EUI-64 Host Address.

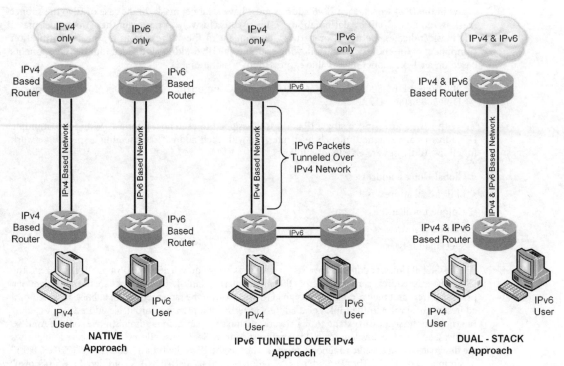

FIGURE 4.1.39 IPv6 implementation approaches.

Stateful autoconfiguration is used when RA does not exist on the network. A DHCPv6 server must be provided in such a case for autoconfiguration to occur.

Implementation of IPv6 in a network can be accomplished in several ways, with the best approach determined by numerous factors. The "native" approach builds a separate IPv6 network from an existing IPv4 network infrastructure. This approach is often used when an IPv6 only network is required; it is a popular method to establish an IPv6 "sandbox" within an organization. The "tunnel" approach builds a separate IPv6 network environment, but utilizes the IPv4 network infrastructure to provide interconnectivity between IPv6 networks by means of a tunnel. This approach is commonly used when an organization's ISP network or carrier transport network does not support IPv6. The "dual-stack" approach maintains one network infrastructure where each host has been implemented with IPv4 and IPv6 IP "stacks" and is often the preferred method of IPv6 implementation. Figure 4.1.39 illustrates the possible conceptual IPv6 implementation methods.

IPv6 is incorporated into all major computer operating systems existing in 2015 such as Windows 7/8/10, OSX versions, and Linux. IPv6 is slowly being incorporated into broadcast equipment as a standard feature. It is expected that IPv6 will be a standard offering especially in broadcast products that are destined for the international and governmental marketplace. Remember, IPv6 is still IP.

4.6.9 Layer 3 Summary

Layer 3 is the primary layer where the IP (IPv4 or IPv6) and the associated family of protocols is implemented. The Layer 3 IP address provides a structured view of the network by the address providing network and host address information by means of a mandatory subnet mask. Recall that the mask may be implied in the case of Classful network addressing. However, as Classful addressing is considered obsolete in favor of Classless addressing, the subnet mask must be specified. Layer 3 provides internetworking between different networks, subnets, or broadcast domains.

4.6.10 The Transport Layer—Layer 4

The Transport Layer of Layer 4 is focused on providing data transfer between a send host and a receive host. The transfer may be performed in a "guaranteed" manner by means of an acknowledgment from a receive host back to the send host or the transfer may be performed in a "nonguaranteed" manner where data are sent from a send host to a receive host without any acknowledgment that the receive host actually received the data sent. The application often determines the method of transfer used, with latency often being the determining factor—especially when real-time data are involved, such as "media" streaming content.

The Transport layer can also provide additional features or services that enhance the efficiency of the data transfer between hosts. Features such as windowing and flow control can often enhance the transfer of data between a send host and a receive host even as network condition changes. In the case of guaranteed data transfer, an acknowledgment and retransmission capability ensures that all intended data are received by the receive host.

TCP (IETF RFC-793) and UDP (IETF RFC-768) are considered the primary protocols utilized by the Transport layer for the transfer of data between hosts.

Transmission Control Protocol or TCP is a guaranteed method of data transfer and the method utilized by the majority of IP network applications. The guaranteed data transfer delivery is achieved by a "three-way handshake" process occurring between the send host and the receive host. Figure 4.1.40 illustrates the handshake process occurring between a send host and a receive host. The send host inmates a communication session with a receive host by sending a "synchronization" message to the receive host. The receive hosts respond with an acknowledgment and their own synchronization message. The send hosts respond with another acknowledgment, which completes the three-way handshake process. Data are now transferred between the hosts in a send/acknowledge process. The transfer efficiency may be improved by use of the windowing features available. Instead of sending a block of data and waiting for an acknowledgment, the send host sends several blocks of data at a time to the receive host where the acknowledgment from the receive host represents several blocks of successfully received data. The amount of data the receive host can receive and process

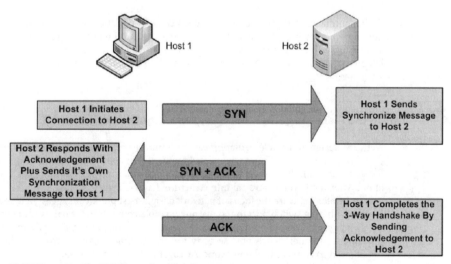

FIGURE 4.1.40 The TCP "3-way handshake" process.

represent the receive host "window." The window information is communicated from the receive host to the send host during the 3-way handshake process.

User Datagram Protocol or UDP is a nonguaranteed method of data transfer utilized where the delay or latency involved in the acknowledgment process is detrimental to application functionality. Most applications that utilize transfer of real-time data utilize UDP. It should be noted that the TCP 3-way handshake is used to establish a UDP data transfer session, but acknowledgment is not utilized after the initial session establishment. Figure 4.1.41 illustrates the UDP data transfer process between a send and a receive host. Once the session is established, data from the send host are sent to the receive host.

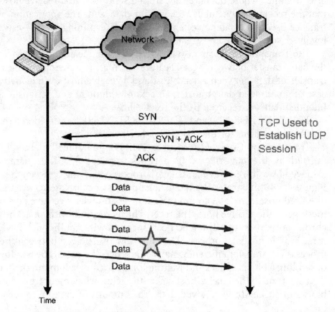

FIGURE 4.1.41 The UDP data transfer process.

In summary, TCP provides a guaranteed transfer of data, but at a cost in terms of latency. UDP does not guarantee the transfer of data between hosts and relies on a reliable network infrastructure and an upper layer to handle any missing data. Thus, how the missing data are compensated for is left up to the Application Layer in many cases. Figure 4.1.42 provides a comparison of TCP and UDP attributes.

4.6.11 Putting the Pieces Together—Building the Segmented Network

Network design and implementation can be found in many forms and configuration. Many factors often come into consideration when determining the best architecture and equipment choices within a specific network. Regardless of the scope of a network several "best practices" come into consideration. After the Layer 1 physical infrastructure, Layer 2 switching and Layer 3 routing make up the core technology and are the focus of network design. Best practices suggest that one maintain a single host per Ethernet switch port to remove any possible contention for bandwidth between the host and the switch port. This action minimizes the "collision domain" and also provides the opportunity to implement Ethernet switch port security features to monitor or control what hosts are allowed or not allowed to connect to the network. At Layer 3 the internetworking often becomes an important consideration. The ability to organize host devices in some logical arrangement by implementation

TCP	UDP
• Connection Oriented	• Connectionless
• Guaranteed Delivery	• Not Guaranteed
• Acknowledgments Sent	• No Acknowledgements
• Reliable, But Higher Latency	• Unreliable, But Low Latency
• Segments & Sequences Data	• No Sequencing
• Resends Dropped Segments	• No Retransmission
• Provides Flow Control	• No Flow Control
• Performs CRC	• Performs CRC
• Uses Port Numbers for Multiplexing	• Uses Port Numbers for Multiplexing

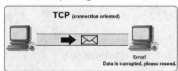
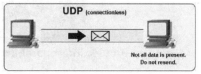

FIGURE 4.1.42 TCP and UDP comparison.

of separate networks is a powerful tool to ensure performance and provide security. The sizing of the "Broadcast Domain" is often a critical aspect in network performance by minimizing the impact of broadcast traffic on a given network.

A frequent question often arises with regard to when should I route and when should I switch?

The answer is simple in general but often more complex in reality. Best practices suggest that you segment a network into separate networks for a specific group of hosts. The grouping of host devices may be based on a geographic location, a company or corporate policy, a state or Federal regulation, or more often based on performance and security enhancement. This grouping also determines the broadcast domain for the network or defines how far a broadcast will travel within a network. A broadcast will not travel past a router interface, thus limiting the scope of reach. Recall that Ethernet was based on a "bus" architecture with all host devices having equal access to the network. It was given that collisions would occur when two or more hosts attempted to transmit at the same time. Collisions were expected to occur and often networks were sized to limit the occurrence collisions so that throughput was not impacted to an excessive amount. It was often felt that 5 Mbps throughput across a 10 Mbps was "good" performance. Today, with the drastic price drops of Ethernet switch ports, each host should be assigned to a specific Ethernet switch port. This approach ensures that the collisions between hosts will not occur and full-duplex communications can occur between the switch port and the host. Switch port security features also allow the switch to identify which host device is connected to the switch by the MAC address. This feature can be used to ensure that a specific host is connected and remains connected to a specific switch port. The individual switch port may in turn be associated with a VLAN implemented at Layer 2. A couple of "best practice" implementation steps should be kept in mind:

- Avoid use of Layer 1 hubs in the network
- Maintain a practice of one host device per switch port

Whereas a hub might be a handy approach to quickly connecting host devices to a network, it has distinct disadvantages. A hub is essentially a multiport distribution amplifier and only operates at Layer 1. There is no intelligence. Thus, all hosts now share the connectivity to the network switch and the autonegotiation process will configure the part to the lowest common value of the host devices. In addition, the legacy CSMA/CD process of Ethernet is enabled, which mitigates host devices transmitting data at the same time. Performance or network throughput is impacted as a result.

Maintaining the practice of one host per switch port provides the best performance, creates a collision domain of one, and allows switch port security features to be effectively implemented. Port security features allow control over what device can be connected to a switch port and also provide notice of when a foreign device is connected to a switch port.

Figure 4.1.43 illustrates when to route and when to switch.

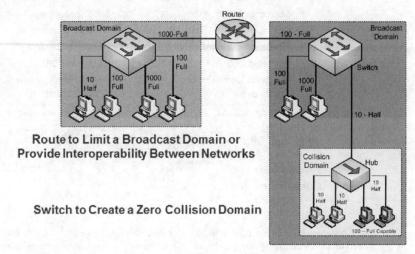

FIGURE 4.1.43 Example of when to route/when to switch. (*Note*: Do not use hubs in your network implementation.)

The use of VLANs in the design of the network architecture can offer several implementation advantages. Host devices can be grouped in a manner of their use or function and are not based on where they may be physically located. Yet, a common physical infrastructure can be utilized to support what are essentially multiple isolated networks. VLANs are constructed between Layer 2 managed Ethernet switch ports by means of tagging an Ethernet frame with the VLAN membership information. In the Cisco environment, a "trunk" port is designated for this purpose while hosts are assigned to an access port and the access port is assigned membership to a specific VLAN. In an HP environment, the same principle applies; however, different terminology is used. The truck port becomes a "tagged" port and the access port becomes the "untagged" port in the HP terminology. You will often find that any industry switch manufacturers will use one of the two terminology approaches to identify the switch port configuration.

It is important to recognize that a VLAN is an independent network, subnet, or broadcast domain. No connectivity exists between VLANs even implemented in the same physical chassis. For connectivity to occur between these separate networks, Layer 3 routing must be implemented. Figure 4.1.44*a* and *b* illustrates several conceptual approaches to VLAN implementation.

4.7 CONCLUSION

The impact of IT on the broadcast plant is here to stay. Information technology is a vast diversified field unto itself, with IP networking often found as the foundation technology or glue that makes other areas of IT integration possible. The understanding of IP networking is an essential foundation to build upon. This brief introduction has just scratched the surface with regard to understanding IP networking. Several suggested references are shown in the Appendix of this chapter to help further

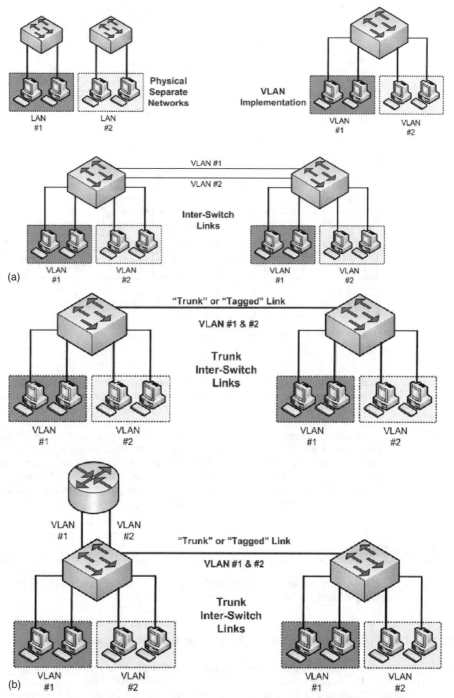

FIGURE 4.1.44 (*a*) Conceptual approaches to VLAN implementation. (*b*) Additional conceptual approaches to VLAN implementation.

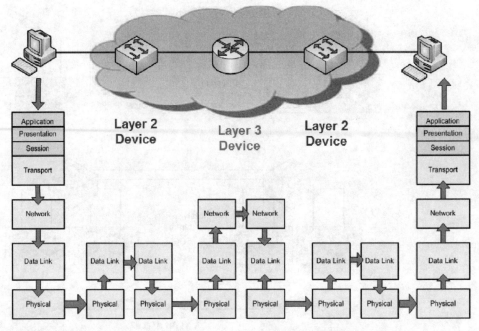

FIGURE 4.1.45 Encapsulation and de-encapsulation within a Layer 2 and Layer 3 conceptual network.

your IP networking knowledge. Figure 4.1.45 is a slightly enhanced version of the two host conceptual network presented at the beginning of this chapter. As the network infrastructure becomes more complex with additional host devices, Layer 2 devices, or Layer 3 devices, do not lose sight of this fundamental conceptual network in your network design and troubleshooting efforts.

IT industry technology advances will likely be an integral part of the broadcast plant ranging from cloud-based services to Software Defined Networking, where application governs the network performance and capabilities. Security is an important and essential aspect of any network design and represents a never-ending assessment, implement, and assessment process. And, of course, the "Internet of Things" or IoT movement will likely engulf all devices in the broadcast facility and redefine what is thought to be a network host device.

A4.1 APPENDIX

The following information is provided to further the reader's understanding of the tutorial information provided previously in this chapter.

A4.1.1 Conceptual Ethernet Switch (VLAN) Configuration Examples

An example of Cisco-to-Cisco implementation is shown in Fig. 4.1.46.
An example of Cisco-to-HP implementation is shown in Fig. 4.1.47.
Caution: Exact configuration commands may vary by the switch model, IOS, and/or software version.

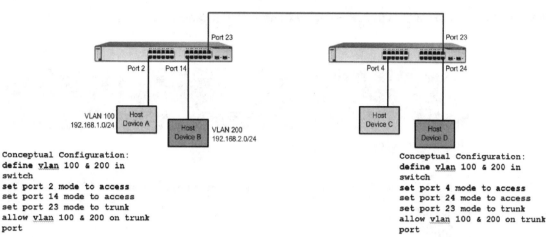

FIGURE 4.1.46 Cisco-to-Cisco example.

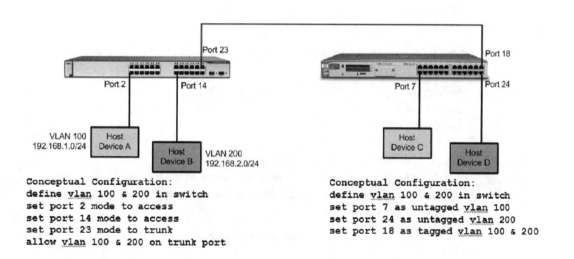

Cisco Terminology	HP Terminology
Access Mode	Untagged
Trunk Mode	Tagged

FIGURE 4.1.47 Cisco-to-HP example.

A4.1.2 Cisco Ethernet Switch Port Security Configuration Examples

A typical Cisco Ethernet switch is shown in Fig. 4.1.48.

FIGURE 4.1.48   Cisco© 2960 Ethernet switch.

A4.1.2.1   Example 1:   Enable port security on Port 1 and set maximum MAC addresses to 1:

Switch#
Switch# config t
Switch(config)# interface Fa/1
Switch(config-if)# switchport port-security
Switch(config-if)# switchport port-security maximum 1
Switch(config-if)# end

A4.1.2.2   Example 2:   Allow only MAC address "e8-80-2e-b5-43-cc" on port 2 and shut down port if a violation occurs:

Switch#
Switch# config t
Switch(config)# interface Fa0/2
Switch(config-if)# switchport port-security
Switch(config-if)# switchport port-security maximum 1
Switch(config-if)# switchport port-security mac-address e8-80-2e-b5-43-cc
Switch(config-if)# switchport port-security violation shutdown
Switch(config-if)# end

Caution: Exact configuration commands may vary by the switch model, IOS, and/or software version.

A4.1.3 Reverse Engineering an IPv4 Address

Reverse engineering an IP address is often a useful tool when troubleshooting or verifying the IP addressing scheme of an unknown network. There are numerous web-based and smartphone calculator applications available to perform this task. Of course, if the network is broken, you might not have access to these tools. The "old fashion" method is a handy technique to know. This technique is known by several names including the "Magic Box," the "Magic Number," or the "Morris-Barker" technique.

The common questions to be answered are: what is the useable IP address range with the IP address? In this example we are given the IP address "128.194.247.54" with a subnet mask of "255.255.255.224." Remember each IP address must have a subnet mask, either implied or explicitly stated as in this example.

Step 1: Examine the first octet of the IP address and determine the address class. The first octet value of "128" indicates that this is a class B IP address, which would have an implied or default mask of "255.255.0.0." It can be seen that this mask value explicitly stated is different and should indicate that this is a subnet of a larger network using VLSM to specify the network and host range.

Step 2: Convert the decimal mask of "255.255.255.224" into CIDR notation to indicate the number of subnet bits. The first three octets are "255" which indicates "8" bits or a total of 24 bits. Converting the decimal value of "224" found in the fourth octet yields "3" bits or a total of "27" bits in the subnet. The fourth octet becomes the focus since the value is not a decimal 255 or "11111111" in binary.

Step 3: Create a reference table of bit position values, mask values, and bit positions on scratch paper or your memory. You can create the table by remembering the number "1" and the simple steps to create the reference tables. Begin with a "1" in the right-most or eighth bit position (least significant). Next, double the value of "1" to create the seventh bit position or "2." And double the second bit position to create the third bit position of "4." Continue this process for all 8-bit positions and you have created the table of all possible decimal bit values for an octet; see Fig. 4.1.49.

128	64	32	16	8	4	2	1	Value

FIGURE 4.1.49 Step 3.

Step 4: Create a second reference table of mask values for each bit position. Begin by taking the first bit position value of "128" from the decimal value table to create the first bit position value of a mask table. Next, take the second bit position value or "64" from the decimal value table and add to the first bit position of the mask table value to create the second bit mask table value or "192." Continue this process for all 8-bit positions yielding the tables shown in Fig. 4.1.50.

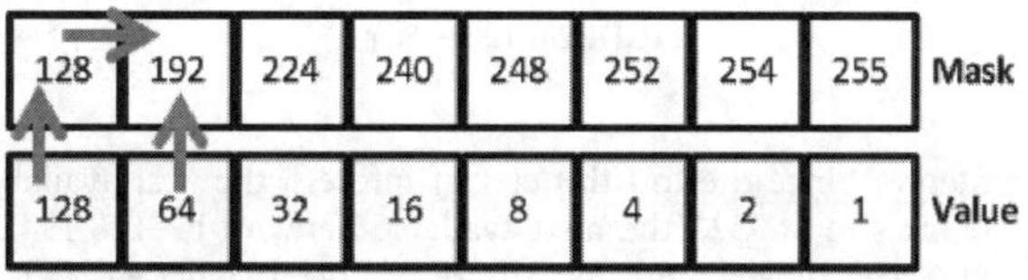

128	192	224	240	248	252	254	255	Mask
128	64	32	16	8	4	2	1	Value

FIGURE 4.1.50 Step 4.

Step 5: You may want to add a third reference table of bit positions from the most significant to least significant position to yield your reference tables; see Fig. 4.1.51.

1	2	3	4	5	6	7	8	Bit Position
128	192	224	240	248	252	254	255	Mask
128	64	32	16	8	4	2	1	Value

FIGURE 4.1.51 Step 5.

Step 6: Remember that the subnet mask requires "3" bits from the fourth octet of the subnet mask, thus identifying the fourth octet as the octet of interest. Draw a line between the third and the fourth bit position; see Fig. 4.1.52.

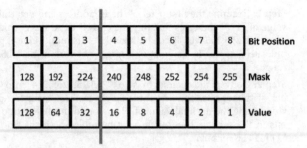

FIGURE 4.1.52 Step 6.

Step 7: Select the decimal value to the immediate left of the drawn line or "32" in this example. The decimal value of "32" becomes the "Magic Number" or the block size of the network in which the IP belongs; see Fig. 4.1.53.

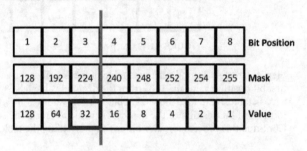

FIGURE 4.1.53 Step 7.

Step 8: Since the fourth octet of interest, the first available network is "128.194.247.0." With a block size of "32," the next available network is "128.194.247.32." The next available network is "128.194.247.64" and thus the IP of "128.194.247.54" falls within the "128.194.247.32" network; see Fig. 4.1.54.

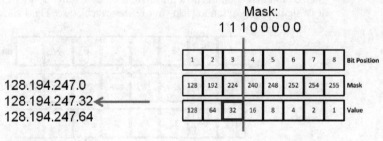

FIGURE 4.1.54 Step 8.

Step 9: With the network identified as "128.194.247.32" we know the first useable IP address is the Network ID plus 1 or "128.194.247.33." Since we know the next network address of "128.194.247.64," we know the Broadcast address of the "128.194.247.32" network to be one host less or "128.194.247.63." With the Broadcast address known, the last useable host is the Broadcast address minus 1 or "128.194.247.62." The IP address has now been re-engineered (Fig. 4.1.55)!

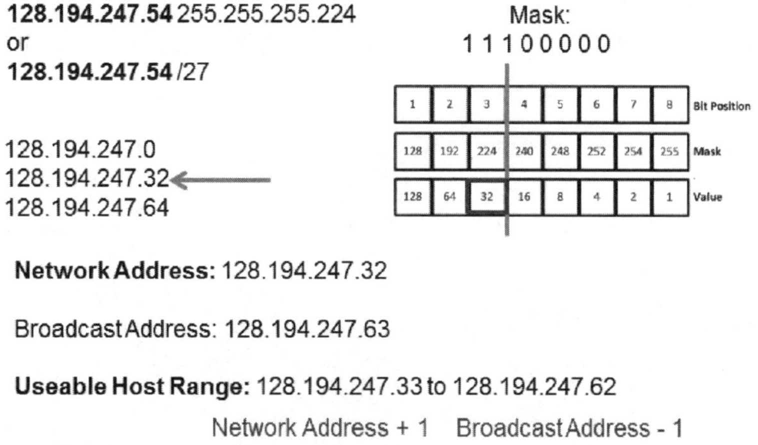

128.194.247.54 255.255.255.224
or
128.194.247.54 /27

Mask:
1 1 1 0 0 0 0 0

Bit Position	1	2	3	4	5	6	7	8
Mask	128	192	224	240	248	252	254	255
Value	128	64	32	16	8	4	2	1

128.194.247.0
128.194.247.32 ←
128.194.247.64

Network Address: 128.194.247.32

Broadcast Address: 128.194.247.63

Useable Host Range: 128.194.247.33 to 128.194.247.62

Network Address + 1   Broadcast Address - 1

FIGURE 4.1.55 Step 8.

A4.1.4   Select IETF RFC References

The IETF web site can be accessed from the following link: https://www.ietf.org/rfc.html.

RFC-768 UDP—User Datagram Protocol
RFC-791 IP—Internet Protocol
RFC-792 ICMP—Internet Control Message Protocol
RFC-793 TCP—Transmission Control Protocol
RFC-826 ARP—Address Resolution Protocol
RFC-1112—IP Multicast
RFC-1208—Glossary of Networking Terms
RFC-1518—Architecture for IP Address Resolution with CIDR
RFC-1723—RIP – Routing Information Protocol
RFC-1883—IPv6 Specification
RFC-2362—Multicast PIM – SM
RFC-3373—Multicast IGMPv3
RFC-3704—Multicast Routing (forwarding)

SUGGESTED FURTHER READING

The following books are suggested for additional information on the concepts presented in this chapter.

Donahue, G. *The Network Warrior*, 2nd ed., O'Reilly Media, Sebastopol, CA, 2011.
Goralski, W. *The Illustrated Network*, Morgan Kaufmann, Burlington, MA, 2009.
Hagen, S. *IPv6 Essentials*, 2nd ed., O'Reilly Media, Sebastopol, CA, 2006.
Kozierok, C. M. *The TCP/IP Guide: A Comprehensive, Illustrated Internet Protocols Reference*, No Starch Press, San Francisco, CA, 2005.
Solomon, M., D. Kim, J. Carrell. *Fundamentals of Communications and Networking*, 2nd ed., Jones & Bartlett Learning, Burlington, MA, 2015.
Spurgeon, C. E., J. Zimmerman. *Ethernet: The Definitive Guide*, 2nd ed., O'Reilly Media, Sebastopol, CA, 2014.

CHAPTER 4.2
NETWORK SYSTEMS

Gary Olson
GHO Group, New York, NY

4.1 INTRODUCTION

The core infrastructure technology of broadcast facilities has changed substantially. The term IP is derived from the original Internet communication protocol, Transmission Control Protocol/Internet Protocol, more commonly known as TCP (Transmission Control Protocol)/IP. When Ethernet was introduced for networks, so was TCP/IP as the transmission protocol. The Information Technology (IT) sector has been using Ethernet switching since 1983 when it first became commercial. As the technology matured and other protocols, such as User Datagram Protocol (UDP), were introduced for large packet transmission, the generic term for networks and computer systems became IP and the terms TCP, UDP, and others now refer only to the transmission protocols.

4.2 NETWORK INFRASTRUCTURE

In broadcasting, the IP network router and switch have in many ways superseded the SDI (serial digital interface) router as the most prominent core infrastructure technology. The IP network has long been the core of the enterprise infrastructure technology and IT services. It has also been a key technology in broadcast, more prevalent in the management of systems and devices. It has now moved front and center handling all aspects of media, from creation to production to distribution.

The design and configuration of network topology for broadcast and production is different than the configuration of network topology in the enterprise. In both cases, there are many broadcast, production, and business layers to the network, as illustrated in Fig. 4.2.1.

There are some terminology challenges to overcome in understanding IP networks. The IP router is a completely different technology than an SDI router. Where the SDI router can "route" any input/source to one or many outputs/destinations in a one way or simplex path, the IP network is fully duplex and all ports can receive and transmit simultaneously to each and all of the connected devices.

When looking at the network topology, there is the architecture of the network and how it connects to the end points, server infrastructure, and storage layers; and then there are the different network protocols to address the different transmission requirements of data, files, and streams. These are business and operational layers that can be either physical, virtual, or both. Each of these layers can have different configuration parameters and addressing schemas. In addition to the routers and switches, a key component of the network is the firewall. The firewall is what manages the rules and policies that govern access to the network from external applications, other networks, and internal applications reaching out to outside networks.

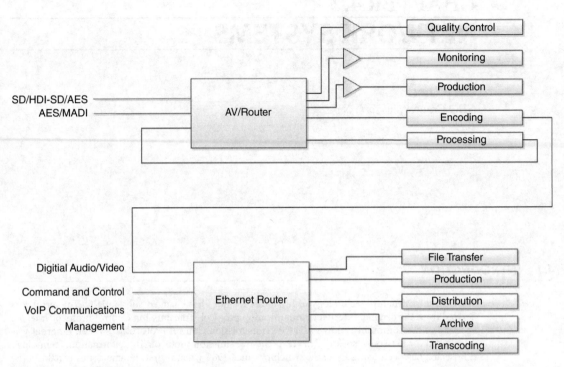

FIGURE 4.2.1 The hybrid network.

4.2.1 OSI Layers

Looking at the construct of network technology, the features and functionalities of routers and switches are identified by layers. The standard that defines the entirety of the hardware and software of the IP network is known as the Open Systems Interconnection (OSI) Basic Reference Model under the joint International Standards Organization (ISO) and International Electrotechnical Commission (IEC). This is a seven-layer network architecture model that is maintained by the ISO under the ISO/IEC 7498-1 Reference. The various layers are detailed in Table 4.2.1.

4.3 NETWORK TOPOLOGY

All devices, applications, and systems on the network require an IP address for them to communicate with each other. The assignment, distribution, and management of IP addresses are done by the Internet Corporation for Assigned Names and Numbers (ICANN). There are different types of IP addresses. There is Point-to-Point Over Ethernet (PPOE), Dynamic Host Configuration Protocol (DHCP), and Static IP addressing. Then there are public and private IP addresses. Public addresses are assigned by ICANN and are typically issued in blocks or ranges. Private IP addresses are restricted to ranges that ICANN has reserved for private networks; these are nonpublic protected ranges that can be used by anyone behind their firewall and are not restricted to any one user group. These private addresses are configured and assigned by the router on each network and based on the requirements of the device can be either DHCP or a static address within the restricted range. One of the roles of the router or the firewall is Network Address Translation (NAT). NAT is needed because private IP addresses are nonroutable on the public Internet, so they must be translated into public IP addresses before they can access the Internet.

TABLE 4.2.1 ISO/IEC Seven-Layer Network Architecture

Layer 1—Physical Layer

This is the electrical and physical specifications, which include:
- Copper and fiber optic cable
- Connectors and connector pin outs
- Line voltage and impedances
- Signal timing
- Hubs, repeaters, and network adapters

Examples of Layer 1:
- RS-232, Full duplex 802.3 (Ethernet), 802.11a/b/g/n/ac (Wi-Fi), MAC/LLC, 802.1Q (VLAN), ATM, FDDI (Fiber Distributed Data Interface), Fibre Channel, Frame Relay, PPP, POTS ("plain old telephone service"), SONET (Synchronous Optical Network), DSL (Digital Subscriber Line), T1, E1, 10BASE-T, 100BASE-TX, 1000BASE-T, and DWDM (dense wavelength division multiplexing)

Layer 2—Data Link Layer

This layer is commonly used for switches and data transport including the protocols used by these devices to communicate with each other:
- Media Access Control (MAC) is a unique ID assigned to each network connection per device and controls their access to data and the permission to transmit it.
- Logical Link Control (LLC): this controls error checking and packet synchronization.
- The point-to-point protocol (PPP) is one part of this layer.

Examples of Layer 2:
- ATM, IEEE 802.2, LLC, L2TP, IEEE 802.3, AppleTalk, X.25 LAPB (Link Access Procedure Balanced)

Layer 3—Network Layer

This is the most commonly known layer. A network defined as an interconnected set of devices each with its own address, which enables the communication of messages and data between devices by the address of the device and finding the best path (route) to deliver the message and data. This is the layer where the network addresses are assigned, managed, and controlled using a number of management protocols.

Examples of Layer 3 addressing and protocols:
- IPv4, IPv6, ICMP (Internet Control Message Protocol), IPsec (Internet Protocol security), IGMP (Internet Group Management Protocol), IPX (Internetwork Packet Exchange), X.25, and ARP (Address Resolution Protocol), maps Layer 3 to Layer 2 addresses

Layer 4—Transport Layer

The transport layer can be compared to how delivery services handle packages. They classify each package and then send it to a destination. The transport layer acknowledges the delivery of the packet and if there are no errors, sends the next packet.
The transport layer is where Transmission Control Protocol (TCP) communicates within the Internet Protocol (IP).

Examples of Layer 5:
- TCP, UDP, SCTP (Stream Control Transmission Protocol), DCCP (Datagram Congestion Control Protocol)

Layer 5—Session Layer

This controls the connections between computers and devices. It establishes, manages, and terminates the connections between the local and remote applications. The session layer manages full duplex, half duplex, and simplex connection on the network.
Within the OSI model, this is the layer responsible for closing sessions, which is part of the TCP, and also for session recovery, which is not typically used in IP.

Examples of Layer 5:
- NetBIOS, SAP (Service Access Point), half duplex, full duplex, simplex, RPC (Remote Procedure Call), SOCKS (Socket Secure), TCP, RTP (Real-time Transport Protocol), PPTP (Point-to-Point Tunneling Protocol)

TABLE 4.2.1 ISO/IEC Seven-Layer Network Architecture (*Continued*)

Layer 6—Presentation Layer

The presentation layer is what ensures that information sent out by the application layer of one network system is readable by the application layer of another network system.

Examples of Layer 6:
- MIME (Multi-Purpose Internet Mail Extensions), SSL (Secure Socket Layer), ASCII, EBCDIC (Extended Binary Coded Decimal Interchange Code), MIDI (Musical Instrument Digital Interface), MPEG

Layer 7—Application Layer 7

This layer is closest to the end user within the OSI layer model. Across the network, both the application layer and the end-user interact directly with the software application. Software that points across the network interacts with the application layer.

These are web services, file transfer, mail and communications to name a few.

Examples of Layer 7:
- NNTP (Network News Transport Protocol), SIP (Session Initiation Protocol), SSI (Simple Sensor Interface), DNS (Domain Name System), FTP (File Transfer Protocol), Gopher, HTTP (Hypertext Transfer Protocol), NFS (network file system), NTP (Network Time Protocol), Dynamic Host Configuration Protocol, SMTP (Simple Mail Transfer Protocol), SNMP, Telnet

Within the network, there are multiple addressing schemas and transmission protocols that are defined above as layers. There is the IPv4 and IPv6 addressing protocol, the MAC address, unicast and multicast; and TCP or UDP. All of these are important to understand.

Internet Protocol—The protocol used to encapsulate data for sending between networks. It has routing functionality that enables internetworking. This is how the Internet was created. IP delivers packets from a source application to a destination application or device based on an IP address embedded in the packet header.

Transmission Control Protocol—The connection protocol that applications use to transmit packets across the Internet. It is used when an application needs to send a large amount of data across the Internet. TCP is optimized for accurate delivery rather than timely delivery and can introduce latency in file deliveries.

User Datagram Protocol—A simpler transmission model than TCP, having removed handshaking, error checking, and other protections found in TCP. Time-sensitive applications may use UDP and apply error checking and correction within the application.

IPv4—Internet Protocol Version 4—A 32-bit address in four 4-byte segments (octets). While this is the most recognizable system, the global proliferation of IP-connected devices has used up the available IPv4 public addresses.

IPv6—Internet Protocol Version 6—A 128-bit address in eight hexadecimal segments, which has a higher number of available addresses. IPv6 was created to resolve the issue that IPv4 had used all available addresses within the 32-bit address range.

MAC Address—A unique ID assigned to every network interface or adapter that is connected on the physical network. The MAC address is typically assigned by the manufacturer of the network interface controller (NIC), stored in the hardware or firmware. The MAC address is usually an encoded version of the manufacturer's registered identification number and in a 48-bit format.

Unicast—When a packet of information is sent from a single source to a specific destination over the network. This is point to point and very bandwidth intensive. This is the most common form of transmission on internal networks (Local Area Network or LANs) and the Internet. All IP networks support unicast, and the applications that are the most familiar are http, smtp, ftp, and telnet. TCP only supports unicast.

Multicast—Multicasting delivers the same packet simultaneously to a group of devices or applications clients. Multicast provides dynamic many-to-many connectivity between a set of senders (at least one) and a group of receivers. The format of IP multicast packets is identical to unicast packets with a special class of destination IP address (class D IPv4 address), which denotes the specific multicast group. Multicast applications must use the UDP transport protocol. Within a multicast environment, the devices and applications receive only the stream of packets they have been configured to by having a unique multicast group address. Routers in a multicast network can learn which subnets have active multicast devices and can minimize unneeded transmission of packets across parts of the network where there are no active multicast devices.

There are different network topologies used for long distances (e.g., ATM [Asynchronous Transfer Mode], SONET, and MPLS [Multiprotocol Label Switching]); these are high-bandwidth network services for interconnecting geographically separated locations. Taking this a bit further, a router is a Layer 3 device; its role and responsibilities are to provide connectivity to the Internet and other networks. It requires a network address from the external network it is connected to, and also provides the network addresses to all devices within the internal network it is managing.

Layer 2 (datalink) and Layer 3 (network) are the most common layers when working with the core network infrastructure for switching and routing. The firewall is the third most common network component that is not identified by which layer it resides on, but is critical to the network architecture.

The IP router is part of Layer 3, the network layer, and is the heart and brains of the network. The network switch is a Layer 2 device that moves data based on the Layer 3 routing tables and the data path that is determined by the Layer 3 router. In a typical network configuration, the router is a separate device that connects to outside routers for Internet and connecting multiple locations together. In larger networks, the more sophisticated devices are Layer 3 switches, which have the feature and functionality of both a switch and router.

The firewall is a hardware device or software technology that protects the network from other networks. The Internet is very much considered a network and is very unsecure. It is the firewall that manages all the incoming and outgoing traffic based on a set of rules. It is firewall technology that is the first line of defense against the proliferation of intrusion and hacking. The network switch handles the distribution of data and controls bandwidth to all the devices on the network. There are managed and unmanaged switches.

- **Unmanaged Switch:** It allows all traffic on the network to be broadcast to all devices, and it becomes the responsibility of the device or application to understand which part of the data stream it needs to perform its function without getting confused or overwhelmed by network traffic.
- **Managed Switch:** Each port can be controlled and, therefore, enable or prevent data from reaching a device or application that does not need it or should not have it. This is accomplished by controlling which IP address or MAC addresses can interact with another IP addresses or MAC addresses.

Once the basic network is created, there are configurations and settings that will optimize it and also be segmented for different systems and file types. This is to maximize network performance, prevent latency, and allow protection of the information. There are two primary types of segments, called Subnets and Virtual Local Area Networks (VLANs). The traffic on these segments can be either fully isolated or enable some amount of cross-over traffic using configurations known as Access Control Lists (ACLs) and Trunking. There are subtle differences between a subnet and a VLAN:

- **Subnet:** The segmentation of a network done at the server or device by using different subnet mask addresses on a single network. In an open network, all subnets on all switch ports and are visible to any device configured for that subnet. Subnets do not manage bandwidth or allow optimization.
- **VLAN:** The partitioning of the network into isolated segments done in the router to function as if they were physically separate switches. VLAN access is controlled at the switch port; if a port is not configured to pass a specific VLAN, then any device assigned an IP address in that VLAN will not be able to access data even if the device is set up with the correct address. Traffic can cross VLANs only if configured in the router. In the IP broadcast environment, VLANs are used to segregate different signal types within the same network (e.g., video, audio, communications, and control).

There are a number of configuration settings within the router and switches to manage priorities (QoS [quality of service]), control the amount of bandwidth, synchronize data movement, and protect the data as they move between systems and devices.

On a broadcast and production network, this is where packet collisions and latency become a major consideration. If there is too much data traffic on an open network, the data packets from one system can interfere with the error-free transmission of another system riding in the same path. In addition, if one system has a large amount of data, it might force a system with a smaller but more critical amount of data to fail to reach its destination at the right time. An example of this would be an encoder is processing a file transfer while an SDI router panel has requested a switch to happen. If the switch does not happen at the right time, there could be "dead air"—that is latency.

This brings us back to the original concept of the many layers in IP network topology as it applies to the broadcast workflows and the processes it supports. In an SDI environment, most of these layers are of different signal types, have their own cable and connector, and are managed by a different type of switching, routing, and distribution architecture. In the IP world, these become VLANs instead of individual cables, and the IP network can be segmented so that each different signal type can coexist in the same routing and switching environment while traveling separately and protected between systems and devices. In most instances, the output of these devices is an NIC port that connects to a port on the network switch. Each port on the switch can be configured individually for the data and/or signal it is transporting, and they can be isolated if necessary. There is now only one cable type, two actually since fiber is used for higher bandwidths and for sending the IP over extended distances.

An example broadcast network infrastructure is illustrated in Fig. 4.2.2.

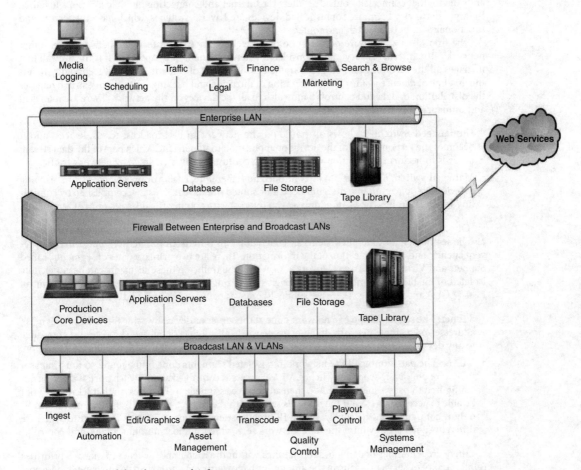

FIGURE 4.2.2 Example broadcast network infrastructure.

In essence, each VLAN can be compared to the unique cables previously used for each of the signal types. Now, each of the signals is in the same switch and groomed into a single IP stream that travels over the same cables/fibers. As it connects to each device, the device understands which VLAN and signal it is taking commands from, sending or receiving files, and what process to execute or file to manage; see Fig. 4.2.3.

VLANs provide a level of security and control over which devices on the network should be exposed to users, or whether these devices only need to interact with other systems. For example, there are devices on the network that do not need Internet access and should not be exposed to it; furthermore, there are users that only need access to specific systems. VLANs also manage access to media and metadata between the business and production layers. The VLAN controls the cross-over into the enterprise network.

4.3.1 System Demands

Broadcast and production systems put high demands on IP networks. Applications that are handling media files and streams are more bandwidth intensive than enterprise applications using documents and spreadsheets. Even large databases are not bandwidth "hogs." Looking forward at next-generation infrastructure requirements, SDI technology is evolving to 1080P and 3 Gb/s and this will enable it to handle new formats with higher bitrates, like 2K, 4K, and 8K. Manufacturers of broadcast and

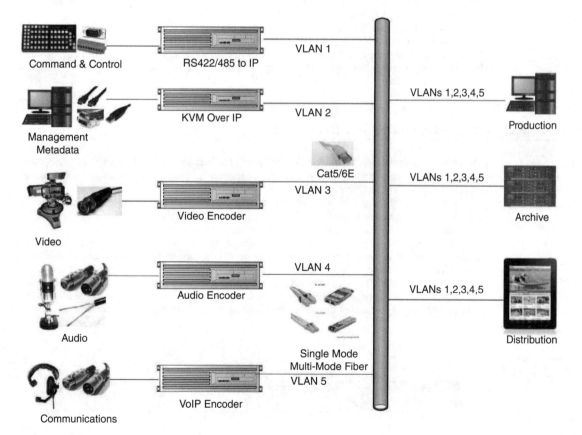

FIGURE 4.2.3 The conversion from cable to VLAN.

production equipment (i.e., audio/video routers, terminal, and distribution systems) are pressed to provide products that will support 3 Gb/s. In the IP world, the next generation is 10 Gb/s, 40 Gb/s, and 100 Gb/s, which in comparison is huge step beyond 3 Gb/s.

Configuring and managing the IP infrastructure is also very different from the SDI plant. An interesting note is that the copper cable distance limitations in IP are the same as SDI. The difference is in the architecture. The design of the SDI infrastructure is based on the core SDI router and point-to-point connections for inputs and outputs of audio and video; then the other signal types each have their own connections and matrices. If there are devices outside the distance limitations, then distribution and equalization amplifiers are needed in addition to fiber optic equipment per device. In the IP architecture, there is a core router and each device gets a single connection that is duplex, plus carries all signals and communications. However, in the IP network architecture, the core router has high-bandwidth connections to satellite switches in proximity to the end devices. This allows a single set of fibers to interconnect the switches and then use copper for the final local connection to the devices. There are fewer long runs and having localized switches makes adding or changing equipment more manageable.

The next generation in networking is Software Defined Networks (SDNs). The complexity of routers and switches has increased to a critical level where the amount of processing, virtualizations, and configurations these devices need to support have added a layer of overhead and in many cases latency to the core functions. To alleviate this burden, dedicated servers tightly integrated to the switch topology became necessary. The next logical step was to "de-couple" the server from the router and switch frames and handle all the processing, routing, and management. In this way, the switches become slaves or clients, with the ports being defined and controlled from the server; the configurations no longer need to reside in each switch.

This will change how the network is designed, where Layer 2 and Layer 3 switches are required, and how the entire LAN, WAN (wide-area network), WLAN (wireless local area network), and VLAN architecture is structured. By using an SDN design, it makes network management less cumbersome. This improves network performance and reduces the requirement to update each device when there are changes or modifications.

4.4 FILE-BASED WORKFLOW ARCHITECTURE

The IP network is the core technology in the IP and file-based workflow infrastructure. More than the SDI router, it transports and manages media files and streams, communication, and command and control of the entire IP-based architecture.

File-based workflow is a broad definition that encompasses all the processes, technologies, and systems that create, convert, produce, edit, manage, store, transport, and deliver media; see Fig. 4.2.4. File-based workflow introduced an entirely new vocabulary and terminologies.

The complete file-based workflow is a combination of processes and technologies. Each process has its own systems. These systems are applications, databases, servers, and storage—all integrated over different networks.

File-based workflow begins at the point of creation or acquisition; see Fig. 4.2.5. Before IP- and file-based workflows, for production there was live, live to tape, and programs produced for postproduction finishing. The concept in the file-based workflow is essentially the same although now it is

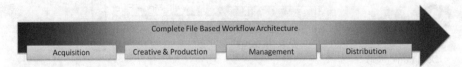

FIGURE 4.2.4 File-based workflow architecture.

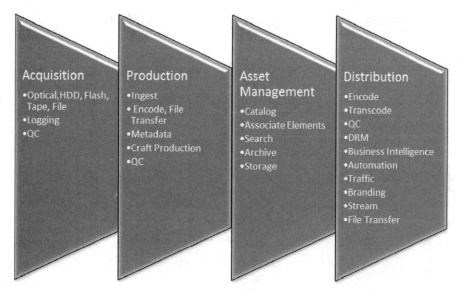

FIGURE 4.2.5 Major steps in the file-based workflow process.

live, live to stream, live to file, and produced for postproduction finishing. One of the benefits that came with file-based workflow is that the content is accessible for editing within seconds of being captured (recorded). One of the most significant changes is in the number of new production processes needed to create content for all the distribution platforms.

All media workflows begin with content. In file-based workflow, content is defined as Essence—the raw video, audio, or graphics—plus the Metadata that describes it. At the beginning of a file-based workflow is acquisition. Here content is captured by cameras and microphones or created by graphics and editing systems. If the content is captured by a camera and microphone, it is encoded to a file and transported over a network into storage. If it is created in either a graphic or editing workstation, it is already in file form and is transported over the network to storage. This process is illustrated in Fig. 4.2.6.

In tape-based production, everything was recorded on tape and the only decision was which tape format to use; every other parameter was defined and built into the tape format. Each tape format had its own machine type; however, the input and outputs were based on a single standard for video, audio, reference, timecode, and control.

This has all changed in file-based workflow. The "machines" are computers that handle the media and use the same servers, storage and network. The production and craft tools are software applications that are not machine- or format-specific. In file-based workflow, the program material or content can be in different file formats, multiple resolutions, and packaged in different wrappers or containers. In the file-based workflow, there are many video file formats with their own codec and each format can have many different bitrates. There is also the audio file format, with different sampling rates and bitrates, and codecs. Then, there is the container or wrapper selection, which takes the audio, video, and metadata and puts it into a single container that can be transported between systems and archived.

4.4.1 Fundamental Technologies

File-based workflow can be broken into technology, systems, and processes. The foundation of file-based workflows is the technologies. The systems are built on the technology and the processes are enabled and created based on the technology and systems.

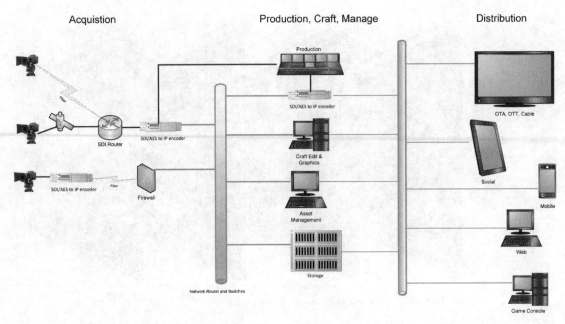

FIGURE 4.2.6 File-based workflow process.

The core technologies that are the foundation in broadcast file-based workflow are the same computer technologies that are core to enterprise IT. These are applications, servers, networks, and storage. And at the most basic level in the same configuration, applications run on servers, servers and applications communicate and transfer files and data across a network, and store the files on a variety of storage devices and in different storage configurations. When these same technologies are used in broadcast, they have different requirements and configurations to handle the heavier loads that media places on them; see Fig. 4.2.7. There are many different products and services within these technologies that make up the entire file-based workflow.

We will explore each of these technology groups and analyze how they differ in requirements and configurations when used in broadcast and production.

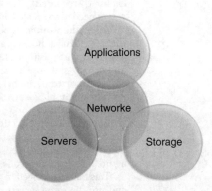

FIGURE 4.2.7 Major elements of a file-based workflow system.

The requirements of applications play a significant role in the specification of the servers. Not unlike enterprise applications, they are typically purpose-specific. There are dedicated and proprietary applications for encoding, transcoding, streaming, capturing, editing, graphics, logging, metadata entry, management, automation, playout, and protection to name a few. These need to integrate with each other and in most cases use other applications called *middleware* for this integration and to transfer and/or move files; and move and copy data sets. Many of the primary applications use databases to manage the media and metadata. There are middleware applications specifically designed to integrate these databases. In addition, in file-based workflows, the business applications are required to integrate with the metadata in broadcast and production databases.

The servers supporting these applications typically are more robust than standard enterprise servers. Media applications require more memory, faster processors with as many cores as possible, and large volume high-speed storage. Not only are the media files themselves large, the media applications are very large and require considerable storage space. The read/write time of the storage is a critical component of file-based workflows. Video and audio processings put considerable loads on the server, and the embedded graphics and audio chipsets are typically not adequate enough to support the applications. There are a variety of hardware solutions available that are installed in these servers to upgrade them to work for media. Some of these are hardware accelerators, graphics cards with on-board processors and their own memory, audio cards that process surround sound, etc., and there are specialty cards for encoding audio and video (SDI and AES), decoding for SDI/AES for playout, fiber channel for storage interface, and a host of others. These separate devices are important as they relieve the main processor and memory to handle the media and execute operations such as rendering, which utilizes all of the system resources.

In the file-based workflow environment, while the applications and servers handle the processes, storage is where the files reside and are accessed. In addition to the server itself having specific requirements, to support file-based workflow, the storage has its own special requirements. In the file-based media workflow, storage plays a number of different roles and for each of these there are a number of different storage types and configurations. When considering the storage architecture, first there are multiple file resolutions and different bitrates for the high, mezzanine, and proxy files. There is online, near-line, offline, and proxy storage; and each of these has their own requirements. In production and broadcast, online storage is for high-availability media access and requires fast disk access and high throughput. Near-line is used for content that needs to be accessible, however, not as highly available. In a news organization, online might be 3–5 days of recent news and near-line 1–2 weeks. Off-line is the archival and preservation copy of the file that is retained "forever" or following retention policies. One of the main considerations in calculating storage is high and mezzanine files can move from online to near-line to offline, but proxies live forever as the search and browse format.

There are different types of storage for each of the storage categories. There is on-board storage, attached storage (USB and Thunderbolt), Network Attached Storage, and Storage Area Network. There is also cloud storage. Online storage is high speed disk or Solid State (SSD). Near-line can be slow speed disk, SSD, or tape (i.e., DLT [digital linear tape], LTO [Linear Tape-Open], etc.). Offline is typically a removable media (i.e., tape or optical disk). Managing storage is important in the file-based workflow. This can be managed by the asset management system.

Some of the first decisions in production are choosing which format, codec, bitrate, and container will be used. The workflow begins with the encoding or transcoding process to capture the content. If the content is captured from a camera, microphone, or from a legacy tape format, it is likely to be analog/SDI/AES, the process for this it is an encode. The encoder parameters are set based on the decisions made on file and container format. If the content is originally created in a workstation or encoded in the field from a remote production and is already in a file form, then the file is transported over the network to a server and ingested to storage. In this scenario, it may need to be transcoded first if it is not in the preferred "house" high-resolution file format. Whether the content is encoded or exists in file form, it needs to be transported into the storage environment, where it becomes accessible for editing, production, distribution, archive, and management.

The production workflow access the content from the shared storage and editing begins. Editing systems use a mezzanine or proxy version of the master and create a timeline of markers that are representative of locations in the master file with the actual content. Once all the editorial decisions are made, the output can be produced.

Once the file is created, it will need metadata to manage it, describe it, and protect it. Metadata is the foundation of the file-based workflow environment. Metadata manages and controls the movement, accessibility, and access permission to the content. Just as important as the file format, equally critical is the metadata schema and asset management tool selection. The asset management tool uses the metadata to register the content and build the rules and policies that will govern it for its entire existence. The metadata schema is the list of unique elements and the data structure that will manage, define, control, move, and protect the content. These are fields within a database. The asset management tool tracks the location of the content, provides the search and browse functionality, enables access permission, and handles the archive and preservation of the content. The asset manager uses metadata to manage the associations between different content elements that may become pieces of a final program. This is known as a parent–child relationship, where the child inherits the metadata properties of the parent in addition to having its own set of properties.

One of the core infrastructure technologies in file-based workflow is the network. The network enables and facilitates the applications to communicate with each other, the command, control, and automation of devices and the transport of files and streams between processes, and how files are transported in and out of the storage environment. There are a number of different network topologies used to connect all the systems, and storage can have its own network topology.

One of the challenges in file-based workflow is creating and managing all the different iterations of the content as it is produced for each delivery platform. It is more than format differences; the creative process or story telling for each platform is different. A program for delivery to large screen, whether it is over the air, cable, satellite, or over the top the program length, resolution, and compression are formatted to the large screen. Content created for web delivery is formatted for the smaller screen and lower bitrates, and following a different business model. Programs created for smartphones and tablets use different codecs, high compression, and have different metadata. There are multiple standards or protocols for each platform. This means a different codec, wrapper, and metadata set for each. All of these versions reside in the storage architecture and need management. The master high resolution is archived and the distribution storage will receive all of the versions with delivery instructions in metadata.

The distribution workflow is responsible for organizing and managing the finished content for delivery to each of the delivery platforms. In a broadcast environment, the playout server will handle the "play to air" content complete with commercials, interstitial material, and network IDs. File delivery is managed differently. The content needs to be formatted to the codec and bitrate for each platform or distribution network, i.e., Flash, HTML5, HTTPLive, ProRes422, and MPG4 H.264/5. The metadata for each platform is a different subset of the master metadata managed by the asset management system.

The entire ecosystem file-based workflow can be characterized as a media management architecture, and the components (applications) that handle the movement of media and metadata in this new architecture are managed and controlled by other systems. The command and control over all

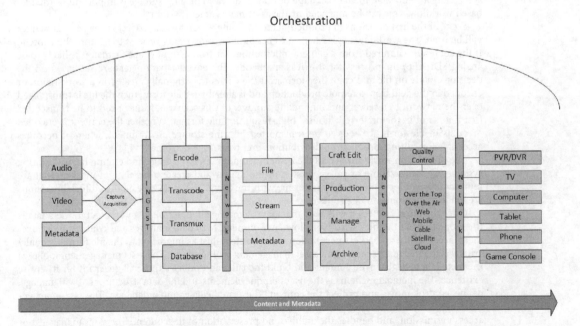

FIGURE 4.2.8 File-based workflow orchestration.

the systems, devices, and applications in the file-based environment require a more comprehensive level of automation. This next generation of automation is called *orchestration*. Orchestration is an overarching automation and management system that integrates with all the devices, applications, and systems within the file-based workflow architecture to provide the scheduling, command, control, process monitoring, and resource allocation; see Fig. 4.2.8.

This chapter discusses the number of different systems that make up the complete file-based workflow. The orchestration system will trigger an encoding or transcoding event based on the input from a scheduling system or interface while managing the resources and handling any conflict resolution of over assigned resources. It monitors the progress of each process and alerts an operator or engineer if a process fails, slows down, or is interrupted. As the asset is required to move between the different systems from ingest to production to distribution, the orchestration system controls the movement, assigns the resources, and communicates with the asset management system for tracking. The orchestration system is a unified interface that enables an operator to monitor all the processes and track the progress of files as they move throughout the entire file-based architecture into distribution.

The introduction of files and streams changed quality control. As files are created and move throughout the file-based ecosystem, checking and confirming the integrity of the file as it transitions between formats, applications, and storage systems is critical. The test and measurement tools to analyze files and streams are substantially different. There are different parameters that need to be monitored and system testing that includes CPU cycles, network bandwidth, and file integrity. It is an essential component of file-based workflow. File checking includes checksum verification, bit rate, frame rate, and GOP (group of pictures) structure to name a few. Stream analysis has a different set of measures such as packet loss, bit error, quantization errors, and multiplexing that begin a long list of analysis points.

It is easy to see the complexity in file-based workflow; other chapters will delve deeper into each of the technologies and systems.

CHAPTER 4.3
TIME AND FREQUENCY TRANSFER OVER ETHERNET USING NTP AND PTP

Nikolaus Kerö
Oregano Systems, Vienna, Austria

4.1 INTRODUCTION

Moving from an SDI (serial digital interface) infrastructure toward a partly or even fully IP-based workflow within a broadcast production environment has a number of significant advantages. Any device attached to an SDI network can lock its local PLL (phase-locked loop) readily enough to the pixel frequency of the SDI data stream while extracting any sync-signal it requires from it at the same time. In contrast to SDI, Ethernet is an inherently asynchronous transmission medium. (For the time being we disregard its extension SyncE—Synchronous Ethernet.) To provide both accurate frequency and phase information to all nodes in an Ethernet network, additional measures have to be taken.

All Ethernet-based clock synchronization methods use the same basic principle; at least one distinct node within a network acts as a time reference. It has to be capable of transmitting messages containing the current value of its local clock. Every device willing to synchronize to this reference node simply denotes the arrival time of such a message on its local clock. As long as the transmission delay over the network is either known or can be calculated (see below), every node is able to adjust its local clock accordingly. Whenever possible, the node acting as a time reference is itself connected to a reliable (time-traceable) time source such as GPS (Global Positioning System). Two different protocols, both relying on this principle, have been standardized worldwide and are widely used to transfer time information: NTP (Network Time Protocol) [1] and PTP (Precision Time Protocol) [2]. In the 1970s, the former had been devised to synchronize computers over wide area networks (i.e., the Internet) while the latter has been published more recently (version 1.0 in 1997 and version 2.0 in 2008, respectively) and was originally intended to be used within local area networks (LANs) where high accuracy is a key requirement for any kind of distributed application.

4.2 PTP—PRECISION TIME PROTOCOL

PTP follows a strict master–slave principle for transmitting time information. For the time being, let us assume that one node has been designated to be the so-called PTP-Master. It continuously transmits synchronization messages (`sync_messages`) to all slave nodes within the respective network (typically at least once every second). The content of these messages is basically the current time of the master. Actually, it should be the very point in time (labeled T_1) at which the master starts sending the message via the physical channel. Every slave with respect to PTP, in turn, denotes the time at which it receives any such `sync_message` on its local time scale (labeled T_2). The difference between these two timestamps is the offset between the two clocks plus the transmission delay of the message via the physical channel:

$$T_2 - T_1 = \text{Transmission}_{\text{Delay}} + \text{Clock}_{\text{Offset}}$$

With this information alone a slave is capable of correcting the frequency offset of its clock with respect to one of the masters by evaluating a sequence of such messages, However, it has to disregard the influence of the transmission delay as it is unknown as of yet.

To calculate the transmission delay, the slave initiates a second time transfer procedure by sending a `del_req_message` and denoting the time when the transmission over the physical medium is initiated (labeled as T_3). The master, in turn, will note the time when it has received such a packet (labeled T_4) and will relay these data back to the querying slave by sending a so-called `del_resp_message`. This measurement cycle is repeated to allow for filtering. The difference of the two timestamps of the `del_req_message` equals the clock offset minus the transmission delay:

$$T_4 - T_3 = \text{Transmission\_Delay} - \text{Clock\_Offset}$$

Now the slave clock is able to calculate both the clock offset and the transmission delay using both timestamp differences:

$$\text{Clock\_Offset} = \frac{(T_2 - T_1) - (T_4 - T_3)}{2}$$

$$\text{Transmission\_Delay} = \frac{(T_2 - T_1) + (T_4 - T_3)}{2}$$

If the master is not able to insert a timestamp into the `sync_message` with sufficient accuracy while actually sending it (details on effects deteriorating the accuracy are given below), it will merely note the time at which the packet is sent over the network by drawing a timestamp from its accurate local clock while actually sending such a message and later on forward this time information by means of a corresponding `follow-up_message` again to all its slaves.

The message flow of the complete IEEE 1588 synchronization process is shown in Fig. 4.3.1 together with the equations required for calculating the delay and the offset.

4.2.1 How to Select a Master Node

IEEE 1588 relies on one single master continuously distributing its time information to all slaves attached to it. In case this master fails or during the start-up phase of a network when no master is decided on, all nodes need to elect a new master among themselves by means of the so-called Best Master Clock Algorithm (BMCA). To this end, every node communicates information about status and quality of its local clock (e.g., stratum level, clock deviation, and the like) to all others using an `announce_message`. Every node will compare the data of all `announce_messages` it has received with each other and, of course with its own dataset. The node with the "best" (most stable) clock is selected and will become the new master of the network. In case of more than one node having clocks with identical qualities, a tie-breaking mechanism ensures convergence of this algorithm, selecting the node with the highest `clockID` to become the new master.

After the BMCA has completed, the newly elected master will continue to send Announce messages aside from the sync messages, but usually at a lower rate. This allows any node entering the network to take over from the current master as long as its local clock is of higher quality. If a reference time source has to be taken offline for repair, it will automatically resume its role after being reinstalled.

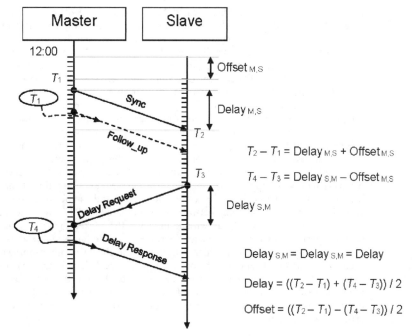

FIGURE 4.3.1 IEEE 1588 message flow.

By default, every node within a network is eligible to take part in the master election process and thus may eventually become a PTP master itself. This is a very effective and desirable way to keep all nodes within a LAN synchronized to each other regardless of which nodes and/or network connections have failed. In industrial automation, distributed control applications have to rely on such a plug-and-play like behavior without the need for preconfiguring a node prior to adding it to a network. For larger networks, where all nodes have to be permanently synchronized not only to each other but even more importantly to an external time reference, node-specific configuration may prove worth the effort. More than one PTP master clock may be added resulting in one acting as a master while the others remain passive merely listening to Announce messages ready to take over in case the current Master fails.

4.2.2 Accuracy of PTP

For the calculation of the delay and the offset described above, we have made a number of implicit assumptions that need to revisited and analyzed. First of all, we presumed that the transmission delay is constant for all messages—and even more critical—identical in both directions (i.e., it takes the same amount of time to transfer a `Sync` message from the master to the slave as it takes for a `delay_request` message to be sent from the slave to the master). Furthermore, we assumed that timestamps can be taken with infinite precision. Unfortunately, none of these assumptions fully hold up against scrutiny. To investigate these possible sources for inaccuracies, we can distinguish between node- and network-related effects.

For a node, IEEE 1588-based synchronization accuracy is dependent on how accurately the four timestamps (T_1 to T_4 in Fig. 4.3.1) can be taken. A purely software-based approach (used in NTP) is an obvious solution. However, if accuracies below a few milliseconds are required, this method can no longer be used because timestamping is subject to a series of highly unpredictable sources of jitter. The timestamp required for transmitting a synchronization packet would be drawn and inserted into the packet in the respective subroutine that assembles the message. Upon completion, the corresponding subsystem is informed that new data are ready to be transmitted over the network. After copying the data into a local buffer memory, it will finally convert it into a serial data stream to be sent over the physical channel. The time span between drawing and inserting the timestamp into the packet and actually sending over the wire is subject to the current loading of the system, the interrupt latencies of the CPU, and the performance of the network subsystem. This time span is usually in the range of tens of microseconds but may very well increase to several milliseconds.

In the receive direction, a similar situation occurs. The CPU is alerted that a packet has been received by means of an interrupt, triggering the related service routine, which, in turn, analyzes the packet, detects that it is a PTP event message, and finally draws a timestamp. Varying interrupt latencies in conjunction with pre-emption mechanisms cause delays varying in the same order of magnitude. Furthermore, the CPU clock may have insufficient resolution; most likely it is driven by a low-grade oscillator with very poor jitter and wander performance.

To obtain the required accuracy, gathering of timestamps has to be accomplished by means of extensive hardware support, the most effective way being to add a timestamping unit into the interface between the physical layer IC (integrate circuit) and the network controller. This module will scan all incoming and outgoing network traffic in search for PTP event messages. Upon detection of such a packet, a timestamp is drawn and linked to the packet. Finally, the underlying software analyzes the packet information, evaluates the respective timestamp, and readjusts the parameters of the control loop for disciplining the highly accurate local clock (usually running off a high-grade crystal oscillator). Fortunately enough, the delay via the physical channel can be considered to be constant. Typically, it varies less than a couple of nanoseconds. Finally, the quantization error introduced by the finite frequency of the local clock should be kept as low as possible, Within modern integrated circuits, a clock frequency in excess of several 100 MHz is feasible, thus reducing the quantization error to less than 3 ns.

4.2.3 Influence of Real World Networks

Within modern Ethernet networks, all nodes are connected with each other via intelligent network devices such as switches or routers, performing Layer-2 and Layer-3 based packet switching, respectively. This technique has proven to be extremely efficient in terms of overall network throughput, allowing every node to operate in full-duplex mode and thus has replaced purely passive network components like hubs. With respect to accurate clock synchronization, such active network components have a severe drawback. As shown above, the accuracy a slave is able to synchronize to its master relies on a known *and* constant transmission time of the sync_message. The time it takes for a standard network switching device to forward a network message is by no means constant. It depends on the current workload, and even more important, on the hardware architecture of the network device itself. If, for example, a master node transmits a sync_message, the network device has to forward this message to all ports within the multicast range indicated in the sync_message. In case the network is unloaded (e.g., no other network traffic has to be handled by the switch), the time span for this forwarding procedure can be assumed to be fairly constant—so long as packet forwarding is performed in hardware within the switch engine. This is true for nearly any Layer-2 network device, but cannot be taken for granted for Layer-3 network devices. Even in this case, the odd sync_message may be delayed unexpectedly long by the switch in the presence of other network traffic, as sparse as it is. Let us assume that a single maximum size packet is processed by the switch while a sync message has to be forwarded at the same time. This causes the sync_message to be delayed up to 12 µs as it has to wait until the complete packet has left the switch.

If the network is loaded, thing become worse; the switch will be busy forwarding other traffic most of the time and, thus, it will delay all PTP event messages for an unknown amount of time. Extensive measurements have proven delay variations may range from several 10 µs up to several milliseconds.

This problem can be mitigated to a certain extent by filtering the measured delay values or even applying nonlinear filtering techniques (e.g., discarding values exceeding a predefined or dynamic limit). As an additional measure, Quality of Service (QoS) techniques can be utilized by configuring a high-priority queue for PTP traffic on every network device and tagging all PTP event messages accordingly. This method does limit the impact of loaded network devices on synchronization accuracy; however, assigning QoS for PTP traffic may very well collide with other requirements within a given network.

4.2.4 Transparent Clocks

IEEE 1588 solved the problem of variations caused by network devices by introducing so-called Transparent Clocks (TCs). These components act as normal network devices treating only PTP event messages (e.g., sync_messages and del_req_messages) in a special manner. A TC comprises an accurate clock, allowing it to measure the time it requires to forward any given PTP event message. A timestamp is drawn from its clock upon reception of such a message and stored locally. If the message is re-transmitted via any other port of the TC, another timestamp is drawn. The first timestamp is retrieved, and the difference between the two timestamps equals the residence time of the packet. This information is either inserted into a correction_field within the sync_message or stored and inserted into the respective field in the follow_up_message or del_resp_message. The former method is referred to as one-step and the latter as a two-step TC.

To enable cascading of TCs, the respective residence times are accumulated, rather than just inserted into the correction field. A schematic of the function of an end-to-end TC is shown in Fig. 4.3.2.

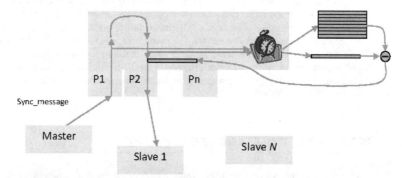

FIGURE 4.3.2 IEEE 1588 Transparent Clock message flow.

Although end-to-end TCs provide a prefect means to cope with packet delay variations within large networks, one limitation still remains uncovered. With an increasing number of slaves, the workload of the master they are attached to will increase as well, as it has to handle a considerable number of del_req_messages aside from having to send sync_messages. This shortcoming can be mitigated by using peer-to-peer TCs or Boundary Clocks (BCs).

Using the value in the correction field of a sync_message a slave can account for the varying delays caused by forwarding the packet within a switch. For obtaining the absolute transmission delay, every slave still has to revert to sending delay_request_messages every so often. If a TC could account for the absolute transmission delay of every incoming message as well, by adding it to the residence time before updating the correction field, any slave has sufficient information to adjust its clock without the need to send delay_request_messages. A peer-to-peer TC does exactly that by measuring the absolute transmission delay on a per-port basis using a similar PTP event message exchange mechanism. A peer_delay_request_message is sent from every port and will be answered by a peer_delay_response together with a peer_delay_response_follow_up message. This message exchange yields the same four timestamps as described above, allowing the peer-to-peer TC to calculate the transmission delay for any of its ports. As this method is used to measure the transmission delay only to the adjacent node, all peer delay messages have to be treated as link-local, i.e., they must not be forwarded by any device. The peer delay mechanism increases the complexity of a TC and all end nodes quite significantly, but it has a couple of undisputable advantages. For large networks, it reduces the overall network load for TCs and the master itself. Furthermore, a peer-to-peer network can change its topology without losing—even temporarily—its accuracy. If a master or any network device fails, sync messages are re-routed autonomously to different paths. If their respective absolute transmission delays are already known (i.e., they have been measured previously), the synchronization accuracy will not deteriorate.

4.2.5 Boundary Clocks

Boundary Clocks are intended to partition time distribution within large networks, effectively reducing the number of messages a single PTP master node has to process. Rather than simply forwarding PTP `sync_messages` from a given master to all ports as TCs do, the content of the `sync_messages` is used to synchronize a local high-accuracy clock of the BC to the master attached to the respective port. Basically, a BC acts as slave synchronizing to the master on this port. All other ports will generate `sync_messages` using the time information of the local clock. To this end, each port of a BC has to be capable of acting both as a PTP master or slave with all ports sharing the internal clock. One port will assume slave role while all other ports will act as PTP masters (or passive master if there is already a better master in this part of the network). Rather than assuming these roles in a predefined way by means of static configuration, the role of every port is determined dynamically by the BC itself.

To accomplish this, it will execute an extended version of the BMCA evaluating clock quality information contained in the `announce_messages` every port is receiving. If more than one port is receiving `announce_messages`, the information on the clock quality is compared and the most accurate clock is selected as a master. The respective port will switch to slave state while all other ports will revert to master state sending `announce_messages` themselves, eventually causing all nodes connected to these ports to switch to slave state. This mechanism supports cascading of BC as well.

Both TCs and BCs will treat any other network traffic according to the respective Layer-2 or Layer-3 forwarding rules.

4.3 IEEE 1588—WHAT THE STANDARD SPECIFIES

First and foremost, the IEEE 1588 Standard defines the actual protocol, explaining the different message types, the various states a node may assume together with the state transition diagrams, and the respective trigger conditions. This is followed by a format specification of all PTP messages, i.e., all data fields with their respective sizes, permitted values or ranges, and default values. Although PTP is designed as a "configuration-less" profile, it provides extensive features for remote configuration and monitoring by means of Management messages. Using a single monitoring device attached to the network, the value of any PTP-related parameter may be read and modified this way for any node. Management messages may even be used to convey user- or application-specific data. Finally, rules for mapping of PTP onto different transport protocols (such as Ethernet Layer 2, IPv4, or IPv6) are explained in the respective annexes.

The IEEE 1588 Standard is deliberately kept very generic to allow adapting it to the requirements and constraints of any application domain. To facilitate such different use cases, the standard has introduced the concept of PTP profiles. Within a well-defined framework, a PTP profile yields its bespoke version. Besides specifying specific subranges and default values for certain or even all PTP parameters (such as messages rates), a profile may allow (or rule out) certain transport mechanism. Furthermore, the capabilities of the network devices with respect to PTP can be specified, together with supported network topologies and sizes.

For telecom applications, for example, the ITU (International Telecommunications Union) has published two profiles (namely, the G.8265 [2] and the G.8275 [3]), the IEEE itself published the C37.238-2011 [4] (a profile for power substation synchronization applications), and the TICTOC working group within the IETF (Internet Engineering Task Force) is working on an Enterprise profile [5]. For broadcasting applications, the SMPTE (Society of Motion Picture Engineers) has published a PTP—namely the ST 2059-2 standard [6].

4.3.1 SMPTE ST2059-2, a PTP Profile for the Broadcasting Industry

In April 2015, SMPTE published ST 2059-2, a standard that describes a PTP profile to be used by the broadcasting industry. This PTP application requested a quite unique new requirement, namely fast locking. It is defined as the time span it takes a slave to remain locked to the master with a maximum offset of less than 1 μs (better 100 ns), less than 5 s after the network cable has been plugged in. To this end, the rates for all PTP event messages have to be chosen sufficiently high. By default, the rate of all event messages is set to 8 per second. Any device has to be capable of handling up to 128 messages per second. The rate for the Announce message is set to a high value as well (4–8 per second) because according to IEEE 1588, a slave has to receive at least two such messages prior to starting the actual synchronization cycle. It should be mentioned that this requirement calls for specific hardware modules accommodating synchronous offset cancellation.

An ST 2059-2 network may be built using PTP-aware network devices; in so doing, BCs as well as TCs may be used in any suitable combination. However, PTP network devices are not mandatory, permitting users to deploy PTP either onto an existing standard network infrastructure or to build mixed networks using PTP network devices at certain critical locations.

The profile allows only Layer-3 communication supporting UDP (User Datagram Protocol), both over IPv4 and IPv6. Transparent Clocks have to support both protocols simultaneously. Boundary Clocks and Ordinary Clocks, however, may support only one protocol at a time.

For Sync, Announce, and follow_up messages multicast has to be supported. A unicast transport mechanism for these messages is optional. delay_request messages may be transmitted either way, allowing implementation of a mixed mode communication scheme using multicast downstream (announce, sync, and follow_up) and unicast in upstream (delay_request and delay_response). To this end, every node has to reply to a PTP event message sent in unicast with a unicast message as well. Management messages may be sent either as unicast or multicast packets. However, to avoid transient traffic congestions in large networks, any reply to a management message has to be sent in unicast mode.

To convey additional information relevant to the broadcasting industry, a special management message has been defined allowing a PTP Master to transmit information such as the locking status of the master, the default system frame rate, information about the local time zone (including daylight saving time and leap seconds), and, of course, information concerning the daily jam time. This message is referred to as Synchronization Metadata.

4.3.2 NTP—The Network Time Protocol

The NTP (defined by RFC 5905 [7]) was designed with rather different objectives in mind. It provides reasonably accurate time information (in the range of milliseconds) for any computer attached to a wide area network (i.e., the Internet) without imposing any constraints or requirements to it other than the capability to handle IP traffic. Reference clocks referred to as NTP servers may reside anywhere in the network. Consequently, NTP has to deal with a number of network-related uncertainties like large PDVs (packet delay variations) or, even worse, unreliable communication paths, resulting in a considerable packet loss [8].

For synchronization, NTP relies on the same principle of exchanging messages containing time information; however, it has a number of distinct differences compared to PTP. First of all, the communication is based on a client-server model and is usually initiated by the node that has to be synchronized (i.e., the NTP client). Consequently, it has to know the IP address (or domain name) of at least one NTP server. Only two messages—a synchronization and a control message—are known to NTP. The former has four timestamp fields, which are, in turn, populated during a synchronization cycle. The first is the Reference Timestamp, denoting the last time at which the local clock of the sending node has to be set or adjusted. It is updated upon sending the synchronization message and is used as an indication of the quality of the local clock. It is followed by the so-called Originate Timestamp and the Receive Timestamp, both of which correlate to T_1 and T_2 in PTP, respectively. Finally, the Transmit Timestamp corresponds to T_3 in PTP, i.e., the send timestamp of the delay_request message.

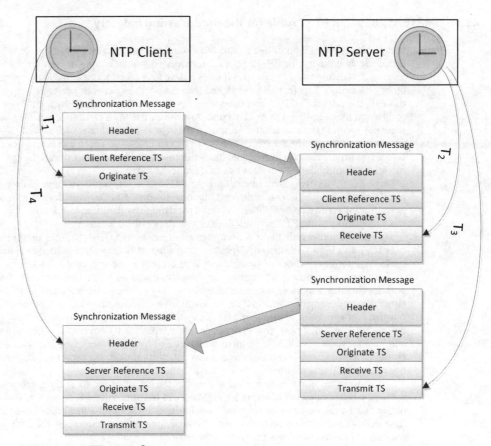

FIGURE 4.3.3 NTP message flow.

Finally, the querying node (i.e., the NTP client) will draw the fourth timestamp upon receiving the synchronization message, which has been bounced back and forth between client and server. Generally, timestamping is done entirely in software, at least on the side of the NTP client. Servers may provide hardware-assisted timestamping to increase the overall accuracy. The message flow with the corresponding timestamps is shown in Fig. 4.3.3.

Aside from basic configuration data, the header of an NTP synchronization message contains information about the precision of the server clock, the maximum error of the local clock with respect to its reference (referred to as *root dispersion*), and, most importantly, the Stratum value.

NTP servers are intended to operate in n multitier configuration. A server directly attached to a time source such as GPS will publish a Stratum value of 1, being the highest level of accuracy. Any server synchronized to a Stratum 1 device will, in turn, advertise its Stratum value as 2 and so on.

IEEE 1588 defines solely a protocol, leaving any implementation-related items unspecified. NTP, on the other hand, specifies many implementation details like the clock synchronization algorithm as well as the related servo architecture. The precision and Stratum information mentioned previously is used to select the best time source from a number of NTP servers a client usually will query. NTP uses the well-known Marzullo function for this purpose [9], which allows selecting the "best" (most accurate) subset of a group of time sources, deliberately discarding inaccurate ones. This procedure yields far better results than using a simple mean value because it disregards excessively faulty servers rather than simply using their data.

A reference implementation of the protocol stack, the ntpd (NTP daemon), is provided as open source software. For all common operating systems it has even been already included as part of the basic distribution. Aside from handling the mere time protocol, ntpd provides the underlying control loop to adjust both the offset and the rate of the local clock accordingly. Over the years, it has been highly optimized to handle high PDV and unreliable connections to NTP servers. The NTP stack is maintained by the Network Time Foundation [9], which is responsible for preparing new releases and verify/test proposals for new functions or enhancements to existing ones [10].

One major difference to PTP has to be considered, especially when relying on both methods within a system where one is used as a fallback solution in case the other method fails. NTP is based on a different time scale. NTP uses a different epoch than PTP; namely January 1, 1900, while PTP has defined January 1, 1970 as its epoch. Rather than using a continuous time representation as PTP does with TAI (International Atomic Time), NTP distributes UTC (Coordinated Universal Time) time. UTC is an inherently discontinuous time scale, because it includes leap seconds, which have to be handled correctly both by NTP clients and servers. It has to be taken into account that time information itself is represented differently as well; NTP uses seconds and fractional seconds rather than seconds and nanoseconds as PTP does.

4.4 SYNCE—SYNCHRONOUS ETHERNET: A SOLUTION TO ALL PROBLEMS?

In the recent past, SyncE has been proposed by the ITU-T in cooperation with the IEEE as an extension to Ethernet. The underlying ITU-T recommendations are G.8261, G.8262, and G.8264. It specifies a way to transfer frequency over Layer 1 effectively generating a fully synchronous network similar to the E1/T1 or SDH (Synchronous Digital Hierarchy) networks previously used in the telecom area. It imposes quite stringent constraints on the network topology, with limited capabilities to extend and reconfigure a network. Needless to mention, all network components have to fully support SyncE, which may lead to significant investments for large networks. Furthermore, SyncE is strictly limited to frequency transfer; PTP is still required to provide proper time information, yet it is capable of increasing the accuracy of PTP considerably.

4.5 GENLOCK OVER IP

If all nodes in a network are tightly synchronized to a reference, their time information can be used to (re)-generate sync-signals. Assuming that each local, high-accuracy clock is operating at a sufficiently high frequency, any rational frequency can be derived from it to be subsequently used to generate sync pulses with the respective pulse widths.

For all common video sync standards, the formulas to calculate the alignment points with respect to the PTP time are defined in the SMPTE ST 2059-1 Standard, which was published in April 2015. This document describes generation of time labels as well.

4.6 CONCLUSIONS

Both NTP and PTP have proven to be viable methods for transferring accurate time information to nodes connected with each other via an asynchronous medium such as Ethernet. By doing so, one of the major shortcomings of Ethernet can be effectively mitigated or even eliminated. They both rely on the same principle of exchanging timestamped messages back and forth, which allow for offset and clock drift correction.

NTP is a ready-to-use protocol with the underlying software being available on any operating system providing network access. It requires no configuration or maintenance effort whatsoever, other than specifying one or more NTP time servers, whose addresses and performance data can be found easily enough on the Internet. NTP synchronizes the system clock by means of software without any requirements on the underlying hardware. It yields accuracies in the range of several milliseconds, which is quite sufficient for most task in a collaborative work flow like joint access to databases, file-based replication, and distribution of data.

Wherever more stringent requirements for synchronization are demanded, PTP should be the preferred choice. Being optimized for highly accurate time transfer over local area networks, it provides viable solutions to cope with both node and network-induced packet delay variations. Accuracies in the submicrosecond range are possible to obtain using dedicated hardware for generating timestamps. PTP can be even tweaked further to provide nanosecond accuracy; however, this requires careful planning of the network and all its components, and selecting appropriate end nodes.

In the foreseeable future, both technologies will most likely coexist, each serving its purpose. NTP continues to be a perfect tool to provide sufficiently accurate time to all PCs in a network, while PTP is capable of synchronizing the nodes to such a high degree that mission-critical tasks like video sync signal generation can safely rely on it.

REFERENCES

[1] IEEE Instrumentation and Measurement Society. "IEEE Standard for a Precision Clock Synchronization Protocol for Networked Measurement and Control Systems," IEEE Standard 1588-2008, IEEE, New York, NY, 2008.

[2] G.8265.1/Y.1365.1. "Precision Time Protocol Telecom Profile for Frequency Synchronization," ITU-T Recommendation, ITU, Geneva, Switzerland, October 2010.

[3] G.8275/Y.1369. "Architecture and Requirements for Packet-based Time and Phase Distribution," ITU-T Recommendation, ITU, Geneva, Switzerland, November 2013.

[4] PC37.238/D6. "IEEE Draft Standard Profile for Use of IEEE Std. 1588 Precision Time Protocol in Power System Applications," IEEE, New York, NY, February 2011.

[5] Internet link: https://tools.ietf.org/html/draft-ietf-tictoc-ptp-enterprise-profile-03, September 18, 2014.

[6] SMPTE ST 2059-2-2015. "Precision Time Protocol SMPTE profile for time and frequency synchronization in a professional broadcast environment," SMPTE, White Plains, NY, 2015.

[7] Mills, DL. *Computer Network Time Synchronization: The Network Time Protocol*, CRC Press, Boca Raton, FL, 2006.

[8] "Network Time Protocol Version 4: Protocol and Algorithms Specification," 2010, http://www.ietf.org/rfc/rfc5905.txt, IETF, Freemont, CA, May 2015.

[9] Marzullo, KA. "Maintaining the Time in a Distributed System: An Example of a Loosely-Coupled Distributed Service," Ph.D. Thesis, Stanford University, Department of Electrical Engineering, February 1984.

[10] http://www.networktimefoundation.org, 6 May 2015.

CHAPTER 4.4
STANDARDS FOR VIDEO TRANSPORT OVER AN IP NETWORK

John Mailhot
Imagine Communications, Frisco, TX

4.1 INTRODUCTION

The current Internet Protocol (IPv4) was developed in the early 1970s as an outgrowth of the original ARPANET (Advanced Research Projects Agency Network) project. The goal of the project was to create a mechanism for making robust connections between computers over large-scale distances using packet-switching technologies. Other protocols and approaches (e.g., UUNET [Unix-to-Unix Network]) had also been developed and even deployed at some scale, but the packet-switching/routing technology underlying the ARPANET and later Internet proved the more successful and widespread approach.

In order to cover long distances, the ARPANET and later Internet used circuit paths provided by telecommunication carriers. These circuits provided strict point-to-point connections, and the Internet "nodes" performed packet switching to determine where to forward the packets on their next hop.

4.2 HISTORICAL VIEW OF TELEVISION TRANSPORT OVER CARRIER NETWORKS

Digital transport of television signals over long distances employed the high-capacity digital circuits of the telecom network. The telecom carrier digital circuits are available in a variety of bandwidth capacities as shown in Table 4.4.1.

It should be noted that historically, T1 circuits were available to almost any destination in the carrier's region—homes, small businesses, and urban areas. T3 and above circuits require optical fiber access to the endpoint from the telecom "central office" that serves the area. Cost of the circuits is also an issue in most applications, and until very recently circuits above DS3 rates were prohibitively expensive for most television transport applications.

A variety of telecom-oriented digital compression methods were developed in the 1970s and 1980s to transport NTSC (525 line interlaced) video signals over DS3 telecom circuits, culminating in the ubiquitous "DV45" codec from Northern Telecom (Nortel). The DV45 transported an NTSC video signal and its associated audio channels over telecom DS3 circuits at a quality level that was deemed usable for high-profile event contribution by the television production industry.

TABLE 4.4.1 Common Telecom Circuit Types and Capacities (North America)

Service	Bit Rate	Configuration	Media
T1	1.544 Mbits/s	24 voice circuits	Twisted pairs (TX, RX)
T3/DS3	44.736 Mbits/s	$28 \times 24 = 672$ voice circuits	Coaxial cables (TX, RX)
OC3	155.52 Mbits/s	$3 \times 28 \times 24 = 2016$ voice circuits	Optical fiber
OC12	622.08 Mbits/s	$12 \times 28 \times 24 = 8064$ voice circuits	Optical fiber
OC48	2488.3 Mbits/s	$48 \times 28 \times 24 = 32{,}256$ voice circuits	Optical fiber
OC192	9953.2 Mbits/s	$192 \times 28 \times 24 = 129{,}024$ voice circuits	Optical fiber
10GBE	10,000 Mbits/s	Ethernet frames	Optical fiber (or cat-7 copper)

The MPEG-2 (ISO/IEC 13818.2, ITU-T Rec. H.262) video compression standard and its related "Transport Stream" (ISO/IEC 13818.1, ITU-T Rec H.222) were published in 1995, and rapidly gained widespread use across many industries. For "delivery-quality" (suitable for final delivery to a consumer) standard-definition television signals, MPEG-2 required about 4 Mbits/s of bandwidth. This was an economic enabler for direct-to-home satellite television, digital cable television, and even the DVB-T and ATSC (Advanced Television Systems Committee) over-the-air digital broadcast systems, all of which were based on MPEG-2 technology.

Professional contribution of television signals (from sporting venues back to broadcast centers, or from television networks to their affiliated stations) required a higher-quality level, leading to the standardization of the MPEG-2 4:2:2 Profile (H.262 Amendment 2, November 1996). This profile was able to deliver contribution-quality standard-definition television signals suitable for production at typical rates between 12 and 20 Mbits/s, and was critical in the shift from analog satellite delivery to digital satellite delivery. It was also widely used for transport over long-haul telecom networks.

4.2.1 The Shift from Circuits to Packets in Carrier Networks

Telecom circuits deliver fixed bandwidths between fixed points—voice circuits between the two callers, higher-rate circuits between their (two) endpoints. Yet, the technology of MPEG-2 allowed several production signals to fit within the DS3 circuit's capacity. Television production also needs to communicate signals between multiple sites—different production centers or newsrooms, for instance.

At about this time (mid 1990s), in response to the increasing market pressure for data circuits of different rates, and data interchange among multiple destinations, the telecom industry was undergoing a packet-switching revolution, moving from providing circuit-based connections to providing packet-based data services. These data services were aggregated to each site over traditional circuit interfaces (typically DS3 and OC3) but inside the network. *Asynchronous Transport Mode* (ATM) packet-switching fabrics were used to route each signal, packet-by-packet, to its specific destination.

The MPEG-2 Transport Stream standard was developed to be compatible with ATM-based packet-switching technology. In particular, the 188-byte length of the TS packet is evenly divisible into four 47-byte ATM cell payloads. The ATM was a popular choice for contribution networking of MPEG-2 TS-based television signals and associated production or control data signals over telecom infrastructure from 1995 through about 2010. Circuits employing this technology inside are still in use as of this writing. ITU J.82 is the definitive document on how to map MPEG-2 transport stream packets into ATM cells.

ATM was a preferred technology for packet-based video signals because it provided methods for quality of service, traffic engineering, and prioritization of traffic classes. In addition to carrying data payloads, telecom carrier core networks built on ATM technology were often used to "emulate" the underlying T1 and T3 circuits across the new packet-based core networks.

Despite the occasional video transport customer wanting to hand off ATM, the bulk of the data circuit business in the telecom customer base is for transporting enterprise data traffic, and this enterprise traffic is almost exclusively IP traffic. As even the telephony traffic moved to IP, and traditional voice circuits went into decline, IP data traffic overtook traditional voice traffic, and most carriers developed and deployed plans to move to packet-based core networks optimized for delivering IP traffic. As a consequence, these larger-scale economics drove the delivery of television signals over the carrier networks to also move to IP.

4.3 MPEG-2 TRANSPORT STREAMS OVER IP NETWORKS

MPEG-2 Transport Streams are a sequence of 188-byte packets, transmitted at a constant rate. The rate varies with the application; ATSC broadcast signals are famously 19.39265846 Mbits/s (an exact number matched to the details of the channel bandwidth and modulation). Similarly, ISDB-Tb and DVB-T/T2 have very exact TS rates based on the modulation details. Satellite-delivered signals also have TS rates that are a function of the occupied bandwidth and modulation details. In all of these cases, the bit rate is constant.

When delivering any constant-bit-rate signal over a packet-based network, several important considerations arise:

- How is the signal formed into IP datagrams for transport?
- How does the receiver recover the original bit-rate accurately?
- How does the receiver know if there have been lost packets or other errors in the transmission (or better yet correct for those errors)?

The mappings described below are based on UDP—the "User Datagram Protocol," which is sometimes also referred to as the "Unreliable Datagram Protocol" because no specific provision is defined to ensure that the packets are delivered, to retransmit them if they are not, or even to measure if they were delivered. In practice, on networks that are well engineered, lost packets are rare (but important to know about).

A key reason for the choice of UDP is the ability to also leverage Multicast Addressing; this allows the stream of packets to be forwarded by the network to a controlled set of destinations, making copies inside the network. The alternative, humorously called "Multi-Unicast," requires the sending device to make separate copies of the same packet with unicast destination addresses for each receiver. Although multicast is not appropriate in every situation (e.g., some long-haul links do not support multicast), multicast addressing is the preferred methodology when copies are potentially required.

When choosing Multicast Addresses and UDP Port Numbers for services, there are some important common practices. The RTP (Real-time Transport Protocol) definitions require port numbers to be even numbers for the media flows (leaving the odd port numbers for control flows). Good practice requires to avoid using port numbers in the "ephemeral" range (49,152–65,535) and also to avoid port numbers that conflict with any well-known protocol port numbers (registered or otherwise) that might be in use. In practice, UDP port numbers between 10,000 and 40,000 are relatively safe to use, subject to keeping in mind the various IANA (Internet Assigned Numbers Authority) registered port numbers in that same range; see the IANA registry for details of the already-reserved port numbers.[1]

Each different stream in the network should have a unique Multicast Group:Port combination. It is also a good practice to ensure that each stream has a unique multicast Group (irrespective of port) as the IGMP processing in Ethernet Switches and IP Routers is based on Group, without regard to port.

---

[1] https://www.iana.org/protocols.

4.3.1 TS-Over-UDP

The first (and still extremely common) method of mapping Transport Stream packets into IP uses UDP datagrams (RFC-768) on top of IPv4 headers (RFC-791). A number of TS packets (typically seven) directly follow. The receiver determines how many TS packets are included by looking at the total packet length in the UDP header; see Table 4.4.2.

TABLE 4.4.2 UDP Header

```
 0                   1                   2                   3
 0 1 2 3 4 5 6 7 8 9 0 1 2 3 4 5 6 7 8 9 0 1 2 3 4 5 6 7 8 9 0 1
+-+-+-+-+-+-+-+-+-+-+-+-+-+-+-+-+-+-+-+-+-+-+-+-+-+-+-+-+-+-+-+-+
|Version|  IHL  |Type of Service|          Total Length         | IPv4
+-+-+-+-+-+-+-+-+-+-+-+-+-+-+-+-+-+-+-+-+-+-+-+-+-+-+-+-+-+-+-+-+ Header
|         Identification        |Flags|     Fragment Offset     |
+-+-+-+-+-+-+-+-+-+-+-+-+-+-+-+-+-+-+-+-+-+-+-+-+-+-+-+-+-+-+-+-+
|  Time to Live |    Protocol   |         Header Checksum       |
+-+-+-+-+-+-+-+-+-+-+-+-+-+-+-+-+-+-+-+-+-+-+-+-+-+-+-+-+-+-+-+-+
|                         Source Address                        |
+-+-+-+-+-+-+-+-+-+-+-+-+-+-+-+-+-+-+-+-+-+-+-+-+-+-+-+-+-+-+-+-+
|                       Destination Address                     |
+-+-+-+-+-+-+-+-+-+-+-+-+-+-+-+-+-+-+-+-+-+-+-+-+-+-+-+-+-+-+-+-+


+-+-+-+-+-+-+-+-+-+-+-+-+-+-+-+-+-+-+-+-+-+-+-+-+-+-+-+-+-+-+-+-+
|      Source Port Number       |    Destination Port Number    | UDP
+-+-+-+-+-+-+-+-+-+-+-+-+-+-+-+-+-+-+-+-+-+-+-+-+-+-+-+-+-+-+-+-+ Header
|    Length (incl UDP header)   |       Checksum (or zero)      |
+-+-+-+-+-+-+-+-+-+-+-+-+-+-+-+-+-+-+-+-+-+-+-+-+-+-+-+-+-+-+-+-+


+-+-+-+-+-+-+-+-+-+-+-+-+-+-+-+-+-+-+-+-+-+-+-+-+-+-+-+-+-+-+-+-+
|     (0x47)    |         Transport Stream Packet               | First
+-+-+-+-+-+-+-+-+-+-+-+-+-+-+-+-+-+-+-+-+-+-+-+-+-+-+-+-+-+-+-+-+ TS
|                                                               | Packet
|         188 Bytes per TS packet = 47 x 32-bit words           |
+-+-+-+-+-+-+-+-+-+-+-+-+-+-+-+-+-+-+-+-+-+-+-+-+-+-+-+-+-+-+-+-+

            (up to seven TS packets total per UDP)

+-+-+-+-+-+-+-+-+-+-+-+-+-+-+-+-+-+-+-+-+-+-+-+-+-+-+-+-+-+-+-+-+
|    (0 x 47)   |         Transport Stream Packet               | Last
+-+-+-+-+-+-+-+-+-+-+-+-+-+-+-+-+-+-+-+-+-+-+-+-+-+-+-+-+-+-+-+-+ TS
|                                                               | Packet
|         188 bytes per TS packet = 47 x 32-bit words           |
+-+-+-+-+-+-+-+-+-+-+-+-+-+-+-+-+-+-+-+-+-+-+-+-+-+-+-+-+-+-+-+-+
```

This basic TS-over-UDP method was widely deployed in digital cable head ends, for transport of TS signals between receivers, scramblers, ad inserters, and modulators. This method has still widespread use, especially in cable head ends, and equipment which implements the RTP-based standards below often (but not always) includes compatibility to receiving this format also.

Receiving devices will typically buffer the data coming in, to remove timing variation (jitter) caused by the network delivery of the IP packets. The size of this buffer directly relates to the amount of jitter the device can effectively remove. The output of this buffer is TS packets at the (reconstructed) original constant rate.

No provision is made in this TS-over-UDP method for detecting if any of the IP datagrams are lost in transit, though downstream devices can detect errors in the MPEG2 TS payload by examining the continuity counters in the MPEG2 TS packets.

4.3.2 TS-Over-RTP (RFC-3550, SMPTE ST2022-2)

The audio/video transport working group within the IETF (Internet Engineering Task Force) developed a generalized set of tools for streaming multimedia content over IP networks. The RTP Protocol (RFC-3550) describes a general framework for streaming audio and video signals and describing their relationships. RFC-2250 describes more specifically the RTP formats for MPEG video, audio, and also for mapping the Transport Stream into RTP. Pro-MPEG Forum (an industry group) developed a "Code of Practice" that further described/clarified the use of RFC-2250/RTP for Transport Streams. This method (Pro-MPEG CoP3r2) gained widespread use in the long-haul transport community (transporting contribution signals over long-haul networks) because it included the ability to detect errors through the RTP sequence numbers, and it also defined a Forward-Error-Correction (FEC) scheme that could correct small numbers of packets lost in transport. When the Pro-MPEG group stopped operating, this work was picked up by the Video Services Forum (VSF), and taken through standardization through SMPTE, leading to the SMPTE ST2022-2 standard for Transport Streams over IP, and the SMPTE ST2022-1 standard for FEC of RTP flows.

The actual packet format looks a lot like the TS-over-UDP method, except there is an RTP header put into the protocol stack. This RTP header is constrained in ST2022-2 to remove some of the variation of possibilities inherent in RFC-2250 in order to promote interoperability between vendors. This history of interoperability of ST2022-2 streams has been quite good; products from different vendors can be reasonably expected to interoperate when using the ST2022-2 standard.

The RTP/SMPTE 2022-2 method is common in the video transport marketplace, and is preferred in many customers for new installations moving forward. Many 2022-2 compliant devices do provide legacy support for the TS-over-UDP format in order to interoperate with legacy equipment, but use of RTP is encouraged when both ends support it.

The RTP/SMPTE ST2022-2 approach has the critical advantage of being able to determine, at the transport layer, whether packets have been lost or reordered by examining the sequence number field in the RTP header. In addition, in situations where there is occasional packet loss, the ST2022-1 FEC system can be applied in order to correct for lost packets.

In important high-availability situations, the ST2022-7 dual-path seamless-reconstruction system can be applied to leverage multiple network paths into one final result. Both the ST2022-1 FEC and ST2022-7 seamless reconstruction systems require the use of 2022-2 as the underlying Transport Stream mapping; see Table 4.4.3.

4.3.3 Forward Error Correction in ST2022-1

IP datagrams traversing a network can be lost for many reasons. Congestion losses (when too much traffic converges on a single queue within a router or switch) are a frequent source of concern, but in managed network situations traffic engineering principles and QoS application can make congestion loss a rarity. Each link in a network has the possibility of bit errors, though. Imagine an Ethernet cable that is really too long (or maybe not well made) operating at gigabit speeds—the opportunity for bit errors is real. Electrostatic discharges and impulse noise from large motors are common sources of electrical interference. Even in optical transport systems, bit errors can occur due to near-fringe fiber spans and dirty or misaligned connectors along the path.

The Ethernet header includes a checksum over the packet, which allows destination devices to know if errors occurred in transit, and most Ethernet equipment provides counters at each interface which disclose the number of header checksum errors. The only recourse in a network interface when it detects corruption via the header checksum is to discard the packet. Since packet bit errors (and also congestion losses) can occur at any link within the path, an FEC system was developed, documented through Pro-MPEG, and later VSF, and then standardized as SMPTE 2022-1. This FEC system is designed as a companion to ST2022-2 RTP packets, but could be applied to most RTP formats.

The ST2022-1 FEC system does not *detect* bit errors; rather, it assumes that bit errors have already manifested as lost packets, and seeks to reconstruct the contents of the packet that was lost. An XOR

TABLE 4.4.3 Table Data Format for TS Over RTP/ST2022-2

```
 0                   1                   2                   3
 0 1 2 3 4 5 6 7 8 9 0 1 2 3 4 5 6 7 8 9 0 1 2 3 4 5 6 7 8 9 0 1
+-+-+-+-+-+-+-+-+-+-+-+-+-+-+-+-+-+-+-+-+-+-+-+-+-+-+-+-+-+-+-+-+
| Version |  IHL  | Type of Service |         Total Length      | IPv4
+-+-+-+-+-+-+-+-+-+-+-+-+-+-+-+-+-+-+-+-+-+-+-+-+-+-+-+-+-+-+-+-+ Header
|         Identification          | Flags |     Fragment Offset |
+-+-+-+-+-+-+-+-+-+-+-+-+-+-+-+-+-+-+-+-+-+-+-+-+-+-+-+-+-+-+-+-+
|  Time to Live |    Protocol     |         Header Checksum     |
+-+-+-+-+-+-+-+-+-+-+-+-+-+-+-+-+-+-+-+-+-+-+-+-+-+-+-+-+-+-+-+-+
|                          Source Address                       |
+-+-+-+-+-+-+-+-+-+-+-+-+-+-+-+-+-+-+-+-+-+-+-+-+-+-+-+-+-+-+-+-+
|                        Destination Address                    |
+-+-+-+-+-+-+-+-+-+-+-+-+-+-+-+-+-+-+-+-+-+-+-+-+-+-+-+-+-+-+-+-+

+-+-+-+-+-+-+-+-+-+-+-+-+-+-+-+-+-+-+-+-+-+-+-+-+-+-+-+-+-+-+-+-+
|      Source Port Number         |     Destination Port Number | UDP
+-+-+-+-+-+-+-+-+-+-+-+-+-+-+-+-+-+-+-+-+-+-+-+-+-+-+-+-+-+-+-+-+ Header
|    Length (incl UDP header)     |    UDP Checksum (or zero)   |
+-+-+-+-+-+-+-+-+-+-+-+-+-+-+-+-+-+-+-+-+-+-+-+-+-+-+-+-+-+-+-+-+

+-+-+-+-+-+-+-+-+-+-+-+-+-+-+-+-+-+-+-+-+-+-+-+-+-+-+-+-+-+-+-+-+
|V=2|0|0|  CC=0  |0|   PT=33     |        sequence number       | RTP
+-+-+-+-+-+-+-+-+-+-+-+-+-+-+-+-+-+-+-+-+-+-+-+-+-+-+-+-+-+-+-+-+ Header
|                  timestamp (90kHz clock sample)               | as in
+-+-+-+-+-+-+-+-+-+-+-+-+-+-+-+-+-+-+-+-+-+-+-+-+-+-+-+-+-+-+-+-+ 2022-2
|             synchronization source (SSRC) identifier          |
+-+-+-+-+-+-+-+-+-+-+-+-+-+-+-+-+-+-+-+-+-+-+-+-+-+-+-+-+-+-+-+-+

+-+-+-+-+-+-+-+-+-+-+-+-+-+-+-+-+-+-+-+-+-+-+-+-+-+-+-+-+-+-+-+-+
|    (0x47)     |          Transport Stream Packet              | First
+-+-+-+-+-+-+-+-+-+-+-+-+-+-+-+-+-+-+-+-+-+-+-+-+-+-+-+-+-+-+-+-+ TS
|                                                               | Packet
|           188 bytes per TS packet = 47 x 32-bit words         |
+-+-+-+-+-+-+-+-+-+-+-+-+-+-+-+-+-+-+-+-+-+-+-+-+-+-+-+-+-+-+-+-+

            (up to seven TS packets total per UDP)

+-+-+-+-+-+-+-+-+-+-+-+-+-+-+-+-+-+-+-+-+-+-+-+-+-+-+-+-+-+-+-+-+
|    (0x47)     |          Transport Stream Packet              | Last
+-+-+-+-+-+-+-+-+-+-+-+-+-+-+-+-+-+-+-+-+-+-+-+-+-+-+-+-+-+-+-+-+ TS
|                                                               | Packet
|           188 Bytes per TS packet = 47 x 32-bit words         |
+-+-+-+-+-+-+-+-+-+-+-+-+-+-+-+-+-+-+-+-+-+-+-+-+-+-+-+-+-+-+-+-+
```

code is used across a sequence of bytes to calculate a "check byte." The process can be reversed to find the original value of any one of the original bytes (assuming you know that one was lost):

$$7B \text{ xor } 5D \text{ xor } 38 \text{ xor } 4C = 52 \text{ (check byte)}$$

$$7B \text{ xor } 5D \text{ xor } 38 \text{ xor } (00) \text{ xor } 52 = 4C \text{ (missing value)}$$

The XOR code is applied in each byte position across several packets, creating a whole "FEC Packet." The contents of the FEC packet can be used to reconstruct any one lost packet within the original group of packets. If more than one packet in a group is lost, then reconstruction is not possible; see Fig. 4.4.1.

STANDARDS FOR VIDEO TRANSPORT OVER AN IP NETWORK **4.85**

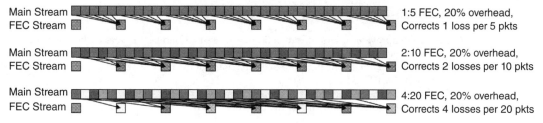

FIGURE 4.4.1 FEC packet correction process.

In practice, error events—even from physical layer issues—tend to occur in bursts; the system described above can only correct a single packet loss. In order to make the system more robust, an interleaver is used; FEC packets are calculated not based on adjacent original packets, but based on original packets distance "L" apart. This allows correction of burst losses of up to "L" packets in sequence, without further increasing overhead. The other key factor is the number of original packets factored into each FEC packet (the "depth" of the interleaver).

These values (L is the length of a burst that can be recovered, D is the depth of the interleaver) are set by the transmitter; the receiver determines them from the headers of the FEC stream. Figure 4.4.2 shows the FEC packets calculated based on L = 4, D = 5. In practice, L = 20, D = 16 (5 percent overhead) is a typical choice for providing a light amount of protection to an important contribution signal. Most receivers provide some statistics on the number of corrected packets and any uncorrected packets.

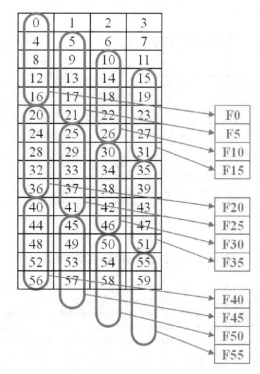

FIGURE 4.4.2 Example FEC packet calculations.

4.3.4 Seamless Reconstruction (ST2022-7) for High Availability

Although FEC provides a level of protection against small numbers of lost packets, often there is a need to provide protection against more colossal failure events such as fiber cuts and network equipment failures (or maintenance). For protection against these types of large-scale events, VSF developed the SMPTE ST2022-7 system.

In the 2022-7 "Seamless Reconstruction" schema, the receiver can reconstruct the original 2022 signal based on packets from either of the two copies transmitted; see Fig. 4.4.3.

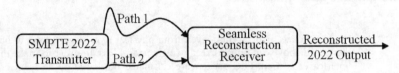

FIGURE 4.4.3 SMPTE 2022-7 seamless reconstruction concept.

The receiver uses an assumption (typically user provided) about the worst-case time skew between the two paths that the signal could take. For long-haul signals these paths might be diversely routed through different carrier networks, while inside a campus these two paths might simply be through disjoint sets of switching/routing equipment; see Fig. 4.4.4. The receiver then uses the sequence numbers from the RTP headers in order to recompose the original stream, which then can be deconstructed without incident.

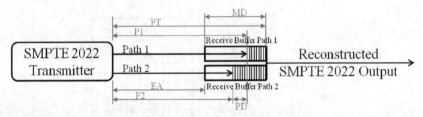

FIGURE 4.4.4 Seamless reconstruction of different signal paths.

The ST2022-7 system has gained significant deployment as a way to insulate feeds against equipment and connectivity failures within the network.

Although one could apply the 2022-1 FEC and the 2022-7 seamless reconstruction technologies to the same original signal, this is not common. Typically, if 2022-7 seamless reconstruction is applied, 2022-1 is not used; however, it is possible and allowed to combine the techniques.

4.3.5 Uncompressed Video on IP Networks (SMPTE ST2022-6)

For most long-haul applications, video compression (MPEG-2, H.264, JPEG2000) is used in order to reduce the bandwidth required from the network. This is a straightforward economic tradeoff between the additional cost of the compression equipment, and the cost of the network service bandwidth, given the acceptability of the quality degradation introduced by the compression in use.

In situations where the signals are inside a building or campus environment, the incremental cost of bandwidth can be quite low—set really by the fixed cost of the switches, routers, and fibers that connect them. With the advent of high-capacity datacenter-grade Ethernet switches at reasonable price points, it becomes feasible to transport pixels (uncompressed video) around the campus over

IP links instead of traditional coaxial cables—particularly when evaluating this in the context of software-based workflow devices such as playout servers, and graphics generators, where the cost includes a Serial Digital Interface (SDI) card. While the economics of such an application are beyond the scope of this chapter, the underlying technology of mapping uncompressed (pixel) payloads into IP is described in SMPTE ST2022-6.

SMPTE ST2022-6 maps SDI television signals into IP packets. The header stack is similar to the one for ST2022-2, but the RTP header has been augmented with an additional "media header" containing some information about the SDI payload.

In the RTP header of 2022-6, the "M" bit is used to signal the last datagram of a video frame. This datagram is stuffed with zero-fill (if needed) after the end of the video frame, to reach the same length as the other datagrams. After the headers, there are 1,376 bytes of data, into which the successive 10-bit words of the SDI structure are sequentially packed; see Table 4.4.4.

TABLE 4.4.4 Data Format for SDI over RTP/ST2022-6

```
 0                   1                   2                   3
 0 1 2 3 4 5 6 7 8 9 0 1 2 3 4 5 6 7 8 9 0 1 2 3 4 5 6 7 8 9 0 1
+-+-+-+-+-+-+-+-+-+-+-+-+-+-+-+-+-+-+-+-+-+-+-+-+-+-+-+-+-+-+-+-+
| Version |  IHL  | Type of Service |         Total Length      | IPv4
+-+-+-+-+-+-+-+-+-+-+-+-+-+-+-+-+-+-+-+-+-+-+-+-+-+-+-+-+-+-+-+-+ Header
|       Identification          | Flags |    Fragment Offset    |
+-+-+-+-+-+-+-+-+-+-+-+-+-+-+-+-+-+-+-+-+-+-+-+-+-+-+-+-+-+-+-+-+
|   Time to Live  |   Protocol  |       Header Checksum         |
+-+-+-+-+-+-+-+-+-+-+-+-+-+-+-+-+-+-+-+-+-+-+-+-+-+-+-+-+-+-+-+-+
|                        Source Address                         |
+-+-+-+-+-+-+-+-+-+-+-+-+-+-+-+-+-+-+-+-+-+-+-+-+-+-+-+-+-+-+-+-+
|                      Destination Address                      |
+-+-+-+-+-+-+-+-+-+-+-+-+-+-+-+-+-+-+-+-+-+-+-+-+-+-+-+-+-+-+-+-+

+-+-+-+-+-+-+-+-+-+-+-+-+-+-+-+-+-+-+-+-+-+-+-+-+-+-+-+-+-+-+-+-+
|      Source Port Number       |    Destination Port Number    | UDP
+-+-+-+-+-+-+-+-+-+-+-+-+-+-+-+-+-+-+-+-+-+-+-+-+-+-+-+-+-+-+-+-+ Header
|   Length (incl UDP header)    |    UDP Checksum (or zero)     |
+-+-+-+-+-+-+-+-+-+-+-+-+-+-+-+-+-+-+-+-+-+-+-+-+-+-+-+-+-+-+-+-+

+-+-+-+-+-+-+-+-+-+-+-+-+-+-+-+-+-+-+-+-+-+-+-+-+-+-+-+-+-+-+-+-+
|V=2|0|0| CC=0 |M|   PT=98     |         sequence number        | RTP
+-+-+-+-+-+-+-+-+-+-+-+-+-+-+-+-+-+-+-+-+-+-+-+-+-+-+-+-+-+-+-+-+ Header
|             timestamp   (27 MHz clock sample)                 | as in
+-+-+-+-+-+-+-+-+-+-+-+-+-+-+-+-+-+-+-+-+-+-+-+-+-+-+-+-+-+-+-+-+ 2022-6
|            synchronization source (SSRC) identifier           |
+-+-+-+-+-+-+-+-+-+-+-+-+-+-+-+-+-+-+-+-+-+-+-+-+-+-+-+-+-+-+-+-+

+-+-+-+-+-+-+-+-+-+-+-+-+-+-+-+-+-+-+-+-+-+-+-+-+-+-+-+-+-+-+-+-+
|Ext |F|VSID | FRCount         | R | S | FEC | CF |   RESERVE  | 2022-6
+-+-+-+-+-+-+-+-+-+-+-+-+-+-+-+-+-+-+-+-+-+-+-+-+-+-+-+-+-+-+-+-+ Payload
|  MAP  |  FRAME   |    FRATE   |   SAMPLE   |   FMT-RESERVE   | Header
+-+-+-+-+-+-+-+-+-+-+-+-+-+-+-+-+-+-+-+-+-+-+-+-+-+-+-+-+-+-+-+-+
          (optional extensions are allowed but not shown here)

+-+-+-+-+-+-+-+-+-+-+-+-+-+-+-+-+-+-+-+-+-+-+-+-+-+-+-+-+-+-+-+-+
| first 10-bit word | next 10-bit word | next 10-bit word   | n | SDI
+-+-+-+-+-+-+-+-+-+-+-+-+-+-+-+-+-+-+-+-+-+-+-+-+-+-+-+-+-+-+-+-+ Data
|ext 10-bit word|  next 10-bit word  | next 10-bit word | next | Packed
+-+-+-+-+-+-+-+-+-+-+-+-+-+-+-+-+-+-+-+-+-+-+-+-+-+-+-+-+-+-+-+-+ Tightly
|                                                               |
| exactly 1376 Bytes of SDI per TS packet = 1100.8 10-bit words |
+-+-+-+-+-+-+-+-+-+-+-+-+-+-+-+-+-+-+-+-+-+-+-+-+-+-+-+-+-+-+-+-+
```

TABLE 4.4.5 Annotated Bibliography of Video-Over-IP Standards

	Official Title	Comments
IANA Port Registry	Service Name and Transport Protocol Port Number Registry	http://www.iana.org/assignments/service-names-port-numbers/service-names-port-numbers.xhtml
ISO/IEC 13818.1 ITU-T Rec. H.222.0	Information technology–Generic coding of moving pictures and associated audio information–Systems	MPEG-2 "Systems" specification. This document defines what an MPEG-2 "Transport Stream" is. ATSC, DVB, and other documents refer back to this definition. The MPEG-2 transport stream is used with many different video and audio payloads.
ISO/IEC 13818.2 ITU-T Rec. H.262	Information technology–Generic coding of moving pictures and associated audio information–Video	This is the defining document of MPEG-2 video compression, including the professional profile extensions. An exceptionally readable document for learning how video compression works.
RFC 768 (IETF)	User Datagram Protocol	Published in 1980 and still the basis of how we transport video signals today. Three pages only.
RFC 791 (IETF)	Internet Protocol–DARPA Internet Program Protocol Specification	Published in 1981 (updating the 1980 original), this is the basic definition of the IPv4 header. Some later RFCs redefine or update a few of the fields, mostly around QoS.
RFC 3550 (IETF) (Obsoletes RFC 1889)	RTP: A Transport Protocol for Real-Time Applications	This replaced/extended the original RFC 1889 definition of RTP, but as used here the differences were minor. This defines the core of RTP as used across a wide variety of video and audio payloads.
RFC 2250 (IETF)	RTP Payload Format for MPEG1/MPEG2 Video	The bulk of this RFC deals with mapping the MPEG-1 and MPEG-2 video elementary stream (PES) formats into RTP; however, it also does define mapping TS into RTP.
RFC 1890 (IETF)	RTP Profile for Audio and Video Conferences with Minimal Control	Section 7, Table 2 of this RFC defines the RTP "Payload Type" numbers for different video and audio formats, including PT = 33 for MPEG-2 TS.
ST2022-2:2007 (SMPTE)	Unidirectional Transport of Constant Bit Rate MPEG-2 Transport Streams on IP Networks	This document references and further constrains RFC2250 (and by extension RFC 1889) RTP format for Transport Streams over IP. It also specifies (by reference to ST2022-1) the FEC system for TS over IP.
ST2022-1:2007 (SMPTE)	Forward Error Correction for Real-Time Video/Audio Transport Over IP Networks	Describes the FEC system which can be applied to 2022-2 TS over RTP streams. This FEC system can protect against packet losses in transport by reconstructing the lost packets. No retransmission is required.
ST2022-7:2013 (SMPTE)	Seamless Protection Switching of SMPTE ST 2022 IP Datagrams	Defines requirements on a pair of 2022-x streams such that a receiver could reconstruct a single stream from any mix of packets received from either stream.
ST2022-6:2012 (SMPTE)	Transport of High Bit Rate Media Signals over IP Networks (HBRMT)	Describes mapping of SDI into a sequence of IP datagrams. Note that an HD flow requires approximately 1.6 Gbits/s of network bandwidth, but is delivered perfectly.
ST2022-5:2012 (SMPTE)	Forward Error Correction for Transport of High Bit Rate Media Signals over IP Networks (HBRMT)	Companion FEC system for ST2022-6. Very similar to ST2022-1.

A companion FEC document (ST2022-5) was produced in parallel with ST2022-6. The method described is almost identical to 2022-1, but allows a larger range of values for the L and D parameters, and corresponds to the packet sizes and additional headers of ST2022-6.

Readers are encouraged to obtain and read the underlying reference documents (ST2022-6 in this case) to learn more about the header structures.

The same data format described above is used for all of the different SDI formats—525, 625, 1080i, 720p, etc. Table 4.4.5 provides an annotated bibliography of video-over-IP standards in current use.

SECTION 5
PRODUCTION SYSTEMS

Section Leader—Andrea Cummis
Managing Partner, AC Video Solutions, Roseland, NJ

Local content is what makes radio and television stations an essential element of their communities. Described variously as "localism," it is what separates broadcasters from the growing list of other content providers that vie for consumers' time. The trust enjoyed by local stations in their service area has typically been earned through countless activities—large and small. Breaking news and coverage of emergency situations immediately come to mind when localism is discussed. But, there are many other ways in which local stations improve their communities through public service programs, fund raising efforts, and direct community involvement.

Local broadcasters are unique in the electronic entertainment and information industry because their focus is on the local area. The connections with their listeners and viewers are real.

Localism is possible because stations invest in people and equipment necessary to tell stories, report the news, and help ensure public safety during an emergency. The starting point for a local presence is the production facility.

The scope of production systems varies from the large and sophisticated complex to the mobile ENG van, and everything in between. This section provides an overview of radio and television production facility design. The chapters that follow serve as a starting point for planning a new facility, or renovating an old one. Indeed, entire books could be written (and have been written) on the subjects covered in this section. The goal here is to give readers a general overview and pointers to additional information and resources. We have combined radio and television applications where it makes sense.

It is important to remember that the technologies used in modern production systems are a moving target. As convincingly documented in the previous section of this Handbook, Information Technology concepts and hardware are rapidly entering the broadcast plant. Nowhere is this trend more prevalent than in production systems.

During the transition to digital video (and DTV), digital systems were introduced one box or system at a time, forming what were called "digital islands." In most production centers today—radio or TV—it is more like "digital continents." One could imagine a point in the not too distant future where we will have "analog islands" in broadcast stations.

As mentioned above, "localism" is a fundamental element of broadcasting. For the broadcast engineer, "never a dull moment" is another.

Chapter 5.1: Production Facility Design
Chapter 5.2: Audio System Interconnections
Chapter 5.3: Audio Monitoring Systems
Chapter 5.4: Remote Audio Broadcasting

Chapter 5.5: Master Control and Centralized Facilities
Chapter 5.6: Video Switchers
Chapter 5.7: Automation Systems
Chapter 5.8: Media Asset Management
Chapter 5.9: Production Intercom Systems
Chapter 5.10: Broadcast Studio Lighting
Chapter 5.11: Cellular/IP ENG Systems

CHAPTER 5.1
PRODUCTION FACILITY DESIGN

Richard G. Cann, Anthony Hoover,

Frederic M. Remley, and Ernst-Joachim Voelker

5.1 INTRODUCTION

Studios are constructed to host and record a wide variety of programs—live and recorded. The production can be relatively simple, as in a fixed-scene news program, or very complex, such as variety show with a live audience—and everything in-between. While fine tuning of the program is carried out in the control room, the overall look and feel of the production is determined by the capabilities of the studio—and the attention to detail during design and construction.

Studios should provide stimulus for the performer. For that, good acoustics, lighting, set dressing, and adequate room for the production are necessary. Television studios have special requirements for lighting and air conditioning. Sets may cover the acoustical wall treatment. Microphones must be used at greater distances to be out of the picture. The view through the control room window is often obscured. These and other restrictions challenge the facility designer to solve challenging, sometimes unique, problems.

5.2 STUDIO DESIGN CONSIDERATIONS

When building a new facility or renovating an old one, it is important to evaluate in detail the suitability of the site for ambient and interfering noise as well as for production and aesthetic requirements. The cost of construction to meet the specified criteria may be much greater at one location than another. For example, the interior noise might be dominated by ground vibration from an adjacent railroad yard. Breaking this vibration path can be very expensive. An industrial site might appear to be an excellent location to avoid the complaints of neighbors late at night, but the facility may not be usable if punch presses are installed next door.

Because the needs and challenges a given facility are often unique, a fresh look must be taken for each studio project. Having said that, there are a number of guidelines that should be followed to minimize surprises later in the project. The following sections provide a starting point for a new or renovated facility, and offer up some important lessons learned from previous experience.

5.2.1 Survey of Studios[1]

In general, professional studio types can be divided into three main groups:

- Broadcasting studios for radio and television
- Recording studios for music, speech, and television and film and dubbing
- Studios in theaters, multipurpose halls, and concert halls

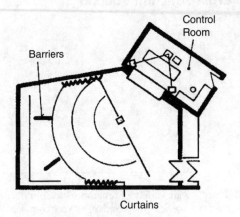

FIGURE 5.1.1 Large music studio.

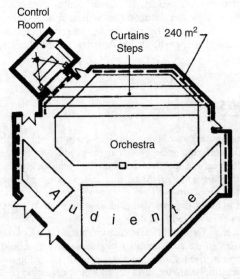

FIGURE 5.1.2 Large music studio with provision for an audience.

Studios are combined with control rooms except for self-operated and disk jockey studios. The control rooms may be used in different ways: for recording the output of the studio, for high-quality monitoring, or for monitoring and recording musical instruments. Figures 5.1.1–5.1.11 show studios in simplified forms with their associated control rooms. Each is described in brief in the following section.

5.2.1.1 Example Studio Designs. Figure 5.1.1 shows a large music studio without an audience for television recording use. This studio could be used for large orchestras, small orchestras, or choirs. Volume is normally 2000 m^3 or more. Reverberation time may be adjusted between 1.5 and 1.2 s, depending on the distribution of draperies and other sound-absorbing materials. Changes of reverberation time are limited to the middle and high frequencies. The maximum sound level within the studio must be taken into account when designing one studio adjacent to another studio.

Figure 5.1.2 shows a large music studio with or without an audience. Such studios may have a volume of about 5000 m^3 and are used for large orchestras, small orchestras, and soloists. When the orchestra is rehearsing in the studio, available volume may amount to 50 m^3 per person, but when an audience is present, this figure falls to about 15 m^3 per person. If audience seating is removed from the studio, reverberation time may rise as high as 2.4 s. With draperies that may be 10–12 meters high and have an area as great as 240 m^2, reverberation time declines to about 1.2 s with audience and orchestra present.

Figure 5.1.3 shows a studio complex with an on-air and production studio. This studio may consist of a control room that works with a talk studio or a multipurpose studio. The adjacent equipment room serves as a control room for separate operation of the multipurpose space. For the differing requirements of speech, choir,

---

[1]This section was adapted from Voelker, E-J., "Studio Production Systems," in *Master Handbook of Audio Production*, J. C. Whitaker (ed.), McGraw-Hill, New York, NY, 2003.

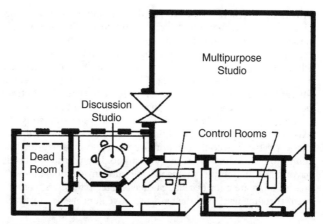

FIGURE 5.1.3 On-air and production studio complex.

orchestra, etc., acoustics are varied with movable wall elements or adjustable absorbent roller blinds. The talk studio is suitable for announcements or for discussion programs. Possible operation of either control room with either studio requires a high level of sound isolation from the windows, doors, and walls.

Figure 5.1.4 shows a drama complex that includes a large studio. A drama complex is normally used for speech recordings. Each room has an individual acoustic condition. The large studio

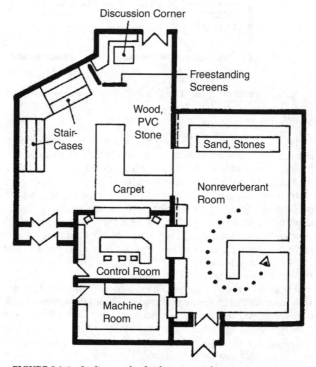

FIGURE 5.1.4 Studio complex for dramatic productions.

provides a reverberant area and, in the corner, a nonreflecting area for intimate speech. Many specific facilities may be provided for sound effects. In some installations, the large studio is divided into three parts by hanging draperies; one end will be reverberant and the other end dead, while the center area simulates a normal living-room condition. This type of studio has the disadvantage that while a recording is being made in one section the other parts cannot be used, even for rehearsals.

The large studio shown in Fig. 5.1.4 is usable for speech recordings with up to about 30 persons, for example, simulating an auditorium presentation. The control room normally occupies a central position and has a view into the individual areas of the complex, as illustrated.

Figure 5.1.5 shows a television studio for live broadcasting. A cyclorama and sets, together with extensive lighting and air-conditioning equipment, are the important features of this studio. In many cases, a catwalk is constructed above the cyclorama. Normally it is not necessary for the control rooms to have a direct view into the studio. In any case, the production control room normally will have a bank of monitors that would obscure any studio window. The acoustic characteristics of such studios often are poor because of unavoidable reflections from the studio sets, the necessary equipment at ceiling level, and the hard reflective floor needed for stable camera movement.

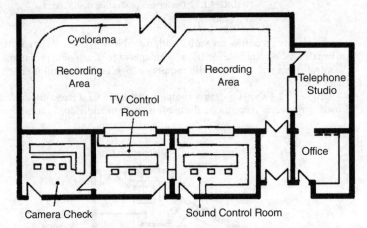

FIGURE 5.1.5 Television studio with separate video and audio control rooms.

Figure 5.1.6 shows a television complex with two studios, each having its own control room but sharing a common equipment room. Each studio includes an announcer booth that opens directly from the control room (Fig. 5.1.7). Double doors and sound locks provide the necessary isolation of the studios from the activities in the control area.

In many cases, a single control room serves for both sound and production control. During production, close communication between the sound engineer and producer is normally more important than the quality of the sound at the monitoring point. The side area is open to give a direct view to the announcer booth, but is acoustically dead to act as a sound lock. The sound engineer can improve listening conditions and separate himself or herself from activities in the control room by closing the movable window between the two areas. The engineer can then listen at a higher level without interfering with other activities.

Figure 5.1.8 shows a news studio complex for live broadcasting. Two control rooms are combined with three studios in the news complex. The single-operator (disk jockey) studio is self-run with its own source and control equipment. It can also work with the other studios—in this case acting as a control room. There are many ways to use the studios—the larger studio having the possibility of including an invited guest to take part in discussions. A certain amount of sound transfer from one area to another may be permissible. The same aim is achieved by windows that link all the areas together. On the other hand, the windows result in undesirable acoustic characteristics with strong

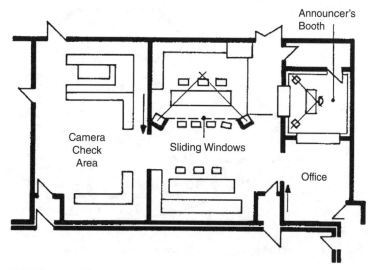

FIGURE 5.1.6 Television complex with two studios.

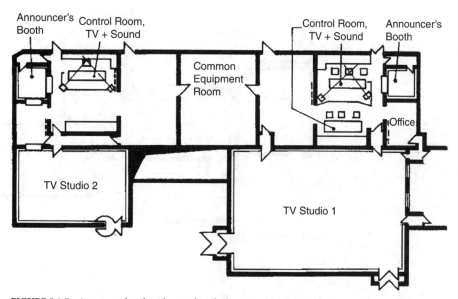

FIGURE 5.1.7 Announcer booth with a combined television and audio control room.

reflections, and they also limit the wall area available for acoustic treatment. In discussions in which microphone distance may be greater because of inexperienced participants, these strong reflections can create unwanted side effects.

Figure 5.1.9 shows a collection of studios with a central control room. The central control room/machine room is a popular concept in a variety of facilities. Studio 1 could be a multipurpose studio for music and speech, while Studios 2, 3, and 4 would be designed primarily for speech-based programs, e.g., news. The control and equipment rooms together form a technical center. This is a disadvantage in that the control room can usually work with only one studio at a time.

5.8   PRODUCTION SYSTEMS

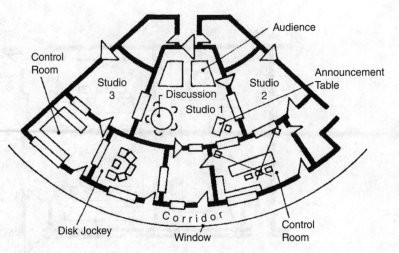

FIGURE 5.1.8   News studio complex for live, on-air broadcasting.

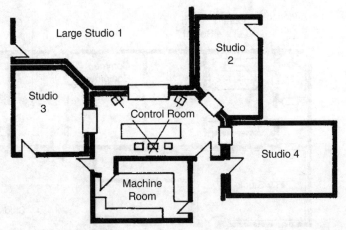

FIGURE 5.1.9   Studios with a central control room.

Figure 5.1.10 shows a small radio station with two disk jockey studios, each of which can act as a control room operating with the common interview studio, with or without an audience. It is typical also to have a small studio for news inserts, weather, and other announcements. Small music groups may be recorded in the larger studio. Most recordings are speech-orientated and include discussions, news, weather, and telephone call-in programs. A certain level of background noise may be permissible and even desirable, but there are certain types of programs for which noise is disturbing. To reduce the necessity to construct expensive, highly insulated studios, close-microphone techniques are normally employed.

Figure 5.1.11 shows another arrangement of studios, control room, and disk jockey studio. For full utilization, one studio may be used for editing and preparation of automated transmissions. During transmission, the control room, disk jockey studio, and weather studio work together.

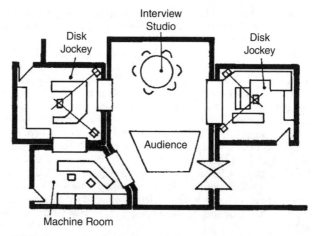

FIGURE 5.1.10 Small radio station studio complex.

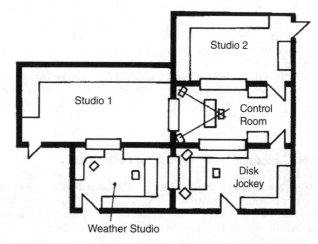

FIGURE 5.1.11 Multiple-studio radio station complex.

5.2.2 Special Considerations for Control Rooms[2]

A comprehensive definition of a television control room should include all spaces that are allocated to control the creative and technical phases of television production. By this definition, video control rooms, audio control rooms, master control rooms, continuity booths, and editing suites would all be included. Each of these spaces has a specific function to perform in the final production. In addition, some aspects of the design will vary depending on the type of facility being considered. For example, the studio control room requirements of a general-purpose production center will differ from those of a control room intended for the evening news program of a local television station. By the same

---

[2]This section adapted from Remley, F. M., "Control Room Design and Layout," in *Television Engineering Handbook*, revised edition, K. Blair Benson and J. C. Whitaker (eds.), McGraw-Hill, New York, NY, 1992.

token, an audio control area dedicated to serve a large, multipurpose studio will have requirements different from the needs of an audio booth assigned to a production or master control operation.

It is important to identify carefully the primary use of a new control room before beginning to design of the room. A variety of design factors are affected by the planned use for the space. Since few architectural designers have experience in control room layout, the responsibility for a careful definition of the form and function usually lies with the future owner, the owner's engineering staff, or a consultant selected by the owner or architect.

Although skilled consultants may be found to assist in control room design, the owner or owner's technical experts must, in all cases, define carefully the specifications for any space that is to serve a control room function. Both architects and consultants will require such guidance.

For a control room associated with a studio or suite of studios, the physical arrangement must take into account the communication and interaction needs of the persons occupying the space. Even though the television facility will be equipped with a complete intercommunications system, visual and spoken cues may also be necessary within a single control area. Hand signals remain an important part of the technique of television production, and visual contact with the director may be important for some control room occupants, such as the audio operator, video switcher, or technical director.

Because of conflicting requirements for listening to program-related audio, the audio operator may be placed in a separate booth adjoining the television studio production control room. By this means, the audio operator will be in close physical proximity to the production area but will not be exposed to the distractions of the video production process. The result will usually be better audio—and more accurate judgment of sound levels, microphone balances, and so on. In addition, the other control room personnel will not be distracted by high loudness levels of the audio monitoring system since the audio operator can make the monitor adjustments independently of those involved in the video portion of the production. Other control room personnel who might be considered for separate booth locations include, of course, an announcer and perhaps a light control director.

Careful consideration of space relationships and of requirements for adjacency of key personnel and equipment will result in reduced loss of valuable production time by permitting immediate, direct coordination.

5.2.3 Criteria for Acceptability of Acoustical Performance[3]

Before beginning the acoustical design of any studio space, one of the first tasks is to establish criteria for its performance. Sometimes it might appear that this is superfluous, that the owner, architect, and engineer have an unwritten understanding of acoustical requirements, and that the design process can begin immediately. However, this is often not the case, and absence of thoughtfully written criteria may lead to fundamental design errors. The development of criteria is very important in defining just how spaces will be used and will ultimately determine just how well the spaces will function acoustically.

For example, if a studio space will be used for both speech and music, basic design decisions will have to be made at the outset. Figure 5.1.12 shows a typical plot of reverberation time at 500 Hz versus room volume for auditoria used for different activities. The preferred reverberation time for music is approximately twice as long as it is for speech. Either the acoustical quality of some activities will have to be compromised by selecting a specific reverberation time, or provision will have to be made for adjusting the reverberation time for each activity.

Though there are generally accepted criteria for the reverberation time of a small auditorium, they are not nearly so clear-cut for recording studios. Some performers insist on live reverberant feedback from the space itself and directly from adjacent performers. Other artists, who work entirely through

---

[3] This section was adapted from Cann, R. G., and A. Hoover, "Criteria for Acceptability of Acoustical Performance," in *Standard Handbook of Audio and Radio Engineering*, J. C. Whitaker and K. Blair Benson (eds.), McGraw-Hill, New York, NY, 2002.

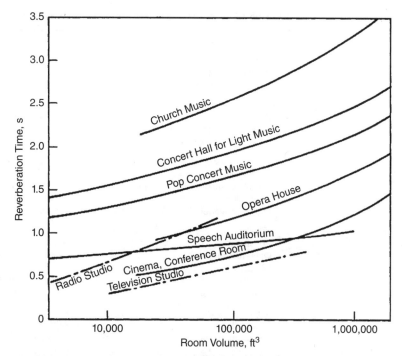

FIGURE 5.1.12 Typical reverberation times for different sizes of rooms and auditoria according to usage.

headphones, are much less concerned about the reverberant quality of a studio. The former scenario requires an architectural solution, while the latter requires none.

The design criteria for a control room often simply invoke, without acoustical reason, some currently fashionable proprietary design concept. It seems that no matter which concept is selected, there are always sound engineers who hate it or love it. And, of course, that varies from one engineer to the next, according to their past experience. Therefore, it is wise to build flexibility into control room design, allowing engineers to produce their own semicustom setup. No one way is best.

5.2.3.1 Background Noise. The background sound level should not interfere with the perception of the desired sound, especially during quiet periods, for music is a rhythm not only of sound but also of silence. It is important not to intrude upon the latter with a rumbling fan or traffic noise. The most frequently used criteria for background noise are the noise-criterion (NC) curves, which are classified according to space usage in Table 5.1.1.

In some circumstances quieter is not necessarily better, for reducing noise levels may reveal other sounds that were previously hidden. Noise levels can often be increased to the NC value specified in Table 5.1.1 without causing noticeable intrusion. This masking of an intruding sound may be less expensive than controlling the sound itself.

It may be necessary to set special criteria for a particular activity, e.g., for a recording studio noise criteria may be set equal to the noise floor of the recording device(s). This level may be above the aural threshold of the performers, with the result that it is possible to identify clearly a source of intruding noise before a recording session begins and yet not have the offending noise be audible on playback. This technique was more applicable in the days of analog tape recording; today, digital systems provide a noise floor that masks very little.

Concern need be given not only to sound levels within the space but also to those outside in the neighborhood. The community may have a specific bylaw limiting noise levels. If not, the criteria

TABLE 5.1.1 Design Goals for Mechanical-System Noise Levels

Type of Area	NC Range
Residences	
Single-family homes	20–30
Apartments and condominiums	25–35
Hotels and motels	
Guest rooms	30–40
Meeting or banquet rooms	30–35
Corridors and lobbies	35–45
Offices	
Boardrooms	20–30
Executive offices	30–40
Open-plan areas	35–40
Public circulation	35–45
Hospitals	
Private rooms	25–35
Operating rooms	30–40
Wards	30–40
Laboratories	30–40
Public circulation	35–45
Churches	
Sanctuaries	20–30
Public circulation	30–40
Schools and universities	
Libraries	30–40
Lecture rooms and classrooms	20–35
Laboratories	35–45
Cafeterias and recreation halls	35–45
Public buildings	
Libraries and museums	30–40
Courtrooms	25–35
Post offices and banks	30–40
Auditoria, theaters, and studios	
Concert halls	15–25
Recording studios	15–25
Multipurpose halls	20–30
TV studios	20–30
Movie theaters	30–35

for maximum noise levels should be based on available annoyance data. It is appropriate to establish these criteria up front in a permit application rather than hope that abutters will not stir up community objections at a later time.

5.2.3.2 Mechanical Systems. Mechanical systems are the source of ventilation for recording studios, control rooms, and associated spaces. The acoustical concerns generally are focused on mechanical systems as a source of noise produced by such items as fans and airflow, and as a path for sound transmission between spaces as through ductwork. A more complete treatment of the mechanical system as a source of noise may be found in [1].

Fans that are required to move air through a ventilation system inherently generate noise. Many factors determine the amount of noise produced, including the type of fan used, the volume of air to be delivered, the static pressure against which the fan is forcing the air, the blade passage, and the efficiency of the fan system.

The most common type of fan used for ventilation systems is the centrifugal airfoil fan, although other types of system are not unusual. Each system tends to produce its own characteristic spectrum of frequencies, but in general fans used for ventilation systems tend to produce more low-frequency noise energy than high-frequency noise energy. In most cases, these fans are contained within a prefabricated housing, which in turn is connected to the supply-air ductwork system and to the return-air ductwork system. It is important to note that the sound generated by a fan propagates as easily through the return-air system as through the supply-air system because the speed of sound is so much faster than the speed of the air within the ductwork.

Noise generated by the fan not only travels down the supply and return ductwork systems but also is radiated off the fan housing. In general, the fan housing is a very poor isolator of sound and for most practical purposes, especially in lower frequencies, can be considered to provide no isolation whatsoever. Therefore, it is good practice to locate the fan assembly well removed from the studio and control room.

5.2.3.3 Turbulent Noise in Ducts. Airflow noise is generated by turbulence within the ductwork and at diffusers and dampers. Air turbulence and, therefore, airflow noise generally increase as the speed of airflow increases. Therefore, it is good practice to keep the speed of airflow low. Several rules for controlling airflow noise include:

- Size ductwork so that the flow of air stays below 2000 feet/min and preferably below 1500 feet/min. The velocity of air in a duct may be calculated by dividing total cubic feet per minute in that duct by the cross-sectional area in square feet of the duct itself.
- Airflow velocities through diffusers should be kept below a maximum of 500 feet/min through all diffusers. For critical applications, lower speeds such as 200–300 feet/min are advisable.
- Air valves and dampers should be located so that the airflow noise that they generate does not contribute to the noise ducted from upstream sources.
- Splits and bends in the ductwork should be smooth. Abrupt corners and bends should be avoided, especially near the fan, near high-airflow-velocity locations, and near diffusers and grilles.

Airflow noise is typically a major component of mid- and high-frequency background noise in studios. However, when there are abrupt bends and turns in ductwork systems, especially with high airflow velocity, a considerable amount of low-frequency energy may be generated that is extremely difficult to control.

5.2.3.4 Attenuation of Noise by Ducts. Various elements within the ducted ventilation system inherently provide some attenuation of the noise as it travels through the ductwork, both down the supply-air system and up through the return-air system. Certain elements, such as internal duct lining and prefabricated silencers, can be added to the system as necessary to increase noise attenuation.

Bare ducts, that is, sheet-metal ducts that lack any added sound-absorptive lining, provide a minimal but measurable amount of attenuation to the noise. The amount of attenuation depends on such factors as width and height dimensions and the length of that section of ductwork under consideration. Such duct attenuation is approximately 0.1 dB/feet of duct length for frequencies of 250 Hz and above regardless of width and height dimensions. For example, the 1000-Hz-band noise level inside the end of a 10-feet length of bare rectangular duct should be approximately 1 dB less than inside at the beginning. For lower frequencies, duct attenuation is approximately 0.2 dB/feet of duct length or even up to 0.3 dB/feet for ductwork as small as 5 to 15 inches in either width or height.

It is important to note that duct attenuation decreases in the lower frequencies because the thin sheet metal of which most ductwork is constructed is a poor barrier for low-frequency sound transmission, and as a result these low frequencies "break out" of the ductwork and into the surrounding space. Thus, it is advisable to reroute ductwork that is known to contain high levels of sound energy, especially low-frequency sound energy, from spaces that require low background noise.

Sheet-metal duct may be lined with sound-absorptive material. This material is generally of about 1½ lb/feet density and of either 1- or 2-inch thickness (the 2-inch thickness generally provides improved duct attenuation, which can be an important consideration, especially for low-frequency noise control), and it often has a mastic facing to reduce shredding and deterioration from high airflow velocities. For lined ductwork, duct attenuation is very much dependent on the width and height dimensions, and on the octave frequency band of interest. Sources such as [1] or specific manufacturers' data should be consulted for a detailed analysis of lined-ductwork attenuation. It should be noted that noise breakout, especially for low frequencies, is not significantly affected by lining.

Bends and elbows in ductwork are not very effective in attenuating low-frequency noise but can provide significant attenuation of higher-frequency noise. Lined elbows and bends provide better high-frequency attenuation than bare elbows and bends. Bare elbows may provide up to 3 dB of attenuation at 2000 Hz and above, and lined elbows can provide between 5 and 10 dB of attenuation in the higher frequencies, depending on the elbow radius and duct diameter.

5.2.3.5 Duct Silencers.

Prefabricated duct silencers generally incorporate a system of parallel sound absorptive baffles between which the air must flow (Fig. 5.1.13). Silencers are available in a wide range of sizes and duct attenuations, which tend to vary with frequency. Manufacturers should be able to provide detailed data on the performance of their silencers. These data should include the octave-band dynamic insertion loss (DIL), the pressure drop across the silencer, and the self-generated noise of the silencer. The effective attenuation of a silencer in any given octave band can change with the airflow velocity through the silencer. This is measured in terms of the DIL, which is duct attenuation in octave bands at different airflow velocities, both positive and negative. Positive DILs rate the effectiveness of a silencer when the noise and the air both flow in the same direction, as in the case of a supply-air system, and negative DILs apply where noise flows in the opposite direction of the airflow as in a return-air system.

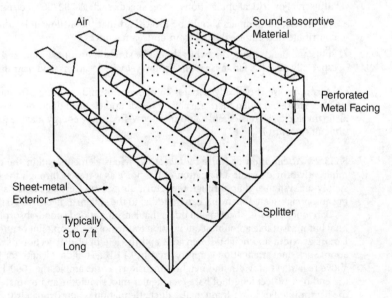

FIGURE 5.1.13 Cutaway view of a typical duct silencer.

Because the baffles in a silencer restrict the flow of air to a certain degree, the silencer can add to the static pressure against which the fan must work, so that the pressure-drop ratings of silencers can become an important consideration. Since the baffles generate a certain amount of turbulence in the airflow, silencers can generate a certain amount of noise. Silencers should be positioned so that the amount of attenuated noise leaving them is still higher than the generated noise of the silencers, which implies that the silencers should be placed relatively close to fans. On the other hand, it is good practice to locate silencers at least five duct diameters downstream of a fan in the supply-air system; otherwise noise generated by turbulent air, especially low-frequency noise, can greatly exceed the rated self-noise of the silencers. Placement of silencers in return-air systems is less critical, but a spacing of at least three duct diameters between fan and silencer is still advisable.

5.2.4 Sound Isolation[4]

Sound may travel to an adjacent space by a multitude of paths. Obviously it can travel through apertures. But if all these holes are sealed, the sound is ultimately received after vibrations have been transmitted through the building structure. Because they travel readily in solid materials, the vibrations may take long, devious paths before arrival. If high noise isolation is required, all paths that the vibration may take need to be interrupted; there is little value in blocking one path when a significant amount of the sound travels through a flanking path.

5.2.4.1 Sound Barriers.
In most building applications involving audio pickup devices a very sizable transmission loss is often required. The performance of the single homogenous wall is inadequate; doubling the mass of the wall gains only 6 dB. Often space or floor-loading restrictions also limit this option. Alternatively, increased performance can be obtained from a two-wall system such as that shown in Fig. 5.1.14, in which one wall is completely separated from the other, over its entire area, by an air gap. Reverberant sound within the interior cavity is absorbed by fiberglass. The coincidence dip is reduced by ensuring that the frequency at which it occurs is different for each wall.

Care must be taken not to inadvertently reduce the design transmission loss (TL) by tying the two walls together with a mechanical connection, such as a wall bolt or perhaps a pipe. Also, care must be taken with air leaks. Electrical outlet boxes on opposite sides of the wall must be staggered by at least 3 feet. Conduit and pipes should pass through the wall at its perimeter. The joints between the wall, ceiling, and floor must be grouted or caulked with an elastomeric compound. Never use foam caulk for acoustical isolation. If an air duct must pass through the wall, special arrangements must be made to ensure that sound does not travel into the duct wall, through the wall, and back out through the duct wall into the adjoining space.

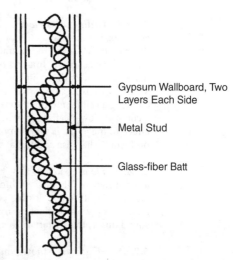

FIGURE 5.1.14 Section of a lightweight double wall.

5.2.4.2 Partial Walls and Barriers.
The effectiveness of sound barriers depends partly on the amount to which sound traveling over the top or around the sides is diffracted. In practice, the effectiveness is limited, even under the most favorable geometric layout, to about 25 dB. The material of construction does not influence a barrier's performance provided that its transmission loss is sufficient to block significant sound from passing through. For many applications, ¼-inch plywood has sufficient mass and may be used effectively, but other materials are often preferred because they meet other criteria such as weatherability or structural strength.

---

[4]This section was adapted from Cann, R. G., and A. Hoover, "Sound Isolation," in *Standard Handbook of Audio and Radio Engineering*, J. C. Whitaker and K. Blair Benson (eds.), McGraw-Hill, New York, NY, 2002.

Where barriers are also used in interior spaces, such as partitions and sets, their performance is degraded by the reflection and reverberation of sound from the adjacent walls and ceiling. In many situations, reverberant sound greatly exceeds the diffracted-sound level.

Where barriers fully or partially enclose a source, some additional performance can be gained by reducing the local reverberant field by applying absorptive materials to the inside surface of the partition and other adjacent surfaces.

5.2.4.3 Doors. The door is potentially the limiting element in noise isolation for an interior wall. Not only may the frame reduce the transmission loss of the mounting wall, but improper sealing of gaskets further reduces performance. Furthermore, a good noise control design does not assume that gaskets will remain good after abuse by substantial traffic. Alternatively, a design with two well-fitting ungasketed doors should be considered. A significant increase in performance can be obtained by placing the doors a minimum of 4 inches apart. Further improvements can be made by increasing the separation between the two to form an *air lock* in which acoustically absorptive materials are applied. The close-separation arrangement is awkward to use since the two doors must open in opposite directions. The air-lock design may be more convenient, although it does occupy more space.

5.2.4.4 Ceilings and Floors. Much of that which applies to walls also applies to ceiling and floor elements, but there are additional precautions to be taken. The addition of carpet to a floor does not decrease the airborne sound transmitted to the space below. However, carpet does reduce the generation of impact noise. Impacts on a hard floor from shoe heels and rumblings from steel-tired hand trucks are readily discernible.

If the transmission loss of the floor structure is found to be inadequate, it can be increased by applying a ceiling mounted to resilient channels, which in turn are fastened to the joists. Better still, a complete lower ceiling may be supported by resilient hangers which penetrate the upper ceiling to attach to the joists. It may be possible to use the cavity for air ducts as long as duct breakout is not a problem. Otherwise additional noise isolation for the duct must be provided.

5.2.4.5 Floating Rooms. Where maximum noise isolation is required and cost is of little concern, a room may be "floated" within a building space on vibration isolation pads. All building services are supplied through flexible connections. All doors are double, with the inner door and frame attached only to the floating room. The cavity around the room and under the floor is filled with sound-absorptive material. This type of construction is little used in commercial applications.

5.2.4.6 Windows. Most window manufacturers can supply TL data on their products. These show that single-glazed windows typically have much less transmission loss than walls, and if they represent more than a small fraction of the wall area, they are usually the controlling element when the composite transmission loss of wall and window is calculated.

Thermal glazing, two glazings with an air space, is also commonly used. However, because the panes are close together, cavity resonance restricts improved performance. But when the panes are separated by several inches and sound-absorptive material is applied around the cavity perimeter, there is a marked improvement in performance.

The local building code may require that a window with a specified sound transmission class (STC) be installed, but for noise control applications STC values are insufficient. The full spectral data should be used in any computations.

5.2.4.7 Vibration. Excessive vibration can cause several problems for audio engineers, including reradiation of vibration-induced noise from walls and direct vibration of microphone elements through mike stands. There exist quite a variety of vibration sources, such as mechanical equipment, automobile and truck traffic, and even pedestrian traffic within a building. Once vibration has entered the building structure, it may be difficult to control, and this problem can be exaggerated by resonances found in all buildings, which commonly occur in the range of 5 to 25 Hz. Thus, it is important to try to determine which sources of vibration may be problematic and to isolate them before vibration enters the building structure.

5.2.4.8 Driving Frequency. The driving frequency of the vibration source is the most important consideration in trying to develop a vibration-isolation system. It is not unusual for a source to have several driving frequencies, but it is the lowest driving frequency that is of primary concern. The lowest driving frequency of most electrical and mechanical equipment can be determined from the lowest rotational or vibrational motion. For example, a fan that operates at 1200 rpm has a lowest driving frequency of 20 Hz (this same fan may have a harmonic at 200 Hz if the fan has 10 blades).

To isolate mechanical equipment, some sort of vibration isolation device is placed between all the supports of the piece of equipment and the building structure, such as underneath each of four legs. The vibration isolation elements generally consist of steel springs, some sort of resilient material such as neoprene rubber, or a combination of the two. As the machine is installed, it compresses the mounts by an amount known as the *static deflection*. When installation is complete, it can also be seen to have its own natural frequency. The natural frequency can be thought of as the frequency in hertz at which the machine would oscillate after it was deflected from rest.

5.2.4.9 Vibration Isolation. It is important to note that several factors can significantly reduce the effectiveness of isolators. Any item that short-circuits the isolator, such as rigid conduit, can carry substantial vibrations to the building structure and seriously degrade the performance of an isolation system. In addition, if the isolators stand on a flexible floor span rather than a basement slab, reduced isolation must be expected. In this special situation more sophisticated methods than those outlined here must be used to ensure adequate performance [2].

REFERENCES

[1] ASHRAE, *ASHRAE Handbook*, American Society of Heating, Refrigerating and Air-Conditioning Engineers, Atlanta, GA, see https://www.ashrae.org/resources—publications/handbook.

[2] ANSI, *American National Standard for Rating Noise with Respect to Speech Interference*, ANSI S3.14-1997, American National Standards Institute, New York, NY, 1997.

BIBLIOGRAPHY

ANSI, *Method for the Measurement of Monosyllabic Word Intelligibility*, ANSI S3.2-1960, rev. 1977, American National Standards Institute, New York, NY, 1976.

Beranek, L. L., *Acoustics*, McGraw-Hill, New York, NY, 1954.

Beranek, L. L., *Noise and Vibration Control*, McGraw-Hill, New York, NY, 1971.

Beranek, L. L., and J. W. Kopec, "Wallace C. Sabine, Acoustical Consultant," *J. Acoust. Soc. Am.*, 69, 1981.

Cann, R. G., and A. Hoover, "Criteria for Acceptability of Acoustical Performance," in *Standard Handbook of Audio and Radio Engineering*, J. C. Whitaker and K. Blair Benson (eds.), McGraw-Hill, New York, NY, 2002.

Cann, R. G., and A. Hoover, "Sound Isolation," in *Standard Handbook of Audio and Radio Engineering*, J. C. Whitaker and K. Blair Benson (eds.), McGraw-Hill, New York, NY, 2002.

Egan, M. D., *Concepts in Architectural Acoustics*, McGraw-Hill, New York, NY, 1972.

Hertz, B. F., "100 Years with Stereo: The Beginning," presented at the 68th Convention of the Audio Engineering Society, Hamburg, Germany, 1981.

Huntington, W. C., R. A. Mickadeit, and W. Cavanaugh, *Building Construction Materials*, 5th ed., Wiley, New York, NY, 1981.

Jones, R. S., *Noise and Vibration Control in Buildings*, McGraw-Hill, New York, NY, 1980.

Kath, U., "Ein Hörspielstudio mit variabler Nachhallzeit," *Rundfunktech. Mitt.*, no. 1, 1972.

Kryter, K. D., *The Effects of Noise on Man*, Academic, New York, NY, 1985.

Lyon, R. H., and R. G. Cann, *Acoustical Scale Modeling*, Grozier Technical Systems, Inc., Brookline, MA.

Marris, C. M., *Handbook of Noise Control*, 2nd ed., McGraw-Hill, New York, NY, 1979.

Marshall, H., and M. Barron, "Spatial Impression Due to Early Lateral Reflections in Concert Halls: The Derivation of the Physical Measure," *JSV*, vol. 77, no. 2, pp. 211–232, 1981.

Moles, A., and F. Trautwein, "Das elecktroakustische Institut Hermann Scherchen," *Gravesaner Blätter*, no. 5, pp. 51–64, 1956.

Morse, P. M., *Vibration and Sound*, American Institute of Physics, New York, NY, 1981.

Noise Rating Curves, IS0 R 1996, 1971.

Remley, F. M., "Control Room Design and Layout," in *Television Engineering Handbook*, revised edition, K. Blair Benson and J. C. Whitaker (eds.), McGraw-Hill, New York, NY, 1992.

Sabine, W. C., *The Collected Papers on Acoustics*, Dover, New York, NY, 1964.

Shankland, R. S., "Architectural Acoustics in America to 1930," *J. Acoust. Soc. Am.*, 61, 1977.

Siebein, G. W., *Project Design Phase Analysis Techniques for Predicting the Acoustical Qualities of Buildings*, research report to the National Science Foundation, grant CEE8307948, Florida Architecture and Building Research Center, Gainesville, FL, 1986.

Talaske, R. H., E. A. Wetherill, and W. J. Cavanaugh (eds.), *Halls for Music Performance Two Decades of Experience, 1962-1982*, American Institute of Physics for the Acoustical Society of America, New York, NY, 1982.

Thiele, R., "Richtungsverteilung und Zeitfolge der Schallrückwürfe in Raumen," *Acustica*, 3, 1953.

Trendelenburg, F., *Einfihrung in die Akustik*, Springer-Verlag, Berlin, p. 123, 1939.

Vermeulen, R., "Akustik und Elektroakustik," *Gravesaner Blätter*, no. 7, 1960.

Vermeulen, R., "Stereo Reverberation," *Philips Tech. Rev.*, 17, 1956.

Voelker, E. J., "Akustik und Aufnahmetechnik im modernen Hörfunk- und Fernseh-studio—Anforderungen im Wandel," *NTG Fachber.*, vol. 56, VDE-Verlag, Berlin, Germany, 1976.

Voelker, E.-J., "Studio Production Systems," in *Master Handbook of Audio Production*, J. C. Whitaker (ed.), McGraw-Hill, New York, NY, 2003.

Voelker, E. J., and M. Brückmann, "Raum- und Bauakustik des neuen Funkhauses des Hessischen Rundfunks in Kassel," *Fortschr. Akustik*, pp. 439–442, VDI-Verlag, Düsseldorf, Germany, 1973.

CHAPTER 5.2
AUDIO SYSTEM INTERCONNECTION

Greg Shay and Martin Sacks
The Telos Alliance, Cleveland, OH

5.1 INTRODUCTION

First, and for a long time, there was just analog.

For the last half of the twentieth century, audio interconnection in broadcast facilities changed little. Broadcast plant wiring was primarily a "one wire per source" (analog) world based on 600 Ω wiring (derived from the telephone industry) until the AES3 digital format was introduced in 1985. AES3, also based on twisted pair wire technology, proved little different than analog—in the minds of most—because the signal density was still low (a single left and right channel in a one pair wire) and AES3 digital did not offer any additional enhancements. MADI (Multichannel Audio Digital Interface) (AES10), introduced in 1991, increased the density to 56 channels in one coaxial cable, but most radio facilities were not able to take advantage of the format because it relied on a type of cabling that is not normally found in radio facilities. As a result, MADI was more frequently implemented in television facilities and recording studios to support large audio mixing desks.

5.1.1 Then Came AoIP

Since 2004, a sea-change has taken place in audio infrastructure for broadcast. Traditional 600 Ω balanced and AES (Audio Engineering Society) wiring has given way to *audio over IP* (AoIP). This is notable because it is the first time that "off-the-shelf" components (not just technology) from a different industry have been installed in technical operations centers (TOCs), control rooms, and studios worldwide. Public and private broadcasters with facilities that range from one to 50+ control rooms are now operating AoIP-enabled console and routing systems. More than 5000 AoIP-enabled consoles and over 75,000 AoIP devices have been installed in broadcast facilities in the last ten years.

AoIP for broadcast is based on the ubiquitous Ethernet standard (IEEE 802.3) and made possible by high performance Ethernet switches created for use by corporations and governments that sought to marry all communications onto a single network, which includes the vitally important telephony function of voice over IP (VoIP). These newer generation Ethernet switches were optimized for VoIP through careful design that optimizes throughput and eliminates port and backplane blocking, the bugaboos of earlier Ethernet switching gear. Any kind of blocking is unacceptable for real-time communications—and absolute deal breakers when moving audio through a network, since dropped audio packets lead to audio interruptions.

However, even with optimized Ethernet switches, many hurdles remained to create an AoIP system that was truly ready for broadcast. The biggest involved creating a timing system to synchronize audio throughout the network. Timing had never been done in Ethernet up to this point and a system to accomplish this was created as the Livewire™ AoIP protocol in 2000.[1]

The first AoIP systems were installed in 2003.[2] While the Telos version of AoIP was the first developed and first to be commonly installed in broadcast facilities, a number of other manufacturers have joined in subsequent years. All modern AoIP variants are based on OSI Layer 3, which has a primary advantage of being able to route audio to and from outside the local area network (LAN). As broadcasters frequently find themselves dealing with sources that originate outside the local facility (interconnecting feeds from multiple cities, for example), Layer 3 capability is considered a minimum requirement of a broadcast audio network.

There are now a number of firms selling different versions of AoIP gear and the range of offerings is quite impressive. In 2015, as this is being written, AoIP is installed in the majority of new and rebuilt facilities. In fact, AoIP has now been codified by the Audio Engineering Society (AES) as standard AES67-2013. This standard is designed to codify profiles that manufacturers can build into their products to allow basic interoperability amongst systems from different providers. Currently, this is limited to audio interoperability, but that will likely change with additional standards, like AES70.

Though early in its life as a standard, AES67 holds great promise because interoperability gives broadcast customers more choices in what they buy. Currently, the primary AoIP formats for broadcast are Livewire and RAVENNA,[3,4] but there is also equipment installed using other implementations. (There is also the AVB standard, based on OSI Layer 2, which—as AoIP systems are all Layer 3 implementations—is beyond the scope of our discussion here.)

From a practical standpoint, the main distinctions among the various AoIP versions have to do with how the audio packets are encoded, how the available streams on the network are advertised and what, if any, capability exists to route GPIO (closures) with the audio streams to provide start/stop capability that is commonly used by broadcasters.

But we are getting ahead of ourselves. We now present an overview of AoIP.

5.2 AUDIO OVER IP—A PRIMER

AoIP offers a revolutionary change in how studios can be built. But at the same time, it is a natural continuation of general trends and what you already know. This section explains the basics and puts AoIP into context.

5.2.1 Why Ethernet?

Ethernet makes overwhelming sense. Today's computers are near universally linked via Ethernet—and telephony is gradually moving that way as well, with VoIP rapidly gaining market share. Even remote-controlled stage lighting is transitioning from the XLR-based DMX protocol to Ethernet. Ethernet cables, plugs, cards, and chips are produced in the hundreds of millions so we get tremendous economy of scale. We get patch bays and cords, testers, and all kinds of "structured wiring" components "off the shelf." Plugs are easy to install and jacks are efficiently small.

But much more important is that Ethernet allows us to combine many channels of digital audio with whatever data transmission we might need on a single cable. This data could be as simple as a start command for an audio player or could be anything that computers and Ethernet do, such as file transfer, e-mail, web communication, etc.

---

[1] Developed by Telos Systems founder Steve Church, along with co-author Greg Shay, and other colleagues at Telos Systems.

[2] Produced by Telos Alliance as the Axia brand.

[3] All product names, trademarks and registered trademarks referred to hereafter are the property of their respective owners.

[4] Developed by ALCNetworX.

Furthermore, we are in the line of future development. Since its invention over 30 years ago, Ethernet has been constantly evolving. It started as a 2 Mbps shared bus over coaxial cable and has grown to today's modern (and very common) 1 Gigabit star and switched system. 10 Gigabit is already widely available on many Ethernet devices and is following the usual curve to low cost as volumes increase. 40 and 100 Gigabit systems were standardized in 2010. While copper is the most common Ethernet connection, fiber is popular as well, and media converters allow the two to be interconnected. Ethernet switches cost $6000 for eight ports several years ago; now high-end 24-port switches cost $500. And they include many advanced features that were unheard of only a few years back.

There are radio links in many varieties, from WiFi for short-range to sophisticated long-range systems.[5] There are satellite links, and LASER links. Ethernet opens the door to a world of options.

Ethernet has proven to be the PC of networking: initially released with only basic capability—low speed and bussed—it has been expanded to today's fast, flexible, switched architectures.

The combination of huge research and development expenditures, open standards, massive economies of scale, technological evolution, and flexible multiservice packet design is hard to beat. Not to mention the surprisingly appropriate name.

5.2.1.1 A Little History. Ethernet was named by its inventor, Robert Metcalf. He had been involved in a radio data network in Hawaii called ALOHA. The first Ethernet was a bussed coax that carried data packets similar to the way ALOHA had sent them—over the "ether."

As to the origin of ether, for many years after James Clerk Maxwell's discovery that a wave equation could describe electromagnetic radiation, the "aluminiferous ether" was thought to be an omnipresent substance capable of carrying electromagnetic waves. In 1887, scientists Albert Michelson and Edward Morley disproved its existence. The ingenious experiment that did so was performed at Case Western Reserve University, in Cleveland, Ohio.

5.2.2 Compared to AES/EBU

As mentioned previously, stereo digital audio transport (also known as AES3) is the main alternative to an Ethernet-based system. Invented in the days of 300-baud modems, it was the first practical answer to connecting digital audio signals. But it's now nearly 20 years old and is showing its age. Compared to AES67 AoIP's computer-friendly, two-way, multichannel + high-speed data capability, AES3 looks pretty feeble with its two-channel and one-way only limitation. Not to mention 50-year-old soldered XLR connectors and lack of significant data capacity. AES3 is a low-volume backwater, with no computer or telephone industry R&D driving costs down and technology forward. Your 300-baud modem has been long retired—that is if you even had one! It's well time to progress to the modern world for studio audio connections as well.

That said, AES3 and AES67 AoIP may comfortably coexist in your facility. You can use AES67 interface end points[6] to interconnect between one and the other. And, in the rare cases that you are using a house sync system for AES3, the AoIP system may be synced to that master timebase as well.

5.2.3 Audio Routing

Low-cost mass-market Ethernet switches offer us something very interesting: Since their function is to direct packets from port to port, we can use them to move our audio signals from whatever source to whatever destinations we want. This means we get a simple, flexible, facility-wide audio routing system, almost for free. Say goodbye to racks of distribution amps or expensive proprietary mainframe audio routers with lots of single channel per pair wiring. An audio source entered into the system from any point becomes available for any number of receiving destinations. This makes audio routing available to any facility that is AoIP enabled.

---

[5] Dragonwave is one supplier.
[6] For example Axia Audio's xNodes.

5.2.4 AoIP and PCs

One of the advantages of an AoIP system is that PC-based and server rack-based audio may be directly connected to the network without soundcards. This means radio station delivery systems can use the Ethernet connection they already have to send and receive audio. Soundcard problems such as noise and in some cases modest audio performance, as well as multiple A/D conversions, are avoided—the audio remains in digital form from the PC's files to the network with no alteration or degradation. Audio received by the PC may have originated from another PC or from a hardware audio node. Audio sent from a PC may be received by other PCs or hardware nodes.

With so much audio in radio and TV stations being either played from computers or recorded into computers, is not this a tremendous advantage? Not only do you save the soundcard (and having to replace the card when PC bus architectures change rendering the old card obsolete) and also the port that it needs at the other end to connect to your console or router. And you can pass control and other information over the same connection. A truly data rich interface.

5.2.5 Support for Surround

Multichannel sound has been the standard in television for many years, arriving with HDTV (high-definition television) and DVDs (digital versatile discs).

While surround for radio broadcast is not mainstream at the time of this writing, it is possible. Were broadcasters interested in moving in this direction, they would find that using AES3 is an impractical and expensive way to handle multichannel audio. The reason is that the 7.1 system needs eight channels: two front, two surround, two rear surround, one center, and one subwoofer. And, if required to keep a separate stereo-mixed version independently, there could be 10 total audio channels. Using the traditional one-wire-per-channel approach, that's a lot of plugs, cables, router cards, and rack space.

For TV facilities, the *status quo* SDI or HD-SDI multiplex for audio transport carries with it the architectural limitation of point to point connection routing.

On the other hand, Ethernet has plenty of bandwidth to carry the multiple channels surround broadcasting requires. All eight channels, plus associated control, can easily be conveyed on one convenient Ethernet cable.

Expanding a traditional console or audio router from stereo to eight channels would be either impossible or very expensive. Not so with Ethernet and AoIP. In fact, there is no additional cost for the core Ethernet switch because the one you need for stereo would also be fine for surround. Audio from PCs and servers can also easily be multichannel.

Designing a facility with AoIP is designing with the future well in mind, and nearly without risk; it already provides the infrastructure for many thousands of broadcast studios worldwide.

5.2.6 Audio Quality

We are often asked, "Is AoIP like audio on the Internet?" Yes and no. While AES67 AoIP uses Internet transport standards that is where any similarity ends. Internet streams are usually compressed for transmission over public links with limited, variable bandwidth and low reliability. AES67 AoIP audio is not compressed—it uses studio-grade 48 kHz/24-bit PCM encoding (and, for the future, is capable of even higher resolution).[7] LANs offer a safe, controlled environment where there is no risk of audio drop-outs from network problems, and plenty of bandwidth to carry many channels of high-quality, uncompressed audio.

Indeed, we often hear from users that they notice an improvement in fidelity when they transition from traditional interconnection systems. This is probably due to the direct connection of PCs

---

[7]For example, Axia xNode audio interfaces have more than 108 dB dynamic range, less than 0.005 percent THD, and headroom to +24 dBu.

(no A/Ds), the 64-bit accumulator in the mixing engines (very precise hardware platform), and to the ultra-careful design of the audio stages of the nodes. Frankly, good audio design is a bit of a lost art—there are devices available that are truly recording-studio quality. In any event, the network takes nothing away from audio performance; it behaves like a very capable wire for our purposes.

5.2.7 Delay

In packet-based systems, delay (latency) is an important issue and certainly has an effect on talent's perception of "quality." Packetizing audio for network transmission necessarily causes delay, and careful design of the system is required to reduce this to acceptable levels. Internet audio delay is often multiple seconds because the receiving PCs need long buffers to ride out network problems and the delays inherent in multiple-hop router paths.

However, with fast Ethernet switching on a local network, it is possible to achieve very low delay. To do this, we must have a synchronization system throughout the network. This also avoids sample or packet slips that cause audio dropouts. Internet streaming does not use this technique, so even if it were to have guaranteed reliable bandwidth, you still could not achieve the very low delay we need for professional studio applications.

The early work that Telos did in DSP based on-air processing was instructive in the design of the Livewire AoIP system. Specifically, the amount of audio delay in the microphone-to-headphones path for live announcers must be carefully considered. Maximum total delay must be held to around 5 ms (or less), otherwise announcers will start to complain of comb-filter or echo problems when listening to themselves through the system. Most implementations of AoIP are created so that an approximate 1 ms latency exists between one part of the system and another in the microphone path, and also in the headphone path, for an approximate total latency of 2 ms—well within the undetectable range as shown in Table 5.2.1.

TABLE 5.2.1 Effects of Various Latency Ranges
(air-talent reactions to delay in a test conducted by Jeff Goode at WFMS in Indianapolis, IN)

Delay	Effect
1–3 ms	Undetectable
3–10 ms	Audible shift in voice character (comb filter effect)
10–30 ms	A slight echo turning to obvious slap at 25–30 ms
30–50 ms	Disturbing echo, disorienting the announcer
>50 ms	Too much delay for live monitoring

Another way to think about delay: audio traveling 1 foot (0.3 meter) in air takes about 1 ms to go this distance. And another data point: A common professional A-to-D or D-to-A converter has about 0.75 ms delay.

But, as is universally the case in engineering, there is a tradeoff—otherwise known as the "if you want the rainbow, you gotta put up with the rain" principle. To have low delay in a packet network, we need to send streams with small packets, each containing only a few accumulated samples, and send them at a rapid rate. Bigger packets would be more efficient, because there would be fewer of them and they would come at slower rate, but they would require longer buffers and thus impose more delay. Big packets would also have the advantage of applying necessary packet header overhead to more samples, using network bandwidth more effectively.

With AoIP, we enjoy our rainbow and avoid the rain by having different stream types: *low latency* streams use small and fast packets, while *standard latency* streams have bigger, slower packets.

Low latency streams typically require dedicated hardware (or at least a very specialized operating system) to achieve the required very low delay for microphone-to-headphone paths. Standard operating systems in PCs are not yet able to handle these small packets flying by so quickly; therefore, they use the standard latency streams, which the AoIP network may receive directly from PCs running delivery software.

The network delay in this case is around 5 ms, and the PC's latency is likely to add perhaps 20–50 ms more. Since PCs are playing files and are not in live paths requiring low latency streams, this is not a problem. Our only concern is how long it takes audio to start after pressing the *On* button, and delays in the range of a few tens of milliseconds are acceptable.

Standard latency streams can also be sent from the network to PCs for listening and recording. Again, in this case the delay of the standard latency stream is not an issue—especially given that PC media players have multiple seconds of buffering. However, off-the-shelf PC hardware with a specialized operating system and software optimized for real time are able to handle the fast streams. Indeed, this approach can be used for a studio mixing and processing engine.

Audio node devices may transmit both stream types and can receive both stream types. There is no inefficiency from having both available, because all streams stop at the Ethernet switch and use no system network bandwidth unless they are subscribed to by a receiver or node. Each receiver takes only the stream needed: the low-delay version if available, or the higher-delay version if not. The selection happens transparently with no user action needed; users simply select the channel they want and audio is delivered by whichever method is appropriate for the equipment they are using.

AES67 AoIP's low-delay streams are also fixed-delay. The delay is constant, regardless of the system size or anything else. In fact, a source being received by multiple subscribers will have a differential delay of typically less than 5 μs—less than one-fourth sample at an audio sample rate of 48 kHz.

5.2.8 Synchronization

For AoIP, we generate a system-wide synchronization clock that is used by all nodes. Within each node, a carefully designed PLL system recovers the synchronization reliably, even in the case of network congestion. For AES67 (as well as the RAVENNA and other AoIP protocols) the network synchronization used is the industry standard IEEE 1588 PTP (Precision Time Protocol), 2008 version.

The IEEE 1588 synchronization standard is widely used across multiple industries to accurately distribute time, with the right equipment down to nanosecond precision.[8] For AoIP, the synchronization accuracy desired is typically some fraction of an audio sample, or ± a handful of microseconds.

A master clock device provides this clock, and in each system there is one master node that sends the clock signal to the network. If it should be disconnected, or stop sending the clock for any reason, another device, another IEEE 1588 grandmaster clock, an IEEE 1588 boundary clock, or in the case of Axia equipment, even one of the networked audio devices, can be configured to automatically and seamlessly take over providing clock.

In cases where an AoIP network's digital sampling system must be synchronized with external systems such as digital video, the two systems may be referenced using the same master clock. For long-distance absolute synchronization reference, all sites can use a GPS-derived master clock.

Note that AoIP sync technology, IEEE 1588, is based on communicating precisely synchronized wall clock time to all locations. Audio and video media sample rates are derived from this common wall clock. Drop frame or vari-speed sample rates do not speed up or slow down the IEEE 1588 clock, which always stays universally constant; the conversion to various media sample rates is done in the devices themselves.

For interchange with AES3 equipment, to avoid passing audio through sample-rate-converters, the AES3 sync generator should use the same master reference as the AoIP IEEE 1588 grandmaster.

---

[8] For additional information see Chapter 8.3, "Time and Frequency Transfer over Ethernet Using NTP and PTP."

5.2.9 Source Advertising

Given the capabilities of an AoIP system to deliver hundreds or even thousands of audio channels, managing this system and *finding* the wanted channels becomes key to getting day-to-day work done. The answer is to use the network to distribute a directory of the multitude of devices and audio channels that are present. This is commonly called *Source Advertising*. Each device announces itself and the audio sources it provides; a list of all devices and audio channels may then be gathered for presentation anywhere in the system.

The AES67 standard defines the format of the AoIP stream itself (the appendix suggests commonly used discovery protocols), but stops short of defining a standard source advertising system, leaving this to be implemented separately.

Multiple vendor source advertising solutions exist. Axia's Livewire audio advertising system multicasts every source's text name and numeric channel number. RAVENNA and Dante both make use of the Apple Bonjour service discovery system, common with many computer network peripherals. A future standard for source advertising, or one of these vendor supplied systems, may become dominant.

5.2.10 Control

Most audio these days needs associated control. A delivery system needs a start input at minimum, but could well benefit from a richer control dialogue, such as "now playing" text that can be sent to the studio mixer and to HDRadio, DRM, and RDS encoders. Satellite receivers have control outputs. Telephone systems need dialing, line status, hold, transfer commands, etc. CD players need a "ready" indication out, and "start" in. Even the simplest source—a microphone—needs to convey on/off status, via the on-air lights. Most conventional controls have been done with primitive GPIO parallel "contact closures."

The same Ethernet network carrying AoIP content is ideally suited to also carry control data, reducing and simplifying cabling, replicating traditional start/stop control. But it continues from there; the network allows sophisticated remote operation of studio equipment, transporting much more advanced information than just simple start commands. For example, Axia can send the song title from a delivery system to a display on a mixing console's fader channel. Control of telephone systems and codecs can follow fader assignment and be accessible from any location.

Individual vendors have each developed integrated, network control-enabled systems that deliver on these promises. The newly issued AES70 standard control protocol for audio equipment (November 2015) has the potential to become universally accepted. More users adopting AoIP will drive demand.

5.3 WHAT CAN YOU DO WITH AOIP?

Imagine everything that you can do with a PC connected to a network: share files, send and receive emails, chat, surf the web, listen to audio, etc. PCs and networks are designed to be general-purpose enablers. There is a similarly wide range of possibilities for audio applications using AoIP. Examples are given below, starting with the most simple, and continuing to the most interesting.

Make a Snake—Concert sound guys need to get a lot of audio from the stage to their mixing consoles in the center of the house. They call the multiconductor cables they traditionally use for this function a "snake." AoIP lets you put such a snake on a diet! A single Ethernet cable connects multiple audio channels. Add a switch at each end and you can have as many nodes as you want. Use Gigabit Ethernet and you can have hundreds of channels. Add fiber optic media converters and cable to extend the distance between units to many kilometers.

Make a High-Performance Sound Card Replacement—AoIP can talk directly to PCs, making the network look like a sound card to delivery systems, editors, etc. Axia's AES67 capable xNodes have excellent audio performance: balanced I/O with better than 108 dB dynamic range, < 0.005 percent distortion, headroom to +24 dBu, etc. They make excellent multichannel "soundcards" for professional applications. You can position the node at a distance from the PC and you get balanced audio on connectors that are a lot more reliable than the mini phone jacks found on traditional soundcards. With the addition of an Ethernet switch you can feed your audio to multiple computers and/or have multiple I/O boxes, which take us to the next application...

Build an Audio Router—A system with AoIP nodes, one or more Ethernet switches, and PC-based routing controller software make an excellent facility-wide audio router. PCs send and receive audio directly to the network without soundcards or audio ports, thus lowering costs and eliminating conversion steps. Processing, mixing, telephone interface, and codec equipment from many vendors is now able to connect directly. To connect conventional analog and AES3 signals to AoIP, interface nodes come in a number of versions. One node operates like a traditional audio router X–Y control panel. But with a difference: audio in and out is available on the same box.

Router control packages are available that make your whole system look like a single entity. You can control which outputs are connected to which inputs just as if the system were a single location box.

Since there is no requirement for a mainframe, the base cost is low; you can make a small system at very reasonable cost and expand it over time. Indeed, the total cost of a large system will be much lower than older approaches due to the use of Ethernet switches at the core that are a fraction of the cost of purpose built broadcast TDM routers. Just as using standard PCs to play audio makes much more sense than a proprietary approach, building routers from common computer industry parts makes similar sense. Indeed, this approach gives you a true "audio network" quite unlike other approaches.

5.3.1 Build a State-of-the-Art Broadcast Studio

Plug an audio processing engine and a control surface into the network and you have a modern radio studio with many advantages over the old way:

- Simplified and unified cabling for audio, control, general data, and telephone.
- No multiple conversions. With most studio audio coming from or going to PCs, audio is kept in the networked digital domain. Audio may be monitored on any PC with a standard media player.
- Integrated data means you are ready for synchronized text and metadata, which is needed for HDRadio or DRM. It will also be possible for audio processor parameters to be controlled depending on source characteristics.
- Tighter integration with delivery systems means that mixing, scheduling, and playing can work together. For example, song titles can appear on the mixer surface, start and other control functions may be conveyed over the network, and logging can confirm that an audio piece was really played on the air.
- Troubleshooting and repair are transformed. Extensive diagnostics are available over the same network that connects the audio. A suspect surface or engine may be swapped by replugging only one Ethernet cable.
- Low-cost power. Computers replaced cart machines because they are a lot more powerful, convenient, reliable, and cheap. The technical side of radio broadcasting is tiny compared the computer and networking industries. We get tremendous value by plugging into the massive R&D and production scale offered by the computer world. Leveraging low-cost mass-produced computer components makes the same sense for studio mixing and audio distribution as it did for cart machine replacement.

For example, an AoIP-based console used in a studio can be plugged into an AoIP capable Ethernet switch along with a converter (node) to allow a handful of peripherals such as microphones,

CD players, a recording device, etc. to be used as local sources, while sources external to the studio—such as satellite network feeds, OB van remotes, and so forth—can interface with a node located in a TOC rack or equipment room.

Certain peripheral equipment connects directly to the network. Audio from, and control of, the delivery PC connects to the network via an Ethernet connection. The network also supports file transfers to the delivery system from a server. The studio operator's surface controls a rack-mount mix engine, which has a single Ethernet connection for both control and audio.

5.3.1.1 Make a Flexible Two-Way Multichannel STL. Studio and transmitter sites may be linked with "Ethernet STLs." AoIP audio nodes provide audio interface to Ethernet point-to-point radios. Today, these are available as off-the-shelf products.[9] In addition to the audio, anything that can be carried over Ethernet can be conveyed over the radio link, such as VoIP telephones, email, file transfers, and transmitter remote control.

5.3.1.2 Create a Facility-Wide Audio Network That Includes Integrated Studio Consoles. Combine all of the above for maximum power, convenience, and flexibility. You get facility-wide audio routing, state-of-the-art studio mixing, a single wiring infrastructure for audio, computer data, control, and telephone.

Audio processors with AES67 ports may easily have multichannel outputs for simultaneous analog FM, HDRadio, and low-delay monitoring feeds. A single Ethernet connection would serve for all needed inputs and outputs. With data capability carried alongside the audio, it would be possible to control processing parameters depending on which audio source is active.

5.3.1.3 Create an Integrated National/Local Radio Network. Imagine all those satellites transmitting IP packets. Now live audio, audio to be stored for later play, and identifying data can be delivered. This transformation of radio networks into something much more interesting, useful, and powerful has been enabled by AoIP, and would not have been possible without it. Include an Internet-based distribution and return path, adding new dimensions. What further transformational changes are possible by leveraging the worldwide IP network infrastructure?

5.3.2 Planning the Network for AoIP

There are multiple ways to build an AoIP network. For many applications a simple one-switch layout will be perfectly sufficient. Others will want to build sophisticated networks to support multiple studios and perhaps hundreds, or thousands, of audio channels. Fortunately, Ethernet scales easily—and therefore so does an AoIP installation.

The following sections provide some examples and ideas to get started.

5.3.2.1 Simple One-Switch Network. Common 1U switches can have as many as 48 ports. At an audio channel capacity of 20 stereo streams in 100BT, or 10× that for GigE, that's a lot of audio!

Let's consider a setup that supports both an on-air studio and adjacent production studio. The switch might be a standard 24-port 100BASE-TX + 2-port 1000BASE fiber version.

An AoIP node with mic-level inputs resides in the on-air studio, and another AoIP node with line inputs is in the rack. The production studio connects with a router-version node, which has one send channel and a selectable receive channel. A network connected GPIO node supplies plenty of inputs and outputs for starting CD players, lighting on-air lamps, remote mic on-off, etc.

The mixing engine connects with a 1000BASE-T copper link to one of the two 1000BASE-T SFP ports with a standard copper transceiver module. The delivery PC connects directly to the audio network with an IP audio software driver. Control for the delivery PC is directly over the

---

[9]Products are available from companies such as Ubiquiti, HP, Cisco, Antaira, Ruckus, Motorola, Dragonwave, MikroTik, and so on.

network. Servers and additional PCs can be connected to the switch for file storage and delivery systems.

Peripherals such as codecs, telephone systems, and satellite receivers may be connected into the network wherever it is convenient. For example, an AoIP-enabled codec and phone system may be attached directly to the switch; many Telos and Omnia devices (as well as those of other manufacturers) can be attached directly with AoIP connection ports. Equipment without AoIP can be interfaced through an audio node, for example, a satellite receiver (although it is worth noting that satellite receivers with direct AoIP outputs are also available).

This could be expanded to two consoles and engines to support two studios since the switch has two 1000BASE-T ports. Or, as described next, an additional 1000BASE-T switch can be used to support as many studios as desired.

5.3.2.2 Hierarchical Multiple-Switch Network.
Moving up a bit in scale, we now examine a layout that is typically used to support everything from a medium size to very large facility. A Gigabit "core" switch resides at the center, and 100/1000 switches are used at the "edge," with one for each studio, or perhaps shared between a pair of smaller studios.

A router-version AoIP node is kept in the central equipment room for test and monitoring. Additional nodes could link audio from non-AoIP studios.

While the mix engines might be plugged into the central switch, keeping them coupled to the individual studio switches helps eliminate single failure points for each studio. If the central switch were to go off line, each studio continues running stand alone.

Gigabit links are used between the edge switches and the center. These could be copper or fiber with a suitable switch using fixed ports or SFP (Small Form-factor Pluggable) modules.

The physical location of the switches is a matter of taste and tradeoff. Putting the edge switches near the studios saves cable runs, but locating all the gear in a central room simplifies engineering activities.

At this time, an appropriate switch for the core typically costs less than US$2000 and the studio edge switches are under US$700. So, the two level hierarchy is a quite reasonable cost option that provides a lot of power, flexibility, and expandability. Dozens of studios and thousands of audio channels are possible.

5.3.2.3 Options for Redundancy.
Ethernet switching has a built-in scheme for redundancy, called *Rapid Spanning Tree* and standardized as 802.1w. Switches with spanning-tree enabled exchange information with each other about the topology of the overall network. There can be redundant backup links that are automatically activated in the case that a main link has failed. Depending on the switch and layout, using Rapid Spanning Tree, it could take as little as a second or as much as a 15 s for a redundant link to be connected.

Link aggregation (sometimes called *port trunking*) is another method. With Spanning Tree, even if there are two links between two switches, only one of them at a time will be active. But it's often better to have both active simultaneously because this gives twice the bandwidth during normal operation and instantaneous backup should one fail. The link aggregation standard is 802.3ad. To be used this has to be specifically enabled on the switch. There are some vendor-specific forms of link aggregation, for example Cisco 'EtherChannel' that can switch traffic away from a down redundant link in a few hundred milliseconds. Link aggregation is supported on some PC network interface cards intended for servers, so it is not only for switch-to-switch links.

For power supply protection, most Ethernet switches offer a redundant power supply option. Nodes from some manufacturers contain multiple network connections and/or multiple power sources to add redundancy to the system.

Network-enabled software can offer levels of supervision and automation to reroute audio during various kinds of failures.

Besides automatic online redundancy, there is also the old standby manual swap-out as a reasonable option. Because RJ-45s are so easy to unplug and replug, and because switches and other AoIP components can be much cheaper than traditional TDM alternatives, you can have spare units on the shelf for fast substitution.

5.3.2.4 Fiber. Fiber optic links can extend the range of Ethernet. Because they are not subject to crosstalk and magnetic interference, fiber optic links can solve problems that might crop up in difficult locations with copper cables.

External media converters can be very simply plugged in line with AoIP nodes and switch ports to convert copper connections to fiber.

Modern Ethernet switches often have the option to plug a media converter directly into a special socket so that fiber may easily be connected from switch to switch. This is useful to make high capacity backbone links without any external boxes.

One typical switch has four "uplink" ports for use with 1000BASE-T SFP (Small Form-factor Pluggable) copper for fiber transceiver modules. The device below it is a typical modern media adapter in the "SFP/mini-GBIC" size—about the same in width and height as an RJ-45 jack.

The SFP modules come in different "flavors," for 1000BASE-SX, 1000BASE-LX, etc., that need to be matched to be compatible with the type of fiber media. Fiber media can be ST multimode fiber for up to 2 km range or SC single mode fiber that can extend up to 75 km. Generally, SX links have a range to 500 meters, LX to 5 km, and LH to 70 km.

5.3.2.5 Radio Links. There are IP capable Ethernet 802.11a/n radios with surprisingly high bandwidth—and at surprisingly low cost. Most of these operate in the unlicensed ISM bands, but with modern spread-spectrum technology and elevated directional antennas, interference does not look to present much problem. Bitrates greater than 1 Gbps and distance to 100 km+ are possible depending on power level, antenna, and terrain. For co-owned stations that are not colocated, these could be an effective way to link studio facilities.

Note: Some of these radio link systems are optimized for speed and are not capable of achieving true wired LAN low error rates, so some caution is in order when using for AoIP.

For studio-to-transmitter link, remote pick-up, and studio-to-studio applications, IP radios offer the capacity for many dozens of uncompressed audio channels, two-way transmission, and the ability to multiplex VoIP telephone, remote control, and general data. When audio and general data are mixed, the Ethernet switch ahead of the wireless link provides the prioritization function. As with all network-connected devices, the far end can be checked with a web browser on a network-attached PC, saving a drive to the transmitter site.

5.3.3 Designing for Security

You will have 100 percent security if you keep the AoIP network completely isolated from any other network, local or wide area. Those very concerned with protecting the studio system may well want to take this approach.

But there are advantages to sharing with or linking to an office or corporate network. You can configure and monitor the system from any connected PC and audio can be monitored on any desktop with access. In this case, separate switches or VLANs can be used to provide isolation. An IP router passes only the correct packets from one to the other and thus provides a firewall function.

The next step up in connectivity would be to have a network linking co-owned or otherwise affiliated stations. In this case, a network engineer is probably in the picture and he/she can take the necessary steps to protect your audio.

Connection to the public Internet brings the advantage that you can monitor and configure from a remote site, but you now have much risk from unwanted intruders, viruses, etc. A qualified network engineer should be consulted to be sure you use an appropriate firewall and other protection.

In AoIP devices, web and Telnet access can be password-protected to provide some measure of security. Stronger security techniques like SSL (Secure Sockets Layer) should be used for devices designed to be exposed to the public Internet without external protection.

Example configurations are illustrated in Figs. 5.2.1 and 5.2.2.

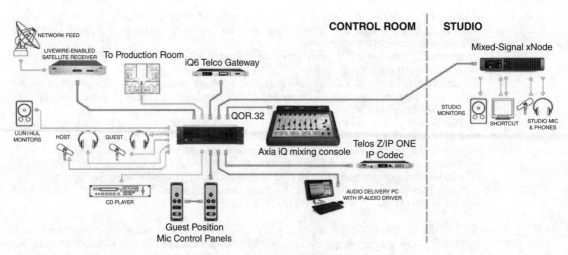

FIGURE 5.2.1 A small AoIP installation, consisting of a single control room with guest studio. (*Courtesy of Axia Audio/The Telos Alliance.*)

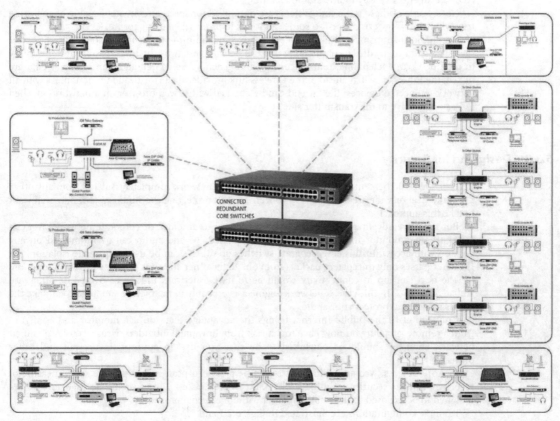

FIGURE 5.2.2 A large, whole-plant network including multiple control rooms, studios, and utility stations. (*Courtesy of Axia Audio/The Telos Alliance.*)

5.4 NETWORK REQUIREMENTS

With AoIP, the network itself brings a large part of the "magic," the special capabilities that are the reason AoIP is valuable. AoIP devices put audio into a compatible format to allow the IP network to do its job. As a result, correct capability and configuration of the network is a requirement to get end to end quality audio.

5.4.1 Quality of Service (QoS)

An important concept in a converged network is Quality of Service. When general data is the only traffic on a network, the concern is only that the available bandwidth is fairly shared among users and that the data eventually gets through. But when studio audio and general data are sharing the same network, all the required steps need to be taken to be sure audio flows reliably.

System-wide QoS is achieved with the following components, each contributing a part of the whole:

- **Ethernet switch**—Allows an entire link to be owned by each node. Isolates traffic by port.
- **Full-duplex links**—Together with switching, eliminates the need for Ethernet's collision mechanisms and permits full bandwidth in each direction.
- **Priority assignment**—Audio is always given priority on a link, even when there is other high-volume nonaudio traffic.
- **IGMP (Internet Group Management Protocol)**—Ensures that multicasts—the audio streams—are only propagated to Ethernet switch ports that are subscribed.
- **Limiting the number of streams on a link**—Nodes have control over both the audio they send and the audio they receive, so they can keep count and limit the number of streams to what a link can safely handle.

The result is rock-solid QoS, combined with the ability to share audio and data on the same or interconnected networks.

5.4.2 The Ethernet Switch for AoIP

AoIP packets include both Ethernet and IP headers. This means that AoIP streams may be either "switched" at layer 2 or "routed" at layer 3. For most installations, it is recommend to use a "managed Layer 2 switch" or a "Layer 3 switch" that includes the required IGMP Querier and snooping functions. IP Routers are able to do layer 2 switching as a subset of their more advanced capabilities, so may also be used.

Requirements for the Ethernet switch are:

- **Sufficient backplane bandwidth**—It is required to be fully "nonblocking" to handle all ports at full capacity.
- **Sufficient frame forwarding rate**—AoIP low latency streams have small packets at a fast rate. The switch needs to handle this in hardware. Software-based routers typically cannot handle the high volume of small packets associated with multiple low latency AoIP streams.
- **Correct handling of IEEE 802.1p/q frame prioritization or DSCP DiffServ**—Audio frames must be given priority without too much delay or jitter. The Ethernet standard specifies 8 levels of priority, but few switches support 8 separate priority queues. Many support only 2 or 4, lumping some of the incoming levels together. Four priority queues are recommended as the minimum for an AoIP system.

- **Support for multicast**—With sufficient address filter entry capacity to cover the total number of audio streams you need. This is important, because when the filter capacity is exhausted, switches forward multicast packets to all ports, subscribed or not. This would cause serious flooding problems. You will probably want a multicast MAC address table size of 512 minimum.
- **IGMP V2 control for multicast**—Traffic must be under IGMP control—strictly no flooding of ports with multicasts under any circumstances, even during switch powering on and initialization.
- **IGMP Querier must hand off correctly** between the higher priority core switch and edge switches. When disconnected from the core, the edge switch must take over becoming IGMP Querier, and when reconnected to the core, the core resuming being the Querier.
- **When planning a combined network, support for port-based and tagged-frame-based VLAN**—Tagged frames is the IEEE 802.1Q standard and allows the switch to determine priority on a frame-by-frame basis. Port-based VLAN can also be useful: it lets you "hardwire" a particular port for a single VLAN, useful to be 100 percent sure an office PC cannot get onto the AoIP audio VLAN.
- **If you will use a separate VLAN for AoIP, the switch needs to have an "IGMP querier" on each VLAN,** which also means that you can assign an individual IP number to each VLAN. This is a somewhat rare capability and its absence disqualifies some switches for combined networks.
- **Management accessible** through the network for remote monitoring.

5.4.3 Some Switches Proven AoIP Capable

There are new switches introduced everyday it seems, with ever increasing performance and falling prices. Having said that, the Cisco Catalyst 6500, 4900, 3750, 3560, and 2960S series switches have been successfully used in hundreds of AoIP installations. Within these series, there are literally hundreds of models that can be used with your networked audio system. Twenty-four 10/100BASE-T ports, forty-eight 10/100BASE-T ports, with or without 1000 Mbps SFP (copper/fiber) transceiver ports—connect them together and have virtually unlimited flexibility and redundancy. All of these units include built-in simple router and IGMP queriers on every VLAN. They also come in a powered-port version that can be used with VoIP phones.

5.4.4 Switch Configuration

Most switches offer three configuration connection options: an RS-232 console port, and either Telnet and/or a web interface over the network itself. Some AoIP equipment vendors offer a configuration "cheat-sheet" that gives the minimum necessary required configuration.

In the AoIP equipment industry, it is not generally expected or required that everyone must be Cisco Certified Network Engineers, so providing what is needed to know about configuring the network switch is part of the service the AoIP vendor can offer the customer.

5.5 NETWORK ENGINEERING FOR AUDIO ENGINEERS

One does not need to know most of the details of Ethernet and IP in order to use AoIP. Just as a beginner can plug analog XLRs successfully together without knowing anything about op-amps, you can connect and use AoIP without knowing details about packets. But just as fixing tricky problems in the analog world calls for higher-level understanding, so does an awareness of AoIP's internal technology help you to solve problems and build complex systems.

The next level of detail is to get familiar with networking technology and terminology—at least enough for you to get a feel for how data networks work and to understand the lingo so you are ready to ask intelligent questions of IT network professionals and vendors.

AoIP is built upon standard components, so if you understand data networking generally, you'll be ready for the specifics of AoIP audio networking. Network engineering is a rich topic, abounding with information and nuance, and in constant flux. Fortunately, AoIP uses only a small subset that is relatively easy to learn and understand. That is mainly because most of the complexity comes with IP routing and wide-area networks such as the Internet—and the system inside a facility does not use much of that, staying only with the much simpler Ethernet LAN level.

If you want to know more, book stores have shelves loaded with networking advice and information. We offer a few starting points in the Resources section. (Steve Church's book is an excellent next destination.)

A good way to think about the new relationship of audio engineers to IT folks is that the IT professionals can ensure the network is correctly designed and configured, and operating properly so data is getting end to end where it should be. But the audio engineer is the one who can say *if it sounds good*! The network carries the audio, but the content of the audio is the *raison d'être*, the very reason we invest in this subject the time and attention to make great sounding audio. *The audience is listening....*

5.5.1 Frequently Asked Questions

There will always be questions. Here are some already heard, and some imagined.

Is AoIP reliable? It's a new technology, isn't it?

AoIP uses the same technology that underlies VoIP telephony. Did you know that the majority of Fortune 100 companies use VoIP? Or that VoIP PBX systems outsell the old kind by a wide margin? With these systems, telephones plug into a standard Ethernet/IP network. Contrast this with traditional PBX phone gear—proprietary devices that required you to purchase phone sets and parts exclusively from the company that built the mainframe. You were locked into a single vendor, because the technology that ran the mainframe was owned by the company that made the gear.

Of course, we would never mix on-air audio and business functions or open ourselves up to hacking. Can I make this a completely separate network?

Yes, we understand and agree. You have a few choices:

- Have a completely separate and isolated network for AoIP. Take advantage of Ethernet, but don't combine any Internet or business functions with studio audio.
- Have two physical networks and link them with an IP router. Correctly configured, the router provides a security barrier.
- Share the network hardware for audio and general functions but isolate AoIP to its own VLAN. Again, an IP router could be used to link the two networks.

Can the network be used for general data functions as well as audio?

Most certainly, you should choose to do so. The Ethernet switch naturally isolates traffic. You may even use one link for both audio and data, since the audio is prioritized. One network link will often be the case when a PC is used—you will sometimes want to download files, receive email, etc., in addition to the audio traffic. Dual NIC cards in the PC is a common solution in the case of separate AoIP and corporate networks.

Do AoIP networks have any single points of failure? Is there a central "brain" I can lose that will take the system down?

AoIP networks are distributed, with no central box. Ethernet networks can be designed any number of ways, including those that are fully redundant and self-healing. Normally, our clients build larger facilities with "edge switches" serving each studio, connected to a redundant core. Each studio is able to operate stand-alone.

So, what about that delay?

For live monitoring, such as when an air talent hears his own microphone in headphones, about 5 ms is the limit before noticeable problems. The AES67 low latency link delay is below 1 ms, so a number of links can be successfully cascaded. To put this in perspective, a normal professional A-to-D or D-to-A converter has about 0.75 ms delay.

Does latency increase whenever you add inputs? In other words, the more sources you add, the higher the delay, right?

No, AES67 AoIP latency remains fixed at the same low value regardless of the channel count. You can run a system with a thousand channels and the latency will be the same as for a single stereo stream. Indeed, the delay is so consistent that channel-to-channel phase shift is typically less than ¼ sample. The total latency of an analog input to analog output "one hop" over the network using the low latency format is about 2.3 ms:

- The time through the A/D and D/A converters is about 600 µs × 2 = 1.3 ms total.
- The network transit time is 1.0 ms.
- To put this into perspective, the analog input to output latency on a self-contained digital console of recent vintage is about 1.75 ms.

The backplane of a modern Ethernet switch can handle full duplex traffic on all ports simultaneously without any packet loss. And since an AoIP device's link never exceeds the port's capacity, we never exceed the switch capacity. The way we prevent port overload is simple: we "own" each port, there is no possibility of contention or audio loss. Every AoIP device is plugged into an unshared port on the switch. Even when all of a node's inputs and outputs are active, we are still well under the bandwidth of the ports, and the switch is completely under control. Because the switch has the backplane capacity to handle all ports fully loaded, the system performance does not change from one channel to thousands of audio channels.

If a node needs both audio and data, such as a PC running an audio editor and a web browser, audio is prioritized and always has precedence. AoIP system vendors have had thousands of hours of testing in our lab with careful logging of packet transmission, not to mention the thousands of consoles that are already in the field. So they can assure you that AoIP works.

How can you promise live audio over Ethernet? Will not it drop out?

No dropouts. Period. The concern about network dropouts comes from many years ago when Ethernet used a shared coax cable and used Ethernet "hubs." In some cases, two devices would grab the bus simultaneously. When this happened, one would back-off and retry a few milliseconds later. These were the famous collisions. But with today's modern switched Ethernet, there is no shared bus—each device completely owns its own full-duplex link. There are never collisions or lost packets as a result of network congestion; it is physically impossible.

But the Internet is a packet network and the quality is not very good for audio.

Right. Internet bandwidth is not guaranteed, so there can be problems when there is not enough. But you completely own and control all the pieces of an AoIP LAN or managed WAN system, there is more than enough bandwidth, and priority is controlled, so performance is fully reliable.

What about compression? I am concerned about codec cascading.

AES67 audio is uncompressed 48 kHz/24-bit, or higher. It would be possible to have compressed audio streams sharing the Ethernet, but compressed audio is not a part of AES67 or studio grade AoIP.

Tell me about "sound card" drivers for workstations.

A sound card driver makes the AoIP network look like a sound card to a PC application, under Windows, Mac OS/X or Linux. Most audio applications should work unmodified. Note that software soundcard drivers are generally not capable of reaching the lowest latency modes of AoIP. This is due to latency limitations of the operating systems, not AoIP itself.

I've got a large facility. How many studios can I interconnect?

There is no practical limit. You may have as many studios and audio channels as your Ethernet switch can support. Switches come in all sizes, some with hundreds of ports. And multiple switches may be cascaded to expand ports. Best practice recommends that you use a switch per studio to isolate any problems to a defined area. These are then interconnected with a core switch. Switches may be physically associated with each studio or may be all in a central location, as you prefer.

What about for smaller stations? This all sounds pretty sophisticated for a simple set-up.

Look at Ethernet for data applications, You have everything from a single PC connected to a printer to a few PCs in a small office tied to the Internet and a couple of printers to huge campus networks with thousands of nodes. This is one of the reasons AoIP went with Ethernet—you can use it for big and small facilities. The technology and economics naturally scale to suit the application size. In fact, small stations may benefit the most as they gain routing capability at a very modest cost. Some consoles even have built-in switches to make interconnection easier.

Are optical audio links supported?

AoIP is fully compatible with copper and fiber connection types. We imagine a common configuration to be switches dedicated to studios with copper connecting nodes, engines, surfaces, etc., in the studio and a fiber backbone connecting the switches to the core in order to share audio among the studios.

What about hooking up over the Internet? With my studio audio in IP form, can I just plug a port from the switch into an Internet router?

This is almost true, but not through the Internet proper. Instead, higher performance Wide Area Network (WAN) services that include QoS and reserved guaranteed bandwidth using MPLS, RSVP, diffserv, IPV6, and other technologies, are beginning to become available. Some large media companies are building their own managed WANs for strategic use. Many cities are investing in metropolitan area fiber high capacity WANs. Using these high performance WANs it is becoming possible to extend AoIP from inside the studio to outside venues, and between facilities.

RESOURCES

The following resources are recommended for additional information on the subjects covered in this chapter.

AoIP For Broadcast:
Steve Church and Skip Pizzi, *Audio Over IP: Building Pro AoIP Systems with Livewire*, Focal Press, 2010, www.focalpress.com/books/details/9780240812441/.
Probably the best single volume written to date on using AoIP in the broadcast environment. Steve and Skip's book has introductory material for networking (including theory) and presents the practical everyday information necessary to actually construct AoIP facilities. Their work goes a long way in demystifying the concepts. While written with a focus on Livewire, the material is also applicable to most other variants of AoIP. It is available in hard copy as well as electronic format.

AES67-2013:
See the AES web site, www.aes.org

Broadcast Vendors/Livewire/RAVENNA:
Axia Audio, www.AxiaAudio.com
Telos Alliance, www.TelosAlliance.com
RAVENNA, www.ravenna.alcnetworx.com/technology/about-ravenna.html

Ethernet:
See the IEEE web site, www.ieee.org
IEEE is the standards body for Ethernet. The documents are now free to download, but will cost you a lot of paper and toner. The basic Ethernet standard is 1268 pages!
C. E. Spurgeon, *Ethernet: The Definitive Guide*, 2nd Edition, O'Reilly & Associates, 2014, http://www.ethermanage.com/books/.
Living up to its title, it is pretty definitive on basic Ethernet topics.

TABLE 5.2.2 AoIP Standards Map

Layer 1	IEEE Ethernet Physical
Layer 2	IEEE Ethernet switching
	IEEE 802.1p/Q prioritization
	IEEE 802.1p multicast management
	IEEE 1588 v2 Precision Time Protocol
Layer 3	IETF IP (Internet Protocol)
Layer 4	IETF RTP (Real-Time Protocol)
	IETF UDP (User Datagram Protocol)
	IETF TCP (Transport Control Protocol)
	IETF IGMP (Internet Group Management Protocol)
Layer 5	IETF DNS (Domain Name Service)
	IETF HTTP/WebIETF ICMP Ping
	IETF SAP/SDP (Session Announcement Protocol/Session Description Protocol)

General Networking and Interest:
See the IETF (Internet Engineering Task Force) web site, www.ietf.org.
IETF is the Internet's main standards organization. Look for the RFC (Requests for Comment) documents to see in detail how the Internet is built.
A. Tannenbaum, *Computer Networks*, Pearson Education/Prentice Hall, 2003.
Our favorite general networking book. Popular college textbook covers it all, including multimedia, with a breezy style and at just the right level of detail: enough to be useful, but not so much as to be overwhelming.
J. Naughton, *A Brief History of the Future*, Overlook Press, 2000.
Not really so interesting for audio and Ethernet, but still worth reading for perspective. This history of the Internet tells how it happened in a friendly—even charming—way. Lots of stories and anecdotes. Of particular note is AT&T repeatedly making clear that digital communication had no future!

Cabling Information and Standards:
J. Abruzzino, *Technician's Handbook to Communications Wiring*, CNC Press, Chantilly VT, 1999.
This book is concise yet contains a lot of great information including proper technique for working with Cat 5 cable and connectors. Small enough to keep with your toolbox.
Cabling Design, www.cabling-design.com; cabling tutorials.
TIA, www.tiaonline.org; standards organization for cables.
Global Engineering, www.global.ihs.com; sells the TIA/EIA cabling standards.

Cable and Contractor Supplies:
AMP/TE Connectivity, www.te.com; RJ plugs and tools.
Belden Cable, www.belden.com; leading cable supplier.
Hubbell Premise Wiring, www.hubbell-premise.com; devices for Cat 5, etc.
Panduit, www.panduit.com; marking and installation products.
Corning Optical Communications, www.corning.com; fiber optic cabling and components.
Siemon, www.siemon.com; punch blocks.

Cable Testers:
Fluke, www.flukenetworks.com; full range of testers.
Agilent, www.agilent.com; top-end test equipment.
ByteBrothers, www.triplett.com/byte-brothers; low-end tester.

Network "Sniffers":
Wireshark, www.wireshark.org

AoIP Network Standards and Resources:
AES67 AoIP operates at both Ethernet and IP network layers, taking advantage of appropriate standards-based resources at each layer (Table 5.2.2).

CHAPTER 5.3
AUDIO MONITORING SYSTEMS

Martin Dyster
The Telos Alliance, Lancaster, PA

5.1 INTRODUCTION

The need to monitor sound levels has always been of critical importance within television broadcasting. Aside from the quality of the image on screen, audio reproduction should be consistent, coherent, and unobtrusive while delivering a compelling listening experience that enhances the viewer's enjoyment. Above all, it should not leave the viewer reaching for the remote control every few minutes because of fluctuations in level or because of problems with intelligibility.

Television sound has evolved slowly over time from mono to stereo and Dolby Surround to Dolby Digital 5.1. The next step will almost certainly include consumer personalization using interactive features that enhance the viewing experience. As a result, broadcast audio is set to become more challenging to produce and deliver. Whether involved in acquisition, production, post production, or delivery, the job of the audio professional is becoming more challenging and the tools required to ensure that "good audio" reaches the home must meet the challenge.

5.1.1 A Short History of the Audio Meter

The BBC Research Department produced its first moving coil "programme" meter in 1932 to coincide with the move to the new purpose-built BBC facility at Broadcasting House. It was the first meter devised with the familiar look of a white scale printed on a black background and was know as the "Smith Scale" after its inventor, Charles Holt-Smith. By 1938, the first PPM (Peak Programme Meter), also devised by BBC Research, came into use at Broadcasting House. It would be instantly recognizable to any sound professionals working in the broadcast industry today and featured the same 1–7 scale commonly in use in many parts of the world.

In the United States, colleagues at Bell Laboratories, CBS, and NBC were developing the "Volume Indicator" designed to help standardize levels across telephone lines. The first *VU meter* as it became known appeared in 1939 and, like the PPM, became an industry standard. Both the PPM and VU meter are still commonplace in professional audio, and despite regional variations, both would still be familiar to their inventors 70 years after the event.

Apart from a pair of expertly "tuned" ears, the moving coil meter became the accepted standard for critical level measurement by the professional and nonprofessional sound mixer alike. The BBC PPM and American VU standards have been adopted and adapted by engineers, content producers, and committees in almost every country across the globe. As technology has improved, bargraph variants have gained acceptance as a cost-effective alternative to moving coil meters in the vast majority of professional audio monitoring applications. As we have moved into the digital age and benefitted from the extended dynamic range that PCM audio enables, metering has adapted and the humble VU and PPM meters no longer provide the resolution required for most modern music and broadcast applications.

In recent years the digital dBFs (Decibels full scale) has emerged as the closest contender we have to an internationally accepted standard. It is now commonplace to find audio mixing consoles, metering, and monitoring devices throughout sound installations with digital dBFs bargraph scales used for critical monitoring applications, and moving-coil meters or emulations still used for main program measurement. Despite a degree of global acceptance, many engineers find the dBFs ambiguous and difficult to equate to their more familiar reference devices.

5.2 AUDIO MONITORING IN BROADCAST

Audio monitoring has been a fundamental requirement of broadcast television production and delivery since the very beginning; however, the complexity of today's systems would have been inconceivable to broadcasters even as recently as the early 1990s.

Today, audio monitoring can be as complex as the multiple bargraphs on the meter bridge of a 48 fader mixing board or as basic as a mono speaker replaying a cue to the talent in their dressing room. The common thread is that the backbone of the monitoring system must deliver the audio that is required to the desired destination reliably, repeatedly, with operational ease and with clarity.

Right across the complete signal chain from the extremities of a remote production at the ballpark through the link back to HQ, Ingest, Post Production, Production, QC, and onward to Transmission, audio monitoring is a critical component in the broadcast workflow. Each discipline and individual operator therein may have fundamentally different requirements ranging from simply listening to program, right through to full compliance checking of the integrity of the broadcast before it hits the airwaves. With such a diverse set of parameters to fulfill there is an equally diverse range of solutions available to meet those needs.

5.2.1 Physical Form

The majority of audio monitoring devices conform to the 19-inch rack-mount standard, whether fitted into desk consoles or equipment racks. Alternative solutions include desktop format, custom enclosures, and portable units—and also software application-based monitoring installed on computers and tablets. As mentioned previously, most devices are rack-mount and generally conform to the majority, if not all, of these common basic criteria:

- Source/input selector
- Signal level indication
- Output volume control
- Loudspeaker/headphone output

Within a broadcast ecosystem monitoring devices normally reside as destinations of the audio and/or video routers, as part of engineering test and measurement functions, or an addendum to the intercom/talkback system where additional facilities are required. Within post production facilities such as edit suites, audio monitoring will usually be connected directly to the system breakout panel or host PC.

5.2.2 Functionality and Operation

Although it might be difficult to claim that there is such a thing as a "typical" broadcast facility, much less a common workflow from acquisition to transmission, it is certainly true that the demands on audio monitoring throughout the sequence have common characteristics.

AUDIO MONITORING SYSTEMS **5.39**

Figure 5.3.1 is indicative of both audio signal flow, and workflow interaction between the various functional elements of a mid-sized broadcaster involved in everything from live sports and news to entertainment programming.

Each area represented and their role requires a different audio monitoring solution, and within that solution, certain specific functions of the monitor will be more pertinent to the role than others. Although not definitive the following descriptions indicate the kind of tasks undertaken by the audio monitoring in each area.

5.2.2.1 External Sources—Remote Broadcasting. A multicamera outside broadcast (OB) truck mimics the function of a complete production studio within the confines of a compact and often physically limited environment. Whether producing surround sound or stereo audio content, the individual members of the production team have specific demands of the audio monitoring system in order to do their job.

5.2.2.1.1 Outside Broadcast Truck. In the case of an OB truck, connectivity can vary a great deal but most modern vehicles use SDI, MADI, or a combination of both to transport audio signals around. Dolby E encoding is still frequently used as a means to maintain metadata and to multiplex audio signals transmitted from the truck.

Production Team—The production team primarily listens to the program output derived by the A1 or sound supervisor. A simple monitoring panel might be used to select the appropriate source and control the level to the control room speakers. Additional functions could include the ability

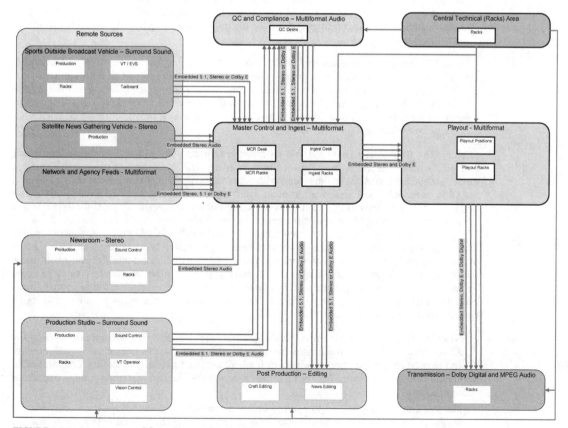

FIGURE 5.3.1 Monitoring workflow schematic for a broadcast facility.

to handle 5.1 audio if the control room is equipped with a complete surround speaker system, and the ability to measure and log loudness is an increasing requirement in this environment. A dedicated loudness meter will probably handle the latter demand.

Aside from program monitoring, members of the production team are likely to use additional audio monitoring to listen to preview and *prehear* circuits. The engineering manager will require a more comprehensive device that provides technical feedback such as signal status, metadata display (if used), and audio delay for sync checks.

VT/EVS—The VT operators require a very specific type of monitor, something that enables them to quickly switch sources from input to output. The signal format supported depends largely on the VT device employed but an increasing number of truck builders rely on MADI for all audio monitoring requirements.

Racks—The rack monitor in an OB truck usually performs a core engineering function and is used to check the integrity of the vehicles audio systems. As a result, the owner will often choose a device with similar capabilities to the engineering (or technical) manager position in the production control area.

Sound Control—When a monitor is deployed within the sound control room of a truck, it is frequently located close to the patch panel and used as a simple signal integrity checker for MADI, AES, and analog circuits.

Tailboard—The tailboard of an OB truck encompasses the connection points to outside facilities such as cameras, microphones, communications, and trunk circuits to other vehicles. The audio monitor chosen for this location frequently combines both basic video and comprehensive audio review capabilities and is used to check signal integrity from both incoming and outgoing circuits.

5.2.2.1.2 Satellite News Gathering Vehicle (SNG). In the majority of situations an SNG truck has quite rudimentary audio system requirements and hence monitoring on board is similarly simplistic. A basic audio monitor might be used to control the main loudspeakers; when the vehicle has more complex needs, a multiinput multifunction device capable of checking signal integrity of a variety of formats is employed. Many SNG trucks still have analog or AES core audio systems but SDI can be cost effective also.

5.2.2.2 *The Broadcast Center.* Audio monitoring requirements within a broadcast center are as diverse and varied as the functions of the various departments therein. Core infrastructure in modern television broadcast facilities tends to be based around SDI with embedded audio; however, MADI is commonly used (described in more detail later) and analog and AES used locally where necessary. Also discussed in the "Connectivity" section of this chapter is IP-based transport, which seems destined to replace baseband infrastructure in the future.

5.2.2.2.1 The News Studio. The news studio in any broadcast facility is a core function where work is fast-paced, often 24/7, and almost invariably, transmission is live to the local area, nation, or the world. Whatever the scale of operation, the workflow will demand intuitive and unobtrusive monitoring that provides seamless access to any audio signal. TV news tends to be created in stereo and so the audio monitoring devices chosen do not need to support surround sound content; however, they may require multichannel support for more than one language or for additional signal content from material gathered in the field.

News Production Team—The news production team listens to the program output derived by the A1 or sound supervisor, usually in stereo. Technical monitoring positions check signal integrity and may be concerned about loudness or intelligibility, and producers handling lines will often use multichannel audio monitoring tools in addition to the intercom system to preview or prehear signals.

News Sound Control—The A1 or sound supervisor monitors audio through the mixing console for the majority of the time; however, additional preview or prehear facilities that can be used in addition to intercom and mixer prefade are common.

5.2.2.2.2 *The Production Studio.* A television production studio succeeds on its versatility. Able to host a late night live discussion program (in stereo) on a Monday night, it must be able to be transformed into the theater required to host a Thursday night prerecord of Saturday nights 5.1 format light entertainment showcase. Where the former might involve a small set with three cameras and a couple of lapel mics, the latter shoot may well require 12 cameras, 20 channels of wireless mics as well as lapels and handhelds plus a PA system for the house band. As a consequence, all aspects of the audio system need to be highly versatile and the audio monitoring system is no exception.

Fundamentally, however, the monitoring requirements of the production studio echo those of the OB truck but with an element of scale dependent on the specific demands and the constraints of the operation in question.

Production Team—As intimated previously, the role of monitoring within the production control room is broadly similar to that of the sister location in the OB truck. The ability to handle surround sound audio will be determined by the studio booking card and/or the broadcasters' transmission format.

VT/EVS—The VT function within a studio is also similar to that of an OB truck; however, the number of operators required will often be scaled to meet production needs. In a dry hire facility, it is not uncommon for desk positions to be able to accommodate different members of the production team from day to day depending on the scale of the shoot, and as a result the monitoring requirement has to be adaptable to suit all eventualities.

Sound Control—When a monitor is deployed within the sound control room of a studio it is frequently located close to the patch panel and used as a simple signal integrity checker for MADI, AES, and analog circuits.

5.2.2.2.3 *Master Control.* The master control room (MCR) function within a broadcast center handles the incoming and outgoing lines to and from the facility and acts as the first and last line of compliance check for any content passing through their monitoring systems. Compliance might include ensuring that outgoing content meets the demands of an international client as much as ensuring that incoming material meets the host's criteria.

As a consequence, the audio monitoring solutions in an MCR tend to be the most comprehensive of all devices, with the ability to handle all relevant signal formats in any configuration (surround, multichannel, Dolby encoded, etc.). Data, metadata, and loudness reporting are also key features alongside down-mixing and the ability to feed external monitoring speakers.

5.2.2.2.4 *Ingest.* The ingest department is responsible for digesting media of any format into the host broadcast systems. Whether recording incoming news feeds onto the server or digitizing from tape archive material, the function also implies checking that all audio content is compliant with house standards. Similar to the specification of audio monitoring equipment in MCR, the ingest monitor function requires versatility and a high degree of technical capability due to the likely variety of formats and information criteria that must be checked during the ingest process.

5.2.2.2.5 *Quality Control.* QC often provides the last check of compliance before a "package" is loaded onto the transmission server for broadcast or sent to a commissioning client. The demands of audio monitoring within QC are therefore very similar to that of ingest, with similar specifications to check for similar criteria.

5.2.2.2.6 *Post Production.* Sound and video editing is an essential piece of the TV broadcast jigsaw. From simple journalist desktop edit stations within the newsroom through to full blown film dubbing suites, a broadcast center may well encompass a broad range of post production functions. Audio monitoring solutions vary accordingly from a simple USB-connected headphone amplifier the journalist uses, a metering bridge in a craft edit suite and the PEC (photo electric cell) monitor panel installed within the mixing console within the film dubbing suite.

Most nonlinear edit suites are built around a networked host PC or Mac that combines video and audio editing capable of handling the majority of projects from short-form to full-length features. The audio system will generally comprise monitoring speakers (stereo or 5.1), a mixing console or control surface, and an audio monitor with quality metering and increasingly, loudness measurement capabilities. Some operators favor on-screen "plug-ins" to display metering and loudness level, but an equal number would choose a hardware solution because of a perceived reliability benefit.

Voice-over is also part of the post production function and a dedicated booth capable of working with the edit suites needs a different kind of audio monitor. A simple headphone amplifier gives the voice artist level control over cue audio from the edit platform, but the same headphone unit may require bi-directional talkback to enable the talent to communicate with the editor. If the voice booth is a shared resource that can be assigned to a number of edit suites then it is likely to be both a router source and destination that can be switched as required. When this is the case, the monitoring system within the booth can become quite complex, having to resolve both video and audio cue feeds, bidirectional intercom, as well as microphone amplification for artists with different vocal characteristics. A common solution is to use embedded audio into and out of the booth with the monitoring system combining some or all of these functions in a single comprehensive device.

5.2.2.2.7 Playout and Transmission. The penultimate and ultimate links in the chain, both playout and transmission, require comprehensive audio monitoring tools that enable operators to ensure that the outgoing content is compliant and correct. Many facilities will be responsible for the delivery of tens or hundreds of channels, and the systems that monitor them are likely to be "probe" devices that automate the process and report dropouts and signal errors via alarm systems. In smaller operations where fewer channels are handled or where premium channels demand, dedicated operators will use audio monitors with similar specifications to those used in MCR, Ingest, and QC.

Playout and transmission facility monitoring may also encompass logging devices capable of recording loudness and peak levels, signal data, and metadata as well as errors against a timeline.

5.3 CONNECTIVITY—SIGNAL TYPES

Audio monitoring devices and systems have evolved as the ecosystems that they support have changed. For example, analog audio prevailed as the signal transport of choice for many decades since the birth of radio and television because digital audio simply did not exist. The introduction of digital audio as a viable emission system within broadcast ushered in an inevitable need for monitoring that could be used to provide confidence checking throughout the broadcast infrastructure in the same way as analog had done for decades. Hybrid system designs where AES and analog coexisted and carried contribution (AES) and communication (analog) circuits became commonplace during the 1990s until the development and mass acceptance of SDI-embedded infrastructure became de facto within television broadcast systems.

Although AES-10 (MADI) is itself an aging technology dating back to 1991, the ability to transport up to 64 channels of 48 kHz audio along a single coax or fiber cable has meant that the format has seen resurgence in recent years in the form of distributed infrastructure and shared resources, supported by the very console industry that championed the standard 25 years ago. With MADI providing an efficient means to trunk audio circuits between intercom, console, and even hybrid video/audio routing systems, MADI monitoring solutions are becoming a common site within fixed and mobile broadcast facilities. Particularly in studio production environments dedicated to news, sports, and entertainment where the proliferation of many diverse audio circuits mean that studio technicians, audio engineers, and creative personnel alike need instant access to a wide variety of audio signals.

As witnessed increasingly in the last couple of years, the race to develop IP-based technology capable of replacing the established yet copper hungry baseband video and audio has reached breathtaking pace. At time of writing audio-over-IP (AoIP) might even be described as *mature* technology with AES-67 compliant protocols such as Livewire, DANTE, and Ravenna commonplace in many broadcast facilities. However, the use of this technology as an infrastructure replacement for MADI, embedded audio, AES, or even analog in TV broadcast systems is virtually nonexistent, aside from a few early adopters and as a means to bridge devices in fixed environments. As a result, the number of "traditional" audio monitoring devices with IP capability is extremely limited; however, this statistic will undoubtedly change as more supporting technology becomes available and the advantages of IP and Ethernet infrastructure is fully realized.

5.3.1 Special and Extended Functionality

As alluded to previously, it is almost impossible to describe a "typical" broadcast audio monitoring tool. There are certainly common features offered by almost all product offerings, from the most basic analog listening bridge right up to the proliferation of mid-price multifunction products offered by the majority of companies currently operating in the audio monitor market.

A handful of established specialist companies have developed product that occupies the rarefied "high-end" of the broadcast audio monitoring market. In this space there exist features and capabilities that command a price premium and are as important to the broadcasters that demand them as are the most basic requirements. Here are a few examples:

- **Audio delay**—The ability to dial in an delay in the monitoring output of the unit in order to compensate for latencies in associated video display devices.
- **Conditional switching**—The audio monitoring system is able to automatically adapt to the incoming signal format. For example, the default audio pair the user wishes to monitor varies depending on whether the incoming SDI signal is SD or HD, and the monitor switches accordingly based on user-defined criteria.
- **Auto down-mix**—When a group of audio signals represent a surround sound (5.1 or 7.1 typically) program, the audio monitor uses a predefined algorithm to down-mix the signals to stereo. This feature is used to ensure that the stereo signal that will ultimately be heard in the home is free from phase issues and lack of intelligibility.
- **Dolby decoding**—Usually implemented using a module manufactured by Dolby, this feature would typically include Dolby E, D, and DD+ decoding as well as the ability to down-mix and display metadata.
- **Loudness measurement**—The ability to measure loudness as well as peak is likely to become a standard feature of all broadcast monitoring equipment over time, but for the moment it is a premium application that features as a part of remarkably few products.
- **SMPTE 2020**—SMPTE 2020 or S2020 is a standard that defines the insertion of Dolby metadata into the ancillary data of an SDI signal. Several broadcast monitors feature the ability to extract this information, display it to the user, and also down-mix any surround sound components to stereo for compliance checking.
- **External speaker outputs**—Most rack-mount broadcast audio monitors have in-built loudspeaker systems; however, it is often the case that the user wishes to listen to the monitored audio through a quality full-range external speaker system and not the physically compromised internal drivers. Some monitors are designed with the ability to mute the internal system and to connect external speakers (stereo and occasionally surround) via a variable line-level connection controlled by the master volume on the front panel.

5.4 THE FUTURE OF AUDIO MONITORING

Broadcast audio monitoring systems are fundamentally a means to check the integrity of the audio signal components of the ecosystem at any point in the chain. When the technology within that ecosystem evolves, the monitoring components have to follow suit and so the evolution of audio monitoring is, to an extent, defined by the way that the broadcast system changes and adapts.

We have discussed already how audio and video over IP is set to become a replacement for the baseband transport mechanisms in use today. As a result, it is somewhat inevitable that traditional rack-mount audio monitors will be required to support whichever protocols emerge from the current battle for wider acceptance.

No less important than the move toward IP and Ethernet-based systems is the growth of ultra hi-definition video formats and the audio definitions and requirements in association. Although audio for UHD is yet to be formally defined, a great deal of discussion has centered on new formats such as *object-based audio*, Dolby ATMOS, and extended surround sound such as NHK's 22.2 channel system. Whatever becomes the new "norm" it is certainly true that audio monitoring will have to embrace it to support those companies involved in production, post, and playout.

A trend in broadcast as in most global industries has been one of rationalization and increased efficiencies. Whether this is achieved by downsizing the workforce, simply improving workflow, or by a combination of both, audio monitoring seems set to play a role. Hand-on rack or desk mount monitoring solutions are designed for manual operation and combine visual and aural perception to tell if a signal is "good." With workforce numbers and skill levels seemingly destined to diminish in the name of economy and efficiency, it seems entirely probable that the next generation of audio monitoring devices will include increasing levels of automated functionality. Such devices will almost certainly deliver hands-off signal checking, possibly combined with autocorrection capability, alarms to tell you when something is wrong, and what action is being taken as well as logging to keep track of what happened, when, and why.

What we might consider to be "traditional" now is unlikely to disappear any time soon; however, we can already witness the introduction of broadcast audio monitoring tools that live on PC, tablet, and handheld platforms. Where signals are physically brought out from the router or matrix to the monitoring panel, probe devices can be embedded within the signal path providing a more efficient way to access content, analyze it, and stream it to a real person or an analytical system designed to monitor automatically.

From the humble beginnings of the moving coil meter to whatever device is destined to become the audio monitor of the future, broadcast audio monitoring has and will continue to evolve and ultimately support the ecosystem in which it has fulfilled a critical yet often understated role.

CHAPTER 5.4
REMOTE AUDIO BROADCASTING

Martin Dyster
The Telos Alliance, Lancaster, PA

5.1 INTRODUCTION

Remote broadcasting, in the context of this chapter, refers to *television outside broadcast* and focuses on the audio components and workflows used in content acquisition away from the broadcast center.

In any genre of television production, the type of programming determines the scale and complexity of the workflow and the system between the camera and the delivery mechanism; remote broadcasting is no exception. The term "remote" or "outside broadcast" can mean anything from a journalist reporting from a war zone using video and audio sent via their smartphone right up to a multicamera shoot at a major international sports event via a fleet of $10m 40 feet HD production trucks. Whatever the scale of production, the end-game is always the same—to deliver a compelling visual and audio experience to the viewer at home.

5.1.1 A Short History of Remote Broadcasting

The history of outside broadcast arguably stretches back before the First World War, far beyond the very earliest days of television and into cinema history when newsreel presentations formed part of the daily schedule at movie houses across America, British Commonwealth Countries, and parts of Europe. Shot on film and distributed by companies like Pathé News (whose first newsreel began in Paris in 1908), the format established a genre of communication that has changed the world forever.

When it comes to "live remote" broadcast, radio was ahead of the curve using leased telephone lines to transmit music shows to U.S. radio listeners in the 1920s; television would not catch up until a decade later. History suggests that the first major outside broadcast was the BBC coverage of the 1937 coronation procession of King George VI in London, England.

Arguably the most famous remote broadcast of all time was the Apollo 11 moon landing, watched live by an estimated half a billion people worldwide. Neil Armstrong's famous sentence achieved iconic status, aided in part by static reportedly masking the word 'a', changing the message from "one small step for a man" to "one small step for man." Neither before nor since has such a minor audio glitch altered the content with such a positive impact.

5.2 NEWS REMOTE BROADCASTING

From local affiliate stations to national and international news networks and agencies, the content that brings the news to life is delivered direct from the field by many varied means. We as viewers have grown accustomed to the fact that when a story breaks anywhere in the world, somebody will have a camera of some description pointing at the action. Almost all smartphones and tablets have a camera and a microphone and the connected world in which we live means that amateur footage can make its way to the news networks and agencies as quickly as the Internet or phone network will allow.

5.2.1 Electronic and Satellite News Gathering (ENG and SNG) Vans

The majority of TV broadcasters that have embedded news departments run satellite news gathering (SNG) vehicle fleets as an integral part of their ENG (electronic news gathering) operation. The SNG truck is a dedicated automobile that combines a compact production system with satellite communications. The specification can vary considerably depending on the type of news production it is designed to report. With IP technology increasingly prevalent in satellite communications and dish size and weight decreasing proportionally, it is possible to build an SNG unit based around vehicles varying in size from an ultracompact car up to a large panel van.

The audio system within an SNG van is generally quite simplistic. The scale of the system varies in accordance with the application, but the general principals are the same regardless of the size, with certain basic functions common to almost all designs. Some or all of the following audio and communications components are likely to be found in a typical van:

- A stereo audio mixer with sufficient channels to handle microphones and any local playback devices that may exist.
- Talent and guest microphone circuits (may be wired or wireless).
- Camera talkback and microphones.
- Audio into DSNG (digital SNG) codec/return audio out (depending on model).
- A simple communications device such as a 4-wire box to handle talkback/IFB between the van and talent, van and camera operator(s), van and Broadcast Center, etc.
- Telephony cellular or IP codecs used for communications links, and also as a means to provide backup contribution circuits in the event of a failure of the satellite link.
- Satellite phone.
- SNG vans may still use traditional POTS or ISDN codecs depending on the infrastructure of the locale in which they are deployed.
- Possibly an audio delay unit and/or effects device.
- Audio in/out of any recording device that may be deployed.
- Loudspeaker monitoring.
- Dedicated audio monitoring device to provide quality metering.
- Tailboard connection panel.

Within compact SNG vans with one or two multiskilled operational personnel, the audio signal flow centers on mixing talent and guest microphones for transmission and the coordination of communications. Larger vehicles might have a dedicated audio or communications engineer/technician whose role is closer to that of a sound supervisor or "A1" in a production truck.

5.2.2 Electronic News Gathering (ENG) Field Operations

Not all news gathering operations benefit from the luxury or comfort of an SNG van. In many locations the only practical method of bringing news from the field to the home is to deploy a journalist or talent working with a single technician or engineer with a camera, microphone(s), portable lighting, and an audio recorder/mixer.

In a simple two-person setup, the audio from the presenter microphone is recorded directly onto a media device within the camera. A more complex recording might have the "wild" sound from the location recorded along with the talent, and then shotgun microphones combined with handheld or clip mics recorded onto separate tracks.

The simplest form of field reporting might be as basic as a report filed via a smartphone camera/mic or from a laptop using Skype.

5.2.3 Sports and Event Remote Broadcasting

The majority of sports and event coverage transmitted from or recorded on location at stadia, arenas, theatres, or auditoriums is produced using one or more outside broadcast (OB) trucks. An OB truck is a highly specialized broadcast production gallery on wheels with many such vehicles deployed around the globe costing anywhere from $0.5m to well over $10m depending on the complexity and application. A typical NFL or English Premier League live shoot will command a fleet of trucks handling all aspects of productions from cameras and sound through to graphics, slo-mo, communications and ultimately transmission out to the wider world.

5.2.3.1 Outside Broadcast Truck. Undoubtedly more complex than the SNG van described earlier in this chapter, the OB truck might be described as a compact facsimile of a traditional static broadcast television production studio complex. The audio systems therein replicate the facilities of a permanent installation; however, the physical limitations of a road-legal vehicle means that certain elements are compromised and, as is often considered the case with broadcast system design and expenditure as a whole, the audio facilities are usually subject to a greater degree of physical compromise than other aspects.

The core of any audio system forming part of a production system is the mixing console. Modern TV mixers combine routing and distributed input/output (I/O) devices as well as the control surface itself. This type of architecture is particularly suited to a mobile environment in that the routing system can be used as to convey the audio signal to and from all contributing, monitoring, and communications equipment whether that signal is sent to the mixer or not. In short, a well-designed OB truck audio system will use the audio console infrastructure as the core router for all audio signals with the exception of the talkback system.

The distributed I/O benefits the truck design by minimizing cable usage and therefore weight, as such systems are connected via MADI or IP. These technologies use fiber, coax, or Ethernet cable to transport dozens of channels of audio where previously a single twin screen cable would have existed for each individual circuit. Another considerable benefit of a distributed architecture is the benefit to riggers. Where miles of multicore cable would have previously been run into a stadium for mics, communications, and cue systems, the same can be replicated using a just few fibers, with the added advantage that distance limitations are virtually nonexistent.

Communication is a critical component of TV production and an intercom or *talkback* system should be a seamless and unobtrusive part of the production process. This assertion is equally true of OB production where the shoot is often live and the pressure intense and immediate. Director talkback provides the majority of instruction and should be loud and clear to the entire crew. The ability to create subdivisions within sound, vision, and engineering to communicate easily with one another is no less essential.

Where the scale of a production requires multiple vehicles, either from the same or different companies, it is often necessary to trunk together the audio, video, control, and data systems so that the combined crews can work as one. MADI and IP-based systems are an ideal means to "join" truck systems together, whether capable of communicating natively via common protocols or simply as a means to bus multiple channels between one truck and another.

Some or all of the following audio and communications components are likely to be found in a typical OB production vehicle:

- A large format audio mixer often capable of handling 5.1 channel surround sound productions. Due to space constraints, models with narrow fader pitch are favored in order to optimize access to the maximum number of channels.

- Distributed I/O for local sources acting independently as a router, as well as providing shared resources for the mixer and audio monitoring system (MADI or IP connectivity, typically).
- Mixer I/O in *stagebox* form for rigging microphone circuits, communications, and cues within the venue (MADI or IP, often via fiber).
- Intercom systems including rack-mount key panels, telephony interfaces, RF talkback, and trunk circuits to/from venue systems and additional vehicles in the fleet.
- Talent and guest microphone circuits (may be wired to stageboxes or wireless).
- Camera talkback and microphones.
- Trunked audio into DSNG codec/return audio out.
- Telephony cellular or IP codecs used for communications links and also as a means to provide backup contribution circuits in the event of a satellite link failure.
- Satellite phones.
- Outboard gear in sound control system.
- Audio monitoring system as required at operational positions.
- Audio in/out of any recording device that may be deployed such as Slo-Mo, servers, etc.
- Audio out of playback devices in sound control such as electronic cart players.
- Loudspeaker monitoring in all operational areas.
- Tailboard connection panel.

5.2.4 Flypack Remote Broadcasting

Longer-term shoots away from the broadcast center are frequently handled using *flypack* systems. Examples of these shoots are major sporting events held over several days or weeks (International Broadcast Centers at Olympic Games, FIFA World Cup, etc.), reality shows filmed on location, and political or news events such as party conferences, rallies, or elections.

A flypack system is a compact version of a complete production facility broken down and shipped via road cases. Systems are often compromised in order to reduce size and weight, and therefore shipping costs; however, the basic components are the same as SNG or OB vehicles. Again, the scale of the system equates to the application, and the choice of equipment is frequently based on practicality rather than "best of breed."

5.3 THE FUTURE OF REMOTE BROADCASTING

At the time of writing, the industry is going through a transition phase with broadcasters looking to IP solutions as the future mechanism for transporting audio, video, and data. While static IP infrastructure solutions are in a state of flux as manufacturers continue developing the next generation of workflow components and promote their initial product launches, the landscape with respect to IP backhaul solutions is more mature.

Audio over IP has been with us for over two decades now and has been used for much of that time in remote radio broadcasting. Audio and video IP solutions are much newer but are increasingly popular as a means to transmit compressed content from remote locations back to the broadcast center. OTT (over the top) coverage for web services and "red button" channels happily use compression systems such as JPEG 2000 delivered via protocols like SMPTE 2022 (S2022/MPEG 2 TS), reducing the cost of production by removing the need to fully crew an event with an OB truck.

Using protocols such as S2022, it will be possible in the near future to broadcast from a remote location back to a static production control room as though it were the studio floor right next door. All cameras, microphone, communications, and data feeds will be transported over IP with or without compression.

CHAPTER 5.5
MASTER CONTROL AND CENTRALIZED FACILITIES

John Luff
HD Consulting, Pittsburgh, PA

5.1 INTRODUCTION

The term *master control* connotes someone sitting at a console deciding what goes on air and personally controlling devices. To be fair, there was a time that such a role existed broadly in the industry. Today, for both technical and economic reasons, the role is quite different. And, it continues to evolve.

5.2 THE FUNCTION OF MASTER CONTROL

To understand master control in the modern sense we first need to understand the flow of information and content within a broadcast facility. Though master control may be the final point in the production of the program stream, it begins in two other departments. Program management decides what programs will air and when. Content might be from live sources, as in networked content, or from content distributed by various *store and forward methods*. That includes satellite delivery of content live (or "day and date" with a short time-to-live), which is then either aired live or recorded for delay broadcast. It also includes delivery via satellite or terrestrial means to local storage platforms used to hold the content until it is staged to playout servers in the station. That is quite often the case with syndicated content.

The other department key to the assembly of the broadcast output is the traffic department. Traffic takes the program schedule, and the interstitials that fill "breaks" and completes an air log showing what will be aired in precisely what order and at what time. It is in the traffic department that orders from the sales department (in the case of commercial stations) are matched up with media to complete the content in the breaks. This also includes promotional materials used to keep the viewer connected to upcoming programming on the station.

The air log is the key piece of information used to assemble the long-form and short-form content into a complete concatenated stream of content, including any graphics content. In early facilities, the log was entirely paper and most likely typed on ancient printing machines that we, of a certain age, called typewriters. Today the air log is electronic and serves as the input document to a process that automates the assembly of the content into the finished stream.

The air log is a representation of a database that includes the identification of the content and the date and time it must air. The fields in the database are customized in each station, but always include a pointer reference to the content itself, usually called a "content identifier." This is the link that connects the traffic database to the inventory of content files for long-form and short-form content. The content files live in a system of hierarchical storage that includes the temporary content present on the storage for syndicated content, the storage for interstitial commercial content, and the air playout server system, which may include near-line and off-line archives of content as well.

5.2.1 The Hardware Environment

Master control systems today are in transition from systems built around discrete hardware components to systems that include networked devices using an IT interconnection topology. The discrete master control facility is rapidly being replaced for many reasons, but due to the number of facilities still based on baseband interconnected components, it is still worthy of description. A typical master control equipment bay is shown in Fig. 5.5.1.

A classic system is built up from playout hardware (servers and potentially video tape recorders), a master control switcher, graphics generators, audio playback devices, and of course an automation system that converts the air log to a native automation playlist and executes each event on air live.

A second paradigm is what is often called CIAB, or "Channel in a Box." This approach uses tightly integrated systems that include each of the above discrete components in a single device, which is then controlled by automation. The CIAB might include playout server capability internally. Some server manufacturers have included the features of a CIAB into a server playout channel. The key point is that a CIAB has many fewer components that must be wired independently and controlled together as a system.

Some CIAB systems also include automation control internally, receiving the air log from a supervisory level of automation control, and executing autonomously. These systems are particularly well suited to installation as a disaster recovery system, which for instance might be located at the station's transmitter site.

Finally, a third approach is to take the functions of a CIAB system and perform all of the work in software in a purely IT-based system, perhaps using virtualization to permit backup and disaster recover scenarios that can be highly flexible. Such a system would include local automation playlist capability, allowing it to run independently. In the event of a failure, a spare server (or another

FIGURE 5.5.1   Equipment rack at WISH-TV, Indianapolis, IN. (*Courtesy: WISH-TV.*)

instance of a virtualized hardware environment) can be quickly attached to the correct playlist and content repositories. This permits flexible operations, like spooling up a channel that needs to run only for a limited time (or part of a day), and then disconnecting the instance. This allows a complex environment to be built without discrete wiring and the need to build complex backup capability out of hardware that sits idle much of the time. The IT nature of this solution makes it possible to think of master control in a "cloud."

5.2.1.1 Baseband Master Control. The discrete environment of a baseband (SDI and HDSDI) is built up from components that are modern equivalents of the devices used in the early days of television. The master control switcher can be thought of as a specialized production switcher, with audio and automation capabilities that are not generally part of a production switcher. In the general case it is a crosspoint switcher with at least two busses, which allows for a program/preset architecture, or transitions between two sources, like dissolves or fade through black. The output of that mixer is fed through at least one layer of keying, often as many as four levels. One of the busses can usually be "squeezed" back to reveal a graphics page underneath, providing an "L shaped" graphic element that can—for instance—contain information on upcoming programs. Usually there is a downstream key over the composite output that allows for logo insertion, or lower third crawls of weather or other alert information.

Sources of dynamic graphic information, like school closings, often come as RSS, ATOM, or other feeds from sources of dynamic content. Those feeds, which are essentially XML-based content delivered asynchronously, allow stations to avoid rekeying changing data, such as stock quotes, or perhaps weather. The ability to interface with such live feeds is not universal in all master control products.

Other graphics are generated in a character generator, or in the more generic case a graphics generator. The interface between the master control switcher and graphics is baseband, often with the key (or alpha) channel separately interfaced. Control over page selection and timing of any graphics transitions must be coordinated between automation and the graphics box. Thus, the interface is quite specific to the devices chosen. It is also affected by the data that comes from the traffic database, which must provide the input to automation to allow timely output of the correct graphics.

Many master control switchers have built-in still storage, which is particularly useful for storage of logos for insertion of "bugs," or for insertion of other information such as ratings. They may also have the ability to accept both SD and HD sources and to convert one to the other. It is also common to put format conversion on the input side of the switcher and keep all feeds native to one format internally. This is at least in part due to the complexity of handling graphics from different formats and defining the behavior of the system to optimize aspect ratio control issues.

There are many different protocols in use for interface between master control switchers and automation. Among the most popular are protocols developed by Grass Valley to support their SD only Master 21 product, and a follow on protocol developed for their HD version, the Master 2100. Other protocols include the Saturn Protocol developed by BTS (now part of Grass Valley), and the Oxtel protocol developed in the UK (now part of Grass Valley after acquisition of both Miranda and Grass Valley by Belden). Each protocol supports some features, but not necessarily all of the features of any one product for obvious competitive reasons. In general it is beyond the capability of any station to modify the interface to support other features since doing that would require access to the manufacturer's embedded control system coding in the master control switcher, not something likely to be available even under license. As a result it becomes a dance when considering interface between automation and master control switchers, with the customer identifying the features they want to support and the automation and master control manufacturer working together to implement the requests if possible.

A second element of baseband master control is the video server. Here the choices are many. It is, however, important to note that control over the video server must be done independently from the master control switcher. Thus, the coordination of playout events becomes a bit more complicated when events must be parsed to different devices for playing content, and getting it to air—perhaps with graphics generated externally, and separately controlled as well. This makes the interface between the traffic system and automation the key link in getting everything to work.

Master control switchers, of course, do much more than process images. They must also deal with sound, which can come from discrete audio inputs, or be interfaced using embedded audio. For

HD feeds it may be necessary to have a master control switcher with surround sound (six-channel) capability. Some master control switchers have the ability to accept compressed Dolby AC3 (up to six-channel) or Dolby E (up to eight-channel) audio. Mix-down from surround to stereo, or mono, is of course a concern for channels that do not distribute surround sound but may receive surround sound sources.

It is important not to leave out audio-over sources, which often match up to graphic elements to provide the matching sound for promotional material (e.g., "coming up at 11..."). Audio-over sources might be either discrete audio sources, mapped to the correct crosspoints, or could be audio files (.wav, MP3, etc.) transferred to local memory. In general, master control switchers have limited storage of audio files. The need may, however, be filled with an external audio record/playback system, which is then controlled by automation through a direct control or perhaps via automation-controlled GPI (general purpose interface, contact closures used for triggering events in a simple and reliable method).

Often master control switchers also provide manual controls to allow the operator to adjust audio levels as necessary, and also may include audio metering in the control panel or separately mounted metering panel. It is clear that with surround sound the audio monitoring and metering can be complex.

If the number of controlled devices seems daunting, consider that one master control switcher panel may address several channels, as in the case of a broadcast station with several sub-channels in the multiplex. Knowing which channel you are controlling is critically important! This brings up the question, "Can one operator with one panel effectively control multiple channels in an automation-assisted, manual control mode? Although it is physically possible, it is clear that a lot is going on. In addition to monitoring audio and video for multiple channels, the operator must watch multiple automation lists to insure all media has successfully cued and is ready to play to air.

There is a tremendous amount going on in master control and many opportunities for problems the master control operator must be ready to solve. With a discrete baseband system like the one being described here, it is clear that it takes a very well trained operator and a system that is fully debugged to make a smooth operation. What is not so clear is that the interface between the master control operator and the traffic department, whether virtual via the log and automation playlist, or real via telephone and other direct communication is the Achilles' heel of ever more complex baseband systems. The operator must be able to fully trust that the log as delivered to automation is without errors, that all of the media is present and quality checked, and that nothing will go wrong. Skepticism is always a healthy way to look at complex system managed by fallible beings.

5.2.1.2 Channel in a Box to the Rescue (?).

The complexity just described was just the kind of prescription for a problem that engineers, and marketing experts, like to have in front of them. The difference between baseband master control and CIAB is principally in that fewer hardware interfaces are required. With fewer boxes to be controlled the automation interface can in principle be simpler. Fewer devices, it is reasoned, will be less costly to install and maintain, and should be inherently more reliable. Over time this has proved to be true.

The simplest of CIAB systems incorporates graphics, switching, and audio and video clip (short) playback into a device that is a hybrid baseband/IT system. Such a system has external video (baseband) inputs, analog or AES digital audio (or embedded) inputs, and perhaps a graphics input (key and fill). But the rest of the processing is done internally, simplifying the interface to automation and allowing for fewer devices that must be maintained. The system can be built in a small, 1 or 2 rack unit, computer, with an interface card to accept the audio and video baseband inputs.

These systems can be treated as black boxes and if failures happen it is easier to replace it than with many individual boxes connected together. In this simple version, no content playout is included, and the box is often called a "branding engine," though it does much more than the name implies.

If one adds file (audio and video) playout to a branding engine, you achieve a further level of integration and reduction in box count that should result in better reliability. Countering this dynamic is the fact that a full CIAB solution is putting a lot of eggs in one basket. For a long time, many stations were reluctant to go to such full integration of many devices. This was in part because some early CIAB solutions were not quite ready for prime time and failures left no escape route. Add to that the high cost of early CIAB solutions and one gets a recipe for disaster (i.e., failures happen more often and the cost savings were minimal). The bad name this gave to some attempts still gets in the way, of course.

But modern CIAB solutions are defined by another important dynamic. As computers become more capable of handling full bandwidth high-speed media, HD streams, we find costs coming down and capabilities going up with increased reliability. The application of GPU (graphics processing unit) technology has materially improved the ability of CIAB systems to fully replace baseband discrete box solutions. Indeed, as processor speed has increased it has become possible for a single server to handle more than one television channel, further improving the economics of implementation.

An additional benefit is the very nature of a system based on common IT hardware. Packaging and software systems are becoming more supportable than low volume broadcast hardware, and the single box nature of such systems makes them easier to swap out in the case of failures. With a full CIAB channel containing local content storage (a few TB), it is quite feasible to use a CIAB as a full disaster recovery solution, deployed in a geographically diverse location, at manageable cost. The software, and commodity hardware nature of CIAB solutions, makes them easy to remotely monitor and control as well.

The tight integration of the graphics and playout systems in a CIAB makes it easy to implement automation internal to the system and maintain tight coupling to insure playlists run unattended most of the time. A typical architecture would include perhaps a few hours to a day of local content storage, with long-term storage in an external share storage platform. The supervisory automation system attending to many locally installed CIAB boxes can manage the movement of content to insure the right files are present in the right channels. A side benefit of this capability is to permit an N for M redundancy scheme where perhaps one spare is available to support five or more playout channels. Swapping a spare into a stream requires moving in the right playout list and beginning to populate the local disk with the appropriate content, perhaps moved at high speed to permit a minimum of switch-over time. While a one for one redundancy approach may be more desirable for some applications where the content is of high dollar value, N for M is a reasonable tradeoff for cost vs. reliability of the complete system.

To this point I have been describing a CIAB that lives in a hybrid world where live content is principally in baseband interconnections. The system is equipped with an interface card to support baseband I/O. There is no doubt that virtually all television plants at the time of this writing are very much of this ilk. But with the growth of OTT (over the top) channels delivered over the Internet, the baseband interface becomes an impediment. If we add only the ability to receive and deliver content as IP streams a further improvement in integration is achieved, and the amount of baseband infrastructure is reduced. OTT channels today are seldom live and as such have little reason to need baseband connections at all. Adding only compression software allows a CIAB system to evolve into a fully standalone channel capable of self-management after delivery of the traffic log.

It is not hard to see where this is going. With the inexorable integration of IT and broadcast infrastructures we are approaching the point that a channel can be defined as a server, or a virtual channel running in a server with other channels. Broadcast stations deliver compressed streams, so why not move the compression into the CIAB and eliminate still more separate boxes. All that holds back this is the ability of channels to receive and process live content. SMPTE and manufacturers are hard at work eliminating this last bottleneck. By the time you read this it is certain that CIAB solutions will be available with no baseband connections, or at the least with baseband relegated to stand-alone interface chassis that convert from baseband to streaming SDI (SMPTE 2022). This is both inevitable and some would say highly desirable since video can then be moved about like any other type of data, taking care to understand the network topology of course.

5.2.2 The Master Control Holy Grail

Assuming we have achieved full integration of an IT-based system using the architecture I have been describing, it is a small step away from our comfortable baseband living room to reach a fully virtualized IT environment that delivers streams of video content indistinguishable from what a television station of 2015 produces. The final step is to abandon single box per channel and accept that modern server implementations can achieve sufficient performance to deliver multiple channels, even from one blade in a modular server architecture. Placed in the same rack room I doubt anyone could tell the difference.

Such an approach would assign a virtualized operating system to each channel, though one to several might actually be implemented in a single compute chassis. Today, the only thing holding this back is the availability of virtualized GPU capability, but by the time this reaches the press, virtualized GPU will have sufficient processing power and speed to support HD video, and in the future 4K video as well.

What have we achieved? Instead of several racks of hardware interconnected systems we can achieve broadcast channels in a completely IT environment that only needs access to sufficient content resources (disk space), and access to the ultimate playout destination, whether OTT, cable, or broadcast release channel.

What will we lose by approaching this paradigm? First we likely lose nothing. If a channel needs to be monitored by an operator, nothing will be different. Attach a user interface (control panel), appropriate monitoring and act like nothing has changed, for at the operations level nothing has changed. Control panels can be built on a touch screen, saving hardware development and maintenance cost. You might even implement a virtual control panel as a web service to enable control from anywhere that is convenient. Couple that with monitoring of the audio and video and run master control from anywhere that is appropriate.

Backup of infrastructure becomes a software management problem. In the case of a cloud-based system, whether private or a commercially run cloud computing environment, disaster planning can be achieved by buying service in a geographically diverse cloud that is self-healing, or in the private cloud putting a second instance in another facility accessible to the delivery network needed.

Lastly take one more step. Move the servers out of the station's rack room into an IT environment that was designed for cooling and maintaining lots of computers. A station group could build a private cloud servicing all of their stations. An IT staff, properly cross-trained in broadcast operations and video/audio technology, could maintain the systems and when necessary swap out hardware. Make the final leap and move that facility to a cloud computing platform and eliminate capital replacement expense, moving to an all operating expense model.

What is holding us back from going this far? First, it is fear of losing control of our critically important real-time final product. There is the illusion in broadcast facilities that because the equipment is in the next room it is somehow more supportable, when the opposite is likely more the case.

Second, it is a lack of understanding of the advantages of commodity hardware and virtualization over purchasing and maintaining purpose-built hardware. Broadcast hardware is very reliable, but training on maintaining it is not easy to acquire. As all broadcast infrastructure moves inexorably toward IT-based equipment, the training in IT becomes ever more important, and the route to getting that training ever more distant for people who entered the business through a "television portal."

Third, management is reluctant to have assets stored in an environment often deemed insecure, and news reports of theft of data over the public networks make those who were not concerned more circumspect. The assumption seems to be that it is more secure in our own facility. But IT security experts are quick to point out that carrier-class cloud computing facilities are better protected than any corporate facility, certainly that of a small broadcaster. Part of this concern is that placing your assets, your crown jewels, in a publically accessible facility makes it easier for competition to see what you are doing. That is simply not true, of course.

Finally, the fear of being first is always difficult to get past. Over time it will become clear that this strategy is successful and those success stories will propagate and begin an inevitable shift to a more flexible infrastructure.

5.2.3 Other Aspects of Master Control

In practically all television stations today the digital stream includes more than one sub-channel. This makes stream assembly and monitoring more complicated, and is one of the principal reasons why all stations today use some form of automation. One operator simply cannot keep track of multiple streams without automation aiding the process. This means multiple automation playlists need to be displayed, most often on a single display that can toggle between views of a single stream or a multistream overview to show the operator a bigger picture of the total operation at the current moment.

5.2.3.1 Off Air Monitoring.
It is essential that master control monitor the outgoing stream as it is sent to the viewer. This presents obvious challenges, for few master control operators have the luxury of seeing the content on a large flat screen with no other visual or aural distractions. But to the largest extent possible it is essential that the viewer get the highest quality possible with captions, ratings information, emergency alerts, and other adjuncts to the program as they are intended. This will require special care and a physical arrangement conducive to hearing and seeing what the viewer will.

Audio is perhaps the hardest part of a television signal to monitor correctly. The difference between the physical environment of master control and that of the home is considerable. Add to the complicated nature of the problem the need to monitor in all audio formats the viewer will have. This includes mono, stereo, and surround sound, along with required descriptive video service (DVS), and potentially a second language program as well and it is clear that real-time monitoring of all formats is just not possible. It is, however, critical that the monitoring environment support all of the above to insure the viewer is well served. Headphones may be necessary (for mono and stereo audio) to overcome the noise in master control and permit quality judgments to be made. There must be surround speakers in any control room that monitors a surround sound service. Clearly, the audio image may be hard to accurately interpret, but it is important that the operator be able to know if there are phasing errors or other problems in the surround sound image.

Video monitoring may be done with multiple individual screens, but today multi-image processors allow a smaller number of monitors to show many more feeds. Those would include each stream as delivered to the transmitter, video from network and other local sources, studio control room output, video servers, and perhaps graphics sources as well. There should be the ability to bring each of the air signals to a larger monitor to view the image as it will be delivered to the home.

Modern flat panel monitoring presents great advantages to master control, including saving space, and permitting more sources to be viewed. But it comes with technology issues that must be understood and accounted for in design. The two most critical are latency and the quality of the picture itself.

Latency arises due to the nature of the processing both in the multi-image processor and the display itself. Both add delay to the video without compensating for the audio in the control room.[1] Compensating for that delay is critical if the operator is to be able to spot lip sync errors in outgoing streams, or source streams as well.

A second problem that is frankly hard to solve is the accuracy and matching of consumer flat panel displays in professional use like master control. This has long-term effects on the quality of the delivered signals, of course, since judging quality requires a monitor one can rely on for accurate rendition of the picture. Second, if monitors fail and need to be replaced, it is often impossible to get the same model again making it at best difficult to accurately adjust the replacement set to match the one it replaces. Make no mistake, this is not a problem we as an industry will solve soon, if ever, since flat panels are designed for the consumer marketplace, not for professional use.[2]

5.2.3.2 Closed Captions and Other Metadata and Content.
Master control must do more than just concatenate programs and interstitials. There are legal requirements to deliver other content with the pictures and sound, including closed captions, descriptive video service, and emergency alert services. In most facilities the caption and DVS content will be time synchronized with the content and embedded in the video/audio stream (SMPTE S292 or S259 most often). In some complex facilities captions are contained in a separate database and must be synchronized at the time of playout. A special case of this is content that is live and the captions must be multiplexed with the content as the program is broadcast.

---

[1] Consumer sets fed by HDMI account for internal differential delay between audio and video if both are delivered to the set on the single interconnect. But use of consumer displays in control rooms normally bypasses that intended use and delivers audio separately to speakers not associated with the display, resulting in lip sync errors if not corrected.

[2] There are a number of manufacturers who make high quality flat panel monitors with accurate and repeatable colorimetry. They are, however, available in only a few sizes, and at high cost.

5.2.3.3 Quality Control.
Master control must also have the ability to monitor both the essence (audio and video) for appropriate technical quality. In most facilities, this is done with waveform monitors that can display overall levels and chrominance information (vector or other display of color). These may be either stand-alone instruments or a "rasterizer" with an output that can be used as a source on the multi-image combiner for the monitor wall.

Other monitoring is also appropriate. For instance, the ability to display closed captions and audio levels (as bar graphs) on picture monitors is extremely important. Audio monitoring should be capable of showing all channels including surround sound, DVS, and other languages if part of the broadcast output.

Audio monitoring must also include loudness[3] as required by the FCC. From a practical standpoint all stations must log the loudness of all content.[4] It is also useful to have live monitoring, in addition to logged loudness values, to enable an operator to correct problems while they are on the air.

Unlike analog television, DTV requires the ability to monitor the syntax of the digital signal in addition to the levels in the audio and video portion of the data stream. This syntax monitoring includes PSIP and other required portions of the ATSC stream.[5]

5.2.3.4 Relationship to Traffic.
As mentioned earlier, master control does not originate the air log, it simply follows the recipe that the traffic department creates. Traffic organized the content, identifying when and how it will air, and the operations department in master control assembles the content according to the recipe. In an ideal world, the log would arrive with no errors, and all content will cue up on command and run successfully to air. The perfect traffic department should be commended, but not expected. Master control often is tasked with correcting problems that arise due to either defects in the air log, or which arise because media (content) is missing or defective. Changes made to the air log must be noted and returned to the traffic department so that the legal records of what was on air are precise.

Loading the log into automation can be either a manual process or one which is automatic, as when the automation system monitors a watch folder for new files and appends new content as received.

After the day is complete, the log must be returned to traffic for reconciliation of the actual aired content with the intended log delivered to master control. In some stations this is a fully manual process. But there are relatively new options that can improve that workflow considerably.

5.2.3.5 Broadcast Exchange Format (BXF).
BXF, SMPTE S2021, is a communications protocol that was developed by manufacturers to simplify the communication of messages between broadcast traffic, automation, and other software systems involved in broadcast workflow. A full description of BXF is beyond the scope of this chapter, but the importance to modern master control is worth noting.

BXF can provide a "live link" allowing traffic software to modify a single event, or append/replace a block of events in an automation playlist without operator intervention. One use is for changes to the log. Traffic can make a change in their system software, and the change is immediately reflected in the automation playlist. This would allow a spot to be replaced with another at the request of the advertiser with little fanfare, even up to nearly the time it is scheduled to air. Operational limits probably should be placed on the ability to change the log close to the time of air to insure content can be located by the automation software and be cued, ready for playback, with sufficient margin for manual intervention if problems arise.

After an event has been aired, BXF can permit the aired automation playlist to be automatically reconciled to the traffic department log. This facilitates efficient workflow and speeds to delivery of affidavits and invoices to the advertiser.

---

[3]See ATSC Recommended Practice A/65, which is referenced in FCC regulations. http://atsc.org/cms/standards/A_85-2013.pdf.

[4]Public stations are currently exempt, but in general full comply.

[5]See Chapter 3.2.

5.3 CENTRALIZING BROADCAST OPERATIONS

The rise of ownership of broadcast stations by large groups, and decline of small broadcast ownership, has put pressure on technology planning and operational costs in a time of falling margins from operations. As cash flow decreases, either sales must increase or operational costs will be further pinched.

Beginning in the mid to late 1990s, groups began to experiment with centralization with an eye to reducing both capital and labor costs. The theory is that the additional cost of interconnection over wide areas, by satellite or terrestrial data transmission, can be covered in reduced capital and operational costs. When true, this can provide multiple benefits. A local station master control facility handling multiple program feeds is shown in Fig. 5.5.2.

Clearly, if profits need to be preserved (or operating margins) and the operational plan makes sense for the mission of the station and the finances as well, then this should be pursued. For many years it was not clear that the cost of interconnection could be kept low enough to avoid swamping the labor savings, especially in small markets where labor cost is modest and personnel perform multiple functions. But as data transmission costs have fallen, the equation has tilted toward successful wide area centralization.

Operational experience has shown that in a well-designed facility a small staff can manage multiple streams of broadcast content. One of the keys to this working well is a very accurate traffic log. If labor is less involved in fixing problems in the log, especially after traffic has gone home for the night, then labor can be applied more efficiently. Indeed, some stations have combined broadcast operations and traffic into one department. In such a case it may well be that the air operation and traffic are in the same room, especially in facilities with little or no live programming.

The conversion from local to shared centralized operations is difficult and time consuming. A well thought-out plan, with contingency plans where barriers might slow or stop progress, is critically important. It is worth remembering that seldom does the staff at the station being moved to a centralized operation choose that option. They resist because their job, and that of their co-workers, is at risk. They may stall, or find reasons why "it can't work," some of which may well be worth exploring in an effort to maximize the result. Suffice it to say there is no single plan that can work in all situations.

FIGURE 5.5.2 The hub master control room at WISH-TV, Indianapolis, IN. (*Courtesy: WISH-TV.*)

5.3.1 Centralization Topology

There are many ways to organize a centralization plan. It is relevant to consider several aspects of the centralized plan. Notably,

- Do the stations retain any capacity for local origination without the central facility?
- Does automation control equipment at both the centralized facility and the stations?
- Is media stored at the central facility, at the station, or both?
- Where is interstitial content ingested into the playout system?
- Is the link to the station bidirectional for media?
- Is the air stream assembled completely at one location?

There is not a single method that can provide the answer. For instance, a group of public broadcasters might centralize all air operations, but leave the ability for each station to cut in during fund raising periods to have local autonomous control over when and what content is played for maximizing viewer pledging. A commercial group might choose to have local ingest of commercials that are locally produced and received, but transfer them via ftp to the central facility for ingest before air. They might also be equipped with localized ingest capability that goes directly into the shared storage at the central facility, giving some local autonomy and reducing the work load at the central facility. That local ingest might happen in the traffic department so operations staff can concentrate on other duties, like news and production of promotional material.

If the goal is to reduce labor to the largest extent possible, it seems intuitive that a fully centralized operation should be the preferred approach. That might depend though on the financial success of each station in the group. Some might be better with more local technical autonomy to allow for more flexibility in decisions that affect revenue-generating capability. For instance, a group might include stations in one of the top 10 markets and some in markets below 50. Choosing a single inflexible model might not be appropriate.

5.3.2 Fully Centralized

The simplest fully centralized model is often referred to as *hub and spoke*. In this model, the centralized facility fully takes over the air operation of each station it services. If there is local content, for instance live news that is produced at the station, it is returned to the network operations center (NOC) for integration into the final composite stream. Local commercial content and promotional material is delivered to the NOC via live video feeds or ftp transfer. This allows a minimum operations staff to remain at the station to provide production for live content and maintenance of infrastructure, including the transmitter and STL.

There are potential benefits to this topology that accrue from the choice to close the local master control. First is reduced capital expense, particularly if the station has little or no high definition capability. Second is labor savings. It is hard to quantify the labor savings in a general case since the labor rates in different markets make calculating the savings highly dependent on the specific station, even within a group. Third is the ability to provide more effective backup from the central NOC since it is more efficient to have redundant components all in one site instead of spread across many locations.

There of course are drawbacks as well. Fully centralizing leaves the station with little or no backup for the possibility of failure of the interconnection link to the NOC. It is also likely that even if some minimal capability is retained at the station, the availability and training of the local staff may make returning to local operations even in an emergency difficult at best. A facility so designed should be planned with some minimum level of disaster recovery, which can be as simple as leaving the old master control in place (assuming it exists), and putting some "evergreen" content on the local server to bridge the transition to disaster recovery operations. Note that these suggestions will apply to all of the topology approaches discussed in this chapter.

This approach was first used in stations owned by Ackerley Broadcasting in the late 1990s for stations in New York and elsewhere. At the time, the technology was not well developed and Ackerley had to break new ground to find hardware and workflow solutions that supported their goals. Ackerley coined the term CentralCasting©.[6]

5.3.3 Distributed Content and Control

Another topology to be considered is one where the content is not all distributed from a single NOC, and air operations are shared between the NOC and the stations. Consider that perhaps a group has a number of affiliates of the same network and chooses to use the same program schedule to support all of them. The common program stream could be created at a central NOC, and the content sent to each station where interstitials are inserted. This would allow local commercials and promotional content to be kept at the station and inserted at the time of air from local servers. The automation can be either fully free-standing, or an instance controlled from the NOC. This allows a minimum local staff needed only for ingest of locally received/generated content to be present, but no master control operator needs to be on site.

This can ease the difficulty of disaster recovery since commercial content can continue to play locally for some time, along with either satellite received network content, or evergreen material. Training for disaster recovery is a necessary element of planning. If not rehearsed regularly, it will be difficult to transition back to any local operation smoothly. This of course assumes there is local staff that can be called in as necessary for the emergency situation. One piece of a technical plan organized this way is that the automation selected should be capable of continuing to control local devices after being disconnected from the WAN that connects to the automation database in the NOC.

Another variant of this might include playing some of the long-form content from servers at the station under the same control of automation centrally monitored and controlled from the NOC. This might be applicable for public television stations, for instance, that receive content locally via file transfer from PBS over a combination satellite/terrestrial link.[7] This avoids unnecessarily transferring content to the NOC that is already resident at the station facility.

5.3.4 Remote Monitoring and Control

A third topology that has been used is perhaps the one with the lowest capital and operating cost. If one simply remote controls the infrastructure in a station from the central NOC, but does not move any high data rate content over the interconnection, then the cost of the WAN connection can be considerably lower. One might use Remote Desk Top or other means to control the station's automation from the central NOC, which clearly would use less bandwidth than moving content over the WAN. This is also the simplest way to provide for disaster recovery, for other than the fact that no one is sitting in the chair in the station's master control nothing dramatic has changed. The same comments about staff training for disaster recovery of course apply as well to this case.

This of course comes with a set of constraints. The WAN connection still must have enough bandwidth to permit remote monitoring. At low bit rate the monitoring may have considerable latency, which makes control over live switching more complicated. The failure of the WAN presents the same challenges as before, and with no eyes on the station locally it may be problematic to leave a transmitter running unattended and unmonitored. This likely means that at least one person needs to be in the building at the station at all times.

---

[6]CentralCasting was copywritten by Ackerley, now part of Sinclair Broadcasting, after acquisition of Fisher Broadcasting.

[7]The transfer of the content from the PBS Station Services Platform (SSP) to local air servers is controlled by local traffic interface to the SSP. The traffic department manager executes the ftp of the content after receipt and confirmation that the content is complete.

This approach, sometimes called the "New York Times Model" as early centralization efforts by the NY Times stations, is inexpensive to deploy, but still relies completely on the infrastructure in place. The automation system may need to be carefully evaluated to insure it is suitable for operation in this way.

5.3.5 Interconnection

Centralized operations can make use of many types of interconnection as appropriate to the topology chosen. Today, almost all NOCs are connected over bidirectional terrestrial data circuits. The bandwidth necessary for a fully centralized operation is easy to calculate The example shown in Table 5.5.1 is not representative of any specific station.

TABLE 5.5.1 Example Bandwidth for Centralized Operation

Service	NOC to Station Bandwidth	Station to NOC Bandwidth
Program delivery	15 Mb	15 Mb
Subchannel 1 delivery	6 MB	
Subchannel 2 delivery	6 MB	
Subchannel 3 delivery	6 MB	
Voice communications	1 Mb	2 Mb
Transmitter remote control	200 kb	200 kb
File transfer	2 Mb	10 Mb
Monitoring	2 Mb	8 Mb
Total bandwidth	38.2 Mb	35.2 Mb

As Table 5.5.1 shows, even in the case of a fully centralized operation with program content returning to the NOC from the station at fairly high data rate for high quality, a single DS3 (45 Mb) circuit would suffice for delivery of mezzanine-level signals for encoding at the station to ATSC compressed and multiplexed streams. Note that this table includes communications and file transfer, and an allocation for full time transmitter remote controls as well.

If a lower data topology is chosen, the data rate might be considerably lower. For instance, using a remote monitoring and control model the total interconnect could probably be accomplished in under 10 Mb (bidirectional).

In the event the WAN is down it is important to have a backup plan, perhaps at lower bandwidth via a VPN over the pubic Internet, to maintain critical services and some portion of the monitoring to allow the NOC to continue reasonably normal operations while the WAN service is being restored.

5.3.6 Traffic and Other Considerations

So far this chapter has related to only centralization of master control. Of course that is where the popular literature points for the maximum "bang for the buck." This is in no small measure due to the potential for savings in labor in master control. But other parts of broadcast operations can also benefit from centralization. Several station groups have chosen to centralize the traffic function, standardizing on a single traffic system implementation and then providing remote access to stations for sales order entry and other functions.

It is also possible to centralize the production of promotional content. This is particularly effective when the group has several stations with the same network affiliation and similar syndicated programming. This can also include centralized graphics production for news, allowing for better labor utilization for both interstitial production and original graphics for news.

5.3.7 Remote Monitoring of Centralized Operations

The purpose of remote monitoring is essentially to extend the view of the control room in the NOC all the way to the remote stations. The design of a successful remote control system always includes the things an operator in the station would be expected to monitor, but may extend beyond that to enable the remote operator to have a good "picture" of what is happening away from what physical access will allow.

That might take several forms. First, a low bit rate camera in operational areas of the station returned to the NOC allows the NOC operators to know when someone is present on the other end of the circuit to help them with troubleshooting. This may represent in some people's view a privacy concern, but it is very helpful when time is short and solutions to problems must come quickly.

Second, an output from the station's routing switcher should be fed into a return feed at moderate bit rate, sufficient to provide a low latency (3–10 frames) copy of feeds throughout the station. This allows the remote operator to look at signals that it is not practical to return full time. One of the choices might be a quad split showing four signals everyone agrees are of particular importance. For instance, in the case of a four channel stream multiplex the four streams should be fed into the quad split,[8] which normally can then be fed to the NOC where it is permanently on one of the monitoring screens. A method for monitoring all of the channels of the multiplex audio is also important, perhaps by selecting one at a time.

If bandwidth permits, the full ATSC multiplexed (off air) stream should be returned to the NOC where it can be not only decoded, but also fed into a stream monitor to check for issues in the encoding system. If that much bandwidth is not available, it is possible to put a stream monitor at the station and return the output to the NOC for evaluation. It is normal for the stream monitor to have a built-in web server to allow control and output screens to be accessed over the WAN from the NOC.

It is also useful to have SNMP monitoring of hardware at the station. The SNMP system might look at the health of local (station) terminal equipment, ATSC encoders, and other critical air chain equipment.

Some manufacturers have designed elegant systems that use simple network management protocol (SNMP), or proprietary software, to enable complex and very user-friendly environments for monitoring broadcast station systems. These are particularly valuable in monitoring stations from a shared NOC. These systems include custom alarms, thumbnails, sampling of audio and video at selected points in the system that are then returned to the monitoring system, and streaming media encoders for returning live feeds as well.

Live operation of a broadcast station, which may be in another time zone far from the NOC, requires a system that presents sound and pictures as close to real time as possible. In the case of remote monitoring and control, if any manual switching is to be done, a delay of even five frames is problematic.

5.3.7.1 Transmitter Controls.
The NOC must also be able to monitor and control the transmitter to meet the requirements of the FCC (on and off). Various schemes exist for extending transmitter remote controls over the WAN, including Remote Desktop Connection, or native protocol from the transmitter remote control system directly to a computer in the NOC. The remote control system should also be configured to send text or email alerts to the NOC in the event the direct connection over the WAN is down.

---

[8]Each channel should be presented on screen with closed captions and audio bar graphs on the quad split.

CHAPTER 5.6
VIDEO SWITCHERS

Brian J. Isaacson
Communitek Video Systems, New York, NY

5.1 INTRODUCTION

In the world of television production and engineering, one of the most relied on components is the video switcher. Video switchers come in multiple types, such as master control, studio production, and field production. In this chapter, we will concentrate on studio and field-based production switchers.

5.2 SWITCHER FEATURES

All switches have what are known as *sources*, otherwise known as *inputs*. The purpose of a video switcher is to allow a given number of inputs to be routed or switched to a designated output, with various transitions and effects. The primary switching occurs on the "Program Bus," and thus the output for that bus is the Program Output. Switchers also have a "Preview Bus," which allows a preselected source to be designated as the next source to be switched on the Program Bus. Whichever input is selected on the Preview Bus will be routed to the Preview output, which is either connected to a monitor or part of a multiviewer screen. Most newer digital switchers have a multiviewer that displays all inputs, preview, and program outputs. A "Take" button on the switcher panel will cause a transition to occur between the preselected preview input and the program input.

Figure 5.6.1 is a schematic block diagram of a typical production switcher.

The sources can be of various video formats, depending on the scaling capabilities of the switcher, and the outputs can also be of various formats depending on whether the switcher has scaling available for the outputs. Scaling is the ability to convert a source from one resolution and frame rate to another.

Ideally, since there are multiple formats in both the HD and SD environments, a studio or field type switcher needs to be flexible enough to input multiple format sources and output multiple formats for distribution or recording. Not all switchers have these capabilities, and thus the user must select a switcher that is compatible with the formats and frame rates of all the required video sources. For practical purposes, we will concentrate on the commercially available HD switchers, and associated broadcast HD formats.

The most common source type for production switchers are cameras. The typical HD camera format in the United States is 1080i or 720p. The most common frame rate for 1080i is 59.94 fps; however, there are instances where the frame rate might be 24 or 23.97 fps. The "i" in 1080i stands for interlace format, and the "p" in 720p stands for progress format. This brings us back to the

5.64 PRODUCTION SYSTEMS

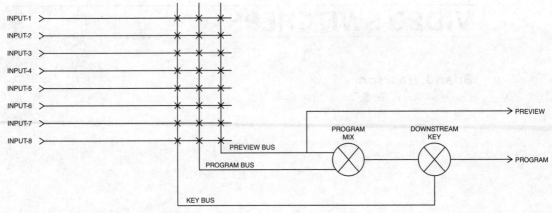

FIGURE 5.6.1 Basic switcher block diagram.

switcher's input scaling capability. Most production switchers will accept 1080i 59.94, or 720p 60 fps, but not necessarily the 23.97 or 24 fps rates. When selecting a switcher for a particular operation or application, it is necessary to know what the desired video source formats are, and the required final output format.

Another common type of video input source is graphics files (see Fig. 5.6.2). Most of the latest generation of switchers will accept direct graphics formats found in computer systems, such as .JPG or .BMP, for example. However, there are times when the required source will be an external computer system. The computer system may output DVI or HDMI. Some legacy computer systems only output VGA, which can be converted to DVI or HDMI using external devices. Likewise, there are some switchers that will accept VGA; however, most newer systems require HDMI or SDI as the source signal. In this case, external sources have to be configured to match the compatible switcher input formats. Most often this will be 1080i as 1920 × 1080 or 720p as 1280 × 720 resolution. With a scaling input processor, a switcher can accept various computer resolution and frame rates.

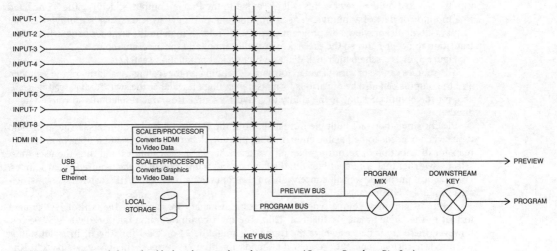

FIGURE 5.6.2 Expanded switcher block with external graphics sources. (*Courtesy Broadcast Pix, Inc.*)

5.2.1 Transitions

The primary function of a switcher, as mentioned previously, is to allow for inputs to be switched or routed to a designated output. The way the switch from one source to another occurs is known as a *transition*. Transitions can be straight cuts, dissolves (also called fades), wipes, 2D effects, 3D effects, key effects, or any combination of the aforementioned.

Straight cuts are just a switch between input sources on the program bus. Dissolves or fades create a blend between sources as one source is switched to another. Wipes are patterns such as horizontal sweep from left to right, or any pattern that the switcher offers as a wipe pattern. 2D effects are digital pushes, zooms, or any digital effect that occurs in the horizontal (X) plane and/or vertical plane (Y) of the image. 3D effects are digital effects that occur in the X, Y, and Z axes (the Z axis is perpendicular to/or at an angle to the plane of view).

5.2.2 Effects

Although dissolves, wipes, and fades can be considered "effects," they are more often termed transitions, and the more significant effects, such as keying, 2D or 3D effects, are more often referred as a "Mix-Effect" (ME). The Mix-Effects bank (or banks) contains a "from" and "to" bus, much like the preview and program bus, except in this case the Mix-Effects bank provides a predetermined effect that will be ported to the final program output. There can be more than one ME prior to the actual Program bus. Some switchers have two MEs or even three MEs, depending on their complexity. In this case, multiple layers of effects can be cascaded, one overlaid into the other, eventually creating a multilayer effect that is switched or transitioned to the program output.

Some typical Mix Effects include the following:

- 2D effects—digital push, zoom, or any digital effect that occurs in the horizontal (X) plane and/or vertical plane (Y) of the image.
- 3D effects—digital effects that occur in the Z axis, or perpendicular to the plane of view.

A sample of common effects is shown in Fig. 5.6.3.

FIGURE 5.6.3 Sample images of 2D and 3D switcher effects.

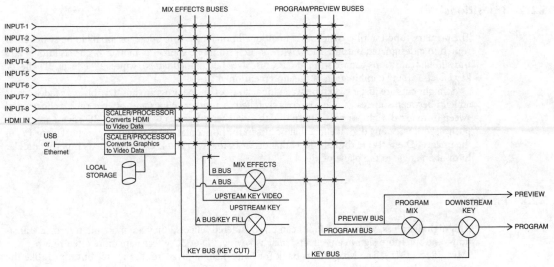

FIGURE 5.6.4 Upstream keying system block diagram.

One of the important effects functions of a video switcher is the ability to "key-in" graphics or titles over a given source or over the program output. A video key cuts a hole into a video source and fills that hole with another source. There are key controls that adjust the "slice level" or the level at which the key kicks-in and starts cutting a hole for the "fill" source.

When the key effect is directly inserted into the program output, it is referred to as a *downstream key*. The downstream key is a video cut-out that appears overlaid into the main program output. This is also sometimes referred as *overlay* instead of keying. Keys can also be "upstream" of the program bus; hence, the term "upstream keyer" (see Fig. 5.6.4). The difference between an upstream key and a downstream key is that the upstream key creates the overlay on a given source, which occurs before it is switched to the program bus, and thus the effect becomes a source that can be switched into the program output, versus a downstream key, which creates an overlay over the program output. Most newer digital switchers allow multiple keyers, and a given number of layers of keys can be combined to produce a multilayered effect.

5.2.3 Chroma Key

Chroma key is another form of keying, which derives the key cutter or hole from a selected color. Typical colors for generating chroma key are green and blue, thus the term "green screen" or "blue screen." If green is the selected color for the chroma key, then anything in the scene or background that has the color green will be keyed-out and replaced with the fill signal. A typical example is a TV news operation that desires to have a person on camera sitting in the foreground, with another video source in the background. The background or fill source can be another camera, graphics, or video file source. Over the years, and with the transition from analog to digital, the chroma key function has become more refined, with the ability to fine-tune the chroma key system. This functionality is also the basis for what is termed the "Virtual Set."

A virtual set is composite of foreground and background components that have computer generated elements, and allows the insertion of camera or other video sources into the "set" (see Fig. 5.6.5). In this case, the composite scene looks like a studio set, without the physical elements normally associated with a studio set.

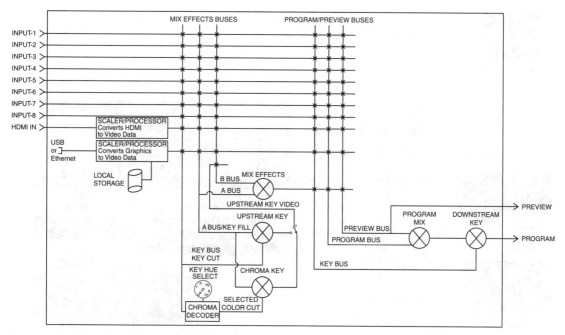

FIGURE 5.6.5 Block diagram of a chroma key system.

5.2.4 Video File Roll-in

Another function of the production switcher is to allow video sources from premade video files to be "rolled-in" or transitioned to, as a source. In the older analog technology, a video source was a video-tape that was physically played back on a VTR, which was connected to the switcher, and selected as a source; hence the term "roll-in." However, with the proliferation of digital technology, most, if not all prerecorded sources are video files that have been captured and edited as a given file type. Once the file is generated by the camera or edit system, it can be ported to the video switcher either by direct file transfer to the switcher or on a video server that provides an HD-SDI digital signal, much like a camera source. See Fig. 5.6.6. File types for video files are typically MPEG2, MPEG4, AVI, or MOV. Switchers generally have specifications for the type of video files that are compatible, including frame rates and resolutions.

5.2.5 Titling

In the past, if you wanted titles superimposed over video, you used a standalone unit known as character generator (CG). Most newer productions switchers have an integrated or optional CG. In many of today's production switchers, and in particular software-based switchers, the titling function is part of the package. Titling is much like keying, utilizing graphics that are both alpha numeric and image based. A typical title operation will place the characters in the lower-thirds of the video field; for example, with a graphic background as a composite graphics that is overlaid with some amount of transparency.

Titles can be placed in any part of the video field, and can incorporate motion, sometimes referred to as animation. For example, see Fig. 5.6.7. Typical motion includes scrolling from bottom to top of the field or vice versa, or a crawl that looks like a ticker tape across the field in a horizontal motion, usually toward the lower third of the field.

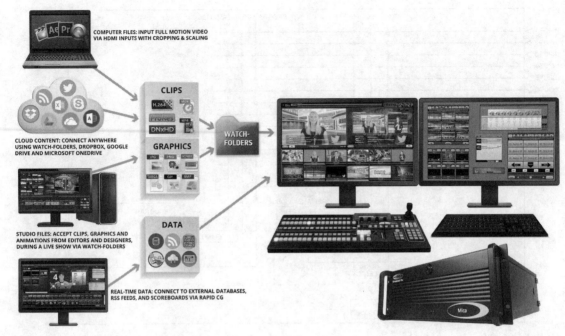

FIGURE 5.6.6 Video file transfer process. (*Courtesy Broadcast Pix, Inc.*)

FIGURE 5.6.7 Tilter images with lower thirds, scrolling, crawling, and multiple graphic elements. (*Courtesy Broadcast Pix, Inc.*)

FIGURE 5.6.8 Large video switcher control panel. (*Courtesy Broadcast Pix, Inc.*)

5.2.6 Control Panels

Most, but not all switchers have a dedicated hardware control panel, sometimes referred to as a control surface. The control panel contains dedicated and/or programmable buttons that permit direct switching of all sources for the program, preview, and Mix Effects buses. Controls for setting the upstream and downstream key levels, chroma key settings, transition rates, wipe pattern, preset effects, and graphic and video file selection, are typical of a dedicated control panel. On most dedicated panels you find a handle called a "T-Bar." The T-Bar allows for manual transitions between preview and program buses or source transitions on a Mix-Effects bank. An example control panel is shown in Fig. 5.6.8.

In software-based switchers, many of these functions are mouse-driven or touch-screen driven. These are known as virtual control panels, and are found on some smaller compute-based switchers.

5.3 THE BIG PICTURE

The video switcher provides a central role in the production of television programs of all kinds—from live sporting events to newscasts and music shows. The switcher has evolved over the past several decades to adapt to new technologies and new levels of program complexity. The sophisticated productions routinely done today would be impossible without advanced switchers. Like audio mixing systems, video switchers have evolved to encompass roles far removed from their original purpose of simply switching input signals. The history of video switchers tells us there is more to come.

CHAPTER 5.7
AUTOMATION SYSTEMS

Gary Olson
GHO Group, New York, NY

5.1 INTRODUCTION

Automation has a long and storied history in broadcasting. An automation system is essentially a machine control system, and can be characterized as a system that controls other devices and systems based on instructions, schedules, rules, and policies. In the complete creative, production, handling, transport, and media management architecture there are very few operations that do not use some form of machine control and automation.

Automation is an integrated and essential technology component in the entire media production and distribution architecture. In many ways, it is one of the more complex systems since there are so many different devices and systems. Automation combines command and control with device and system management.

The automation system is responsible for some core features and functionalities; "automatic" operation and control is just one of them. The automation system needs to be aware of all the resources it will be managing. It needs an interface to communicate with each of these resources. Notably:

- Are they online and available?
- Are there enough channels, playout devices or router I/O to handle its instructions?
- What policies and rules does it need to obey as it makes these decisions?

The automation system will schedule these resources, so it needs to be able to do conflict management and resolution. What's that? If there are 10 events that need to occur but only five available devices, there is a conflict. The automation system needs to have a set of rules to decide which event takes priority. The automation needs to report if an event does not happen and if there is additional error information to capture that in a log.

5.1.1 A Bit of History

Automation has actually been around since 1959 when NBC first used it to delay the recording and playout of programs for different time zone feeds. However, playout automation as we know it today was for all practical purposes introduced in the 1970s. It was a little cumbersome and it was triggered via relay closure.

The automation was limited to triggering a tape machine to start or stop—there was no cueing; programs were stacked on tapes and the operator had the rundown sheets or "playlists." Traffic was a printout and the schedule was entered by hand. Early automation was intended to resolve some of the "On-Air" mistakes that required "make-goods" (missed aired spots) and therefore revenue loss. In one case study, a cable operator felt the need to automate their commercial insertion process in master control. On a tour of their master control facility, there were operators changing tapes for the next set of commercials. As they were putting tapes in, they tried to put a tape in a machine that already had one in it. It took a while to realize the problem. Automation would not completely solve this issue, but having many breaks on a single tape, and the automation to run them, would greatly reduce errors. Traffic logs were hand-written notes taken by the master control operator and transcribed when delivered back to the traffic department. The master control operator was responsible for starting audio carts, setting up the graphics in the character generator, and keying it in. There was no automation here.

The machine control was a time clock with pegs that tripped a lever. This was set up by an operator working off a hand-written or typed schedule and as the clock hit the correct time, the peg would throw the lever that tripped a relay. This was all hard wired to either the record or play button of tape machines. Automation began taking on a larger role as the master control switch could also be triggered. It could not be programmed, only triggered. So, the operator still had to cue the correct source and the switch had a GPI (general purpose interface) input that was simple contact closure. When the clock triggered the GPI, it controlled the TAKE button and took the next source to air, essentially off the preview buss. Any keying was manual. Audio carts were triggered to start, not stop. There was no cueing; that was all done manually.

One of the next iterations of automation was using audible tones in the program where the breaks should be, with space left in the program for inserting the commercials and interstitials. These tones, DTMF (dual tone multifrequency), were adopted from the telephone industry. Broadcasters found that these audible tones could be placed in the audio track before and after the program and could be decoded to trigger a relay. This could start the tape machine and tell the master control switch to switch to the next source. Now all the operator had to do was make sure the correct commercials and interstitials were loaded into tape machines. The content for each program break was edited together on a single tape. There was no switching between tapes.

Cable operators had the toughest challenges. They were the first multichannel distribution systems. They had to insert commercials and interstitials across multiple channels at the same time. Now it started getting complicated. They had multiple schedules and prerecorded commercials and interstitials. Each channel had a few machines assigned to it. Some for program and some for breaks. This added up to a lot of machines and it was hugely impractical to have operators running up and down equipment aisles pressing buttons. This is where DTMF tones came in. The networks were able to remotely manage the breaks from their broadcast center to the local stations and then cable operators by placing the tones where they wanted the breaks and putting black or a network ID as a place holder if the local station or cable operator missed the break.

As serial RS-232 and parallel machine control entered the picture, automation could now do a few more tasks than just start a tape machine. Jumping forward, RS-422 was introduced and quickly adopted by SMPTE as the standard for machine control protocol. It introduced a whole new world of machine control and automation into the broadcast and production process. The SMPTE RS-422 protocol enabled complete control of a device including the integration of timecode, making it frame-accurate and fully synchronous. This protocol is a point-to-point or device-to-device physical connection from the controller. SMPTE RS-422 was a standard communication protocol that enabled controllers to interface with different types of machines. The command set for individual device types can be different; however, the communication between the controller and the device uses a standard protocol.

Automation stayed in the master control and program origination areas. As the networks needed to originate multiple programs while cable and satellite services expanded, automation was challenged. Similar to what has more recently occurred with asset management systems, organizations started building their own automation systems as there were very few vendors in the automation space to begin with and fewer doing multichannel.

Automation needs instruction and for program origination that's where traffic came in. Actually, traffic was one of the original cloud services before "cloud" even existed. Companies like BMS, BIAS, Columbine, JDS2000 (Encoda), and VCI collected the traffic information from program providers and then sent the aggregated data over modems from a central service to mini mainframes located at each station. These computers had printers attached to receive the schedules. There were machine controllers as the automation and after the commercials and interstitials were stacked on tapes, and this automation would receive the DTMF tones and trigger the next element to play.

5.1.1.1 Case Study. In the early days of cable systems, a regional cable operator running about 10 channels was using a tape machine sequencer to run their playback machines. A sequencer was an early machine controller than handled up to three tape machines. They received the schedules on a mini mainframe and printed it for the master control operator to use when they programmed the sequence. The traffic department was responsible for consolidating all the ad spots and interstitials together for each break. The operator programmed the sequencer with time clock based on where the breaks were. The sequencer had the audio and video outputs from each machine and would switch to the next machine once the machine was given the play command. At the end of the sequence, the operator would change tapes. All breaks happened at a specific time for a specific duration.

In a desire to reduce the number of make-goods, they wanted to implement an automation system. At the time there were two vendors in the market. (Neither one exists anymore.) They opted for the third alternative, commissioning a programmer to develop a custom automation program that would be developed for them on a new computer system. The programmer did not know broadcast operations or have any experience with machine control. At the time his background was accounting and he had recently installed a new computer for their traffic department. Clearly, he had the right knowledge base to build a new automation system with a new interface for machine control interfacing with a traffic system. Or not. It was not pretty. It was not ultimately delivered as promised.

Custom has many drawbacks: sustainability, extensibility, and maintenance are some. It's more complicated when the programmer is unfamiliar with the use of case and devices they need to control. In the end, the cable operator went with one of the two products in the market.

5.2 BUT ENOUGH HISTORY

Automation has continued to evolve as the device interfaces and protocols have become more robust. GPI/O and relay had limited functionality, RS-232 serial enabled more control, and RS-422 introduced timecode and frame-accurate triggering and cueing. Cueing is BIG. The automation system can instruct the tape machine to go to a specific timecode position and then play that program or interstitial. Now breaks did not need to be edited together. All the interstitials could be stacked on a tape or tapes and the automation systems given the in and out points. As the RS-422 protocol became the control standard, automation could now have greater control over the switcher, tape machines, and graphics system. The automation system began to control routers, graphic devices, master control switchers and playout VTRs, and now servers.

First, the master control room has evolved into a multichannel program origination center. Even a single station has multiple program feeds and in addition to having their own program content, each feed has its own branding and interstitials. This means each feed has a playout channel, graphics, commercials, interstitials, and needs it's own master control switcher to control and manage playout.

Multichannel origination added many layers of complexity, and this is where automation plays a significant role. The obvious benefit of automation was the reduction in "On Air" errors. That being said, automation is essential for multichannel origination, considering the number of sources, playlists, and different branding—it would be difficult if not impossible without automation. New protocols allowed the automation to control more devices.

File-based program origination introduced a new set of challenges. It needed its own automation system and had to interface with the traffic system. Now, there is a tape-based automation control and file-based automation control running off the same playlist. The next significant change in program origination was multiplatform. This added encoders, transcoders, and file transfers and more work for the automation system.

While automation had its origins in program delay and playout, it quickly moved into production—beginning in the edit suite. As computer-controlled editing technology was introduced into the edit process, machine control took on new responsibilities. In an interesting segregation of terminology, playout retained the terminology "automation." The editing community adopted the term "machine control" as the way edit systems integrated with devices, leaving automation as a function of master control. They are the same thing.

The creative process creates the Edit Decision List (EDL), which identifies the entry and exit points, source materials, transitions, and effects of the audio and video. Once the EDL is finalized, the edit system was told to execute the list or to *conform* it. As nonlinear editing (NLE) entered the picture, so did more automated processes. Files moving between systems using "watch folders" is automation. When it is time to output or render the finished program, the NLE using the EDL executes the rendering process. This is the same as conforming. The rendering process is an automation process, following an instruction set to execute an operation.

The next process to get automation was the production studio. The production switcher has GPI/O and triggers the digital effects or image store, possibly the character generator and it tells the router which sources to feed which inputs. How about camera robotics or multiviewer? Or the router? Isn't a router salvo automation?

There are many new devices that all use automation to perform their operations: encoders, asset managers, file transport, playout servers, studio in a box, and channel in a box. The QC processes for files and streams are automated processes.

In the media handling and management workflow there are very few operations that do not use some form of machine control and automation. There are a number of reasons to use automation for different operations and processes in the broadcast and production workflow. One of the first thoughts that come to mind is increased efficiency and reducing the number of manual operations. While that is one, it is not typically the main reason.

Broadcast and production devices and systems have become IP appliances or software applications. They are servers operating over a network and in the IP and file-based architecture, devices and systems have a reason to communicate with each other. Even those devices that are not IP in their operation (i.e., production switch, SDI router, CCU, and lighting console) all use an IP management interface and are network-attached. Given all this, the complexity of applications, systems, and device control has made it very difficult for manual operations. In addition, the number of concurrent processes that occur in all workflows almost requires an automation system to control and manage them. Furthermore, there are many processes that run unattended (and need automation to manage them).

The communication protocols have changed as well; instead of RS-232, RS-485, RS-422, and GPI/O, there is IP. The management or control interface between devices is an IP connection, even if the device is not fully IP.

Automation needs to communicate to all the devices it will be controlling. One interesting development or complication is at the time of this writing, there are no standards in machine communication over IP. There are programming languages and descriptive instructions but no standards. XML and JSON are the most common languages; however, each vendor and service provider has their own instruction set called an API or Application Programming Interface. They will sometimes provide a Software Development Kit (SDK), which is the tool they will allow to build the interface for systems to communicate. EBU in a joint task force is making progress with a common command protocol, called Framework for Interoperability of Media Services (FIMS). Their goal is to have interoperability at the control level of devices.

Staying with program origination a while longer, in addition to multiple channels or program streams there are different platforms and within these platforms different formats and bitrates. Commercial and interstitial integration is different. Also, the commercials and interstitials are different by markets segment. This adds more complexity to the automation process. Sending commands to a cloud-based service to move files, play content, and track resources is a lot of programming.

At the end of the day, automation is still a software application. As of this writing, cloud-based master control or program origination was being introduced. Automation takes on a new set of responsibilities and interfaces. The automation system has to track where the content is, transport it between systems and locations, and trigger processes based on multiple playlists. The automation system needs to know what format and bitrate each channel and platform requires, what metadata needs to be sent with the content, and what metadata gets sent to the Electronic Program Guide aggregator.

5.3 ORCHESTRATION—THE NEXT GENERATION OF AUTOMATION

Miriam Webster defines orchestration as "the arrangement of a musical composition for performance by an orchestra."

The computer industry describes it as "the automated arrangement, coordination, and management of complex computer systems, middleware, and services."

Actually they are not very different, but what does any of this have to do with broadcast? Quite a lot! As previously mentioned, automation has evolved as broadcast and production systems have become more complex and sophisticated. Each aspect of production and broadcast uses automation tools to create, manage, transport, distribute, protect, and archive program content. File-based workflows are dependent on automated processes to manage and move files throughout the entire media architecture.

It is cumbersome and difficult to program each system's automation. At the same time, it is difficult to communicate with each system to trigger a sequence of processes.

The Enterprise IT world realized that the interaction required between software applications and the movement of files between systems and processes needed to be managed, controlled, monitored, and access-protected. Business Process Management tools or BPM was introduced as a way to resolve this problem. BPM is a layer of middleware applications that sits between other applications (in the middle) to integrate them or facilitate file movement and access. Another process to manage multiple systems and devices is called the Enterprise Service Bus or ESB.

ESB is an application that connects to other applications and creates a common platform for control, monitoring, and management of the connected applications. This is analogous to using a browser application to control devices in a card frame where the backplane is the common bus for communication.

As cloud services entered the picture and hosting application development in the cloud became commonplace, the need developed to have a single application to control and manage all these different and disparate processes and file movement.

Enter Orchestration, the next tier or generation in automation for the broadcast and production industry. Instead of each process having its own control, monitoring, and management layer, orchestration is a single tool that unifies the control of multiple systems and devices into a single control application. This application manages and controls each of the processes by sending commands to each application and subsystem, reading and reporting on the status of each process in a clean and easy way to monitor user interface.

As broadcast and production technology has become more application- and server-centric, the need has also increased for new applications and middleware tools to handle it. The stream- and file-based workflow are not single thread processes. There are multiple devices and applications at each stage of production and broadcast. There are multiple encoders, processing different formats and bitrates with files moving down one path, streams another. Asset management associates metadata with media and allows business applications access for protection and monetization. Files move between production, edit, and distribution systems. While many vendors offer "end to end" solutions, there really are no "beginning to end" products or services.

It is too complex to try and manage all this using separate dashboards (user interface) for each application. Orchestration tools streamline this and adds the "single screen" interface to monitor all the processes.

The broadcast vendors have realized this and are now including in their offerings orchestration products.[1] The asset management companies are now supporting device control, and the automation vendors are expanding beyond master control. Each of these needs to be able to connect to the various systems. This is where FIMS can play a major role. Otherwise, connecting systems becomes a major software development exercise.

---

[1] Imagine Communications has Magellan and Zenium, Sony has the Media Backbone, Evertz has Magnum, and IRM and Grass Valley has iControl.

In the evolution to IP, the master control room has become more of a Network Operations Center (NOC) and orchestration is a needed and critical component. The following list captures some of the responsibilities that the orchestration technology must monitor and manage:

- Make sure devices are to the correct parameters (i.e., encoders and transcoders)
- Configure each device based on traffic and schedule data
- Manage the transfer or transport of the encoded file and/or stream to each stage in the process
- Assign resources to each process
    a) Fiber or satellite circuit
    b) Encode chain
    c) Storage location
    d) Play out chain
    c) QC chain
- Monitor each of the processes to address any conflicts such as over-allocating resources
- Monitor network utilization
- Balance the distribution of processes across applications to optimize performance
- Monitor systems status for any alerts and alarms
- Monitor automated QC processes

5.4 SUMMARY

The need for orchestration evolved as a natural next generation of automation technology. As devices, applications, and systems all communicated using a command and control layer within the IP infrastructure, orchestration enabled integration with each other. Orchestration is also critical as more applications are becoming cloud-based. The integration of cloud services into the production environment will depend on orchestration tools to manage and control the processes in a seamless way.

We will continue to see more orchestration products enter the marketplace and also watch as the solution providers understand the importance and value of looking at the entire architecture from a command and control perspective. They will either develop their own products or create software interfaces that will work with another vendor's orchestration tool to manage their products.

CHAPTER 5.8
MEDIA ASSET MANAGEMENT

Sam Bogoch
Axle Video LLC, Boston, MA

5.1 INTRODUCTION

There are few tasks more frustrating than staring at an unsorted mass of media files, and being asked to find relevant material for a task at hand; Media Asset Management (MAM) systems are used to solve this problem. MAMs are databases, usually based on IT-standard SQL technology, that contain entries for all the media files for a given enterprise, department, or workflow. By using a database, it's possible to assign additional keywords and metadata to individual assets to make it easier to find them later. In addition, MAMs generally include technology to make low-resolution proxy media that can be viewed from desktop and mobile devices. When users are watching the media, they can add timeline-based comments and mark in/out points, further streamlining later use of the media in nonlinear editing (NLE) applications.

MAM is an obscure undertaking. It's only evolved into a field in its own right over the last decade, after spending years in the shadow of the more widely known DAM (digital asset management) and ECM (enterprise content management) fields. ECM covers true enterprise software for warehousing all the files typical of a large organization—things like Word, Powerpoint, and Excel files, for instance. Its cousin DAM emerged over the last 25 years to do many of the same things, but focused on files from creative applications like Adobe Photoshop, Illustrator, InDesign, and Quark XPress. MAMs are like DAMs, but aimed at video, audio, and accompanying image files.

MAM is a relatively new category mostly because the storage capacities and network speeds needed to natively support working video catalogs and archives have started to become available after 2000. But there are clear trends ensuring that MAM is growing faster than its better-established cousins. Mostly, it's because video is becoming a mainstream format, rather than an obscure offshoot of digital technology that required proprietary file types and storage formats. Examples of this mainstreaming can be seen as widely as in the case of YouTube, Facebook Video, and Vimeo, which are global web-based MAMs of a sort.

Due to the popularity of and immediate accessibility of video editing software, the perceived role of MAM software has been a secondary one even within the video field. But this too is changing. The big driver here is the sheer amount of HD video being captured, edited, and published. Hundreds of hours of video footage are being uploaded to YouTube every minute. Video has become a key battleground between online technology companies. Meanwhile, Microsoft and Google are building video-specific capabilities into their cloud offerings as a way to catch up with Amazon's early lead in commercial cloud computing. And, of course, mobile network providers and media companies are increasingly duking it out to provide not only the most video content, but also uniquely differentiated video content to their subscribers. With all these forces at work, it is inevitable that the role of managing all this content—media asset management—will become more mainstream, even central, over time.

5.1.1 How MAM Is Used

The uses of MAM technology can probably best be understood as falling into five categories:

- Browsing and searching media
- Tagging and adding metadata
- Collaboration within a local or distributed team
- Archiving and repurposing media for playout, web, and mobile
- Workflow automation and process orchestration

Compared to some of the more traditional solution categories ("playout" and "nonlinear editing," for instance) media asset management has a relatively short history and is likely still early in the process of evolving into what it will eventually become. This is because, until broadcast systems went all-digital and centralized on shared digital storage in the years 2000–2010, there was not much room or even need for a media asset management system. It would have been completely impractical to take every videotape or cartridge off the shelf, digitize it, create a low version proxy of it, take the further time to review this proxy and add metadata to it, and so forth just to have a complete catalog. But because broadcast and postproduction facilities have passed the tipping point on their transition to digital, and can store most of their media on large shared systems—whether network-attached storage (NAS), storage-area network (SAN), or a mix of both—it is now not only practical, but compelling, to have a software system comb through the media, make low resolution versions of them, grab as much metadata as possible from the files, and allow users to add further metadata, locators, sub clipping in/out points, etc.

Without such a system, you would just have an ever-growing, undifferentiated pool of files with similar-sounding filenames. With it, you have a MAM.

5.2 A BRIEF HISTORY OF MODERN MAM

The field of MAM is a diverse one today, and the entrants are almost too numerous to cover here. Rather than drill down into details of all the products on the market today, I am going to provide some key criteria and categories that can help distinguish between the types of MAM product that exist. Using these criteria, we can start to understand how the use of these products, which are still developing and likely to keep evolving in the coming years. There are four main distinctions I will draw: between work-in-progress and archival systems; between passive and active systems; between heavy-client and browser-based user interfaces; and between on-premise and cloud systems.

Before we get into the distinctions, I should spend a little time outlining the general history of the MAM marketplace. Probably the first widely used MAM system, though the term didn't exist back then, was the Media Manager software provided as part of Avid's Unity hardware/software workgroup solution in 2002–2007. The hardware part of the solution was pretty straightforward, at least in hindsight—a FibreChannel SAN with drives formatted in RAID 0 and 1 layouts. But the idea of a system that included software for searching and managing this storage was pretty revolutionary at the time. It was the first mainstream attempt at a work-in-progress MAM, which would track video and audio assets even while they were being edited, rather than waiting for finished media to be output and archived.

This early implementation was quite forward-looking, as well—for one thing it took advantage of browser technology on the client side, so that searching and browsing media could be accomplished on any workstation or laptop with a browser, rather than requiring installation of client software. On the back end, it used a standard SQL engine, rather than a proprietary database engine. The product was quite popular, used by hundreds of TV stations and media companies. However, as its uses grew, customers encountered limitations in these early tools. They wanted interactive features that were difficult or impossible to implement in the browsers of the time, and they wanted the system to handle hundreds of thousands (or millions) of constantly changing assets. Other, powerful MAM systems were starting to become available around this time, making the limitations of the first-generation system more apparent.

For these reasons, Avid undertook the development of a next-generation, production-oriented MAM, called Interplay. It was a powerful multiapplication suite which offered tight integration with Avid's NLEs, as well as lesser degrees of integration with other editors in the future, and no less than three additional client applications: Interplay Access (for administrators) and Interplay Assist and iNEWS Instinct, which also served as a bridge to Avid's newsroom products. The back end ran on industry standard Wintel systems and consisted of an object-oriented database acquired with AlienBrain, a Munich-based development team with a DAM background, along with additional dedicated systems for data transfer, media transcoding, and system monitoring. Later, even more back-end services were added to the system such as a streaming server for H.264 proxies, and a mixdown server capable of rendering out flattened clips from simple AAF sequences. A complete Interplay system, of which over 2000 have been sold, typically runs on 6–20 servers and is connected to editing storage as well as anywhere from 5 to 300 workstations.

This general approach—of a number of dedicated servers, often one or more per function, running on industry-standard rack-mount hardware—was also taken by Interplay's main competitors as well as a number of other vendors in the sprawling MAM marketplace that evolved over time. There were, and are, a number of systems that specialized in different niche capabilities or geographic customer bases; each of these occupied a small slice of the global marketplace, which at this writing is approaching 7000 networked systems in total.

5.2.1 The New Wave

Around the same time (2009–2012), a new wave of MAM software began to emerge, characterized by browser user interfaces rather than Mac and Windows client applications, shorter learning curves, and a focus on review and approval applications rather than in-depth production tools. Most of these are cloud-based; one entrant, axle Video, took a contrasting approach of providing an on-premise solution, connected to main storage, which nonetheless offers 'radical simplicity' in its user interface and deployment. These vendors have continued to rapidly evolved their products and add functionality over time, so that they are beginning to rival legacy systems in their capabilities. However, their more modern front ends and Service-Oriented Architecture (SOA)/Representational State Transfer (REST) back-end architectures may give them an advantage in speed of development and integrations. An array of legacy systems have not stood still, however, and are all starting to offer more modern browser front ends, though typically with layers of integration underneath that increase complexity and cost.

The most recent wave of cloud-based browser applications (2012–2015) can best be described as even lighter-weight; these are focused on offering great user experiences for review and approval, and are all cloud based, while some shared storage vendors are beginning to offer some basic media management functionality with their storage systems (much as Avid had done a decade earlier with Unity and MediaBrowse).

5.3 FOUR WAYS TO CATEGORIZE MAM

5.3.1 Work in Progress versus Archival

Some MAMs focus on media that has just been ingested and is being actively edited. These, known as work in progress MAMs, tend to offer integration with production tools such as FCP, Adobe, and Avid, and support for a wide range of file types and workflows. Others, however, focus on finished media and its use in playout, send-to-web, and send-to-mobile applications. These, referred to as archival MAMs, will often have support for more sophisticated metadata schemas, and librarian-quality cataloging and search tools as well as integration with CDNs and rights management software.

5.3.2 Active versus Passive

Nearly all MAMs have historically taken an active approach to managing media, where the MAM controls the location of the media and there is a rigorous check-in/check-out workflow process for each asset. This can be real impediment to adoption in creative environments; however, freelancers and others on the wider team may want to continue to working with assets located, and named, as they see fit. In contrast, a few MAMs have taken a less-is-more, "passive" approach, where the MAM simply catalogs the contents of one or more storage volumes or folder structures. The biggest advantage of such a system is that it requires no fundamental changes in the workflow of the team using the media, and can also be more immune to upgrade and migration problems in editing or other client-side software where version-specific plug-ins are often required by the active MAM approach. Since over 90 percent of media storage today is unmanaged by any sort of MAM, passive approaches would appear to hold great promise in providing most of the benefits of MAM without nearly as many steps in implementation and training.

5.3.3 Native Client versus Browser Based

The design of MAM systems is fairly evenly split between those which opt for a browser interface, favoring the ease of client deployment over the loss of native-application speed and functionality once installed, and those which offer a native (i.e., installable) client for Mac, Windows, and sometimes Linux. The recent surge in importance of mobile devices also means that "native" can include iOS and Android applications as well, though these are in many cases being implemented for MAM systems that were once browser-only.

5.3.4 On Premise versus Cloud

An on-premise MAM system typically includes the MAM software itself, as well as the CPU hardware to run it and the storage it catalogs, installed at the customer's corporate or departmental headquarters. Access can generally be from local or remote locations, depending on the firewall settings chosen by the customer. A cloud MAM, on the other hand, will generally have both the MAM software, and the processing and storage hardware all located in cloud facilities. Here too, distinctions are being blurred, as on-premise MAMs are increasingly integrating with cloud storage, and cloud systems are beginning to be deployed on premise in "private cloud" configurations.

5.4 XML—A KEY INTERCHANGE FORMAT

One important ingredient in the maturing of the MAM marketplace is the ability of customers and their integrators to easily transition teams from one media asset management system to another. While MAMs often employ similar technology suites, for instance transcoders to create their H.264 proxies of video, or SQL databases to store the metadata, they differ widely in the exact details of how this is implemented. As a result, it is almost never possible to continuously access the database of another MAM while doing a migration. Instead, a popular approach has emerged which consists of doing an XML export of the metadata for all the assets in the database.

XML is a specific kind of formatted text file, in which fields and their contents are laid out logically and can be generated and read ("parsed") by software systems. By using XML batch exports from one MAM system and imports to another, it is often possible to take a team to a new system in a matter of days, rather than weeks or months as was traditionally the case. This is particularly true in the case of passive MAMs, where the locations of the media files are visible not only to the MAM system, but also directly to the users of that media. With such systems, a migration becomes a much less risky process, as ongoing media access is never in doubt, even while the outgoing MAM is being decommissioned and the new MAM system put online.

5.5 WORKFLOW AUTOMATION AND PROCESS ORCHESTRATION—A PARADOX

One of the main benefits of enterprise MAM systems—their ability to automate and orchestrate at scale—has also been a limitation. This is because the scripting tools and enterprise service buses provided by each MAM vendor have often been unique to that system, and even in a few cases completely proprietary. Even when open systems standards from the IT world have been employed as part of MAMs, they have generally been unsuccessful in driving adoption across multiple vendors in the broadcast space. As a result, pockets of expertise have sprung up only at the vendors and a handful of technically ambitious customer sites. At this point, the broader problem of industry-wide workflow/process integration remains to be solved. If and when such open solutions become available, they are likely to trigger significant economies of scale and efficiencies throughout the broadcast industry.

5.6 CONCLUSIONS

Media Asset Management is a dynamic, rapidly evolving field where a variety of technical approaches and even business models are contending to begin managing the flood of video content that is growing at an unprecedented rate. One can even see global systems such as YouTube, Facebook Video, and Vimeo as examples of MAMs. There has been a proliferation of vendors in the space, many of which offer departmental or enterprise systems. Cloud and lightweight on-premise systems are beginning to fill the needs of smaller teams and freelancers, which is the highest-growth area in the industry. Meanwhile, systems are becoming both more powerful, and far easier to use, over time. One thing is certain—due to the rapid pace of technical development and market evolution, change will be a constant.

CHAPTER 5.9
PRODUCTION INTERCOM SYSTEMS

Vinnie Macri
Communications Specialist, Millington, NJ

5.1 INTRODUCTION

"Intercom" is a general term to describe a voice communication system or device that allows people in different locations to speak to each other. An apartment building door intercom comes to mind. While this might be a universal term, it is better to understand any television broadcast voice communication system as a "production intercom." *Production* is the definitive means to the end of why an intercom is necessary as there are many elements that must be coordinated to bring about a successful television show.

There are several types of production intercom subsystems, including wired party-line systems, wireless party-line systems, matrix systems, and interfaces. Over the years, production workflows have grown larger and, accordingly, the means to communicate have expanded. The solutions found today are diverse—keeping up with the times; so, this would naturally include virtual communications found on mobile devices and desktop and portable PCs, including tablets and work stations. What used to be limiting and costly telephone and ISDN dial-ups have been replaced with universal IP network connections where everything is accessible.

Wired party-line systems have until recently been analog systems. The analog systems have been in use since the late 1960s and early 1970s, and yet today permeate all segments of the production market. These systems are also referred to as "talkback" systems and two-wire (TW) systems. There are currently many applications within small regional broadcast centers, remote field production, and ENG/SNG work where analog party-line is relevant. Party-line workflow is still the primary means of communications in all shows—the director calls the show and many listen. Where we find the use of wired analog party-line stations declining is when users have the need for wireless functionality.

The digital wired party-line offerings are very different and offer many attributes one would expect from a digital platform; however, their functionality remains basically that of their analog counterparts.

Wireless party-line systems have also been analog for many years and can still be found and purchased, although use of the radio spectrum they occupy is in a state of change. These analog wireless systems use UHF frequencies of 470–698 MHz, part of which have been selected by the FCC to be auctioned. Despite the efficiency of the UHF band, long-term use may be in decline. There are many digital radio choices that are now outside the UHF bands.

Matrix intercom systems intended for use in production environments have been available for almost as long as the analog party-line. Today, these systems have evolved and offer much higher density input/output (I/O) than almost all analog and digital audio formats, while still maintaining legacy hardware. The benefits of the matrix intercom include the many functions that can be realized through computer control of matrix cross-points. Matrix intercoms offer superior audio quality in comparison to analog party-line intercom systems. As the communications needs of production facilities become greater and more complex, a matrix intercom offers the user great power and flexibility.

5.2 ANALOG PARTY-LINE/TW INTERCOMS, WIRED

It is worth nothing that every intercom circuit starts out as a four-wire circuit, i.e., headset earphone/microphone, or separate microphone/loudspeaker. (Note: In "carbon mic" circuits, one side of the headset earphone and microphone are connected together to become a "three-wire" connection, but for practical purposes it is treated as a four-wire circuit.)

From a technical point of view an analog party-line or two-wire system is a communications system where the path is the same for both talk and listen. The name *party-line* (PL) came from the original telephone systems where more than one subscriber shared the same line and could hear and join all conversations at once. Therefore, party-line type intercoms are always full-duplex and are commonly nonprivate. As a group communication tool a party-line allows a group of people with mutual workflow functions to intercommunicate with each other all the time. An example in broadcast would be a camera PL. Most users only converse on one or two channels, receiving cues from the directors and talking within their own group.

5.2.1 Wired Party-Line System Components

Party-line intercoms consist of the following hardware components:

- Power supplies or master stations—A master station is a multichannel user station that includes power. These multichannel stations are usually used by directors and producers.
- User stations, such as multichannel remote main stations, user belt-packs, and user speaker stations.
- Interconnecting cable, headsets, and/or panel microphones, sometimes push-to-talk microphones.

5.2.1.1 Power Supplies. The power supply or master station generates the DC power (typically 28–30 V DC), for the entire system. In analog party-line systems the DC voltage is applied to the XLR3 pin 2 audio conductor. The power supply includes a 200 Ω system termination for the audio channel. In some RTS models an impedance generator exists with DC the source, and in other RTS and Clear-Com systems, there is a simple resistor and capacitor in series as the termination circuit. This audio termination circuit is what allows many intercom user stations to be connected to a single party line. Each channel of the system would have its own terminator. The termination prevents drastic changes in the impedance of the PL channel if remote stations or belt packs were added or removed from the PL line. An unterminated line will cause excessive levels, possible oscillation of line drivers, and squealing in the headsets. An intercom line with double or multiple terminations will cause low levels and the inability to null the headsets. All unused intercom lines must also be terminated.

Typical connection schemes are illustrated in Fig. 5.9.1.

A main station or power supply is the heart of a party-line intercom system. It has special features that are not found in traditional designs. It must supply low-noise 30 V DC to multiple intercom lines. It must continue to operate in adverse conditions such as low AC line voltage, momentary shorts on the DC power lines to the stations, and excessive peak loads during power-on conditions.

5.2.1.2 Overall User Station Descriptions. Wired party-line intercoms can be described as a distributed amplifier. The distributed amplifier is built into the various user stations. The distributed amplifier permits each user to adjust their listening level. User stations also include a microphone preamplifier, a headset or speaker line amplifier, volume control(s), talk switch(es),

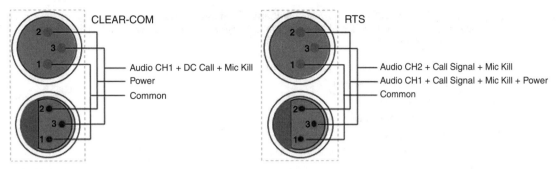

FIGURE 5.9.1 Two-wire intercom pin-outs.

and call switch(es). Some user stations may have visual signal circuitry (call light indicators), status indicators, (e.g., an LED to signify power on the line plus power fault), and channel selectors. In many user stations, switches such as talk buttons are programmable, i.e., they can be momentary or latching.

Party-line intercom systems use a single audio pair (i.e., a shielded twisted pair) to connect user stations to an intercom master station/power supply via daisy-chain wiring. Passive splitter wiring can also be used to expand the system. The circuitry in the user stations allows each user to hear everything on the channel except for their own voice. (A slight amount of the user's voice, called *sidetone*, is retained so the user knows the unit is working.) Most of all user stations have an adjustment for more or less sidetone.

The user stations connect to the power supply and intercom line with a two-conductor shielded cable and/or a standard microphone cable terminated with three pin XLR type connectors. A person uses a headset or microphone/loudspeaker. For a given channel or channels, the user stations are connected to each other in parallel and/or from a passive multidistribution center.

Clear-Com systems use a female 4- or 5-pin XLR connector as a headset connector and a male 4- or 5-pin XLR connector for the user station intercom line. RTS uses a male 4- or 5-pin XLR connector on their headsets and a female 4- or 5-pin XLR connector for the user station intercom line.

5.2.1.3 Belt Pack Headset User Station Description. Belt pack stations have a Talk switch/button to turn the microphone on/off, and a headset listen volume control (typically a rotary knob, sometimes recessed). They may also have a call indicator and a call send button. Two channel belt packs include a channel selector switch and usually have two talk buttons, two listen volume controls, and two status indicators to tell which talk button is engaged. Two channel belt pack units have intercom line(s) and loop thru (XLR-6 for Clear-Com/XLR-3 for RTS) connectors and a headset connector (XLR-4 or XLR-5). Remember the XLR genders are reverse between Clear-Com and RTS. Some stations have an independent program input and some stations are programmable for left ear/right ear listen options. More sophisticated settings include binaural functionality.

5.2.1.4 Speaker User Station Description. Speaker stations provide intercom communication capability in places where wearing a headset may not be feasible (although some speaker stations are supplied with a headset connector). Remote speaker stations permit the user to select between one or two channels of party-line communications, with the ability to talk and/or listen on the selected channel. The user can listen via the integrated speaker, or may use a headset or telephone-style handset—and can talk via a headset mic, a telephone handset, or a push-to-talk microphone. Remote speaker stations may be desk/console mounted, or optionally mounted in standard NEMA sized electrical back boxes for in-wall use and/or can be mounted as a portable table-top speaker stations.

5.2.1.5 Master and Remote Master Stations.
A master station or remote master/main station allows a user access to talk and listen on multiple party-line intercom channels. These stations have extra features for special tasks such as:

- Program interrupt function for IFB (interrupted fold back)
- Stage announce (SA) output
- Relay closures
- "Hot" microphone output
- Multiple program (PGM) source inputs
- Link functions
- All-talk capabilities
- Nulling of the hybrid circuitry per channel
- System remote microphone kill switch

Master stations can send and receive call light signals on any channel. Master stations allow simultaneous monitoring of any channel, any combination of channels, or all the channels. In addition, some master stations can monitor a program source.

5.2.1.6 Signaling.
Call lights are very useful to augment communications in high noise environments, backstage/green room signaling, in theatrical under stage areas, and for various industrial applications.

Clear-Com uses a DC voltage to activate call lights; RTS uses an in-audible 20 kHz tone. Call signals can also be used to trigger general-purpose input output (GPIO) for two-way radios and other devices. Devices are available that will detect both types of call light signals and generate a relay closure. Auxiliary "flasher" devices are available that provide visual and/or audible call signal indication whenever a call signal is present on the intercom line. These devices are wired directly in-line to the party-line intercom channel.

5.2.1.7 Remote mic-Kill.
Remote mic-kill is a useful feature of many two-wire party-line intercom systems. This feature is valuable when an unattended user station with an activated microphone is introducing unnecessary noise into a line. Extraneous noise picked up by an open unattended microphone obviously raises the noise floor and can be heard by everybody on the intercom channel, making communication difficult. Clear-Com systems generate remote mic-kill by momentarily interrupting the DC on an intercom line. RTS TW systems use an inaudible 24 kHz tone to shut off microphones.

5.2.1.8 Stage Announce Output.
Some main and remote stations are equipped with a three-pin XLR male connector on the rear panel to feed announcements into a studio PA or dressing rooms. Pressing the Announce or SA button on the front panel places line-level audio from the selected headset or panel microphone to the rear panel connector. Optionally, pressing the Announce button can also disconnect the selected headset or panel microphone from the intercom line(s). This option may be controlled by an Interrupt Announce option switch or internal jumper. Simultaneously, if a program audio feed to the Announce Output is enabled, it can be interrupted, (dimmed or muted) by the announcement. Program audio feed to the Announce Output is usually selected by setting jumpers.

5.2.1.9 About Headsets and Microphones for Analog PL.
Intercom stations, whether they be analog or digital party-line or more sophisticated matrix user key panels, have been generally designed for use with dynamic headsets. A dynamic headset typically incorporates a noise cancelling dynamic microphone with an impedance range of about 150–500 Ω depending on the manufacturer. An electret microphone headset is useable with an analog party-line intercom and may be self-sensing (i.e., internal circuitry senses an electret element) or may have to be selected for use.

Intercom station panel microphones are typically electret design and have an impedance of 300–2000 Ω, and designed to be phantom-powered with a voltage range of 1.5–15 V DC. These panel microphones are typically gooseneck, and are offered in various lengths and with differing connectors (e.g., ¼ inch TRS and/or DIN screw in).

Push-to-talk handsets and telephone-type handsets are useable in any party-line intercom system but more practical to users who cannot use a headset because of their workflow.

Headphone impedances in headsets range from 50 to 1000 Ω. The use of a 50 Ω headphone must have a headphone amplifier capable of powering the line. Lower impedance loudspeakers will draw more current from the amplifier. This becomes more important when we talk about IFB earpieces that may have extended audio bandwidth that includes low frequency content that excites the loudspeaker to pull more current. While low impedance headsets are a good choice to provide enough SPL (sound pressure level) to overcome the interference from loud environments, caution should be used to avoid distortion and most important ear damage. One option for high noise environments are headphones that offer an acoustic isolation of 20 dB or more to protect the user. Headphones from manufacturers are typically labeled *lightweight* and *full cushion*, and offered as single-sided for one earphone or dual-sided with two headphones. It is important to note many user stations including matrix key panels and wireless user stations have a headphone impedance range from 25 to 600 Ω.

5.2.1.10 Source Assignment. Analog party-line systems are, by "older" design, contingent on the cable that has a particular channel on it. So, if a cable has channel A on it and you need to be on channel B, you have to find that cable and repatch—unless you have a multichannel user station or source assignment panel.

Source assignment panels are manually operated matrix switches. One manufacturer offers a slider matrix source assign feature in their 4-channel master station. A source assign panel is usually a separate piece of hardware. Applications include any facility with at least four channels of party-line and twelve or more "drops" or "busses" or "circuits," and where the assignment setups change on a regular basis.

Inputs to the source assign panel are sensibly called "sources" and are typically the party-line channels from a main or master station or party-line interfaces from a matrix or similar system. The "destinations" are the party-line circuits or busses themselves, i.e., cabled outputs from the source assign panel that are "wet" (have supply voltage, 24–30 V DC). Cables are typically three-pin XLR microphone cables. Remote intercom stations, belt packs, or other interfaces that can also be groups of such devices connected together are patched to various circuits/busses. The source assignment panel offers flexibility in assigning intercom channels to intercom busses and provides the ability to easily reassign channels to busses on the fly. With the advent of digital party-line systems, the need for source assignment hardware has been eliminated as all channels now are found on the one cable and channel assignment is selected inside the user stations.

5.2.1.11 Interfaces to Analog Two-Wire Party-Line Systems. Interfacing generally involves interconnection of separate production communications systems like a party-line two-wire system with a matrix four-wire system and/or an incompatible communication device, such as a two-way radio or telephone or a camera to a production intercom system. Interface challenges include conversion (e.g., two-wire to four-wire), level matching (to prevent squealing in 2W-to-2W connections), and signaling.

The most common interface method to wired analog party-line systems is to convert the two-wire signals to four-wire signals. As described earlier, two-wire refers to talk and listen on one path; a four-wire intercom signal is a separate talk and separate listen path. In an audio world these would be known as an input and output, and in a radio/telephony world these might be considered transmit and receive. When we place ourselves in a four-wire world, we can easily interface to those incompatible communication devices listed above.

The best means for conversion is the 2W-to-4W converter also known as a "hybrid." Almost all hybrid devices allow for separate send and receive level adjustments. Some hybrids will convert party-line call signals to serial data for recognition in some third-party communication platforms, such as a matrix intercom.

In 2W-to-4W interfaces, it is necessary to null (minimize) the rejection when an external analog party-line is placed in the matrix environment. Ideally, there should be no portion of the Talk signal

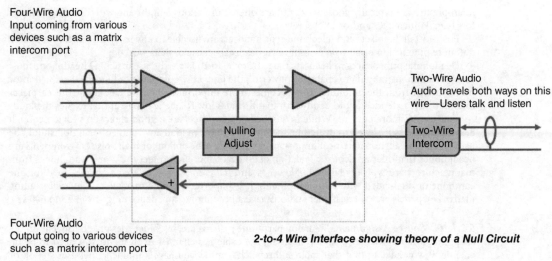

FIGURE 5.9.2 2W-to-4W interfacing.

in the Listen signal. There are many factors that must be accounted for when interfacing an analog party-line to a matrix; these are explained in manuals of the various hybrid manufacturers. The basic concept is to subtract the Talk signal by adding an inverse polarity copy of the Talk to the Listen. The effectiveness of the cancellation is driven by many factors, including the characteristics of the 2W line (capacitance/inductance/resistance) and amplitude of the 4W line, to name a few. Modern hybrids offer an automatic nulling function to simplify this procedure. See Fig. 5.9.2.

It is important to note that every time communication goes through a 2W-to-4W conversion (hybrid) the signal is degraded somewhat. The more hybrid conversions a signal goes through, the more the signal is degraded. Therefore, in any intercom system, the fewer hybrids a signal must go through, the cleaner the system.

5.2.1.12 *Cameras and Analog PL Systems.* The best way to interface a camera intercom to analog party-line systems is, without exception, to access the camera CCU intercom circuit as a four-wire circuit, and then use a 2W-to-4W interface to convert to a party-line. The most difficult party-line interface problem is when the camera intercom is not available as a four-wire circuit. A four-wire connection is frequently not available on low-cost cameras.

If a four-wire connection to the camera intercom is not available, almost always the next best choice is to interface to the headset connection on the camera CCU/DCU. (The headset connection must be a three- or four-wire circuit.)

It is always best to interface each camera individually. Trying to combine the two-wire party-line circuits the CCU/DCU provides into a single party-line and then use an interface to convert the camera party-line to Clear-Com or RTS party-line is the least effective way to interface multiple cameras. There are several reasons.

- First, the more cameras you connect in parallel, the worse the party-line impedance characteristics become, therefore a good null is difficult, if not impossible, to obtain.
- Second, if the impedance characteristics of a single camera are worse than the other cameras (which is not uncommon), the overall party-line impedance characteristics are degraded to the level of the worst camera.

Note that many camera CCUs/DCUs have the intercom connection on a three-pin connector, or on three pins of a multipin connector (frequently along with the "tally" controls).

The pin out is frequently identified as: "common," "+," "–." Many people assume that this indicates a three-wire intercom circuit, with one pin "send", one pin "receive", and one pin "common" to both. This is not true. The "+" and "–" pins are a two-wire party-line, with the "common" simply being a shield/ground. There are no known exceptions to this.

These problems can be eliminated completely by bypassing the camera altogether. Run a microphone cable with the camera cable, and plug a belt pack in at the end. You see this commonly done at a remote production such as a news shoot or sporting event when there is a set on the field.

5.2.1.13 Two-way Radio Interfacing to Analog PL.
The most effective radio interfaces are those that provide complete electrical isolation between the intercom system and the radio system.

There are three common methods in connecting two-way radios and 2W intercom. The first is with interfaces that are offered by several companies whereby the intercom connects to the handheld radio. When used with remote radios, the radio connected to the interface acts as if it were the base station. A connection is made via the headphone and microphone connectors on the handheld and a relay is provided. A relay delivers the required "push-to-talk" transmitter "keying" of the radio. The shortcoming of this method is that every radio is different, even models from the same manufacturer. This means that custom cables have to be constructed for each radio.

The most difficult setting is getting your particular radio to key properly. Every radio has a different method of externally keying its transmitter. To determine how to set the jumpers and wire the connector, you need to know what type of external microphone is used in your radio. For some walkie-talkies or two-way radios, a dedicated power path may be needed to activate their transmit function.

Figure 5.9.3 shows some typical examples.

These simple radio interfacing devices typically offer a 3pin XLR 2W intercom connection that goes directly to your party-line and a "D-type" multipin connector to attach to the custom cable you will have to make. Some models have a timed relay so as to maintain a longer key depending, once again, on the radio. The 2W intercom typically fires the relay from either the call on the user station, or user stations can be programmed to signal a call-on-talk. This is a more simple method as it eliminates a two-keystroke call then talk. These units feature front panel or internal Transmit and Receive level controls for correct level matching. The relay contacts, in conjunction with the custom-made cable, allow for correct transmitter keying to virtually any type of radio. Figure 5.9.4 shows one implementation.

There are rental companies who specialize in communications for television broadcast who have built two-way radio to intercom interfaces. You will find these types of interfaces used mostly in remote sports production for such applications as airship communications and within the TV compound at large events. These devices include both two-wire interconnects and four-wire interconnects to intercom matrices and other communication devices. The audio is the simplest and most straight forward part of interfacing. In these devices, to adapt to a wide range of applications, the transmitter keying can be done in various ways, including:

- By operating the PTT switch usually found on the interface
- By call light signals on the two-wire party-line intercom
- By voice-activation on the four-wire audio input (VOX)
- Via an external contact closure

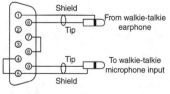
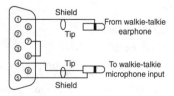
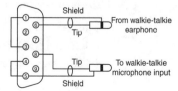

FIGURE 5.9.3 Radio transmitter keying examples. (*Courtesy: Clear-Com.*)

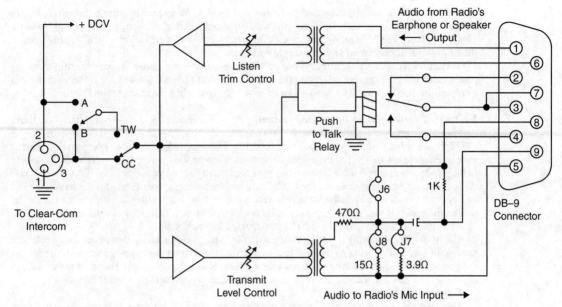

FIGURE 5.9.4 Intercom/radio interface implementation. (*Courtesy: Clear-Com.*)

Lastly, there are products referred to as *gateways* that not only interface production intercom two-wire and four-wire but also interconnect disparate land mobile radios (LMR) and digital mobile radios (DMR) together. Some gateways include the signaling communications protocol Session Initiation Protocol (SIP) for interoperability to VoIP telephones and SIP radios. These products are available from various manufacturers that serve the public safety market.

5.2.2 Advantages to PL/TW Systems

The advantages of party-line systems include:

- Relatively easy setup
- Simple operation
- Use of common cabling (standard microphone cables/shielded twisted pair)
- Portability
- Clear and reliable communications

There are practical implementation issues that need to be considered, including cabling. The cable size must be increased for:

- Increasing the number of user stations
- Increasing the length of cable run
- Use of high current-drawing user stations (loudspeaker stations)

Where voltage drops are excessive, it may be more practical to add a remote power supply than to increase wire gauge.

When considering how to install and wire an analog party-line intercom system, several factors must be taken into account, including:

- The number of user stations
- The length of the cable runs
- Whether single or multiple channels are required

If multichannel stations are connected with multipair cable, then crosstalk may become an important issue.

Cable capacitance, resistance, and crosstalk affect analog party-line intercom system performance.

Very long cable runs and wire specification—gauge AWG, shield DC resistance, capacitance (conductor to other conductor and to shield)—limit the number of user stations on a channel circuit. Different types of user stations have different current draws relative to call lights and number of talkers latched on, so one must estimate the total system current draw "worse case" so as to not cause a PSU fault. A system with cumulative cables (i.e., all runs, adding up to 10,000 feet or more) will have a reduction in high frequency response due to cable capacitance. Both resistance and capacitance affect crosstalk.

Excessive resistance in the conductors of the cable results in a loss of sidetone null at remote stations, and some overall loss of level. Excessive resistance in the ground conductor or shield greatly increases crosstalk between channels. This can significantly affect the performance of multichannel systems.

5.2.2.1 Crosstalk. There are many causes to excessive crosstalk on a PL system, including the following:

- High DC resistance in ground return.
  Solution: Use heavier cable; add additional conductor(s) to ground return.
- Multichannel cable pairs are not individually shielded.
  Solution: Replace cable with individually shield pairs.
- Headset cables are not wired properly or shielded properly.
  Solution: Correct wiring. Use headsets with properly shielded wiring.

Make sure the mic pairs are individually shielded. Recommended shielding practices are illustrated in Fig. 5.9.5.

5.2.2.2 Other Practical Information. In addition to the above guidelines, consider the following implementation rules:

- Keep cable runs under 500 feet. If a longer cable run is unavoidable and approaches 1000 feet or more, make sure the appropriate long line option switches or jumpers are set in the stations. For permanent installation runs longer than 500 feet, do not use wire smaller than 20 gauge.
- Use a cable whose common shield has a low DC resistance.
- Connect unused cable wires of a multipair cable to the Pin 1 shield.
- The Pin 1 ground connection of each XLR connector must be isolated from the chassis. Pin 1 should not be connected to the shell of the XLR connector.
- Individually shielded multipair cable is acceptable for use in multichannel systems. For crosstalk considerations, the shields must be tied together on both ends of the cable to produce the lowest possible DC resistance path for the ground return. Note that all multipair cables must have individually shielded pairs.
- Do not run TW intercom system cables along the same duct ways and pathway areas as digital equipment, high current primary power conductor, motors, transformers, transmitters, and lighting dimmers as these cables may cause electrical interference. The dynamic microphone in a

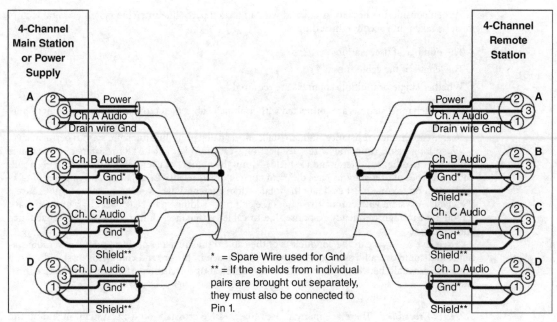

FIGURE 5.9.5 Recommended shielding practices.

headset can pick up stray magnetic fields and introduce unwanted hum and noise into a system. Do not place the headset on or near other equipment that has strong magnetic fields.

5.3 DIGITAL PARTYLINE SYSTEMS

At the time of this writing, there are two digital party-line platforms on the market, and they differ greatly from each other. The one common characteristic between these systems and the analog party-line is that they offer two- and four-channel master/remote main rack-mount stations, wall-mount and desktop speaker stations, and two-channel belt pack headset stations.

One system is based on an AES protocol and the wiring concepts are similar in nature to the analog party-line system, i.e., power to end points via standard shielded twisted pair mic cable. The other is IP-based and useable on standard IP networks, and also with power to end points via standard shielded twisted pair mic cable. The AES system maximum number of channels to an end point is 2 on a mic cable. The IP-based system can generate up to 12 party-line channels in one system of which all channels are available to an end point on a mic cable. With IP use to a great extent acceptable, this system makes sense for those looking to a network solution. IP production intercom connectivity and solutions are discussed in the Ethernet/IP section below.

5.3.1 Digital Party-Line #1

An analog wired party-line powers the line with DC voltage applied to pin 2; this digital party-line applies a 30–48 V DC phantom power to both pins 2 and 3 (with the shield as return) to power belt packs and speaker stations. The DC voltage is generated from a power supply or a master station. The audio is a modified AES-3 format and all signaling, such as remote mic-kill and call signaling, is transmitted as user bits in the modified AES-3 bit stream. The digital audio is two full duplex channels, each at 16-bit resolution sampled at 48 kHz. The system can only be cabled with 110 Ω cable supporting AES standards, and distance limitations need to be considered. Active splitters are available for branch circuits. A system interface is available for integrating digital party-lines in matrix intercom environments—four two-channel CAT5 matrix ports to four phantom-powered belt pack lines. Channels are configured via an audio assignment software package. User stations include a 2-channel belt pack/headset station, wall mount two-channel remote station, and two-channel remote desktop station.

5.3.2 Digital Party-Line #2

This system is based today on four-channel master stations that can be linked together locally or networked by standard Ethernet to form a 12-channel party-line system. Each Main station provides at least four channels of full-duplex intercom communication, plus a separate program audio feed to a combination of at least 20 belt pack, wall, desktop, or similar user stations. Y-split and star-split cable configurations are made without the use of active splitter devices. The intercom audio is immune to electromagnetic- and RF-induced interference on the cables.

The intercom main station has two power line modem circuits to supply power to user stations, with each circuit providing 59 V DC power to both pins 2 and 3 on the line. The audio data and control data are embedded onto each of the powered circuits. Connected user stations are automatically detected and configured with the system. Three rear-panel slots in the main station are provided to accommodate any mix of optional two-wire, four-wire, Ethernet, or optical interface modules. These interfaces allow the main station to connect with a variety of digital matrix, wireless, or analog wired party-line intercom systems, as well as other audio communication circuits. Two-channel user belt stations are able to select among, monitor, and communicate on any two of those channels at any given time, and also receive and monitor a separate level-controllable program audio feed. Remote and speaker stations are four-channel devices and have the same programmable features. Remote and speaker stations have the capability to be locally powered via an external power supply or a Power-over-Ethernet (PoE) connection, and can be placed on a network.

5.4 WIRELESS PRODUCTION INTERCOMS

5.4.1 The Rules

The use of radio equipment is subject to regulations in each country. A given device may not be allowed to cause harmful interference to other authorized users. The device must accept any interference caused by other users. To comply with FCC part 15 rules in the United States, radio equipment must only be used in systems that have been FCC certified.

RF equipment must be installed by qualified professional personnel. It is the responsibility of the installer to insure that only approved equipment/systems are deployed and that effective radiated power does not exceed permissible limits established by the country's regulatory agency in which it is used.

In most applications, the wireless intercom is an extension of the wired system, used by those staff members who require mobility for safety or convenience.

Wireless intercoms are typically used in a group/party-line communication workflow as detailed in the analog party-line section above. They are seldom used as point-to-point communications; however, there are wireless systems available that have these capabilities.

Systems are generally comprised of a "base station" and a limited number of user wireless belt pack/headset stations that can work with and are "paired" to the base station.

With most platforms, multiple base stations can be collocated to work as one system, increasing the number of wireless belt pack users for the application. Depending on the platform (UHF/2.4 GHz, DECT, etc.), this workflow typically requires a higher degree of engineering, as explained below.

Wireless communication is full-duplex; however, some systems offer different operational modes whereby the system can be used in a half-duplex manner, allowing more wireless users on the system. Modern digital radio wireless systems include wired intercom features such as wired call signals and remote microphone kill. These modern digital systems also offer multichannel wireless user belt packs with two, four, and up to six channels of party-line communications.

Advantages of wireless intercom systems include:

- Greater freedom of movement for the user.
- Simple installation; avoidance of cabling problems common with a wired intercom.
- Reduction of cable "trip hazards" in the performance/work space.

Disadvantages of wireless intercom systems include:

- Sometimes limited range (some wireless systems have a shorter range).
- Possible interference with or, more often, from other radio equipment or other radio intercoms.
- Noise or dead spots (places where it does not work, especially in nondiversity systems).
- Limited number of operating user stations at the same time and place, due to the limited number of radio channels (frequencies).

Today's professional wireless intercoms being used in television production applications are offered in four frequency spectrums (ranges):

- UHF—470–698 MHz
- UHF—900 MHz
- DECT—1.9 GHz
- 2.4 GHz

Comparatively, available UHF products are typically limited in features, require FCC licensing, may need frequent maintenance, and often suffer from interference from other UHF systems. One common solution to the interference issue is to perform complicated frequency coordination for all wireless systems at a given venue. Because of the need to share the UHF radio spectrum with other users (wireless microphones and wireless IFB/in-the-ear monitoring systems), frequency coordination often requires a reduction of the available useable frequencies (channels) for many users during a given event or show. In addition, several years ago the FCC made public its plan to proceed with the reallocation of the UHF TV frequency spectrum. This change also had an effect on the part of the UHF used by wireless intercom and other UHF wireless devices, such as talent microphones. While the useable frequencies available for intercom and other UHF wireless products were not completely eliminated, these changes reduced the frequencies available for those products and will likely create further crowding of the airwaves. The advantage of these systems is that they offer an extended audio bandwidth of 8 kHz, which makes them sound very natural, and affords a user the ability to use the system for longer time without fatigue.

One of the biggest innovations incorporated in non-UHF (470–698 MHz) wireless intercom systems is the all-digital radio. A significant feature of some digital wireless systems (UHF, 900 MHz and 2.4 GHz) is they use transceivers that require no FCC end user license for operation. These systems never need frequency coordination and do not suffer from the interference that analog radio signals incur regularly. In addition, the wireless systems allow for many more simultaneous users than ever possible in UHF (470–698 MHz) systems.

5.4.2 Interfacing to Wireless PL Systems

5.4.2.1 2W/4W Interfacing. Wireless intercom systems are commonly equipped on a per channel basis with a two-wire loop-thru (three-pin XLRM&F) interface to analog and digital party-line systems and four-wire I/O (typically an RJ45) for connection to matrix intercoms and other third-party audio devices. The two-wire interfaces generally have options for interfacing to both RTS TW and Clear-Com party-line systems. Send and receive audio trims are incorporated for both 2W and 4W I/O. Some base stations include a nulling feature for the two-wire circuit.

5.4.2.2 Audio PGM/AUX. A program or AUX input, or both connectors (XLR3F), is generally available at the wireless base station to feed analog program line-level audio (or any audio input) to the wireless belt packs. In some systems, the audio is selectable at the belt pack. The audio input is set for line-level signals.

5.4.2.3 Stage Announce. Similar to the analog and digital wired party-line systems, most wireless base stations include the stage announce feature and output connection. These outputs are always analog line-level and often do not have an output level trim control. In operation, the audio from a belt pack, which is the headset microphone signal, or audio from the base station microphone appears on an output XLR3 male connector on the base station. Audio from this connector is typically wired to a paging system or series of paging systems for the benefit of anyone on a studio set, scene shop, or other work areas. A typical application is the stage manager announcing to everyone how much time remains until a show is on the air. A relay can be associated with the SA output so that the activation of the SA will key a two-way radio or turn on an on-air light.

5.4.2.4 Wireless ISO/Talk-around. In systems that include wireless ISO/talk-around, all wireless users on the same base station can talk with each other, isolated from either two- or four-wire intercom channels. ISO function can be restricted to specific users. Some systems call this "Wireless Talk Around."

5.4.2.5 Collocated Systems. It is often necessary to locate several wireless systems together in a venue for a large production. There are some important guidelines that must be followed to ensure success.

Avoid harmful desensing of receivers. Desensing occurs when one or more transmitters are broadcasting while one or more closely located receivers are trying to "listen". Desensing happens even when the transmitter and receiver in question are not on, or even near, the same frequency. This effect is extraordinarily harmful to wireless systems and must be avoided at all costs.

Some digital wireless systems do not require intensive antenna distribution systems (splitters/combiners). These systems rely on individual distributed antennas. For systems that use individual antennas per base station, the antennas of each individual system should be separated from every other system's antennas for best collocated performance. There are antenna distribution systems available for these systems by third-party providers. These antenna distribution systems carry FCC licensing.

5.4.2.6 Antenna Placement. One of the most important factors in a successful RF system is the placement of the antennas relative to the desired coverage area. Digital wireless intercoms (900 MHz, 2.4 GHz, and DECT systems) use a dual antenna diversity system. It is critically important that both antennas be connected and properly placed at all times for best RF performance. Unlike older analog VHF and UHF wireless intercom systems that had dedicated transmit and receive antennas, each digital antenna is both a transmit and a receive antenna. This is very important to remember when setting up antennas because it means that both antennas are necessary and equally important. Both antennas must be used and properly placed at all times or RF performance will suffer.

Some important points to remember when placing antennas include the following:

- Every antenna has a certain pattern of coverage for which it is useful. The patterns of both antennas need to overlap in the desired coverage area to ensure best RF results.
  - Do not point directional antennas in two different directions.
  - Do not separate omnidirectional antennas too far away from each other.

- Higher is almost always better when placing antennas.
  - Maintaining a direct line of sight from the base station antenna to the belt pack is the best possible antenna scenario.
  - The minimum acceptable application of this rule is to get the base station antennas above head level.
  - In many cases, the best execution is to get the base station antennas well above the desired coverage area and point antennas directly down at the coverage area.
- Centrally position omnidirectional antennas in the middle of the desired coverage area.
- Position directional antennas on the edge of the desired coverage area and point them across the area to be covered.
  - Always make sure the patterns overlap.
- Always keep antennas away from:
  - Large metal objects; stay at least two feet away.
  - Large containers of liquid. Most liquids are intense RF absorbers.
  - Confined spaces. Do not set up antennas in rooms or areas that are closed in with very few RF enter/exit points. Wide open spaces are good.

5.5 IFB

IFB is an abbreviation for *Interrupted Fold-Back*. It is a communication circuit feature or a separate system that interfaces with the intercom system. It is often used as a wired subsystem to the intercom. "Wired" describes the talent listening devices are hardwired to the IFB system, usually with a simple mic cable. Wireless is described below.

The term "fold back" refers to sending "program" audio, or some other audio mix, back to announcers (talent), while they are on the air. Doing so allows announcers to monitor themselves, other announcers, videotapes of commercials, or some mix of sources, while they are on the air. This is typically found in television news and live broadcast event. It varies somewhat in remote operations.

Announcers typically wear a small earpiece driven by a headphone amplifier so they can hear the selected fold back audio mix. When a director wants to give directions and/or cues to an announcer on air, or to announce changes in the program, the director must "interrupt" the fold back. To do this, the director uses a channel specifically set up to interrupt the fold back audio.

IFB "communications" are one-way only from an access location to the selected talent position. (At an optional split-feed receiver, the program is dipped only on the cue side; the other side has continuous program with no cue.)

Wireless ITE (In-the-Ear) monitoring systems, typically used and seen in use with musicians on stage, can be used with IFB systems. Care must be taken in understanding that the IFB system may supply voltage to the output connector feeding the talent wired receiver/headphone amp. This voltage must be eliminated (dried) when connecting to third-party listening devices such as ITE wireless systems. A wireless IFB system employs a transmitter whose source audio is the IFB output of a matrix or two-wire intercom system. The talent wears a wireless receiver to hear the IFB. The IFB itself is created upstream of the wireless transmitter in either the two-wire or matrix intercom system.

With the widespread use of IP circuits and cellular, IFB functions are facilitated using cellular mobile devices and/or Internet/network connectivity. This is most prevalent in remote news gathering applications. In these cases, the IFB signals are still generated and created in a wired intercom system.

A central IFB device connects all control panels (destination buttons on a panel) used by directors or producers to cue talent, talent receivers, and intercom together. This unit contains the circuitry for interrupting program via the Talent Control Station and is where a variable audio program dip-level would be sent.

IFB central electronics typically supply two or four independent IFB outputs. Each circuit consists of one or two audio sources, sometimes called program or mix-minus feeds (see below), and one or two-channel split-feed IFB outputs that are connected to listening devices called "talent boxes."

In wired systems, Talent Amplifiers or Talent boxes are self-contained belt pack units that drive talent earpieces or headsets. These portable belt pack units interconnect with a standard microphone cables. With wireless systems, the talent box is a wireless belt pack receiver. The talent amplifier is typically used by on-air personnel, camera, and production staff. The talent amplifier often contains a source selection switch, along with one or two output level controls.

5.5.1 Mix-Minus/Clean Feeds/SAR (Select Audio Return)

In the past, radio and TV talent could be sent out into the field with a link back to the studio and the ability to monitor the station's off-air signal. The fact that they heard their own voices in headphones or earpieces was not distracting, as there was essentially no delay between speaking and hearing.

Today's remote feeds are usually done through digital audio codecs for news gathering, Internet broadcasts, and podcasts—and there can be extensive processing time for audio in both directions. Your brain can only handle a relatively short echo of your voice while speaking. Therefore, it has become absolutely necessary to keep the remote audio from being sent back to the remote site. This is done with mix-minus. It is called mix-minus because it is the mix of all the sources you want to feed the talent, minus the talent; or a mix of all needed audio, except from the remote.

It is usually easy to create one mix-minus feed on a broadcast console. Most consoles have an audition or auxiliary bus (some use the term $N - 1$), which allows the creation of an additional mix of certain console sources without affecting the main or program mix. If a telephone hybrid is fed this mix, and a codec feed is required later, the console must be reconfigured for each mix-minus. This can be confusing and cause program delay. Also, if the hybrid and codec need to be on-air simultaneously, a single mix-minus is insufficient. Newer consoles provide for more than one mix-minus feed.

Some modern digital audio consoles have an $N - 1$ (mix-minus) function that allows for setting up complex conference calls. All subscribers who are off-air (fader down) can talk to each other. If one of them is put on air (fader up), the corresponding path will be removed from the conference. When the fader is closed that person is once again added to the conference.

5.6 MATRIX INTERCOMS

Today's matrix production intercom systems utilize time division multiplexing (TDM) technology to provide interconnection of intercom stations, including the ability to provide custom-mixed intercom signals to specific intercom stations and other connected devices within the system. The utilization of TDM technology in mixing signals and setting custom volume levels for system listeners permits relatively large intercom systems to be constructed. TDM is a form of multiplexing many signals, transferred as subchannels, onto one channel, but are physically taking turns on the channel. Today, nearly all matrix intercoms are based on TDM or similar technology.

Time division multiplexing is not discussed in this chapter. (Numerous sources for this information can be found on the web.)

The best method to understanding matrix production intercom is from an audio perspective. These platforms are referred to as full summing, nonblocking matrices. This refers to the capabilities that any input or number of inputs could be routed to any output or any number of outputs with full cross point level control/adjustment. The term cross point refers to a one-way audio path from one port's input to another port's output. Cross points exist between every pair of ports in the system, and are connected and disconnected as needed to provide communication paths between system ports on the TDM backplane. The outstanding feature here is the cross point level control/adjustment, which is not typical of traditional audio routers. Clearly, when you have many people talking, the ability to adjust listening levels (cross points) is critical to production cueing and directions. This becomes especially important with multitalent IFB work and group/conference communication. A feature of audio matrices, and production intercom matrices included, is the function of audio conversion. Because the I/O to modern intercom matrices include various analog and digital audio formats, the circuit I/O cards convert whatever audio format is being generated into one of the many digital formats used by other circuit cards. You may have analog program audio input put on a MADI stream to feed the audio console, as an example.

Configuration software is used to create and manage communication pathways between devices including matrices, interface cards, interface modules, and user panels.

For example, software can be used to:

- Create individual, point-to-point talk and listen paths between members of a matrix intercom system.
- Create groups and one-to-many calls (fixed groups).
- Create many-to-many party lines.
- Create isolate or ISO—a temporary private discussion amongst two parties.
- Manage telephones—single key to answer an incoming telephone call, or to make an outgoing call, (requires a telephone interface).
- Store complete system setups (configurations).
- Interface with telephones, two-way radios, camera intercoms, and more.
- Bring an outside audio source, such as program source, into the system's audio stream.
- Interrupt the program audio for announcements (IFBs).
- Use the GPIO facility to activate an applause light in a studio, a lock on a door, or other control functions each time an external device such as a switch is triggered.
- Control, monitor, or run diagnostics on a matrix system remotely from anywhere in the world, if the matrix is set up on a network.
- Link matrices across cities, nations, and continents.

5.6.1 Matrix Intercom System Components

A complete matrix system comprises the configuration software, a central hardware matrix (which includes various client cards or circuit cards used for I/O), user panels, and other remote devices (interfaces, four-wire equipment, and digital equipment) connected to the matrix. All components within matrices are typically hot swappable. In addition, all cards can be used with various card-frame sizes offered from each manufacturer. The following sections give a brief overview of the matrix system components.

5.6.1.1 Central Processing Unit (CPU). The central processing unit (CPU) provides the serial data and an Ethernet connection to the PC hosting the software. The CPU also coordinates the data flow for other features of the system, such as general-purpose outputs and inputs. The CPU stores complete system configurations in its memory. Each matrix system contains at least one CPU, located either on its own removable CPU card, or in internal circuitry. Most if not all matrices are supplied with two CPU cards (in a master and slave relationship), ensuring fail-safe operation.

5.6.1.2 Internal Circuitry. The matrix internal circuitry controls the operation of the panels and interfaces connected to it. User key panels and interfaces connect to the internal circuitry through an RJ-45 connector, or port, on the rear panel of the matrix. With analog I/O, the internal circuitry sends balanced duplex audio and RS-422 serial data signals in Clear-Com systems and RS-485 serial data signals for RTS systems to and from connected user key panels. These serial control data protocols are embedded in digital I/O circuits.

5.6.1.3 Power Supplies. Each matrix typically can run with two independent power supplies. These may be connected to a main and backup power source for redundancy. If one of these power supplies fails, the second supply automatically takes over. In the event of a complete power failure, a matrix may automatically restore itself to complete functioning by retrieving configuration information from its nonvolatile operational memory.

5.6.1.4 Rear-panel Connectors. A matrix connects to remote devices such as user key panels, interfaces, general-purpose inputs and outputs, local area networks (LANs), the computer that hosts the configuration software, and other matrices through its rear-panel connectors. RJ45 connectors are found on analog four-wire, IP circuit cards, and Ethernet for PC control, while digital connections may include BNC or fiber or both for MADI I/O and BNC for AES I/O. Typically D-sub connectors are used for external functions like GPIOs.

5.6.1.5 Client Cards. There are a wide variety of different client cards available for matrix systems. These client cards are the I/O and interfacing connections to various user intercom stations and third-party sources and destinations. These client cards offer various audio format conversions as well. Client cards are described in terms of "ports." Ports refer to the number of connections available to external devices from the matrix. Typically, a port is assigned a function within the configuration software. This function specifies what type of device is connected to the port such as a user key panel, an interface, or a four-wire. Software is used to set the parameters of the function for that port. See Fig. 5.9.6.

5.6.2 About Client Cards

Analog client cards afford a standard balanced line-level audio input and output per port. Some analog client cards also carry the manufacturers' protocol of serial data communications used for signaling (tally) and control of a user panel, as mentioned earlier. The ports on analog client cards are typically referred to as four-wire ports. Four-wire ports can be used as "split" ports, meaning that the input may be used for a different purpose than the output purpose. These client cards are available with 8- and/or 16-port connections and are typically supplied with an RJ45, but other break-out connector types may be found. Quoted specifications state nominal levels of 0–8 dBu and maximum levels of +18 to +20 dBu. A port is typically made of ¾ pairs—one pair for audio in, one pair for audio out, and one/two pair for serial data. See Fig. 5.9.7.

AES client cards afford a single or two-channel bidirectional digital connection. These client cards carry the manufacturer's protocol embedded in the user-bits of the digital signal for connecting user control key panels. AES client cards that are not used for user panels are useful for bringing digital audio signals in and out of the matrix environment, and may be connected to AES audio routers or digital console environments fitted with AES I/O. IFB program source and/or mix-minus feeds to IFB destinations are typical uses. These client cards are sometimes offered as unbalanced BNC AES format (75 Ω) or similar with RJ45 connections.

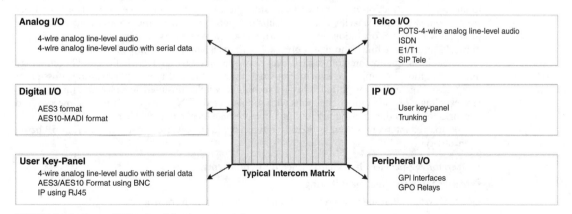

FIGURE 5.9.6 Types of I/O and peripherals to a matrix intercom.

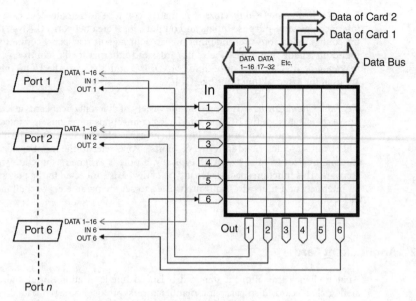

FIGURE 5.9.7 Matrix ports analog example.

MADI (Multichannel Audio Digital Interface) client cards may be used to connect any AES-10 compliant device over coaxial or fiber connections in and out of the matrix. Some manufacturer's MADI circuit cards include the serial data protocol for serial pass-through port connections, making them useable to connect user control intercom key panels. These systems allow individually configurable audio parameters of each MADI channel. Because MADI is a high density audio format (32/48/56/64 channels) using a single cable, it is useful as a connection to interface program audio and IFB signals to and from the intercom system. MADI is also used for trunking matrix frames together. Trunking will be discussed later in this chapter.

5.6.3 Communicate Over IP

The days when real-time audio and IT networks made poor bed fellows are slowly but surely coming to an end. The attraction of IT networks as an audio carrier has never been clearer: low cost per connection, very high capacity, and an established global infrastructure. Additionally, commodity products with proven operation over an enormous array of media types—including copper, 3G/4G networks, microwave links, and fiber optic cable—have also made it a compelling option.

Transmitting quality intercom signals over networks is not without its challenges. IT technology has not traditionally supported synchronous real-time traffic, and typically uses retransmission to compensate for underlying unreliability—a technique wholly unsuited to voice communications. However, higher bandwidth connections with sophisticated VPN management tools allow standard networks to rise to the challenge of routing intercom on a local and global basis.

For matrix intercom systems, there are typically four scenarios that benefit from IP-based connectivity:

- Operator key-panel-to-host matrix (including soft clients)
- Matrix-to-matrix intelligent linking
- "Glue" systems that interconnect traditional systems over IP infrastructure
- Matrix-to-digital telephony gateways

Currently, there are three IP connectivity solutions available in matrix intercom systems. One system supports both an eight-channel AVB (Audio Video Bridging) card alongside a more mature eight-channel VoIP card. AVB is an extension to the Ethernet standard which is designed to support very high quality, low latency audio and video.[1] Disappointingly, AVB has had poor uptake in the IT world, and its compatible kit is very limited. It can coexist with existing Ethernet LAN equipment, but requires all intermediate infrastructures to be AVB-compliant to support AVB transmission. It is not clear if or when mainstream IT suppliers will start to support this standard, as the market is a relatively niche market, and standard LAN technology is most often sufficient.

Another system offers an IP card utilizing the Dante protocol for very high quality audio on high bandwidth LANs with the promise for RTP support in the future, although the scope and timescales are not defined. Dante is a combination of software, hardware, and network protocols that deliver uncompressed, multichannel, low-latency digital audio over a standard Ethernet network. There is a commitment to provide Open Control and support standard audio over IP/Ethernet networks. This card currently supports a maximum of 64 IP channel connections per card sold in 16-channel licenses. Currently, their VoIP card (8- and 16-channel) remains available as a solution for remote connectivity.[2]

Another system offers a client card for the system that provides up to 32 channels of securely encoded G.722 audio over standard LAN, WAN, and Internet infrastructure. The card supports suitable optimizations for each connection type. System intelligent linking is peer-to-peer and does not require an online computer management system. The card itself supports $N + 1$ card redundancy as an option. G.722 audio encoding represents an excellent compromise between network bandwidth, audio quality, and transmission latency.

5.6.4 GPIO (General-Purpose Input/Output)

5.6.4.1 Opto-Isolated Inputs. GPIs can be set up as remotely controlled key-panel keys to activate intercom ports, party-lines, GPOs, etc. within the intercom system. The positive and negative input and common connections may be provided from a remote source such as a GPO from an external device. A GPI can be used to tally (call) a user key panel, typically associated with a tie-line or external source and/or to trigger an event within the matrix.

5.6.5 Relay Outputs

Intercom matrices provide a number of independent single-pole, double-throw (SPDT) relays that are directly controlled by the CPU card in the matrix frame by software assignment. There is usually DC isolation between relay device(s) and the central matrix. An LED indicator may be provided for each relay that lights when the relay is active. Logic control of relays is provided by associating a relay with any other label in a system using the configuration system programming software. The GPOs are typically, but not always, assigned for activation from key-panel keys. Activating a Talk or Listen label with a relay associated to it is a common use. They can be used to control lighting or to key remote transmitters, paging systems, camera tally, etc. There are special classifications of labels for relays called control labels in most matrix intercom systems.

5.6.6 Other Connectivity

Some systems offer an array of native audio carrier types (Fiber, E1/T1), in addition to IP that allows the system to negotiate interface failover and trunk forwarding. (See Trunking below.)

---

[1] See http://en.m.wikipedia.org/wiki/Audio_Video_Bridging
[2] For more information, see http://en.wikipedia.org/wiki/Dante_(networking)

5.6.6.1 Trunking. When resources in disparate matrix systems need to communicate with resources in another system or group of matrix frames, "trunking" is applied. Trunking uses proprietary protocols and algorithms, and the support for trunking is tightly integrated into all manufacturers' matrices. Most manufacturers' trunking schemes are referred to as "intelligent," meaning it functions automatically, establishing the trunk, continuously monitors and reports on status of trunk utilization, and releasing the trunk when the conversation is over.

Multiple calls to one destination use only one trunk line. Also, calls from one source to multiple destinations will use only one trunk line. This enables an efficient use of available audio bandwidth.

Matrices may be dispersed within a building or separated by countries. Different manufacturers offer different trunking methodologies. They are thru the use of: four-wire circuit cards, AES and MADI digital circuit cards, E1/T1 with E and M signaling, IP circuit cards, and optical fiber circuit cards. Some of the high density circuit cards like MADI and IP may be partitioned to allow multiuse, such as X number of ports for user stations and X number of ports for trunk circuits, allowing flexibility in design and affordability.

Trunk lines may support not only point-to-point calls but also conference calls and group calls.

Some matrices offer an external trunk controller unit to manage the trunk lines that others have this built into the system. Most matrices allow trunk line routes to be routed to the final destination even if there is no direct link. In some systems, if a direct trunk line connection exists then this is used. If there is no physical connection, the software knows that there is a possible path using another matrix as a hub connection.

Figure 5.9.8 shows an example system.

In general, as one plans for expansion, you need, as a maximum, as many trunk lines as there will be different conversations.

5.6.7 User/Subscriber Control Key Panels

User panels are varied and so complex in feature sets that it is beyond the scope of this chapter to explain every programmable feature from every manufacturer. We shall only describe most of the common attributes of control key panels.

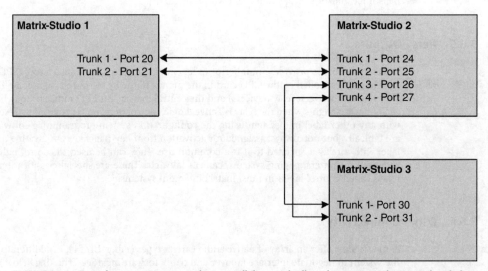

FIGURE 5.9.8 Example system: A user in studio 1 can talk/listen and call another user in studio 3 even though there is no physical direct trunk line connection between the studio 1 and studio 3 matrixes. The software would route the audio through studio 2's matrix.

PRODUCTION INTERCOM SYSTEMS

Control key panels are the human interface hardware in intercom systems used by production staff to communicate while performing their function (creating a show). In most applications it is the only piece of equipment between the production people and the show. Control key panels are populated with communication paths to the various individual personnel, and/or groups of people with similar tasks (like cameras). These communication paths are accessed on the key panels by the use of "keys." Thus the name key panel!

Keys refer to the number of control positions on the face of the unit. A typical key would include the following:

- A display (sometimes LCD, sometimes OLED, depending on the manufacturer) and panel series from a manufacturer showing a destination.
- A talk button or lever function (usually down for talk); a listen button or lever function (usually up for listen).
- A listen level control (sometimes a rotary encoder or up/down button).

User panels are fully compatible with digital matrix systems but not interoperable between manufacturers. They are available in 8-key, 12-key, 16-key, 24-key, and 32-key pushbutton, rotary, and lever key formats. See Fig. 5.9.9.

Some of the more common front panel controls include the following:

- **Mic On**—The Mic On/Off button turns the currently selected microphone (gooseneck microphone or headset microphone) on or off.
- **Shift Page**—Certain devices include multiple pages of key assignments. Pressing and releasing the Shift Page button, toggles between the main page and the currently selected shift page.
- **Headset Select**—The Headset Select button enables you to select the panel headset for audio output. Typically, the panel microphone is deselected, if active.
- **Menu**—Depending on the manufacturer, detailed listing of panel functions are configured in Menu mode.
- **Loudspeaker Main Levels (Volume) Control**—The main levels (volume) control comprises a rotary encoder, sometimes with push-switch action and a tri-color loudspeaker volume indicator LED or bar graph display. Some key panels include an auxiliary levels (volume) control to adjust optional rear panel inputs or optional external loudspeaker.
- **Dial Pad**—The dial pad is commonly used to access some menu pages directly (as a shortcut) and to enter dial codes, to dial out thru an external telephone interface.

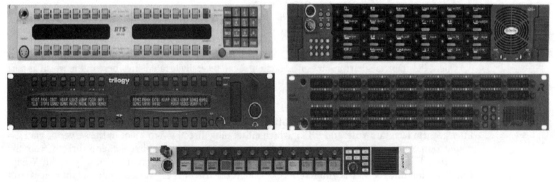

FIGURE 5.9.9 Example key panels: 32-key lever, 24-key rotary, and a 16-key pushbutton.

- **Status LEDs (Tallies)**—Status LEDs (tallies) indicate the status of a key, audio route, or menu option. Depending on the manufacturer these indicators can be multicolored. Indications include but are not limited to:
  - A listen path (audio route) is active.
  - A talk and listen path (audio route) is active.
  - Key is either not configured or the menu option is not selectable.
  - Either user action is required, or there is an incoming call or call signal.
  - Key is configured as a talk key, or a menu option is selectable.
  - Key is configured as a listen key, or menu option is selectable
  - Key is configured as a talk and listen key, or menu option is selectable.
  - Menu mode is active.

5.7 VIRTUAL INTERCOMS

The concept of virtual intercom was inevitable with the proliferation of the use of IT networks, and necessary to support this challenging paradigm shift to ensure a smooth transition into the new world of broadcast-IT.

This idea replaces the matrix intercom hardware frame we talked about earlier with software that runs on standard computers and servers. The architecture for these systems is basically the same for all manufacturer offerings. There is a server or multiple servers that run administration software, and there are clients (end points or users, which otherwise would have been hardware key panels). This administration application configures the system settings and includes:

- Adding, editing, and removing users.
- Optimizing audio settings and network settings, including codec selection, sample rate, and jitter buffer size.

Each server controls the audio routing. Basically the system administrators configure, manage, and monitor the system. These servers communicate to all clients, (virtual stations), regardless of location, via standard Ethernet routers and Internet connections. These systems can provide a number of talk paths over a single Internet connection.

Client software runs on standard computers. Some manufacturer client software also runs on tablets and smartphones, which eliminates the mouse and keyboard. Features of virtual intercoms include:

- Ad-hoc communication, instantly initiate multiparty conferences and text messages
- Drag-and-drop functionality to begin conferencing with one or many participants
- Presence awareness, with visibility of participant availability
- Multiple conference modes, including: talk only, talk and listen, or listen-only modes, and party-lines
- Text message capabilities

Client PCs/tablets/smartphones or "soft clients" can be easily located anywhere in the world as part of a network infrastructure where there are multiple facilities, control rooms, studios, edit bays, audio sweetening, multiple wired or wireless IFB audio channels, ENG, SNG, and full production trucks.

Virtual intercom servers also feature what we'll call Interface Gateway nodes. These can be four-wire interfaces via PCI cards or USB audio interfaces, as well as third-party SIP-to-telephone interfaces (to facilitate VoIP and PBX), which either plug directly in to a slot in the server or via USB connection. Some manufacturers have the ability to interface to these virtual clients directly from the respective IP circuit card in the hardware matrix platform (as mentioned previously in the "Client Card" section under "Matrix Intercoms"). Through the use of these various gateway interfaces, soft clients can communicate with hardware matrix key panels and once introduced to the matrix, can be programmed to any and all elements within the matrix environment.

CHAPTER 5.10
BROADCAST STUDIO LIGHTING

Frank Marsico
Shadowstone Inc., Clifton, NJ

5.1 INTRODUCTION

The broadcast medium is a development of the twentieth century and has grown into a media form that most of the Earth's population has heard or viewed over wireless airwaves, across cable lines, and /or the internet. Broadcast content is created today at many varied locations but the majority of programming is still created in the studio.

Studio space is probably the most costly area per square foot of the entire facility and all properties of the studio must contribute to the final broadcast product. Focusing on the audio and visual aspects of the production, skills in both the engineering and creative disciplines are utilized. Not unlike high technology manufacturing facilities of computer chip or mobile phone companies, television programs are also manufactured in a regulated environment, and a quality lighting system is one of the basic requisites to generate acceptable and competitive images. A studio lighting system is defined as the essential components for controlled illumination of a person or group of people, scenic pieces, or a generally defined area such as a cyclorama, green screen, or multiuse portion of the studio. This chapter is not about how to light video productions but contains specific information regarding studio lighting system components and other related studio infrastructure elements. With that said, and before beginning, it is helpful to have an overview of selected lighting principles and practices to better understand the day to day requirements of a studio lighting system.

5.2 THREE POINT LIGHTING

Since the dawn of mankind we have had one primary light source—the sun. In our effort to light a subject naturally we would also establish a single primary source, which is referred to as the key light. In video lighting, this single main source provides viewers a subconscious comfort level with the lighting environment. Once it is decided where to place the Key light and at what angle and intensity, then other lighting will be positioned. (See Fig. 5.10.1.) The typical angle of the Key light from the camera lens line to the subject, ranges between 30° and 40°.

Small point sources create hard edge definition and shadows. The sun is again our reference and performs as a relative single point source projecting defined edges and is a good model for the key light.

Note that certain information is lost in the conversion from color images and charts to black-and-white for publication. The color version is available in the online and electronic versions.

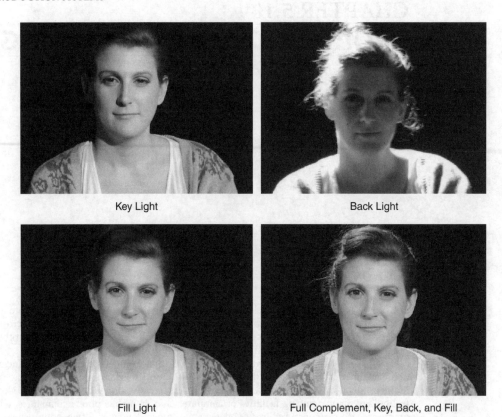

FIGURE 5.10.1   Primary lighting elements. (Black-and-white reproduction of a color illustration.)

Complementary to the key light is the fill light which minimizes and softens the shadows of the single key light source. Large light sources, such as the sky on an overcast cloudy day or large aperture instruments produce a softened shadow projection by means of diffused and indirect light rays.

Supplemental to key and fill is the back light, sometimes referred to as a "hair" or "rim" light and is necessary to facilitate the illusion of three dimensions in a two-dimensional medium. The back light delivers a crisp bright edge that separates the subject from the background allowing the viewer to perceive picture depth. The backlight must be properly focused to avoid lens flare.

The key, fill, and back lights are the most important part of a basic lighting setup. A fourth light also important to quality imaging is the set or background light. Even if the background is a black curtain or flat muslin cyclorama (cyc) cloth, it too could be better rendered with a texture provided by the set light. These elements are illustrated in Fig. 5.10.2.

There are formulas used regarding the intensity ratio of key-to-fill-to-back lights but they must be adaptable based upon the talent's characteristics. If a 3-to-1-to-3 key-to-fill-to-back light ratio is employed but today's subject has thinning or no hair we have to be able to adjust accordingly. The lighting system must allow for this type of tweaking without disrupting other positioned lights.

5.3   LIGHT SOURCE INFORMATION

Visible light is within a narrow band of the electromagnetic spectrum between 400 and 700 nm. UV (ultraviolet) light centers at 350 nm. The spectrum is illustrated in Fig. 5.10.3, with details provided in Table 5.10.1.

BROADCAST STUDIO LIGHTING

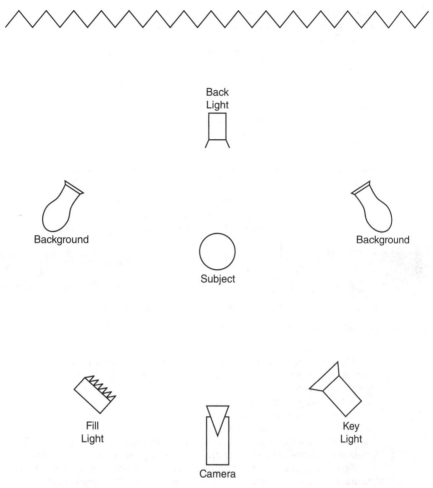

FIGURE 5.10.2 Key, back, fill, and set lighting model.

The most common studio light sources are thermal incandescent tungsten–halogen (T–H), which emit approximately 10 percent of their energy as visible light. Conventional T–H sources have been used for many years but are slowly being replaced by more energy efficient sources such as solid state LEDs (light emitting diodes), compact fluorescent lamps (CFLs), and gas discharge plasma lamps (LEP).

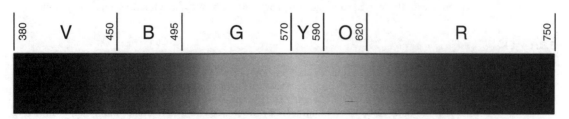

FIGURE 5.10.3 The visible light spectrum (Bruno [1]). (Black-and-white reproduction of a color illustration.)

TABLE 5.10.1 Characteristics of Common Colors
(Note: in the online version of this Handbook, hyperlinks are provided to additional information.)

Color	Wavelength (nm)	Frequency (THz)	PHOTON ENERGY (eV)
Violet	380–450	668–789	2.75–3.26
Blue	450–495	606–668	2.50–2.75
Green	495–570	526–606	2.17–2.50
Yellow	570–590	508–526	2.10–2.17
Orange	590–620	484–508	2.00–2.10
Red	620–750	400–484	1.65–2.00

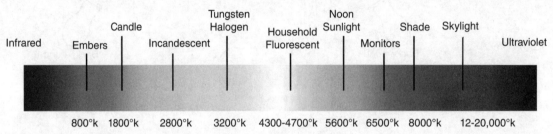

FIGURE 5.10.4 Various light sources color temperature in Kelvin degrees (Tiffen [2]). (Black-and-white reproduction of a color illustration.)

Light sources emit color based upon the principle that a black body radiator produces light of specific color that depends upon the temperature of the radiator. The unit of measure of color temperature is degrees Kelvin, which is one of the seven base units in the International System of Units (SI) and is assigned the unit symbol "K." (See Fig. 5.10.4.)

Color temperature benchmarks for photographic reproduction (film and video) are 3200 K, which is based on tungsten film emulsion, and 5600 K from daylight film. These color temperature standards remain constant for studio (tungsten) and outdoor (daylight) location production lighting even though variations have become more common and attainable due to the flexibility of solid-state lighting.

Color temperature of incandescent sources drops (warmer) when reducing power (voltage) through a dimming system. Studios sometimes set the overall color temperature to approximately 2900 K or lower rather than tungsten film's 3200 K, allowing a degree of latitude when dimming the conventional T–H lamps.

The inverse square law states: Light energy that is measured twice as far from the previous measured distance is expanded over four times the area delivering one-fourth the light. (See Fig. 5.10.5.) This becomes important when determining which size instrument to use at specific distances.

A source and its associated fixture's projected light quality may be evaluated on multiple index scales such as the Color Rendering Index (CRI), Color Quality Scale (CQS), and Television Lighting Consistency Index (TLCI).

All are designed to grade color characteristics on a scale of 1–100, with 100 being the highest level. Light sources with elevated values, usually above 90 are better suited for video and film production.

5.3.1 CRI

CRI is a measure of a light source's ability to show object colors realistically or naturally compared to a familiar reference source, either incandescent light or daylight. The CRI is calculated from the differences in the chromaticity' of eight CIE (International Commission on Illumination) standard color samples (CIE 1995) when illuminated by the source and by a reference illuminant of the same correlated color temperature (CCT). The CRI of a light source does not indicate the color of the light source. See Fig. 5.10.6.

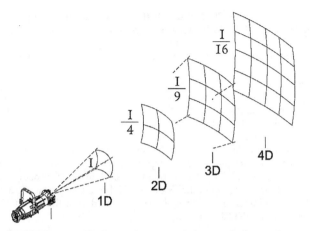

FIGURE 5.10.5 The Inverse Square Law: light twice the distance from the previous measured distance is one-fourth the previous light output.

Type of Light Source	CRI
Incandescent	100
Tri-phosphor Fluorescent	85
Metal Halide	85
Fluorescent	70-50
Coated Mercury	49
High Pressure Sodium	24
Clear Mercury	17
Low Pressure Sodium	5

(a)   (b)

FIGURE 5.10.6 CRI: (*a*) light sources chart, (*b*) test color chart (EYE Lighting [3]). (Black-and-white reproduction of a color illustration.)

5.3.2 CQS

The National Institute of Standards and Technology (NIST) developed the Color Quality Scale (CQS) based on the CRI test sample method of measuring a light source and a reference source to a standardized set of color chips. The primary difference is that the CRI 8 color chips are more pastel in nature while the CQS set of 15 chips are based on more saturated color (Fig. 5.10.7).

5.3.3 TLCI

The Television Lighting Consistency Index (TLCI) 2012 was developed by the European Broadcasting Union (EBU). The following text is from the EBU website (https://tech.ebu.ch/tlci-2012):

FIGURE 5.10.7 CQS test color chip chart—primary colors (N.I.S.T. [4]). (Black-and-white reproduction of a color illustration.)

"In November 2012, the EBU Technical Committee approved a Recommendation designed to give technical aid to broadcasters who intend to assess new lighting equipment or to reassess the colorimetric quality of lighting in their television production environment. The TLCI mimics a complete television camera and display, using only those specific features of cameras and displays which affect color performance. The TLCI is realized in practice using software rather than real television hardware. The only hardware that is required is a spectroradiometer to measure the spectral power distribution of the test luminaire, and a computer on which to run the software analysis program to perform the calculations."

Additionally, the following is from http://www3.ebu.ch/contents/news/2013/06/a-warm-welcome-for-the-ebus-led.html:

"On June 24, 2013, a workshop jointly organized by the EBU and the Guild of Television Cameramen brought industry experts together at BSkyB's Sky Studios near London to learn about the EBU's new LED light assessment tools. Former BBC broadcast engineer and color sciences expert Alan Roberts presented the Television Lighting Consistency Index. The TLCI was developed by the EBU to address many of the issues associated with the use of LED lighting in film and television production." (TLCI [5]).

5.4 THE LIGHTING SYSTEM

With the preceding lighting techniques and light source information as background we may better understand why the lighting system must allow lighting operators to perform certain tasks without hindrance as well as enhance and maintain image quality and productivity.

Assuming the studio is used exclusively for broadcast program production, without interruption, and common to both new construction and converted or renovated spaces the following are the primary lighting system components:

- Suspension—rigging and grids
- Cyclorama, curtains, and track
- Hard cyc, green screen
- Electrical power and distribution
- Network control signal distribution and DMX
- Dimming
- Lighting control
- Lighting instruments and accessories

5.4.1 Suspension—Rigging and Grids

A fundamental component to any lighting system is the suspension method used to secure lighting instrument hanging positions. Lights are mounted overhead, allowing operators to utilize a multi-camera recording approach. Lights and cables are to be up and out of the way so the studio floor is clear, allowing camera and talent movement. As with many other elements, the type of suspension system employed should be determined by the studio's use. The following are general types of permanent suspension systems:

- Fixed, nonmoving
- Moving, manually operated
- Moving, motorized

Types are sometimes mixed within a facility to meet production and budget requirements.

5.4.1.1 Fixed, Nonmoving: Pipe Grid.
A fixed pipe grid suspension type is seen in many broadcast studios due to cost and the fact that most studio ceiling heights are less than 20 feet. This type of grid consists of pipes that are arranged in either a parallel or crisscross pattern and suspended from the ceiling by anchors or clamps. The industry standard 4-feet × 4-feet pipe grid spacing provides effective instrument and distribution positioning. An optimum height for a pipe grid is approximately 14 feet × 16 feet to accommodate 10 feet 0 inch scenic heights and standard ladder sizes. A common grid pattern is shown in Fig. 5.10.8.

If the height is considerably lower, the spacing pattern could be reduced from the 4 feet × 4 feet size to a 3 feet 6 inches × 3 feet 6 inches or tighter patterns to accommodate close-fitting instrument positioning and focus angle. Alternatively as the grid height is raised, spacing may expand. See Fig. 5.10.9.

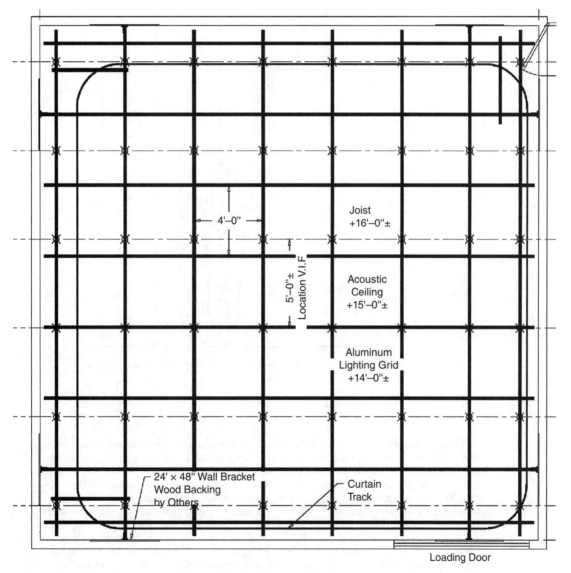

FIGURE 5.10.8 Example of a 4-feet × 4-feet fixed pipe grid, 14 feet 0 inch height with overhead anchor points indicated (X).

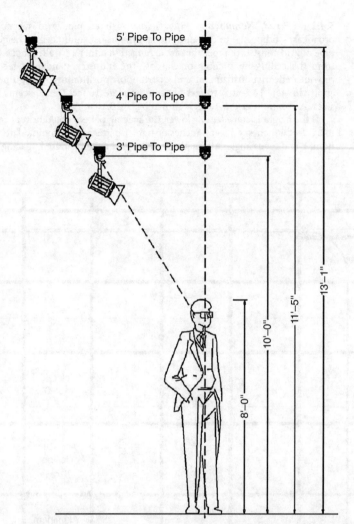

FIGURE 5.10.9 Pipe grid spacing calculation example with fixture angles at different heights.

In most fixed pipe grid studios, schedule 40 black steel pipe has been utilized; however, recently there has been an increased use of T6061 aluminum pipe due to the cost parity, technical equalities, and weight savings. Schedule 40 steel pipe has a wall thickness of 0.145 inch and weighs 2.72 lb per foot versus aluminum pipe schedule 80 with a wall thickness of 0.20 inch and a weight of 1.25 lb per foot. The weight difference will reduce the overall structural load; as an example, if aluminum pipe is utilized in a 1200 feet² studio, there could be a reduction of approximately 1000 lb less weight than if steel pipe were installed. Both types are available in black as well as other finishes. Aluminum pipe deflection concerns are minimized when it is realized that hanging points from beams and ceilings are based on 5 feet to 6 feet center spacing, creating smaller gaps between points plus the added support from rigid crossover clamps. Aluminum became economically practical over the past years as steel exports to the far east grew due to demand, and domestic cost was raised as a result.

5.4.1.2 Fixed, Nonmoving: Channel or Strut Grids. Channel type grids are often installed for special lighting applications, as well as when the highest position to mount the grid is approximately 10 feet or less, making it necessary to optimize every inch of height. By changing the top hanging adapter, lighting instruments are snugged up to the channel, optimizing height by saving the space a typical pipe mounted C-clamp device requires. Channel grids are also arranged in parallel or crisscross patterns similar to the pipe grid style.

Note that whether hung pipe or strut channel, all lighting fixture maintenance and changes require the use of a ladder or lift, which can be inconvenient and problematic when working around scenic elements. A solution to using a ladder when refocusing lights on a fixed grid is to employ pole or motor operated type lights as discussed later in the lighting instrument section.

5.4.1.3 Fixed, Nonmoving: Low Ceiling Fixed Frame Method. There is an alternative approach to pipe or channel for fixing lights in medium to low-ceiling environments when the ceiling layout is either a 2 feet × 2 feet or 2 feet × 4 feet hung frame or a level drywall configuration design more akin to an office than a studio. Common approaches are;

- Video-centric fluorescent fixtures in a 2 feet × 2 feet or other frame to fit the ceiling tile
- Recessed video fixtures with a trim frame for dry wall ceiling types
- 2 feet × 2 feet or 2 feet × 4 feet composite lay-in tile cut to size and strong enough to bolt small or mid-sized instruments
- 2 feet × 2 feet shaped metal panel domes that accommodate fixture focusing without restriction

These methods may require added support due to the load increase but allow instrument bottoms to be flush with the ceiling line similar to standard recessed office lighting maximizing height and flexibility. Figure 5.10.10 shows an example.

5.4.1.4 Fixed, Nonmoving: Catwalk. Catwalks are derived from theaters and motion picture studios and require added building height to accommodate overhead clearance and the additional structural support steel. Also, a fixed grid may be applied between catwalk members allowing instruments to be mounted in numerous positions when considering railings, positions under the walkways, and the overhead possibilities. However, due to the higher costs of structural steel, the added overhead space to heat and air condition, plus the practice that most broadcast productions hang lights in a single position for long periods without moving, catwalks are generally considered impractical in most broadcast studios (but are often seen in multiuse rental facilities).

FIGURE 5.10.10 Fixed frame ceiling fixture, T-Series. (*Courtesy: Brightline.*)

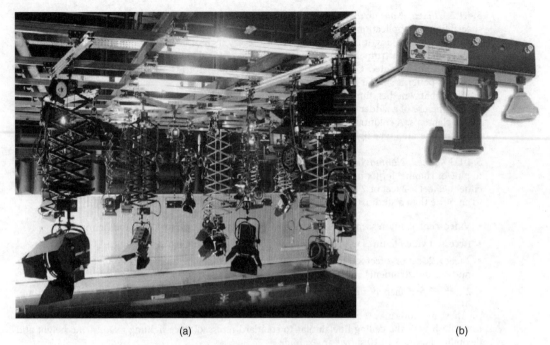

FIGURE 5.10.11 Track and trolley approach: (*a*) studio with track and trolley system, (*b*) trolley device. (*Courtesy: DeSisti Lighting.*)

5.4.1.5 Moving: Manually Operated Track and Trolley. A semi-moveable system is the track and trolley style that employs track suspended from the overhead slab or beams in a regular grid pattern, or one that best suits the scenic layout. The lighting instruments are mounted to a trolley device (see Fig. 5.10.11) that fits onto the track and allows side-to-side movement. The trolley is equipped with nylon or rubber wheels and then drawn side-to-side by a pole with an end hook that latches into the trolley's steel loop, as shown. A rotating cup with an inner post twists to lock the trolley in position.

5.4.1.6 Moving: Manually Operated Winch and Counterweight. Manually operated winch or counterweight systems have been used on live production stages for many years and put into studio operations where dramatic or entertainment events are regularly produced. Counterweight rigging offers versatile performance capabilities for scenic and rapid lighting change turnarounds (Fig. 5.10.12). Video production studios that employ counterweight systems are generally designed for recorded theatrical production events rather than relatively static broadcast applications. The counterweight system requires the user to correctly balance the arbor load, which in turn affords variable raise and lower speeds.

Manual winch systems wind up rope or wire rope cable by a hand crank and stores the cable on a winch drum (Fig. 5.10.13). Winches are typically less expensive to install than counterweight systems but still require the overhead loft blocks and as with the catwalk system. Unless there is a specific production need for this flexibility, counterweight and winch systems are often seen in multiuse rental spaces rather than broadcast studios.

5.4.1.7 Motorized Rigging. There has been an increase of motorized rigging installations as labor costs increase and productivity is continually optimized. Benefits include a simple and repeatable operation, managed by a single operator, and an efficient way to maintain lighting instruments and other mountable production equipment. Overhead motorized systems require added structural steel, a control desk, and a well thought-out plan to accommodate future use.

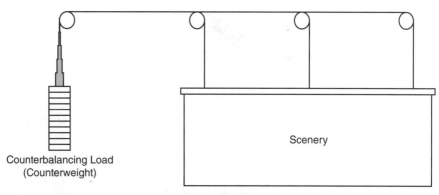

FIGURE 5.10.12 Counterweight system. (*Courtesy: J.R. Clancy.*)

5.4.1.8 Motorized: Self Climbing Hoists. Self-climbers are designed as an independent self-contained rigging device. Included with the housing is a motor, gearbox, electrical distribution, raise/lower cables with safety switches built-in as well as DMX data and A/V lines with connectors, and a batten or track to mount lighting instruments. The motor is sized to lift a limited load by winding wire cable on and off a drum without the use of counterweights. The hoist is hung from prearranged overhead points and incorporates self-leveling compensating brackets. An example is shown in Fig. 5.10.14.

Electrical circuit and data connections are made to an overhead-mounted distribution box run to the hoist body by loose folded flat cable or on flip-flop style trays. With proper planning, self-climbing hoists are a cost-effective approach for studios as they streamline structural steel requirements and eliminate the need for a catwalk. A drawback to self-climbers is the limitation of providing a way to securely bridge between battens located in adjacent rows, which may limit instrument placement.

5.4.1.9 Motorized: Electric Winch Hoists. Electric winch hoists are available that are mounted to studio walls, ceilings, or overhead grids. The hoists are controlled by computers or custom consoles and perform accurate and repeatable pipe batten raise and lower actions. A line-shaft hoist hung from the structural steel has a separate drum for each lift line, resulting in a vertical load. These systems require added structural support, and typically there is an overhead grid (loft) or catwalk above for maintenance access.

5.4.1.10 Other Rigging Options. In addition to raising and lowering battens that support multiple lights or devices,

FIGURE 5.10.13 Manual winch device. (*Courtesy: Thern Stage Equipment.*)

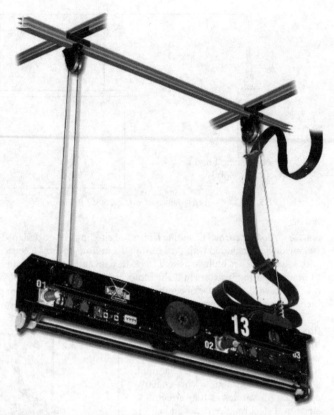

FIGURE 5.10.14 Self-climbing hoist. (*Courtesy: DeSisti Lighting.*)

there are methods to suspend and raise/lower individual lights. One such device is the motorized pantograph, which carries power and data within tubular channels that collapse or expand as the device is raised and lowered (Fig. 5.10.15). These devices can be stationary, manual, or motorized travel modes.

Contrasted to the sophisticated pantograph or motorized single point apparatuses are manual mode only extension rods, telescoping hanger that adjust instruments vertically so that illumination points may be moved closer to the subject by a crew member on a ladder or lift. See Fig. 10.5.16.

5.4.2 Cyclorama, Curtains, and Curtain Track

Broadcast studio curtains or cyclorama ("cyc," pronounced *sike*) are similar to theatrical cycloramas in that they may serve as set masking or as a primary scenic element filling empty spaces as a background cover-all. Studio cycs are hung from a track or multiple tracks along the walls, often covering the entire studio perimeter.

Curtains have weighted bottoms to keep the fabric taut top to bottom and are hung from wheeled carriers approximately 12 inches apart that should smoothly ride in the level track. At each studio corner the track is curved, allowing for even lighting projection—and the wider the curve, the smoother the lighting effect.

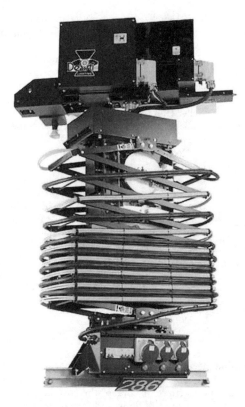

FIGURE 5.10.15 Pantograph system. (*Courtesy: DeSisti Lighting.*)

FIGURE 5.10.16 Telescoping hangers. (*Courtesy Matthews Studio Equipment.*)

Cyc curtains may be visually altered by lighting changes of color, intensity, or projections. The various fabric types used as cyc curtains include the following:

- Cotton Sharkstooth Scrim with tooth-like openings
- Cotton Leno filled scrim similar to Sharkstooth but with closed openings
- 60 percent reflectance gray muslin
- Bleached white or raw muslin
- Different weighted velours, available in a variety of colors but most often black or blue

Other types of fabrics or painted scenic drops may also be used. The type of cyc curtain material selected should be dependent on the overall studio use as well as size. Larger studios may have bleached white muslin backgrounds for rich lighting coloration, while smaller studios may utilize dark colored (often black) velour in order to project an undefined endless space behind the subjects, increasing viewer depth perception and sense of size.

The cyc track is typically attached to a fixed grid or independent mounting points and defines the available production space, as all other visual set pieces are located in front of the track-mounted hung curtains. Cyc track is often a single run, but paralleled double or triple tracks are also possible

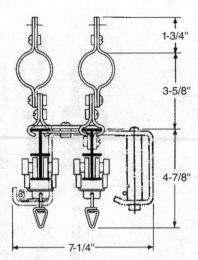

FIGURE 5.10.17 Double curtain track attached to pipe grid. (*Courtesy: Automatic Devices Company.*)

with switches to accommodate positioning of curtains from front to back or possibly bisecting the production space. See Fig. 5.10.17.

In a studio that is built for a specific, defined production, curtains may not be necessary if hard scenic elements like LED video tiles, monitors, or translucent and colored Plexiglas panels are fully covering the studio walls. However, having a black velour curtain on track available adds a flexibility that may come in handy for an unexpected shot.

Note that fabrics used in public spaces are required by law in many states and cities to be certified as flame retardant according to standards developed by the National Fire Protection Association (NFPA). When purchasing curtains they must have a certificate attached, with other available documentation stating adherence to the fire safety code. Fabrics that are labeled Flame Retardant (FR) are topically treated and are certified for one year and require annual testing. Fabric that is certified as "IFR" (Inherently Fire Retardant) or "PFR" (Permanently Fire Retardant) has been woven from fibers that are noncombustible for the life of the fabric and will not dissipate after cleaning.

5.4.3 Hard Cyc/Green Screen

A hard cyc is a continuous background piece made of rigid material that serves the same purpose as cyc curtains discussed in the previous section. The added benefit of a hard cyc is the ability to paint it different colors as needed and to curve or "sweep" the cove between floor and walls making it one continuous surface. If the entire wall, floor, and sweep are uniformly finished, a hard cyc provides the option of an 'infinity' space effect. The cyc surface finish must be level, even in texture and color for the infinity effect to be attained, as well as for chroma key functionality. See Fig. 5.10.18.

Chroma key is the technique of digitally joining two images together when removing one color from one image revealing the second image behind it. The choice of color to be removed is typically blue or green and is determined by the dominant colors of the subject. If for instance the subject contains an abundance of blue then a green finished screen could be used to key out the blue, thus the term "green screen." Selected other colors work for chroma key effect but green and blue are the most in use, with green being brighter and therefore requires less light than blue.

A major concern of chroma key is to have not only an even background color and surface but also to have even lighting approximately within one-half stop overall. The use of dimmers is not advised on chroma key screens as they cause a color temperature shift, which reduces the purity of the green or blue backing. Recommended cyc illumination level should be approximately one stop less than the talent key light. Flesh skin tones are projected equally with green or blue screens when using modern digital keys.

Hard cycs are commercially available in ready to assemble modules and may be built-into the building structure or free-standing from the walls, or as a hybrid combination of both. There are also skilled craft persons who make hard cycs by hand, sometimes sculpting them around room windows and columns. It is recommended that with either method, a crew that has experience with installing a hard cyc is called on to guarantee the correct leveling and smoothness of the surface and structure. See Fig. 5.10.19.

If the studio will not be producing full body emersion views (i.e., talent in frame head-to-toe touching the floor), chroma key curtains or flat painted walls will also work for "ankles-up" or sitting subjects upper body shots.

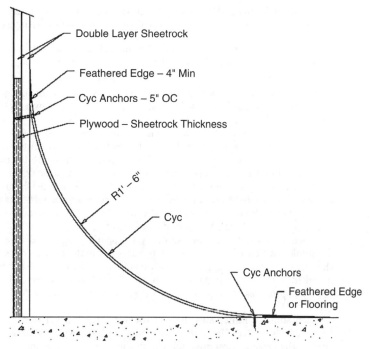

FIGURE 5.10.18 Example of hard cyc wall to floor sweep module.

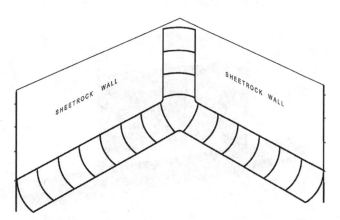

FIGURE 5.10.19 Hard cyc modular layout example. (*Courtesy: Pro Cyc.*)

5.4.4 Electrical Distribution and Power Capacity

Studio electrical and control systems distribution are at a crossroads today with the increased use of fluorescent and digital solid-state energy efficient lighting instruments versus the conventional incandescent filament type of sources that have been in use for many years. Due to this technology changeover, two approaches to electrical distribution systems are in use:

- Conventional—Based on electronic dimming that regulates the voltage and in-turn controls illumination intensity of higher wattage filament type tungsten–halogen source instruments, but may also 'phase dim' fluorescent and even some LED.
- Digital—Where lower wattage instruments receive a constant voltage that is simply "On" or "Off" and the intensity and other fixture attributes are controlled by a digital signal on a separate line.

5.4.4.1 Conventional Approach. The conventional approach in the United States utilizes 120 VAC dimmers that reduce the voltage to the instrument in order to reduce the illumination level. If the light is to be discreetly controlled, this method requires each instrument be fed by a circuit that begins at an individual dimmer with a typical 10 or 20 A capacity per dimmer. This is referred to as a "dimmer-per-circuit" approach, where all grid or other receptacles in the studio lighting dimming scheme require a dedicated dimmer. The dimmers are typically 20 A as that is the most cost-effective mass produced dimmer capacity of 120 V sizes. In large studios there may be a need for 50 A circuits that is determined by the height or desired effect.

Power is brought from the dimmer on properly gauge sized wires and distributed throughout the studio, particularly at grid height, by means of outlet troughs or boxes that attach to pipe, walls, channel track, moving rigging devices, or other surfaces. The distribution devices may have flush-mounted connectors or pigtails, usually 18 inches long, for each circuit with the preferred connector. Typical approaches are shown in Fig. 5.10.20. A selection of standard studio connectors is shown in Fig. 5.10.21.

Each studio dimmer must have its own neutral run between the dimmer and the receptacle wherever it is located. The distribution trough or box device will have 2-wires plus a ground for each circuit. The ground wires are often run to the housing of the box or trough and a single ground is then run from the box to a properly grounded terminal.

As the majority of lighting instruments are hung from the overhead grid out of the way of cameras and set pieces, an appropriate number of receptacles on the grid are required. An often-used formula for the quantity of outlets is one 20 A receptacle per 20 feet^2 of studio space. Or a calculation of

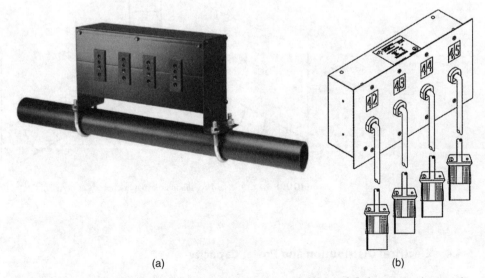

(a)   (b)

FIGURE 5.10.20 Lighting electrical connector approaches: (*a*) pipe-mounted box with flush connectors (*courtesy: Lex Products*), (*b*) recessed box with pigtails (*courtesy: Electronic Theatre Controls*).

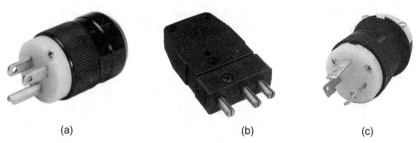

FIGURE 5.10.21 Common connector types: (*a*) parallel blade (often referred to as Edison), (*b*) stage pin, (*c*) grounded twist-lock.

the useable net production area, the square footage inside the curtain tracks, is also applied with one circuit for each 16 feet^2 of the net square footage amount. This approach reflects the common 4 feet × 4 feet fixed pipe grid layout. It can be problematic if too few outlets are available and only available across the studio grid in an opposite corner. An entertainment-oriented studio with multiple scenery possibilities would require more outlets than a news studio with generally fixed sets and defined areas.

These calculations do not include dedicated outlets for cyc lighting from above or from the floor. If overhead cyc lights or cyc ground-row fixtures are anticipated, an ample amount of circuits are necessary in order to satisfy the cyc lights single or multiple color compartments. The cyc lighting circuit layout is dependent on height and length of the cyc as well as the selected fixture's unique distribution output. The cyc circuits are in addition to the production lighting circuit layout.

If lights are to be mounted on floor stands or in set pieces, they will require dimmed circuits in wall outlet boxes, either surface or recessed mounted that are positioned 24 inches to 30 inches above the floor to allow for the pigtail length and the connected instrument's cable. Receptacles are sometimes mounted in scenery for added convenience and for use with practical on-set fixtures, and then are attached by an extension cable to a dimmed or nondimmed wall box outlet or possibly a floor receptacle.

Other floor- or grid-mounted devices such as foggers, smoke machines, discharge lamps, followspots, or motor-operated lights or machinery will require nondimmable power. A recommended amount of nondim circuits is approximately 20 percent of dimmed circuits, but more importantly is dependent on the studio's use. When nondim outlets are provided they typically are fed from a standard contractor's electrical panel with a protective circuit breaker for each outlet. There are also modular dimmer systems that allow for the swapping of modules from a dimmed to a nondim type within the cabinet providing a flexibility to change the location of nondim circuits as needed. All outlets in the studio should be numbered with a corresponding dimmer or nondim identification number.

The estimate for the amount of power required for this approach is based on the number of dimmers multiplied by their amperage, and the same for the nondim circuits. This gives us total amperage if all circuits are raised to full output. It is very rare a studio uses the full capacity of all sources and it is common practice to apply a diversity factor to that total calculated amperage. This factor assumes we will work with only a certain amount of circuits at any one time. For instance, if we have 96 channels of 2400 W 20 A circuits and we determine that we need eighteen 20 A nondim circuits, the total power capacity of the studio would be (1920 + 360) = 2280 A. Using a diversity factor of 60 percent calculates to 1368 amps anticipated usage. In a three-phase 120 VAC system, this equates to 456 A per phase. Then it is common practice to provide protection for that amount rounded to the next larger size standard main breaker, in this example 500 A three phase.

5.4.4.2 Digital Lighting Approach. With the refinement of video acceptable LED and fluorescent lighting sources, a major change to studio systems is now possible in the area of the power required and its distribution. Instead of the high power demand due to the tungsten filament source with wattage ranges from 100 to 10,000, the digital or fluorescent source instruments are available with equivalent optics and beam projections at wattages from 30 to 1000.

LED, also referred to as solid-state sources, generally operates with one less zero than tungsten–halogen instruments. A 1000 W fixture becomes a 100 W instrument, a 2 K becomes 200 W and so on. This is a simplistic viewpoint but as a rule of thumb is somewhat accurate depending on fixture type and manufacturer. The point is that today studio lighting systems may have a power distribution layout with not only an overall lower power requirement but also fewer outlets. Considering 20 A outlets may now possibly feed up to 10 instruments rather than the one 10 or 20 A dimmer required for a standard 1000 W light. This is also possible because fluorescent and LED fixtures may be dimmed by means of a separate line transmitting a digital signal and is not dependent on lowering large amounts of voltage from a dedicated dimmer. The power source to the fixture is "On" or "Off," nondimmable and constant. This changes the studio power distribution requirement to standard circuit breaker panels or controllable relays.

At this time in the new technology, there is not a tried formula for the number of circuits needed for a "standard" studio distribution. With lower wattage instruments, the plan would be based on convenience and access to receptacles, which may be grouped to a quantity of outlets per 20 A circuit. So instead of the conventional formula of one 20 A circuit per 20 $feet^2$ of studio space, we would calculate one receptacle per 20 $feet^2$ and one 20 A circuit per three receptacles. This translates into a 66 percent reduction in circuits and the corresponding power required, and allows for more LED or fluorescent instruments into that area than the conventional approach.

The outlets could be supplemented with boxes or plug strips to provide added receptacles if additional instruments are needed in the future. There is also a discussion about the energy an LED continues to draw after the controller is lowered to zero output as the LED driver still searches for a signal. Controllable hardware has been developed that is added to or in place of breakers that stop the electrical flow to the LED source for an absolute zero amount of current, thus maximizing the energy savings and reduces LED degradation.

At this writing it is a new evolving technology with the best solutions and practices not all known due to frequently changing developments.

5.4.5 Network Data and DMX Distribution

The need for an industry standardized dimming system control signal protocol became evident when manufacturers developed their own proprietary signals incompatible with other's similar apparatuses. In studios whether controlling a dimmer rack, a motorized intelligent moving light rig, or multicolor LED solid-state lighting instruments, DMX 512 was developed in 1986 as an agreed-upon communications protocol standard by manufacturers of different lighting devices.

DMX 512 is a digital multiplex signal that uses a serial communication framework (RS-485) and a particular data packet system to control up to 512 separate addresses or channels on the serial bus with 8-bit commands. At a data level, the DMX 512 controllers send asynchronous data at 250 kbaud; 1 start bit, 8 data bits, 2 stop bits, and no parity checking. Some devices will combine two channels, providing a multiplexed 16-bit data packet if necessary (USITT [6]).

The original DMX standard was known as "Digital Multiplex with 512 Pieces of Information" and was revised in 1990, then in 1998 for ANSI standard recognition and revised again in 2008.

DMX 512 1990 specifies that system shall use five-pin XLR style electrical connectors (XLR-5), with female connectors used on transmitting (OUT) ports and five-pin XLR male connectors on receiving (IN) ports. Additionally the use of three-pin XLR connectors was prohibited. After ANSI approval in 2004 the DMX standard became known as DMX 512-A. Typical connectors are shown in Fig. 5.10.22.

One dimmer, whether stand-alone or within a pack or rack, typically requires a single channel address channel to raise or lower light output intensity, and each intensity level may have 256 steps divided over a range of 0–100 percent. The DMX 512 signal is a set of 512 separate channels that are constantly being updated and repeats all 512 intensities at a frequency up to 44 times per second.

One DMX network segment of 512 channels is referred to as a DMX Universe and different sized controllers are able to communicate with just a single universe while larger consoles are built to support two or more universes.

FIGURE 5.10.22 DMX connectors: (*a*) five-pin XLR female, DMX IN; (*b*) five-pin XLR male, DMX OUT.

DMX wiring consists of a shielded twisted pair with an impedance of 120 Ω and a termination resistor at the extreme end of the line furthest from the control console to absorb signal reflection. Specification requires two twisted pair data paths but only defines the use of one of the pairs. DMX 512-A allows the use of eight-pin modular (8P8C or RJ-45) connectors for fixed installations where regular plugging and unplugging of equipment is not required.

A DMX network consists of a network master DMX 512 controller and one or more slave devices. A lighting console is frequently the control master for a network of slave devices such as LED, fluorescent or intelligent instruments, dimmers, and fog or haze machines. The controller generates the DMX signal and each slave device is assigned a "slot" number within its universe 1–512. The slot number is commonly referred to as the DMX "address" and some slave devices require multiple addresses since they have several different functions beyond intensity. Only the control master transmits over the network and all slaves receive DMX data sent to each unique address. The address is stored inside the device and is set by different methods such as an LED menu display, DIP switches, Rotary Dial, or directly from Software.

Slave devices typically have a DMX 512 "IN" XLR connector and usually an "OUT" (or "THRU") XLR connector. The controller, which has only an OUT connector, is connected to the IN connector of the first slave device. A second cable then links the OUT or THRU connector of the first device to the IN connector of the next slave device and so on, creating a daisy chain system. Figure 5.10.23 shows a straightforward network consisting of a controller and three slaves. Note the final device in the chain is terminated.

The DMX 512-A standard as written supports 32 devices on one network at a distance of up to 4000 feet; however, most manufacturers recommend runs of no longer than 1200 feet total (300 feet between devices). Similar to other digital systems the signal quality degrades over long distance runs and employing a repeater or optically isolated splitter is recommended. The opto has an IN terminal or connector and multiple OUT lines. The isolated splitter also protects failure on one output line from effecting instruments or devices on another OUT line. Additionally, if there is an electrical fault in one device, the isolated splitter protects the other DMX connected equipment from damage. Each splitter output may drive up to 1000 of 120 Ohm data cable similar to a control console.

For short cable runs of less than about 150 feet with only a few devices, it is possible to operate without termination. As the cable length and/or the number of devices increases following the specification for termination and correct cable impedance should be adhered to.

DMX has also expanded to uses in nontheatrical architectural lighting, in applications ranging from holiday lights to electronic billboards.

Implementing a DMX network requires forethought and professional planning to insure correct hardware choices, reliable data flow, convenient access, and correct implementation by the installing integrator or contractor.

The next generation of lighting data protocols include RDM (Remote Device Management) for two-way communication between RDM-compliant devices over a standard DMX line and ACN (Architecture for Control Networks) designed for use on Ethernet networks focused primarily toward theatrical applications.

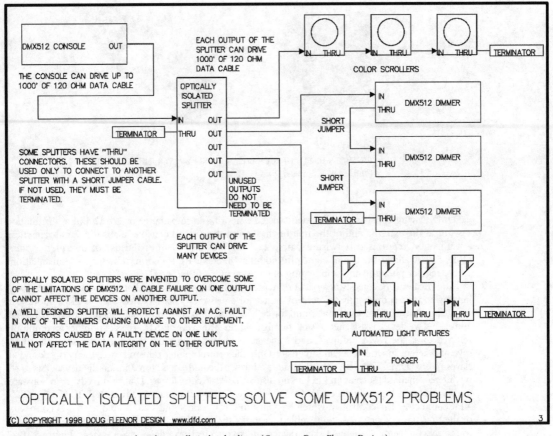

FIGURE 5.10.23 DMX network with optically isolated splitter. (*Courtesy: Doug Fleenor Design.*)

5.4.6 Dimming

Fixture dimming is the heart of a lighting system. The ability to instantaneously raise and lower multiple fixture intensities for composition or effect, or change colors by intensity variances on the colored gels, provides the lighting designer with countless ways to affect viewer's focus and the overall visual composition.

Dimmer modules are typically based on SCR (silicon controlled rectifier) or SSR (solid-state relay) devices to manage voltage and, correspondingly, the output of the light source. In broadcast studios each dimmer typically utilizes toroidal chokes to regulate the noise and rise time of the dimmer, therefore filtering possible filament vibration and RF interference. The modules are located in a manufacturer's custom cabinet with an electronics module that manages the data traffic flow and may report on the system's status. The internal modules may be quickly replaced if faulty or to swap types of modules for specific functions. However, since most modules house two 20 A dimmers, both circuits will be affected by the change of type of module.

If there are a large quantity of circuits, and therefore dimmers, they are typically housed in a rack of 48 dual dimmer modules for a total of 96 dimmers per rack. There are also smaller racks that may house 24 modules for a total of 48 dimmers, or 12 modules for 24 dimmers. When working in quantities less than 24 dimmers they are usually referred to as "packs."

The dimmer packaging provides for two distinct system approaches.

- Centralized—the system employs racks located in one area.
- Distributed—dimmers are dispersed and located in the studio where they will be used.

Centralized dimmers are in a clean and dry room with uninterrupted air flow. The racks are electrically fed individually or through connected buss bars located within each housing, or through added auxiliary cabinets. Air conditioning of the room is normally calculated as 5 percent of the total dimmer capacity since most professional grade dimmers are more than 95 percent efficient. The racks utilize fans to move air within the cabinet, so locating them outside the studio is advisable to avoid unwanted audible noise. Ideally the dimmers would be located in a room near the studio to minimize the wire run distance and minimize voltage drop. All circuit wiring is run from the outlet, through the selected type of distribution trough or box in conduits to the dimmer room. Local code will determine the conduit sizes and quantity of wires permissible within the conduit.

Distributed dimmers are manufactured with outlets as part of the device and may be packaged as multiple dimming units in one housing. In this approach the modules have a high level of noise suppression or utilize a different quieting technology. Sine wave dimmers that alter the output without changing the waveform are often specified in concert halls and other audio sensitive venues. Distributed dimmers may be configured in packs, as well as configured within a trough type housing or strip, and often mounted directly to rigging devices. Distributed dimming offers advantages such as the ability to add or change dimmer positions on the fly, expand existing centralized systems without a full changeover, and accommodate the new solid-state low wattage technology as a supplement to instruments that only require a DMX control signal to affect light output.

The disadvantages of the distributed dimming approach are that nondim constant power must be available throughout the studio and, depending on the dimming device, a three-phase outlet may be required. Also, a control wire must be run to each dimmer device and DMX addressing needs to be maintained across the control network when moving the dimmers from their initial position.

Distributed dimming devices are shown in Fig. 5.10.24.

5.4.7 Lighting Control

Typically, there are various objects to be illuminated in a single composition requiring discreet lighting control of all lighting instruments. The lighting control console—also called a light board, a console, board, or lighting desk—is the brain of the studio lighting system and its function is to set and simultaneously manage the intensity, contrast, and other features of all lighting instruments and related devices. Lighting consoles today employ electronics, computerization, and utilize DMX 512 with other electronic signal protocols to communicate with dimmers and controllable devices such as LEDs, intelligent lights, video walls, audio MIDI signals, and system effects devices.

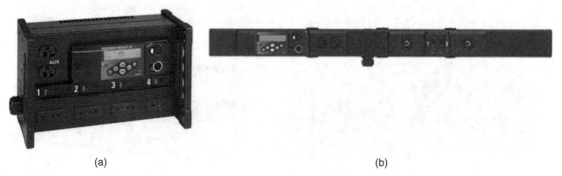

FIGURE 5.10.24 Distributed dimming components: (*a*) dimmer module, (*b*) dimmer bar. (*Courtesy: Electronic Theatre Controls.*)

The particular dimmer or intelligent light function's DMX address is assigned to a control channel and is accessed through the console. Multiple addresses may be assigned the same channel number through a patching procedure, which makes those addressed items respond to that specific channel's command.

Physically, lighting controllers in their simplest form have a numbered arrangement of potentiometers that as they slide up or down, a numbered scale between 1 and 100 the intensity raises or lowers accordingly. Whenever there are two rows of faders where the second row is the same numbered channel sequence as the first, it is intended to crossfade between the two scenes, changing the lighting from one preset look to another. Switching between each scene allows the lighting designer to set up the next look while the current scene is being played.

Prior to the computerization of consoles, lighting controllers were custom manufactured to the specified quantity of presets and a 2-, 5-, or 10-preset board with nondim off/on buttons were found in many broadcast studios. Today, two scene preset consoles often include added functionality such as recording a single slider "submaster," an effects package, the option of turning the second group of faders into submasters, a timer for crossfades, and often an LED readout screen.

Other common faders or buttons on consoles include:

- Grand Master fader, which controls all channels.
- Two Cross-Faders, "A" and "B" or "X" and "Y" that fade between the lights "on" scene and the subsequent assigned scene.
- *Blackout* switch or button that immediately brings all instruments to zero output and when released or switched, back to their previous state.
- Chase, where channels are sequenced to fade in or bump "on" repeatedly in a designed pattern similar to a building marquee.

Consoles range from handheld sizes to boards that have the capacity for thousands of channels as well as wheels and dials to command intelligent instruments, other motorized devices, and multiple monitors. See Fig. 5.10.25. The cost of the console is determined by many factors, from chip technology and the platform used to the number of universes or channels it may output and the board's feature set. However, the basic function of raising and lowering intensity remains the same and selecting a console becomes a choice of operator preferences, cost, and what best serves the types of productions to be produced.

Alternative controllers such as PC or MAC based, tablets, and hand-held or pocket-size units are expanding in use to satisfy the increased demand for familiar user interfaces, simple access, and the economic restraint of reduced staff. Other control applications may use a group of wall-mounted buttons or faders and require minimal operation once presets are established on a standard lighting board. A tablet-based approach is shown in Fig. 5.10.26.

5.4.8 Lighting Sources

Throughout this chapter we have used the words *fixture*, *instrument*, *luminaire*, and *light* interchangeably to describe studio lighting equipment. A more specific definition of a studio lighting instrument

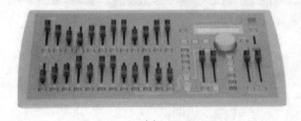

(a)   (b)

FIGURE 5.10.25 Example lighting consoles: (*a*) Electronics Theatre Controls Smartfade console, (*b*) Gio console. (*Courtesy: ETC.*)

FIGURE 5.10.26 Luminaire tablet console software app. (*Courtesy: Synthe FX.*) (Black-and-white reproduction of a color illustration.)

is: an apparatus from which light is produced and the projection may be modified for specific production illumination applications. Lighting instruments are similar to a doctor's or musician's tools and clearly becomes an "instrument" when utilized by a skilled lighting professional.

The selection of studio lighting instruments offered today is greater than any time in the past primarily due to the many types of light sources implemented by lighting fixture manufacturers.

The following sections discuss current studio lighting illumination sources (lamps).

5.4.8.1 Incandescent Tungsten-Halogen Lamps (T-H). *Incandescent* means to glow or cause to glow with heat. T–H lamps emit light as electrical current is passed through the tungsten filament that incandesces creating light as a by-product of the heating process (Fig. 5.10.27). The incandescent lamp converts less than 5 percent of the energy burned into visible light. Luminous efficiency of a source is expressed as lumens-per-watt which tells us how many lumens of visible light the source generates from a watt of heat. A typical incandescent source produces approximately 16 lumens-per-watt, contrasted to the 60 lumens-per-watt of a compact fluorescent lamp, and over 50 lumens-per-watt of LEDs.

5.4.8.2 LED—Solid-State Light Source. An LED is a two-lead semiconductor source, specifically a diode that emits light when stimulated (Fig. 5.10.28). When an electrical current is applied to the leads, energy is released as light in a process known as *electroluminescence*. Other properties of the light, such as color, are controlled by additional elements of the semiconductor's structure or externally applied caps or phosphors.

The intensity of LEDs may be controlled by electronic low-voltage dimmers (ELV) or directly from a DMX control console communicating with the LED driver without the need of an intermediary dimming device.

LED sources offer advantages over incandescent lamps such as extreme longer life, less energy demand, lighter weight fixtures, and less copper use, and also eliminate the need for resistance, sine-wave, or solid-state on–off switching dimming systems. Unlike the T–H lamp, LED color temperature does not shift with power level variations.

LED fixtures do often require a cooling fan and possible audible noise must be considered in the fixtures evaluation.

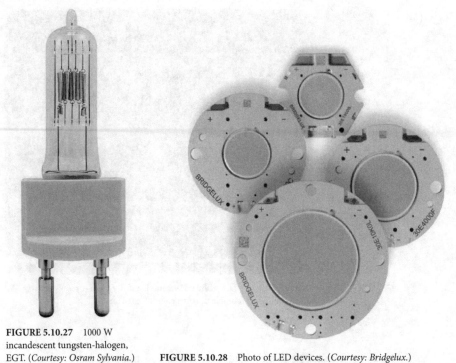

FIGURE 5.10.27 1000 W incandescent tungsten-halogen, EGT. (*Courtesy: Osram Sylvania.*)

FIGURE 5.10.28 Photo of LED devices. (*Courtesy: Bridgelux.*)

5.4.8.3 Light Emitting Plasma™ (LEP). The Light Emitting Plasma™ source utilizes radio frequency (RF) power to excite a plasma material inside a quartz envelope creating a 90% optical efficient point source (Fig. 5.10.29). The LEP emits full color spectrum high lumen density light with a CRI of up to 98.

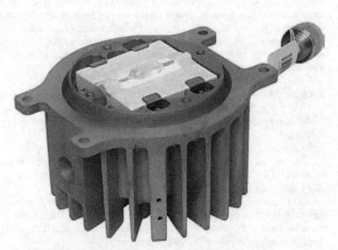

FIGURE 5.10.29 Photo of a plasma lamp. (*Courtesy: Luxim Corp.*)

The difference between LED and LEP technology is that LEP light sources utilize a solid-state device to generate RF energy that stimulates a plasma emitter while LEDs generate light output using the solid-state device. A single LEP lamp, the size of a tic-tac mint, can replace a large HID lamp or a cluster of high powered LEDs.

LEP-sourced instruments are often used in high speed recording applications due to their high frequency range. A LEP drawback is that they require a warmup time at initialization and restrike.

5.4.8.4 Tubular Fluorescent. A fluorescent lamp or fluorescent tube is a low pressure gas-discharge lamp that uses the fluorescence process to produce visible light. After a ballast-assisted ignition the controlled electrical current excites gases contained in the tube and radiates ultraviolet light, which in turn causes the phosphor coating on the inner wall of the tube to glow. This method of converting energy to light may produce more than 100 lumens-per-watt of light, far surpassing the efficacy of an incandescent filament and other types of sources.

With fluorescent studio lighting lamp wattage sizes ranging from 10 to 100 W, fluorescent sources are extremely energy efficient. In fluorescent fixture design it is common practice to group more than one lamp to an individual fixture in order to increase the projected light and beam spread. The total wattage is typically less than 300 W per fixture in larger array instruments. Due to the necessary ballast to regulate the lamp current, fluorescent lamp fixtures typically are more costly than incandescent units.

5.4.8.5 Compact Fluorescent Lamps (CFLs). CFLs are a subset of tubular fluorescent lamps employing similar low pressure gas-discharge technology requiring a ballast for ignition and stabilization; however, the difference is that the CFLs are curved, folded, or spiral shaped to fit into existing A-lamp socketed fixtures or in a different U-shaped format sized to fit into 2 feet × 2 feet tiled ceiling fixtures. Most broadcast lighting instruments utilize the 55 W biax CFL. Both the tubular and compact fluorescent lamps are available in 3200 K, 5500 K, and other color temperatures.

5.4.8.6 Remote Phosphor LED. A remote phosphor source is a three-component LED-based lamp that produces white light with no visible point source as a phosphor composite layered onto a substrate and separated, by a space referred to as the mixing chamber, from blue LEDs, creating a low glare consistent light without hot spots. See Fig. 5.10.30. By placing different panels with different phosphor mixes in the holder away from the blue LEDs, the color, color-temperature and/or the CRI of the fixture may be changed to suit the application with less color shift than other sources (LED Magazine [8]).

A drawback of the remote phosphor source is its unidirectional beam, which may require various shaping devices to contain the light from unwanted spread. However, due to the color accuracy and the LED efficiency, instrument availability using remote phosphor technology has rapidly grown.

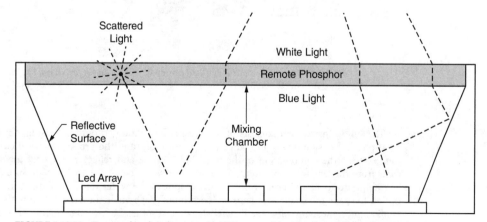

FIGURE 5.10.30 Remote phosphor diagram with LED array, mixing chamber, and phosphor panel.

5.5 LIGHTING INSTRUMENTS

Broadcast lighting instruments may be organized into three general categories: spotlights, floodlights, and supplementary lights.

5.5.1 Spotlights

Spotlight is a term commonly used for a luminaire that projects a beam of light onto a subject and casts a single hard-edge shadow. In certain spot fixtures the shadow edge may be softened by a treated lens, filters, or other intermediary diffusion. Referring to the earlier discussion of three-point lighting, spot instruments are typically used for key and back lighting and therefore are the most widely used instruments in the studio. The method of controlling the light beam differentiates spot instrument types.

5.5.1.1 Spotlights—Fresnel. This spotlight uses a lens named after French physicist Augustin-Jean Fresnel who developed the stepped lens for long-throw lighthouse use in 1822 allowing the beam to be seen many miles away as the lens collimates the light beam. See Fig. 5.10.31.

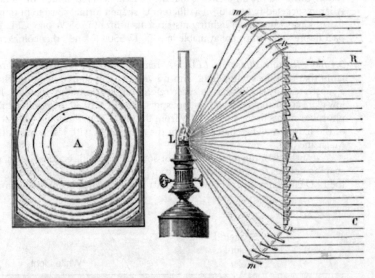

FIGURE 5.10.31 Fresnel lens illustration from *Natural Philosophy for General Readers and Young Persons* (Ganot [8]).

The studio instrument consists of the Fresnel lens mounted onto a housing, a reflector usually of a spherical design, the light source (lamp) assembly, and the proper electrical components. See Fig. 5.10.32. The combination of the fixed light source and reflector assembly travels toward and away from the lens focal spot to increase or decrease the beam size from a tight spot to a wide field. The ability to focus the beam from a narrow spot to the wider flood positions combined with the option to shape the beam with flaps or "barndoors" have made the Fresnel fixture the workhorse of studio lighting instruments. The Fresnel also casts a softer shadow edge than many other spotlights, allowing for blending from fixture to fixture as well as the ability to be used as an area wash light.

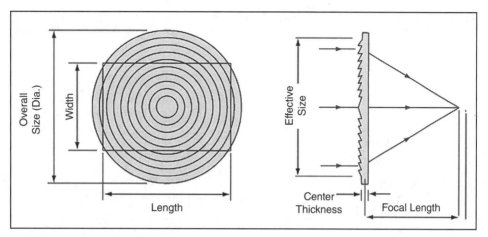

FIGURE 5.10.32 Fresnel lens front and side. (*Illustration Courtesy: Edmund Optics Inc.*)

Ears on the lens holder or lens door allow for a frame to be secured in front of the lens, which may hold plastic film (gel) to color or diffuse the light or wire screens to dim the light without changing color temperature when tungsten–halogen lamps are the source.

Today Fresnel instruments are available with incandescent tungsten–halogen, LED, and LEP Plasma sources.

5.5.1.1.1 Pole-Operated Instruments. With the Pole-Op feature added, many fixtures may be panned, tilted, focused, and accommodate adjusting barndoors from the studio floor. See Fig. 5.10.33. Three different colored cups are each connected to twist and maneuver a specific fixture function. A post positioned within the hollow cup is latched by the pole's hook, much like the pole with a hook used to open and close commercial awnings. The pole has a handle crank at the other end and a sleeve around the outer tube for the operator's hand to grip, allowing the core of the pole to rotate in turn, moving the unit as needed.

5.5.1.1.2 ERS Spotlight. ERS stands for Ellipsoidal Reflector Spotlight and is a type of stage and studio lighting instrument based on the ellipsoidal reflector used to gather and direct the light beam. See Fig. 5.10.34. The ERS is also referred to by the manufacturer's product names as "Leko," an acronym for the inventors Joe Levy and Edward Kook, or "Source Four" after the Electronic Theatre Controls company lamp that utilizes four compact filaments.

The optics of an ERS instrument are similar to those of a slide projector (Fink [7]). The lens train contains one or two lenses and the distance between the lenses, as well as distance from the lenses to the reflector, determines the width and focus of the light beam; adjusting the lens tube in or out from

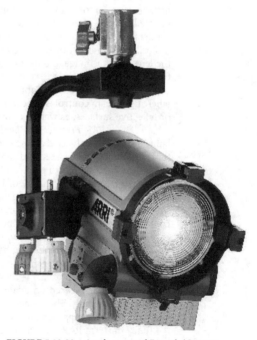

FIGURE 5.10.33 A pole-operated Fresnel. (*Courtesy: Arriflex.*)

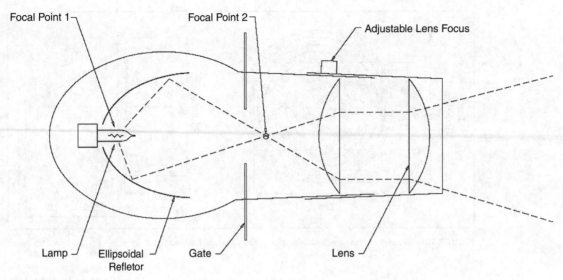

FIGURE 5.10.34 ERS optics.

the reflector alters the projected light beam focus from soft to hard edge. Incandescent, LED, or plasma light sources are used in ERS as a tight point source for light gathering and beam control.

The focal point of the lens system is in front of the aperture, or gate, so that images or patterned metal inserted into a slot at that point will be projected, appearing backward and upside down. The inserted images, often referred to as a "gobo," vary in material based on the source. Four adjustable shutters are mounted at the focal point, allowing for beam shaping and sizing. Accessories such as motorized gobo rotators, dichroic filters, or a circular beam-controlling iris also may be mounted at the lens gate for image and beam control, and light effects. "Zoom" type ERS instruments can vary the size of the beam as well as the focus.

Ellipsoidal reflector spotlights come in many sizes and configurations and are used in applications where tight beam control is needed in particular to project a pattern or image, cut the light off scenery or reflective surfaces, as well as to highlight an object or actor. See Fig. 5.10.35.

5.5.1.1.3 PAR Fixtures Spotlight. PAR is an acronym for the Parabolic Aluminized Reflector lamp and is primarily used as a sealed beam automobile headlight. The PAR lamp is a lens and reflector combination that is fixed and cannot be altered relative to the filament. The PAR lamp is mounted into a "can" and by rotating the entire lamp the elliptical beam can be moved from a horizontal to a vertical orientation. A variety of lamps are available in very narrow, narrow, medium, wide, and extra wide beam spreads and by changing the lamps the beam spread may be adjusted. PARs were the standard lamp used in many broadcast location events since they were relatively inexpensive and could project a considerable light beam from a distance. See Fig. 5.10.36.

These types of lamps/instruments come in varying diameters, the most common being designated PAR56 and PAR64, where the number indicates the diameter of the PAR lamp source in eighths of an inch. A PAR56 is a seven inch diameter lamp and the PAR 64 is eight inches in diameter.

A different PAR approach is the ETC PAR unit, which uses the same filament lamp as the ETC Source Four ERS and accommodates beam spread changes by changing lens as they are not welded to the reflector assembly. The distance and position of the lamp filament relative to the lens remains constant in this novel approach and is efficient as the lenses are less costly than PAR sealed beam lamps.

Today the term PAR refers more to traditional applications of the incandescent PAR lamp, including color or white light washes for set lighting or for special effects, as RGB LED PARs may fulfill those functions without requiring a colored gel placed in front of the PAR can holder.

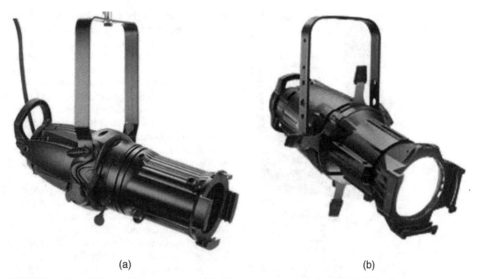

(a)   (b)

FIGURE 5.10.35 ERS spotlight instruments: (*a*) LEKO (*courtesy: Strand Lighting*), (*b*) Source Four (*courtesy: Electronic Theatre Controls*).

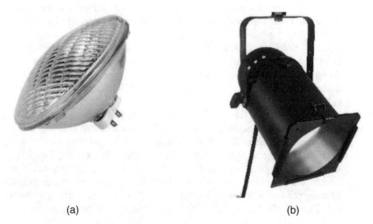

(a)   (b)

FIGURE 5.10.36 PAR instruments: (*a*) PAR sealed beam device (*courtesy: General Electric*), (*b*) PAR 64 housing (*courtesy: Altman Lighting Inc.*).

5.5.2 Flood Lights

Flood lights are instruments that project a relatively unfocused beam of light onto a subject with either a soft or hard edge shadow. Flood instruments are typically used for fill lighting to contrast with spotlights. Floods are characteristically without a lens, and project a wide beam based on the relationship of the source to the reflector and the reflector's size and shape.

5.5.2.1 Softlights—Indirect Beam. Studio softlights are named after the near shadowless light projected from an indirect beam produced from a source that is not in a direct line of projection to the subject. The softlights source, out of the subjects view, is directed toward an unseen primary reflector and then on to a secondary larger reflector, which in turn reflects the light toward the subject.

FIGURE 5.10.37 Baby softlight. (*Courtesy: Mole-Richardson.*)

The light quality of the softlight eliminates harsh and sharp edged shadows and blends evenly with other instruments. Softlights are an ideal fill light in our three point lighting scheme and sometimes are used as an overall light onto a set so talent may walk to different positions and be evenly illuminated. See Fig. 5.10.37.

The softlight is an inefficient instrument due to the indirect beam that causes a considerable amount of light loss. The projected light is also characterized by a "quicker" light fall off than other instruments, making it an inappropriate choice for long throw applications. There are incandescent linear lamp types as well as LED softlights that use the indirect source method. Accessories such as eggcrates, color frames, and louvers are available that shape and limit the beam.

In video or film, as well as still photographic applications, the light source—whether it be a strobe type or constant—may be focused onto a reflective board or matte surface for a very wide uncontrolled indirect softlight effect. This approach is often referred to as "bounce" light as the light is imagined to be bounced back off the reflective surface. In this use, the fixture may be a spotlight or almost any other type regardless of the source as the "bounce" light illuminates the image not the primary source unit.

Portrait photographers and videographers often utilize umbrellas with the inner side covered in a matte white or a highly reflective material for a soft effect. Many actors who are constantly under the lights often request to be lit with softlight as it helps to minimize facial blemishes and wrinkles.

5.5.2.2 Softlights—Direct Beam. Direct beam softlights are instruments that have the source of light in direct view of the subject yet still project a softer beam and a softer shadow than a spotlight. Different sources such as tubular or compact fluorescent and remote phosphor types were originally designed to approximate a full size aperture of light similar to an indirect softlight but with the source lamps in view.

5.5.2.2.1 Direct Beam—Fluorescent. Through the fluorescent process of phosphor energized inner walls, a soft beam is created as the light's center is along the length of the tube making an extended projection. The tubular shape of the lamp emits a constant glow that is 360° around the center so it projects light evenly and is directly reflected. See Fig. 5.10.38. It is a broader and longer source than the plasma point source and incandescent filament, and due to those projection qualities fluorescent instruments have been adopted as fill lights as well as a fixture of choice on green screens. Fluorescent lamps require a ballast and may be dimmed through two-wire phase dimmers as well as DALI and DMX protocols, and utilize similar accessories as the indirect softlight.

FIGURE 5.10.38 A 4-bank select fluorescent direct beam device. (*Courtesy: Kino Flo.*)

Louvers or screens of differing thicknesses are often placed over the face of fluorescent fixtures to limit peripheral light, thus focusing the resultant light beam on the subject. Fluorescent instruments also may have a diffusion type hood placed over the lamps affecting the beam quality softness.

5.5.2.2.2 Direct Beam—Remote Phosphor. Remote phosphor units use the phosphor panel as the instrument's primary projection point. The blue LED emitters located behind the panel out of view generate the initial light and the front mounted phosphor panel becomes the instrument's source. See Fig. 5.10.39. This is similar to placing a material on the face of a spotlight amending the source from the lamp to the diffusion.

Depending on the size of the panel, a wide and diffuse beam is created similar to the indirect beam softlight, satisfying the requirement of the three point lighting's fill light. The continuous linear spectrum provides an enhanced color quality and renders superior flesh tones. With the wide even beam angle approaching 160°, cycloramas and green screens are also lit by remote phosphor instruments. Eggcrates, color frame, and louver accessories are available to alter the light beam similar to the indirect softlight. An example is shown in Fig. 5.10.40.

FIGURE 5.10.39 Direct beam remote phosphor light, Cineo LS unit. (*Courtesy: Cineo Lighting.*)

FIGURE 5.10.40 Direct beam remote phosphor light BBS Area 48. (*Courtesy: BBS Lighting.*)

5.5.2.2.3 Direct Beam—LED Panels. In broadcast studio lighting the initial LED fixture configuration was the 1 foot × 1 foot panel by Litepanels. In a panel light construction LED emitters are arranged in condensed rows and columns across the face of the unit, creating a square pane of light. This collection of LED cells projects a beam with the angle determined by the lens degree of each emitter. Typically, lens are 20° and 50° optics producing a spot field in the former and flood in the latter. There is also the choice of 5600 K daylight or 3200 K tungsten or a mix of both for a bicolor variable color temperature fixture, the first of its kind in studio lighting. Intensity and color temperature control is by either a direct wired handheld remote, a DMX interface, or rear-mounted manual control knobs.

As panels were the first LED studio types on the scene, they gained rapid market acceptance and only since traditional fixtures such as LED Fresnels and LED softlights were developed has LED panel usage been reduced. However, LED panel lights are still produced in great variations around the world and used as a low wattage fill-type light with better color rendering and more output than previous generations. Panels and many other forms of LED fixtures are used as location battery powered units due to their low power consumption which is ideal for location and ENG applications.

5.5.2.2.4 Direct Beam—Incandescent Scoops and Broads. Scoops and broads are classic historical instruments using a tungsten filament burning lamp that fulfilled many needs for cyc and fill-type lighting prior to the use of energy efficient LEDs and fluorescent lighting. The lamp is directly exposed to the subject in both the broad and scoop, supplying a somewhat even field but with a brighter center tapering to the beam edge. As with most direct exposed source instruments, a hard edged shadow is produced and often requires a diffusion material placed in front of the aperture to soften the beam.

Scoops, as the name implies, are elliptically shaped and formed by an aluminum spinning process. The scoop appears as a large bowl with a hole for the socket at the apex of the ellipse for a 500–2000 lamp wattage range and are typically fitted with accessory ears on the face for the color filter frame. See Fig. 5.10.41.

Broads are rectangular shaped housing fixtures utilizing a double ended linear lamp source across the center of the reflector, emphasizing the rectangular shaped beam they project (Fig. 5.10.42). Like the scoop, the broad has a hot center due to the filaments' direct projection onto the subject and is often utilized in tight spaces due to its varying sizes from small to large.

FIGURE 5.10.41 Altman lighting 16 inches scoop. (*Courtesy: Altman Stage Lighting.*)

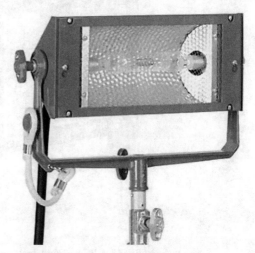

FIGURE 5.10.42 Broad Nook light. (*Courtesy: Mole-Richardson.*)

Broads have been used as cyc and set lights as they are formed similar to the shape of horizontal scenery and cycloramas. Broads are also referred to as "nook" lights and are available in a wattage range from 300 to 2000.

5.5.3 Supplementary Lights

Supplementary lights is a general term used here to include motorized robotic and studio specific application instruments. The following light types have no unifying feature other than they are specific to a particular application when used in the studio.

5.5.3.1 Incandescent Cyclorama Lights. Cyclorama lights are almost exclusively used to project light across a hard or soft material background. Whichever material the cyc is constructed of the cyc light must project a flat even field. The lighted cyc is frequently intended as "skylight" with ERS fixtures overlaying cloud-shaped patterns, creating an immediately recognizable natural backdrop. Cyc lighting is the most efficient way to show time of day and other weather-related projected effects.

The incandescent cyc light instrument is similar to the broad fixture in that it utilizes a linear double-ended lamp, and its housing is rectangular in shape reflecting a typical cyc shape. Certain cyc lights are different from the broad, as the lamp is located in an asymmetrical reflector that directs more light in one direction; the light is usually mounted overhead of the cyclorama and the beam must reach the bottom of the cyc. This approach allows for the units to be hung at even intervals, depending on the height and width of the cyc. Cyc lights based on the asymmetrical reflector are often referred to by the manufacturer's names Farcyc, Skycyc, or Iris.

An alternate incandescent type cyc light is a strip light characterized by a continuous line of compartments with units fastened one to another with no space in-between. Either type of incandescent cyc light is based on a compartmentalized approach with each compartment dedicated to a specific color, or no color, that accommodates color mixing.

5.5.3.2 *Incandescent Groundrow Cyc Light.*   Groundrow lights are, as the name implies, lights that are located on the floor and light the cyc or background from bottom to top. The configuration of the groundrow is similar to the overhead cyc light, spaced at intervals or evenly across the cyc as a striplight. The groundrow is often used when a sunrise effect is called for, with the fixture hidden behind scenery or some other visual element.

As with the other incandescent cyc lights, the groundrow employs a linear lamp in a rectangular-shaped housing.

5.5.3.3 *LED Cyclorama Lights.*   Cyc lighting has benefitted from the introduction of professional level solid-state lighting as the original LEDs were colored red, green, or blue. These cyc lights are referred to as RGB type, often with a "W" or an "A" suffix to indicate the additional colors white and/or amber as added color choices. The LED offers a tremendous energy saving, considering 1000 W linear lamps are used in incandescent cyc lights. The striplight configuration of LED cyc lights come in different sectional lengths from 1 foot to 8 feet long. Each color may be addressed by a DMX signal allowing discreet control for intensity and color mixing. An example is shown in Fig. 5.10.43.

The LED cyc light may be used as an overhead or groundrow type, and as the evolution of LED optics advances, units will be better able to light tall cycloramas evenly.

5.5.3.4 *A Note About Green Screen Lighting.*   Related to cycloramas, green screens also require even lighting across the entire area within the frame. The overhead cyc lights have been used for green screen lighting but recently dedicated white light fluorescent, remote phosphor, or LED type fixtures are found to be better choices due to even beam spread, blending, and color consistency features. Cyc lighting often calls for color mixing whereas green screen lighting is concerned specifically

FIGURE 5.10.43   Chroma-Q Color Force cyc lights. (*Courtesy: Chroma-Q.*) (Black-and-white reproduction of a color illustration.)

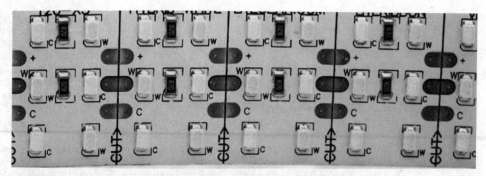

FIGURE 5.10.44 Lite Gear LED VHO tape. (*Courtesy: Lite Gear.*)

with white light projection. Additional green or blue colored LEDs or fluorescent tubes may be added to green (or blue) screen lighting but general use and emphasis is on high CRI white light projection.

5.5.3.5 LED Tape. LED tape is a malleable strip of uniformly spaced light emitting diodes mounted on a tape-like thin strip that has a bonding agent backing and operates on 12 or 24 V of direct current from a driver. See Fig. 5.10.44. An LED controller is necessary to adjust color and/or intensity. Initially, LED tape was used for general lighting applications such as lighting counter tops mounted under cabinets, behind monitors or pictures and for scenic lighting and effects.

As the LED light quality has improved, LED tape is now used as a replacement for certain talent fixtures and has bred a new fixture type. Users may customize the tape in length, color, beam spread, and emitter density. Available LED tapes include single white light ranging from 2700 to 6500 K in color temperature or a select single RGB color or combined RGB multicolor. Emitter pattern addressing may be in as many possible configurations as desired, if discreet control is available. USB strip type lights operate on the standard 5-volt direct current used by USB devices.

5.5.3.6 Robotic Motorized Fixtures. Motorized fixtures are mentioned here as they have been used in select broadcast venues, but typically are used for theatrical shows live or recorded, live events, and concert lighting where the lights movements enhance the programs experience or serve as multi-positioning beams for specific purposes. Motorized fixtures are generally classified as spot or wash types with many variations of wattage, lamp source type, and available features.

Traditional broadcast applications do not justify the specialized crew and expense required to set up, maintain, control, and operate robotic type fixture unless it is determined that a specific situation is best lit by a remote-controlled fixture. With that said, the next generation of LED solid-state source motorized fixtures lessen the operating expense of lamp replacement, and as fixture and console setup becomes less arduous, the tradeoff costs will turn out to be minimal and may eventually reach an even balance.

5.6 ACCESSORY HARDWARE

Lighting accessory hardware may be categorized into two groups: light mounting and light control.

5.6.1 Light Mounting

Every studio needs or eventually will need to position a light away from the grid. The light may be stand-mounted or hung from the grid by means of a telescopic hanger or extension rod. Examples are shown in Fig. 5.10.45.

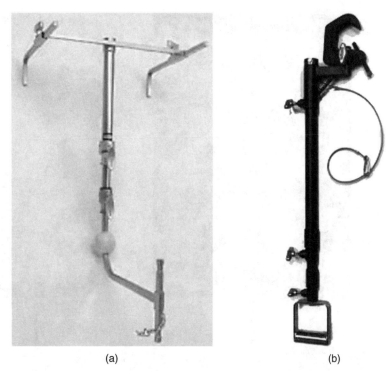

FIGURE 5.10.45 Light mounting devices: (*a*) MSE baby trombone (*courtesy: Matthews Studio Equipment*), (*b*) adjustable hanger (*courtesy: American Grip*).

Stands are available in a large array of configurations depending on height to be attained, the collapsed size for storage, the weight of the unit to be mounted on the stand, and if the stand needs wheels for portable positioning.

Manually operated adjustable telescopic hangers are utilized to lower the light beneath the grid, reducing the angle of the light or bringing a light closer to the subject. Most mid-height and higher gridded studios will have a complement of hangers for these purposes. Additionally, the instrument may need to be perched on a scenic element by means of a base plate or "pigeon." Grip equipment hardware solutions include many types of clamps, brackets, and heads that are often named after the inventor or the image the piece recalls. The world of grip equipment has many mounting devices for almost every circumstance, and somehow keeps coming up with new solutions for light mounting challenges.

5.6.2 Light Control

Not all light intensity and other characteristics are controlled through a dimmer or DMX system. Mechanical dimming by means of wire screens, or scrims, was introduced in the motion picture industry due to color temperature change when incandescent lamps are dimmed by lowering the voltage. With a wire full single scrim you would reduce the light output almost a full f-stop, depending on the manufacturer, without compromising the source's color temperature. A double scrim is double the density of a single. Examples are shown in Fig. 5.10.46.

Other control equipment includes reflectors of different colors and types of mirrors to redirect light.

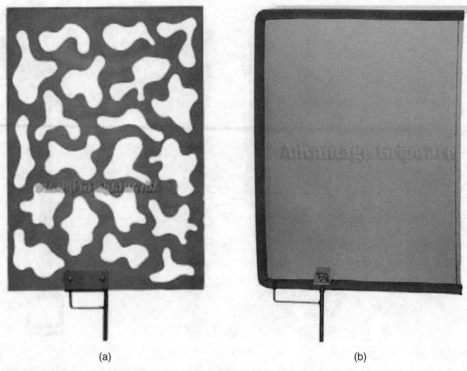

FIGURE 5.10.46 Light control devices: (*a*) wooden Cucolori (*courtesy: Advantage Grip*), (*b*) open end scrim; note the green edging for single density (*courtesy: Advantage Grip*).

A range of fabric products are available such as large "overheads" and "butterflies" for diffusing light, and the same materials are also sewn onto frames to affect the light quality when mounted closer to the instrument. "Flags" or "cutters" are used to cut the light off a subject and are fabricated with a black material.

5.6.3 Other Information

In order to have an appropriate overall viewpoint it is important to review other related studio infrastructure elements that affect system design choices. These include:

- Building and interior construction
- Building power
- Studio size and location
- Studio ceiling construction
- Floors
- Walls
- Studio doors
- HVAC (air conditioning)
- Acoustical considerations
- Adjacent area uses

5.7 SUMMARY

In this chapter, we have highlighted that a television lighting system has many interrelated sub-systems and components. From the fixed pipe grid suspension technique to track considerations, DMX control, new LED instruments, and accessory items we have reviewed various elements that could each require its own chapter.

Today, with many new technologies being introduced and put into use we are in a unique time simultaneously testing and clarifying the results to arrive at a better understanding of the latest developed lighting inventions' uses and limitations. There will always be advancements in lighting, many that will find their way into the studio.

As you begin planning a studio lighting system we encourage you to continue this research as well as to enlist a qualified lighting systems professional early in the development stage.

The following References and Resources are related to this chapter and available for your further use.

BIBLIOGRAPHY

Bennett, A., "Recommended Practice for DMX512, A Guide for Users and Installers," incorporating USITT DMX512-A and remote device management, RDM, Eastbourne, UK, PLASA, 2008.

Gillette, J. M., and M. J. McNamara, *Designing with Light: An Introduction to Stage Lighting*, McGraw-Hill Companies, Dubuque, IA, 2013.

Jones, G.A., D.H. Layer, and T.G. Osenkowsky, *National Association of Broadcasters Engineering Handbook*, "Lighting for Television," authors Bill Marshall and Harvey Marshall, Taylor and Francis.

Kane, R., and H. Sell, *Revolution in Lamps, A Chronicle of 50 Years of Progress*, Fairmont Press, Lilburn, GA, 2001.

Mobsby, N., *Lighting Systems for TV Studios*, Entertainment Technology Press, Royston, UK, 2001.

REFERENCES

[1] Bruno, T. J., and P.D.N. Svoronos, *CRC Handbook of Fundamental Spectroscopic Correlation Charts*, CRC Press, 2005.

[2] Tiffen, Lowel Light: http://lowel.tiffen.com/edu/color_temperature_and_rendering_demystified.html.

[3] Courtesy of EYE Lighting International of North America, Inc. a division of Iwasaki Electric of Japan.

[4] National Institute of Standards and Technology (NIST) is an agency of the U.S. Department of Commerce. The Physical Measurement Laboratory (PML) develops and disseminates national standards.

[5] European Broadcasting Union, The EBU (UER in French) is the world's leading alliance of public service media with 73 Members in 56 countries across Europe and beyond. https://tech.ebu.ch/tlci-2012.

[6] United States Institute for Theatre Technology DMX512 FAQ.

[7] Fink, D.G., and H.W. Beaty, *Standard Handbook for Electrical Engineers*, McGraw-Hill, New York, 1978.

[8] Ganot, A., *Natural Philosophy for General Readers and Young Persons*, D. Appleton & Co., New York, 1872, p. 328.

[9] *LEDs Magazine*, "Remote-phosphor technology can deliver a more uniform and attractive light output from LED lamps." September 2012.

RESOURCES

United States Institute for Theater Technology, http:// www.usitt.org/ usittHome.html

Bennett, A., *Recommended Practice for DMX512: A Guide for Users and Installers*, incorporating USITT DMX512—a and remote device management, RDM, Eastbourne, UK, PLASA, 2008.

http://www.lutron.com/en-US/Education-Training/Documents/DMX%20webinar_7-29-2010.pdf

http://www.elationlighting.com/pdffiles/dmx-101-handbook.pdf

Stage Lighting for Students www.stagelightingprimer.com, by Jeffrey E. Salzberg (jeff@jeffsalzberg.com) with Judy Kupferman (judithku@post.bgu.ac.il)

https://www.energystar.gov/index.cfm?c=cfls.pr_cfls_about

The color version of illustrations contained in this chapter are available in the online and electronic editions.

5.8 ORGANIZATIONS

5.8.1 USITT

The United States Institute of Theatre Technology (USITT) supports, develops, and promotes a wide variety of standards for the theatrical and entertainment industry. In 1986, USITT developed the DMX 512 protocol as a simple, flexible, and reliable standard for lighting control. In 1998, USITT transferred maintenance of the DMX 512 protocol to the Technical Standards Program of ESTA. The standard is constantly revised and updated as technology continues to advance.

5.8.2 PLASA

The leading international membership body for those who supply technologies and services to the event, entertainment, and installation industries. As a proactive trade association, it looks after the interests of its members and seeks to influence business practices and skills development across the industry.

5.8.3 ESTA (Now Incorporated in PLASA)

The Entertainment Services and Technology Association (ESTA) is a nonprofit trade association representing the entertainment technology industry. ESTA promotes professionalism and growth in the industry and provides a forum where interested parties can come together to exchange ideas and information, create standards and recommended practices, and address issues of training and certification.

5.8.4 ANSI

The American National Standards Institute (ANSI) is an organization composed of representatives from industry and government that collectively determine standards for the electronics industry as well as many other fields, such as chemical and nuclear engineering, health and safety, and construction. ANSI also represents the United States in setting international standards. New electronic equipment and methods must undergo extensive testing to obtain ANSI approval. In 2004, ANSI approved the DMX 512 standard, and has since approved several other related standards including Remote Device Management (RDM) and Architecture for Control Networks (ACN).

5.8.5 EIA/TIA

The Electronics Industry Alliance (EIA) is a trade organization composed of representatives from electronics manufacturing firms across the United States. EIA began in 1924 as the Radio Manufacturers Association (RMA), and has grown to include manufacturers of televisions, semiconductors, computers, and networking devices. The group sets standards for its members, helps write ANSI standards, and lobbies for legislation favorable to growth of the computer and electronics industry. The EIA is composed of several subgroups including the Telecommunications Industry Association (TIA). The EIA/TIA-485 standard is the communication basis for DMX 512.

5.8.6 Industry Manufacturer's Information

Advantage Gripware Inc., www.advantagegrip.com
Altman Lighting Company, www.altmanlighting.com/
American Grip, www.americangrip.com
Arriflex, www.arri.com
BBS Lighting, www.bbslighting.com
Benjamin Electric, www.benjaminelectric.com
Brightline, www.brightlines.com
Chroma Q, www.chroma-q.com
DeSisti Lighting, www.desisti.it
Doug Fleenor Designs, www.dfd.com
Elation Lighting, www.elationlighting.com
Electronic Theatre Controls, www.etcconnect.com
Eye Lighting, www.eyelighting.com
J.R. Clancy, www.jrclancy.com
Lee Filters, www.leefilters.com
Lex Products, www.lexproducts.com
Lite Gear, www.litegear.com
Litepanels, www.litepanels.com
Luxim Corporation, www.luxim.com
Matthews Studio Equipment, www.msegrip.com
Mole Richardson Co., www.mole.com
Osram Sylvania, www.sylvania.com
Prime Time Lighting, www.primetimelighting.com
Pro Cyc, www.procyc.com
Rosco, www.rosco.com
SSRC, www.ssrconline.com
Strand Lighting, www.strandlighting.com
Synthe FX, www.synthe-fx.com
The Light Source, www.thelightsource.com
Thern Inc., www.thern.com
Tiffen, www.tiffen.com
Videssence, www.videssence.com

CHAPTER 5.11
CELLULAR/IP ENG SYSTEMS

Joseph J. Giardina and Herbert Squire
DSI RF Systems, Somerset, NJ

5.1 INTRODUCTION

It all started with the phrase "Film at Eleven!" Every station wanted to cover local news but had to cope with the mechanics of processing 16 mm news film, or editing on color reversal film; and experiencing broken splices in a newscast was very common. The processing cycle often took an hour or more, which effectively prohibited a news operation from airing late breaking stories.

5.1.1 An Abbreviated History of ENG

The first real electronic news breakthrough came with the introduction of the U-matic tape deck and the birth of Electronic News Gathering (ENG). There was no more film processing, no more broken splices, and the news could air within minutes of the tape's return to the studio. But the U-matic revolution still had some major shortcomings; anyone then working in the industry will remember the heavy tape decks, big, heavy battery belts for power and, yes, those unstable time bases. The ability to air news events faster and more reliably offset these minor disadvantages.

ENG is a very evolutionary business, so the next and obvious progression in news gathering would be the ability to air live broadcasts. Thus was born the ENG truck with its telescoping mast and antenna, in-truck editing, and support electronics (Fig. 5.11.1). The news room edit bay had moved to the street. Now, newscasters could broadcast news events live whenever the ENG truck was within range of the central receive site, or they could transmit an edited news story back to the station whenever the truck came within range of a central receive site. This greatly reduced the time to air for a breaking news story.

A major shortcoming remains, however—connectivity. If the ENG truck can't see the receive site there is no live news.

Although, broadcasters can install more central receive sites to expand the geographical news coverage area, as the news collection infrastructure grows, multiple receive sites mean added maintenance, signal interconnects, and site costs.

A more flexible, yet sometimes more costly solution, was the introduction of the satellite news truck with a portable deployable Ku band antenna and uplink electronics. A station could now cover stories anywhere as long as the SNG truck's antenna had a clear path to the satellite in the southern sky. The typical SNG truck presents its own set of problems; it's a significantly larger vehicle than the normal ENG van, and it needs a larger staging location and time to deploy its stabilization system. This limits its usefulness for everyday news events.

Whether it's an ENG or an SNG truck, the central receive site or satellite downlink can typically accept only one channel (story) at a time. However, new technology that uses multiple sectored antennas at a central receive site to accept multiple feeds is evolving.

PRODUCTION SYSTEMS

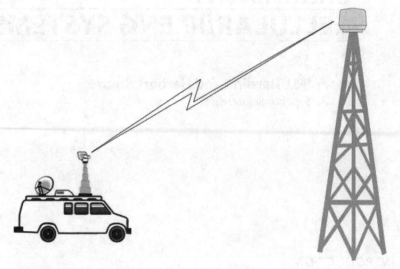

FIGURE 5.11.1 Traditional ENG system.

So, while news gathering has progressed from 16 mm film to ENG, with its dated point-to-point microwave technology (and inherent line-of-sight and channel capacity limitations), to SNG with the ability to broadcast from anywhere, the broadcaster still faces some serious challenges in today's evolving news environment.

The demands of social media, a 24-h news cycle, and the real-time nature of cell phones, force news operations to provide instantaneous event coverage to remain competitive in the now international news arena. Therefore, an obvious need exists for a new, more robust and more flexible transport system. Individual stations and networks can continue to expand their infrastructure with more receive sites and added ENG and SNG vehicles, but the fundamental point-to-point system model problems remain—limited channel capacity and the need for line-of-site connectivity. You cannot deliver the story from behind an obstruction.

Today, several implementations using cellular and WiFi systems enable instant news reporting. The available system types use different coding, and either single embedded or bonded modems to provide wider and more instantaneous ENG coverage.

5.2 THE CELLULAR REVOLUTION

The majority of the available cellular-based solutions use bonded modem technology to achieve the bandwidth required for their systems to transmit quality HD signals. A bonded system distributes the encoded video and audio across the multiple modems using numerous cellular channels on the same or competing cellular carriers, the belief being that this approach provides additional bandwidth and system reliability.

This chapter focuses on the more efficient, more cost-effective, and smaller alternative to the bonded modem technique—the single modem model. This is an alternative system architecture that uses a Digital Signal Processing (DSP)-based coding engine with a single highly tuned cellular modem and gain antenna to achieve superior upstream data transmission.

Before looking at an in-depth comparison of new systems, we must ask why not just use a cell phone? Many new cell phones offer 1080i video capability, better camera imagery with each new model, and, in a large part of this country and the world, LTE connectivity.

But even with the improved imaging and higher-speed connections cell phones offer, adapting them for use as professional news gathering exposes several critical shortcomings, as one might expect when attempting to retask consumer-oriented devices to broadcast news use. Streaming video with large data buffers is okay for watching time-shifted video offerings or a FaceTime call, but these transmissions often suffer transmission anomalies related to poor connectivity or coding that are all perfectly acceptable to the consumer.

Aside from cellular phones, some major communications companies are offering WiFi hot spots but these come with their own set of system issues, for example:

- Does a usable hot spot exist near the breaking news story? Reporters cannot waste time trying to log onto a hot spot from the local coffee shop while the story is breaking around them.
- Does the hot spot have the necessary capacity to get the story out? A story cannot hinge on that one last person logging on and bogging down the system.
- Most WiFi hot spots are undermanaged or not managed at all, which eliminates any significant Quality of Service (QOS). Hot Spots that are not professionally managed can result in interrupted streams because interfering data traffic is taking channel capacity from the story.

These shortcomings are story killers.

The hallmarks of a professional newscast are good imagery, reliable connectivity, usable communications, and low latency. So the question becomes: How can one compress the features of a typical ENG van or SNG truck into a package small enough to be portable and yet capable of transmitting usable—if not superior—news coverage?

The ideal replacement must be small enough either to mount on the news camera or to fit in a belt pack. In order to provide maximum operational flexibility, it must be either self-powered or use the camera battery.

5.3 THE IDEAL ENG DEVICE

Any new device should serve as a possible replacement to, or augmentation of, the ENG or SNG truck model. Such a system should incorporate not only the current system capabilities, but must also include the desirable features that are now missing from the existing point-to-point ENG or SNG model.

Remember, the traditional ENG/SNG system bottleneck is the transport construct. Even if a news organization expands its market coverage with added central receive sites, the system is one-way with limited bandwidth and, most importantly, it relies on point-to-point communication with high-power radios and large antennas. To add to these shortcomings, only a limited number of ENG channels exist in each TV market. Many news operations have only a primary channel and must rely on sharing the remaining channels on an ad hoc basis.

Up to now, we have used the terms "News Operations" and "ENG" to refer to an FCC-licensed TV station using the Broadcast Auxiliary Service (BAS) that allocates the seven 12 MHz-wide digital ENG channels from 2025 to 2110 MHz. However, both these terms have become synonymous with news-gathering operations in general, and they are no longer the sole purview of the television news operation or of a truck alone. Many organizations that serve nonbroadcast news—for example, cable, social media, and Internet—now provide a significant amount of daily news coverage.

Because the FCC Rules and Regulations forbid their use of the BAS microwave channels, these nonbroadcast affiliated news operations suffer a major disadvantage. Some stringers operate under a mutual agreement with a specific TV station (this is not practical if the news event is competitive) or they license their ENG equipment under the FCC Rules for Common Carrier operation. This approach often needs a separate point-to-point system because the Common Carrier channels operate in the 6 GHz band and are not supported by the normal station ENG central receive sites.

In addition to the 2 GHz ENG bands, several TV stations conduct ENG operations in the 7 GHz remote pickup frequencies. In fact, many existing antennas have dual elements to accommodate either band.

An ideal replacement system should give anyone the opportunity to transmit news events from anywhere without having to commit a significant amount of capital into a receive site infrastructure.

Earlier in this chapter mention was made of the use of a public network, cellular WiFi, or both, which have many access nodes (cell sites or WiFi hot spots) and viable bandwidth.

Here, we should note that some stations can look at a transition from a station-owned transmission network to a public network and the requisite loss of system control as a major paradigm shift.

While such a transition may seem almost repugnant to a news organization, the reality is that almost all TV operations use the public network for information transfer. A station's studio-to-transmitter link is most likely a fiber from the local telco, the link to a distant news bureau or even that SNG feed that has come through a third-party satellite company. Granted, that usually these are established links properly engineered for reliable failsafe operation. However, the fact remains that the station or news operation has relinquished transmission transport to a reliable third party.

By embracing the highly dependable public cellular network, a news operation expands its coverage area geographically from local to global; literally, news coverage is possible anywhere there is adequate cell or specialized WiFi service.

The use of the public networks in conjunction with an IP-based coding system is possible as a result of the global cellular modem standardization that permits a user to switch systems by changing a SIM card. Cell service is available in almost every market, and most countries have superior cellular and Internet systems that allow efficient H.264 transmission.

A public network appears to be the ideal solution, but that raises more questions:

- Is the large and costly ENG or SNG truck with its generators, editing equipment, etc., still needed?
- Is the new cellular or WiFi encoder small enough to be portable?
- Is the new cellular or WiFi system cost-effective?
- Is it easy to deploy?
- Will the new system address the traditional ENG/SNG shortcomings?
- Is it a single person device?

These questions directly impact any news group's finances and operations. The new ENG system must be a properly engineered arrangement of complementary technologies that benefit all users and can deliver professional news stories. The most critical design criteria must address the fact that news personnel are in the field to cover the story, not to spend time setting up equipment.

If we set aside the existing historic ENG system model to design a new portable ENG system, then we must consider certain fundamental design and operational considerations:

- A low glass-to-glass[1] latency (<3 s).
- Efficient coding to produce the lowest possible bit rate to work in congested cell sites during a breaking story.
- High video/audio fidelity; it must be better than cell phones and two-way radios, both of which may severely distort the nonvoice segments of the IFB audio.
- Dual IFB channels—one for the talent and a second for the camera person.
- Robustness; it has to work almost all the time.
- Small and lightweight.
- Long battery life.
- Easy integration into existing news systems.
- Minimal operating controls for a fast learning curve.
- Fast deployment—turn it on and go.
- Self-contained—no extra wires other than video, audio, and battery.

---

[1]The time for the coded stream to transit from the camera lens to the monitor in the news room.

- Lightweight—no added weight burden to the camera person.
- The ability to work with smaller form factor cameras.
- Safe operation—minimal SAR (Specific Absorption Rate)[2]
- Rugged and able to withstand the rigorous field conditions experienced by news teams.

This expansive feature list reflects the need to identify the basic requirements drawn from the existing truck-based ENG model, which has taken decades to evolve and adapt to new field requirements. These specifications will provide the most usable and cost-effective portable ENG system.

Portable ENG is now so important that many news operations are willing to deploy systems that lack some of these key criteria (e.g., no IFB, bulky physical implementation, poor coding, and sub-par video performance) in order to provide faster news coverage and compete with the various nonbroadcast news entities.

5.4   IP-BASED ENG SYSTEMS

A new and innovative news system must adhere to accepted IP protocols to insure adaptability to such public transmission networks as the cellular phone network or WiFi hot spots. These systems use IP-based coding, which calls for a different and sometimes unique approach to the transmission protocols traditionally used with ENG and SNG systems. Typical ENG systems use DVB-T transmission with bandwidths ranging from roughly 6 to 26 Mbit/s, which is well beyond the capacity of a single cellular channel or a typical WiFi hot spot. Many ENG radios have as a standard input SDI or HDSDI, a few have HDMI, and many accept only ASI *versus* an IP-based input. Thus, any shift to cellular or WiFi ENG requires a transport system compatible with the existing IP-based transport.

To meet the target of using one cellular modem, a new IP-based ENG system needs a higher degree of compression for efficient and reliable cellular transmission. Fortunately, the H.264 (MPEG-4 AVC) protocol compresses the video and audio content from the field camera at substantially lower bit rates than previous standards like MPEG-2. The H.264 standard also provides for multiple profiles and several adjustable parameters to tailor the signal stream to the cellular or WiFi transmission system. Efficient audio coding within the H.264 system (e.g., Advanced Audio Coding, AAC) provides robust audio even with a less-than-desired picture. The audio compression rate is not a big contributor to the overall data stream, even at a 16 kbit/s sampling rate. So, in almost all instances, the audio always gets through.

A major shortcoming of the legacy ENG system is one-way communications; there is no way to implement usable IFB communications or system control from the studio to the field reporter. The traditional IFB workaround for this problem was off-air receivers; the field reporter could simply wait for the cue from the anchor who was using the program audio. With stereo TV, many stations used the Pro channel to act as a single IFB channel. However, this was not practical when more than one ENG truck was in the field during a newscast. The Pro channel disappeared with the implementation of HDTV, and the off-air reception became unusable because of the ATSC system's inherent multisecond coding delay. For IFB communications, a vast majority of news operations were forced to switch to either licensed remote pickup frequencies or cell phones.

An IP-based system allows duplex audio transmissions. Ideally, the audio system provides dual IFB channels, with audio quality far superior to that of a cell phone: one channel for the field reporter and a second for the camera person, which allows the field personnel to immediately connect to the director or producer at the receive site.

A very important factor in deciding the overall system delay or the "Glass-to-Glass" latency is the efficiency quotient of the signal compression, which in turn determines the time needed to encode and decode the signal stream (latency). An efficient and low-latency encoding/decoding system does not significantly increase the added delays experienced in the signal transport. System latency is an additive condition in which all processing and transmission contribute to the overall delay.

---

[2]SAR is a measure of the rate of absorption of RF energy in the body.

The International Telecommunication Union (ITU) G.114 Standard recommends that to satisfy most telephone listeners, the audio latency should not exceed 200 ms; which is simply not achievable when coupling an IP-based coding system to a public network. However, to be truly usable and effective, a duplex IFB communication system's overall audio delay must not exceed 1.5 s (the latency typically experienced with SNG). News personnel are quite adroit in dealing with these delays, and the resultant verbal interchange is acceptable.

When used for ENG, the system latency most affects the audio as the field reporter needs timely cues and instructions, and the ability to conduct a conversation with the anchor at the studio. Audio delays that exceed 1.5 s disturb viewers and are almost unmanageable for reporter-to-anchor conversations.

The most critical aspect of a usable IP-based ENG system is how the H.264 protocol affects the total system latency coupled with the efficient use of the available cellular bandwidth.

One seemingly obvious approach is the use of compression software in conjunction with bonded modems to achieve a significant amount of data bandwidth. This approach has many negative aspects when used for an RF-based transmission system. Its principal drawback is an unacceptable latency rate of multiple seconds to often tens of seconds, which makes it unusable for interactive field reporting. Most software coding programs call for a stream restart to make significant adjustments to the H.264 parameters. The high latency values and high data bandwidth requirements, coupled with stream restarts, make software coding less than desirable for a mobile RF-based system.

Fortunately, a hardware application-specific DSP engine provides several key advantages over software compression. A DSP coding engine has very low latency and less overall transmission bandwidth, which results in a more robust signal path. A very important advantage is the ability to make live system adjustments without interrupting the story. Many software-based coding schemes call for a reboot of the stream for any coding changes to take effect. This causes unacceptable story interruptions. Lastly, the DSP engine minimizes user adjustments. This results in faster setup and use.

Therefore, the ideal ENG system will implement a hardware DSP engine running the H.264 protocol as the preferred compression method for reliable RF transmission.

Now that we have quantified the ideal ENG replacement hardware and software, how do we make it insure that the breaking news story gets on the air? After all, the business of news is covering the news.

Let's summarize the requirements for a practical IP-based ENG transmission system:

- Low latency under 1.5 s from the lens to the monitor (glass-to-glass)
- Efficient H.264 coding, preferably under 4 Mbit/s for a usable HD transmission
- Dual duplex high-fidelity IFB channels
- The ability to connect to cellular or WiFi networks
- Dynamic system adjustment during streaming without interruption

5.4.1 System Design Considerations

Before continuing with a discussion of cellular ENG, we need to understand some of the variables inherent in with cellular systems that do not exist with normal ENG trucks. Some of these variables are not a major concern for private WiFi networks since the host controls the system bandwidth, so we will concentrate on cellular networks and possible private WiFi networks.

Even with efficient program content compression, the major limitation when streaming over a cellular network is the signal-to-noise ratio (SNR) of the transmission channel, which is one of the most determinant deciders of the available bandwidth allowed by the cell provider.

Several key variables affect data bandwidth, including the following:

- Terrain—can sometimes be corrected by moving the device
- Device RF power—controlled by the cell site and cannot be adjusted by the user
- Atmosphere—variable
- Device antenna—often fixed by the device manufacturer, but can be the only parameter controllable by the user

A given cellular device (e.g., a phone or USB modem) has a known fixed antenna designed by the device manufacturer to produce adequate signal performance in a typical cellular coverage area.

Cellular sites monitor the received signal using a variety of proprietary measurement techniques and algorithms and instruct the transmitting modem to adjust its output power to maintain an optimum connection. Since the cell site equipment is optimized and fixed, the received carrier level at the site is dependent on the terrain, the distance to the cell site, and the transmitting modem and antenna on the ENG unit. A cell site will instruct a robust signal to throttle down its power to maintain the optimal SNR and the highest bandwidth, while it instructs the lower-powered modem to boost its output power to maximum to maintain the mediocre bandwidth with the lower SNR. In essence, the cellular system is one in which the dominant RF signal will command the best bandwidth and, therefore, the highest bit rate. This is very important as it affects the success of a given field transmission.

There are several ways to address the cellular data bandwidth issue: efficient content coding, improved antenna performance, or the use of both. Using Forward Error Correction (FEC) provides a significant improvement, but adds a large amount of latency to the live stream. For this reason alone, FEC is not a viable system for ENG use.

The cellular bandwidth is controlled by the cell company, by the FCC, and by the laws of physics, which limit the SNR for specific bandwidth and bit rates,

$$SNR = 10\log\left(\frac{Bandwidth}{Bitrate}\right)$$

where SNR = signal-to-noise ratio.

The Shannon channel capacity, shown mathematically below, dictates that as the SNR decreases the bit rate must also decrease,

$$C = B\log_2\left(1 + \frac{S}{N}\right)$$

where[3] C = bits per second, B = frequency in hertz, and S/N = signal to noise.

So, an increase in the SNR results in an increase in the data rate. This provides a more robust and reliable transmission channel. Remember, the same laws that dictate the bit rate also prescribe that any effort to send bits faster than the Shannon channel capacity will increase the error rate.

Unfortunately, there is no appealing the Laws of Physics!

The major domestic carriers promise spectacular data speeds, which often apply to the cell's download capability, not to the upstream channel, which often has a lower bandwidth. Because the portable ENG system relies almost solely on upstream transmission, this channel's bandwidth is the controlling factor for successful portable ENG.

Anecdotal data shows that most 4G cellular systems can typically deliver a megabit and higher of upstream bandwidth with a given modem. This is not true in all cases, or in those instances of significant channel usage; namely, a breaking news event or a feature news event concurrently with the public trying to upload their social media data. The bandwidth then takes a nosedive to the hundreds of kilobits per second.

This problem seriously impacts the viability of a cellular-based device.

5.4.2 Modem Configuration Options

Presently, there are two competing technologies for cellular ENG coverage: bonded and single modem devices. Both systems employ IP-based coding to interface to the cellular network. A more detailed explanation of IP-based coding will be discussed later in this chapter, but for now we must first look at the differences between the two techniques.

---

[3] If the logarithm is taken in base 2.

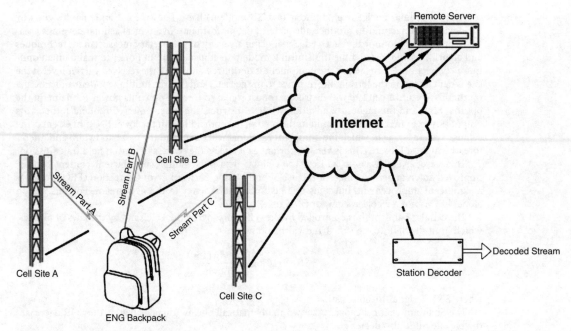

FIGURE 5.11.2 Bonded modem system.

5.4.2.1 Bonded Modem System.

As its name implies, a bonded modem system uses a number of cellular modems, each of which establishes a data channel with a cell site. The IP stream is divided among these modems and is transmitted, often to a server, which then recovers and reassembles the multiple streams into one program stream for application to the decoder. This technique is illustrated in Fig. 5.11.2.

Bonded modem systems do appear to offer one advantage: the potential to use different cellular suppliers at the same time to increase the appearance of reliability. After all, it seems to make sense that if any modem on given cellular system goes off line, then the remaining modems on the other cellular system should pick up the slack. Regrettably, they rarely work this way. The modems are already using the maximum channel capacity based on their weak RF performance (we will discuss this later) and the coding methods often used with bonded modems are software based and are not very efficient. Therefore, they need more bandwidth than a DSP-based coding engine, which only exacerbates the situation.

Bonded modems appear to be a practical solution, but they do experience some significant technical shortcomings that do not permit this technique to be considered as the most reliable portable ENG device. Specifically, bonded systems have the following disadvantages:

- Inefficient internal strip antennas that often result in "negative" gain![4]
- Multiple modems tend to draw more power, which drains the system batteries faster.
- Low RF output power so there is less SNR and reduced bandwidth.
- Mechanical instability; the modems may easily jar loose.
- Excessive SAR levels that can exceed the mandated FCC limits.

---

[4]An AT&T technical brief indicates that it is difficult to get even 75 percent efficiency from an embedded antenna, which equates to a power loss of −1.2 dB. Any wasted power directly relates to decreased bandwidth as the signal level decreases into the cell site.

- They use noncoherent sources such as dongle modems operating in close proximity, which desensitize the adjacent modems and contribute to potential adverse antenna pattern distortion.
- High "glass-to-glass" latency, which is often measured in tens of seconds for stream reassembly.
- Reliance on the news camera battery to supply the significant system power.
- Heavy backpacks, bulky equipment cases, or large camera-mounted devices, which offset their portability.

One of the most significant drawbacks of the bonded modem solution is that the modems are incapable of transmitting the power required to achieve the maximum Shannon channel capacity of the cell site. Anecdotal and empirical data shows a less than stellar upstream data rate unless the device is relatively close to the cell site, has good line of sight, and experiences no cell congestion. Add to this the potential antenna pattern distortion caused by multiple modem antennas, and the antenna pattern and gain are unknown.

The alternative is to use a single modem device. How can one modem be better than four, five, or even, in some instances, ten separate modems?

5.4.2.2 Single Modem System. A single embedded modem, illustrated in Fig. 5.11.3, eliminates many bonded system shortcomings and offers:

- Extended-range specifications
- Rugged construction
- External antenna connection
- 100 Mbit/s download and 50 Mbit/s upload capable
- An optimized RF path and a known gain antenna
- Receiver equalization to improve performance in noisy environments
- SAR compliance
- Maximizes Shannon channel capacity

Couple the embedded modem to an external gain antenna and the system now has predictable and stable RF gain, there is finite control of the relationship of the antenna location *versus* the user's head location (very important for SAR compliance), and there is a good ground plane for excellent RF performance.

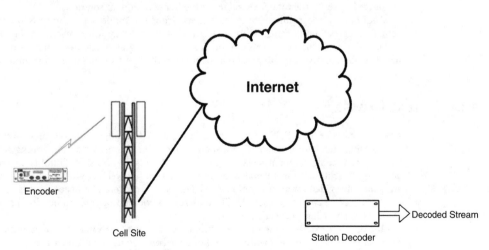

FIGURE 5.11.3 Single modem system.

One very important benefit is that a single robust modem with a gain antenna and matched impedances results in the optimal RF system. As the Shannon channel capacity formula shows, a correctly designed antenna system leads to the highest SNR and so the highest number of bits transmitted.

5.4.3 Cellular Transmission Systems

So one would think that just getting a router, plugging in a USB cellular modem dongle, hooking up an H.264 encoder, and transmitting some live news would be possible. Unfortunately, it is not that simple or easy.

As was mentioned earlier, the ideal ENG unit should be, among other things, small, self-contained, lightweight, easy to use, and ruggedized for field news. You would think that, in essence, this describes a cell phone; it's small, it's cellular, and has everything built-in. But this is not the case. A cell phone is optimized for phone calls, text messaging, and emails—all activities that may interfere with a data transmission. Couple this with a cell phone's less-than-stellar antenna system, and we quickly eliminate cell phones as efficient and reliable data streaming devices.

But consider, people use cell phones for data services all the time, so why cannot they be used to send live video to the studio? A cell phone's data stream is internally buffered and communicates with a server that can request a resend of corrupted or missing packets. The packets are often buffered internally within the phone and sent to the phone's application as a pseudo-uninterrupted data stream. Regrettably, a live video and audio stream cannot resend corrupted or missing packets and maintain the low latency and throughput needed for live news coverage.

Fortunately, reliable cellular connections are possible with the use of proper hardware and coding designs.

All cellular phones or modems are RF radiators; therefore, SAR is a very important safety consideration. The user must fully understand the implications and dangers of excessive SAR levels for personnel safety and potential organizational liability. SAR is the measure of the radio frequency energy absorbed by the user's body when using any RF device; in this situation, a mobile phone or modem. The FCC limits SAR for public exposure, from cellular telephones and modems, to a level of 1.6 W/kg. The typical cellular data modem transmits at 0.466 mW/g and must conform to the governmental limits.

A bonded modem system uses multiple USB-stick modems in proximity to each other. This can contribute additive levels of SAR energy depending on their physical location, proximity to each other, and other criteria. The authors are unaware of any definitive SAR testing that has been conducted to verify the total SAR values of the bonded modem solutions.

Additionally, the resultant antenna pattern from the modem proximity is unknown and can actually create antenna transmission patterns that are substantially worse than the patterns from a single cellular modem. The types of modems and the integrated antennae vary by manufacturer so a user has no definitive data on potential coverage or SAR levels using bonded solutions.

5.4.4 What About WiFi?

It seems that a cellular ENG unit using an integral WiFi modem is a logical answer to link to the studio using high bandwidth WiFi hot spot. Public WiFi networks are proliferating across the country and many cable companies claiming millions of hot spots. Most WiFi hot spots are excellent for accessing the Internet for normal web browsing and other web activities, but they fall far short for streaming live video for a variety of reasons. Furthermore, overall WiFi coverage is not on a par with cellular coverage and the present WiFi standards do not allow a graceful handoff between hot spots if the crew is moving to cover the story.

The upload capacity to the Internet is unknown because a vast majority of public access locations often suffer from user overloading. The hot spot connectivity could be a DSL circuit with typical 750 kbps or lower throughput for its connection to the Internet. Couple this with unknown number of users, and the upload capacity is often overloaded. This eliminates the local hot spot as a usable ENG connection. In addition, these systems do not often factor quality of service into the network design. People using these networks have to accept poor coverage, slow connections, etc.

Imagine, the news crew is covering a breaking story and that one extra customer gets their coffee and decides to upload a large data file; this bogs down the system enough to drop the news crew's connection. This is one of the most serious negative aspects that public WiFi hot spots presents to reliable ENG coverage.

Now, consider a private WiFi access point, which is a viable and valuable tool to transmit from known locations. It is a controlled medium that guarantees the connection to the Internet or the studio with a known bandwidth. It is designed by the news organization to cover a specific area with a known WiFi signal for their exclusive use.

Private WiFi access points are categorized as either "Direct Mode" in which the WiFi access point connects directly to the local decoder, or "Internet Mode" in which the system transports the received stream to the decoder at the studio using the Internet via a cable modem or fiber network. Empirical data shows that private WiFi networks need different attention to the H.264 coding because of network jitter, the fluctuation in the timing of packet flow, and packet loss.

Experimentation shows that a properly designed WiFi access point can communicate with the field encoder at distances of three city blocks (about 1500 feet) in a very high RF environment, with more distance and less RF congestion. This is quite suitable for covering news events at fixed locations like a courthouse, or for transitory events like a parade or concert.

A private WiFi network design requires the proper gain antenna, a professional access point, and consideration of the band saturation at the chosen location. Often, the 2.4 GHz band is unusable in most public locations due to user concentration and the limited number of available channels. Therefore, the authors have dismissed the 2.4 GHz band and will refer to only the 5.8 GHz WiFi band as the standard for ENG news transmissions.

It may appear obvious, but the most important part of a local WiFi access point system design is the radio device. Today, almost every Ethernet router or switch that connects to an IP network has a built-in WiFi for general home and business. However, most WiFi device designs are centered on small to moderate-sized environment with integrated antennas. These devices have RF front ends with poor sensitivity and selectivity, and lack the key features for ENG use, namely:

- Remote system control
- Remote spectrum analysis
- High power output
- Nonintegrated antenna pattern and gain flexibility

The supplied antennas often have unity or very low gain figures. This type of antenna is not conducive to a professional installation. The access point is located outdoors in most remote locations and needs to be weatherproof and environmentally rated for extreme weather conditions.

A more professional and reliable approach is to design the access point using a carrier-class radio and antenna, as illustrated in the following section.

5.4.4.1 Implementation Example. A typical access point is the Bullet M series offered by Ubiquiti Networks. This is a robust and weatherproof outdoor access point that attaches directly to a 5.8 GHz antenna and offers carrier-class speeds up to 100+ Mbit/s real TCP/IP throughput. Alternatively, the Ubiquiti Networks NanoStation series offers an integrated 10.4–11.2 dBi gain antenna and radio, all in a weatherproof case. It also offers carrier-class speeds up to 150 Mbit/s real TCP/IP throughput. Both these devices have demonstrated their capabilities in many field tests conducted by the authors at high-capacity venues.

We set the WiFi link, which operates in the half-duplex mode, throughput at 10 to 20 Mbit/s to minimize errors and to maximize the reception. An efficient IP ENG system will produce a coded stream around 5–6 Mbit/s for a near-perfect picture which is well within the link's capabilities at these lower throughput.

5.4.5 General Considerations

Regardless of the chosen access point manufacturer, remember that a private WiFi hot spot is designed for a station's needs, not for those of the general public. Public WiFi hot spots, even ones employing a carrier-class WiFi radio, often incorporate a medium gain omnidirectional antenna to provide service over the desired area.

The location of a private ENG hot spot dictates the design where the antenna is directional to shape the beam to the needed news coverage area. This effectively extends the news coverage, along the antenna's main axis, for the location because of the higher gain of a typical directional antenna. This directionality minimizes interference from other users. Typical directional WiFi antennas include corner reflectors that provide approximately 10 dBi gain with a 10° beamwidth, to panel antennas exhibiting 12 dBi gain with a 90° beamwidth. See Fig. 5.11.4.

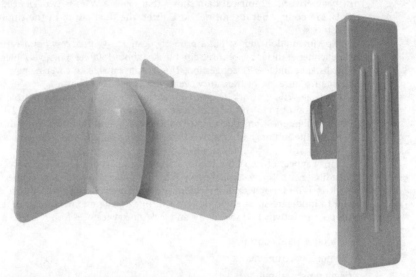

FIGURE 5.11.4 Antenna types: (left) typical corner reflector, (right) typical panel antenna.

5.4.6 Portable ENG System Economics

Thus far, this chapter has dealt with a practical ENG implementation using a single cellular or WiFi modem, but a user looking to implement a new cellular ENG system cannot rely solely on the technical aspects of the new system. An equally important criterion is the financial impact of the system.

Bonded modem solutions have a unique set of financial variants when compared to a single modem solution: initial capital costs and the recurring operational costs associated with maintaining multiple cellular modems.

The capital costs for some of the bonded solution hardware are becoming more reasonably priced, with some bonded solutions available only on a lease that often incorporates the cellular data into the monthly charge. However, the aggregated data usage costs for the multiplicity of modems can seriously impact an operation's bottom line. As an example, a typical single modem device might cost the user $80.00 a month while a bonded modem device, using an average of six modems, would total $480 a month. Whether the bonded device is leased or purchased, the estimated recurring costs can run from $5,760/year for a purchased unit to $36,000/year for a leased system based on typical usage.

Alternatively, using the same anticipated costs per modem, a single modem device would cost $960/year with one modem slot used and $1920/year if using both modem slots.

A single modem device incorporates a second modem slot to choose between two preselected cellular carriers. This gives the user coverage flexibility in a wider geographical area. An important reminder is that only one modem is on line at a time; therefore, data usage charges are shared between the two modems.

Ultimately, a news organization will not deploy a new cellular ENG system that has a poor return-on-investment performance.

5.5 GETTING THE STORY

An equally important aspect of the IP-based ENG is the field implementation of a new system. Common sense dictates that a simple and easy to use system is more likely to be deployed, while a more complex one is used less.

The authors' opinion is that splitting the stream across numerous less-than-optimum RF paths is not the answer. We also know that there is a finite amount of data bandwidth available at any given cell site.[5] So, we must consider how the optimized single modem solution helps deliver the news in almost any situation by providing a simple and easier to use cellular device. We are excluding private WiFi systems as these are controllable systems with known characteristics.

For the purposes of this example, we have categorized news stories as:

- Breaking news
  - Competing organizations covering the same story
  - Getting the story on the air quickly
  - Frequency congestion (cellular or microwave ENG)
  - Need for reliable studio and field communications
  - Frequently not enough bits to go around

- Feature News
  - Often not a time-critical story
  - Often the sole news entity at the story
  - Usually there are enough bits to get the story

Every news story dictates the location and the location dictates the cellular system site.

The coverage of breaking news is the most stringent news requirement and has the highest demand on a specific cell site. This is due to the traffic from competing news organizations and the general public. After all, everyone is vying to get their video on line and capture their "15 min of fame."[6]

When covering breaking news, the field reporter is locked into several conditions that dictate the success or failure of the coverage: the location, the competition, and the cell site. Since these criteria are site-determined, the reporter has only one tool available to insure that they get the story: a superior ENG device.

As we have shown earlier, the reporter has little control over the complex variables experienced in the field when covering a story. Even with a single modem solution and a superior RF antenna system, sometimes situations will dictate that the cell site needed for the all-important story is less than optimal.

Remember that the cell site is somewhat Darwinian in nature inasmuch as, with all other data modems vying for attention, the single modem ENG device has the advantage of being able to transmit more RF power than the bonded modem devices; thus, it obtains a better bandwidth than the competing devices.

---

[5]The cell site backhaul is often the limiting factor in achieving usable Internet bandwidth.

[6]Attributed to photographer Nat Finkelstein in response to Andy Warhol's comment "*...everyone wants to be famous...*" to which Finkelstein replied, "*Yeah, for about fifteen minutes, Andy.*"

Take for instance, this projected scenario; the reporter is at the story, there are a dozen other news entities there with bonded modem devices. We already know from Shannon that our single-modem device, with its superior RF antenna, will garner the most available bandwidth, but this is for the RF path only. There is one other bandwidth-determining factor; i.e., the backhaul link from the cell site to the Internet. The cell provider may manage this return link by limiting the maximum bandwidth to and from the cellular device by the total number of users accessing the site. The user cannot predetermine this data restriction as it is solely dependent on the cell provider and the cell site router settings. So even with an adequate RF bandwidth channel, that congested site may allow only a limited IP bandwidth. In these instances, the reporter needs some H.264 "magic" in the ENG device to insure coverage especially on a competitive breaking news story. The ideal system must allow the user to make adjustments to the coding parameters to get the story on the air.

The DSP hardware coding engine results in extremely low system latency, which permits making real-time adjustments to the stream without stopping the coding process. Ideally, the transmission should just change as the newscast is being streamed so that the viewer will not see any stream breakup or frozen frames. This is critical to presenting a live news cast with no transmission interruptions.[7]

5.6 FINDING A PRACTICAL SOLUTION

Through the application of some H.264 "magic," the story can still get through while maintaining the transmission without interruption. This answer is a complex set of algorithms identified by the term Robustivity™ the definition of which is: "The ability to resist the uncertainties, pitfalls, and other data integrity issues when traversing the Internet by using a series of algorithms that optimize the IP stream to account for different bandwidths and the known empirical variables in the public cellular system and the Internet." Simply put, the system can adjust the operational parameters in real time and maintain invisibility.

This means that almost anyone can, with enough time and effort, adjust the encoder to produce the best video stream from a remote location given a set transmission path. This assumes that they have sufficient knowledge of and experience with MPEG technology.

However, cellular, and sometimes WiFi, systems present an ever-changing transmission path because of RF conditions and Internet processing and routing issues. Furthermore, our experimentation has shown that although cellular and WiFi transmissions call for a unique set of adjustment parameters, the general techniques are the same; only the parameters for the different services change.

As we mentioned earlier, one of the key aspects of an IP-based ENG system is ease of use and rapid deployment, but these benefits disappear if someone has to optimize the coding for every story.

The solution is to establish a set of predetermined adjustments derived from empirical analysis and make these adjustments en masse and at the push of a button. Wait a minute, that's just adaptive coding, right? Not really; adaptive coding often refers to coding methods typically associated with lossless data compression. A typical DSP coding engine uses adaptive coding for both the video and the audio stream.

Our initial step was to identify the most egregious offender—the transmission path. For these purposes, the transmission path includes all aspects of conveying the stream from the field encoder through to the decoder: RF, Internet, and local WAN and LAN systems. Once we know the cellular and WiFi transmission system characteristics, we can adjust the H.264 feature set to produce the best possible picture for a given condition. The development of such a system is highly subjective and often requires years of MPEG coding expertise and experience with RF transmission systems to be able to determine what parameters need enhancement and at what cost to other aspects of the video transmission.

---

[7]The authors applied these findings and their knowledge to a new single modem cellular system development which became known as the NewsShark Plus. This device takes advantage of the cellular benefits while providing a methodology for dealing with the technical transmission issues beyond the news operator's control.

Further empirical data indicated that an external antenna system and embedded modem could actually realize over 10 Mbps in several cellular nodes. Armed with this information and with testing of the CODEC pair, the authors decided that the highest practical bandwidth would be 6.3 Mbps, which is the point at which the coding showed no appreciable improvements. Our next step was to determine the lowest possible usable bandwidth, as this is the more realistic value needed for congested events. This effort involved not only the H.264 adjustments but changes to the coding engine's source code to produce a low-bitrate, usable high-definition stream.

In closed loop tests, this properly tuned coding scheme, coupled with the dedicated DSP engine and optimized RF system, can promise breaking news video and audio at streaming data rates as low as 256 kbps.

At this rate, the video does exhibit some minor motion artifacts but delivers clear and undistorted audio. So, overall it is far superior to other transmission methods currently available. The result is that the news entity covering the story gets the breaking news on the air first and in HD not SD.

Knowing the bandwidth limitations, we were now faced with applying this to the transmission system.

The wireless 4G/LTE cellular system is far from being a dedicated channel for a specific video and audio transport stream. We have mentioned that mobile networks have operating characteristics that are inherently different from conventional wired networks. While the rated uplink and downlink speeds are more than adequate for public data feeds (e.g., web surfing, photo and file transmissions, etc.), the actual error-free throughput is limited by existing traffic at the local cell site at any given time. Factored into the equation is the Carrier-to-Interference-plus-Noise Ratio (CINR) and the Received Signal Strength Indicator (RSSI) at the originating cell site. The system jitter, the fluctuation in the timing of packet flow, and the packet loss that is caused by wireless interference, degrading signal integrity, and network congestion create artifacts and dropouts in both the video and audio streams, all of which contribute to the quality of the transmitted stream.

In early in 2010,[8] we studied and tested several video codecs, a few of which worked very well in dedicated fixed microwave links. We discovered that the system throughput was consistent using the 3G/EVDO wireless services generally in use then at the time these codecs worked well. If the maximum upload speed of the network connection dropped, the codec had to be stopped manually, the operating settings readjusted, and the codec restarted.

Hopefully, these new settings would compensate for the reduced link throughput. Frequently, there were multiple start and stop sequences performed in order to find a workable combination. The decoded video at the "studio" end of the link was the only place a user could see the results of the settings changes. This back and forth technical parameter configuration work was very time-consuming. We quickly decided, based on our news operational experience, that trying to cover a breaking news story under these circumstances would not be practical. In fact, it would be a product killer, since no one would take the time to start and stop the stream while trying to correct for the ever-changing cellular parameters. We then realized that the recalibrated connectivity would remain adequate at the start of the actual news report and could suddenly drop even lower during the actual broadcast. Conversely, the throughput could increase, but there was no way of knowing that this improvement had taken place. Also, without a suitable IFB system there is no adequate method to convey the system anomalies to the field personnel.

We decided after many more hours of field testing, that the only useful way to solve this major problem was to develop a codec data link with an interface that would monitor the video quality as received at the decoder and send a data stream back to the encoder that would change the encoder settings to compensate for the transmission errors and remain on line during those configuration changes. Our design philosophy was to create an ENG News appliance that was simple to use and therefore, by definition, the studio viewer and not the field personnel could make all needed adjustments. This was the point at which the concept of Robustivity™ entered the picture. Luckily, the NewsShark Plus system incorporates a duplex data path, for the IFB system, that could now be shared as the return data path for the studio-based system corrections.

---

[8]During the initial development of the NewsShark Plus.

At this juncture, the tests and experimentation were conducted toward providing a suitable standard definition NTSC low-resolution (352 × 240 lines) video stream using the then available 3G/EVDO type systems. Since on a very good link the uplink throughput only averaged around 700 kbps, when the network throughput dropped down to 100 kbps or less during heavy traffic and/or low RSSI with marginal signal strength in certain locations, the video quality became very marginal. Even in good RSSI locations, there were many times that the actual network throughput dropped to 50 kbps and lower for short high-network-traffic periods. At this point we discovered that even the program audio and the IFB cue circuits would become very intermittent, thus creating a communications link definitely not ready for prime time.

In the long term, 3G/EVDO technology was not truly viable for reliable news gathering under many of those adverse transmission conditions. Also, because an increasing number of news operations incorporated high definition cameras, the low bandwidths associated with the 3G/EVDO systems would not suffice, possibly not even for NTSC. Fortunately, newer wireless transmission technology was just around the corner.

In October of 2008, Sprint introduced its WiMAX (first 4G type) network. By early 2010, its growth and coverage area still left much to be desired. Under good conditions, the typical uplink maximum throughput ran around 1.2 Mbps. Operating in the 2.5 GHz band, many locations had an additional RSSI loss beginning in the spring, when leaves appeared on the trees and the line-of-sight path attenuation increased substantially. A number of locations that we tested and that worked well in February ceased to work by late March and early April. In most of the country, there was no service available outside some major cities and their close-in suburbs. By July 2012, Sprint began to implement its newer 4G/LTE service, and began to phase out the existing WiMAX service, which was scheduled to be totally shut down in November 2015.

The good news in wireless network connectivity was, that by late 2010, Verizon began to offer LTE service. AT&T started their LTE by the second half of 2011. With other carriers adding 4G and/or LTE service, the practicality of having an uplink throughput averaging 5 Mbit/s or better has become a reality.

Knowing that these bandwidths were coming available, and using prototype embedded modems, we could now simulate a transmission path that mimicked our closed loop tests of 256 kbit/s to 6.3 Mbit/s.

Hundreds of hours of transmission tests were made under varying conditions. These included numerous decoder parameters from many encoder locations both fixed and mobile. The encoder settings were adjusted to compensate for the errors and to restore valid video throughput at a workable lower data rate. The preliminary empirical data showed that the corrections could be keyed to the RF system data bandwidth, which made a convenient pointer for all adjustments.

Once we had a workable number of settings from the high throughput to the lowest practical throughput, each setting was tested using a closed-circuit network emulator so that the final throughput settings and parameters could be tested and the image quality optimized, thereby simulating the anticipated 4G services being deployed by the major carriers.

Research prescribed the Robustivity™ settings for both cellular and WiFi transmission systems are documented in Table 5.11.1.

Even with the significant wireless bandwidth network improvements, Robustivity™ still remains an extremely important factor to keep "making ends meet" on a live broadcast link.

TABLE 5.11.1 Settings as a Function of Transmission Type

Transmission Type	Number of Settings	Minimum Bandwidth (kbs)	Maximum Bandwidth (Mbs)
Cellular	20	350	2.1
Direct WiFi	18	700	6.3
Internet WiFi	25	350	6.3

At this point, the new 4G/LTE services allow high definition video 720P (1280 × 720 lines) to become operational. The 4G/LTE wireless networks, which are several times higher in throughput than the older 3G/EVDO, still exhibit the same type of transmission anomalies as the older technology.

As with the 3G/EVDO Robustivity™ implementation, we conducted many more hours of real-world testing to determine the appropriate settings for each of the individual parameters between high and low throughput conditions with the newer 4G/LTE technology with HD video. While there are still network connectivity issues that will cause glitching and short-term loss of signal under adverse conditions, the probability of these problems occurring is less with 4G/LTE links than the earlier 3G/EVDO technology.

5.7 NO MORE "FILM AT ELEVEN!"

News coverage started with a camera operator armed with a 16 mm film camera and an audio technician, and was within the purview of a limited number of entities capable of shouldering the enormous cost bringing news to the TV public. It has now evolved to a single person with a handheld HD camera and a robust single modem system that allows live news coverage from almost anywhere.

Major TV networks can deliver late-breaking news from almost any spot on the globe while any town's local TV channel can now cover at Little League game that, in the past, was too cost prohibitive or just impossible.

SECTION 6
FACILITY ISSUES

Section Leader—Jerry C. Whitaker, Editor-in-Chief

Building and maintaining a broadcast facility is a challenging task, but also one that holds the potential for great rewards. The sense of accomplishment that results from designing, building, and then maintaining a facility can be immense. In order to achieve that level of satisfaction with the project, however, it is necessary to focus on details during planning and execution.

There are a number of key elements that go into a successful, modern broadcast facility. Step one is identifying the functional (or user) requirements. There is an old saying, "If you do not know where you are going, any road will take you there." This applies to facility design as well. A project executed without clearly defining the user requirements is in danger of failure from the start. Some projects are small and simple and can be implemented with little advance planning because the objective is obvious. Such projects might include replacing an old piece of equipment with a new one. For projects with a broader scope, however, each of the facility constituencies and stakeholders must be engaged in the planning stages.

Once a plan is set, the installation phase can begin, usually with a whole new set of players. Managing outside vendors is no easy task as worker and equipment availabilities can vary depending on the workload in the area, or even just at the building itself. Competition for resources can be a very real challenge.

Facility planning does not end with the opening of the competed broadcast plant. Fine-tuning may be needed as users find issues that were not sufficiently addressed during planning, or simply never occurred to anybody. Ongoing maintenance begins almost as soon as the facility is put into service. Keeping ahead of developing problems is the hallmark of a good broadcast engineer. Routine maintenance should not be left to chance, or deferred until real problems surface.

This section includes seven chapters that describe many of the important elements of broadcast facility planning, installation, and maintenance. Each is written with an eye toward the broadcast engineer and his/her role in station operation. There are many things that a broadcast engineer does very well. There are also areas where the engineer may say, "Let us bring in an expert." There was a time when the engineering department of a radio or TV station could perform most needed work without bringing in outside expertise. With the complexity of broadcast systems today, those times have long since passed. As the role of the broadcast engineer changes, so does the range of expertise needed to do the job.

Chapter 6.1: Broadcast Facility Design

Chapter 6.2: Wire Management

Chapter 6.3: Equipment Rack Enclosures and Devices

Chapter 6.4: Broadcast Systems Cooling and Environmental Management

Chapter 6.5: Facility Ground System

Chapter 6.6: AC Power Systems

Chapter 6.7: Transmission System Maintenance

CHAPTER 6.1
BROADCAST FACILITY DESIGN

Gene DeSantis

6.1 INTRODUCTION[1]

Planning how a broadcast facility will function is a difficult exercise, and one that will have far-reaching effects. The success of any project—whether an upgrade to an existing facility or a "green-field" project—depends in large part on the planning that goes into the effort before any concrete is poured or any cables are pulled.

Every project begins with a set of specifications, which must be carefully considered and drafted. Specifications are a compilation of knowledge about how something should be done. An engineer condenses years of personal experience, and that of others, into the specification. The more detailed the specifications, the higher the probability that the job will be done right.

6.2 CONSTRUCTION CONSIDERATIONS

Even when a facility is upgraded, there is typically a demolition phase where old equipment is removed and the space is cleared. Demolition and construction of existing structures may, therefore, have to be planned and specified. Electrical power, lighting, and air conditioning requirements must be identified and layout drawings prepared for use by the electrical and mechanical engineering consultants and the architect.

Most facility design projects begin with a "big picture" view and work toward the details. As illustrated in Fig. 6.1.1, the big picture is the overall scope of the facility—what it needs to provide to make the station efficient and productive. From that point, individual teams may focus on specific rooms or functions and drill down toward documenting of the details that make up the big picture.

During preparation of final construction documents, the architect and the system engineer can confirm the layout of technical equipment wire ways, including access to flooring, conduits, trenches, and overhead raceways. At this point, the system engineer also provides layouts of cable runs and connections. This makes equipment installation and future changes much easier. An overhead cable routing plan is shown in Fig. 6.1.2.

When it is necessary to install ac power cables in conduit, follow National Electrical Code (NEC) requirements for conduit fill and the number of pull boxes. More pull boxes or larger conduit is required in conduit runs that have many bends. Specify direct-burial-type cable when the conduit

---

[1] This chapter was adapted from: *Standard Handbook of Video and Television Engineering*, 4th Edition, Jerry C. Whitaker, Ed., Chapter 9.4, McGraw-Hill, New York, NY, 2003.

6.4   FACILITY ISSUES

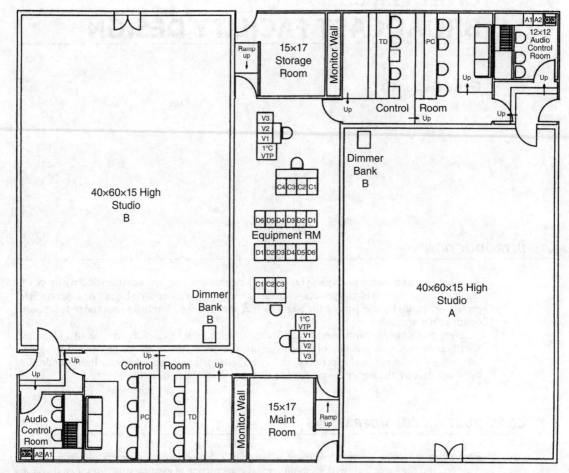

FIGURE 6.1.1   "Big picture" view of a new broadcast facility.

or cable trays are underground, and where there is a possibility of standing water. Conduit and cable trays should be designed to accommodate the minimum bend radius requirements for the cables being used.

Debur and remove all sharp edges and splinters from installed conduit. Remove construction debris from inside the lines to prevent damage to the cable jacket during pulling. Cover openings to the conduit to prevent contamination or damage from other construction activities. If the cable is damaged during pulling, moisture could enter the cable and ultimately result in failure.

6.2.1   Component Selection and Installation

Equipment selection is normally based on the function it will perform. User input about operational ease and flexibility of certain models is also important. However, to ensure that the most cost-effective choice is made, consider certain technical issues before making a decision. The system engineer should research, test (when required), and provide the technical input needed for selecting hardware and software.

Equipment features and functional capabilities are probably the main concerns of the users and management. Technical performance data and specifications are important considerations that

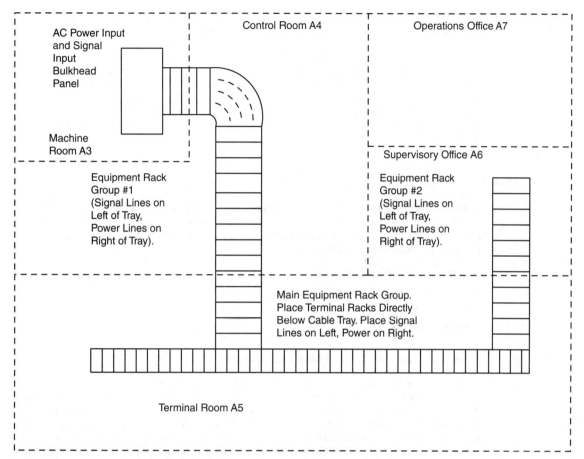

FIGURE 6.1.2 Cable wire routing plan for a new facility.

should be contributed by the system engineer during the selection process. For a piece of equipment to qualify, technical specifications must be checked to ensure that they meet set standards for the overall system. Newly introduced products should be tested and compared before a decision is made. The experienced system engineer can test and measure the equipment to evaluate its performance.

A simple visual inspection inside a piece of equipment by an experienced technician or engineer can uncover possible weaknesses, design flaws, and problem areas that may affect reliability or make maintenance difficult. It is therefore advisable to request a sample of the equipment for evaluation before committing to its use.

The availability of replacement parts is another important consideration when specifying products. A business that depends on its equipment functioning to specifications requires that the service technician be able to make repairs when needed in a timely manner. Learn about manufacturer replacement parts policies and their reliability. When possible, select equipment that uses standard off-the-shelf components that are available from multiple sources. Avoid equipment that incorporates custom components that are available only from the equipment manufacturer. This will make it easier to acquire replacements, and the cost of the parts will—most likely—be less.

When possible, specify products manufactured by the same company. Avoid mixing brands. Maintenance technicians will more easily become familiar with equipment maintenance, and experience gained while repairing one piece of hardware can be directly applied to another of the same model. Service manuals published by the same manufacturer will be similar, and therefore easier to understand and use to locate the information or diagram needed for a repair.

6.6 FACILITY ISSUES

Commonality of replacement parts will keep the parts inventory requirements and the inventory cost low. Because the technical staff will be dealing with the manufacturer on a regular basis, familiarity with the company's representatives makes it easier to get technical support quickly.

Sometimes components are selected that are not really compatible. The responsibility then falls on the system engineer to devise a fix to make the component compatible with the rest of the system. The component may have been originally selected because of its low price, but additional components, engineering, and labor costs often offset the expected savings. Extra wiring and components can also clutter the equipment enclosure, hampering access to the equipment inside. Nonstandard mounting facilities on equipment can add unnecessary cost and can result in a less than elegant solution.

6.2.2 Design Mockups

Design renderings (drawings created by an artist or drafter to show a realistic flat or perspective view of a design) or full-color 3-D models can be generated for viewing from different perspectives. A color and materials presentation board is often prepared as well for review by decision-makers. The presentation may include the following:

- Artist or computer renderings
- Color chips
- Wood types
- Work surface laminates
- Metal samples
- Samples of carpeting, furniture fabrics, and wall coverings

Several different combinations may be prepared for consideration.

When a drawing cannot be interpreted easily by the owner and/or staff, a scale model or full-size mockup of the facility (or portions of it) can be constructed. This will help familiarize them with the design, allowing informed decisions and changes to be made. Models can also be used to present a design concept to company executives. Models can provide a cost-effective way to evaluate new ideas. Inexpensive materials can be formed to represent racks, consoles, or equipment. The more detail provided in the model, the better.

Full-scale mockups, like models, can be built using any combination of construction materials. Stiff foam board is a relatively easy and inexpensive material to use to prepare full-size models. Pieces can be cut to any shape and joined to form 3-D models of racks, consoles, and equipment. (See Fig. 6.1.3.) Actual-size drawings of equipment outlines, or more detailed representations, can then be pasted in place on the surfaces of the mockup.

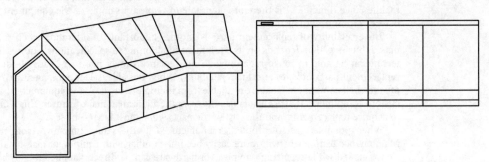

FIGURE 6.1.3 Equipment rack and console mockup for planning purposes.

6.2.3 Technical Documentation

Engineering documentation describes the practices and procedures used within the industry to specify a design and communicate the design requirements to technicians and contractors. Documentation preparation should include, but not be limited to, the generation of technical system flow diagrams, material and parts lists, custom item fabrication drawings, and rack and console elevations. An example drawing is shown in Fig. 6.1.4. The required documents include the following:

- Documentation schedule
- Signal flow diagram
- Equipment schedule
- Cable schedule
- Patch panel assignment schedule
- Rack elevation drawing
- Construction detail drawing
- Console fabrication mechanical drawing
- Duct and conduit layout drawing
- Single-line electrical flow diagram

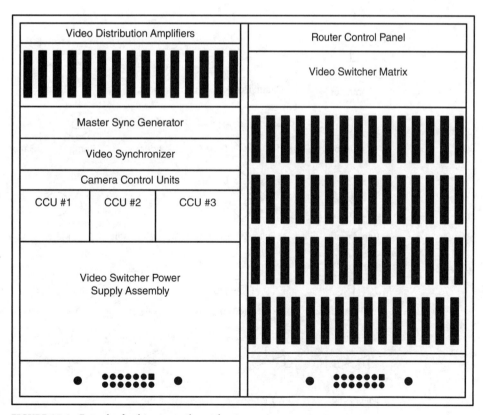

FIGURE 6.1.4 Example of rack equipment layout drawing.

6.8 FACILITY ISSUES

The documentation schedule provides a means of keeping track of the project's paperwork. During engineering design, drawings are reviewed, and changes are made. A system for efficiently handling changes is essential, especially on big projects that require a large amount of documentation.

Completed drawings are submitted for management approval. A set of originals is signed by the engineers and managers who are authorized to check the drawings for correctness and to approve the plans.

6.2.3.1 Symbols. Because there are only limited informal industry standards for the design of electronic component symbols to represent equipment and other elements in a system, custom symbols are usually created by the designer; each organization typically develops its own symbols. The symbols that exist apply to component-level devices, such as integrated circuits, resistors, and diodes. Some common symbols apply to system-level components, such as amplifiers. (See Fig. 6.1.5.)

The proliferation of manufacturers and equipment types makes it impractical to develop a complete library; but, by following basic rules for symbol design, new component symbols can be produced easily as they are added to the system. For small systems built with a few simple components, all of the input and output signals can be included on one symbol. However, when the system uses complex equipment with many inputs and outputs with different types of signals, it is usually necessary to draw different diagrams for each type of signal. For this reason, each component requires a set of symbols, with a separate symbol assigned for each signal type, showing its inputs and outputs.

If abbreviations are used, be consistent from one drawing to the next, and develop a dictionary of abbreviations for the drawing set. Include the dictionary with the documentation.

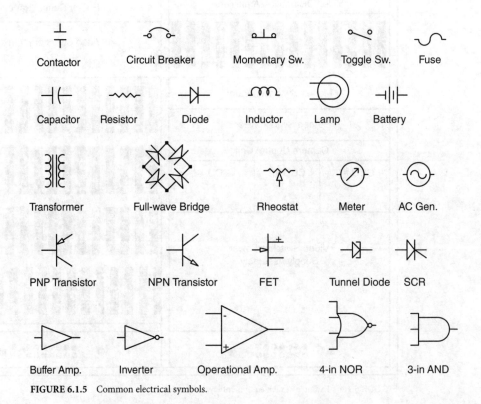

FIGURE 6.1.5 Common electrical symbols.

6.2.3.2 Cross-Referencing Documentation.
In order to tie all of the documentation together and to enable fabricators and installers to understand the relationships between the drawings, the documents should include reference designations common to the project. That way, items on one type of document can be located on another type. For example, the location of a piece of equipment can be indicated at its symbol on the flow diagram so that the technician can identify it on the rack elevation drawing and in the actual rack.

A flow diagram is used by the installation technicians to assemble and wire the system components together. All necessary information must be included to avoid confusion and delays. When designing a symbol to represent a component in a flow diagram, include all of the necessary information to identify, locate, and wire that component into the system. The information should include the following:

- Generic description of the component or its abbreviation. When no abbreviation exists, create one. Include it in the project manual reference section and in the notes on the drawing.
- Model number of the component.
- Manufacturer of the component.
- All input and output connections with their respective name and/or number.

6.2.4 Working with the Contractors

The system engineer must provide support and guidance to contractors during the procurement, construction, installation, testing, and acceptance phases of a project. The system engineer can assist in ordering equipment and can help coordinate the move to a new or renovated facility. This can be critical if a great deal of existing equipment is being relocated. In the case of new equipment, the system engineer's knowledge of prices, features, and delivery times is invaluable to the facility owner. The steps to assure quality workmanship from contractors on a job include the following:

- Clarify details.
- Clarify misunderstandings about the requirements.
- Resolve problems that may arise.
- Educate contractors about the requirements of the project.
- Assure that the work conforms to the specifications.
- Evaluate and approve change requests.
- Provide technical support to the contractors when needed.

BIBLIOGRAPHY

DeSantis, G., J. C. Whitaker, and C.R. Paulson, *Interconnecting Electronic Systems*, CRC Press, Boca Raton, FL, 1992.

Whitaker, J. C., *Facility Design Handbook*, CRC Press, Boca Raton, FL, 2000.

CHAPTER 6.2
WIRE MANAGEMENT

Fred Baumgartner
KMGH TV, Scripps, Denver, CO/Nautel Television

6.1 INTRODUCTION

Done smartly, connection changes are fast, easy, and reliable. You can identify any wire mid-span and describe a wiring change from the comfort of your laptop.

Broadcast engineers manage people, technology, facilities…and miles of interconnecting optical fiber and signaling wire. Data centers and broadcast equipment rooms may both have racks of equipment, CRAC (Computer Room Air Conditioning) units for cooling, uninterruptable power system (UPS) units, wire trays, and "computer" (raised) flooring; but they are managed entirely differently. Data centers are generally purpose-built and have an end-of-life point; typically about seven years out, or 35,000 hours of beneficial use (the same as some popular transmitter tubes). The data center space is then typically stripped to the concrete and rebuilt. Sometimes racks, flooring, some pieces of electrical distribution, interconnected bulkheads, and room lighting are saved and reused. Usually, cooling and UPS components are reconditioned or replaced. Broadcast equipment rooms, however, are forever. Consider how many AM transmitter buildings that are still in use as they approach the century mark. Visit one of these historical sites and one can see the tale of continual infrastructure and equipment upgrades written into the building itself.

Poor wire management is costly in terms of labor and reliability. It is a practical reality that most stations have had a series of owners and through periods of cost cutting, missing management or mismanagement, a broadcast facility can slip into a state where the documentation is inadequate to maintain the facility properly and efficiently, and abandoned wiring and equipment make what should be simple work into a risky and time-consuming task. Well-managed facilities are a thing of beauty and it is obvious to virtually anyone looking at it. Some owners actively enforce good engineering practice and upon acquisition of a station may even completely rewire a facility to bring it into alignment with corporate standards simply because that is the most cost-effective path.

Wire management in broadcasting is straightforwardly being able to add new connections, remove old connections, and to quickly learn what each connection's role is in the operation of the station. The challenge is one of variety, scale, and change; the stronger these forces, the more difficult and important connectivity management is. Variety is an especially strong force in broadcasting where, unlike most data centers, the same subsystem is *not* repeated over and over.

Anything that reduces the number of trips between rooms to discover information allows reuse, reduces time to repair, and is faster and cheaper to administer, install, remove, and saves a significant amount of labor; it makes a major positive impact on the cost of operating a broadcast facility.

6.12 FACILITY ISSUES

6.2 LABELS

There are a number of types of wire labels. Each is suitable for faculties of a certain size and purpose. While there are means of printing directly on wire, even at regular intervals, labeling each end with a machine printed "wrap around" self-adhesive label is the economical norm.

Some electrical supply houses and all telecom suppliers have a variety of labels, printers, pre-printed labels, and wire tied labels and tags, as well as an assortment of wire management hardware that makes for a neat and orderly installation. Convenient resupply is a consideration, as is cost and quality of the selected label.

The location of the label needs to be where it can be conveniently read, ideally without torqueing the wire and connector. There are labels that are designed to be loose enough that they can be moved along the wire and rotated for easy reading. If the label is too close to the termination or where wiring is dense, it can only be read with a dental inspection mirror or it may require disconnection. If the label is too far back, it might land in a bundle that would need to be broken into to read. If one is prefabricating cables, the ideal distance might not be obvious, so placing redundant labels at about 3 and 12 inches from the termination (usually a removable connector) is a one method.

6.3 DOCUMENTATION

There are two customary schools of wire documentation. Rarely is one method spot-on for every purpose, but clearly one is best for a particular installation. A potential hazard is starting with a plan that becomes too small if the facility grows, sometimes unexpectedly (consider the station that suddenly becomes home to a cluster). The worst case though is where the documentation database is destroyed or lost. Groups usually require that the documentation is backed up off site.

6.3.1 Ascension Numbering

With ascension numbering, each wire is numbered in an incremental manner. A database ties the number to the information about the wire. For example:

1. Signal type (e.g., tri-level sync)
2. Source (e.g., basement equipment room, rack #3)
3. Destination (e.g., Studio A equipment room, rack #1)
4. System (e.g., part of basement sync generator)
5. Subsystem (e.g., tri-level reference for studio A)
6. Purpose (e.g., feeds DA for studio cameras)
7. Wire color/type (e.g., black RG-59)
8. Route Info (e.g., second conduit blue room, hallway ceiling…)
9. Dependencies (e.g., studio A reference also feeds edit bay 4)
10. Length/footage marks (useful for ID and if an extension or repurposing of a wire is anticipated)
11. Bulkhead or punch-block interconnect position
12. Installer (who, or if an integrator, what project it is tied to)
13. Date—this can provide an amazing amount of useful context
14. More—there is a surprising amount of information that can be useful

This method is entirely dependent on having an accessible and secure database, usually a simple spreadsheet. *Document control* is critical as there can be only one version of the database. In a typical broadcast facility this is usually accomplished by enforcing a policy where only one person can check out the document at a time. This works well in a small shop with good communication. Larger facilities where many people are making entries will need to use collaborative software that allows several people to use the same database at the same time and ideally tracks authors and changes to make corrections and reversals easier.

Periodically printing the database or making it accessible to tablets and work stations is ideal. Updating and correcting the information about a wire is fast and easy with this method. Repurposing requires little effort or research.

Repeating the number on the label makes it easy to read from any angle. Preprinted labels can be used.

There is a natural temptation to encode the source and destination in the wire number. For example "0122333-1455666" where 01 is the room of the source and 14 is the destination room, 22 and 55 are the respective racks and 333 and 666 are the sequential wire number, which results in a 14-digit number for a mid-sized facility. This requires a bigger label, is harder to read, write down and remember. It is harder to administer, search, and sort. There are more opportunities for error, even without the dyslexic issues common to the engineering mind. Compare that to the three- to five-digit number that a simple sequence generates that also tacitly encodes the date and order of installation. The simple sequence is probably more efficient, useful, and economical.

It is also hugely beneficial to be able to search and sort the database. The database can be linked to drawings and other documentation.

Recording the footage markings (most cable has a footage number every foot) at both ends of the cable makes it easy to calculate the length in a spreadsheet. Finding a cable in a bundle or finding out what a given cable in a bundle does is simply a matter of searching the database for those wires that include that value. Knowing the color, type of cable, and the footage number should be all one needs to identify the wire anywhere in its span. Given that most wire is labeled with a four- to eight-digit footage number, there are thousands to a hundred-million possible numbers and maybe only a few thousand wires in a facility. In most cases, any given wire footage number is unique or one of a very few possibilities. While a simple sort and search of the database should identify the cable in a few seconds, a simple formulae or query in more elaborate databases is even better. Simply type in a footage number and go directly to the record. Being able to identify wires mid-span helps greatly with removing abandoned wiring, or fixing an accidental break.

Ascension numbering is also a bit more IP friendly where a growing number of connections are peer-to-peer, bidirectional (like an IP router) and thus the concept of source and destination fits poorly.

Backing up the database is critical and should be part of the disaster recovery planning.

There can be a significant label cost savings with this method. One can order preprinted labels in economical bulk, usually with two or four copies of each number. No printer, ribbons, special label stock, or software is needed and most importantly there is no labor to operate the printer...entering data into a spreadsheet is usually faster and easier than most label printers.

6.3.2 Source/Destination

Source/destination numbering typically uses a label large enough to describe in some suitable detail the source and destination of each wire, and a few characters about its purpose. The first line generally identifies the room and rack the signal comes from, the next line the room and rack it goes to, and the third line identifies some useful information like the subsystem or the type of signal. Generally, printing a fourth line results in an unreadable label as the print gets too small to read on typical wire. On thin wires such as mini-coax, audio, or fiber, even a second line can produce an unreadable result unless the label is mounted as a tag. Tags protruding from a wire do not pull well and can tear off.

Source/destination labels are large, take time to make, and are prone to error. Corrections require cutting off or covering the old labels. Multiline cable labels usually require rotation (sometimes dislodging or breaking the wire in the process), a dental mirror, or disconnection to read all of the lines of text.

Source/destination documentation is ideal for small systems and stations. News gathering vehicles, stand-alone radio stations, and A/V installations are well served by this method. This is especially good where maintaining a database document and keeping it available to whoever needs the information is difficult. With this method, anyone can walk into a radio station, or any operator can find their way around a news truck without having to find a document to know what a given wire does.

Many stations use source/destination documentation as a matter of tradition. If the station was built before accessible and secure databases were available and wires were few, it most likely is documented in this manner. It is hard to justify redocumenting the facility when there are decades of institutional momentum. Usually, the time to change documentation systems is on the eve of a major project.

Likewise, inconsistent engineering oversight or a series of subsystem installations by a procession of different integrators, make source/destination wiring documentation the only practical method. Inconsistent engineering oversight can be a fact of life in broadcasting as sales and consolidations, as well as the use of contractors and the politics of stations and groups, can put the station at risk of losing any information that is not printed on the wiring itself…and that is a very good reason to document in this manner.

6.3.3 Self-Documenting

Not every wire needs to be documented and some systems, in particular IT systems, tend to self-document. Consider IP routers. In a redundant system, each connection back to the router is made with "A" colored wire and the redundant connection with "B" colored wire. If a bulkhead interconnect is used, it is generally easy to trace a wire from a server to the bulkhead because they are in the same rack. The bulkhead replicates the series of router ports, and if one needs to identify a given physical connection, one need only to identify the router port it was plugged into and the router control software makes that an easy task. Within a given VLAN (virtual local area network), all physical connections can be equal. In this circumstance, the same pattern is duplicated over and over; the ports are visible without needing wire identification and the paths from bulkheads to equipment are short.

There is also little point in labeling or documenting utilities like power cables to dumb power strips and interconnect wires where both ends are clearly visible and close to each other. Managed outlets and A-B power requires some labeling and/or color coding to assure that equipment is plugged into the correct outlets.

Looping reference between video processing cards in the same frame hardly needs documentation, unless the drawings or other pieces of the plant documentation will not support wires without numbers. For short interconnect wires, the use of light weight mini-coax, etc., where diminutive pull strength or length is involved, not only conserves space and reduces bend radius issues, but identifies the wire as self-documenting.

6.3.4 Convenience Labeling

Ideally, one should be able to remove a piece of equipment, remove all the connecting cables, and replace it, secure in the knowledge of which wire connects where. The previous methods of wire management do not always address this well. Even if it is a just marking the cables with a Sharpie˚, this saves intuiting its proper home from the wire label text, or in the case of ascension numbering, looking it up.

A hand-written wire-tie tag label installed at a convenient point at the back of the machine is ideal. Wire-tie tags will not pull well (as they stick up from the wire and can cut one's hands if one tries) so they are not good for most wire labeling; however, they rotate easily for viewing. Matching sticker labels on the equipment is called for when there is some ambiguity or difficulty conveniently seeing the markings on the equipment.

6.4 PHYSICAL LAYER

Arguably, nylon wire ties are hugely overused and underused in broadcast plants. It is sometimes hard to remember when removing abandoned wires that the engineer that wire-tied the cables to a wire tray or into a now arbitrary bundle, did so with good intention and maybe some excusable lack of knowledge. Likewise, the technician that did not take the time to wire tie and secure a "rat's nest" has not finished the work. Wire ties are easy to over tension, impairing operation of the cable. There are special tensioning tools, but as a practical matter, their use day-to-day is inconvenient and probably does not really produce significant results. Wire ties have their place; unbundled wiring is undesirable in many locations. Wire ties should have their excess length trimmed to prevent an eye hazard, cuts to hands, or entanglements.

Velcro™ wire bundling presents less stress on the wires and is "green" in that it can be reused. To do the same with wire-ties, one needs to carry replacement ties and at least a pair of cutters.

Figure 6.2.1 shows an installation using wire management brackets as a replacement for wire ties.

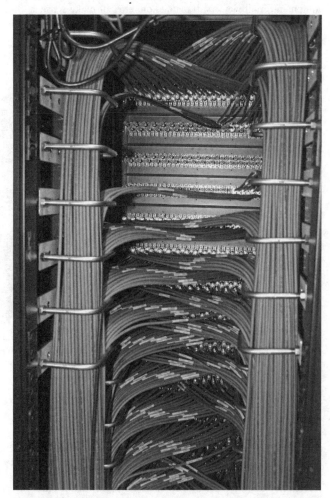

FIGURE 6.2.1 This patch bay is made easy to maintain through the use of wire management brackets as opposed to wire ties. (*Courtesy: Tony Roccanova 5280 Broadcast, Denver.*)

6.16 FACILITY ISSUES

Bulkhead interconnects allow the preinstallation of reusable wiring between two points. In the ascension numbering scheme, each "tie" is given a number and that number is inherited by the equipment cross-connects. If the cross-connect is short and obvious, only convenience (so it can be reconnected easily) labels are required. In the source/destination scheme, there is no perfect way of dealing with bulkheads. Most often each bulkhead port is treated as a source on one end, a destination on the other, and the third line of label information is reserved for a note on the remaining information (as visibility into the wire's full span is lost at the interconnects). Bulkheads approach permanent infrastructure and are treated in that manner. The use of conduit, large wire-tied bundles, placing tie line bundles outside wire trays, and the like takes advantage of the idea that one is not going to remove or add to that wiring in its lifetime. While older coax probably is not useful in the digital world, there are a number of those nearly one hundred-year-old AM plants where audio and control still go through connected bulkheads that were installed when the building was constructed. Bulkhead interconnects are a near necessity between equipment rooms, studios, edit bays, etc. They are desirable inside an equipment room for utilities such as Ethernet, control, reference, and any reasonably predictable path that is interconnect-intensive. Bulkheads between routers and production and on-air switching equipment, multiviewers, transmission, etc., all make sense as they follow the layout of the room. The best wire, even if it is expensive and hard to work with, is justified. Where Plenum-rated wire is needed between rooms, interconnects and dedicated tie-lines are a good investment.

A common bulkhead installation is shown in Fig. 6.2.2.

FIGURE 6.2.2 A bulkhead in a studio, edit bay or other special place allows permanent interconnects to be installed between locations and reused through several generations of equipment. This bulkhead serves a studio now on its third news set since the bulkheads were installed.

6.4.1 Color

Some facilities are all about wire color. Assigning a color (very common for IP systems with redundant switches and A/B power systems) to a particular purpose can make sense at any level. For example, one facility uses red-wire to indicate adult services to prevent accidental airing on the wrong service channel. Another uses color to indicate which room's wire is coming into the central equipment room. Simply mixing up colors with no particular pattern makes tracing and removing wire easier. It is often difficult to keep enough of every color and type on hand, so one is often tempted to use an improper color.

Specialty jumpers or cross-connects, in particular null or crossover cables, need to be well identified to keep them separate from the general population. With discipline, color can be part of that process.

6.4.2 Service Loops and Slack

Probably, the best way to deal with excess cable length is not to have much. Fiber, which is little thicker than hair if the outer jackets are removed, often uses a small closed storage cabinet of a few square inches to wind up the reasonable excess fiber. Most fiber is ordered preconnectorized and tested, so some slack is almost necessary. Coax and copper Ethernet cables do not have similar options for excess wire. Looping slack into a wire tray uses three times the space and tangles easily with other wires. One cannot pull a wire that is looped to remove it without catching other wires and creating knots that are hard to remove.

Some cables just cannot be shortened reasonably. Fiber and special cables with preconnectorized terminations do not lend themselves to trimming. Outside of the wire trays and paths, one can take up slack by forming a coil or using a wire reel and stabilizing it with wire ties (never too tight). Small quantities can be hidden in racks. If raised flooring is used with a wire tray, use the wire tray for runs and outside the tray for reels or rolled up slack. If the height of the computer floor is small, or there is no ground-mounted wire tray, the entire area is active wire space and any impediment is a point of entanglement and kinking, so placing unused slack into this environment is not desirable. Studios often use common garden hose holders found at any hardware store for operating cables and excess camera cable that is seldom needed in its full length. There are professional wire reels for cable that is redeployed often.

Some slack is necessary to allow the regular movement and servicing of equipment. Equipment that can be pulled out on rack rails, or can be partially removed should be able to do such without having to disconnect its interconnections. Likewise, there should be enough slack to allow a replacement with a reasonably different apron layout. Folding arm assemblies are available for servers and other deep rail-mounted equipment to which the connecting wire is attached.

The use of service loops at a broadcast facility is shown in Fig. 6.2.3.

Broadcast facilities often need to use deep racks and mount equipment from both sides. The space between racks becomes an important asset for wire management. Ideally, the covers for the inter-rack space are easily removable. They may be necessary for cooling, and so should be replaced when wiring is complete. Wiring directly from one rack to another is generally considered bad practice, but may be a necessity if the alternatives are overloaded. The rack to rack wire prohibition is often a religious one.

A common approach to inter-rack wiring is shown in Fig. 6.2.4.

6.4.3 Installation and Dress

Most of wire management takes place at installation. All wire and fiber have a minimum bend radius and maximum pull strength, easily found in the wire's specifications. Long wire pulls might exceed the maximum recommended force for a given wire. Likewise, dress that requires a sharp bend might be expedient and even pretty, but might also degrade the wire's performance to the point of unreliability if the minimum bend radius is exceeded. Strain relief is necessary in some cases to prevent overbending wires, or minimize fatigue on wires that are often flexed.

A kink changes the electrical characteristics of a cable, most notably at high frequency and data speeds and creates a point of physical weakness. If the insulation is damaged, corrosion might set in. Wiring that runs through poor environments, especially conduit and underground runs, will fail quickly if the insulation fails.

6.4.3.1 Special Wire. There are nearly an infinite variety of interconnect materials available. *Plenum-rated* cable is required in most places where air is ducted for heating and cooling. *Flooded* cables

6.18 FACILITY ISSUES

FIGURE 6.2.3 Servers with proper service loops are easy to pull out to work on. (*Courtesy: Tony Roccanova 5280 Broadcast, Denver.*)

contain gelatin like compounds that heal breaks in the cable's coverings, even underground. Tiny flexible cables are perfect for trucks and tight, short distance connections. Flexible cables are for constant movement. Strong, stiff, and low loss cables are available for long tie runs. Ideally, wires on equipment like cameras and steerable antennas should be easily replaced as even wire made to flex will wear out before stationary wire does.

6.4.3.2 Wire Tray and Race Way. Common wire tray can be welded wire mesh for above and between racks and down hallways, or rolled aluminum or galvanized steel open top ducts that can be mounted below computer tile floors and through mechanical tunnels. Cables should rest in the tray without tension. Grounding is often bonded to the trays. Adding and removing wire is straight

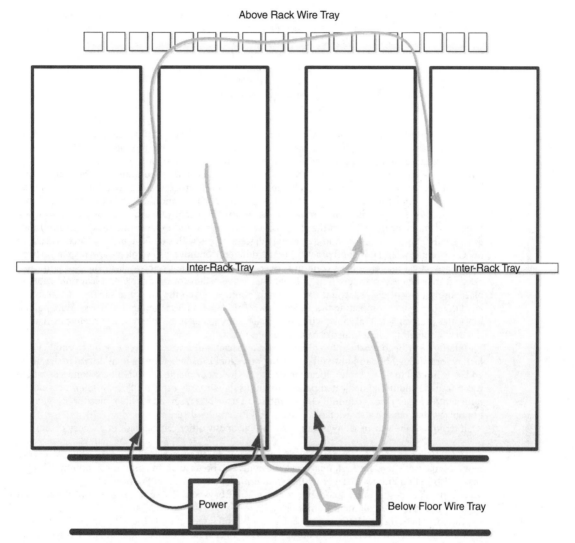

FIGURE 6.2.4 In general, inter-rack wiring goes up into above rack wire trays, or below the computer floor, or both. One scheme is to put east-west wires below the floor and north-south wires above the racks. Generally wiring directly from one rack to another is considered bad practice, unless a dedicated tray is provided. Even then, it is considered bad form to pass through a rack to reach one that is not adjacent.

forward, provided the tray is not overfilled and the wiring is free of wire ties and loops. Intersections, where a north-south wire tray crosses an east-west wire tray, it is acceptable to do such on the same level if the loading is light—no more than 40 percent of each tray's capacity. If the combined loading exceeds the 80 percent mark, the intersection needs to be on two levels (an overpass). If not, the intersection becomes overloaded and presents a blockage to both trays.

Raceway (colloquially a "tangle box") is a long, narrow, electrical junction box ("J-box"), almost always painted gray, with hinged doors or panels intended to connect a series of primary power electrical boxes. It is also a desirable option for low voltage wiring in rough or insecure environments, such as a shared transmitter room. Removing locks or screws is necessary to open most boxes. Knock outs are often provided for conduit connections.

Wire trays and raceways should be filled to about the one-third point. They are overloaded when they are two-thirds filled. By definition, when the wire falls out of the tray, they are critically overloaded and no longer efficiently serve their purpose. Wire in quantity is heavy. Wire trays do break up, collapse, and can present a safety hazard when overloaded.

As a practical matter, if it cannot be fixed, one can fabricate the necessary components to keep wiring from falling out of overloaded trays.

6.4.3.3 Conduit and Interduct. Conduit is simply metal or plastic pipe, commonly one to four inches in diameter. Although larger sizes are available, the cost and tooling (especially concrete coring tools and conduit benders) needed to work with six inch and larger conduit might be excessively expensive. Four inch is a very manageable size as longer, smaller conduits with heavier loading become hard to add or remove wiring. Conduit (metal) is often required between floors and between certain fire control zones. The unused space at the end of each conduit, ideally more than half the area, should be stuffed with a fireproof, rodent, and rot resistant material such as fiberglass wool. Wire needs to be protected from the sharp edges at the entry and exit of conduit. Overloaded conduits are difficult to maintain, and damaging adjacent cables with excessive pulling forces is easy to do. Code requires (and it is good practice) to use different conduit for low voltage and primary power.

As a practical matter, once a conduit is overloaded, one cannot remove dead cable or add cable safely. It may be best to simply find or build alternative routes for new cable and leave old cable in place forever. Use of dish soap and water often is not ideal, given the mess and the need to clean out all of the soap before completing the project (it will "glue" the cables together as it dries). Wire pulling lubricants are available that make wire pulls easier. Care needs to be taken to use chemicals that do not dry to something that impedes cable maintenance.

Interduct is ribbed plastic tubing that creates a special, safe route, from one point to another in a facility for delicate fiber and low voltage wiring. Interduct is usually only an inch or two in diameter, and so is limited to a few cables. It can be tied to a wire tray, outside and often below the tray where it can provide additional dedicated paths without using wire tray capacity. Several interducts can be pulled into a larger ridged conduit. Orange interduct usually is reserved for fiber, blue for low voltage; gray, black, and red are also available (even at some chain hardware stores).

Interduct is one way of safely isolating the on-air or any other critical path in a legacy plant.

6.4.3.4 IDFs and MDFs. Facilities likely have wiring closets for servicing phones, cable TV, alarms, news rooms, etc. These are as standard as bathrooms and break areas in commercial building design. Main Distribution Frames (MDFs) are tied to Intermediate Distribution Frames (IDFs), which in turn serve individual desks, offices, conference rooms, edit bays, etc. These terms are from the telecom industry and refer to equipment frames or racks.

The paths between the IDFs and the MDFs are by and large conduit. The MDF might be the main equipment room, although most buildings provide a separate room for equipment that attends to office equipment, local IP switches and servers, telephone (if not VoIP phones), building alarms, station monitoring feeds, CATV, etc. If the building and office systems are under broadcast engineering control, the documentation can be combined. In any station, there is significant crossover between the broadcast systems and the other facility systems. What General Manager does not have a time-code clock and router head, or some variation of that? What broadcast engineering department does not spend time in the electrical closets?

6.5 RINGING OUT THE PLANT

Serious facilities might employ third parties to verify the configuration, labeling, and documentation, and measure the electrical characteristics of the wiring—usually before accepting work from a contractor. Sweeping the cabling can reveal manufacturing defects, kinks, and other damage and certify its capacity for reliably carrying high speed signals at low signal levels for the required distances. If the project is big enough, there will likely be issues. Some of these will not be an issue for some time, and because wiring is a passive component, it tends to be among the last items suspected in an outage or failure.

Cable manufacturers (and sometimes installers) test the cable's performance before installation. Tolerances are tight and manufacturing complex. Not every reel that comes off the line is good. If the facility is mission critical, onsite tests are also critical.

6.5.1 Challenges

There are several challenges in wire management that have difficult, time-consuming, sometimes expensive, and often unsatisfying solutions. As electronics become more compact, interconnect density has to be answered with either more physical connections or interconnection that moves more information on fewer physical connections. Consider that the first generation of 4K video is best carried on four separate coax connections. Countering this is video on IP and IP on fiber, which minimizes the number of physical connections for the information carried.

Break-out cables (where a high density connector is "fanned-out" to more common but space inefficient connectors such as XLRs and BNCs) exist because there is already a density issue . . . the necessary connectors will not fit on the equipment. Break-out cables tend to be thin and delicate compared to the ordinary interconnect cables. Securing the interconnect and dressing it while leaving it serviceable and modifiable is a challenge without a great answer, but there are suitable fixtures for sale and ones that can be easily crafted that allow wires to be dressed and tied down in an orderly manner.

"Black-boxes," those small converters and distribution devices that are powered by wall-warts, are often installed with no physical support other than the wiring. This works well enough if the connectors, in particular the power connection, are locking . . . up to a point. Rack-mounted trays and do-it-yourself supports with ties and some means of mounting the wall-warts in such a way that they are secure and out of the way is probably the best that can be done. Often these same "black-box" functions are available in card frames with well-designed power, monitoring, and cooling. Occasionally, a function is only really available in a "black-box," but most often when these appear as part of a station's infrastructure it is driven by the low entry price and speed of installation.

It is probably the temporary installs that present the biggest challenge. Without good engineering practice, what begins as a temporary fix sometimes becomes a key piece of a long lived critical system. The implications only begin with the wire management concerns.

6.6 CLEANING AND REMOVING

Abandoned wiring is like cholesterol in a broadcast facility. In a data center, abandoned wiring is rarer because new wiring is rarer, and in any case will be removed when the data center is rebuilt. Because broadcast facilities see constant change, wiring is frequently retired. Some transitions, like the over-the-air 2009 digital conversion, orphan much of the plant's wiring in a short period of time and likely involved a fair amount of temporary wiring. Because broadcast plants tend to have long life cycles, dead wiring can build up.

As wire trays and computer floor load, the old wire, mostly on the bottom, is weighed down by new wire and can only be removed a few feet at a time, one piece at a time. Meanwhile, the new wire spills out of the trays that are critically overloaded, stressing connections and patience. If the documentation is weak, there is no way to identify a wire mid-span and there may not be an easy way to restore a circuit that breaks while wire maintenance is performed.

Broadcast engineers shy away from the clean up because it is physically demanding, results in cuts, bruises, and sore muscles. There is habitually little reward in that management often reacts to the heart attacks, but not the cholesterol that caused it. There are always more urgent and glamorous projects to do. See Figs. 6.2.5 and 6.2.6.

FIGURE 6.2.5 The process of removing abandoned wiring from overloaded trays is not pretty, but it is simple. At each step free the old wire by hand and cut it back, recycling the spent cable.

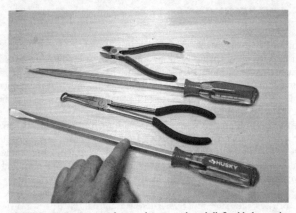

FIGURE 6.2.6 Oversized screwdrivers with a dull flat blade can be used to pry wires without damage. A special plier with provisions for grasping and fishing a wire out from deep in a tight bundle is especially valuable, but hard to find.

FIGURE 6.2.7 There are too many opportunities for pictures like this. This one is particularly interesting in that this connection serves a mission critical purpose and while it looks like it has been pulled apart, in fact, the few connections you see being made are the necessary ones. It is also interesting in that not even tape or wire nuts were used; clearly a years-old temporary connection made in haste and forgotten.

The process is simple. Carefully remove as much dead wire as you can. Some will want to remove the entire length of a wire and on occasion that works. Most often there is an overloaded crossover point, wire tied cables, splices in mid-span, or looped slack in the wire tray to make that practical or safe. In this case, cut the wire; pull the tail out of the tray so that the removal process can be continued latter as the wire pack loosens. Severely overloaded wiring might have to be removed a few feet at a time.

A nonsecure splice in a wiring tray is shown in Fig. 6.2.7.

It is best to disconnect wiring a few days before removing it so that it can be restored if it proves to be active. Use a sharpie to mark the disconnected cable with enough information that it can be restored if needed.

One plan is to simply embark on a program to remove an hour's worth of dead wiring each day, preferably not in prime times. The best tools are the very large flat blade screwdrivers about two-feet long and big and heavy needle-nose pliers with a hook in the beak. Use the screw drivers to off load pressure on lower wiring in a tray and the pliers to pull the wire out so it can be cut.

6.7 BOTTOM LINE

Wire management is what broadcast engineer's do. There is a lot to know as a broadcast engineer and wire management is often undervalued. Broadcasting is often about being fast and cheap, but the equipment room is forever. Until about 1975, wire management for most broadcasters was not all that important simply because there was not that much of it. By the turn of the century, many stations had more wiring than the networks first did. Getting to a workable wire management scheme is more than wire-ties and labels; it is finding or fabricating what is needed and often it is creative solutions to unglamorous problems. The most obvious sign of good engineering is good wire management. Good engineering practice pays back in reliability and efficiency, and poor engineering practice continually costs in failures and not so obvious inefficiencies.

Here is a set of starter "rules":

- No splices in wire trays or below computer floor, and especially no wire nuts or nonlocking connectors.
- Avoid wire ties in wire runs.

- Avoid tied bundles unless the bundle will always be used as a bundle, as in bulkhead-to-bulkhead wiring.
- Avoid placing bundles in wire trays.
- Minimize slack and avoid looping slack in wire trays.
- Label almost everything.
- Avoid dangling "black-boxes."
- Secure wall warts.
- Avoid paths between adjacent racks that go above or below the rack by making provisions for an inter-rack wire tray when appropriate.

6.8 FOR FURTHER INFORMATION

Belden's Steve Lampen's guide to wire is the quintessential guide for installation, explaining fully the use of specialized wire and the correct methods of pulling and otherwise installing wire. See http://belden.com/resourcecenter/documents/.

CHAPTER 6.3
EQUIPMENT RACK ENCLOSURES AND DEVICES

Jerry C. Whitaker, Editor-in-Chief

6.1 INTRODUCTION

In a professional broadcast facility, most equipment will have to be rack-mountable. To assemble the equipment in racks, the installer needs to know the exact physical location of each piece of hardware, and all information necessary to assemble and wire the equipment. This includes the placement of terminal blocks, power wiring, cooling devices, and all signal cables within the rack.

6.2 RACK/EQUIPMENT LAYOUT

Equipment locations can be shown on a rack elevation form. Other forms and drawings can specify terminal block wiring, ac power connections, patch panel assignments, and signal cable connections. An example of an equipment location drawing is given in Fig. 6.3.1.

Drawings showing the details of assembly, mounting hardware, and power wiring generally will not change from rack to rack. Therefore, they can be standardized for all racks to avoid having to repeat this part of the design process. Exceptions can be shown on a separate detailed drawing. This approach is illustrated in Fig. 6.3.2.

When more than one rack is to be assembled side by side, it is normal practice to show the entire row on one drawing. The relationship of all of the equipment in adjacent racks can then be easily seen on the drawing (Fig. 6.3.3).

A large equipment bay for a television station is shown in Fig. 6.3.4. Equipment placement typically follows signal flow. Slide-out keyboards are helpful for equipment configuration, operation, and maintenance.

6.3 INDUSTRY STANDARD EQUIPMENT ENCLOSURES

The modular equipment enclosure, frame, or equipment rack is one of the most convenient and commonly used methods for assembling the equipment and components that make up a technical facility. The ANSI/EIAJ RS-310-D standard for racks provides the basic dimensions and specifications for racks, panels, and associated hardware. Other specifications, such as the European International Electrotechnical Commission (IEC) Publication Number 60297 and German Industrial Standard DINJ 41494, have matching dimensions and specifications.

6.26 FACILITY ISSUES

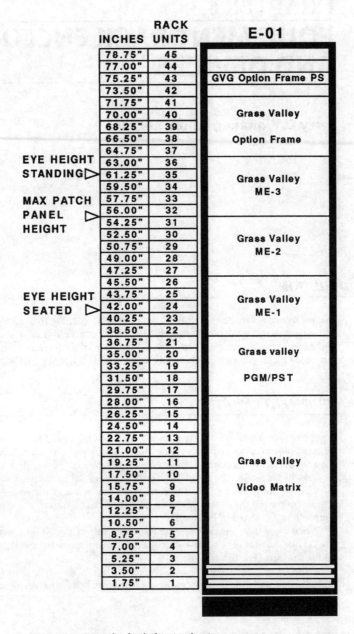

FIGURE 6.3.1 Example of rack elevation drawing.

Applicable standards for equipment racks include the following:

- UL-listed type 12 enclosures
- NEMA type 12 enclosures
- NEMAJ type 4 enclosures
- IEC 297-2 specifications

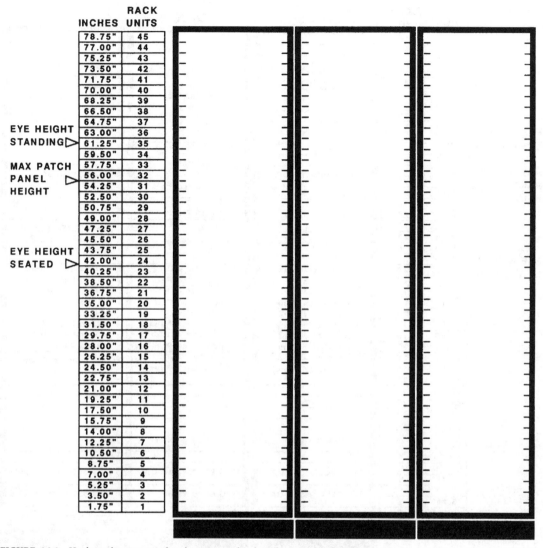

FIGURE 6.3.2 Hardware location template for a series of rack enclosures being installed at a facility. Deviations from the standard template are shown as drawing details.

- IEC 297-3 specifications
- IP 55/NEMA type 12/13 enclosures
- DINJ 41494

The chassis of most of the electronic equipment used for professional audio/video applications have front panel dimensions that conform to the EIA specifications for mounting in standard modular equipment enclosures. Figure 6.4.5 shows the standard RS-310-D rack-mounting hole dimensions.

Blank panels, drawers, shelves, guides, and other accessories are designed and built to conform to the EIA standards. Rack-mounted hardware for interconnecting and supporting the wiring is

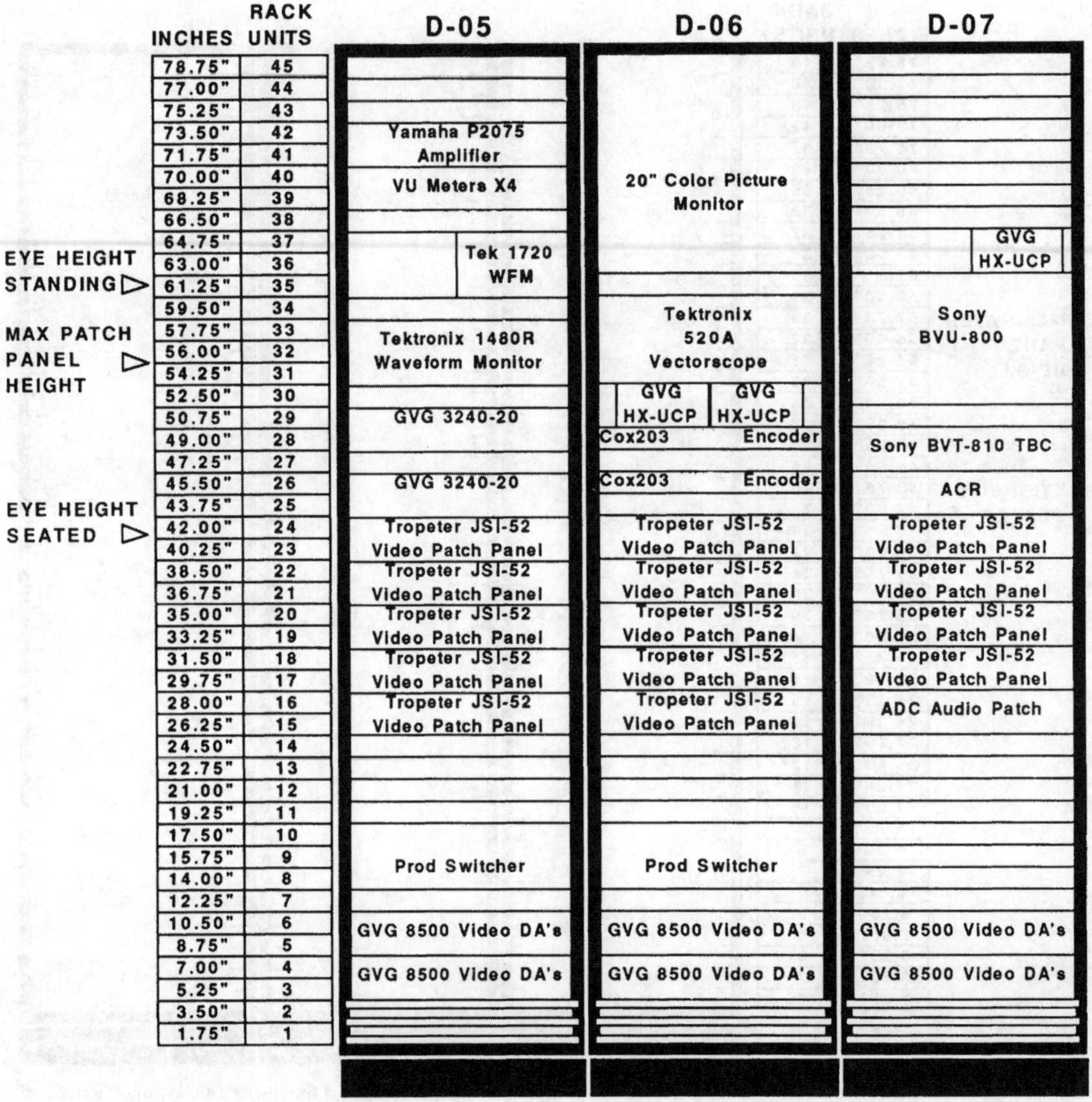

6.28 FACILITY ISSUES

	INCHES	RACK UNITS	D-05	D-06	D-07
	78.75"	45			
	77.00"	44			
	75.25"	43			
	73.50"	42	Yamaha P2075 Amplifier		
	71.75"	41			
	70.00"	40	VU Meters X4	20" Color Picture Monitor	
	68.25"	39			
	66.50"	38			
	64.75"	37			GVG HX-UCP
EYE HEIGHT STANDING ▷	63.00"	36	Tek 1720 WFM		
	61.25"	35			
	59.50"	34			
MAX PATCH	57.75"	33		Tektronix 520A Vectorscope	Sony BVU-800
PANEL ▷	56.00"	32	Tektronix 1480R Waveform Monitor		
HEIGHT	54.25"	31			
	52.50"	30		GVG HX-UCP	GVG HX-UCP	
	50.75"	29	GVG 3240-20	Cox203	Encoder	
	49.00"	28			Sony BVT-810 TBC
	47.25"	27			
	45.50"	26	GVG 3240-20	Cox203	Encoder	ACR
	43.75"	25			
EYE HEIGHT	42.00"	24	Tropeter JSI-52 Video Patch Panel	Tropeter JSI-52 Video Patch Panel	Tropeter JSI-52 Video Patch Panel
SEATED ▷	40.25"	23			
	38.50"	22	Tropeter JSI-52 Video Patch Panel	Tropeter JSI-52 Video Patch Panel	Tropeter JSI-52 Video Patch Panel
	36.75"	21			
	35.00"	20	Tropeter JSI-52 Video Patch Panel	Tropeter JSI-52 Video Patch Panel	Tropeter JSI-52 Video Patch Panel
	33.25"	19			
	31.50"	18	Tropeter JSI-52 Video Patch Panel	Tropeter JSI-52 Video Patch Panel	Tropeter JSI-52 Video Patch Panel
	29.75"	17			
	28.00"	16	Tropeter JSI-52 Video Patch Panel	Tropeter JSI-52 Video Patch Panel	ADC Audio Patch
	26.25"	15			
	24.50"	14			
	22.75"	13			
	21.00"	12			
	19.25"	11			
	17.50"	10			
	15.75"	9	Prod Switcher	Prod Switcher	
	14.00"	8			
	12.25"	7			
	10.50"	6	GVG 8500 Video DA's	GVG 8500 Video DA's	GVG 8500 Video DA's
	8.75"	5			
	7.00"	4	GVG 8500 Video DA's	GVG 8500 Video DA's	GVG 8500 Video DA's
	5.25"	3			
	3.50"	2			
	1.75"	1			

FIGURE 6.3.3 Equipment rack drawing for a group of enclosures, showing the overall assembly.

also available. Figure 6.3.6 shows some of the hardware available for use with standard equipment enclosures.

6.3.1 Types of Rack Enclosures

There are two main types of racks. The first is the floor-mounted open frame, shown in Fig. 6.3.7. This EIA standard equipment enclosure consists of two vertical channels (with mounting holes), separated at the top and bottom by support channels. The frame is supported in the free-standing mode by a large base, which provides front-to-back stability. The rack can also be permanently secured to the floor by bolts, eliminating the need for a base. Equipment is mounted directly to the vertical members and is accessible from the front, side, and rear.

EQUIPMENT RACK ENCLOSURES AND DEVICES **6.29**

FIGURE 6.3.4 Television station machine room. (*Courtesy: WISH-TV.*)

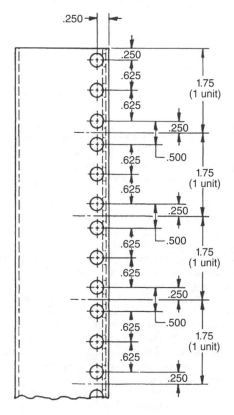

FIGURE 6.3.5 Standard equipment-mounting dimensions for RS-310-D rack enclosures.

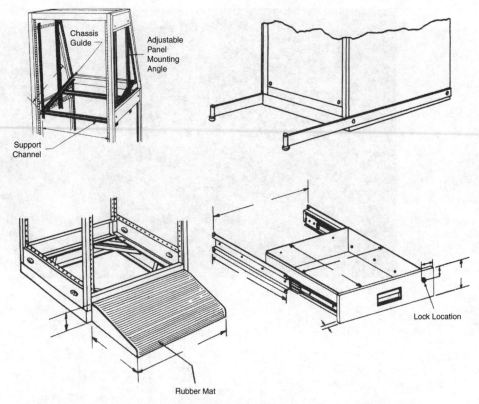

FIGURE 6.3.6 Common accessories available for use with rack enclosures. (*Courtesy: Emcor Products.*)

The second type of rack is a box frame that is free-standing, with front and, optionally, rear equipment-mounting hardware. This is illustrated in Fig. 6.3.8. The frame can be completely enclosed by installing optional side, rear, top, and bottom panels or access doors. Horizontal brackets mounted on the left and right sides of the frame increase rigidity and provide support for vertical mounting angles and other accessories.

Standard racks are available in widths of 19, 24, and 30 inches. The preferred, and most widely used, width is 19 inches (482.6 mm). The racks are designed to hold equipment and panels that have vertical heights of 1.75 inch (44.45 mm) or more, in increments of 1.75 inch. One *rack unit* (RU) is defined as 1.75 inch (44.45 mm). The rack height is usually specified in rack units. Holes or slots along the left and right edges of the equipment front panel or support panel are provided for screws to fasten the unit to specific mounting holes in the rack.

The mounting hole locations are defined by EIA specification so that equipment mounts vertically only at specific heights—in 1.75-inch increments—within the enclosure.

The simplest vertical mounting angles are "L"-shaped and have holes uniformly distributed along their entire length through both surfaces (see Fig. 6.3.9). The holes on one surface are used to attach the angle vertically to the horizontal side members of the rack frame. The horizontal members usually allow the vertical mounting angle to be positioned at different depths from front to rear in the rack. The holes on the other surface of the mounting angle are used for securing the equipment. Mounting holes are arranged in groups of three, centered on each 1.75-inch rack-unit interval.

Rack angles of a more complex design can be used when necessary to mount accessories, such as drawers, shelves, and guides that do not use the front-mounting angles and, therefore, must be secured by an alternate means. Figure 6.3.10 shows one common rack angle of this type.

EQUIPMENT RACK ENCLOSURES AND DEVICES **6.31**

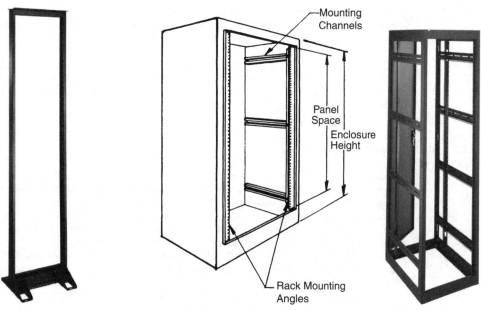

FIGURE 6.3.7 Standard open-frame equipment rack. (*Courtesy Middle Atlantic Products.*)

FIGURE 6.3.8 Standard enclosed equipment rack. (*Photo courtesy Middle Atlantic Products.*)

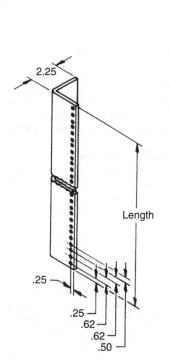

FIGURE 6.3.9 "L" equipment mounting bracket, used to support heavy instruments in a rack.

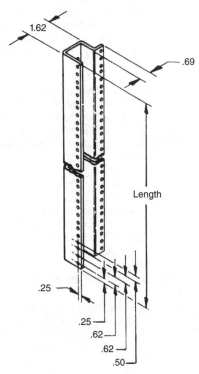

FIGURE 6.3.10 Rack angle brace used for mounting shelves, drawers, and other accessory hardware in an enclosure.

6.32 FACILITY ISSUES

To secure equipment to vertical supports or other mounting hardware, 10-32 UNF-2B threaded clip nut fasteners are placed at the appropriate clearance holes in the mounting angles for each piece of equipment being mounted. Some rack enclosures provide 10-32 threaded mounting holes on the vertical support channels.

The manufacturers of equipment enclosures offer a wide variety of accessory hardware to adapt their products to varied equipment-mounting requirements. A variety of paint colors and laminate finishes are available. Vertical rack-mount support channels are available unpainted with zinc plating to provide a common ground return for the equipment chassis.

6.3.2 Rack Configuration Options

Groups of rack frames can be placed side by side in a number of custom configurations. The most common has racks arranged in a row, bolted together without panels between adjacent frames. Side panels can be mounted on each end of the row, resulting in one long enclosure. Side panels can be installed between frames in a row, if necessary, to provide electromagnetic shielding, a heat barrier, or a physical barrier. A rear door, with or without ventilation perforations or louver slots, can be installed to protect the rear wiring and provide a finished appearance. A front door can be used if access to the equipment front panel controls is not necessary. Clear or darkened Plexiglas[1] doors can be used to allow viewing of meters or other display devices, or to showcase some aspect of the technology used in the rack.

A top panel is recommended to protect the equipment inside from falling debris and dust. Bottom panels are usually not installed unless bottom shielding is required.

Standard equipment racks provide flexibility because equipment can be mounted at any height in the rack. Many different-shaped frames are also available. These shapes conform to the same equipment-mounting and -mating dimensions, which permit assembling different frames together. Shapes are available with sloping fronts and various wedge shapes are common. Racks can be angled with respect to each other by inserting wedge-shaped frames as intermediaries between adjacent frames. Complex consoles for housing control panels and monitoring equipment can be assembled by bolting together the differently shaped frames. With these options, a complex console shape can be assembled to meet functional and human factor requirements. Figure 6.3.11 illustrates several of the stock configurations.

Although control consoles can be assembled from standard components that conform to standard enclosure dimensions, in many instances, custom-made consoles are desirable. These are helpful to achieve a more efficient layout for controls, or to develop a more sophisticated appearance within the control room environment. An elaborate video production/control arrangement is shown in Fig. 6.3.12.

Some equipment enclosure manufacturers offer an intermediate step between a stock rack and a custom-made console. By using off-the-shelf rack elements, the customer can specify the exact size and configuration required. After the dimensions have been provided to the manufacturer, the individual supporting rails and frames are cut to specification, and the unit is assembled.

Rack accessories are available specifically for cable management and high density cable terminations. An example is shown in Fig. 6.3.13.

6.3.3 Selecting an Equipment Rack

When selecting the model of rack that will be used in a facility, the physical dimensions and weight of equipment to be mounted will be needed. Specify racks with enough depth to accommodate the deepest piece of equipment that will be installed. At the same time, allow ventilating air to flow freely past and through the equipment. Allow additional clearance at the rear of the equipment

---

[1]Registered trademark

FIGURE 6.3.11 A selection of stock equipment enclosures designed for specialized installations. (*Courtesy Middle Atlantic Products.*)

FIGURE 6.3.12 Video production/control center. (*Courtesy Middle Atlantic Products.*)

chassis for connectors, and allow enough space for the minimum bend radius of the largest cable. Additional depth may also be required for cable bundles that must pass behind a deep piece of equipment.

Select a rack model that has sufficient strength to support the full array of anticipated equipment. Also, allow a margin of error for future modification and expansion.

Select paint and laminate colors and textures for the rack assemblies and hardware. This information should be included in the specifications for the racks.

FIGURE 6.3.13 Rack wire management and cable termination. (*Courtesy: WISH-TV.*)

6.3.4 Equipment Rack Layout

When specifying the location of equipment within racks and consoles, give careful consideration to the following factors:

- Physical equipment size and weight
- Power consumption
- Ventilation needs
- Mechanical noise

Human factors also must be considered. Equipment placement should be governed by the operational use of the equipment. Human factors that need consideration include:

- Accessibility to controls
- Height with respect to the operating position
- Line of sight to controls, meters, and display devices, from the operator's point of view
- Reflections on display devices from room lighting or windows
- Noise generated by the equipment

Do not completely fill a given rack with equipment. From a practical point of view, leaving blank spaces will allow for future equipment expansion and replacement.

Provide storage spaces in the racks, if required. Rack-mountable shelves and drawers are available in different sizes for this purpose.

Avoid cable clutter by providing easy access to wiring and connections. This will make installation, maintenance, and modifications easier throughout the life of the system.

Place the tops of jack fields at or just below eye level. Jack field labels must be readable. The average eye level of males is 65.4 inches (1660 mm) and females 61.5 inches (1560 mm). Lining up the tops of jack fields that are mounted horizontally across several racks will create a neat appearance. If room is available, place blank panels between patch panels to space them vertically and to allow room for access from the front and rear. Keep the field as confined as possible to allow the use of the shortest possible patch cords.

Provide a pair of rear vertical mounting angles for supporting heavy or deep equipment. Mount heavy equipment in the lower part of the rack to facilitate easier installation and replacement. One exception might be a piece of equipment that generates excessive heat. Mounting it at the top of the rack will allow the heat to escape by convection, without heating other equipment (power supplies are a good example).

6.3.4.1 *Cooling Considerations.* It is a normal practice to cool the room in which technical equipment is installed. At the same time, comfort of the personnel in the room must be ensured and usually takes precedence over the comfort of the equipment. Additional steps should be taken to control heat build-up and hot spots within equipment racks and consoles. Use all possible heat-removing techniques within the racks before installing fans for that purpose. Fans cost money, consume power, take up space, are noisy, and will eventually fail. Dust drawn through the fan will collect on something. If that something is a filter, it must be cleaned or replaced periodically. If the dust collects on equipment, overheating may occur. Some steps that can be taken by the system engineer in the design phase include the following:

- Limit the density of heat-producing equipment installed in each rack.
- Leave adequate space for the free movement of air around the equipment. This will help the normal convection flow of air upward as it is heated by the equipment.
- Specify perforated or louvered blank panels above or below heat-producing equipment. A perforated or louvered rear door may also be installed to improve air flow into and out of the rack.
- When alternative equivalent products are available, select equipment that generates the least amount of heat. This will usually result from lower power consumption—a desirable feature.

- If a choice exists among equivalent units, select the one that does not require a built-in fan. Units without fans may be of a low power consumption design, which implies (but does not guarantee) good engineering practice.
- Balance heat loads by placing high heat-producing equipment in another rack to eliminate hot spots.
- Place equipment in a separate air-conditioned equipment room to reduce the heat load in occupied control rooms.
- Remove the outer cabinets of equipment or modify mounting shelves and chassis to improve air flow through the equipment. Consult the original equipment manufacturer, however, before operating a piece of hardware with the cover removed. The cover is often used to channel cooling air throughout the instrument, or to provide necessary electrical shielding.
- When specifying new equipment designs, describe the environment in which the equipment will be required to operate. Stipulate the maximum temperature that can be tolerated.
- If a forced-air design is necessary, pressurize each rack with filtered cooling air, which is brought in at the bottom of the rack and allowed to flow out only at the top of the rack.
- When forced-air cooling is used, provide a means of adjusting the air flow into each rack to balance the volume of air moving through the enclosures. This will control the amount of cooling and concentrate it in the racks where it is needed most. Adjust the air flow to the minimum required to properly cool the equipment. This will minimize the wind noise produced by air being forced through openings in the equipment.
- Install air directors, baffles, or vanes to direct the air flow within the rack. This strategy works for controlling convection and forced-air flow.
- Provide a minimum of 3 feet (1 m) clearance at the rear of equipment racks. Besides enabling the enclosure door to swing fully open, this will facilitate efficient cooling and easy equipment installation and maintenance.

If required in a given installation, cooling fans and devices are available for equipment racks. Common types are shown in Fig. 6.3.14.

6.4 RACK GROUNDING

Equipment racks and peripheral hardware must be properly grounded for reliable operation. Single-point grounding is the basis of any properly designed technical system ground network. Noise and spurious signals should have only one path to the facility ground. Single-point grounds can be described as *star* systems, whereby radial elements circle out from a central hub. The equipment grounds are connected to a *main ground point*, which is then tied to the facility ground system. Multiple ground systems of this type can be cascaded as needed to form a *star-of-stars*. The object is to ensure that each piece of equipment has one ground reference.

The design of a ground system must be considered as an integrated package. Proper procedures must be used at all points in the system. It takes only one improperly connected piece of equipment to upset an otherwise perfect ground system. The problems generated by a single grounding error can vary from trivial to significant, depending on where in the system the error exists and whether high RF fields are present. This consideration naturally leads to the concept of ground-system maintenance for a facility. Any time new equipment is installed or old equipment is removed from service, give careful attention to the possible effects that such work will have on the ground system.

6.4.1 Grounding Equipment Racks

The installation and wiring of equipment racks must be planned carefully to avoid problems during day-to-day operations. Consult the rack manufacturer for recommendations with regard to grounding and ac power distribution.

FIGURE 6.3.14 Rack accessories used for cooling equipment enclosures. (*Courtesy Middle Atlantic Products.*)

Figure 6.3.15 shows one possible approach. Bond adjacent racks together with 3/8 to 1/2 inch-diameter bolts. Clean the contacting surfaces by sanding down to bare metal. Use lock washers on both ends of the bolts. Bond racks together using at least six bolts per side (three bolts for each vertical rail). Run a ground strap from the *main facility ground point* and bond the strap to the base of each rack. Secure the strap at the same location for each rack used. A vertical ground bus may be installed in the rack if needed. In addition, it may be helpful to mount a vertical ac strip inside each rack to power the equipment. (See Fig. 6.3.16.) Install an ac receptacle box at the bottom of each rack, or some other form of power distribution (a number of commercial options are available).

Power equipment using standard three-prong grounding ac plugs. Do not defeat the safety ground connection. Equipment manufacturers use this ground to drain transient energy. Defeating the green wire ground violates building codes and is dangerous.

Mount equipment in the rack using normal metal mounting screws. If the location is in a high-RF field, it may be appropriate to clean the rack rails and equipment panel connection points to ensure a good electrical bond. In a high-RF field, detection of RF energy can occur at the junctions between equipment chassis and the rack.

6.5 COMPUTER FLOORS

Many large technical centers are built on raised "computer floors." Such floors are convenient because they provide a means to route wiring between various racks, and they provide room for air conditioning ducts that deliver fresh air to the base of each rack. A computer modular floor is also a form of ground plane. Grid patterns on 2 feet centers are common. The basic open grid electrically functions as a continuous ground plane for frequencies below approximately 20 MHz. To be effective, the floor junctions must be bonded together. The mating pieces should be plated to prevent corrosion, oxidation, and electrolytic (galvanic) action.

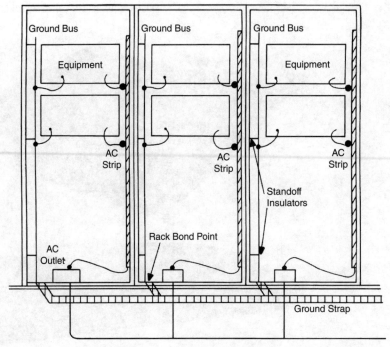

FIGURE 6.3.15 One approach for grounding equipment racks. To make assembly of multiple racks easier, position the ground connections and ac receptacles at the same location in all racks.

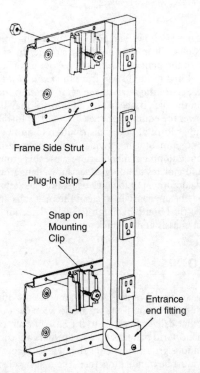

FIGURE 6.3.16 Detail of ac line power strip rack attachment.

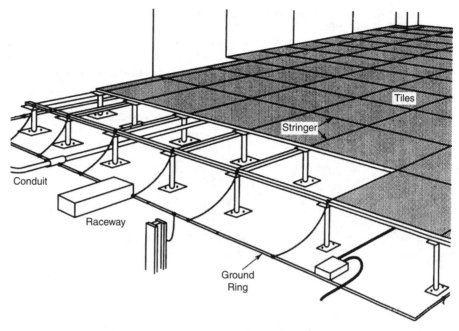

FIGURE 6.3.17 Grounding system for a raised computer floor.

Floor tiles are typically backed with metal to meet fire safety requirements. Some tiles are of all-metal construction. The tiles, combined with the grounded grid structure, provide for effective *electrostatic discharge* (ESD) protection in equipment rooms.

Figure 6.3.17 illustrates the interconnection of a grounded raised computer floor. Note that the grid structure is connected along each side of the room to a ground ring. The ground ring, in turn, is bonded to the room's main ground conductor. Note also that all cabling enters and leaves the facility in one area, along one wall, forming a bulkhead panel for the room.

REFERENCE

[1] Federal Information Processing Standards Publication No. 94, *Guideline on Electrical Power for ADP Installations*, U.S. Department of Commerce, National Bureau of Standards, Washington, D.C., 1983.

BIBLIOGRAPHY

Benson, K. B., and J. Whitaker, *Television and Audio Handbook for Engineers and Technicians*, McGraw-Hill, New York, N.Y., 1989.

Block, R., "How to Ground Guy Anchors and Install Bulkhead Panels," *Mobile Radio Technology*, Intertec Publishing, Overland Park, Kan., February 1986.

Fardo, S., and D. Patrick, *Electrical Power Systems Technology*, Prentice-Hall, Englewood Cliffs, N.J., 1985.

Hill, M., "Computer Power Protection," *Broadcast Engineering*, Intertec Publishing, Overland Park, Kan., April 1987.

Lanphere, J., "Establishing a Clean Ground," *Sound & Video Contractor*, Intertec Publishing, Overland Park, Kan., August 1987.

Lawrie, R., *Electrical Systems for Computer Installations*, McGraw-Hill, New York, N.Y., 1988.

Morrison, R., and W. Lewis, *Grounding and Shielding in Facilities*, John Wiley & Sons, New York, N.Y., 1990.

Mullinack, Howard G., "Grounding for Safety and Performance," *Broadcast Engineering*, Intertec Publishing, Overland Park, Kan., October 1986.

Whitaker, J. C., *AC Power Systems*, 3rd ed., CRC Press, Boca Raton, FL, 2007.

Whitaker, J. C., *Maintaining Electronic Systems*, CRC Press, Boca Raton, FL, 1992.

CHAPTER 6.4
BROADCAST SYSTEMS COOLING AND ENVIRONMENTAL MANAGEMENT

Fred Baumgartner
KMGH TV, Scripps, Denver, CO/Nautel Television

6.1 INTRODUCTION

For many broadcasters, electrical power for equipment is a significant expense. The unfortunate part is that most of this power becomes waste heat that costs even more to remove. It is a special challenge to broadcast engineers to minimize this waste or put it to good use. The cooling for studios, transmitters, and technical space is one of the critical systems that are part of the broad knowledge a broadcast engineer must have. Your HVAC (heating, ventilation, and air conditioning) partner probably does not have many clients whose business is so completely dependent on their HVAC and specialized cooling systems. This is important as the highest value a broadcast engineer can bring to this equation is to work with the HVAC vendors to meet the unique and unusual needs of a broadcast facility.

Managing waste heat is an essential part of the design at every level of electronics. Even in the smallest integrated circuits, a thermal conductive path to removed heat has to be provided. Heat is by far the most significant ordinary stress that limits the life expectancy of equipment. Hot electrolytic capacitors "dry out" sooner, solid state junctions fail more rapidly when hot, and even passive components like resistors and connectors experience shorter lives under environmental stress. Power devices—all varieties of electron tubes and transistors—generally will fail catastrophically and very quickly when subjected to a cooling system failure. All reasonably well-designed high power devices will shut down when cooling fails, but they may not respond fast enough when there is a catastrophic cooling failure or a failure of the sensing system. Disabling overtemperature safety systems or waiting to replace a bad sensor is often expedient, but frequently a very expensive misjudgment.

6.2 COOLING SYSTEM DESIGN CONSIDERATIONS

Cooling technology is advancing at all levels of scale. Solid state coolers, convective tubes, solar powered HVAC, and opportunities that are unique to a location and climate may all play into a broadcast facilities heating and cooling systems.

6.2.1 Studios

Studios present a unique design challenge. Not only must the HVAC system be very quiet, but the duct work should not carry sound from other rooms into the studio. *Low velocity—high volume* HVAC systems reduce the noise from diffuser turbulence, and lower fan speeds are quieter. Low velocities are also needed to avoid having set pieces like plants from blowing around. Isolating the HVAC system to a single studio prevents sharing sound from other rooms through the ducts, minimizes the number of penetrations through otherwise sound proof walls, and inherently results in zone control; but special care is needed to prevent sound from being generated in or entering via the outdoor unit. Long circuitous flexible duct with integral sound absorbing material is a key component of most designs, especially since there are no sharp corners (as typical of rigid ducts) to make noise. Special sound dampening chambers and diffusers are also used. Comfort and building codes require a certain percentage of outside fresh air, so it is inevitable that some ducting to the outside is needed. All of these costs more, requires more maintenance, and takes more space than a typical HVAC system. It is also crucial that the people working on the system understand its unique purpose and requirements. Unfortunately, the easiest way to improve performance is often to remove or disable specialized sound proofing, flow modifiers, and most dangerously special controls needed for the survival of delicate high power equipment.

Legacy studios that were built with the capacity to cool inefficient incandescent lamps with a much higher light level than is now common can be improved dramatically by reducing the capacity and thus the noise of a legacy system. With each rebuild cycle, some attention should be paid to downsizing to the new right size as the opportunity presents itself.

6.2.2 Equipment Rooms

The expertise to build data centers is a different, specialized, and a concerted subset of HVAC technology. Broadcast facility equipment rooms can reach the point where that level of expertise is needed. As a practical matter, most single station level facilities can be serviced by a reasonably effective general HVAC vendor with guidance, and selective use of computer room techniques.

Computer Room Air Conditioning (CRAC, pronounced "crack") units are the norm for data centers and modern broadcast equipment rooms. CRAC units are stand-alone systems with an inside unit and an outside heat exchanger connected with a pair of pipes that carry the cooling fluid in a closed circuit. The general configuration is an $N + 1$ design, where there is one more CRAC unit than is necessary. This allows a unit to be taken off line for repairs or maintenance.

Liquid cooled computer room equipment for broadcast and media that need cold water rather than air is rare at this writing although some multicore-based computers have small internal fluid cooling systems. In major data centers (that are generally located where power is affordable and cooling plentiful) direct liquid cooled equipment is fast becoming the norm because of the physical density it allows. In very dense data centers, dangerously hot air is exhausted by some equipment to the extent that it limits human access to some areas. In time, broadcast engineers may see more liquid cooling in equipment rooms as the computer industry moves in this direction.

Most equipment rooms are set up with some cooling scheme; although many legacy equipment rooms might use ducting and "interesting" approaches to direct cooling air to where it is needed. A common design scheme is *hot row/cold row*. This takes advantage of equipment that takes in cold air from the front and delivers hot air to the rear. Rows alternate, so the cold row sees the fronts of equipment on both sides of the row, and the hot row sees the backs of equipment on either side of the row. One set of rows might have "meat freezer strip" curtains and closed ducting to remove the hot air directly and efficiently. Air that "blows by" without cooling anything is inefficient. Figure 6.4.1 illustrates the concept.

It is common practice to use the space under the computer floor and/or above a suspended ceiling as a plenum. This requires fire resistant plenum wire that should not emit toxic gasses when heated. Furthermore, there may not be enough useful area to handle the capacity of air required. Over time, wire is added and the heat load increases, so initial designs need to be conservative. Plenum systems

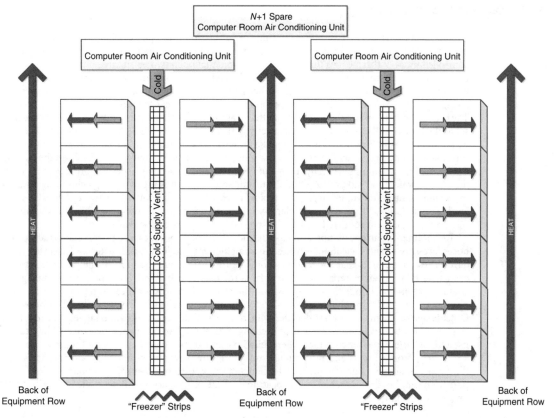

FIGURE 6.4.1 In the hot row/cold row configuration, cold air is forced below the computer floor and vented into racks and the cold row. Racks face into the cold row, where air moves through the rack and is exhausted from the rear. Heat is removed from the back of the racks either by ceiling vents or simply the plenum formed by the racks. Some form of baffling is desirable to keep cold air from directly short circuiting to the hot air intakes from isle to isle and above racks.

present a particular fire risk that needs to be considered. Computer floor needs sufficient openings below racks and usually benefits from integral vents in selected tiles located in the center of aisles, which is the norm for hot row/cold row designs.

While computer centers usually periodically employ specialty cleaners for computer floor space and facilities are routinely completely stripped and rebuilt on a seven to ten year basis, broadcast facilities generally grow and change organically over decades so that computer floor space is often clogged with abandoned wiring, hardware, and dirt. Likewise, a 6-inch floor to tile distance is not enough air space in most reasonable calculations, which usually result around 12 or 18 inches.

In broadcast equipment rooms, there may be no real scheme at all. This is not entirely bad or inappropriate, given the light weight nature of most broadcast gear compared to IT equipment. If the heat load is low, a couple of CRAC units located around the room taking in air from near the ceiling and putting out cold air near the floor (downward or under floor delivery) or vice versa (upward or displacement delivery) at some significant velocity will do what is needed, albeit somewhat inefficiently with significant blow by. Legacy broadcast facilities often lack the space or the opportunity to put in a contemporary computer room cooling system. Uneven cooling might be uncomfortable, but is often tolerable as long as there is enough cooling where it is needed.

6.2.3 Outdoor Heat Exchangers

Outside heat exchangers are composed of thin copper or aluminum pipes welded or soldered to cooling fins. Generally the fluid is a distilled water and glycol solution that will not freeze. Fans, usually controlled by 24 V thermostats that actuate contactors that open two or all three phases of the typical three phase power when the fluid is cool, pull air through the cooling fins. Ideally each fan has an independent control transformer, contactor, and thermostat. Sharing a transformer is common but presents a single point of failure. Thermostats are typically staggered so that they turn on in order as needed. The order should be routinely varied so that the wear on the fans is spread evenly over time as the fans set to the lowest temperature will see more activity than those set to the higher temperatures. The fans will unintentionally suck in debris and clog the air passages over time. Cottonwood trees going to seed can clog a heat exchanger within hours. Leaks generally start slow, but will eventually drain the cooling fluid. Cooling fluid pumps generally suffer from bearing failure. Routine inspections for clogging, leaks, fluid levels, water flow and pressure, and the telltale sounds and heating of bearing failure will preempt most failures; however, failures will occur typically every two to three years for each unit. Ideally, some level of redundancy is desired in particular for transmitters and equipment rooms where the station can stay on the air, even at some reduced level of power, until repairs can be made.

Replacement controls, pumps, and fans are something most broadcast engineers can install. Water hoses, pressure washers, cleaning brushes, fin straighteners, and leak repair materials (these are standard HVAC hand tools that allow you to comb out bent fins) are the tools of the trade. Most heat exchanger failures can be repaired within a day if the parts are on hand. Freezing a heat exchanger is a fatal event and seldom is the damage within the range where it is practical to repair (typically soldering the ruptured tubing). Obtaining and replacing a heat exchanger typically takes days or weeks. Outside heat exchangers can last decades with proper maintenance, and are relatively affordable. It is a good idea to have more outside heat exchanger capacity than one needs, designed to be flexibly useful in a variety of configurations where failed pieces can be isolated and spares brought into use.

Security from vandals and protection from falling tower ice can be serious concerns, as can be the supply of ambient cooling air. Heat exchangers stuck in sunny corners with poor air flow work poorly on very hot days. Ice shields and locating the heat exchangers on different sides of the building improves reliability.

6.2.4 Indoor Heat Exchangers

Some transmitters use distilled water or a specialized (at least mineral) oil that typically serves as an electrical insulator for their cooling circuit. Generally, this fluid in turn transfers its heat to the fluid used in the outside heat exchanger, typically with the addition of a *shell and tube* heat exchanger (one pipe inside another where the inner pipe's heat is transferred through the metal pipe wall to water surrounding it inside the outer pipe) and an additional pump or redundant pair of pumps.

In other installations, building-supplied chilled water might be the destination for the heat. This is typical in tall buildings where transmitters are located on the top floors, and occasionally a building might recover transmitter heat for use in heating office or other space. Distributed Heating and Cooling (DHC) systems can make efficient use of waste heat in HVAC *district networks*.

6.2.5 High Availability

Most engineering approaches *high availability* design for critical systems with some level of redundancy. Fluid pumps can be run in parallel separated by check valves that prevent water from being pushed backward through a failed or idle pump. Some designs have a series of valves that allow bringing on line or isolating individual pumps and heat exchangers. Advanced systems use electronically controlled valves available via remote control or under a *monitoring and control* (M&C) system.

Almost all transmitter facilities have dual air conditioning systems, each one capable of adequate cooling if operated alone. Most operators routinely or automatically rotate which unit leads and which unit lags. Starting two HVAC units at the same time puts a lot of unnecessary stress on electrical systems, so ideally one stabilizes before the other starts. This can be a manual process where the two isolated thermostats settings are alternated during each routine visit to the site. Typically the leading unit comes on at 72°F and off at 68°F, while the lagging unit comes on at 74°F and off at 70°F. A third thermostat might trigger an alarm at 78°F. Key to reliability is that each system should be as isolated and independent as possible; different controls, power sources as far apart (different transformers and entries, if practical), located on opposite sides of the building (one side might pile up with snow while the other blows clear, ice from the tower might take out two units located side by side, etc.).

The third level of redundancy might be a fan and damper system that activates when all else fails. This can bring in all sorts of dirt and insects, and on hot days has limited utility. Typically these are set for around 80°F (after the alarm is sent) and ideally have a differential thermostat that stops the fans from operating when the outside temperature is warmer than the inside temperature. If it is 110°F outside, it makes little sense to activate the tertiary fans until the room temperature is at least as hot, though there is some benefit in circulation. In unmanned sites these can be a single fan and damper, or as elaborate as an intake, circulation, and exhaust fan with filters. A conservative rule of thumb is to have 1000 cubic feet/min of fan capacity for each 1000 W of heat load, at 1000 feet above sea level, for 110°F days with 125°F equipment maximum operating temperature.

Some, in particular large HVAC systems, do not recover quickly from power bumps, nor do they come on line quickly. A "screw" chiller may take 20-min to come on line. The tertiary fans may be necessary to bridge this time. A useful test is to chart how quickly a room heats up without HVAC. A tight transmission room might heat up by as much as 2°F/min. It is not uncommon for a site to have on the order of 20-min of operational life after an HVAC failure. Knowing that number is important to any recovery plans, and to determine how much additional cooling might be needed to meet operational expectations.

A variation of the tertiary fan system is to set up an air cooled transmitter that is normally vented into a closed room to vent outside and open dampers to bring in filtered fresh air when all else fails. Some air cooled transmitter sites operate in this mode constantly, with or without temperature controlled dampers. It is simple, inexpensive, but the total cost of ownership (TCO) might be high as it is difficult to keep clean and maintained, and the thermal swings often dramatically reduce the life expectancy of equipment. It is also often uncomfortable and always dirty, which also reduces equipment life.

The final level of HVAC security is to keep enough box fans and the like (livestock fans for large installations work well) that can be placed in doors and any other opening that can be made to the outside world and to keep air circulating inside. A simple stock of fans, small leak fixes, a fresh change of water and spare pumps, etc., will keep you on the air until a permanent fix can be made ... if you can get to the site in a timely manner. More than one fan has sat on a chair pointed at a flakey exciter or other overheating piece of equipment.

Heat exchangers must simply be segmented and redundant in such a way that a bad or leaking unit can be worked around. In some cases, water from a ground water pump, building, or city water supply can be used for cooling in an emergency, if permitted.

Air conditioners with insufficient charge or in very cold outside air, might freeze water that collects on the cooling coils. *Freeze-stats* detect this condition and typically employ one of several techniques to thaw the coils. For high availability (HA) designs, equipment that will operate in temperature and humidity extremes without freezing, that have fast recovery times, and avoid a common point of failure are critical. Likewise freeze-stats that can be reset remotely and even bypassed upon observed failure are of great value.

6.2.6 Site Heat

There are times when transmitters are off-line either intentionally or because of failure. A similar situation might exist at a studio or equipment room. Cooling fluids (in particular spare water), some maintenance materials (cleaning supplies, paint), and any domestic water (toilets, sinks) will freeze if the space becomes too cold. Some equipment will not function properly if too cold. For this reason, some source of heat is needed at many transmitter sites, albeit rarely. Electric heat is cheap to install and nearly maintenance free, so it is almost always available even after years of sitting idle. Unfortunately, if the transmitter is down because the power and maybe even the backup power have failed, the space might freeze. If the transmitter waste heat is used to heat offices or studios, the problem is extended.

Because these events can cascade . . . the power goes off, the spare water and interior distilled water loops freeze, then the power comes back and the water melts and floods critical components . . . a secondary heat source might be called for. The simplest is a domestic gas (usually LPG, liquid propane gas) space heating "stove." A simple (non-battery) thermostat and an integral millivolt pilot light controlled main gas valve, with a carefully placed exhaust flue (that will not blow out the pilot in a wind storm), can keep the site from freezing for days with just a couple of portable, easy to refill, LPG tanks kept in a safe place (outside). You will want enough LPG tank space so that on the coldest day there is still enough LPG evaporated to keep the stove running for a reasonable period of time. Propane gas is heavier than air, so care has to be taken to prevent gas from collecting from a small leak in low places like basements and sump pits. Natural gas is lighter than air so has the opposite issue, collecting in high places that are not ventilated. Ideally a cage prevents the stove from inadvertently collecting manuals, rags and keeping the stove safely away from any other flammable material.

6.3 TRANSMITTERS

Transmitters are either liquid or air cooled. Liquid cooling is by far the most efficient medium available and higher power devices may be impractical to cool in any other manner. Baring leaks, it is also very clean as the cooling fluid can be routed outside to heat exchangers either directly, or in the case of oil or distilled water (both used where insulation of high voltages is required) an intermediate heat exchanger is used to keep the outside loop separate from the inside loop. The outside loop needs to be freeze-proof, so water and glycol is usually the fluid of choice. The inside loop should not be exposed to freezing temperatures, but likely needs to be cleaner and as mentioned, even nonconductive. There is a catastrophic failure mode where the outside water circulates too fast (or continues to circulate when the transmitter is off) on cold days and freezes the inside water that needs to be addressed. Maintaining the nonconductive condition of fluids that necessitate this requires periodic maintenance. The process and chemistry is always designed by the transmitter manufacturer.

RF dummy loads can be water or air cooled. The higher frequency, higher power levels generally require a well-designed water cooled load composed of a glass tube with deposited resistive material covering the element. Dummy loads have a relatively short life expectancy if water is circulated constantly as the resistive material is abraded away by moving water. Eventually, the resistance increases and the power capacity drops. When the mismatch becomes obvious, the element needs to be replaced. Air cooled dummy loads tend to collect dust if the fan runs constantly. In the worst case, the dust burns when powered up. Water cooled dummy loads also allow direct and very accurate power measurements, either by using a distilled water source of known volume heated by the transmitter output over a specific time interval, or the more complicated calculations where the flow rate, DT (temperature change between input and output), and specific gravity of the cooling fluid is known. Each transmitter manufacturer has a preferred power measurement procedure.

Water systems will eventually leak. Water sensors are useful if not essential. A drainage system and cleanup materials (at least a bucket, mop, wet-dry vacuum, and small utility pumps and hoses) are needed. Some components are designed to be easily removed and replaced in liquid cooled systems without drips and leaks. An amplifier drawer, one of several solid state amplifier drawers in a given transmitter, can typically be removed while the rest of the transmitter is operating. Drip proof connectors seal the cooling system when the component is removed. Typically, each amplifier drawer has a set of individual valves and switches that allow the drawer to be isolated. Typically, the cooling system valves are turned off before removing the drawer to limit leaking should a drip-free connector not seal properly and the power to the amplifier is shut down. Most manufacturers ship a set of hoses with connectors to allow the drawer to be drained before shipping, and typically drawers are repaired by returning them to the manufacturer. The cooling, power, RF, and control connectors all have to align very closely for the drawer to be easily removed and replaced. There also has to be some process and protections to prevent an amplifier drawer from operating when there is not sufficient cooling.

Water systems are also subject to electrolysis. Dissimilar metals will set up a situation where metal migrates within the water. Impellers, pipes, valves, and transmitter parts can react with each other. The transmitter manufacturer will specify what materials the cooling system can be made of and may even specify a treatment or cathodic protection system that may include a sacrificial anode. Failing to address this issue in water systems can quickly lead to the failure of fittings and components. An experienced HVAC partner is especially important if you have exposure to electrolysis.

Dirt in the cooling fluid is abrasive to cooling system components. Very fine filters are used to screen out all but the smallest particles. It does not take too many particles to clog such a fine filter, and that results in decreased flow and increased back pressure. Components that are deteriorating in the cooling system are usually seen first as dirt in the filters. Cooling systems that use distilled water or oil as an insulating material often see leakage current (or "body" current) go up as the deteriorating component contaminates the fluid.

Closed liquid cooling systems typically use a pressure tank not unlike the ones popular for home pumped water systems to deal with expansion of the warm fluid. Most pressure tanks have a bladder and specify nitrogen as the gas that is compressed by expansion in the tank. A common Schrader valve might invite you to use a bicycle pump to refill the tank, and some manufacturers allow this. Others require replacement of the pressure tank when the gas bleeds out. A healthy tank bleeds out rarely, maybe every two to ten years. Each manufacturer has a procedure to add fluid to the system and to bleed out any air bubbles that inadvertently wind up in the system (which often use special automatic bleeder valves). There are valves to isolate, bleed, and pump in fluids as required. All of this is easy for anyone familiar with electrical circuits to understand intuitively, but there are subtleties to any system that only reading the manual and experience will help with. The fluid replacement schedule for closed systems is often over several years.

Open fluid systems usually have a reservoir tank located at the high point of the system where replacement fluid can be poured in and expansion simply pushes fluid into that reservoir. Being open to the air does create degradation and contamination of the fluid over time, so open systems typically call for flush and refills on a more frequent, often yearly basis.

Proper disposal of cooling fluid varies with location and chemical makeup. Mineral oils are usually fairly benign and disposal might be simple. Glycol can be disposed of at most auto garages; however, the one or two year old, almost immaculately clean solution, in hundred gallon quantities, can be reused in tractors and other machinery. You can usually give this away to good use. Truly toxic cooling fluids (such as PCBs) are now rarely used if at all. Again the transmitter manufacturer will specify disposal and replacement. Glycol formulations vary, and while you can usually buy what you need at a local store, it is important to get a formulation that fits the vendor's recommendations and not mix formulations. It is a good practice to stick with and stock up on a brand that meets the requirements. You should always have at least one water change worth of water and glycol on hand. Water can be delivered in five gallon water-cooler bottles. Glycol might take several shopping trips. Typically, you use something like 50-gallons of distilled water to flush a system and 25-gallons of distilled water and another 25-gallons of glycol to refill such a system. The empty water bottles can be used to bring the used glycol solution to recycle, but you have to have a dozen or so empties on hand to start that process.

Operating pressures and flow speeds vary widely with the transmitter manufacturer specifying the optimum parameters for any given configuration, climate and altitude. Seldom can one improve on the specifications, and in most cases, operating out of specifications will void warranty provisions and reduce mean time between failures (MTBF).

Air cooled systems leaks are not messy, and systems can be very simple to maintain and reliable. Air cooling is useful for lower power operation like radio and low power TV. A closed air system dumps transmitter heat into the transmitter room where it is cyclically removed by the HVAC system. This is clean and simple and presents a minimum of thermal cycling stress on equipment. In a typical AM or FM radio and VHF TV installation, the efficiencies of the transmitters are fairly high and thus the waste heat is often on the order of a few kilowatts or so . . . no different than running a hair dryer or two or three in an air conditioned room. In most cases, the cost of power is such that the best total cost of ownership occurs with a closed air system.

Still, venting a transmitter to the outside is clearly the most efficient configuration; except that the additional thermal stress, dirt, filters, and insect expense often cost more than the power saving. More complex systems that vent a transmitter to satellite antennas and back (this requires a sealed vestibule at the dish and significant ductwork and generally additional fan capacity) are very efficient and clean. Open systems see the most stress at ambient temperature extremes. It is uncomfortable and stressful to bring subfreezing or very hot air into the transmitter room. Humid or polluted air also corrodes components faster.

Smart systems that take advantage of adequately filtered outside air when the temperature and humidity are favorable but revert to an essentially closed system when it is too hot, cold, or humid, are complex and more costly at installation, but in many cases more than pay for themselves over a period of time. Using the waste heat effectively can make this a very "green" installation.

Air cooled transmitter fans are sized for a certain amount of duct resistance, often assuming almost no resistance as would be the case if vented directly into a closed room. If long, twisting, or constricted duct runs are needed, or higher altitudes and filters are involved; the fan size needs to be increased or additional fan capacity added. Most air cooled transmitters see the fan as a single point of failure, and they use pressure sensors and or differential thermostats to determine when to back off power or shut down for a cooling failure. Overheated tubes have short lives and usually destroy their sockets, in particular the finger stock that contacts the tube.

Convection air cooling, where no fans are used, works for some small transmitters, translators, and repeaters. Simply allowing convection by not blocking or crowding the cooling heat sinks is adequate. In exceptionally hot environments, convection cooled equipment needs to back off power or simply go off line to prevent damage. Over time, dust buildup on the heat sinks needs to be cleaned off to keep cooling efficiently.

Vapor phase cooling for transmitters is rare these days, reserved only for the highest of power levels at lower altitudes. Basically these are international shortwave stations approaching 10 MW, or older transmitters. The concept is that the state change from water to steam carries significantly more heat that circulating water. The complexity is that the steam needs to be recovered, condensed, and cooled for reuse, and that whole loop needs to be kept clean to avoid buildup of minerals on hot surfaces. These are exceptionally dangerous and maintenance-intensive systems and thus unpopular.

6.4 SATELLITE ANTENNAS

Satellite antennas lose efficiency when covered with snow or ice. Smaller dishes in some climates can be covered by a membrane, and at least one manufacturer vibrates this *radome* to keep ice from sticking. Electrical heaters and gas-fired heaters for large antennas can be used under the control of an icing sensor. Icing sensors discern the combination of water (usually detected by a warmed water sensor—a printed circuit board with traces that interleave in such a way that they conduct a small current when wetted by rain or melting snow) and freezing temperatures. These usually come on at the first sign of icing conditions and stay on for a while after the threat goes away. Most are adjustable. These should be mounted conveniently (there is no real benefit to having the sensor high in the air) so that the sensor can be cleaned and tested periodically (an ice cube from a drink and freeze spray pointed at the thermal sensor). It is surprisingly easy to fool an icing sensor, so some systems are activated when satellite receivers report an increasing EbNo with freezing temperatures, or the sensors used to activate sidewalk melting systems activate. A system that activates too late might be overwhelmed, so much so that some facilities activate the heaters when conditions are forecast.

While typically a do-it-yourself (DIY) project, satellite antennas and concrete that might freeze and block an entry door can be a great place to dispose of waste transmitter heat. Overheating a dish can cause the dish to warp, so constant heating in places where ambient temperatures peak is not good. An indicator, something as simple as an outdoor light that comes on when the heaters are active can save you a big power bill from a stuck sensor, and keeps you in touch with the appropriate on and off cycles. The backup is a long handled broom to clear the antenna, and in some cases, a supply of deicing liquid and sprayers similar to that used for aircraft, that can recover a critical antenna in a severe storm or a system failure.

6.5 OPERATING PARAMETERS

Efficient HVAC system design and operation requires attention to a number of parameters. In some cases, a compromise must be made in one parameter in order to accommodate another parameter.

6.5.1 Altitude

As the air thins with increasing altitude, it becomes less able to carry away heat. Generators, HVAC equipment, transformers, and fans and thus transmitters all must be *derated* for altitude. If a device relies on air to cool it, and that means motors, power conductors, thermal breakers, computers, and even light dimmers, they are generally designed to operate at less than 3300 feet above sea level. While this varies with the design of each device, a ball park derating factor of 15 percent at 5000 feet and 25 percent at 10,000 feet is useful as a sanity check. Good engineering practice is generally to operate devices at a maximum of 80 percent of capacity. At 10,000 feet, following this rule of thumb, we'd be operating devices at 55 percent of sea level rated capacity. Instead, we might replace standard fans with higher capacity fans in transmitters and servers, and operate building space at lower temperatures, which in turn requires larger cooling units. High voltage arcs initiate easier at altitude, so additional insulation (with even poorer cooling characteristics) and spacing might be required. The National Electrical Code (NEC) does not address the topic directly with guidelines, but power wire, thermal breakers, and other devices rely on ambient air to remove heat, so the NEC does note that wiring at altitude requires additional capacity.

Working above 10,000 feet presents its own dangers and limitations. Transmitter manufacturers all make provisions for operation at altitude, even if that is to say the transmitter should not be used above a given altitude.

6.5.2 ΔT

A closed cooling circuit's efficiency can be easily monitored by comparing the input and output temperatures. The difference in temperature between the cool and warm water remains nearly constant despite the ambient temperature. As the efficiency of the system drops, the ΔT increases. Many monitoring and control systems allow this to be calculated and displayed and alarmed. It is a far more valuable metric than input or output temperature.

In a *constant flow* system, one with fixed speed pumps and fans, the ΔT remains reasonably constant while the overall temperature of the system tracks the ambient temperature of the heat exchanger. If the efficiency of the system is controlled by activating more or fewer heat exchanger fans as is usually the case, the ΔT will lightly track the number of active fans.

Variable flow systems use pumps and fans whose speed can be controlled to maintain a more constant cooling fluid temperature and save power by running pumps and fans at lower speeds when ambient temperatures are low. The ΔT remains reasonably stable over a wide range of *normal* ambient temperatures, but the system can smartly avoid pushing subfreezing fluids into a transmitter or crank up the flow and fans when it is hot and additional efficiency is needed. Monitoring the ΔT over time or a more sophisticated means of measuring cooling system efficiency by factoring in flow variations can tell you when maintenance is required.

6.5.3 Power

Technical sites usually have tiers of power starting with mains power, then generator-backed power, and uninterruptable (UPS) battery power, which might be small to keep monitoring and control equipment up and running so as not to lose site visibility or large enough to carry some or all of the transmitter capacity while the generator starts. HVAC gear generally operates at the generator level, and may represent the largest electrical load for the generator. In some cases, the generator is not sized to carry the full site load, so backing off to a smaller auxiliary transmitter and smaller HVAC and cooling units requires some sophisticated control and/or procedures.

6.5.4 Size

A ton of air conditioning is roughly the amount of heat 2000 pounds of ice (a short ton) absorbs in a 24-hour day. Practically, a ton of air conditioning will remove 12,000 BTUs per hour, or the heat generated by 3500 W of electricity. Domestic air conditioners and smaller transmitter sites run from 1 to 5 tons, and require from 3 to 20 kW. Larger transmitter facilities, large studios, and data centers typically require multiple units in the 10-30 ton range.

Your HVAC partner has sophisticated tools to size equipment. The ideal design for broadcast is a system where one HVAC unit can carry the load on the hottest day at 80 percent capacity and the generator can carry both HVAC units and either the main or auxiliary transmitter.

Figure 6.4.2 illustrates heat transfer principles.

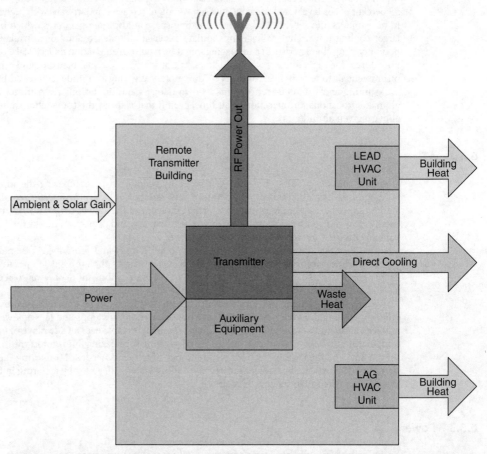

FIGURE 6.4.2 The heat transfer equation for a simple transmitter site is straight forward. The AC power used in the building is transferred directly outside as part of the transmitters cooling system, radiated by the antenna or radiated into the room. All power feeding auxiliary equipment as well as any solar gain and ambient heat that comes into the building from the outside adds to the overall heat load of the building. The building heat is removed by redundant air conditioning units. The amount of cooling has to exceed the heat load of the building under the worst case scenario, which may include operating an auxiliary transmitter as well as the main on a hot day, or with a failed HVAC unit.

6.5.5 Alternatives

Rarely is geothermal an option, but evaporative cooling might be an option where there is water, heat, and dry air. Evaporative cooling should have the humid parts of the system isolated from equipment.

Mountain tops often get snow, but little else in the way of water. At least one site, the Mount Farnsworth complex that serves Salt Lake City, uses waste heat to melt thousands of gallons worth of snow over the winter to supply the facilities water needs for the year. The site is staffed continuously, and has indoor plumbing.

Humidity is often an enemy, so most HVAC equipment will remove excess water from the air. Too little humidity is also a challenge especially at altitude and staffed sites. Arcing is more of a threat in dry air and some materials simply dry out and fail if the humidity is too low.

6.5.6 Noise

HVAC equipment and generators are notorious for causing neighborhood noise complaints. Generators can be *blimped* and heat exchangers can be configured with more but lower speed units and thus quieter fans (budgets and space permitting). This may be governed by an association or local approval process. Rooftops are popular places for HVAC gear for both security from vandals and to keep noise from ground level.

6.5.7 Transmission Lines

Transmission line power-carrying capacity is limited by either flash-over voltages or heat conduction capacity. In air filled lines, the center conductor has to dissipate more heat than the outer conductor (same current, but in a smaller wire), which it radiates to the outer conductor and then into the ambient air. Some designs use a denser gas than air or nitrogen, or operate with higher than the normal 3–5 pounds per square inch of pressure to improve the conduction of heat from the center conductor.

The use of denser or high pressure gas for cooling some RF components like filters is rare, but not unheard of. High pressure and some toxic, dense gases, require special handling.

6.5.8 Tuning

HVAC professionals use a number of tools to visualize heat distribution. IR (infrared) thermometers and Forward Looking Infrared (FLIRs) cameras measure temperature variations so that air and water flow can be tweaked. Adjusting or "tuning" the system to minimize hot and cold spots, balance loads and in particular to deal with areas of concentrated high heat load is part of the process of delivering an efficient and reliable HVAC system. Routinely tuning the system and in the case of equipment, in particular transmitters, documenting the heat distribution in a manner that can be compared at each inspection is a major means of preventing catastrophic failures. Virtually any motor, power distribution, or RF transmission system failure first shows up as a "hot spot," often months or years in advance.

6.5.9 Safety

HVAC equipment usually has rotating machinery, belts, and pulleys, as well as extreme temperatures and toxic materials that can harm or kill. Obviously, obey signage and do not alter safety devices and guards. All HVAC gear has easy to reach power isolation switches that should be locked out or tagged out for maintenance. Do not work on energized equipment as HVAC equipment can come on line unpredictably and dangerously. "Boiler" rooms need to be restricted areas.

Scalding from hot fluids and steam can be a deadly danger. Likewise any fluid under pressure presents a risk of explosion. Heating fuels present a fire hazard. Heating fuels and vandalism at remote sites is a special concern.

There are plenty of opportunities to use mouth pressure to start siphons or blow out amplifier cooling fluids. Avoid getting glycol, dirty water, and other chemicals in your mouth. If you do poison yourself, there are instructions on most glycol and other containers.

6.5.10 Fire

Ideally the HVAC system shuts down in case of fire to reduce the amount of oxygen available to feed the fire. A common worst case scenario is where an otherwise small fire heats the room and the HVAC system calls for more air in an attempt to cool the space. If a tertiary fan system is used, the heat eventually causes the fans to come on, feeding massive amounts of fresh air into the space, creating a blast furnace effect. More than one transmitter facility has been destroyed by what would have been a small fire with limited damage when the tertiary fans came on. Fire control systems should shut down power and HVAC when activated, even if there is no fire suppression system.

6.5.11 HVAC Partners

It rarely makes sense for broadcast engineers to keep the tools or nurture all of the knowledge necessary to do the station's HVAC maintenance beyond a certain point. The broadcast engineer might replace fuses, transformers, thermostats, filters, pumps, and do cleaning and the like. Quarterly to annual maintenance and inspection should be done by a professional, ideally with a long and deep relationship. It is useful to have an HVAC partner that knows what to expect at a transmitter site and what parts, tools, and supplies need to be brought to avoid having to make more than one trip. Likewise, they need keys and access and the trust that they can safely go to the site on their own.

Transmitter and studio facilities have fairly unique HVAC requirements. This relationship works best when information and understanding is heavily exchanged. Like broadcast engineering, there is a lot to know about HVAC, and lots of room for innovation.

6.5.12 Monitoring and Control

CRAC units in particular are SNMP enabled with MIBs that take the measure of hundreds of parameters and provide for dozens of adjustments. Likewise, tools like controllable power strips and remotely resettable breakers allow a lot of visibility into a system's operation. Sites that are unstaffed or have little technical support on site can benefit from HVAC systems designed around systems with advanced monitoring and control features.

Lower level alarming is a near necessity; even if all it is, is an over temperature alarm. It is important to know as early as possible when there is a developing problem, especially at an unstaffed location.

A useful device in some circumstances for independent monitoring if not adjusting the temperature as you arrive at a site is an Internet-ready thermostat. Do keep in mind the security issues this presents.

6.5.13 Stress

Periodically, there will be an extraordinarily hot day where you are asked to run on generator or the power grid sags and your HVAC vendor is overloaded keeping critical care facilities up and running. If you have the capacity and kept up with the maintenance, your station should be there for the community. Maybe the biggest part of broadcast engineering is doing all the things necessary for the critical systems to work well enough under extraordinary stress.

CHAPTER 6.5
FACILITY GROUND SYSTEM

Jerry C. Whitaker, Editor-in-Chief

6.1 INTRODUCTION

The attention given to the design and installation of a facility ground system is a key element in the day-to-day reliability of the plant. A well-designed and -installed ground network is invisible to the engineering staff. A marginal ground system, however, will cause problems on a regular basis. Grounding schemes can range from simple to complex, but any system serves three primary purposes:

- Provides for operator safety.
- Protects electronic equipment from damage caused by transient disturbances.
- Diverts stray radio frequency energy from sensitive audio, video, control, and computer equipment.

Many different approaches can be taken to designing a facility ground system, but the goal is the same: establish a low-resistance, low-inductance path to noise and surge energy. The type of system used is determined by a number of factors, not the least of which is cost. Some ground system choices are determined by the practical realities of the facility. The options for a studio in a high-rise building are different than for a stand-alone structure in a rural area. Some of the more common approaches will be discussed in this chapter.

6.2 THE GROUNDING ELECTRODE

The process of connecting the grounding system to earth is called *earthing*, and consists of immersing a metal electrode or system of electrodes into the earth [1]. The conductor that connects the grounding system to earth is called the *grounding electrode*. The function of the grounding electrode conductor is to keep the entire ground system at earth potential.

The basic measure of effectiveness of an earth electrode system is the earth electrode resistance, which is the resistance, in ohms, between the point of connection and a distant point on the earth called "remote earth." Remote earth, about 25 feet from the driven electrode, is the point where earth electrode resistance does not increase appreciably when this distance is increased. Earth electrode resistance consists of the sum of the resistance of the metal electrode (negligible) plus the contact resistance between the electrode and the soil (negligible) plus the soil resistance itself. Thus, for all practical purposes, earth electrode resistance equals the soil resistance. The soil resistance is nonlinear, with most of the earth resistance contained within several feet of the electrode. Furthermore, current flows only through the electrolyte portion of the soil, not the soil itself. Thus, soil resistance varies as the electrolyte content (moisture and salts) of the soil varies. Without electrolyte, soil resistance would be infinite.

Earth electrodes may be *made electrodes, natural electrodes,* or special-purpose electrodes. Made electrodes include driven rods, buried conductors, ground mats, buried plates, and ground rings. The electrode selected is a function of the type of soil and the available depth. Driven electrodes are used where bedrock is 10 feet or more below the surface. Mats or buried conductors are used for lesser depths. Buried plates are not widely used because of the higher cost when compared to rods. Ground rings employ equally spaced driven electrodes interconnected with buried conductors. Ground rings are used around large buildings and in areas having high soil resistivity. Natural electrodes include buried water pipe electrodes and concrete-encased electrodes.

The grounding electrode is the primary element of any ground system. The electrode can take many forms. In all cases, its purpose is to interface the electrode (a conductor) with the earth (a semiconductor).

6.2.1 Grounding Interface

The grounding electrode (or ground rod) interacts with the earth to create a hemisphere-shaped volume, as illustrated in Fig. 6.5.1. The size of this volume is related to the size of the grounding electrode. The length of the electrode has a much greater effect than the diameter. Studies have demonstrated that the earth-to-electrode resistance from a driven ground rod increases exponentially with the distance from that rod. At a given point, the change becomes insignificant. It has been found that for maximum effectiveness of the earth-to-electrode interface, each ground rod requires a hemisphere-shaped volume with a diameter that is approximately 2.2 times the rod length [2].

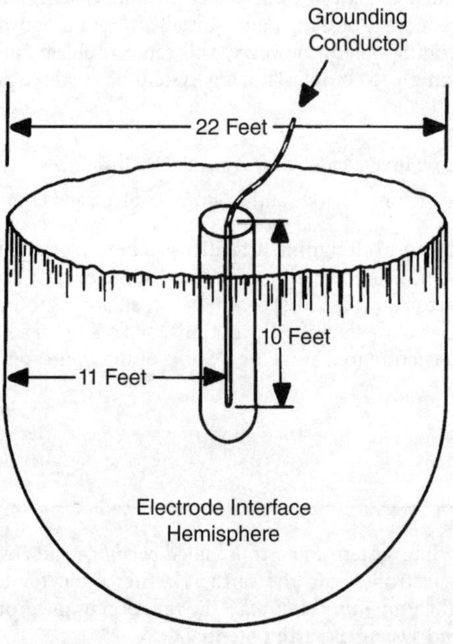

The constraints of economics and available real estate place practical limitations on the installation of a ground system. It is important, however, to keep the 2.2 rule in mind because it allows the facility design engineer to use the available resources to the best advantage. Figure 6.5.2 illustrates the effects of locating ground rods too close (less than 2.2 times the rod length). An overlap area is created that effectively wastes some of the earth-to-electrode capabilities of the two ground rods. Research has shown, for example, that two 10-feet ground rods driven only 1 feet apart provide about the same resistivity as a single 10-feet rod.

Ground rods come in many sizes and lengths. The more popular sizes are ½, 5/8, ¾, and 1 inch. The ½-inch size is available in steel with stainless-clad, galvanized, or copper-clad rods. All-stainless-steel rods also are available. Ground rods can be purchased in unthreaded or threaded (sectional) lengths. The sectional sizes are typically 9/16- or ½-inch rolled threads. Couplers are made from the same materials as the rods. Couplers can be used to join 8- or 10-feet-length rods together. A 40-feet ground rod, for example, is driven one 10-feet section at a time.

FIGURE 6.5.1 The effective earth-interface hemisphere resulting from a single driven ground rod. The 90 percent effective area of the rod extends to a radius of approximately 1.1 times the length of the rod. (*After* [3].)

The type and size of ground rod used is determined by how many sections are to be connected and how hard or rocky the soil is. Copper-clad 5/8-inch × 10-feet rods are probably the most popular. Copper cladding is designed to prevent rust. The copper is not primarily to provide better conductivity. Although the copper certainly provides a better conductor interface to earth, the steel that it covers is also an excellent conductor when

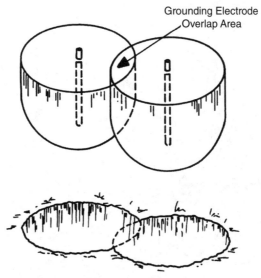

FIGURE 6.5.2 The effect of overlapping earth interface hemispheres by placing two ground rods at a spacing less than 2.2 times the length of either rod. The overlap area represents wasted earth-to-grounding electrode interface capability. (*After* [3].)

compared with ground conductivity. The thickness of the cladding is important only insofar as rust protection is concerned.

Wide variations in soil resistivity can be found within a given geographic area, as documented in Table 6.5.1. The range of values shown results from differences in moisture content, mineral content, and temperature.

Temperature is a major concern in shallow grounding systems because it has a significant effect on soil resistivity [4]. During winter months, the ground system resistance can rise to unacceptable levels because of freezing of liquid water in the soil. The same shallow grounding system can also suffer from high resistance in the summer as moisture is evaporated from soil. It is advisable to determine the natural frost line and moisture profile for an area before attempting design of a ground system.

After the soil resistivity for a site is known, calculations can be made to determine the effectiveness of a variety of ground system configurations. Equations for several driven rod and radial cable configurations are given in [4], which—after the solid resistivity is known—can be used for the purpose of estimating total system resistance. Generally, driven rod systems are appropriate where soil resistivity continues to improve with depth or where temperature extremes indicate seasonal frozen or dry soil conditions. Figure 6.5.3 shows a typical soil resistivity map for the United States.

TABLE 6.5.1 Typical Resistivity of Common Soil Types

Type of Soil	Resistivity in Ω/cm		
	Average	Minimum	Maximum
Filled land, ashes, salt marsh	2400	600	7000
Top soils, loam	4100	340	16,000
Hybrid soils	6000	1000	135,000
Sand and gravel	90,000	60,000	460,000

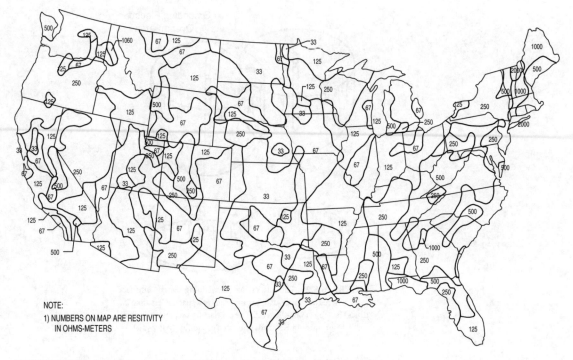

FIGURE 6.5.3 Soil resistivity map for the United States. (*From* [4]. *Used with permission.*)

6.2.2 Chemical Ground Rods

A chemically activated ground system is an alternative to the conventional ground rod. The idea behind the chemical ground rod is to increase the earth-to-electrode interface by conditioning the soil surrounding the rod. Experts have known for many years that the addition of ordinary table salt (NaCl) to soil will reduce the resistivity of the earth-to-ground electrode interface. Salting the area surrounding a ground rod (or group of rods) follows a predictable life-cycle pattern, as illustrated in Fig. 6.5.4. Subsequent salt applications are rarely as effective as the initial salting.

Various approaches have been tried over the years to solve this problem. One such product is shown in Fig. 6.5.5. This chemically activated grounding electrode consists of a 2-½-inch-diameter copper pipe filled with rock salt. Breathing holes are provided on the top of the assembly, and seepage holes are located at the bottom. The theory of operation is simple. Moisture is absorbed from the air (when available) and is then absorbed by the salt. This creates a solution that seeps out of the base of the device and conditions the soil in the immediate vicinity of the rod.

Another approach is shown in Fig. 6.5.6. This device incorporates a number of ports (holes) in the assembly. Moisture from the soil (and rain) is absorbed through the ports. The metallic salts subsequently absorb the moisture, forming a saturated solution that seeps out of the ports and into the earth-to-electrode hemisphere.

Implementations of chemical ground-rod systems vary depending on the application. Figure 6.5.7 illustrates a counterpoise ground consisting of multiple leaching apertures connected in a spoke fashion to a central hub. The system is serviceable in that additional salt compound can be added to the hub at required intervals to maintain the effectiveness of the ground. Figure 6.5.8 shows a counterpoise system made up of individual chemical ground rods interconnected with radial wires buried below the surface.

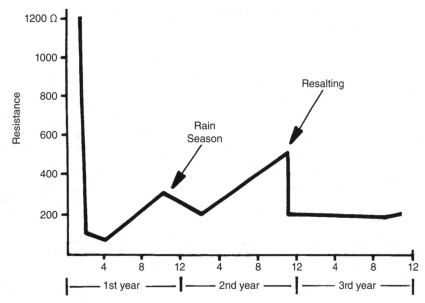

FIGURE 6.5.4 The effect of soil salting on ground-rod resistance with time. The expected resalting period, shown here as 2 years, varies depending on the local soil conditions and the amount of moisture present. (*After* [3].)

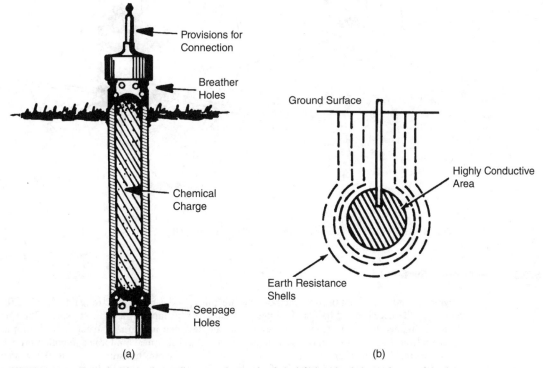

FIGURE 6.5.5 An air-breathing chemically activated ground rod: (*a*, left) breather holes at the top of the device permit moisture penetration into the chemical charge section of the rod; (*b*, right) a salt solution seeps out of the bottom of the unit to form a conductive shell. (*After* [3].)

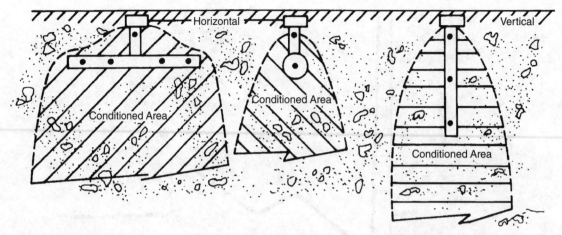

FIGURE 6.5.6 Variations on the chemically activated ground rod. Multiple holes are provided on the ground-rod assembly to increase the effective earth-to-electrode interface. Note that chemical rods can be produced in a variety of configurations. (*After* [3].)

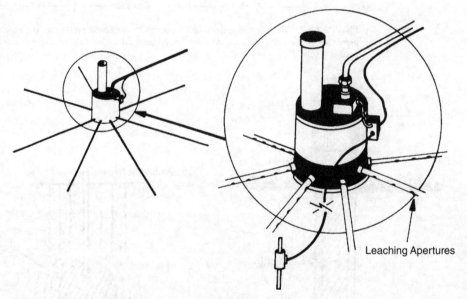

FIGURE 6.5.7 Hub and spoke counterpoise ground system. (*After* [3].)

6.2.3 Ufer Ground System

Driving ground rods is not the only method of achieving a good earth-to-electrode interface [2]. The concept of the Ufer ground has gained interest because of its simplicity and effectiveness. The Ufer approach (named for its developer), however, must be designed into a new structure. It cannot be added on later. The Ufer ground takes advantage of the natural chemical- and water-retention properties of concrete to provide an earth ground. Concrete typically retains moisture for 15 to 30 days after a rain. The material has a ready supply of ions to conduct current because of its moisture-retention properties, mineral content, and inherent pH. The large mass of any concrete foundation provides a good interface to ground.

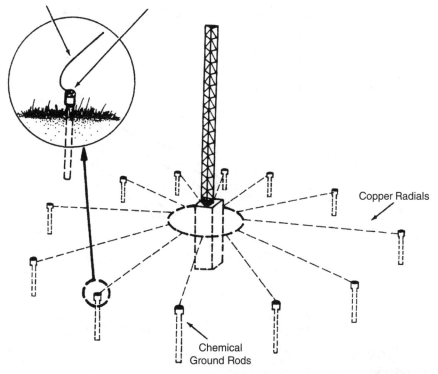

FIGURE 6.5.8 Tower grounding scheme using buried copper radials and chemical ground rods. (*After* [3].)

A Ufer system, in its simplest form, is made by routing a solid-copper wire (no. 4 gauge or larger) within the foundation footing forms before concrete is poured. Figure 6.5.9 shows one such installation. The length of the conductor run within the concrete determines the effectiveness of the system.

As an alternative, steel reinforcement bars (rebar) can be welded together to provide a rigid, conductive structure. A ground lug is provided to tie equipment to the ground system in the foundation. The rebar must be welded, not tied, together. If it is only tied, the resulting poor connections between rods can result in arcing during a current surge. This can lead to deterioration of the concrete in the affected areas.

The design of a Ufer ground is not to be taken lightly. Improper installation can result in a ground system that is subject to problems. The grounding electrodes must be kept a minimum of 3 in. from the bottom and sides of the concrete to avoid the possibility of foundation damage during a large lightning surge. If an electrode is placed too near the edge of the concrete, a surge could turn the water inside the concrete to steam and break the foundation apart.

The construction of the Ufer ground is influenced by the following elements:

- Type of concrete (density, resistivity, and other factors)
- Water content of the concrete
- How much of the buried concrete surface area is in contact with the ground
- Ground resistivity
- Typical ground water content
- Size and length of the ground rod

Before implementing a Ufer ground system, consult a qualified contractor. Because the Ufer ground system will be the primary grounding element for the facility, it must be done correctly.

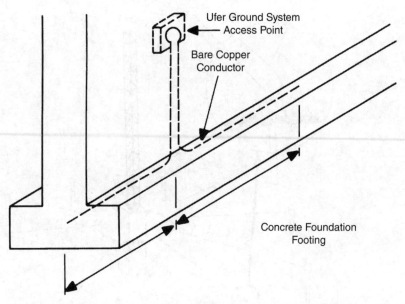

FIGURE 6.5.9 The basic concept of a Ufer ground system, which relies on the moisture-retentive properties of concrete to provide a large earth-to-electrode interface. Design of such a system is critical. Do not attempt to build a Ufer ground without the assistance of an experienced contractor. (*After* [2].)

6.2.4 Bonding Ground-System Elements

A ground system is only as good as the methods used to interconnect the component parts [2]. Do not use soldered-only connections outside the equipment building. Crimped/brazed and exothermic (Cadwelded[1]) connections are preferred. To make a proper bond, all metal surfaces must be cleaned, any finish removed to bare metal, and surface preparation compound applied. Protect all connections from moisture by appropriate means, usually sealing compound and heat-shrink tubing.

It is not uncommon for an untrained installer to use soft solder to connect the elements of a ground system. Such a system will be problematic. Soft-soldered connections cannot stand up to the acid and mechanical stress imposed by the soil. The most common method of connecting the components of a ground system is silver soldering. The process requires the use of brazing equipment, which may be unfamiliar to many facility engineers. The process uses a high-temperature/high-conductivity solder to complete the bonding process. For most grounding systems, however, the best approach to bonding is the exothermic process.

6.2.4.1 Exothermic Bonding Exothermic bonding is the preferred method of connecting the elements of a ground system [2]. Molten copper is used to melt connections together, forming a permanent bond. This process is particularly useful in joining dissimilar metals. The completed connection will not loosen or corrode and will carry as much current as the cable connected to it. Figure 6.5.10 illustrates the bonding that results from the exothermic process.

The bond is accomplished by dumping powdered metals (copper oxide and aluminum) from a container into a graphite crucible and igniting the material by means of a flint lighter. Reduction of the copper oxide by the aluminum produces molten copper and aluminum oxide slag. The molten copper flows over the conductors, bonding them together. A variety of special-purpose molds are available to join different-size cables and copper strap.

---

[1]Cadweld is a registered trademark of Erico Corp.

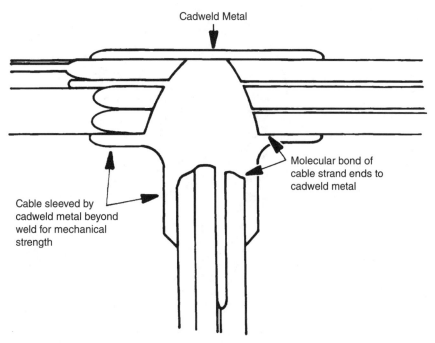

FIGURE 6.5.10 The exothermic bonding process. (*After* [2].)

6.2.5 Ground-System Inductance

Conductors interconnecting sections or components of an earth ground system must be kept as short as possible to be effective [2]. The inductance of a conductor is a major factor in its characteristic impedance to surge energy. Any sharp bends in the conductor will increase its inductance and further decrease the effectiveness of the wire. Bends in ground conductors should be gradual. Because of the fast rise time of most lightning discharges and power-line transients, when planning a facility ground system view the project from an RF standpoint.

The effective resistance offered by a conductor to radio frequencies is considerably higher than the ohmic resistance measured with direct currents. This is because of an action known as the *skin effect*, which causes the currents to be concentrated in certain parts of the conductor and leaves the remainder of the cross section to contribute little toward carrying the applied current.

When a conductor carries an alternating current, a magnetic field is produced that surrounds the wire. This field continually is expanding and contracting as the ac current wave increases from zero to its maximum positive value and back to zero, then through its negative half-cycle. The changing magnetic lines of force cutting the conductor induce a voltage in the conductor in a direction that tends to retard the normal flow of current in the wire. This effect is more pronounced at the center of the conductor. Thus, current within the conductor tends to flow more easily toward the surface of the wire. The higher the frequency, or the faster the rise time of the applied waveform, the greater the tendency for current to flow at the surface. Figure 6.5.11 shows the distribution of current in a radial conductor.

When a circuit is operating at high frequencies, the skin effect causes the current to be redistributed over the conductor cross section in such a way as to make most of the current flow where it is encircled by the smallest number of flux lines. This general principle controls the distribution of current regardless of the shape of the conductor involved.

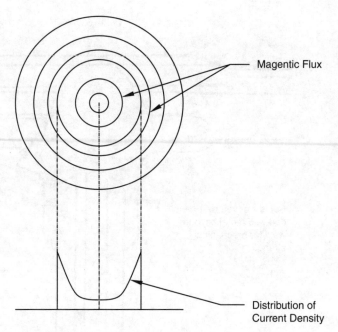

FIGURE 6.5.11 Skin effect on an isolated round conductor carrying a moderately high-frequency signal.

6.3 GROUND SYSTEM OPTIONS

The first determining factor for a ground system is typically the available real estate—or in some cases the lack of available real estate.

Figure 6.5.12 shows a building ground system using a combination of ground rods and buried bare-copper radial wires. This design is appropriate when the building is large or when it is located in a suburban area. Most newer office buildings have ground systems designed into them. If a comprehensive building ground system is provided, use it. For older structures (constructed of wood or brick), a separate ground system may be required.

Figure 6.5.13 shows another approach in which a perimeter ground strap is buried around the building and ground rods are driven into the earth at regular intervals (2.2 times the rod length). The ground ring consists of a one-piece copper conductor that is bonded to each ground rod.

If a transmission or microwave tower is located at the site, connect the tower ground system to the main ground point via a copper strap. The width of the strap should be at least 1 percent of the length and, in any event, not less than 3-inch wide. The building ground system is not a substitute for a tower ground system, no matter the size of the tower. The two systems are treated as independent elements, except for the point at which they interconnect.

Tie the utility company power system ground rod to the main facility ground point per requirements of the local electrical code. Do not consider the building ground system to be a substitute for the utility company ground rod. The utility rod is important for safety reasons and must not be disconnected or moved. Do not remove any existing earth ground connections to the power line neutral connection. Doing so may violate local electrical code.

Bury all elements of the ground system to reduce the inductance of the overall network. Do not make sharp turns or bends in the interconnecting wires. Straight, direct wiring practices reduce the overall inductance of the system and increase its effectiveness in shunting fast rise-time surges to earth.

In most areas, soil conductivity is high enough to permit rods to be connected with no. 6 or larger bare-copper wire. In areas of sandy soil, use copper strap. A wire buried in low-conductivity, sandy

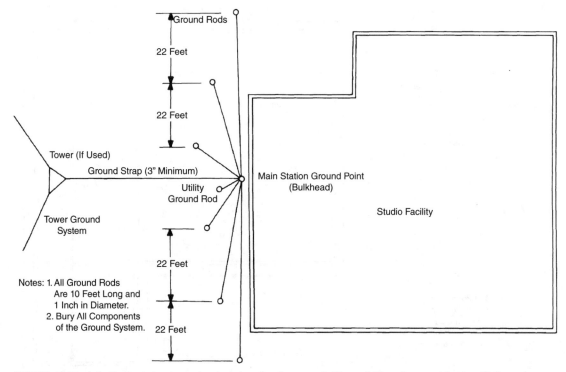

FIGURE 6.5.12 A facility ground system using the hub-and-spoke approach. The available real estate at the site will dictate the exact configuration of the ground system. Consult the local electrical code for utility ground rod requirements.

soil tends to be inductive and less effective in dealing with fast-rise-time current surges. Again, make the width of the ground strap at least 1 percent of its overall length. Connect buried elements of the system, as shown in Fig. 6.5.14.

6.3.1 Bulkhead Panel

The *bulkhead panel* is the cornerstone of an effective facility ground system. The concept of the bulkhead is simple: establish one reference point to which all cables entering and leaving the equipment building are grounded and to which all transient-suppression devices are mounted. The panel size depends on the spacing, number, and dimensions of the coaxial lines, power cables, and other conduit entering or leaving the building.

To provide a weatherproof point for mounting transient-suppression devices, the basic bulkhead can be modified to accept a subpanel, as shown in Fig. 6.5.15. The subpanel is attached so that it protrudes through an opening in the wall and creates a secondary plate on which transient suppressors are mounted and grounded. A typical cable/suppressor-mounting arrangement for a communications site is shown in Fig. 6.5.16.

Because the bulkhead panel establishes the central grounding point for all equipment within the building, it must be tied to a low-resistance (and low-inductance) perimeter ground system. The bulkhead establishes the main facility ground point, from which all grounds inside the building are referenced. The recommended approach is to simply extend the bulkhead panel down the outside of the building, below grade, to the perimeter ground system. See Fig. 6.5.17.

If cables are used to ground the bulkhead panel, secure the interconnection to the outside ground system along the bottom section of the panel using a wide copper strap.

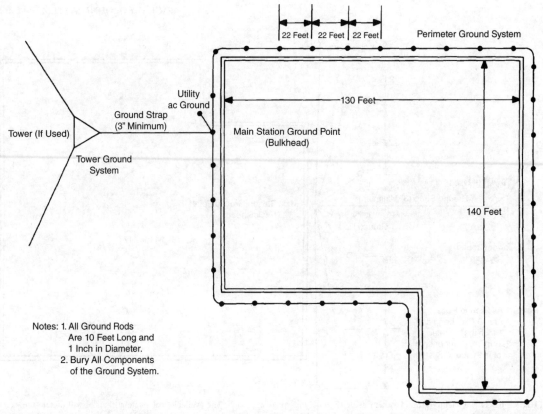

FIGURE 6.5.13 Facility ground using a perimeter ground-rod system. This approach works well for buildings with limited available real estate. Consult the local electrical code for utility ground rod requirements.

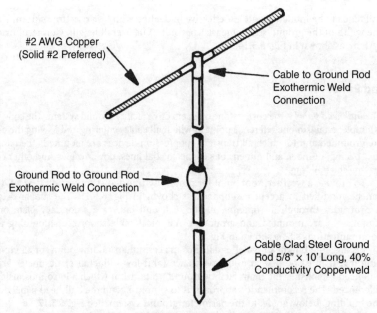

FIGURE 6.5.14 Preferred bonding method for below-grade elements of the ground system.

FACILITY GROUND SYSTEM

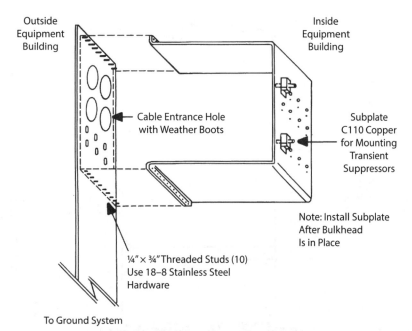

FIGURE 6.5.15 The addition of a subpanel to a bulkhead as a means of providing a mounting surface for transient-suppression components.

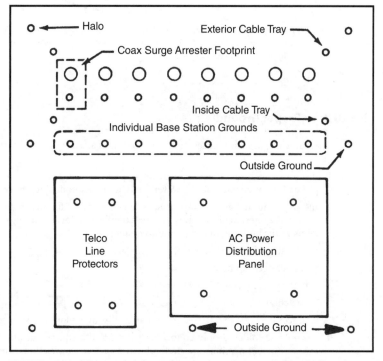

FIGURE 6.5.16 Mounting-hole layout for a communications site bulkhead subpanel.

6.66 FACILITY ISSUES

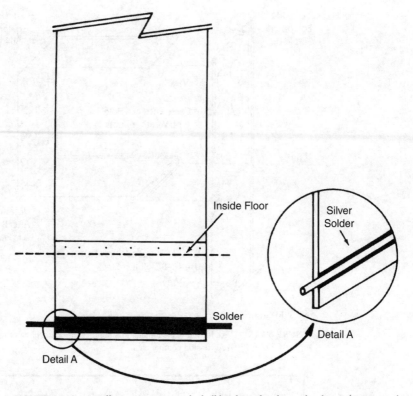

FIGURE 6.5.17 An effective way to ground a bulkhead panel and provide a low-inductance path for surge currents stripped from cables entering and leaving the facility. The panel extends along the building exterior to below grade.

6.3.2 Checklist for Proper Grounding

A methodical approach is necessary in the design of a facility ground system. Consider the following points:

1. Install a bulkhead panel to provide mechanical support, electric grounding, and lightning protection for coaxial cables, power feeds, and telephone lines entering the equipment building.
2. Install an internal ground bus using a star-of-stars configuration. Do not allow ground loops to exist in the internal ground bus. Always comply with local building codes. Connect the following items to the building internal ground system:
   - Chassis racks and cabinets of all hardware
   - Auxiliary equipment
   - Battery charger
   - Switchboard
3. Connect the bulkhead to the ground array through a low-inductance, low-resistance bond.
4. Do not use soldered-only connections outside the equipment building. Crimped, brazed, and exothermic connections are preferable. Protect all connections from moisture by appropriate means.

5. Enlist the assistance of a contractor experienced in ground system design.
6. Follow the local electrical code.

A facility ground system is essential to proper operation of a technical facility. It is too important to be left to chance. Once installed, the ground system is very difficult to expand or otherwise change. Money spent to do it right the first time is a good investment. Consider securing the help of an experienced designer and contractor before proceeding.

REFERENCES

[1] DeWitt, W. E., "Facility Grounding Practices," in *The Electronics Handbook*, Jerry C. Whitaker (ed.), CRC Press, Boca Raton, FL, pp. 2218–2228, 1996.

[2] Block, R., "The Grounds for Lightning and EMP Protection," PolyPhaser Corporation, Gardnerville, NV, 1987.

[3] Carpenter, R. B., "Improved Grounding Methods for Broadcasters," *Proceedings, SBE National Convention*, Society of Broadcast Engineers, IN, 1987.

[4] Lobnitz, E. A., "Lightning Protection for Tower Structures," in *NAB Engineering Handbook*, 9th ed., Jerry C. Whitaker (ed.), National Association of Broadcasters, Washington, D.C., 1998.

BIBLIOGRAPHY

Benson, K. B., and J. C. Whitaker, *Television and Audio Handbook for Engineers and Technicians*, McGraw-Hill, New York, NY, 1989.

Block, R., "How to Ground Guy Anchors and Install Bulkhead Panels," *Mobile Radio Technology*, Intertec Publishing, Overland Park, Kan., February 1986.

Lanphere, J., "Establishing a Clean Ground," *Sound & Video Contractor*, Intertec Publishing, Overland Park, Kan., August 1987.

Lawrie, R., *Electrical Systems for Computer Installations*, McGraw-Hill, New York, N.Y., 1988.

Little, R., "Surge Tolerance: How Does Your Site Rate?," *Mobile Radio Technology*, Intertec Publishing, Overland Park, Kan., June 1988.

Morrison, R., and W. Lewis, *Grounding and Shielding in Facilities*, John Wiley & Sons, New York, N.Y., 1990.

Mullinack, H. G., "Grounding for Safety and Performance," *Broadcast Engineering*, Intertec Publishing, Overland Park, Kan., October 1986.

Schneider, J., "Surge Protection and Grounding Methods for AM Broadcast Transmitter Sites," *Proceedings of the SBE National Convention*, Society of Broadcast Engineers, Indianapolis, IN, 1987.

Schwarz, S. J., "Analytical Expression for Resistance of Grounding Systems," *AIEE Transactions*, Vol. 73, Part III B, pp. 1011–1016, 1954.

Technical Reports LEA-9-1, LEA-0-10 and LEA-1-8, Lightning Elimination Associates, Santa Fe Springs, CA.

Whitaker, J. C., *AC Power Systems*, 3rd ed., CRC Press, Boca Raton, FL, 2007.

Whitaker, J. C., *Maintaining Electronic Systems*, CRC Press, Boca Raton, FL, 1992.

CHAPTER 6.6
AC POWER SYSTEMS

Jerry C. Whitaker, Editor-in-Chief

6.1 INTRODUCTION

Broadcast stations today invariably operate 24 hours a day, 365 days a year. Radio and television stations have a long history of providing reliable service even during the most trying of circumstances. Indeed, the public has long depended on broadcasters to bring them information—sometimes critical, life-saving information—during natural disasters. Reliability does not just happen, of course, it needs to be designed into a facility. And it is not cheap. However, reliability is a key element in making broadcasting the indispensable service that it is.

This chapter will explore some of the ways the ac power supply to the station can be maintained at a high level of reliability. All ac wiring within a facility should be performed by an experienced electrical contractor, and always fully within the local electrical code. Experts are available that can help design a reliable, fault-tolerant system that will keep the station running even during worst case scenarios.

6.2 DESIGNING FOR FAULT-TOLERANCE

To achieve high levels of power system reliability—with the ultimate goal being 24-hour-per-day availability, 365 days per year—some form of power system redundancy is required, regardless of how reliable the individual power system components may be [1]. Redundancy, if properly implemented, also provides power-distribution flexibility. By providing more than one path for power flow to the load, the key elements of a system can be shifted from one device or branch to another as required for load balancing, system renovations or alterations, or equipment failure isolation. Redundancy also provides a level of fault tolerance. Fault tolerance can be divided into three basic categories:

- Rapid recovery from failures
- Protection against "slow" power system failures, where there is enough warning of the condition to allow intervention
- Protection against "fast" power system failures, where no warning of the power failure is given

As with many corrective and preventive measures, the increasing costs must be weighed against the benefits.

6.2.1 Power-Distribution Options

There are a number of building blocks that form what can be described as an assured, reliable, clean power source for broadcast systems, peripherals, and other critical loads [2]. They include:

- Properly sized utility and service entry gear (step-down transformer, main disconnect, and panel board, switchboard, or switchgear)
- Lightning protection
- Efficient facility power distribution
- Proper grounding
- Power conditioning equipment
- Critical load air-conditioning
- Batteries for dc backup power
- Emergency engine-generator
- Critical load power-distribution network
- Emergency readiness planning

A power system to support a critical load cannot be said to be reliable unless all these components are operating as intended, not only during normal operation, but especially during an emergency.

It is easy to become complacent during periods when everything is functioning properly, because this is the usual mode of operation. An absence of contingency plans for dealing with an emergency situation, and a lack of understanding of how the entire system works, thus, can lead to catastrophic shutdowns when an emergency situation arises. Proper training, and periodic reinforcing, is an essential component of a reliable system.

6.2.2 Plant Configuration

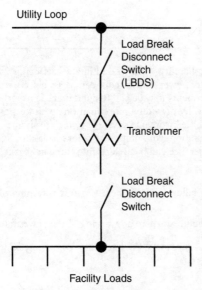

FIGURE 6.6.1 Simplified service entrance system. (*From* [2]. *Used with permission.*)

There are any number of hardware configurations that will provide redundancy and reliability for a critical load. Each situation is unique and requires an individual assessment of the options and—more importantly—the risks. The realities of economics dictate that cost is always a factor. Through proper design, however, the expense usually can be held within an acceptable range.

Design for reliability begins at the utility service entrance [2]. The common arrangement shown in Fig. 6.6.1 is vulnerable to interruptions from faults at the transformer and associated switching devices in the circuit. Furthermore, service entrance maintenance would require a plant shutdown. In Fig. 6.6.2, redundancy has been provided that will prevent the loss of power should one of the devices in the line fail. Because the two transformers are located in separate physical enclosures, maintenance can be performed on one leg without dropping power to the facility.

Of equal importance is the method of distributing power *within* a facility to achieve maximum reliability. This task is more difficult when dealing with a facility stretching over several buildings, where—instead of being concentrated in a single room or floor—the critical loads may be in a number of distant locations. Figure 6.6.3

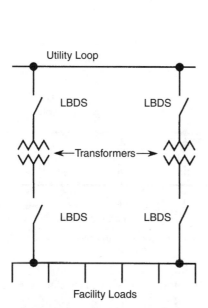
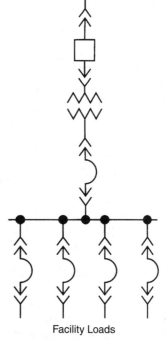

FIGURE 6.6.2 Fault-tolerant service entrance system. (*From* [2]. *Used with permission.*)

FIGURE 6.6.3 Secondary plant distribution using a simple radial configuration. (*From* [2]. *Used with permission.*)

illustrates power distribution through the facility using a simple radial system. An incoming line supplies the main and line feeders via a service entrance transformer. This system is suitable for a single building. It is simple, reliable, and low in cost. However, such a system must be shut down for routine maintenance, and it is vulnerable to single-point failure. Figure 6.6.4 illustrates a distributed and redundant power-distribution system that permits transferring loads as required to patch around a fault condition. This configuration also allows portions of the system to be de-energized for maintenance or upgrades without dropping the entire facility. Note the loop arrangement and associated switches that permit optimum flexibility during normal and fault operating conditions.

6.2.3 Critical System Bus

Many facilities do not require the operation of all equipment during a power outage [2]. Rather than use one large standby power system, key pieces of equipment can be protected with smaller, dedicated backup power systems. This separate power supply is used to provide ac to critical loads, thus keeping the protected systems up and running. Unnecessary loads are dropped in the event of a power failure.

A standby system built on the critical load principle can be a cost-effective answer to the power-failure threat. The first step in implementing a critical load bus is to accurately determine the power requirements for the most important equipment.

When planning a critical load bus, be certain to identify accurately which loads are critical, and which can be dropped in the event of a commercial power failure. If air conditioning is interrupted but the hardware in the central equipment racks continues to run, temperatures will rise quickly to the point at which system components may be damaged or the hardware automatically shuts down.

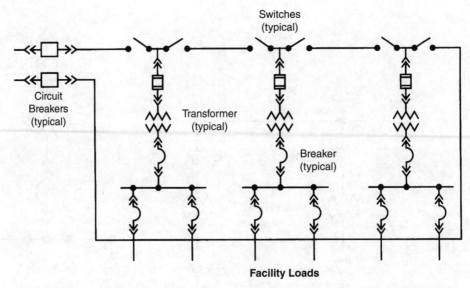

FIGURE 6.6.4 A redundant, fault-tolerant secondary plant distribution system. (*From* [2]. *Used with permission.*)

It may therefore be necessary to require at least some cooling fans, chillers, and heat-exchange pumps to run without interruption.

6.3 PLANT MAINTENANCE

Maintenance of the facility electrical system is a key part of any serious energy-management effort. Perform the following steps on a regular basis:

- Measure the current drawn on distribution cables. Document the measurements so that a history of power demand can be compiled.
- Check terminal and splice connections to make sure they are tight.
- Check power-system cables for excessive heating.
- Check cables for insulation problems.
- Clean switchboard and circuit-breaker panels.
- Measure the phase-to-phase load balance at the utility service entrance. Load imbalance can result in inefficient use of ac power.
- Measure and chart the power factor of the load.

Develop and post a simplified one-line schematic of the entire power network as well as other building systems, including heating, air conditioning, security, and alarm functions. A *mimic board* is helpful in this process. Construct the mimic board control panel so that it depicts the entire ac power-distribution system. The board may have active indicators that show what loads or circuit breakers are turned on or off, what functions have been disabled, and key operating parameters, including input voltage, load current, and total kVA demand. Safety considerations require that systems not be activated from the mimic board. Permit systems to be energized only at the apparatus.

Environmental control systems should be monitored closely. Air conditioning, heating, and ventilation systems often represent a significant portion of the power load of a facility. Data-logging equipment with process control capability can be of considerable help in monitoring the condition of the equipment. The logger can be programmed to record all pertinent values periodically, and to report abnormal conditions.

6.3.1 Switchgear Maintenance

All too often, ac power switchgear is installed at a facility and forgotten—until a problem occurs. A careless approach to regular inspection and cleaning of switchgear has resulted in numerous failures, including destructive fires. The most serious fault in any switchgear assembly is arcing involving the main power bus. Protective devices may fail to open, or open only after a considerable delay. The arcing damage to busbars and enclosures can be significant. Fire often ensues, compounding the damage.

Moisture, combined with dust and dirt, is the greatest deteriorating factor insofar as insulation is concerned. Dust and/or moisture are thought to account for as much as half of switchgear failures. Initial leakage paths across the surface of bus supports result in flashover and sustained arcing. Contact overheating is another common cause of switchgear failure. Improper circuit-breaker installation or loose connections can result in localized overheating and arcing.

An arcing fault is destructive because of the high temperatures present (more than 6000°F). An arc is not a stationary event. Because of the ionization of gases and the presence of vaporized metal, an arc can travel along bare busbars, spreading the damage and sometimes bypassing open circuit breakers. It has been observed that most faults in three-phase systems involve all phases. The initial fault that triggers the event may involve only one phase, but because of the traveling nature of an arc, damage quickly spreads to the other lines.

Preventing switchgear failure is a complicated discipline, but consider the following general guidelines:

- Install insulated busbars for both medium-voltage and low-voltage switchgear. Each phase of the bus and all connections should be enclosed completely by insulation with electrical, mechanical, thermal, and flame-retardant characteristics suitable for the application.
- Establish a comprehensive preventive maintenance program for the facility. Keep all switchboard hardware clean from dust and dirt. Periodically check connection points for physical integrity.
- Maintain control over environmental conditions. Switchgear exposed to contaminants, corrosive gases, moist air, or high ambient temperatures may be subject to catastrophic failure. Conditions favorable to moisture condensation are particularly perilous, especially when dust and dirt are present.
- Accurately select overcurrent trip settings, and check them on a regular basis. Adjust the trip points of protection devices to be as low as possible, consistent with reliable operation.
- Divide switchgear into compartments that isolate different circuit elements. Consider adding vertical barriers to bus compartments to prevent the spread of arcing and fire.
- Install ground-fault protection devices at appropriate points in the power-distribution system
- Adhere to all applicable building codes.
- Seek advice and support of a qualified electrical contractor for installation, upgrades, and ongoing maintenance.

6.4 STANDBY POWER SYSTEMS

When utility company power problems are discussed, most people immediately think of blackouts. The lights go out, and everything stops. With the facility down and in the dark, there is nothing to do but sit and wait until the utility company finds the problem and corrects it. This process generally takes only a few minutes. There are times, however, when it can take hours. In some remote locations, it can even take days.

6.74 FACILITY ISSUES

Blackouts are, without a doubt, the most troublesome utility company problem that a facility will have to deal with. Statistics show that power failures are, generally speaking, a rare occurrence in most areas of the country. They are also usually short in duration. However, for a 24/7 operation, any disruption in the ac power supply is a problem.

6.4.1 Blackout Effects

A facility that is down for even 5 min can suffer a significant loss of productivity that may take hours or days to rebuild. Many broadcast transmitter sites are located in remote, rural areas or on mountaintops. Neither of these kinds of locations are well-known for their power reliability. It is not uncommon in mountainous areas for utility company service to be out for extended periods after a major storm. Few operators are willing to take such risks with their business. Most choose to install standby power systems at appropriate points in the equipment chain.

The cost of standby power for a facility can be substantial, and an examination of the possible alternatives should be conducted before any decision on equipment is made. Management must clearly define the direct and indirect costs and weigh them appropriately. Include the following items in the cost-vs.-risk analysis:

- Standby power-system equipment purchase and installation costs, fuel costs, and maintenance expenses.
- Exposure of the system to utility company power failure.
- Alternative operating methods available to the facility.
- Direct and indirect costs of lost uptime because of blackout conditions.

A distinction must be made between *emergency* and *standby* power sources. Strictly speaking, emergency systems supply circuits legally designated as being essential for safety to life and property. Standby power systems are used to protect a facility against the loss of productivity resulting from a utility company power outage.

6.4.2 Standby Power Options

To ensure the continuity of ac power, some broadcast facilities depend on either two separate utility services, or one utility service plus on-site generation. Because of the growing complexity of electrical systems, attention must be given to power-supply reliability.

The engine-generator shown in Fig. 6.6.5 is the classic standby power system. An automatic transfer switch monitors the ac voltage coming from the utility company line for power failure conditions. Upon detection of an outage for a predetermined period of time (generally 1-10 s), the standby generator is started; after the generator is up to speed, the load is transferred from the utility to the local generator. Upon return of the utility feed, the load is switched back, and the generator is stopped. This basic type of system is used widely in industry and provides economical protection against prolonged power outages (several min or more).

The transfer device shown in Fig. 6.6.5 is a contactor-type, break-before-make unit. By replacing the simple transfer device shown with an automatic overlap (static) transfer switch, as shown in Fig. 6.6.6, additional functionality can be gained. The overlap transfer switch permits the on-site generator to be synchronized with the load, making a clean switch from one energy source to another. This functionality offers the following benefits:

- Switching back to the utility feed from the generator can be accomplished without interruption in service.
- The load can be cleanly switched from the utility to the generator in anticipation of utility line problems (such as an approaching severe storm).
- The load can be switched to and from the generator to accomplish *load shedding* objectives.

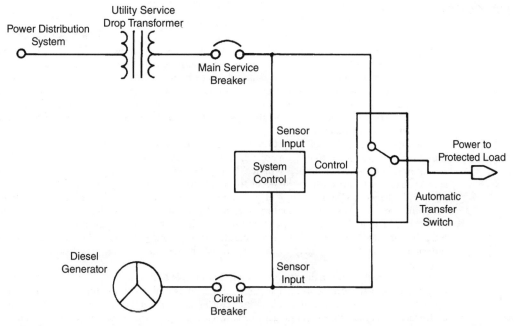

FIGURE 6.6.5 The classic standby power system using an engine-generator set. This system protects a facility from prolonged utility company power failures.

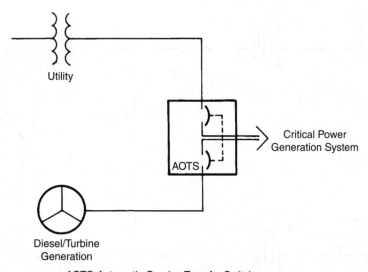

FIGURE 6.6.6 The use of a static transfer switch to transfer the load from the utility company to the onsite generator.

6.4.3 Dual Feeder System

In some areas, usually metropolitan centers, two utility company power drops can be brought into a facility as a means of providing a source of standby power. As shown in Fig. 6.6.7, two separate utility service drops—from separate power-distribution systems—are brought into the plant, and an automatic transfer switch changes the load to the backup line in the event of a main-line failure. The dual feeder system provides an advantage over the auxiliary diesel arrangement in that power transfer from main to standby can be made in a fraction of a second if a static transfer switch is used. Time delays are involved in the diesel generator system that limit its usefulness to power failures lasting more than several minutes.

The dual feeder system of protection is based on the assumption that each of the service drops brought into the facility is routed via different paths. This being the case, the likelihood of a failure on both power lines simultaneously is remote. The dual feeder system will not, however, protect against area-wide power failures, which can occur from time to time.

The dual feeder system is limited primarily to urban areas. Rural or mountainous regions generally are not equipped for dual redundant utility company operation. Even in urban areas, the cost of bringing a second power line into a facility can be high, particularly if special lines must be installed for the feed. If two separate utility services are available at or near the site, redundant feeds generally may be less expensive than engine-driven generators of equivalent capacity.

Figure 6.6.8 illustrates a dual feeder system that utilizes both utility inputs simultaneously at the facility. Notice that during normal operation, both ac lines feed loads, and the "tie" circuit breaker is open. In the event of a loss of either line, the circuit-breaker switches reconfigure the load to place the entire facility on the single remaining ac feed. Switching is performed automatically; manual control is provided in the event of a planned shutdown on one of the lines.

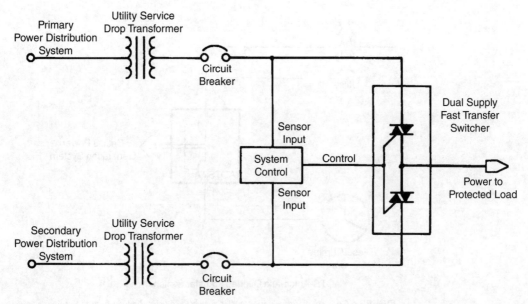

FIGURE 6.6.7 The dual utility feeder system of ac power loss protection. An automatic transfer switch changes the load from the main utility line to the standby line in the event of a power interruption.

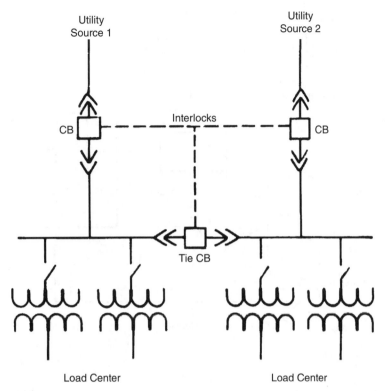

FIGURE 6.6.8 A dual utility feeder system with interlocked circuit breakers. (*From* [2]. *Used with permission.*)

6.4.4 Peak Power Shaving

Figure 6.6.9 illustrates the use of a backup diesel generator for both standby power and *peak power shaving* applications. Commercial power customers often can realize substantial savings on utility company bills by reducing their energy demand during certain hours of the day. An automatic overlap transfer switch is used to change the load from the utility company system to the local diesel generator. The changeover is accomplished by a static transfer switch that does not disturb the operation of load equipment. This application of a standby generator can provide financial return to the facility, whether or not the unit is ever needed to carry the load through a commercial power failure.

Peak power shaving is not something most stations think about during their normal operations. However, in some areas of the country utility companies occasionally request large users to cut back on their power usage, and provide financial incentives to do so. And, since good engineering practice calls for regular testing of backup systems, peak power shaving can sometimes result in a win-win situation for the station and the utility.

6.4.5 Advanced System Protection

A more sophisticated power-control system is shown in Fig. 6.6.10, where a dual feeder supply is coupled with a motor-generator set to provide clean, undisturbed ac power to the load. The motor-generator set will smooth over the transition from the main utility feed to the standby, often

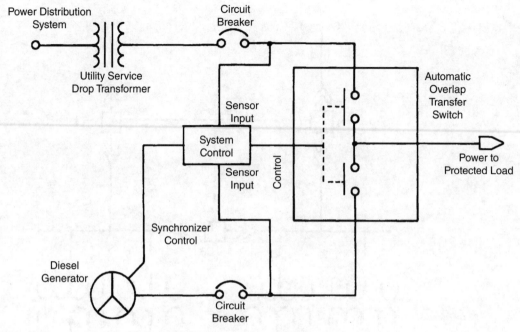

FIGURE 6.6.9 The use of a diesel generator for standby power and peak power shaving applications.

making a commercial power failure unnoticed by on-site personnel. A conventional motor-generator typically will give up to 0.5 s (or more) of power fail ride-through, more than enough to accomplish a transfer from one utility feed to the other. This standby power system is further refined in the application illustrated in Fig. 6.6.11, where a diesel generator has been added to the system. With the automatic overlap transfer switch shown at the generator output, this arrangement also can be used for peak demand power shaving.

6.4.6 Choosing a Generator

Engine-generator sets are available for power levels ranging from less than 1 kVA to several thousand kVA or more. Machines also can be paralleled to provide greater capacity. Engine-generator sets typically are classified by the type of power plant used:

- **Diesel**—Advantages: rugged and dependable, low fuel costs, low fire and/or explosion hazard. Disadvantages: somewhat more costly than other engines, heavier in smaller sizes.
- **Natural and liquefied petroleum gas**—Advantages: quick starting after long shutdown periods, long life, low maintenance. Disadvantage: availability of natural gas during area-wide power failure may be subject to question.
- **Gasoline**—Advantages: rapid starting, low initial cost. Disadvantages: greater hazard associated with storing and handling gasoline, generally shorter mean time between overhaul.
- **Gas turbine**—Advantages: smaller and lighter than piston engines of comparable horsepower, rooftop installations practical, rapid response to load changes. Disadvantages: longer time required to start and reach operating speed, sensitive to high input air temperature.

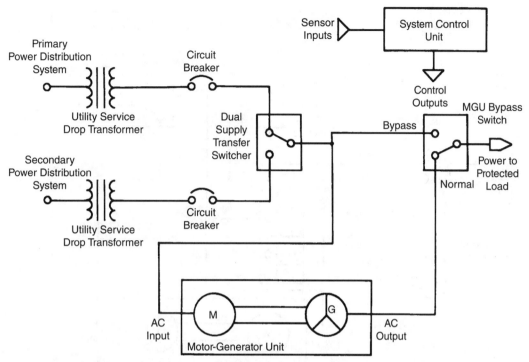

FIGURE 6.6.10 A dual feeder standby power system using a motor-generator set to provide power fail ride-through and transient-disturbance protection. Switching circuits allow the motor-generator set to be bypassed, if necessary.

The type of power plant chosen usually is determined primarily by the environment in which the system will be operated and by the cost of ownership. For example, a standby generator located in an urban area office complex may be best suited to the use of an engine powered by natural gas, because of the problems inherent in storing large amounts of fuel. State or local building codes can place expensive restrictions on fuel-storage tanks and make the use of a gasoline- or diesel-powered engine impractical. The use of propane usually is restricted to rural areas. The availability of propane during periods of bad weather (when most power failures occur) also must be considered.

The generator rating for a standby power system should be chosen carefully and take into consideration the anticipated future growth of the plant. It is good practice to install a standby power system rated for at least 25 percent greater output than the current peak facility load. This headroom gives a margin of safety for the standby equipment and allows for future expansion of the facility without overloading the system.

An engine-driven standby generator typically incorporates automatic starting controls, a battery charger, and automatic transfer switch. (See Fig. 6.6.12.) Control circuits monitor the utility supply and start the engine when there is a failure or a sustained voltage drop on the ac supply. The switch transfers the load as soon as the generator reaches operating voltage and frequency. Upon restoration of the utility supply, the switch returns the load and initiates engine shutdown. The automatic transfer switch must meet demanding requirements, including:

- Carrying the full rated current continuously
- Withstanding fault currents without contact separation
- Handling high inrush currents
- Withstanding many interruptions at full load without damage

6.80 FACILITY ISSUES

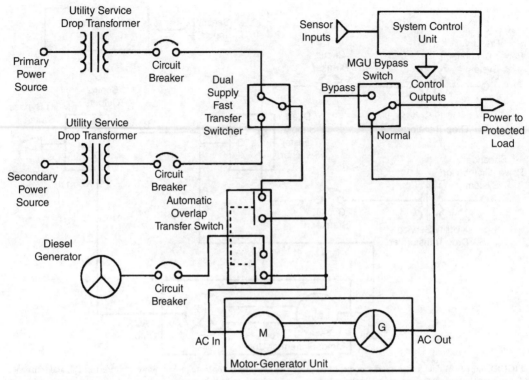

FIGURE 6.6.11 A premium power supply backup and conditioning system using dual utility feeds, a diesel generator, and a motor-generator set.

The nature of most power outages requires a sophisticated monitoring system for the engine-generator set. Most power failures occur during periods of bad weather. Most standby generators are unattended. More often than not, the standby system will start, run, and shut down without any human intervention or supervision. For reliable operation, the monitoring system must check the status of the machine continually to ensure that all parameters are within normal limits. Time-delay periods usually are provided by the controller that require an outage to last from 5 to 10 s before the generator is started and the load is transferred. This prevents false starts that needlessly exercise the system. A time delay of 5–30 min usually is allowed between the restoration of utility power and return of the load. This delay permits the utility ac lines to stabilize before the load is reapplied.

Automatic starting and synchronizing controls are used for multiple-engine-generator installations. The output of two or three smaller units can be combined to feed the load. This capability offers additional protection for the facility in the event of a failure in any one machine. As the load at the facility increases, additional engine-generator systems can be installed on the standby power bus.

6.4.6.1 Generator Types. Generators for standby power applications can be induction or synchronous machines. Most engine-generator systems in use today are of the synchronous type because of the versatility, reliability, and capability of operating independently that this approach provides [2]. Most modern synchronous generators are of the *revolving field alternator* design. Essentially, this means that the armature windings are held stationary and the field is rotated. Therefore, generated power can be taken directly from the stationary armature windings. Revolving armature alternators are less popular because the generated output power must be derived via slip rings and brushes.

The exact value of the ac voltage produced by a synchronous machine is controlled by varying the current in the dc field windings, while frequency is controlled by the speed of rotation. Power output is controlled by the torque applied to the generator shaft by the driving engine. In this manner, the synchronous generator offers precise control over the power it can produce.

Practically all modern synchronous generators use a brushless exciter. The exciter is a small ac generator on the main shaft; the ac voltage produced is rectified by a three-phase rotating rectifier assembly also on the shaft. The dc voltage thus obtained is applied to the main generator field, which is also on the main shaft. A voltage regulator is provided to control the exciter field current, and in this manner, the field voltage can be precisely controlled, resulting in a stable output voltage.

The frequency of the ac current produced is dependent on two factors: the number of poles built into the machine, and the speed of rotation (rpm). Because the output frequency must normally be maintained within strict limits (60 or 50 Hz), control of the generator speed is essential. This is accomplished by providing precise rpm control of the *prime mover*, which is performed by a governor.

There are many types of governors; however, for auxiliary power applications, the *isochronous governor* is commonly used. The isochronous governor controls the speed of the engine so that it remains constant from no-load to full load, assuring a constant ac power output frequency from the generator. A modern system consists of two primary components: an electronic speed control and an actuator that adjusts the speed of the engine. The electronic speed control senses the speed of the machine and provides a feedback signal to the mechanical/hydraulic actuator, which in turn positions the engine throttle or fuel control to maintain accurate engine rpm.

The National Electrical Code provides guidance for safe and proper installation of on-site engine-generator systems. Local codes may vary and must be reviewed during early design stages.

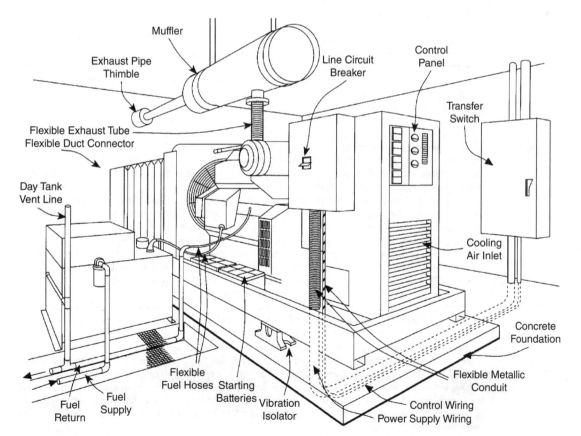

FIGURE 6.6.12 Typical configuration of an engine-generator set. (*From* [1]. *Used with permission.*)

6.4.7 UPS Systems

An uninterruptible power system is an elegant solution to power outage concerns. The output of the UPS inverter can be a sine wave or pseudo sine wave. When shopping for a UPS system, consider the following:

- Power reserve capacity for future growth of the facility.
- Inverter current surge capability (if the system will be driving inductive loads, such as motors).
- Output voltage and frequency stability over time and with varying loads.
- Required battery supply voltage and current. Battery costs vary greatly, depending on the type of units needed.
- Type of UPS system (forward-transfer type or reverse-transfer type) required by the particular application. Some sensitive loads may not tolerate even brief interruptions of the ac power source.
- Inverter efficiency at typical load levels. Some inverters have good efficiency ratings when loaded at 90 percent of capacity, but poor efficiency when lightly loaded.
- Size and environmental requirements of the UPS system. High-power UPS equipment requires a large amount of space for the inverter/control equipment and batteries. Battery banks often require special ventilation and ambient temperature control.

6.4.8 Standby Power-System Noise

Noise produced by backup power systems can be a serious problem if not addressed properly. Standby generators, motor-generator sets, and UPS systems produce noise that can disturb building occupants and irritate neighbors and/or landlords.

The noise associated with electrical generation usually is related to the drive mechanism, most commonly an internal combustion engine. The amplitude of the noise produced is directly related to the size of the engine-generator set. First consider whether noise reduction is a necessity. Many building owners have elected to tolerate the noise produced by a standby power generator because its use is limited to emergency situations. During a crisis, when the normal source of power is unavailable, most people will tolerate noise associated with a standby generator.

If the decision is made that building occupants can live with the noise of the generator, care must be taken in scheduling the required testing and exercising of the unit. Whether testing occurs monthly or weekly, it should be done on a regular schedule.

If it has been determined that the noise should be controlled, or at least minimized, the easiest way to achieve this objective is to physically separate the machine from occupied areas. This may be easier said than done. Because engine noise is predominantly low-frequency in character, walls and floor/ceiling construction used to contain the noise must be massive. Lightweight construction, even though it may involve several layers of resiliently mounted drywall, is ineffective in reducing low-frequency noise. Exhaust noise is a major component of engine noise but, fortunately, it is easier to control. When selecting an engine-generator set, select the highest-quality exhaust muffler available. Such units often are identified as "hospital-grade" mufflers.

Engine-generator sets also produce significant vibration. The machine should be mounted securely to a slab-on-grade or an isolated basement floor, or it should be installed on vibration isolation mounts. Such mounts usually are specified by the manufacturer.

Because a UPS system or motor-generator set is a source of continuous power, it must run continuously. Noise must be adequately controlled. Physical separation is the easiest and most effective method of shielding occupied areas from noise. Enclosure of UPS equipment usually is required, but noise control is significantly easier than for an engine-generator because of the lower noise levels involved. Nevertheless, the low-frequency fundamental of a UPS system is difficult to contain adequately; massive constructions may be necessary. Vibration control also is required for most UPS and motor-generator gear.

6.4.9 Batteries

Batteries are the lifeblood of most UPS systems, and the initial power source used to start generators. Important characteristics include the following:

- Charge capacity—how long the battery will operate at rated voltage?
- Weight
- Charging characteristics
- Durability/ruggedness

Additional features that add to the utility of the battery include:

- Built-in status/temperature/charge indicator and/or data output port
- Built-in overtemperature/overcurrent protection with auto-reset capabilities
- Environmental friendliness

The last point deserves some attention. Many battery types must be recycled or disposed of through some prescribed means. Proper disposal of a battery at the end of its useful life is, thus, an important consideration. Be sure to check the original packaging for disposal instructions. Failure to follow the proper procedures could have serious consequences.

Research has brought about a number of different battery chemistries, each offering distinct advantages [3]. Today's most common and promising rechargeable chemistries include the following:

- *Nickel cadmium* (NiCd)—Used for portable radios, cellular phones, video cameras, laptop computers, and power tools. NiCds have good load characteristics, are economically priced, and are simple to use.
- *Lithium ion* (Li-ion)—Now commonly available and typically used for video cameras and laptop computers. This battery promises to replace some NiCds for high energy-density applications.
- *Sealed lead acid* (SLA)—Used for uninterruptible power systems, video cameras, and other demanding applications where the energy-to-weight ratio is not critical and low battery cost is desirable.
- *Nickel metal hydride* (NiMH)—Used for cellular phones, video cameras, and laptop computers where high energy is of importance and cost is secondary.
- *Lithium polymer* (Li-polymer)—This battery has the highest energy density and lowest self-discharge of common battery types, but its load characteristics typically suit only low current applications.
- *Reusable alkaline*—Used for light duty applications. Because of its low self-discharge, this battery is suitable for portable entertainment devices and other noncritical appliances that are used occasionally.

No single battery offers all the answers; rather, each chemistry is based on a number of compromises.

A battery, of course, is only as good as its charger. Common attributes for the current generation of charging systems include quick-charge capability and automatic battery condition analysis and subsequent *intelligent* charging.

6.4.9.1 Terms.
The following terms are commonly used to specify and characterize batteries:

- **Energy density**—The storage capacity of a battery measured in *watt-hours per kilogram* (Wh/kg).
- **Cycle life**—The typical number of charge-discharge cycles for a given battery before the capacity decreases from the nominal 100 percent to approximately 80 percent, depending on the application.

- **Fast-charge time**—The time required to fully charge an empty battery.
- **Self-discharge**—The discharge rate when the battery is not in use.
- **Cell voltage**—The output voltage of the basic battery element. The cell voltage multiplied by the number of cells provides the battery terminal voltage.
- **Load current**—The maximum recommended current the battery can provide.
- **Current rate**—The *C*-rate is a unit by which charge and discharge times are scaled. If discharged at 1*C*, a 100 Ah battery provides a current of 100 A; if discharged at 0.5*C*, the available current is 50 A.
- **Exercise requirement**—This parameter indicates the frequency that the battery needs to be exercised to achieve maximum service life.

6.4.9.2 Sealed Lead-Acid Battery. The lead-acid battery is a commonly used chemistry. The *flooded* version is found in automobiles and large UPS battery banks. Most smaller, portable systems use the *sealed* version, also referred to as SLA or *gelcell*.

The lead-acid chemistry is commonly used when high power is required, weight is not critical, and cost must be kept low [3]. The typical current range of a medium-sized SLA device is 2-50 Ah. Because of its minimal maintenance requirements and predictable storage characteristics, the SLA has found wide acceptance in the UPS industry, especially for *point-of-application* systems.

The SLA is not subject to memory. No harm is done by leaving the battery on float charge for a prolonged time. On the negative side, the SLA does not lend itself well to fast charging. Typical charge times are 8-16 h. The SLA must be stored in a charged state because a discharged SLA will sulfate. If left discharged, a recharge may be difficult or even impossible.

Unlike the common NiCd, the SLA prefers a shallow discharge. A full discharge reduces the number of times the battery can be recharged, similar to a mechanical device that wears down when placed under stress. In fact, each discharge-charge cycle reduces (slightly) the storage capacity of the battery. This wear-down characteristic also applies to other chemistries, including the NiMH.

The charge algorithm of the SLA differs from that of other batteries in that a *voltage-limit* rather than *current-limit* is used. Typically, a multistage charger applies three charge stages consisting of a *constant-current charge*, *topping-charge*, and *float-charge*. (See Fig. 6.6.13.) During the constant-current stage, the battery charges to 70 percent in about 5 h; the remaining 30 percent is completed by the topping-charge. The slow topping-charge, lasting another 5 h, is essential for the performance of the battery. If not provided, the SLA eventually loses the ability to accept a full charge and the

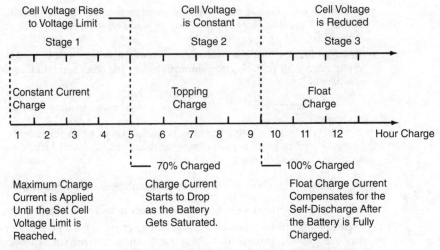

FIGURE 6.6.13 The charge states of an SLA battery. (*From* [3]. *Used with permission.*)

storage capacity of the battery is reduced. The third stage is the float-charge that compensates for self-discharge after the battery has been fully charged.

During the "constant current charge," the SLA battery is charged at a high current, limited by the charger itself. After the voltage limit is reached, the topping charge begins and the current starts to gradually decrease. Full-charge is reached when the current drops to a preset level or reaches a low-end plateau.

REFERENCES

[1] Gruzs, T. M., "High Availability, Fault-Tolerant AC Power Distribution Systems for Critical Loads, *Proceedings, Power Quality Solutions/Alternative Energy*, Intertec International, Ventura, CA, pp. 20–22, September 1996.

[2] DeDad, J. A., "Considerations in Designing a Reliable Power Distribution System," in *Practical Guide to Power Distribution for Information Technology Equipment*, PRIMEDIA Intertec, Overland Park, KS, pp. 4–8, 1997.

[3] Buchmann, I., "Batteries," in *The Electronics Handbook*, Jerry C. Whitaker (ed.), p. 1058, CRC Press, Boca Raton, FL, 1996.

BIBLIOGRAPHY

Angevine, E., "Controlling Generator and UPS Noise," *Broadcast Engineering*, PRIMEDIA Intertec, Overland Park, Kan., March 1989.

Baietto, R., "How to Calculate the Proper Size of UPS Devices," *Microservice Management*, PRIMEDIA Intertec, Overland Park, Kan., March 1989.

Fardo, S., and D. Patrick, *Electrical Power Systems Technology*, Prentice-Hall, Englewood Cliffs, NJ, 1985.

Federal Information Processing Standards Publication No. 94, *Guideline on Electrical Power for ADP Installations*, U.S. Department of Commerce, National Bureau of Standards, Washington, DC, 1983.

Highnote, R. L., *The IFM Handbook of Practical Energy Management*, Institute for Management, Old Saybrook, CT, 1979.

Hill, M., "Computer Power Protection," *Broadcast Engineering*, Intertec Publishing, Overland Park, Kan., April 1987.

Lawrie, R., *Electrical Systems for Computer Installations*, McGraw-Hill, New York, NY, 1988.

Smith, M., "Planning for Standby AC Power," *Broadcast Engineering*, PRIMEDIA Intertec, Overland Park, KS, March 1989.

Whitaker, J. C., *AC Power Systems*, 3rd ed., CRC Press, Boca Raton, FL, 2007.

Whitaker, J. C., *Maintaining Electronic Systems*, CRC Press, Boca Raton, FL, 1992.

Whitaker, J. C., *The Electronics Handbook*, 2nd ed., CRC Press, Boca Raton, FL, 2005.

CHAPTER 6.7
TRANSMISSION SYSTEM MAINTENANCE

Steve Fluker
WFTV/WRDQ Cox Media Group, Orlando, FL

6.1 INTRODUCTION

Your transmitter/tower site is a very important part of your broadcast operation. A major failure there can cripple your business. Unfortunately, these sites often do not get the attention they should. Too many times they are out of sight, out of mind. With shrinking engineering staff, sometimes being responsible for multiple stations and tower sites—not to mention many of these towers are located far away from the studio—it can become very difficult to fit these inspections into your regular schedule.

Newer transmitters and equipment are very reliable and typically just run, so unless there is a problem at the site, maintenance tends to get put on the back burner and is often forgotten. Sometimes months can slip by between inspections, and in many cases, 6 months, a year, and even more can go by without even realizing it. Another issue in today's broadcasting business is that with most studio equipment becoming computerized, we are seeing managers favoring IT professionals as our engineering managers. That makes sense for the studio; however, transmitter site equipment can be intimidating for those whose training has primarily been in computers. There is no shame to it, but they may not know quite how to treat the equipment there, and may even avoid going.

This chapter will give you an outline of what should be done at your transmitter sites to keep them running efficiently and reliably, and hopefully keep you out of the manager's hot-seat when something goes wrong that could have been prevented or predicted months in advance (which would have given the business office time to prepare for upcoming expense).

6.2 TRANSMITTER SITE VISITS

The first thing I want to stress here before we move on is safety. Tower sites can be very dangerous places to work because of extreme high voltages and RF radiation hazards. Your regular inspections can be done by you alone, but NEVER work on the inside or high voltage areas of your transmitter or transmitter site alone. Make sure your manager knows that this is a safety issue. If you are the only engineer for the station and will be doing work inside of a transmitter, make sure someone else goes out with you. This person does not have to be an engineer, just someone who can be there to help should you get hurt.

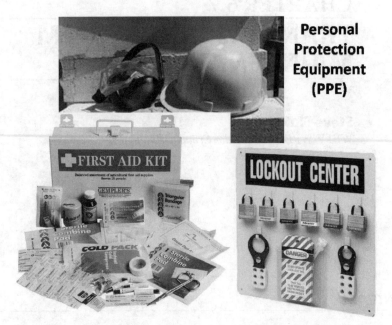

FIGURE 6.7.1 Safety equipment and devices that should be at all transmitter sites.

When working on a transmitter be sure you turn the power off and use "Lock out Tag out" devices to prevent someone else from turning the power back on. Typical safety equipment is shown in Fig. 6.7.1.

Each tower site should have proper personal protection equipment (PPE). This includes a hard hat, eye goggles, gloves, and ear protection. A good first aid kit is also advisable. Safety should be your first concern. If you are working on an AM tower site, you should also have a set of lineman's gloves to protect your hands from RF burns from exposed tuning devices.

With Internet or data links being installed at more and more sites, we are now able to see almost any parameter of our transmitters, power meters, STLs, tower lights, etc. through an IP connection. This is a great benefit and can really help you watch what is going on out there, but it can also give you a false sense of security in thinking that such remote inspections can replace actual "in person" visits. Yes, this kind of connectivity can provide alarms and daily checks, but it is still very important to personally visit your sites on a regular basis to look for things you cannot see with your IP connection. You need to use your eyes to visually inspect the site for problems like rust on a tower, damage to a building, dirt on air filters, etc. Use your nose to smell anything burning, use your ears to hear failing bearings on blower motors or cooling fans on equipment, and touch and feel to detect something heating up, or possibly the air in the building feeling humid or warm indicating an HVAC issue. The importance of regular in-person visits is visually demonstrated in the photo in Fig. 6.7.2.

To keep from letting time slip away, put your visits on your calendar as you would any other important meeting or event. I recommend a visit every other week, but knowing how tight an engineer's schedule is today, at least twice a month. If you are responsible for more than one transmitter site, you need to schedule visits to each of your sites, and not one site 1 month, and another the next. An example schedule is shown in Fig. 6.7.3.

A good way to make sure you check all of the major systems at your tower site is to create a log book to use for each of your visits to the site. You can create these logs using spreadsheets. See Fig. 6.7.4 for examples of transmitter log sheets.

The logs should include any meter readings on your major pieces of equipment. These items will be discussed in more detail later. Some items might not have metering on them but rather status indicators (or maybe nothing at all). For these items a simple check box on the log will be sufficient.

FIGURE 6.7.2 Neglected transmitter site equipment.

June 2015						
Sunday	Monday	Tuesday	Wednesday	Thursday	Friday	Saturday
	1	2	3 Meet with GM	4 Transmitter Site Visit	5	6
7	8 Engineering Dept Meeting	9	10 Dept Head Meeting	11	12 Capital Budget Meetings	13
14	15 Engineering Dept Meeting	16 Back up automation files	17	18 SBE Webinar	19 Transmitter Site Visit	20
21 Father's Day	22 Engineering Dept Meeting	23	24 Meet with GM	25	26 New Phone System Arrives	27
28	29 Engineering Dept Meeting	30 Transmitter Site Visit				

FIGURE 6.7.3 Schedule transmitter site inspections on your calendar.

6.90 FACILITY ISSUES

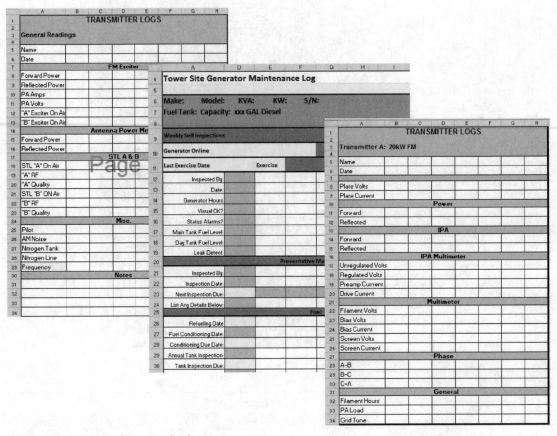

FIGURE 6.7.4 Examples of transmitter log sheets.

The purpose is to remind you to take a look and not forget that it exists. One such item could be your tower lights. Under normal operating conditions you can just check that you have no alarms on your controller box.

When you create logs, try to arrange them so that you walk around your entire building both inside and out. Figure 6.7.5 is an example of a tower site and the path you might take when inspecting the site and filling out logs.

At the bottom of each log you should include a few lines, or have a separate blank page for hand-written comments. Include things on this page that are not normal inspection items, such as "changed a transmitter tube," "added water to a water-cooled transmitter," "cleaned air filters," etc. I suggest copying these lines into a computer document so that you can do key word searches. This will help in the future when you have a problem and remembered this happened before. A quick search may help find the problem and fix it.

6.3 TRANSMITTERS

Perhaps the most important piece of equipment at your tower site is the transmitter. There are many types of transmitters: AM, FM, TV, tube type, solid state, air-cooled, and liquid-cooled. Some are small and mount in a rack taking only a few rack spaces, while others are large with multiple

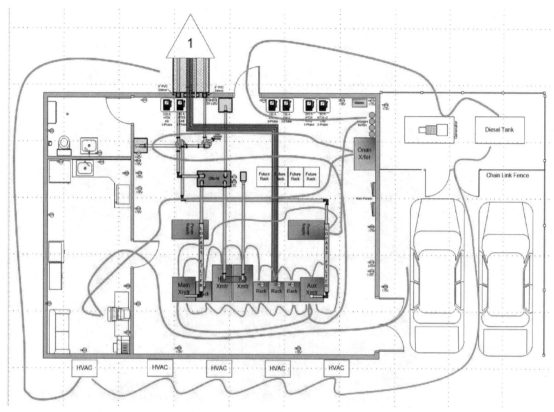

FIGURE 6.7.5 Example of how you may arrange your site inspection to cover the entire building.

cabinets and possibly separate power supplies. Most of these transmitters will have multiple meters on the front, and many times more meters behind the front panel doors. Some of these meters have multiple functions with a selector switch. Include all of these meter readings on your inspection log. See Fig. 6.7.6.

As you log the metering readings week after week you will begin to remember what each value should be, even before you look at it. This is one of the main purposes of this exercise. You will hopefully notice that you get the same readings each time. A value that begins to change over time could be an indication of a problem. See Fig. 6.7.7.

In this example, you will notice that the screen voltage has been increasing on each visit while other readings on the transmitter all remained the same. This particular transmitter's power output is adjusted by raising and lowering the screen voltage. The transmitter also has an automatic power feature to maintain 100 percent power, so what we are seeing here is that the power output has been dropping and the transmitter has been compensating for this drop by turning up the screen. This is most likely an indication that the tube is starting to fail. If you were just monitoring the transmitter by remote control, you probably would have been looking only at the main power output, which would have appeared as if nothing was wrong. This problem probably would have been missed until the screen supply's power reached its maximum, and then the output power would have begun to fall. At this point a tube change would need to happen immediately, catching your business office off guard on a very large expense. Finding this problem early allows you to alert your boss early so there is time to prepare for this hit to the budget. If you see something changing and do not know the reason, you can call the manufacturer's technical support line for help and possibly fix a problem

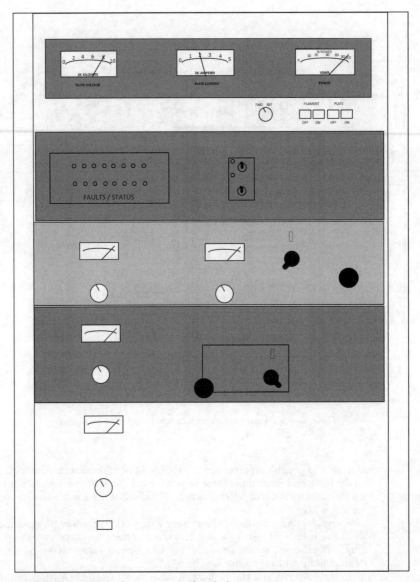

FIGURE 6.7.6 Typical radio transmitter with multiple meters.

before it actually was noticed or took you off the air. This example shows the importance of regular inspections and logging.

After you have written down the meter readings you next need to make sure that they match the values on your remote control. You will most likely only have plate (PA) voltage, plate (PA) current, and power output on the remote control, but these must match the actual values on the transmitter. Recalibrate the remote control as needed using the process specified in the user manual.

Some newer transmitters may have a more sophisticated computer monitoring system. Such systems will most likely have too many parameters to write down in a log. The system may feature a menu type arrangement or offer various tabs or pages of readings. See the example shown in Fig. 6.7.8.

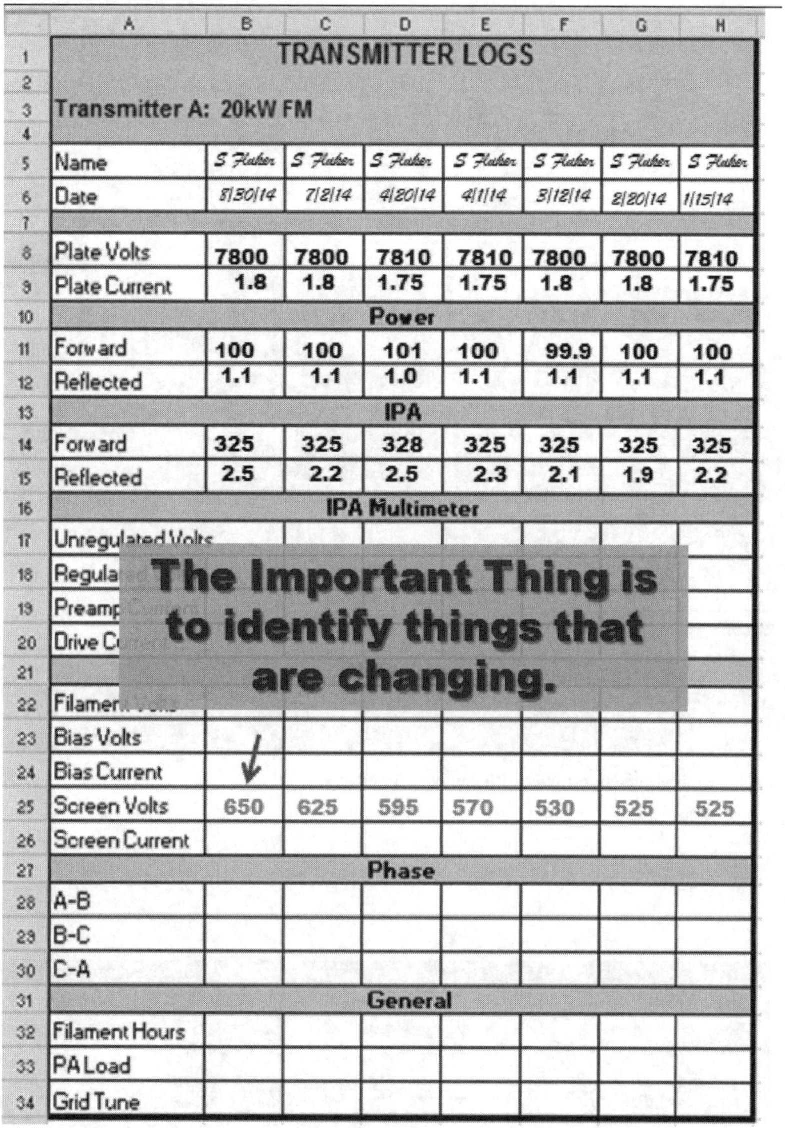

TRANSMISSION SYSTEM MAINTENANCE

TRANSMITTER LOGS

Transmitter A: 20kW FM

	B	C	D	E	F	G	H
Name	S Fluker	S Fluker	S Fluker	S Fluker	S Fluker	S Fluker	S Fluker
Date	8/30/14	7/2/14	4/20/14	4/1/14	3/12/14	2/20/14	1/15/14
Plate Volts	7800	7800	7810	7810	7800	7800	7810
Plate Current	1.8	1.8	1.75	1.75	1.8	1.8	1.75
Power							
Forward	100	100	101	100	99.9	100	100
Reflected	1.1	1.1	1.0	1.1	1.1	1.1	1.1
IPA							
Forward	325	325	328	325	325	325	325
Reflected	2.5	2.2	2.5	2.3	2.1	1.9	2.2
IPA Multimeter							
Unregulated Volts							
Regulated Volts							
Preamp							
Drive Current							
Filament Volts							
Bias Volts							
Bias Current							
Screen Volts	650	625	595	570	530	525	525
Screen Current							
Phase							
A-B							
B-C							
C-A							
General							
Filament Hours							
PA Load							
Grid Tune							

The Important Thing is to identify things that are changing.

FIGURE 6.7.7 Log showing a changing transmitter parameter.

The best thing to do in this case is to pick a few important readings on each tab. This will guide you through each page where an alarm may be spotted—even if it is not one of your normally logged items, such as shown in Fig. 6.7.9.

If your transmitter has the capability of printing or storing a full set of meter readings, you should do this on each visit. If you can do it remotely at your studio, you might want to do this once a month. Logs can be saved in PDF files to stay paperless. These files do not replace your written logs, though. You still need to do a full visual inspection and hand-write logs so that you can spot developing problems. The saved logs are mainly to help you troubleshoot in case a problem does happen by giving

6.94 FACILITY ISSUES

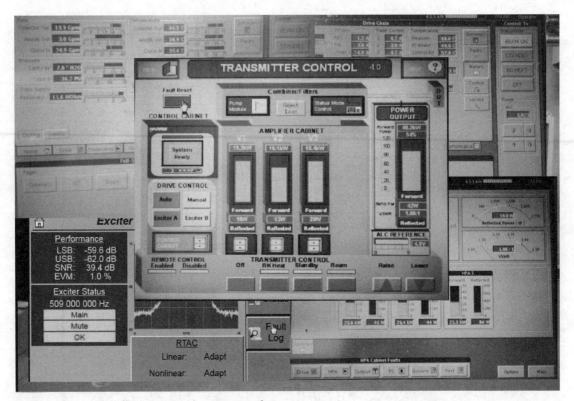

FIGURE 6.7.8 Example of a transmitter with computer interfaces.

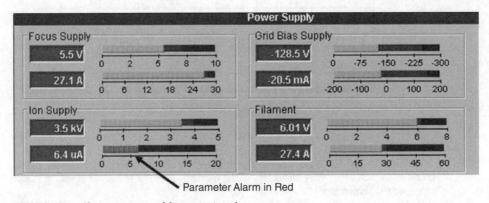

FIGURE 6.7.9 Alarm seen on one of the monitoring tabs.

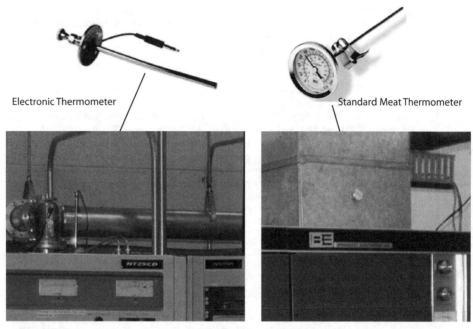

FIGURE 6.7.10 Types of stack temperature thermometers and possible locations.

you a set of logs when everything was operating normally, and to provide data to the manufacturer should you need help with troubleshooting a failure.

After checking the metering on the transmitter, take a look at the temperature of the air coming out of the top of the transmitter. This is called the stack temperature. It is more important to monitor this temperature on a tube transmitter than a solid state, but it can still be a good indication of trouble in both cases. You can use either an electronic thermometer—such as the one pictured in Fig. 6.7.10, which connects to the remote control—or a standard meat-type thermometer. If the transmitter air output is open and the air is blowing into the building, secure the thermometer to the top of the transmitter with a tie wrap or bracket. The temperature on the thermometer could vary dramatically if it rolls around on top. If the transmitter has duct work to exhaust the hot air outside of the building, poke a hole in the duct and push the thermometer into it.

Check and log this temperature. As long as the room temperature remains stable, the stack temperature should also remain the same. If the room temperature changes, the stack will change accordingly. Basically, the difference between the temperature input and output of the transmitter should remain constant. A rise in the stack temperature is an indication of a problem with transmitter efficiency. It could be as simple as adjusting the tuning, or it could be something more serious such as a reduced air flow or a component failure in the PA.

If you have never tuned a transmitter, contact the manufacturer for instructions, or ask another engineer to help you. Tuning the transmitter is not difficult, but should be done under direction the first time and until you are comfortable with it yourself.

Next check the air filter on the back of the transmitter. If it is dirty, replace it, or if it is a washable type, clean using a vacuum or wash it. If you do not have facilities to clean the filter at the tower site, keep a second filter on hand to swap out so you can take the dirty one out for cleaning.

On the topic of cleaning, inspect the inside of the transmitter and clean it periodically. The frequency of the cleaning depends on how clean the building is kept. If the hot air of your transmitter is ducted to the outside, you will save on your air conditioning bills; however, you will need to have some kind of a vent to bring in outside air to replace what you are pumping out of the building (or you could actually create a vacuum inside the building that can starve the transmitter of air flow). The outside air will bring in more dirt and will require more frequent cleaning of the inside of the transmitter. A cleaner method is to use a closed air system in which the stack of the transmitter is blown into the building. It does require a larger air conditioning system for cooling, but it will keep the air cleaner and reduce the number of times you will need to clean the inside of the transmitter. It will also extend the life of the transmitter.

If you have additional transmitters such as one for HD Radio, or you have a backup transmitter, inspect them regularly as well. Turn on your backup transmitter during each visit and keep a log on it. Before turning on your auxiliary transmitter, make sure the test load is ready with the fan turned on, or if it is water-cooled, make sure water is flowing properly through the load. For air-cooled loads, listen to the fan blower for proper sounds. Note any sounds that could be bearing issues or anything rattling inside. Correct as necessary.

A quick note on the test load: be sure there is an interlock between the load and the transmitter, possibly through a transmitter coax switch controller. Should the load fail, you want the interlock to automatically switch the transmitter off before damage occurs.

6.4 TRANSMISSION LINE

Check transmission lines for heating and proper pressurization. Most high power radio and TV stations use a type of transmission line that requires pressurization. As you can see from the pictures in Fig. 6.7.11, these lines have a copper center conductor suspended inside of an outer line with Teflon rings or spacers. The rest of the area inside the line is just air. This line is run up the tower, exposed to outside weather elements and subject to humidity and moisture condensation. Water build-up inside the line can cause arcing, which can lead to a burn out in the line.

The inner section of a transmission line that failed due to arcing is shown in Fig. 6.7.12.

To prevent moisture buildup, the inside of the transmission line must be pressurized. The line can either be pressurized with dry nitrogen or by use of a dehydrator, which is a pressure pump that first dries the air. Having this positive pressure inside the line prevents moisture from developing. Examples of these two systems are shown in Fig. 6.7.13.

Of the two methods, nitrogen is the better option because the dehydrator is just pumping air into the line, which has oxygen in it. Oxygen will eventually cause oxidation on the inside of the copper line, which will reduce its conductivity and can eventually lead to problems. Nitrogen is oxygen-free

FIGURE 6.7.11 Two common types of transmission line: (left) flexible Heliax cable, and (right) rigid line.

FIGURE 6.7.12 A burned inner section of transmission line.

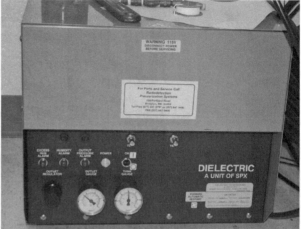

FIGURE 6.7.13 Transmission line pressurization methods: (left) nitrogen tanks and (right) dehydrator system.

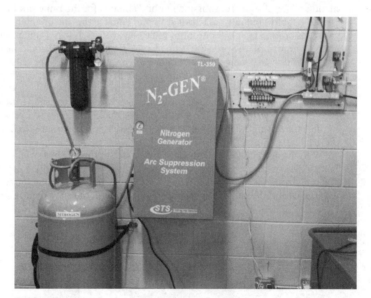

FIGURE 6.7.14 Nitrogen generator system.

and will help extend the life of the line. Dehydrators can still be a favorite choice to eliminate the monthly rental costs of the nitrogen tanks, and also, if you develop a leak in your line, it can become expensive to replace nitrogen tanks frequently. Another option is to purchase a nitrogen generator (Fig. 6.7.14). This approach will be a little more expensive upfront than a dehydrator, but will give you the best of both worlds in the long run.

6.4.1 Maintenance Considerations

For all three pressurization methods, it is important to check the pressure gauges and log the readings. You should have a gauge on the transmission line itself to measure the pressure inside the line.

FIGURE 6.7.15 Tap on the pressure gauges to be certain they are not stuck.

A typical reading may be around 5 or 6 psi. Some engineers prefer to run the pressure lower in the area of 3 psi, and in cases where you are running the RF power in the line near the maximum capacity, you will probably want to use nitrogen and run the pressure up a bit higher. Check with the manufacturer of your transmission line and antenna for their maximum pressure specification so that you do not blow out any gaskets or seals and damage the line or antenna. When reading the pressure gauge, be sure to tap on the front of it. See Fig. 6.7.15. It is not uncommon for these gauges to stick and you may think the pressure is fine when in reality the line could be empty.

In addition to a gauge, there should be some type of pressure sensor on the line that can trigger an alarm on the remote control if the pressure drops too low. Better yet, install a pressure sensor that will give you an actual pressure readout on your remote control, as shown in Fig. 6.7.16. That way if you get an alarm, you can see the actual pressure and determine if the problem is an immediate emergency, or if you can wait until the next business day to go add pressure.

If you are using a dehydrator or nitrogen generator, make sure to check and log the "hours" meter on the device (Fig. 6.7.17.)

With nitrogen tanks, you can catch a new leak in the line by how fast the tank pressure goes down. With a dehydrator or nitrogen generator, the only way to identify a new leak is by logging the hours and calculating the average daily run time. When you see a jump in the run time, you know something is going wrong. These devices typically run very well, but may require some maintenance. There will

FIGURE 6.7.16 Electronic pressure gauges can allow direct psi readings on the remote control.

FIGURE 6.7.17 Hours meter on a nitrogen generator.

be a dryer module (Fig. 6.7.18) on these units that will need periodic replacement when they have removed too much moisture. The blue crystals will start to turn lighter in color and then pinkish.

The seals on a dehydrator can also wear out and may need to be replaced from time to time. Check the overall operation of the dehydrator each time you visit the site and repair as necessary.

6.5 GENERATOR MAINTENANCE

Another very important part of the transmitter site is the backup generator. It is another item that we tend to take for granted until you have a power outage and the generator fails. Some simple maintenance steps can keep the generator running well, and tip you off when something is starting to go wrong.

Hopefully your generator has a built in timer that starts it up and runs it for about 20 min every week. You should also have status indicator on the remote control to let you know when the generator starts up, and another status to tell you when the generator transfers and is actually online. I know of a station that did not have this status, and a fuse on a utility power pole blew. It was the line that fed their building only, so no neighbors had a failure or reported it. The generator ran fine until it ran out of fuel; then they were off the air. A simple status light would have tipped them off that the generator had been running for a long time.

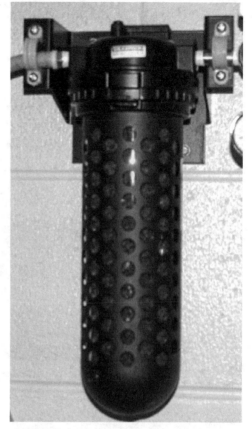

FIGURE 6.7.18 Dryer module on a nitrogen generator.

Even if you have a timer, you should still start up the generator when you visit the site. You want to be able to listen while the generator starts to be sure it is fast and strong. If it struggles to start, it could be an indication of the battery starting to fail. Once it is running at full speed, listen for any strange sounds that you're not used to. (Be sure to wear ear protection when close to the generator.) Write down any meter readings that you can find on the generator in the inspection log. Write down the generator run hours on each inspection too. If you have a visit where you do not run the generator, compare this run time to the last visit to be sure the automatic exercise routine actually worked. A typical generator control/meter panel is shown in Fig. 6.7.19.

Look for a switch marked Auto/Manual, as shown in Fig. 6.7.20, and be sure it is in the Auto mode. In manual, the generator will not start up if the power fails. When a service company is working on your generator, they will typically put this switch to manual, and sometimes will forget to switch it back. Do not just check after servicing though, get in the habit of checking it every time you are at the site. Also look at the status lights on the generator and log problems, if any.

Next, take a look at the transfer panel and write down any readings on it. Check the status lights, as illustrated in Fig. 6.7.21. The type of indications will vary from manufacturer to manufacturer, but may be similar to the photo.

Usually you should see the "Normal" light lit, indicating the system is on utility power. The white "Available" light means that you have normal power from the utility company. When you start up the generator, you should see the Orange "Emergency Available" light come on. This tells you that the

FIGURE 6.7.19 Typical generator metering to log.

FIGURE 6.7.20 Example of a manual/auto switch on the generator.

generator power is getting to the transfer switch. If this light does not come on, you either have a burned out indicator light, or there really is not any power coming from the generator. Check the generator's main breaker. If this trips, the generator will run, but there will be no power from it. Figure 6.7.22 shows the position of a main circuit breaker on a common type of generator.

This is another important reason to make regular visits to the transmission site. You might see the generator starting up every week, but if this breaker is tripped, you will be off the air during a power outage. Make a notation in the log if this breaker does trip. It can happen sometimes, but if it continues to happen, you need to have a service company come out and inspect the generator.

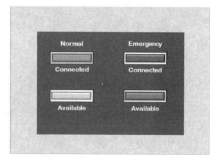

FIGURE 6.7.21 Power status lights on a generator transfer panel.

FIGURE 6.7.22 Main circuit breaker on a generator.

The next step is a visual inspection of the generator and fuel tanks. Walk around and look for any oil, coolant, or other fluid leaks. Check the fuel lines for leaks or rust and make any necessary repairs. A leak problem is shown in Fig. 6.7.23.

Check the fuel level in the fuel tanks. Like pressure gauges, do not trust the fuel gauges. You can tap the gauge to be sure it is not stuck, and you might want to keep a measuring stick at the site and manually check the fuel level with the stick from time to time. A fuel gauge is shown in Fig. 6.7.24.

Check the diesel fuel tanks too. Just a visual inspection is all that is necessary. Check the fuel lines for any rust or leaks and treat or repair as necessary. If you have a double walled tank as shown in Fig. 6.7.25, look for a leak-detect gauge. These tanks are actually a fuel tank inside of a second tank, so that if the inner tank leaks, the fuel is contained in the outer tank to prevent a hazardous fuel spill. The leak gauge shows the fuel level in the outer tank. It should be zero. If you see a level on this gauge,

FIGURE 6.7.23 Coolant leak at a transmitter plant.

FIGURE 6.7.24 Generator fuel tank gauge.

have it checked immediately. Also check the level of the main tank. If there's a leak, that level should have gone down. If it has not, it is possible that water has gotten into the outer tank. Either way, have it checked immediately.

Single walled tanks, as in Fig. 6.7.26, do not have the containment tank and so any leak will be a spill. These tanks, by code in most areas, must have a containment wall around them. Check this wall for any cracks or breaks in the seal. Also, if outside, rain will be retained as well and should be drained so there enough capacity to contain a spill.

On the subject of tanks, fuel can have a shelf life and go bad. While a lot of stations have small fuel tanks for just a couple of days use and go through it just with the weekly tests, other stations can have large tanks of 2500 gallons or more to remain on the air for several days in an emergency. Under good times this fuel can sit there for years. Also, some people like to top off the tank annually, so you may have some pretty old fuel down in the tank. Old fuel can accumulate dirt and even some algae in the tank. It is a good idea to have the fuel tested periodically. Check with your fuel provider for a treatment that can be put in the fuel to prolong its life. If the fuel is really old, you may consider

FIGURE 6.7.25 Double-walled fuel tank (left) and leak detect gauge (right).

FIGURE 6.7.26 Single-walled fuel tank with containment wall for spills.

having a company come in and pump out the tank, running it through a filtering system and returning it to the tank.

Another fuel issue is getting water in the tank. You can perform a simple check for water by putting a special paste, such as Kolor Kut on the measuring stick and dipping it in the fuel. This is a brown paste that will turn a bright red if it touches water. If you find water in the tank, call your fuel supplier for service immediately.

Generators with large fuel tanks located a distance away, may use a "day tank" that sits right next to the generator. This will provide enough fuel for a short period of runtime. It will help the generator's fuel pump work easier by only pumping through a short fuel line. (See Fig. 6.7.27.)

FIGURE 6.7.27 Generator day tank and fuel gauge.

When the tank level is low, a pump inside will transfer fuel from the larger tank. There should be a fuel gauge on the day tank. Locate this and write down the level in the maintenance log. If the pump fails, the generator will run until the tank is empty, even though you may still have a large supply of fuel.

About once a quarter it is a good idea to actually transfer the generator to the load of the building. The generator could run every week just fine, but running the building is a different story. When you switch, be sure your generator does not struggle. Under full load look for a gauge that will indicate the amount of power load being used. Some generators may have a "percent power load" meter. You can use this to determine if the generator can handle the full load of the building, including back up transmitters, or if it can only handle a partial load. You can also "underload" a generator, or run it with too small of a load. Generators run better under a heavy load approaching its maximum capacity. Most manufacturers recommend a load of 70–80 percent of capacity. If the load of the building is light—for example, the generator is a 100 kW generator and the building only draws 40 kW—you may want to consider hiring a service company to perform a "load bank test" annually. They will drive in with a large load on a trailer and hook it to your generator to run it at full capacity for five to eight hours. A generator running under too low of a load will have low combustion pressure, causing carbon build-up and soot that can clog the piston rings. A load bank test will help bring up the exhaust temperatures and help to clean out the cylinders and improve the efficiency. If the generator is running with a marginal load, try turning on back up transmitters and air conditioners. If this brings the load up to the 70 percent or higher level, you can perform your own version of a load bank test.

Generators are like your car in that they need regular and routine servicing and preventative maintenance. I recommend service at least twice a year—more often if you are in an area that has frequent power failures, or after a long-term power failure. Make sure the battery is checked and replaced as necessary. I have found replacement after 2 years is good idea. You can check the battery charger as well to be sure the voltage is correct and that it is taking a charge. A generator failure can create a panic and many of the problems could have been avoided with proper service and repairs.

Cost Savings Tip: Some power companies across the country will have a charge on your bill based on the peak power, or highest amount of power you used at any single time during the month.

This fee can be added to the bill for several months after this peak usage. Testing your back-up transmitter a couple of times a month for 15 min at a time may be enough to increase this peak power fee. If the generator can support the load of your main and backup transmitters together, you may consider running the building on the generator power during your inspection. It may reduce this fee on the power bill.

6.6 SUPPORTING EQUIPMENT INSPECTION

Your transmitter is not the only piece of equipment inside your building that you need to inspect. Make sure you inspect the supporting equipment too. Be sure to check and log the parameters of the exciter(s). Like with the transmitter, the exciter may be an older type unit with a few meter readings, or a newer model with multiple computer pages and a lot of readings. Pick and choose a few readings that will help you monitor the health of the unit.

Also check the STL receiver(s). This could be a microwave system or an IP type of receiver/decoder. For traditional microwave receivers, write down whatever meter readings are available, especially the signal level (Fig. 6.7.28). If you notice a drop in signal level, move on to the rest of your inspections and recheck the signal level several times again before you leave. Certain atmospheric conditions can cause the signal to fade so you need to determine if this signal loss is caused by a fade, or if there is another problem. If the signal does not come back, have the STL dish or antenna checked for alignment. If wind blows the antenna out of alignment, atmospheric fades could cause a total loss of signal taking you off the air.

FIGURE 6.7.28   STL signal strength readout.

As an engineer, you are required to maintain the licensed carrier frequency within the following tolerances:

- AM stations must be within ±20 Hz.
- FM stations must be within ±2000 Hz. The 19 kHz stereo pilot frequency must be within ±2 Hz.
- The DTV pilot signal must be within ±1000 Hz.

The FCC no longer specifies how often you need to measure the operating frequency, just as often as necessary to maintain compliance. You should have a spectrum analyzer or frequency counter to measure the frequency. If the transmitter is very stable, which most newer models are, I would suggest checking the operating frequency at least once a quarter. Check more often if you are noticing a drift in the frequency. In some regions, you may be able to find an independent contractor who can perform the measurements and provide you with a report. This is money well spent as it takes the burden off of you and is looked upon favorably by an FCC inspector.

Monitor and log any and all power meters in the system, including forward and reflected power to the antenna. In a combined transmitter system, monitor and log the forward and reflected power out of each transmitter to the combiner. Finally, in a combined system, log the power being sent to the reject load. A comprehensive power monitoring setup is shown in Fig. 6.7.29.

Next check the equipment racks. Open the back door (if there is one) and look inside. Use your nose to smell anything burning. Look at cable harnesses to be sure they are still supported and not stressed on connectors. Check for dirty air filters on the equipment in the rack. If there is dust and dirt inside the rack, take a moment to clean it out.

Up until now, we have discussed general inspections for all types of facilities. Next, we will discuss unique items that will apply to only AM, FM, or TV stations.

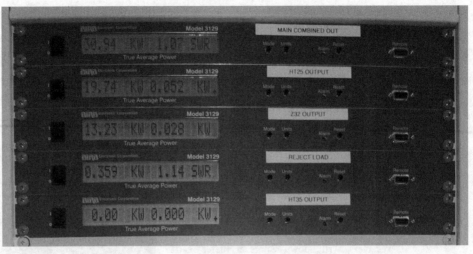

FIGURE 6.7.29 Power meters setup.

6.6.1 FM Radio Station

Check the modulation monitor for proper modulation and SCA parameters, and adjust as necessary. Check the stereo pilot level. It should be between 8 and 10 percent; is typically around 9 percent.

If you are running subcarriers (SCAs), inspect the SCA generators for proper modulation and deviation, and verify that you have the right SCA injection level. This injection may vary depending on how many subcarriers are running. These may include the 57 kHz RDS data carrier, and possibly music or data services on the 67 and 92 kHz channels. The sum total of the SCA injections should not exceed 20 percent of overall modulation.

Also check overall modulation. It should not exceed 100 percent unless you are running subcarriers. In that case, take the sum total injection percentage of all of the subcarriers (up to 20 percent), divide by two, and add that to the normal 100 percent modulation, and that will be the limit. For example:

Normal modulation (includes pilot)	100 percent
57 kHz RDS carrier	9 percent
67 kHz SCA	9 percent
Total SCA injection	18 percent
Half of SCA injection	9 percent
Total allowable modulation	109 percent

FM transmitters can be a single box, or may include two or more transmitters combined together. If you have two transmitters combined together, the transmitters must be in phase with each other and the power levels must match. If either of these parameters is off, you will lose some of the power to a reject load instead of going to the antenna. Check with the manufacturer of the transmitters and combiner to learn how to adjust the phase between the two transmitters. It is typically just a pot adjusted with a screwdriver, or it could be a line section after the exciter called a "trombone section" (see Fig. 6.7.30). This allows you to lengthen or shorten a section of coaxial cable from the output of the exciter to one of the two transmitters to provide fine matching phase adjustments. Between this adjustment and matching the power output of the two transmitters, you should be able to have very little or no power feeding the reject load.

An HD Radio signal will complicate things slightly. The HD signal is actually a separate carrier from the main channel. Think of it as two separate radio stations operating on the same center frequency. There will be a main analog carrier, plus a second HD carrier, which will have the main carrier portion suppressed to ride on the outside of the analog carrier. See Fig. 6.7.31.

This carrier will be achieved in one of three methods. Either the signal will be combined in the exciter and passed through a single transmitter (common amplification), through two transmitters with their outputs combined (mid-, high-, or split-level combining), or with two transmitters feeding two separate antennas (dual antenna or interleaved antenna). With mid-, split-, or high-level combining, you will have a combiner similar to analog dual transmitters; however, you will not be able to eliminate all of the reject load power. If you are not familiar with proper tuning, ask for help or training. For dual antennas or interleaved antennas, make sure you do not have excessive induced power coming back down either line into the transmitters. A problem with a reject load could result in excessive power back into a transmitter in the form of reflected power. The transmitter may react and shut down; however, the problem will actually be with the second transmitter, which may not shut down. Carefully monitor the forward and reflected powers of the entire combining system and watch for changes. Address this immediately or damage could result.

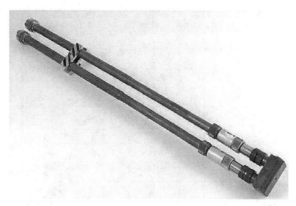

FIGURE 6.7.30 Tuning section for coax line called a "trombone section."

6.6.2 TV Station

High power TV transmitters are typically liquid-cooled. They use either glycol or distilled water for cooling; some transmitters use a combination of both methods. These cooling systems called heat exchangers require additional maintenance. The glycol or water levels must be checked on each visit. If the tank level is low, add more to it. A pump is used to circulate the coolant throughout the transmitter, and possibly the IPA. Check and log all of the flow pressure readings. If you have a main and alternate pump, switch pumps every few visits to confirm that both are working. There should also be pressure gauges feeding each transmitter cabinet. Fluid-cooled systems typically use fluid-cooled dummy and reject loads, and each of these should have pressure gauges. Write the value of each of these gauges in the inspection log. (See Fig. 6.7.32.)

You may also find gauges inside of the transmitter itself (see the example in Fig. 6.7.33). Caution: only open portions of the transmitter that are safe and do not expose high voltages. Log the pressure value and be sure that it is within the manufacturer's specifications. Adjust as necessary.

Water filters may also be included inside the transmitter (Fig. 6.7.34). These should also be checked and replaced per the manufacturer's recommendations, typically about once a year.

If using glycol, check with the transmitter manufacturer for guidelines of how often the glycol should be changed. Typically this will be about every 5 years. At that time you can choose to either replace it, or test it. Test kits are available if you want to do it yourself, but it may be better to send a sample to a lab for analysis. You may find that the glycol is still good and its life can be extended.

It is also important to monitor and log the temperature of the cooling system. The radiator and cooling fans of the heat exchanger are typically outside, so the temperature will vary depending on local weather conditions. Figure 6.7.35 shows the outside cooling portion of a typical heat exchanger. Check the cooling fans to be sure they are operating properly. In a larger system like the one shown, you most likely will not see all of the fans running at once; however, in very hot conditions, or if the

6.108 FACILITY ISSUES

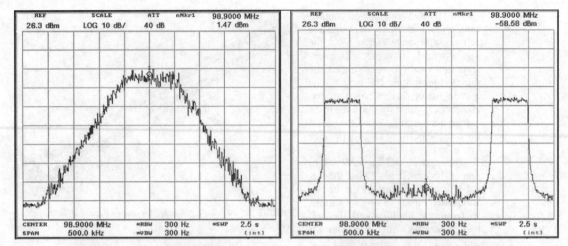

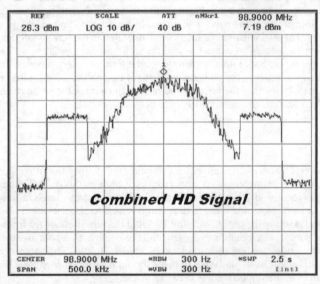

FIGURE 6.7.31 Components of an HD radio carrier.

backup transmitter is running into a dummy load on the same cooling system, you may. The fans are usually controlled by a thermostat, and as the temperature rises, more fans will come online.

The air intake of the heat exchanger should be checked for dirt or debris. In the system pictured, the intake is on the bottom. If the coils or fins on the bottom become clogged, it will restrict the air flow and a rise in the coolant temperature will be observed. This can simply be cleaned off with a hose and nozzle for pressure. Fall leaves can commonly clog the intake.

FIGURE 6.7.32 Temperature and pressure gauges in a liquid-cooled transmitter.

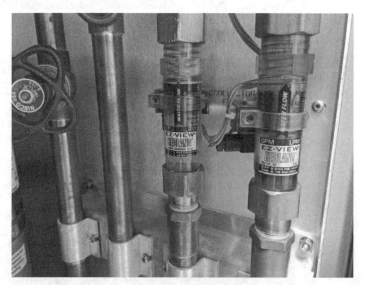

FIGURE 6.7.33 Flow gauges inside a transmitter.

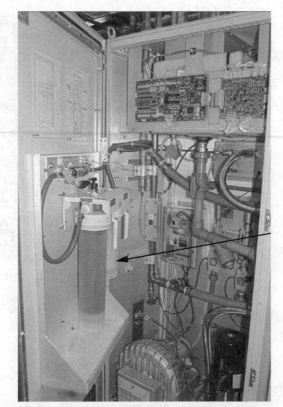

FIGURE 6.7.34 Water filter and deionizing filter inside a transmitter.

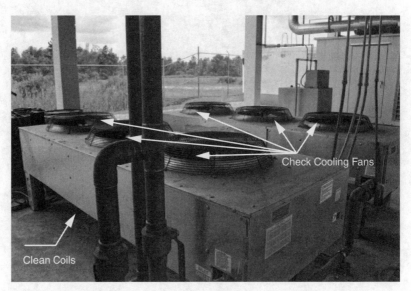

FIGURE 6.7.35 Typical heat exchanger cooling system.

6.6.3 AM Station

AM radio stations have a set of challenges of their own as many of them have directional antenna patterns that must be maintained. Another difference between AM radio stations and FM or TV stations is the type of antenna that they use. TV and FM antennas are mounted on a tower, many times at the top. For an AM station, the tower itself is the antenna. Since the tower is hot, it cannot be grounded directly so it typically sits on an insulator. See Fig. 6.7.36.

Caution: the tower is energized when on the air and you could receive a severe RF burn—or worse—by touching it. To work on an AM tower you must first turn off the transmitter, and be sure the tower is not energized. This is still not enough. Again this is an antenna, and a large tuned one at that. To make the tower safe to work on, short out the base of the antenna as shown in Fig. 6.7.37. Again, be certain that the transmitter is not turned on and connected to the tower before you apply the short. You will not be shorting the tower on a routine site inspection; however, there is one item at the base of the tower you should check. Notice the arc gap in the figure. Because this tower is "floating" and not grounded, this gap is close and grounded on the nontower side and will provide a discharge path for energy buildups such as lightning, thereby protecting the transmitting equipment. Take a look at this gap to be sure it is in good shape. Some pitting is normal, but if the balls are really eaten up from lightning, or even melted, it is time to replace them.

An AM tower is matched to the transmission line with an *antenna tuning unit*, or ATU, which is located at the base of the tower. See Fig. 6.7.38. The ATU can be in a cabinet as in the photo below, or it can actually be housed in a small building, called a doghouse. As with the tower, when you open this tuning unit, high voltage is exposed in the coils and capacitors, and care should be taken not to touch anything without turning off the transmitter and shorting out the tower. The one thing you

FM antenna mounted on the tower

Tower is the antenna for an AM station

FIGURE 6.7.36 Difference between FM and AM antennas.

6.112 FACILITY ISSUES

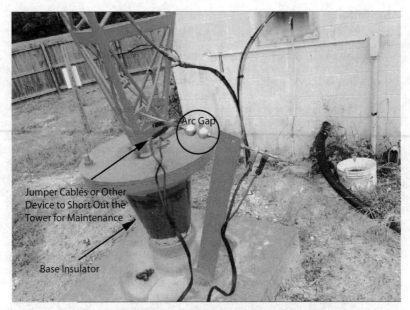

FIGURE 6.7.37 The base of an AM tower and how to short it out for maintenance.

FIGURE 6.7.38 Typical AM antenna tuning unit.

should log during your inspection is the base current at the tower. This is the meter shown in the figure. The base current is a licensed value and must be within tolerance. If you have a directional AM array, you will need to check the base current on each tower for both the day and night patterns.

A nondirectional AM station will simply have a transmitter with a transmission line connected to the ATU, whose output connects to the tower. A directional AM system is a bit more complicated. A directional will have two or more towers, each fed with power from the transmitter. The power levels will vary from tower to tower, and each tower will be fed with a phase variation. This power and phase differential, along with the physical spacing of the towers, produces a designed and predicted signal pattern to put a stronger signal over populated areas, while sending a lower power level in directions of other cochannel stations to protect them from interference. A piece of equipment called a *phasor*

TRANSMISSION SYSTEM MAINTENANCE **6.113**

FIGURE 6.7.39 AM phasor cabinet.

is used to split and distribute the power to the different towers. A typical phasor cabinet is shown in Fig. 6.7.39. This particular phasor controls a four-tower directional antenna array. Each tower has tuning adjustments to increase or decrease the power sent to the tower, and an adjustment to vary the phase of the signal sent to the tower. As the station engineer, you are responsible for the proper tuning of this device. There are several meters and monitors used to keep the pattern in line with the FCC assigned license parameters. First is the "common point" measurement (Fig. 6.7.40).

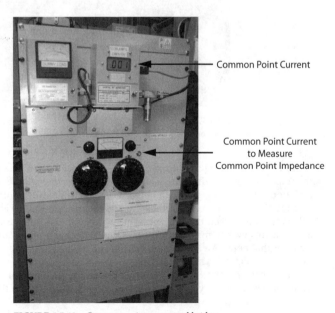

FIGURE 6.7.40 Common point meter and bridge.

The common point meter is located at the point where the transmitter power enters the phasor cabinet. It measures the overall current from the transmitter before it is split and distributed to the individual towers. Log this value in the transmitter logs. This reading is tweaked by adjusting the power output of the transmitter.

Also shown in Fig. 6.7.40 is the common point bridge. This instrument measures the common point impedance of the antenna system and, again, is a licensed value. Not all AM stations have this device on line all the time. Many have a portable version that can be inserted into the transmission system. If you are unfamiliar with the use of the common point bridge, find someone who knows how to use it to teach you. It is not difficult, but is beyond the scope of this chapter.

The next piece of equipment you need to be familiar with is the antenna monitor (Fig. 6.7.41). With this instrument, you will set the value for the main reference tower, and it will give you the power and phase ratio measured for each additional tower. Again, these values are assigned and licensed to each individual radio station and you are responsible for maintaining these parameters by adjusting the phasor. Again, if you have never adjusted an AM array, do not attempt to do this on your own. Seek help from another engineer with directional experience. Turning the controls on the phasor cabinet can get you into serious trouble. For example, if you have a five-tower array and the power ratio on one tower is too high, you might think you can simply use the power adjustment for this tower on the phasor cabinet to lower the power. Unfortunately, if you lower the power to this tower, you will be increasing power to the other towers as the transmitter overall power and common point current will not be changing. The power has to go somewhere. So, any change you make with any tower will have an effect on every other tower in the array, and in turn will cause the RF pattern to change. It is not difficult once you learn how to adjust the array. Some things that can cause a change in the tower parameters can include aging parts, such as capacitors, oxidation of coils, and even weather and ground conditions. Many AM stations are built in swampy and wetlands. If the land dries, the signal reflections may change. Snow on the ground can also cause temporary changes in the array parameters. Log the values for each tower on each site inspection. If the station is a DA-2, which means you have one pattern in the daytime and a second pattern at night, you will need to check and log the values for each pattern.

FIGURE 6.7.41 Typical AM antenna monitor, which measures the phase and ratio for each tower in a directional array.

You will also be required to verify that the signal pattern is correct. This is done by measuring the "monitor points." These are geographic locations that will be listed on the station license, with directions on how to get to them. You need to drive to every point listed on the license and measure the signal strength at that location, and compare it to the value on the license. The measured value must not exceed the licensed value at any point. A value too low is not in violation, but if it is extremely low, you may want to check your antenna monitor and make adjustments to raise the signal at that point. Measurements are made through the use of an AM field strength meter, such as the one in Fig. 6.7.42.

In the past, monitor points were required to be checked monthly. This is another rule that has been relaxed; however, you are responsible to know at any given time that the points are within the

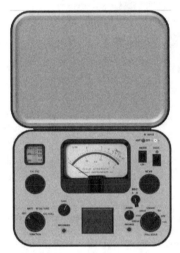

FIGURE 6.7.42 AM field strength meter.

FIGURE 6.7.43 RF contactor on an AM antenna system.

licensed parameters. Once-a-quarter measurements are strongly recommended—more often if you have an antenna system that is not stable.

Finally, AM directional arrays will usually have some type of switch contactor (Fig. 6.7.43).

Take a look at the fingerstock of the switch. Hopefully, your system is set up to turn the transmitter off before the switching takes place. Switching hot could cause arcing and damage to the fingerstock. Age can cause this device to fail as well. Do a visual inspection. If damaged, schedule a time when you can take the station off the air to replace the defective component(s). Remember there is exposed high RF voltages and current on these contactors; do not touch.

6.7 HVAC AND ELECTRICAL SYSTEMS

The air conditioners at a transmitter site work much harder than in most other buildings, especially if you use a closed air system (exhausting the transmitter stack air back into the building instead of ducting it outside). Because of the heat load of the transmitter, the AC units will most likely run 24/7, 365 days a year. Even on very cold days there may be enough heat inside the building for the AC units to continue to run. The HVAC system should be split into multiple units. If using two units, both should have the capacity to handle the entire load of the building should one fail. A better plan is to use four or five units. If sized properly, all of the units could run at all times; however, a failure in any one or two could be compensated by the other units running harder. A five unit installation is shown in Fig. 6.7.44.

FIGURE 6.7.44 Transmitter room with five HVAC units for redundancy.

During the site inspection, check each of the units to be sure they are blowing cold air. You can feel it with your hand, or use an infrared thermometer to check the air temperature. If one unit is putting out warm air, it may just be the thermostat, or it could be a unit failure. Do a visual inspection to be sure the coils are not iced up.

Also do a visual inspection of the condenser units outside to be sure they are clean, and are not making excessive noise (Fig. 6.7.45). If you notice a problem such as squealing bearings or belts, or damage to either the inside or outside units, call for service immediately. The HVAC units should be serviced on a regular basis. I suggest setting up a service contract with a reputable local company. You can talk with this company to set up the proper preventive maintenance schedule for your building and type of HVAC units.

FIGURE 6.7.45 Inspect the outside HVAC units as well as the general building perimeter.

While checking the outside units, take a walk around the perimeter of the building for a visual inspection. Clean out any vegetation growing up the side of the building. Look for foundation cracks, water damage around the building, etc. Turn on the outside lights and be sure they are all working. The last thing you want is to get called out for a problem at night and discover the outside lighting is not working.

6.7.1 Electrical Systems

The building electrical system should not be forgotten either. Visually inspect any surge protectors. Larger protectors will have status lights on them. If it shows any failures, call an electrician for repairs or replacement. A typical system is shown in Fig. 6.7.46.

FIGURE 6.7.46 Building surge protector.

Transmitters and large HVAC systems demand a lot of power and current. This could cause circuit breakers and wiring to run hotter than what you might find in a normal office building. About once a quarter it is a good idea to run your hand down the breaker panels, and touch other electrical panels feeling them for heat. This is another good job for an infrared thermometer. If you notice a panel running warmer than normal, call an electrician to check it out. It is also a good idea to schedule an infrared study of the electrical system annually, or at least once every other year. The electrician will come out with an IR camera and remove the breaker panel covers to take images (Fig. 6.7.47).

As you can see from the images in Fig. 6.7.47, a hot breaker will show up brightly. The problem may simply be that a lug on the breaker has come loose. I have seen electricians tighten these loose lugs and within minutes, the breaker cools down. Not all bright breakers are actually a problem. It could just be a breaker running close to its capacity. You can see in the images that even slight increases in temperature can light up in an infrared picture. If you do have a breaker running near capacity, you may want to consider splitting the load on into two separate circuits.

Any UPS in the building should also be inspected. If you have a large unit that carries your entire site including the transmitter, have a local company do a full inspection at least once a year (Fig. 6.7.48). They will be able to measure the capacity of each battery in the system and give you a report on any problems found. Batteries should be replaced every 3–5 years or when directed by the

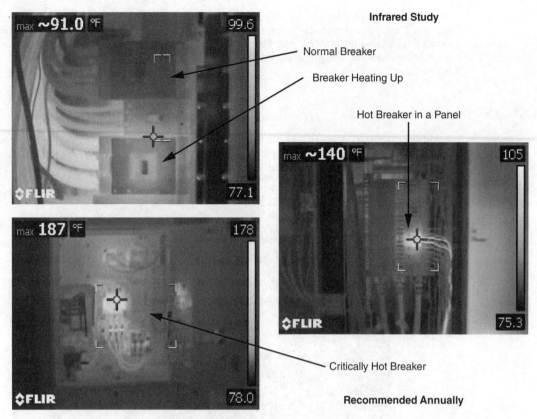

FIGURE 6.7.47 Infrared photos take of breaker panels at a tower site.

FIGURE 6.7.48 Large UPS to power a transmitter and supporting equipment.

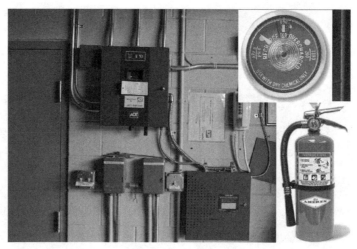

FIGURE 6.7.49 Typical fire systems at a transmitter site.

electrician. Even smaller units should be inspected regularly. You do not want to have a failure of a UPS be responsible for taking you off the air.

The fire systems in the building also require inspections by a qualified fire company. Figure 6.7.49 shows some of the items that should be inspected. Check your local fire code for how often these systems need to be inspected. Fire extinguishers should be inspected and tagged annually, but a monthly inspection by you is advised:

- Make sure you can get to the extinguisher quickly—that nothing is blocking it.
- Check the pressure gauge.
- Check the hose and nozzle for any breaks or cracks.
- Check to be sure the pin is in place properly.
- Check the handle and be sure it is not broken.
- Make sure there are no dents in the body of the extinguisher.
- Check for rust or chemical deposits.
- Check the tag to be sure it is not outdated.

6.8 TOWER MAINTENANCE

Your responsibility for maintaining the tower will vary depending on whether your company owns the tower or rents space on it. If you rent, the bulk of the responsibility will fall on the tower owner and not you; however, you should still do a visual inspection of the tower and report problems you see to the owner. If the owner misses a problem and the tower fails, you are off the air, so be an extra set of eyes for them and avoid surprises and disasters. You will still most likely be responsible for your own transmitting antenna(s), STL antenna(s), and transmission line(s) mounted on the tower. Hire a qualified tower company to do a climb and inspection of your equipment on the tower and make repairs as necessary. Annually is recommended.

If your company owns the tower, you have a lot more responsibilities. First and foremost, you are responsible for the tower lights. Most tower light systems now have monitoring equipment that will

FIGURE 6.7.50 Inspect the base of the tower and associated equipment.

trigger an alarm for the remote control in the event of a failure. Without this monitoring equipment you must do a visual inspection of the lights at least once a day. If a failure occurs, you must notify the FAA to issue a NOTAM[1] immediately. For a Red Light system, you only need to notify the FAA if any of the red flashing beacons goes out. A failure in a side light must still be repaired as soon as possible, but does not require a NOTAM to be filed. For a white strobe system, any failure must be reported.

A quarterly tower inspection is required and completed from the ground; no climbing is required. First check the tower lights. Turn them on and physically check for proper operation. Do not rely on the alarms on the box for this. View the base of the tower and inspect for rust, faded paint, proper posting of tower registration, and RF warning signs. See Fig. 6.7.50. Check the cement foundation for any cracks or problems. Small cracks can just be logged and monitored over time. Larger cracks, especially in colder environments, should be repaired and sealed to prevent water from freezing in them, worsening the problem.

Check copper grounding wires or copper straps. Copper theft is common and can compromise the safety of the tower site, and the protection of your equipment (Fig. 6.7.51).

Next, go out to each of the guy anchor points and take a pair of binoculars with you (Fig. 6.7.52). The guy anchors should be clean and free from rust. Inspect the cement foundation for cracks and take appropriate actions as in the tower foundation. The large nuts on the tower base plates should have cotter pins on them. Be sure they are in good shape. Check the wires themselves, especially where they attach to the anchor plates. Use the binoculars to look up the guy wires and inspect the tower. Look at the antenna(s) and line(s). If you notice any problems, call a tower crew to perform the appropriate repairs. Continue to each guy wire anchor and do the same inspection.

Fencing is required only around AM towers to prevent accidental contact. I recommend fencing around any tower and around each guy wire anchor. Too many towers have come down from mowing equipment catching the line. It can also help with curious bystanders, and to discourage climbers. At an AM site, you must either fence every tower, or have a fence line around the entire parameter of the property. Again I would suggest fences around each tower for safety reasons. The fencing must be locked and secured. Make sure there is no vegetation growing inside the fences. Any growth should be

---

[1]Notice to Airmen.

FIGURE 6.7.51 Photo of cut and stolen copper ground wires.

FIGURE 6.7.52 Guy wire anchor.

removed before it gets out of hand as in the photo shown in Fig. 6.7.53. Trees growing up through the wires can bring down a tower.

Again, these inspections should be done every quarter. In addition to the quarterly inspections, it is recommended that you also have a qualified tower crew climb the tower on an annual basis for a full inspection. This crew should provide you with a written report including photos of any issues found on the tower. These items should be corrected as soon as possible, even if they

FIGURE 6.7.53 Overgrown guy wire anchor should be cleared out immediately.

are not yet critical. Putting them off will only result in a major budget buster years down the line as more issues accumulate.

The final recommendation for your inspection is to do a basic cleaning of the building. Sweep the building and clean and dust as necessary. A clean building will reduce the number of times you will need to open and clean the inside of the transmitter. It will also extend the life of the equipment.

So as you can see, there is a lot to check at the tower site. Regular and routine visits will help your sites run smoothly for years to come.

SECTION 7
BROADCAST MANAGEMENT

Section Leader—Wayne M. Pecena
CPBE, CBNE, Texas A&M University—KAMU, College Station, TX

Broadcast engineering management is often a challenging area for many technically astute broadcast engineers as they move from the workbench and transmitter plant to an engineering management and leadership role. Their achievements quickly change from personal accomplishments to those of the overall department, good or bad. New areas of responsibility come into play as well as responsibility for a wide scope of functional areas.

The first chapter of this section examines many of the diverse aspects of engineering management beginning with leadership and management skill development. Mike Seaver defines the common Director of Engineering (DOE) roles and responsibilities and differentiating between management and leadership traits. Disciplines in which the Engineering Manager is often responsible for are presented. Specific roles are detailed in budget planning, staffing assessments, and staffing oversight, maintaining essential services, and industry trends as one looks forward to the future. The chapter concludes with the importance of maintaining a healthy interdepartmental exchange practice within your organization.

Project Management (PM) ability is yet another critical aspect of the successful Engineering Manager's skill set. The broadcast technical plant is often very complex and involves numerous hardware and software sub-systems integrated as one to deliver the desired capability and reliability. Jerry Whitaker and Robert Mancini review the art and process of project management. Successful planning is considered the cornerstone of successful PM where failure is often attributed to errors in goal setting, whether setting the bar too low or setting the goals too high. Elements of the Project Management process are examined to aid the Broadcast Engineering Manager in formulating a PM plan of their own to fit their specific environment. Project Management is often considered an inexact science; however, the application of a structured and documented process provides an enhanced success rate.

Systems engineering is considered a multi-discipline approach to implementing a successful system. System engineering regardless of the system size is a major aspect of the broadcast technology manager's required skillset. The broadcast plant is commonly comprised of a diverse collection of specialty hardware and software manufactured by different companies and interconnected via various standard and even proprietary interface schemes. Gene DeSantis provides an introduction to the phases of systems engineering and looks at the specifics of each phase in detail. The process begins with design and development, and moves to performance standards or specification development. Detailed budget preparation follows with an emphasis on budget or cost control tracking as completing a project within budget is often an important gauge of a manger's overall performance. The chapter wraps up with a look at proper staffing approaches and overall staff organization.

Local broadcast content whether radio or TV is often in the strongest demand during a disaster event. Thomas Osenkowsky authors a chapter on disaster planning and recovery by first defining a "disaster." Anticipation of possible disasters is often a critical step in the overall preparation process to be prepared.

Areas prone to significant weather events such as flooding, tornadoes, or hurricanes often have advance notice of an approaching event. Even with advance warning, successful disaster plans must be ready to implement with implementation supplies on-hand. Preparation also includes staff reassignments and maintaining relationships with local emergency authorities. These are important aspects as the broadcast station may be the only medium for local information dissemination. Assessment checklists are provided to ascertain vulnerabilities and the aid in the development of your own mitigation plans. Specific aspects related to the broadcast plant such as generator refueling, alternate STL options, data backups, and transmitter plant redundancies are included. The chapter ends with a disaster preparation checklist that may be utilized as is or modified to suit the needs of your broadcast facility in times of disasters.

The broadcast transmitter is certainly a critical system in the overall broadcast technical plant. The transmitter plant presents a challenging environment containing lethal voltages even where solid-state amplification is exclusively utilized. High RF levels and numerous other factors provide an environment where safety is paramount. Jerry Whitaker explores many of the areas of safety that the broadcast engineer should be aware of and respect. Electric shock and the impact upon the human body is the first area outlined as well as practical precautions when working in a high voltage environment. RF considerations are addressed and applied through an awareness of operating hazards and guidelines one should follow in high power broadcast transmitter environments. The impact of these hazards is emphasized by recommended first-aid procedures to follow in the treatment of electrical shock and burn incidents. RF burns often occur, but burns from heat should not be overlooked. OSHA aspects and common infractions are reviewed as well as nonionizing radiation guidelines are presented including liability aspects that is the reality of today's litigation focused society.

Technical staffing is often a challenge of many fields of technology. Existing technical staffs are in constant need of continuing education to simply keep up with often rapid changing technology. The impact of information technology in the broadcast plant only accelerates the rapid technology change often found. Society of Broadcast Engineers (SBE) leaders Ralph Hogan, Wayne Pecena, and SBE Executive Director John Poray introduce the SBE organization—providing fellowship, certification, professional development, and numerous support programs to the broadcast and media community. Certification and professional development are considered two of the cornerstone of member programs offered to broadcast and media processionals. Certification programs range from the entry level to the professional level in radio, TV, and networking technology. Industry salary surveys have consistently shown those holding SBE certifications receive higher pay levels than their noncertified peers. Continuous learning is a trait of a successful technology professional. SBE professional development offerings include online webinars, more in-depth SBE university online classes, and the renowned Ennes Workshop held each year in partnership with the National Association of Broadcasters (NAB) conference. In addition, several regional Ennes Workshops are held throughout the United States each year. The SBE Engineering Leadership Workshop is a unique annual event that provides a hands-on personal approach to preparing the broadcast engineer for management and leadership responsibilities. If you are not already a member of the SBE, consider the benefits available to members and join today. If you are a member, Thank You and be sure to take of the many member benefits offered by your society!

Jerry Whitaker and Richard Chernock, PhD, wrap up this section and the handbook with a look at the future of broadcast TV transmission and what ATSC 3.0 brings to the industry at a critical time in the future of over-the-air broadcast TV. An understanding of the development process and the structure of the Technology Group on ATSC 3.0 introduces the reader to the key technologies and capabilities of the layered model design of ATSC 3.0. The future may be challenged by spectrum availability and available new generation receivers. However, ATSC 3.0 promises to bring a flexible, extensible, and scalable transmission system to the TV broadcast industry.

Chapter 7.1: Management and Leadership
Chapter 7.2: Project Management
Chapter 7.3: Systems Engineering
Chapter 7.4: Disaster Planning and Recovery for Broadcast Facilities
Chapter 7.5: Safety Issues
Chapter 7.6: About the Society of Broadcast Engineers
Chapter 7.7: Looking Toward the Future—ATSC 3.0

CHAPTER 7.1
MANAGEMENT AND LEADERSHIP

Mike Seaver
CBT, Seaver Management and Consulting, Quincy, IL

7.1 INTRODUCTION

In previous days, the Chief Engineer was almost always the person who could best repair the transmitter and equipment associated with maintaining the essential systems of a broadcast facility. The early broadcast engineer had most often been trained in the military or some similar manner of technical education. Often these individuals also were amateur radio operators and were quite capable of performing repairs on almost anything. In addition, regulations set forth by the Federal Communications Commission required stations to be maintained and operated by someone who had successfully obtained the FCC First Class Radio Telephone Operators License. This rigorous exam was rather difficult, thus limiting the recipients to a relatively small, elite field of participants. The FCC requirement that all operational broadcast transmitter facilities be staffed at all times during operation by a licensed engineer created a natural environment for engineers to thrive and exist in almost complete vacuum. They were mandated by the government, and as long as they did a respectable job of keeping the station legal and on the air it really did not matter how involved they were in the daily operation of the station's business.

7.2 SCOPE OF MANAGEMENT AND LEADERSHIP

With the development of more stable equipment, the abolition of the requirement that the transmitter site(s) be staffed around the clock by licensed engineers, and the eventual elimination of the requirement of the FCC First Class Radio Telephone Operator License, the role of the engineer, particularly the Chief Engineer, changed dramatically. Whether this change was good or bad will be left for another debate and forum. But nevertheless, the role of the Chief Engineer became less a technical position and required a new set of management and leadership-based skills. Often, the people serving in the capacity of Chief Engineer were ill-trained and equipped in this new skill set, which involved areas of expertise that were unfamiliar to the engineering discipline. Instead of just being an equipment repairman, the Chief Engineer must now be an expert in:

- Budget planning and compliance
- Strategic system planning and development
- Personnel staffing, scheduling, and oversight
- Physical facility planning and maintenance
- Staff training

- Maintenance of essential operational services
- Interdepartmental exchanges
- Industry trends and projected changes
- Legal and legislative mandates and trends

In addition to the above skills, the Chief Engineer, now actually a "Director of Engineering," must be able to adequately converse with other personnel from multiple disciplines such as business, financial, and product production, along with being able to maintain all the equipment. Quite a paradigm shift from being the person best at transmitter repair.

To begin, we must establish a working definition of both "Management" and "Leadership."

- Management is "the conducting or supervision of something or someone," "the judicious means to accomplish an end."
- Leadership can be defined as "to guide in a way especially by going in advance; to direct on a course or action."

In this chapter on *Management and Leadership* we shall be using these definitions.

The foundation of engineering, or any department for that matter, begins with the project we like least—documentation. A well-documented department will allow for a smoother, more reliable, consistent operation, and time-efficient in the long term.

It is best to write a statement of purpose for your engineering effort. What is the basic purpose of the department? Whom do you serve, what is your responsibility to the operation of the business? What is the chain of command, and to whom do you answer? Next, what actions and/or services are the department responsible to provide or perform? What is required to make these actions or services a reality?

A Mission Statement giving a short but concise statement of the very purpose and goal of the Engineering Department should be written. An example could be:

"...finding better ways to serve through efficient work methods..."

Along with a written statement of purpose and Mission Statement for your department, you should have a well-documented job description for the Chief Engineer or Director of Engineering. This should come from either your Human Resources department or from the General Management, but the Director of Engineering should have input to the content also. It should outline the duties, responsibilities, and authority. This job description should be discussed thoroughly with the appropriate management personnel, and you should have it reviewed and approved by all those involved with the implementation.

Well written yet concise job descriptions should be established for each engineering position. Each description should include the title of the job, to whom the position reports, the duties for which that position is responsible, expected outcomes, authorities and responsibilities, description of the knowledge level and skills required to perform the duties of the position, and the expected end results. Job descriptions should avoid using extremely specific details, except where the tasks at hand are extremely specific. Clear, unambiguous language should be used to describe each area of the job description. Avoid being narrative in the composition of the duties. The purpose of the job description is to allow the employee to perform the prescribed duties without being "micromanaged."

Duties should set forth the major responsibilities. How these duties are to be carried out should be included in concise, understandable, and unambiguous language. The end results should be identified in understandable, unambiguous terms.

Decision-making authority should be identified in clear, understandable terms. The employee should understand the limits of their decision-making authority to determine what action(s) they can take. You can include examples to illustrate:

"...if the widget requires a new AA type battery you have the authority to install one from inventory without permission. If the device requires a part that needs to be ordered you must request permission to order from your supervisor..."

A well-written job description will include the education level, skill set required to perform, and differentiate between what is required and what is desired. The job description should outline any unusual or exceptional factors affecting job performance. It should also outline any unusual or erratic scheduling of work.

To summarize, the job description should outline the major responsibilities of the job, details of how the job is to be accomplished, the knowledge, skills, and abilities required to perform the duties, and the authorities and limitations assigned to the position. A well-written, concise job description should allow for a smooth, efficient work staff who knows their authority and their limitations. Each job description should be thoroughly reviewed with each employee at the time of hire and reviewed as prescribed by the Human Resources Department or as outlined in the company policy or policies. Copies of the job description should be given to each employee, and the Departmental Manager should keep a copy for each position.

Although your company will most likely have a written inventory of each piece of equipment or property for which engineering is responsible, a new, complete inventory should be conducted. Each item of property should be identified using a unique inventory number that will lend itself to scheduling and tracking methods. This will be quite a cumbersome and time-consuming task but will prove to be well worth the investment in time and work. It will pay dividends in scheduling tasks, preparing budgets, and performing quality assurance.

The inventory should include as much information as is necessary to give each item a unique identity. A workable system might include identifying each item into a generic category, using as many subcategories as necessary. An example might be, "10 Series Portable News Cameras." This category ("10" series; portable news cameras) can be further delineated by adding a suffix of types to identify the individual piece of equipment; for example, "10/001."

An inventory system such as this will allow for the recording of vital information concerning each device. Information about this individual item should include at least: (1) the brand, (2) model, (3) serial number, (4) date of purchase, (5) date put into service, (6) department responsible for use (i.e., News, Creative Service, Sales, etc.), (7) accessories, (8) vendor information, (9) vendor contact, (10) service contract information (if applicable), (11) warranty information, (12) (expected life of service), (13) any other unique identifying factor for this individual device or property. A finished inventory of this product might look like this:

Inventory. # 10/001 – Portable News Camera	
Manufacturer:	Panasonic
Model:	PS 2a
S/N:	123456789
DOP:	01/02/2015
DOS:	02/01/2015
Dept.:	News
Accessories:	Case, Canon Lens, Anton Bauer power supply, two chargers
Vendor:	Acme Products, 2525 Desert Rd., Lupton, AZ 86508 (928.555.1212)
Warranty:	2 years parts/1 year labor from factory
Expected life:	5 years

While some inventory categories will contain many individual items, others—such as towers or buildings—might only contain one or two items. But these items should have their own individual category for quality assurance purposes. A unique numbering system will allow ease in scheduling preventive maintenance, tracking repairs performed, and evaluating individual equipment status for budget planning purposes. The time and effort required to initiate a system such as this will be well worth the effort as each item is tracked for efficient use and utilization. While this is only one example of sequential numbering, about any system that allows for easy tracking will suffice. But at least this basic information concerning each device should be included in any inventory list. This list should be easily available to all engineers.

The unique inventory will allow quality assurance tracking of each specific device in a controllable manner. From this inventory, a comprehensive repair and maintenance history for each system/device/property can be developed and tracked using an action-by-action reporting system.

KXYZ
WORK ORDER

Date: ___/___/___ Requested by: _____

Work requested: _____

_____

Date received: ___/___/___   Date serviced: ___/___/___

Serviced by: _____

Inventory number: _____ Location: _____

Work performed:

_____

_____

_____

_____

Material required:

_____

_____

Date of completion: ____/____/____

FIGURE 7.1.1 Example work order.

The work order form used to request and document all service work, be it repair, calibration, or preventive maintenance, should become the standard working tool of the Engineering Department. This document (see Fig. 7.1.1) will serve to track all work activities and production of the Engineering Department. In addition, this form will provide the necessary documentation during budget procedures to justify replacements for defective equipment as that equipment becomes obsolete. It will also provide numerous feedback data as to the performance of each device, user knowledge and requirements for training, inventory requirements for spare materials and parts, frequency of scheduled preventive maintenance, along with the overall cost of operation.

By recording the date which the work order was written, the date the work order was received by engineering, the date it was serviced, and the date of completion, a time-budget for many repairs/services can be estimated. While repairs will vary in the amount of time required due to unforeseen variables, it will provide an approximation of each service request.

The name of the person requesting the work is valuable since often questions concerning operation procedures and personal observation of the user can be of great importance in determining how to solve the problem. By isolating the user, if the problem is operational in nature, additional trainings can be suggested and arranged to improve on the efficient use of any system or device.

The work requested should be concise, detailed as necessary to implement corrective action. The name of the person performing the service, along with the inventory control data, time, and material, is crucial in keeping track of quality assurance information.

Remember, these illustrations are not designed to be a working form but to illustrate some of the data to begin prudent record-keeping. Using a spreadsheet is prudent, easier to use, and will require much less effort on the part of users. Employ the method best suited to your level of comfort and individual operation. These records are "fluid" in that they can, and should be modified as tasks differ and the need for quality data change.

7.2.1 Budgeting

One of the major roles of engineering should be that of budgeting. For our discussion, we shall consider three types of budgets: (1) revenue budgets, (2) operational budgets, and (3) capital budgets. We will begin our discussion with the revenue budget. While budgets are often referred to as "estimates," or by more cynical people as "guestimates," they are—in fact—a dynamic tool to try, as closely as possible, to determine how much revenue will be available by which to build and project a business model for a period of time. In many aspects, a budget is a "best guess" scenario but is based on many factors that are not just "guesses," but rather historical projections of past revenue/expense models. Of these budgets, the revenue budget is by far the most important. It has been said that a budget is based on the best, most accurately determined projection of revenue. If that projected revenue does not materialize, the rest of the budget is useless. While not quite that stark, the principle is accurate; if the money the plan is based on does not materialize, the plan, as outlined, will not work.

Many of the revenue models are beyond the realm and responsibilities of the Engineering Department and will be determined by those better qualified in determining the business methodologies. But the Engineering Department can, and should, be part of the revenue plan. In business there are two categories—income and outgo. Income that Engineering can contribute to offset their expense, often referred to as "burden" (to revenue), should be from sources such as income from vertical real estate (more commonly referred to as tower space rental), tower site utilization rental, or other sources of utilization of data sale revenue. Cable and satellite retransmissions are often negotiated with the aid of the Engineering Department. Do not limit your imagination when it comes to revenue generation by the Engineering Department. A word of caution, if "tower site utilization" includes renting some land to outside entities be absolutely certain, they will in no manner interfere with operation of the transmitter site.

While revenue generation will never be, nor should be, the largest budget of an Engineering Department, it can serve to reduce the burden on the company's budget. The Engineering Department can demonstrate that they can be part of the income to the station, not only a burden.

The operational budget is the most fluid and dynamic and the most likely to affect the overall operation of the Engineering Department. In general terms, the operational budget usually consists of items and materials that have no long-term investment values and are considered to be expendable things. Items included in an operation budget salaries, benefits, and vacations. The salary is really a compilation of the hours worked, pay scale, the cost of benefits, vacations, healthcare, and other associated expenses. These costs can be determined with the assistance of the Human Resources or the Business Manager. As more and more data is collected using the quality assurance/work order information, the costs of these expenses can be tracked with more efficiency. For example, how much does it cost to perform certain tasks? This information can easily be gleaned from the work order information. As time passes and more data is collected, it becomes easier to predict what the expense of salary/benefits is going to cost.

Another operational expense is the cost of space required to house the Engineering Department. This space is usually based on the cost to "operate" this area and includes things such as lighting, cooling/heating, upkeep/maintenance, taxes, insurance, building depreciation, and any other fixed expenses associated with how much of the total cost of the building the Engineering Department is consuming. The Engineering Department should review its requirements on a regular basis to determine if all the space is really necessary, if more is required, or if the current floor space is satisfactory. Often this "fixed" income can be offset in creative ways. An example would be renting space used for storage to house ancillary services or other revenue-producing services. Although reductions in personnel are often the first to occur when a reduction in "fixed" expense becomes necessary, you should never discount the option of reducing the footprint of the Engineering Department nor the ability to offset any reduction in operational monies by reducing other fixed costs. Other fixed assets such as furniture and other house-wide systems are also considered in operational budgets.

One of the major expectations and goals of an Engineering Department is to maintain the services essential to generate the revenue necessary to support the operational effort in a profitable way. Having the essential material readily available at all times, along with the qualified personnel to perform the work, is part of the operational budget. Perhaps one of the biggest expenses to the operational budget is the maintenance of materials. Storage of material affects the operational budget by requiring more "space" to store that material. Material sitting on a shelf somewhere awaiting use is static, costing revenue while not producing any. This balance is often difficult to predict and can be time-consuming to enact. Again, the work order information plays a great role in determining what materials will be needed to be inventoried and which can be ordered on an as-needed, just-in-time basis. If storage is available and the usage demonstrated, quantity discounts can be considered; but if the need is not there, it is not a good idea to order more just because of a quantity discount. The determination of what material to keep in inventory is best enhanced by utilization of information contained on the work order.

If your department is allotted a vehicle, maintenance, licensing, insurance, and expendables such as fuel should also be considered in an operational budget. Depending on the arrangement your company has (ownership, rental, lease, etc.), you will need to take mileage into consideration. Be sure to discuss this with your Business Manager to gain a better understanding of how to address this expense in the operational budget. Often the stations' vehicular maintenance will be the responsibility of the Engineering Department. While changing the oil, aligning the front end, or replacing the alternator is not within the skill set of engineers, you should have a qualified service provider on standby to be able to provide routine or corrective vehicular maintenance procedures. A good idea might be to work with the sales department to secure a "trade" account which would assist in keeping those expenses within a reasonable amount.

Building and system maintenance, including HVAC, electrical, plumbing, structural maintenance/repairs, snow removal, parking lot maintenance, grounds maintenance (these are but a few of the tasks required in building maintenance) should also be arranged to be covered by qualified personnel that, again, could be a corporate trade or other arrangement.

The Business Manager should become your "go to" person for all budget-related information. He or she is the person most likely to have a view of the "big picture" of the money situation of the business. The Business Manager can help you gain a better understanding of how to plan your cash flow for operational expenses. The Business Manager has the overview of the entire financial plan for the station and can be a most valuable resource in determining how and when to plan on expenses. The Business Manager also knows when the money is most available or be able to assist in creative ways to accomplish your goals. We will discuss the role(s) of the Business Manager more in a later part of this chapter.

By far the more expensive and extensive part of the budget process will be that of the capital budget. Capital can be, in the simplest form, described as outlay of money spent to acquire something of intrinsic, lasting value to the company. A typical example would be a station vehicle. For the outlay of cash you acquire something that has value that could be converted to some diminishing return of cash. While the cost of a vehicle is substantial, its useful life is projected for several years. Usually your company will determine at which level of expense capital will be determined. That is a question to be asked to the Business Manager or someone who administers those decisions. For the sake of discussion here we will use an imaginary level of items costing over $200 and having a life usage of 5 years. Some obvious examples would be computer-based weather systems, transmitter, news gathering equipment, and various business machines. There will always be exceptions to whatever rule your company uses. Software might be an example, as could other subscription services or equipment. But for discussion purposes, we will consider the gross definition of something exceeding $200 in cost having a life-span of 5 years or more (as was stated, this will vary from business to business; check with the Business Manager for your specifics).

The first step in preparing the capital budget is to determine what is needed. Simple as this sounds there are many things that need to be prioritized. It would be judicious to break things into as many categories of needs; building/facilities, equipment/services needed to maintain or enhance product delivery, expansion, and maintenance.

The capital equipment budget, sometimes referred to as the "cap x" budget process, begins by identifying each item in each category, then stating why you wish to place that item on the budget. There are really only a few reasons: (1) it is new, (2) it is going to replace an aging device or system, or (3) it will expand or upgrade existing systems. Remember, any revenue spent should yield some type of financial gain, be it direct revenue or enhancing the revenue through increased efficiency (demonstrable), or increased revenue potential. It can also be achieved by demonstrating the proposed item will reduce expense to a level that it becomes "profitable" to invest.

Dividing the proposed purchase into (1) *mandatory*, (2) *necessary*, or (3) *desirable* categories will assist in laying out the budget proposal. "Mandatory" describes the reason in terms such as being legally required to maintain operation (by not spending the capital the product will not be able to be delivered or produced in a documented manner) or revenue will be directly compromised without the expenditure of capital. An example of "mandatory" would be licensing; without a license, be it with the FCC or software vendor, operation of the business would not exist. It is easy to visualize almost everything you need as being classified "mandatory" but in reality, that category should be used only for those things that the station could not operate without. This does not mean that you should not use the category, just that you should be careful as to what you categorize as "mandatory."

The next category, "necessary," is where most items should be categorized. As the word denotes, the item/system you are requesting is considered to be necessary for optimum operation of the business. "Necessary" is a broad category, including items such as business equipment, phone systems, transmitters (unless the current device is totally undependable to the point that it is always "down"), studio equipment, test equipment; in short, any item or device that will in anyway maintain or enhance revenue. If, for example, a certain acquisition device is requiring repair beyond a reasonable limit and is causing News to miss an important story due to equipment malfunction, the replacement should be included as "necessary" to maintain essential services.

The final category is that of "desirable" to enhancing revenue potential. This can be a difficult category in that you must be able to demonstrate why a certain device or item will increase the efficiency of operation to the point that it actually increases the existing revenue generated by a device beyond the level of the expense of purchasing the new system or item. An example of this would be the desire to replace 1 5/8 inch transmission line to the antenna with 3 inch line. While the expense of new line is substantial, the increased efficiency and thus the saving in electricity required to operate the transmitter at an increased level would, over the life of several years, reduce the electric consumption bill by X percent, thus paying for the upgrade in line. In addition, a smaller, more efficient transmitter can be employed to achieve the identical desired area of coverage. It is the cost-saving ideas such as these that will help establish Engineering as part of the revenue team.

Let us next delineate the reasons for spending capital; they are (1) replacement, (2) expansion of services, and (3) enhancement of demonstrated efficiency of services. Each category should be measured by approximately how much, and how the expenditure of this money will increase the revenue at the bottom line. As much as we do not like to consider the bottom line, it is the measure of why any business spends money. The old adage of "it takes money to make money," while being correct, also implies that the expenditure of that money must be a wise investment.

The final category is one that can be somewhat difficult to justify—that of increasing revenue efficiency. By upgrading a system or device to a more efficient level should be weighed against the current cost of the existing method. In what manner will it reduce revenue expenditures? How will it affect the ability to make the product more salable? Will it affect staffing levels? If it is just "new" and "better," that is hardly a valid reason to expend the revenue.

Things to consider in preparing the capital budget should be the urgency of the expenditure, the cost-to-revenue ratio, and the recoverability of resources. Realizing the scope of all the areas that fall within the responsibility of the Engineering Department, submitting the Capital Equipment Budget is a major function within a well-managed organization. A final note: as you plan the capital expenses, keep in mind that it will be impossible for the company to purchase each and every item. They will need to be prioritized, even over the span of several years. This should be incorporated into the overall budget, with indications of when the item should be included in the expense budget.

7.2.2 Strategic System Planning and Development

More than anyone else in the station, the Director of Engineering should know the technical trends in broadcasting. Engineering should be aware of cutting-edge equipment and methods. The Engineering Manager should spend sufficient time reading the latest professional journals to keep up with current research related to broadcasting, and spend time networking with other engineers, equipment manufacturers, and broadcast academics. This should include belonging to and participating in the Society of Broadcast Engineers, and taking any continuing educational courses that might assist in the professional development of the Engineering Department.

Courses offered through SBE University should also be considered. Written by professionals in the field, they provide progressive and in-depth materials from which one can glean extensive knowledge. These courses are relatively inexpensive and many employers will cover the cost.

Attendance to events such as the National Association of Broadcaster's convention is also a great place to learn about cutting-edge technologies. With newer, more efficient technologies coming about all the time and more efficient products coming to market almost daily, a show such as the NAB fall spring event is something to encourage your management to enable your participation. Again, the Engineering Manager should be the person at the station who is the resource for better technology and methodology to make the station more competitive and profitable. Being on the leading edge of system planning and implementation is vital to the efficient operation of any Engineering Department.

7.2.3 Staffing Oversight/Personnel Management

One of the duties of engineering management is successful and efficient oversight and management of the Engineering Department. Corporate structure varies widely in the areas of the oversight of and duties of engineers. The title of "engineer" is often used to designate personnel charged with operational duties not specifically associated with the normal tasks of plant maintenance and installation. In certain environments, this often includes a responsibility to operate the production/distribution equipment to deliver the product to air. This includes, but in no way is limited to, those personnel who operate the audio mixers, control the various methods of playing commercial and other materials via various kinds of record/playback equipment, acquire programming via Internet or satellite, and control various video equipment used in delivery of the on air product.

The management of these personnel, and those charged with the maintenance and installation of equipment and systems, is similar. First, the Engineering Manager should be well acquainted with the company's personnel policies. If any uncertainty exists, it must be resolved with the director of personnel or person responsible for assuring federal, state, and local regulations governing employees. If the employees are members of an organized labor union, you must be intimately familiar with those requirements as well. Be as familiar with regulations governing overtime, work rules, and any other rules or regulations governing employees. Company policy should always be followed. Reviews of employees must be performed honestly and without prejudice, and employees must be treated fairly and equally at all times and in every circumstance.

7.2.4 Maintaining Essential Services

The genesis of broadcast engineering—the maintaining of essential services—will most likely be the central function of the Engineering Department. For the purpose of this discussion, we are going to separate the maintenance of broadcast systems from that of plant/facility maintenance.

The unique inventory, conducted during the initial phase of development of the Engineering Department, will provide much of the information needed to establish and operate the engineering maintenance organization. Perhaps the first and most important action is to be certain that technical information concerning each device be in an equipment library. Unless cost prohibitive or space restricted, a master copy of each should be established in a central location with satellite information in areas where it is easily available for technicians to access. As an example, a complete copy of the service and operational manuals for the transmitter should be at both the technical library location and at the transmitter location. The rationale for this is quite obvious and the same principle applies to all other systems.

As noted, you should establish the library to include both the service/maintenance manual(s) and any and all operational instructions/manuals. Operation manuals should also be conveniently available to the personnel operating equipment. It is equally important to have available manufacturer operational instructions as well as the manufacturing maintenance instructions. Usually, these can be negotiated into the purchase of systems, or a reduced price can be arranged. The operation manuals provide the service personnel with the exact sequences of operation. While usually written in a less technical manner, these are the things the personnel out in the field using the equipment need to know. This information can be quite useful in resolving operational issues. While most engineering technicians understand the operations of equipment, having the factory operational specifications will eliminate any question regarding the capability of the system to perform in a manner in which it was designed. They will also serve a purpose as training is required. Do not underrate the value of having a library of operational manuals in your department.

It goes without saying that a library of complete service/maintenance/parts manuals should be established. Any and all updates, revisions, or additions, including manufacturer modifications, should be meticulously included. This will ensure that your technicians always have the most accurate, up-to-date information with which to work. Again, these manuals should be, if possible, in both a central library and at the location where the equipment is located. The library should be organized so time is not wasted looking for needed information. As this material becomes worn or illegible, it should be replaced. Along with the manuals, some method of "note taking" should be made available such as spiral notebooks in which technicians can make repair notes to assist with future repairs. The service repair forms can also serve to accomplish this by writing repair notes/shortcuts on the reverse side. These forms, as was observed earlier, should be kept in a location convenient to the technicians.

As with constructing a capital budget, priorities of repair should be established dependent on the urgency of need. When performing maintenance on systems, essentials should take priority. Often, this is where the leadership of the Engineering Manager will be called upon. While almost everything in the broadcast business is an urgent need, it must be determined which is the overriding need and it is incumbent upon the manager to be able to demonstrate that priority. The overall knowledge of the Engineering Manager concerning operational priorities will be essential in resolving service priorities. The urgency perceived needs to be weighed against the urgency concerning the overall product. Often this will require the Engineering Manager to be able to speak to the News department in terms they can relate to or to the technical personnel in terms to which they can relate. Be prepared to justify the priority to the general management as it would relate to the overall continued service of the station.

Finally, do not hesitate to involve the operator or operators reporting or noticing the problem. Asking the right questions and listening to their explanations can go a long way in resolving a maintenance issue. The cause of the problem is not as important as the resolution. If the operator did something incorrect, a matter of fact "mini training" will often bring about a satisfactory resolution and prevent further problems.

Replacement parts and materials should be kept in inventory if possible. A review of the work order forms can assist in determining what material to keep in the maintenance inventory. These parts and material should be kept in a convenient location and available to all service technicians. An inventory method should be developed to maintain adequate tracking of materials dispersed. Any form of control that is suited to the operation of your Engineering Department is acceptable. This will assist in maintaining adequate levels of parts on hand and prevent material from disappearing from stock.

While replacement parts and materials are usually available from the manufacturer, often the manufacturer acquires their material from general vendors. Usually, these identical parts and materials can be purchased directly from the parts manufacturer or one of their wholesale vendors at a cost less than they can be acquired through the manufacturer. This is not always the case, but when it is, and the part is identical, considerable reduction in repair expense can be realized. The Engineering Manager should develop an extensive library of parts vendors (this can be easily done online), establish a relationship with them, and become familiar with their entire product line. It is always wise to solicit competitive bids on many of the more expensive items, if possible. Realize that certain proprietary parts will only be available from the manufacturer. Also be aware of lead times involved in materials. If time is of essence, be sure that the vendor can and will meet the expected delivery schedule.

If systems require a service contract, be sure that is negotiated into the original acquisition costs. If you are overseeing the service contract, be sure to treat and track the service as you would noncontracted items. Develop a service log and be sure it is meticulously kept. Even simpler, over-the-phone or internet repairs or actions should be documented. This system of documentation will again pay dividends when budget time comes. Developing good rapport with the contract service vendor is paramount to ensuring that the station receives both good service and receives full value of the cost of the contract. Often, it takes leadership from the Engineering Manager to bring contracted services to a speedy and concise resolution. Again, a good service log or some easy to track method of keeping service information that was performed by some outside party will be of great value to the overall station operation and will aid in any and all future negotiations with the vendors. Ensuring the smooth, efficient, cost-effective, delivery of repair and maintenance, along with effective and complete tracking is the goal of, and the responsibility of the Engineering Manager. Doing this will ensure a smooth and efficient operation.

7.2.5 Physical Facility Planning and Oversight

As with technical equipment, the Engineering Manager is often responsible for the maintenance and oversight of the physical property including the buildings and associated systems such as HVAC, plumbing, electrical, structural matters, outdoor property maintenance, and operational budgeting for all the above listed items (and more). In addition, the Engineering Manager is often charged with planning and development of any and all new structures. Although this seems out of the scope of "broadcast engineering" in today's broadcast culture, it is becoming more the norm. Not only does the Engineering Manager need to be an expert in information technology and electronics, he or she are now required to have knowledge of buildings, building systems, and all other "nonbroadcast" systems and items. While this may seem a bit daunting to the average engineer, it is not out of the realm of abilities of an Engineering Manager who applies good engineering principles and leadership to the tasks.

As with the technical inventory, a complete listing of all properties for which you are responsible should be documented. An outline of requirements to maintain each should be developed, time lines for performance established, and time and materials—along with levels of skill and expertise required to perform these tasks—should be developed and documented. While building management may not be the forte of the broadcast engineer, the principles remain the same.

A frank discussion with management should be conducted to establish what level of building management will be expected. An analysis of the required time and the necessary skill set to perform each duty should also be discussed. Expecting an IT technician to be experienced in repairing plumbing problems, installing hardware, performing lawn care and shrub maintenance, snow removal, auto mechanics, or roof maintenance would be unrealistic. That discussion will need to be arranged and some mutual understanding of compromises that will be involved.

Proper facility maintenance will require the Engineering Manager to establish networking rapport with professionals who specialize in those trades. Outside contractors, service facilities, and free-lance professionals will need to be identified. When utilizing outside contractors, be certain of their liability insurance, omission error insurance, any licenses that might be required, and local reputation of those with whom you are considering using as a contractor.

As with other systems and devices, special budgets should be developed using many of the same criteria such as legal violations, urgency, and other prioritizing information. A special budget for building and property maintenance might be considered.

Networking will be of great value in performing the tasks of a building manager. Get acquainted with various trade groups, service providers, building associations, local building and code officials; anyone associated with the maintenance of facilities or grounds. Having them as resource contacts will be worth the investment in time to establish those relationships. Be sure to be aware of and obtain all necessary governmental permits as required by local, state, and federal governmental authorities.

Always be evaluating by prudent record keeping, such as you do for technical and contractual work records, to determine that the station is getting the best value for its money.

Building and facility maintenance need not be intimidating to the Engineering Manager.

7.2.6 Strategic Building Development

This is an area in which input from the Engineering Manager will be of great value. New construction, be it for technical areas or business areas require input and coordination from someone knowledgeable in building techniques, practices, codes, and construction/material vendors. Most likely, an architectural firm will be retained to coordinate any new construction. But prior to that phase of building development, a purpose of—and requirements of—any expansion or new construction should be well documented.

First we will focus on the development phase. All new development will most likely be technical in orientation. This is the forte of the Engineering Department. Also, since Engineering is mostly responsible for facility maintenance, planning of any new construction, expansion, or development will require a high level of coordination involving Engineering expertise. While future construction and planning is usually a group effort, you must be certain that Engineering is at the forefront.

Once again the Engineering Manager should develop a statement of purpose for the project, including the goal and broad specifications required to achieve the desired results. Some timeline should be established, usually requiring group input, and a budget developed. Because of the technical nature of most construction and expansion projects, it would seem only natural that the Engineering Manager be the project coordinator. The knowledge of broadcast operations and needs, along with familiarity with the various materials involved (including those necessary for physical construction), is a valuable resource to bring the project on line with operational efficiency—and on-budget. Engineering's ability to make quick, accurate decisions should be a valuable asset to the operation. It is incumbent that Engineering Management be quickly involved in any proposed project.

Of course, the continual forward-looking engineer will be always evaluating the station's needs and seeking ways to improve overall efficiency, which leads to greater profits.

7.2.7 Training and Education

The engineering staff know how the equipment and systems operate and how to derive greatest efficiency. The department responsible for equipment maintenance and repair will most likely note many repair requests that are not caused by some malfunction of the device, but rather by some operator error or misunderstanding of the operational parameters. While it is tempting to become belligerent or callous about such operational problems, that accomplishes little to resolve operational issues, to reduce the amount of repair required, or increase profitability. By using the work order/repair form developed, you can isolate problems such as operational errors and identify who may need additional education. This information, of course, must be shared with the appropriate departmental personnel. It is the Engineering Manager's responsibility to illustrate how further training will serve to enhance the operation and increase employee confidence and satisfaction. An explanation will be needed on how to improve performance, which will also enhance the product they are trying to provide.

Training offerings concerning new equipment should be a priority. When new or improved devices are implemented, all personnel who will be involved with the use of, or application of that equipment or device need to be trained in proper operation, limits of, and user maintenance of the device or system. Engineering is the logical source of that knowledge and information. But, to properly train and educate, the Engineering Manager must be able to explain the procedure in terms and language applicable to the end user. Extreme technical terms that are understandable to the engineer but would mean little to the operator should be avoided. Training should cover not only what the various switches do but also working examples of how to gain the greatest optimization from the system or device. Practical tips from a user standpoint are of great value in making the intended operator at ease with the device, and will help them do a more efficient job. While it can be difficult to maintain the proper attitude, it is also imperative to remember that there is no such thing as a stupid question. Answer all questions sincerely using language that is not threatening.

Training of both the Engineering Manager and the engineering staff is vital to the successful operation of an Engineering Department. It goes without saying that when hiring personnel they should have the education required to perform the tasks at hand. But even the most highly educated person has a need for continuing education to maintain skill levels and to acquire new skills. Participation in professional engineering groups such as SBE should be encouraged and, budget permitting, the company should assist in any financial burden. But even if the company does not participate financially, the employee should be encouraged to participate. Various seminars and trainings, often free or at some nominal cost, may be offered in your area. Encourage attendance. This is an opportunity to exercise leadership by management and staff attending together. All must continue to grow in their knowledge of the requirements of the field, changes, trends, and new systems and/or devices.

SBE University should be utilized, when possible, both for your staff and for yourself. Conference calls, on-line trainings, and individual studies are great ways to maintain and increase skill levels. Local colleges are a good resource for both the Engineering Manager and staff. A well-trained staff will not only increase the efficient operation of the department, it will also lower the overall cost and serve to increase the profitability of the station.

7.2.8 Interdepartmental Cooperation

In the days that the Chief Engineer simply was charged with maintaining equipment, the inability to communicate effectively with other members of the management "team" was not very important. Today, as one of the management team, it is not only desirable, but crucial. This often requires the Engineering Manager to acquire a new language; in fact, several new languages. The Engineering Manger must be able to speak in language understood by the on-air personnel (news, creative service, operations, etc.), the sales department, the business department, and general management. Remember, few, if any of these speak the more comfortable engineering language. The Engineering Manager must learn how to speak their language, if not fluently, at least to the point of being understood.

Perhaps one or the most important languages needed is that of the Business Manager. The Engineering Manager needs to understand the business world and the language associated within it. While the Engineering Manager does not need to become an accountant, he/she does need to understand what information is important to them, why, and what it really means. How does it express the overall profitability and operation of the station? It is incumbent to be sure everything is understood, and one must remain persistent until at least a remedial understanding of their language has been attained. Likewise, they do not speak "engineering" and most likely it is totally unfamiliar to them. It becomes the Engineering Manager's job to explain what is being said in terms they can comprehend. In that manner you both become bilingual.

The Business Manager is one of the key people in accomplishing the goals of the Engineering Department. It is the Business Manager who knows the money associated with the business. They know the overall picture, including how much revenue is available, when it is available, and can assist in developing priorities of engineering needs. When the Business Manager understands the Engineering Department and the Engineering Manager understands what information is vital to the Business Manager, a line of communication can be established that will enhance engineering's performance and effectiveness.

One of the more difficult languages to learn is the General Manager's. The General Manager, or GM for short, usually has a background in revenue acquisition. After all, without income all the plans in the world are just ink stains on paper. But without good, reliable product for the station to sell, their job is extremely difficult. A frank discussion with the GM about the station's goals from a financial standpoint, the product content, and the technical requirements will assist in establishing good communications with the GM. The Engineering Manager must learn the language the GM is accustomed to speaking and, again, must be able to explain in understandable terms how the Engineering Department can aid and assist in the overall goals. A good line of open communication with the GM can, and will, make it easier for the GM to understand why engineering does what they do and how it assists in the overall goal of the station. It allows the Engineering Manager to become part of the solution, not viewed as just a "burden" to the financial operation. Keeping the GM informed of the "state of the station" will assist in planning and implementing the overall station goals and strategy. Suggestions on how to improve the product from an engineering view will then be received in a more considerate manner.

Dealing with on-air personnel can be a bit daunting sometimes but it is important to know what is important to them. They have their own language, the Engineering Manager needs to acquire skills in its use but as with other departments they need to be taught a bit of engineering language. The same applies to sales and any other department that is part of the station. The Engineering Manager must become a team member and player—part of the solution.

7.3 SUMMARY

In the distant past, the Chief Engineer, now often referred to as the Director of Engineering (or Engineering Manager), was only concerned with the legal operation of the station and the maintenance of associated equipment. In today's world of broadcasting, the Engineering Manager must walk in multiple shoes. Basic knowledge of the field of engineering must be supplemented with continuing education—not only in the broadcast engineering field, but in all the disciplines for which the Engineering Manager is responsible. The Engineering Manager must be multilingual and able to communicate within other disciplines. The ability to network effectively is mandatory.

Good and accurate record keeping and constant review of methodology must be done to maintain the high level of demands placed on the Engineering Department. A well-documented department is the basis. Record keeping is essential to maintaining an efficient and profitable operation. Through this, the Engineering Manager will be able to operate a highly successful department.

CHAPTER 7.2
PROJECT AND SYSTEMS MANAGEMENT

Jerry C. Whitaker, Editor-in-Chief

Robert Mancini
Mancini Enterprises, Whittier, CA

7.1 INTRODUCTION

The subject area encompassed by "project and systems management" is broad in scope. It includes, but is not limited to, writing, organization, people management, project management, and problem-solving. Within each of these broad groups additional distinctions can be identified; indeed, entire books have been written on these subjects. Beyond a deep-dive into these and other topic areas, there is a need to integrate separate disciplines into a cohesive program and/or process.

The lessons of project and system management can be applied across a wide variety of project types and organizational structures. This being the case, guidelines, suggestions, rules, and all the other elements that make up a process rarely are applied strictly to an organizational structure. Instead, that structure is imparted on individuals to execute and maintain. It is important, therefore, for those individuals to understand the process and have ownership of it. Furthermore, management needs to make the project a priority. If management is not supportive of the project (which includes allocation of appropriate resources and personnel), then it is likely to fail.

While an ad-hoc approach to a given project can be successful, the likelihood of success may be reduced due to the informal nature of the activity. Equally important, lessons learned during execution of one project may be lost and forgotten when the project is completed. This is important but usually overlooked. What are the good and bad things we learned during this project? How can they help us improve future projects and processes? Documentation and follow-up are key.

One benefit of defining a process is that it compels the leadership to look at the big picture and to try to anticipate unforeseen challenges.

Documentation skills come into play here as well. Documentation also, of course, touches on many areas of business and plays a major role in the success of a project or product.

7.2 PLAN FOR SUCCESS

Well thought-out and documented plans increase the likelihood of a successful project. A structural process that cannot be implemented in a given organization is certain to fail, and the individuals tasked with carrying out that process may fail as well. Success has many fathers; failure is an orphan. This well-known truism has been proved right countless of times in any number of organizations over a long period of time.

While the reasons for failure vary from one situation to the next, certain common threads tend to emerge, which include:

- **The goals were set too high.** In business, like most everything else, you cannot always get what you want. The needs of the organization must be balanced with the realities of resources, time, and capabilities. It is always a good practice to challenge individuals and organizations to produce their best; however, setting goals so high they are generally believed to be unachievable often results in team members simply giving up when the impossibility of the task ahead becomes clear.
- **The goals were set too low.** If a project is completed but the end result is inadequate to meet the need, then the effort can result in failure, or at least lost time as the project is rescoped and restarted. Individuals like knowing they are a part of something big, something important to the organization. A small project with only a minimal chance of having a positive impact tends to encourage lackluster participation and effort on the part of contributors.

Between these two extremes, naturally, there is a sweet spot where the organization and individuals within it are challenged with achievable goals and given the resources necessary to accomplish the task at hand. Although there are numerous factors involved, elements of successful projects may be generally summarized as follows and illustrated in Fig. 7.2.1:

- **Adequate resources**. A project starved of resources is in trouble from the start. Typically, such resources translate into available personnel and/or money. The two are usually interrelated, of course. Other types of resource limitations include insufficient time made available for focused work, restrictions on travel, and so on. The resources allocated to a project say something about the importance of the project to those tasked with carrying it out. If management does not think a project is important, then the employees are unlikely to put much effort into it.
- **Adequate time allocated**. The time needed to complete a project is closely related with the resources applied to complete the project. Adding resources usually shortens time, while reducing resources usually lengthens time. In some situations, the timeline is fixed in that a project needs to be completed by a specified date (e.g., promised date for completion of a new facility). In other situations, the timeline is flexible, with no firm end date (although there is often a goal). As a practical matter, projects with no end date tend not to end. An end date is important—but as before with goals, the date needs to be realistic. Personnel working on the project need to understand what the deadlines are, and why they are important to the success of the business.
- **The right people in the room**. This challenge can be a tough one to solve. It is related to resources, but has a unique dimension as well. In any organization, there are key individuals that either direct projects or make decisions about projects. As such, it is important for those persons to be involved in the projects and processes that will fulfill the goals of the organization. Getting a slice of their time may be a challenge. If a key person is not involved, the project will tend to move on but with the risk that when the decision-maker becomes involved it may be so late in the process as to derail or seriously delay the effort. Perhaps more importantly, a key person not invited early-on might not be able to contribute to discussions, resulting in the group going down rabbit trails that might have been avoided if the key person was able to provide wisdom and experience to the group up-front. One approach that often works well is to identify key points in the process where review takes place. This permits the key players to focus on other tasks most of the time, but step into a project at predetermined points to provide input or to suggest changes.
- **Well thought-out design**. An organizational structure can have considerable inertia. Once a project has begun it usually rolls forward. Occasionally, after some work in a particular direction, it may become clear to decision-makers that the original concept and/or technology were flawed, and the best approach is to go in an entirely new direction. This can be wrenching to individuals who have put months into a project only to have it stuck on the shelf. The decision to kill a project or take a radical new approach is difficult to make for a number of reasons, not the least of which is the knowledge of wasted resources. Still, cutting the losses may be the best approach, as the only thing worse than stopping a project in mid-stream is to complete the project only to find that it is not what the business really needs.

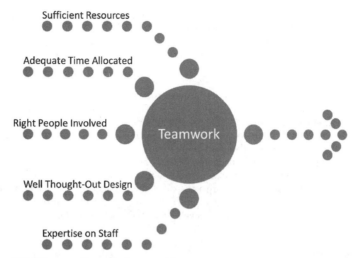

FIGURE 7.2.1 Elements of a successful project.

- **Realistic view of expertise**. Some concepts are exciting and hold considerable promise for the business. The problem is, they can also be very difficult to accomplish and may require expertise that just does not exist within the organization. As noted with goals, stretching the capabilities of an organization is generally good, but being unrealistic about what can be accomplished is not. There are a number of possible solutions to such a challenge. One would be to partner with another company or organization that has the needed expertise, with each party bringing something different to the table. Business issues, naturally, come into play here; still, it may be an area worth exploring.
- **Building the team**. Teamwork is essential to a successful project—and one element of a successful project is the energy and enthusiasm behind it. A lack of energy behind a project is often a result of a combination of the challenges previously outlined. Most organizations can identify one or more projects that have plodded along to completion, but that generated very little energy, interest, or enthusiasm on the part of contributors. The end results are, predictably, uninspired as well. One important task of management in any organization is to motivate individuals. Lacking motivation, people do not produce their best work.

Just as individuals can be set up to succeed, or not, projects and processes can also be set up to succeed or fail. The role of management is to foster the former while preventing the latter.

Timing is another component in the success of a project. Some elements of timing are within the control of the organization and are set in the project timeline as discussed previously. Other elements of timing are outside the control of the organization. For most things, there is a window of opportunity. The window may be large or small, and it can be very difficult to predict opportunities that are 1, 2, or 3 years out.

7.3 ELEMENTS OF PROCESS

Process, within the scope of this chapter, focuses on the steps and structures needed to accomplish a set of stated goals. These steps include developing an organizational structure, coordinating the activities of participants, monitoring progress, documenting results, planning for unforeseen problems, and reporting the results of the work. The process developed for one group within a particular organization is often transferrable to another group working on a related (or even unrelated) project. Such repurposing of management structures is helpful in that it reduces the time needed to begin work and tends to refine the individual process steps. Improvements in the process can be identified through documentation of things that worked well, and documentation of things that did not work as intended.

Process involves looking at the big picture and identifying the key steps necessary to get from here to there. The best structure is often a loose one, where guidelines and guideposts are established at key points along the way, but not so much detail and structure that it inhibits progress and creativity in the face of unforeseen events.

Invariably, documentation comes into play at all steps in a given process. Communication of ideas, problems, and solutions is essential to keep all members of the team on the same page, and top management advised of the status of important projects.

It is difficult to thrive in the business world today without effective communications skills. Foremost among these is the ability to clearly communicate ideas in written form. As email steadily replaces the telephone as the primary business communications tool, good writing techniques have never been more important. Writing impacts all facets of business.

Likewise, meetings are a critical element in any process. Meetings can serve as an opportunity to develop new ideas and concepts. They also give contributors a common vision of the task at hand. And, critically, they serve as vehicle to make key decisions.

7.3.1 Meetings

It is easy to poke fun at meetings. Sometimes they can be long, unproductive, and leave the participants wondering whether it was worth the time spent. However, meetings are critical to bringing together ideas from different perspectives. The key is to have *effective* meetings. While this may be easier said than done, some guiding principles will help to produce the desired result.

Meetings are important for an organization because products and processes are more complex today than ever before. There was a time not too long ago when a small group, or even a single person, could be assigned responsibility to develop, for example, a facility upgrade. With the group located in the same building, meetings could be informal and done with little or no planning. Today, there are very few enterprises that can secure their future based on the vision of a handful of people who happen to be in the same geographic location. Meetings have become a necessity for most companies because the business itself is complex and often needs to address different goals.

While it is satisfying for an individual or small group of workers to develop a new facility or upgrade on their own, the chances of success are limited because of the limited input such a process allows. (It is fair to point out there are exceptions to this rule.)

In any event, companies utilize meetings to bring together—in a structured manner—ideas from people who have different perspectives, backgrounds, goals, and objectives. Meetings also offer the opportunity for individuals to see the bigger picture they might otherwise miss because they are focused on their particular discipline, and not necessarily on the overall needs of the company.

Historically, when people think of meetings, they think of face-to-face meetings. That mindset has changed considerably over the years thanks to improved communications tools. Organizations with widely spread employees may use teleconferences for most meetings, coming together in person only at critical decision points. Among the driving factors for this approach are the improved efficiency of setting up and conducting the meetings, and cost savings. Travel can be extraordinarily expensive. The time required for travel is another matter entirely.

To be effective, meetings require a spirit of cooperation and the willingness to give on points that may be important to one person or group but for the greater good of the organization are not practical or appropriate. Productive meetings begin with a positive attitude and the acknowledgement by all participants that no one person has all the answers. Cooperation and compromise are the keys to success.

The typical committee structure is illustrated in Fig. 7.2.2. The chairperson is the administrative leader of the meeting. This person is charged with the responsibility to conduct the meeting, but moreover to manage the necessary work before and after the meeting. Planning for a meeting involves, at minimum, setting the agenda. Some agendas are simple; others are long and complex. More complex agendas require input from members of the committee, department heads, or others that will participate in the meeting.

The chairperson may choose to appoint a vice-chair. In the case of the chairperson's absence, the vice-chair will temporarily assume the responsibilities of the chair. In addition, during the meeting there may be times where the chairperson will find the need to hand the gavel (and management of the meeting) to the vice-chair in order to make a personal statement on the subject being discussed.

The vice-chair may also be assigned certain responsibilities by the chair, which may change from time to time. For example, the vice-chair may assume responsibility for organizing the agenda for upcoming meetings, and for following-up on action items after the meeting.

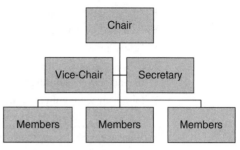

FIGURE 7.2.2 Typical committee structure.

An effective chairperson will look for ways to move the discussion and the work forward, even in difficult situations where there exist entrenched positions. It is important for each viewpoint to be aired—each person given his or her opportunity to make a particular case. If strong opposing views are held by a small group of members, it may be productive to assign the task of coming up with an acceptable compromise to those individuals. If the members are willing to understand the viewpoints of others and compromise for the good of the organization, then progress can be made.

Sometimes the best approach to a problem is to defer a decision until later in the day or until the next meeting. A great deal of discussion takes place in the hallways outside meetings. Such private encounters can be quite productive in moving past sticking points.

In order to make sure major projects maintain forward momentum, the chairperson will want to assign action items to specific persons or groups. The action items need to be focused and should include a stated completion date, such as prior to the next meeting of the group.

The chairperson is responsible for calling meetings at intervals necessary to handle the assigned work. There may be a fixed pattern to meetings (e.g., every Tuesday at 9:00 a.m.) or a flexible schedule, often dictated by key milestones set out in the Work Plan.

Meetings may be in-person, by telephone, or a combination of in-person and telephone. Using a Web-sharing service where committee members can view a shared screen can be very effective in conveying ideas and reaching agreement on a particular strategy. Web-based desktop sharing is also quite useful for document editing by a group of experts.

Sufficient advance notice of upcoming meetings is important to ensure good participation. Getting committee members to agree on a meeting schedule can be a challenge, however, given the complexities of travel and fixed or unpredictable events. Setting up a meeting cadence of once a week or twice a month—ideally at the same time of day—is often helpful in solving this challenge. Scheduling meetings well in advance can also be effective in keeping attendance high. It is not uncommon to schedule meetings out many months in advance, with the understanding that changes may need to be made based on unforeseen events.

When scheduling meetings, efforts should be made to avoid date and/or time conflicts with other company or related industry meetings. However, it is understood that this may not always be possible. The availability of meeting facilities must also be taken into account for in-person meetings.

Whatever process is used for meetings, the rules should be clear to all participants well in advance of the meeting. It is easy for a committee to get distracted with procedural issues that have little or no benefit to the task at hand.

As a practical matter, a good meeting is one that produces useful results. More generally speaking, however, the best meetings typically have the following attributes:

- Agenda distributed well in advance of the meeting
- Appropriate documents distributed well in advance of the meeting
- All key stakeholders show up
- The meeting begins on time
- Presentations and briefings are clear and to the point

- Where specific actions are required, they are clearly outlined to the committee members
- Decisions are made as needed
- Action items are recorded, assigned to specific members with agreed-upon due dates
- Concise minutes of the meeting are recorded
- The meeting ends on time
- Follow-up actions after the meeting are completed

When done right, meetings can serve as an essential element of organizational development and even be an enjoyable experience for those involved.

7.4 CONCLUSION

Process development is an inexact science. However, by documenting the lessons learned on past projects, the knowledge base of the organization grows. This learn-by-doing approach implies that a process developed for a given project may change over time as unexpected problems are encountered and solved.

A company of any substantial size will invariably have a formal structure. Ideas can flow from the top down, or from the bottom up. In the ideal case, ideas will flow both ways. Likewise, ideas can come from inside the organization or from outside. Here again, under the best-case scenario they come from all available sources. It is the responsibility of the organization to facilitate effective communications.

BIBLIOGRAPHY

Whitaker, J. C., and R. K. Mancini, *Technical Documentation and Process*, CRC Press, Boca Raton, FL, 2013.

CHAPTER 7.3
SYSTEMS ENGINEERING

Gene DeSantis

7.1 INTRODUCTION[1]

System design is carried out in a series of steps that lead to an operational unit. Appropriate research and preliminary design work is completed in the first phase of the project, the *design development* phase. It is the intent of this phase to fully delineate all requirements of the project and to identify any constraints. Based on initial concepts and information, the design requirements are modified until all concerned parties are satisfied and approval is given for the final design work to proceed. As illustrated in Fig. 7.3.1, the first objective of this phase is to answer the following questions:

- What are the functional requirements of the project?
- What are the physical requirements of the project?
- What are the performance requirements of the project?
- Are there any constraints limiting design decisions?

A subset of the constraints question leads to the following questions:

- Will existing equipment be used?
- Is the existing equipment acceptable?
- Will this be a new facility or a renovation?
- Will this be a retrofit or upgrade to an existing system?
- Will this be a stand-alone system?

One of the systems engineer's most important contributions is the ability to identify and meet the needs of the customer and do it within the project budget. Based on the customer's initial concepts and any subsequent equipment utilization research conducted by the systems engineer, the desired capabilities are identified as precisely as possible. Design parameters and objectives are defined and reviewed. Functional efficiency is maximized to allow operation by a minimum number of personnel. Future needs and system expansion are also investigated at this time.

After the customer approves the equipment list, preliminary system plans are drawn up for review and further development. If architectural drawings of the facility are available, they can be used as a starting point for laying out an equipment floor plan. The systems engineer uses this floor plan to be certain that adequate space is provided for present and future equipment, as well as adequate

---

[1] This chapter was adapted from: Whitaker, J. C., G. DeSantis, and C. R. Paulson, *Interconnecting Electronic Systems*, CRC Press, Boca Raton, FL, 1992.

7.24 BROADCAST MANAGEMENT

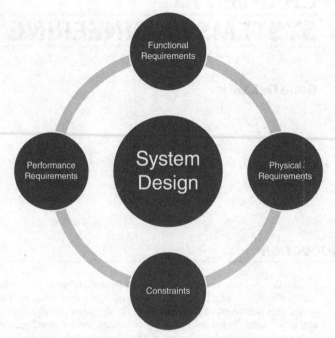

FIGURE 7.3.1 Typical phases of a system design project.

clearance for maintenance and convenient operation. Equipment identification is then added to the architect's drawings.

Documentation should include, but not be limited to, the generation of a list of major equipment including:

- Equipment manufacturers, models, and prices
- Technical system functional block diagrams
- Custom item descriptions
- Rack and console elevations
- Equipment floor plans

The preliminary drawings and other supporting documents are prepared to record design decisions and to illustrate the design concepts to the customer. Renderings, scale models, or full-size mock ups may also be needed to better illustrate, clarify, or test design ideas.

Ideas and concepts have to be exchanged and understood by all concerned parties. Good communication skills are essential. The bulk of the creative work is carried out in the design development phase. The physical layout—the look and feel—and the functionality of the facility will all have been decided and agreed upon by the completion of this phase of the project. If the design concepts appear feasible, and the cost is within the anticipated budget, management can authorize work to proceed on the final detailed design.

7.2 ELECTRONIC SYSTEM DESIGN

Performance standards and specifications have to be established up front in a technical facility project. This will determine the performance level of equipment that will be acceptable for use in the system and affect the size of the budget. Signal quality, stability, reliability, and accuracy are examples of the kinds of parameters that have to be specified. The systems engineer has to confirm whether selected equipment conforms to the standards.

At this point, it must be determined what functions each component in the system will be required to fulfill and how each will function together with other components in the system. The management and operation staff usually knows what they would like the system to do and how they can best accomplish it. They have probably selected equipment that they think will do the job. With a familiarity of the capabilities of different equipment, the systems engineer should be able to contribute to this function-definition stage of the process. Questions that need to be answered include:

- What functions must be available to the operators?
- What functions are secondary and therefore not necessary?
- What level of automation should be required to perform a function?
- How accessible should the controls be?

Overengineering or overdesign must be avoided. This serious and costly mistake can be made by engineers and company staff when planning technical system requirements. A staff member may, for example, ask for a seemingly simple feature or capability without fully understanding its complexity or the additional cost burden it may impose on a project. Other portions of the system may have to be compromised to implement the additional feature. An experienced systems engineer will be able to spot this and determine if the tradeoffs, added engineering time, and cost are really justified.

When existing equipment is going to be used, it will be necessary to make an inventory list. This list will be the starting point for developing a final equipment list. Usually, confronted with a mixture of acceptable and unacceptable equipment, the systems engineer must sort out what meets current standards and what should be replaced. Then, after soliciting input from facility technical personnel, the systems engineer develops a summary of equipment needs, including future acquisitions. One of the systems engineer's most important contributions is the ability to identify and meet these needs within the facility budget.

A list of major equipment is prepared. The systems engineer selects the equipment based on experience with the products, and on customer preferences. Often some existing equipment may be reused. A number of considerations are discussed with the facility customer to arrive at the best product selection. Some of the major points include:

- Budget restrictions
- Space limitations
- Performance requirements
- Ease of operation
- Flexibility of use
- Functions and features
- Past performance history
- Manufacturer support

The goal of the systems engineer is the design of equipment to meet the functional requirements of a project efficiently and economically. Simplified block diagrams for the video, audio, control, data, and communication systems are drawn. They are discussed with the customer and presented for approval.

7.2.1 Detailed Design

With the research and preliminary design development completed, the details of the design must now be concluded. The design engineer prepares complete detailed documentation and specifications necessary for the fabrication and installation of the technical systems, including all major and minor components. Drawings must show the final configuration and the relationship of each component to other elements of the system, as well as how they will interface with other building services, such as air conditioning and electrical power. The documentation must communicate design requirements to the other design professionals, including the construction and installation contractors.

In this phase, the systems engineer develops final, detailed flow diagrams and schematics that show the interconnection of all equipment. Cable interconnection information for each type of signal is taken from the flow diagrams and recorded on the cable schedule. Cable paths are measured and timing calculations performed. Timed cable lengths (used for video and other special services) are entered onto the cable schedule.

The flow diagram is a schematic drawing used to show the interconnections between all equipment that will be installed. It is different from a block diagram in that it contains much more detail. Every wire and cable must be included on the drawings.

The starting point for preparing a flow diagram can vary depending on the information available from the design development phase of the project and on the similarity of the project to previous projects. If a similar system has been designed in the past, the diagrams from that project can be modified to include the equipment and functionality required for the new system. New models of the equipment can be shown in place of their counterparts on the diagram, and minor wiring changes can be made to reflect the new equipment connections and changes in functional requirements. This method is efficient and easy to complete.

If the facility requirements do not fit any previously completed design, the block diagram and equipment list are used as a starting point. Essentially, the block diagram is expanded and details added to show all of the equipment and their interconnections and to show any details necessary to describe the installation and wiring completely.

An additional design feature that might be desirable for specific applications is the ability to easily disconnect a rack assembly from the system and relocate it. This would be the case if the system were to be prebuilt at a systems integration facility and later moved and installed at the client's site. When this is a requirement, the interconnecting cable harnessing scheme must be well planned in advance and identified on the drawings and cable schedules.

Special custom items are defined and designed. Detailed schematics and assembly diagrams are drawn. Parts lists and specifications are finalized, and all necessary details worked out for these items. Mechanical fabrication drawings are prepared for consoles and other custom-built cabinetry.

The systems engineer provides layouts of cable runs and connections to the architect. Such detailed documentation simplifies equipment installation and facilitates future changes in the system. During preparation of final construction documents, the architect and the systems engineer can firm up the layout of the technical equipment wire ways, including access to flooring, conduits, trenches, and overhead wire ways.

Dimensioned floor plans and elevation drawings are required to show placement of equipment, lighting, electrical cable ways, duct, and conduit, as well as heating, ventilation, and air conditioning (HVAC) ducting. Requirements for special construction, electrical, lighting, HVAC, finishes, and acoustical treatments must be prepared and submitted to the architect for inclusion in the architectural drawings and specifications. This type of information, along with cooling and electrical power requirements, also must be provided to the mechanical and electrical engineering consultants (if used on the project) so that they can begin their design calculations.

Equipment heat loads are calculated and submitted to the HVAC consultant. Steps are taken when locating equipment to avoid any excessive heat buildup within the equipment enclosures, while maintaining a comfortable environment for the operators.

Electrical power loads are calculated and submitted to the electrical consultant and steps taken to provide for sufficient power and proper phase balance.

7.2.2 Budget Requirements Analysis

The need for a project may originate with customers, management, operations staff, technicians, or engineers. In any case, some sort of logical reasoning or a specific production requirement will justify the cost. On small projects, like the addition of a single piece of equipment, a specific amount of money is required to make the purchase and cover installation costs. When the need may justify a large project, it is not always immediately apparent how much the project will cost to complete. The project has to be analyzed by dividing it up into its constituent elements. As shown in Fig. 7.3.2 these elements include:

- Equipment and parts
- Materials
- Resources (including money and time needed to complete the project)

FIGURE 7.3.2 Primary elements of budget requirements analysis.

An executive summary or capital project budget request containing a detailed breakdown of these elements can provide the information needed by management to determine the return on investment and to make an informed decision on whether or not to authorize the project.

A capital project budget request containing the minimum information might consist of the following items:

- Project name. Use a name that describes the result of the project, such as control room upgrade.
- Project number (if required). A large organization that does many projects will use a project numbering system or a budget code assigned by the accounting department.
- Project description. A brief description of what the project will accomplish, such as "design the technical system upgrade for the renovation of production control room 2."
- Initiation date. The date the request will be submitted.
- Completion date. The date the project will be completed.
- Justification. The reason the project is needed.
- Material cost breakdown. A list of equipment, parts, and materials required for construction, fabrication, and installation of the equipment.
- Total material cost.
- Labor cost breakdown. A list of personnel required to complete the project, their hourly pay rates, the number of hours they will spend on the project, and the total cost for each.
- Total labor cost.
- Total project cost. The sum of material and labor costs.
- Payment schedule. Estimation of individual amounts that will have to be paid out during the course of the project and the approximate dates each will be payable.
- Preparer name and the date prepared.
- Approval name(s) signature(s) and date(s) approved.

More detailed analysis, such as return on investment, can be carried out by an engineer, but financial analysis should be left to the accountants who have access to company financial data.

7.2.3 Feasibility Study and Technology Assessment

Where it is required that an attempt be made to implement new technology and where a determination must be made as to whether certain equipment can perform a desired function, it will be necessary to conduct a feasibility study. The systems engineer may be called upon to assess the state-of-the-art to develop a new application. An executive summary or a more detailed report of evaluation test results may be required, in addition to a budget request, to help management make a decision.

7.2.4 Project Tracking and Control

A project team member may be called upon by the project manager to report the status of the work during the course of the project. A standardized project status report form can provide consistent and complete information to the project manager. The purpose is to supply information to the project manager regarding work completed and money spent on resources and materials.

A project status report containing the minimum information might include the following items:

- Project number
- Date prepared
- Project name
- Project description
- Start date
- Completion date (the date this part of the project was completed)
- Total material cost
- Labor cost breakdown
- Preparer's name

After part or all of a project design has been approved and money allocated to build it, any changes may increase or decrease the cost. Factors that affect the cost include:

- Components and material
- Resources, such as labor and special tools or construction equipment
- Costs incurred because of manufacturing or construction delays

Management will want to know about such changes and will want to control them. For this reason, a method of reporting changes to management and soliciting approval to proceed with the change may have to be instituted. The best way to do this is with a change order request or change order. A change order includes a brief description of the change, the reason for the change, and a summary of the effect it will have on costs and on the project schedule.

Management will exercise its authority to approve or disapprove each change based on its understanding of the cost and benefits and the perceived need for the modification of the original plan. Therefore, it is important that the systems engineer provide as much information and explanation as may be necessary to make the proposed change clear and understandable to management.

A change order form containing the minimum information might contain the following items:

- Project number
- Date prepared
- Project name
- Labor cost breakdown
- Preparer's name

- Description of the change
- Reason for the change
- Equipment and materials to be added or deleted
- Material costs or savings
- Labor costs or savings
- Total cost of this change (increase or decrease)
- Impact on the schedule

7.3 PROGRAM MANAGEMENT

Systems engineering is the management function that controls the total system development effort for the purpose of achieving an optimum balance of all system elements.[2] It is a process that transforms an operational need into a description of system parameters and integrates those parameters to optimize the overall system effectiveness.

Systems engineering is both a technical process and a management process. Both processes must be applied throughout a program if it is to be successful. The persons who plan and carry out a project constitute the project team. The makeup of a project team will vary depending on the size of the company and the complexity of the project. It is up to the management to provide the necessary human resources to complete the project. A typical organizational structure is shown in Fig. 7.3.3.

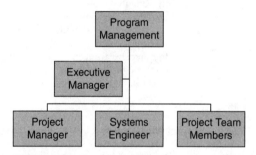

FIGURE 7.3.3 Elements of a program management system.

7.3.1 Executive Manager

The executive manager is the person who can authorize that a project be undertaken. This person can allocate funds and delegate authority to others to accomplish the task. The ultimate responsibility for project success is in the hands of the executive manager. The executive manager assigns group responsibilities, coordinates activities between groups, and resolves group conflicts. The executive manager establishes policy, provides broad guidelines, approves the project master plan, resolves conflicts, and assures project compliance with commitments.

Executive management delegates the project management functions and assigns authority to qualified professionals, allocates a capital budget for the project, supports the project team, and establishes and maintains a healthy relationship with project team members.

Management has the responsibility to provide clear information and goals—up front—based on their needs and initial research. Before initiating a project, the company executive should be familiar with daily operation of the facility and analyze how the company works, how jobs are done by the staff, and what tools are needed to accomplish the work. Some points that may need to be considered by an executive before initiating a project include the following:

- What is the current capital budget for equipment?
- Why does the staff currently use specific equipment?

---

[2] Hoban, F. T., and W. M. Lawbaugh, *Readings in Systems Engineering Management*, National Aeronautics and Space Administration, Science and Technical Information Program, Washington, D.C., p. 9, 1993.

- What function of the equipment is the weakest within the organization?
- What functions are needed but cannot be accomplished with current equipment?
- Is the staff satisfied with current hardware?
- Are there any reliability problems or functional weaknesses?
- What is the maintenance budget and is it expected to remain steady?
- How soon must the changes be implemented?
- What is expected from the project team?

Only after answering the appropriate questions will the executive manager be ready to bring in expert project management and engineering assistance. Unless the manager has made a systematic effort to evaluate all of the obvious points about the facility requirements, the not so obvious points may be overlooked. Overall requirements must be broken down into their component parts. Do not try to tackle ideas that have too many branches. Keep the planning as basic as possible. If the company executive does not make a concerted effort to investigate the needs and problems of a facility thoroughly before consulting experts, the expert advice will be shallow and incomplete, no matter how good the engineer.

Engineers work with the information they are given. They put together plans, recommendations, budgets, schedules, purchases, hardware, and installation specifications based on the information they receive from interviewing management and staff. If the management and staff have failed to go through the planning, reflection, and refinement cycle before those interviews, the company will likely waste time and money.

7.3.2 Project Manager

Project management is an outgrowth of the need to accomplish large complex projects in the shortest possible time, within the anticipated cost, and with the required performance and reliability. Project management is based on the realization that modern organizations may be so complex as to preclude effective management using traditional organizational structures and relationships. Project management can be applied to any undertaking that has a specific end objective.

The project manager must be a competent systems engineer, accountant, and manager. As systems engineer, there must be understanding of analysis, simulation, modeling, and reliability and testing techniques. There must be awareness of state-of-the-art technologies and their limitations. As accountant, there must be awareness of the financial implications of planned decisions and knowledge of how to control them. As manager, the planning and control of schedules is an important part of controlling the costs of a project and completing it on time. Also, as manager, there must be the skills necessary to communicate clearly and convincingly with subordinates and superiors to make them aware of problems and their solutions.

The project manager is the person who has the authority to carry out a project. This person has been given the legitimate right to direct the efforts of the project team members. The manager's power comes from the acceptance and respect accorded by superiors and subordinates. The project manager has the power to act and is committed to group goals.

The project manager is responsible for getting the project completed properly, on schedule, and within budget, by utilizing whatever resources are necessary to accomplish the goal in the most efficient manner. The manager provides project schedule, financial, and technical requirement direction and evaluates and reports on project performance. This requires planning, organizing, staffing, directing, and controlling all aspects of the project.

In this leadership role, the project manager is required to perform many tasks including the following:

- Assemble the project organization.
- Develop the project plan.

- Publish the project plan.
- Set measurable and attainable project objectives.
- Set attainable performance standards.
- Determine which scheduling tools (PERT, CPM, Gantt, etc.) are right for the project.
- Using the available scheduling tools, develop and coordinate the project plan, which includes the budget, resources, and the project schedule.
- Develop the project schedule.
- Develop the project budget.
- Manage the budget.
- Recruit personnel for the project.
- Select subcontractors.
- Assign work, responsibility, and authority so that team members can make maximum use of their abilities.
- Estimate, allocate, coordinate, and control project resources.
- Deal with specifications and resource needs that are unrealistic.
- Decide on the right level of administrative and computer support.
- Train project members on how to fulfill their duties and responsibilities.
- Supervise project members, giving them day-to-day instructions, guidance, and discipline as required to fulfill their duties and responsibilities.
- Design and implement reporting and briefing information systems or documents that respond to project needs.
- Control the project.

Some basic project management practices can improve the chances for success. Consider the following:

- Secure the necessary commitments from top management to make the project a success.
- Set up an action plan that will be easily adopted by management.
- Use a work breakdown structure that is comprehensive and easy to use.
- Establish accounting practices that help, not hinder, successful completion of the project.
- Prepare project team job descriptions properly up front to eliminate conflict later on.
- Select project team members appropriately the first time.
- After the project is under way, follow these steps:
- Manage the project, but make the oversight reasonable and predictable.
- Get team members to accept and participate in the plans.
- Motivate project team members for best performance.
- Coordinate activities so they are carried out in relation to their importance with a minimum of conflict.
- Monitor and minimize interdepartmental conflicts.
- Spot problems and take corrective action.
- Discover the strengths and weaknesses in project team members and manage them to get the desired results.
- Help team members solve their own problems.
- Exchange information with subordinates, associates, superiors, and others about plans, progress, and problems.
- Make the best of available resources.

- Measure project performance.
- Determine, through formal and informal reports, the degree to which progress is being made.
- Determine causes of and possible ways to act upon significant deviations from planned performance.
- Take action to correct an unfavorable trend or to take advantage of an unusually favorable trend.
- Look for areas where improvements can be made.
- Develop more effective and economical methods of managing.
- Remain flexible.
- Avoid activity traps.
- Practice effective time management.

When dealing with subordinates, each person must:

- Know what is to be done, preferably in terms of an end product.
- Have a clear understanding of the authority and its limits for each individual.
- Know what the relationship with other people is.
- Know what constitutes a job well done in terms of specific results.
- Know when and what is being done exceptionally well.
- Be shown concrete evidence that there are rewards for work well done and for work exceptionally well done.
- Know where and when expectations are not being met.
- Be made aware of what can and should be done to correct unsatisfactory results.
- Feel that the supervisor has an interest in each person as an individual.
- Feel that the supervisor both believes in each person and is anxious for individual success and progress.

By fostering a good relationship with associates, the manager will have less difficulty communicating with them. The fastest, most effective communication takes place among people with common points of view.

The competent project manager watches what is going on in great detail and can, therefore, perceive problems long before they flow through the paper system. Personal contact is faster than filing out forms. A project manager who spends much of the time roaming through the work place usually stays on top of issues.

7.3.3 Systems Engineer

The term systems engineer means different things to different people. The systems engineer is distinguished from the engineering specialist, who is concerned with only one aspect of a well-defined engineering discipline, in that the systems engineer must be able to adapt to the requirements of almost any type of system. The systems engineer provides the employer with a wealth of experience gained from many successful approaches to technical problems developed through hands-on exposure to a variety of situations. This person is a professional with knowledge and experience, possessing skills in a specialized and learned field or fields. The systems engineer is an expert in these fields; highly trained in analyzing problems and developing solutions that satisfy management objectives. The systems engineer takes data from the overall development process and, in return, provides data in the form of requirements and analysis results to the process.

Education in electronics theory is a prerequisite for designing systems that employ electronic components. As a graduate engineer, the systems engineer has the education required to design electronic systems correctly. Mathematics skill acquired in engineering school is one of the tools used by the systems engineer to formulate solutions to design problems and analyze test results. Knowledge of testing techniques and theory enables this individual to specify system components and performance and to measure the results. Drafting and writing skills are required for efficient preparation of the necessary documentation needed to communicate the design to technicians and contractors who will have to build and install the system.

A competent systems engineer has a wealth of technical information that can be used to speed up the design process and help in making cost-effective decisions. If necessary information is not at hand, the systems engineer knows where to find it. The experienced systems engineer is familiar with proper fabrication, construction, installation, and wiring techniques and can spot and correct improper work.

Training in personnel relations, a part of the engineering curriculum, helps the systems engineer communicate and negotiate professionally with subordinates and management.

Small in-house projects can be completed on an informal basis and, indeed, this is probably the normal routine where the projects are simple and uncomplicated. In a large project, however, the systems engineer's involvement usually begins with preliminary planning and continues through fabrication, implementation, and testing. The degree to which program objectives are achieved is an important measure of the systems engineer's contribution.

During the design process the systems engineer:

- Concentrates on results and focuses work according to management objectives.
- Receives input from management and staff.
- Researches the project and develops a workable design.
- Assures balanced influence of all required design specialties.
- Conducts design reviews.
- Performs tradeoff analyses.
- Assists in verifying system performance.
- Resolves technical problems related to the design, interface between system components, and integration of the system into any facility.

Aside from designing a system, the systems engineer has to answer any questions and resolve problems that may arise during fabrication and installation of the hardware. Quality and workmanship of the installation must be monitored. The hardware and software will have to be tested and calibrated upon completion. This, too, is the concern of the systems engineer. During the production or fabrication phase, systems engineering is concerned with verifying system capability, verifying system performance, and maintaining the system baseline.

Depending on the complexity of the new installation, the systems engineer may have to provide orientation and operating instruction to the users. During the operational support phase, the systems engineer:

- Receives input from users.
- Evaluates proposed changes to the system.
- Facilitates the effective incorporation of changes, modifications and updates.

Depending on the size of the project and the management organization, the systems engineer's duties will vary. In some cases the systems engineer may have to assume the responsibilities of planning and managing smaller projects.

7.3.4 Other Project Team Members

Other key members of the project team, where building construction may be involved, include the following:

- Architect, responsible for design of any structures.
- Electrical engineer, responsible for power system design if not handled by the systems engineer.

- Mechanical engineer, responsible for HVAC, plumbing, and related designs.
- Structural engineer, responsible for concrete and steel structures.
- Construction contractors, responsible for executing the plans developed by the architect, mechanical engineer, and structural engineer.
- Other outside contractors, responsible for certain specialized custom items which cannot be developed or fabricated internally or by any of the other contractors.

BIBLIOGRAPHY

Defense Systems Management, *Systems Engineering Management Guide*, Defense Systems Management College, Fort Belvoir, VA, 1983.

Delatore, J. P., E. M. Prell, and M. K. Vora, "Translating Customer Needs into Product Specifications," *Quality Progress*, 22(1), January 1989.

Finkelstein, L., "Systems Theory," *IEE Proceedings*, Pt. A, 135(6), pp. 401–403, 1988.

Hoban, F. T., and W. M. Lawbaugh, *Readings in Systems Engineering*, National Aeronautics and Space Administrator, Science and Technical Information Program, Washington, DC, 1993.

Shinners, S. M., *A Guide to Systems Engineering and Management*, Lexington Books, Lexington, MA, 1976.

Tuxal, J. G., *Introductory System Engineering*, McGraw-Hill, New York, NY, 1972.

Whitaker, J. C., G. DeSantis, and C. R. Paulson, *Interconnecting Electronic Systems*, CRC Press, Boca Raton, FL, 1992.

Whitaker, Jerry C., and Robert K. Mancini, *Technical Documentation and Process*, CRC Press, Boca Raton, FL, 2013.

CHAPTER 7.4
DISASTER PLANNING AND RECOVERY FOR BROADCAST FACILITIES

Thomas G. Osenkowsky
Radio Engineering Consultant, Brookfield, CT

7.1 INTRODUCTION

A simple definition of disaster can refer to any event which prevents an establishment from conducting business in their usual manner. A simple example would be a blown fuse in an automation system. A more serious example would be a tower failure. Both of these will prevent normal programming from airing. Broadcasters can best deal with disasters by anticipating, planning, and practicing emergency options. Anticipate what failures can affect business, plan an alternative, and practice the routine to ensure a smooth, transparent transition to normal operation.

7.2 ANTICIPATION

Some events are predictable such as rising waters in flood prone regions or high winds in regions with significant hurricane or tornado history. Proper planning dictates locating studios and equipment above the highest expected flood water level and constructing a building with hurricane resistant buttressing. Utility apparatus such as electric panels, Uninterruptible Power Supply (UPS), transfer switches, and stored items of a critical nature should be installed well above flooding levels. Careful examination of a facility can yield interesting results. If the studios are in leased space, consider what is above you. What risks can you expect from other tenants? Is there piping in the ceilings that, if ruptured, could affect your operations? Events such as an overflowing sink in a hair salon above a studio can cause major damage. A tenant who uses a horizontal fire sprinkler as a coat hanger hook can cause flooding of a number of rooms in a building.

A fire, collapse, bomb scare, or other event on another floor in your building or nearby building can force an evacuation or power shutdown of your facility. These scenarios are not under the broadcaster's control but can have major consequences when they occur.

A critical component of most broadcast facilities is data. This includes, but is not limited to, the music and commercial spot library, traffic and music logs, automation and network configuration files, and client and billing information. These are in addition to the operating system, word processing and spreadsheet programs, automation system programs, and traffic/music scheduling programs. Hardware and software must be treated as equals for one cannot function without the other for many systems.

7.3 PREPARATION

Being prepared to deal with disaster-related events offers the best opportunity to restore as normal operation as possible given unforeseen circumstances. If the event affects a wide area, many stations will report on rescue and recovery efforts. Personnel may have their duties temporarily reassigned. Account executives may serve as field reporters, announcers may serve in news gathering roles, and so on. Authorities should be consulted to determine if any special credentials are required for station and contract personnel to travel in affected areas should a curfew be enacted. Contract personnel refer to snow plow and fuel delivery operators, tower crews, electricians, carpenters, and plumbers employed by the broadcaster.

Establishing and maintaining good relations with local authorities is very important. They depend on the broadcaster to relay important information pertaining to road closures, shelter, school, and other safety-related issues. They will also be more receptive to the station's needs pertaining to getting on the air and having access to facilities if a cooperative relationship exists before any major incident occurs.

7.4 ASSESSMENT

Assessing vulnerabilities and solutions should be an ongoing practice. Employees, facilities, and needs constantly change, as do risks. This simple checklist, while universal in nature, should be tailored to individual needs:

- ✓ Are there trees that can affect power lines, guy wires, facility access, or buildings?
- ✓ Are your facilities secure from employee and nonemployee acts?
- ✓ Do your facilities have generators? Are they routinely tested? How long will their fuel supply last?
- ✓ How reliable is your studio-to-transmitter link? What are the contingency plans should it fail?
- ✓ Are your facilities fully insured? Do you have business interruption insurance?
- ✓ Is your data routinely backed up and stored in an accessible location?
- ✓ Are your backup facilities maintained and tested? Do you have sufficient spares?

Trees can come into contact with tower guy wires with disastrous results. They can fall across roads blocking access to facilities. This is especially true for remote transmitter sites. During an emergency it may be necessary to broadcast from the transmitter site. Access by employees, delivery and emergency vehicles should be free and clear of any obstacles. Ice falling from towers can damage antennas on the tower, transmission lines, ATUs, and transmitter buildings as well as any vehicles parked nearby. Proper buttressing techniques should be employed to minimize potential damage to equipment and persons. This is especially important in earthquake prone areas where transmitters, equipment racks, and other equipment needs to be tethered in a manner designed to minimize movement and damage. Strategically placed concrete pilings can prevent vehicular contact with fuel tanks, guy anchors, generators, and other outdoor equipment. Fences can deter all-terrain vehicles from entering a tower field and damaging the ground system and/or transmission lines.

While rare, acts of employee sabotage do occur. These may include, but are not limited to, theft, damaging equipment, alteration or destruction of data, and physical harm to other persons or their property. All facilities should be examined for single point of failure. An example would be restricting access to a server room. Nonemployee acts can be theft of ground system copper from transmitter sites and transmission lines from towers. Burglarizing remote transmitter facilities has also been a problem for broadcasters. An evaluation of company property by a security expert is a wise investment. Such experts offer advice on securing facilities but do not sell equipment or installation services. Enacting policies such as prohibiting carrying firearms or weapons on company property can enhance security and put minds at ease. The use of swipe cards, RFID, or keyless entry to the studio can reduce the possibility of access by unauthorized persons. Increasing security may also qualify for reduced insurance premiums.

Standby power generators are key equipment when commercial power fails. Depending on size, load, and fuel tank status, they can power a facility for a period of time. Ensuring your relationship with the fuel supplier is strong maintaining an adequate amount of fuel is essential. If a storm is predicted, it is wise to top off the tank. Access may not be possible if roads are blocked. The capacity of a generator should be carefully calculated. The chosen value should consider present and future load needs. These needs must include any devices required by law. An example would be a water pump if water is not provided by the municipality. Other examples include hallway and restroom lighting, as well as studio equipment and lighting, and heating/air conditioning if capacity permits or is required by law.

The size of the fuel tank should take into account how long an extreme event dictates its operation; that is, a prolonged commercial power outage. A UPS should be tested to ensure it is compatible with generator power. Some UPS units are sensitive to generator frequency and voltage variations that may exceed factory limit settings, causing them to remain on battery and eventually shut down despite being powered by the generator.

There are many variations of studio-to-transmitter (STL) links. These range from radio circuits leased from a telephone company or fiber optic supplier, wireless methods, or via the Internet. Whatever the vehicle, your program signal must be delivered from the studio to the transmitter. If your STL fails, you must have a backup or relocate the studio to a location where a viable link can be established. This may be at the transmitter site itself. You will also need a source for program material which may not be readily available (satellite, stored on automation system, etc.). Some radio transmitters have a silence sense feature that uses material stored on a USB device should there be loss of incoming audio. This material should be kept current by the programming department.

Insurance is a broad ranging topic. A broadcaster may be covered by several different underwriters. Inland property coverage insures common items such as desktop computers, office furniture and equipment, etc. Equipment specific to broadcasting such as the tower, transmission and sample lines, ATU, isocoupler, STL equipment, transmitter, remote control, audio processor, and the like are covered by a marine policy. Liability coverage protects against the words and actions of an employee. This is important for anything causable stated over the air. Business interruption insurance may compensate for commercials or programs that failed to be broadcast while the station was off the air. It is difficult to assess any long-term effects where listeners or viewers chose other stations while your station was off the air and did not return. This can also be detrimental during a ratings sweep. While some items and their installation and testing may be covered by insurance, other costs such as road improvement to a tower site to permit safe access by delivery vehicles, cement mixers, and the like may not. Every broadcaster should conduct a comprehensive review of their facilities in a "what if" scenario to determine if they are sufficiently covered for any losses. This process must be ongoing as equipment, facilities, and costs frequently change.

Backing up data should be part of the station's daily routine. Word processing, spreadsheet, database and all business, traffic, music, and configuration files should be stored on a file server that is automatically backed up each day. The type of backup must be accessible at all times. A USB drive that is placed in a bank safe deposit box will be secure; however, it is not accessible when the bank is closed. Online backup is one option and again accessibility is of prime importance since backed up data may need to be accessed at any time. The reliability of the company providing the backup service is critical. If they go out of business or suffer equipment malfunction, your data is inaccessible. Multiple backups are another option, that is, online plus a USB drive maintained off-premises by a trusted employee.

Having a backup studio, transmitter, and/or antenna is critical but only valuable if they have been maintained and routinely tested. Realizing a critical failure in time of need devalues the equipment since it cannot serve its intended purpose. Every studio should have the ability to serve not only its intended purpose but as backup for others. In other words, a control room for one station should be able to serve in that role or as a production room for other stations under the same roof. Should one studio fail it can be backed up by another while repairs are undertaken. Maintaining a portable audio mixer that can be pressed into service at the main studio or other location can be a valuable asset in time of need. While it may seem like an expenditure with low rate of return, it will pay dividends should the regular studio become unusable. At the very least you are on the air. The same can be said for wisely chosen spare parts.

Some multiuser transmitter sites have an $n + 1$ transmitter. This frequency agile transmitter can serve as backup for multiple stations. While this is efficient from an economical and space standpoint, it should be very clear in written form which station has priority in the event of multiple failures and for what period of time.

7.5 WHO IS IN CHARGE?

Every facility should have a "go to" person who is responsible for directing operations during an emergency or crisis. This may not necessarily be the General Manager. It should be someone who is familiar with the facilities, what equipment is capable of performing what task and understands how to accomplish tasks in an efficient manner. This person should be responsible for ensuring the generator(s) is adequately fueled, repairs or reconstruction is underway and on schedule, coordinating with other departments in the station as well as local officials, contractors and other personnel. Any contractors hired by the station should be insured and licensed, if required. Establishing priorities is the first task in an emergency. Some routines may not be possible or required and personnel may have their duties reassigned to accomplish the tasks at hand. Every situation is dynamic and there is a limit to what can be practiced during normal times. This is why planning and preparing for the worst is important.

7.6 PERSONNEL MATTERS

Another concern that must be addressed is personal. Everyday tasks that are performed by an individual should be shared with others. More than one employee should know how to schedule traffic in the event the Traffic Manager is unavailable. This is critical to the business and daily operations. The same applies to the Business Manager who sends out invoices and pays bills, Program Director who manages the schedule, and Sales Manager(s) who establishes account lists and manages remotes and promotions.

Engineering should have documented all studio and transmitter site wiring, satellite receiver relay interfaces with automation system and automation system configuration files. The individual charged with IT duties (Chief Engineer, etc.) should have a master list of all login information for company-related computer web sites and access. Examples of these are music research sites, record company music download sites, news data and audio sites, personalized automation login screens, company email login password, transmitter remote control login, VPN or remote login and audio processor password protection, if used.

7.7 COMMON PRECAUTIONS

Closely examine the electrical distribution in your studios. Critical equipment should be powered by separate circuit breakers and their outlets not accessible in the common area. Studios have shut down due to space heaters being plugged into outlet strips in the rear of equipment racks, a shorted vacuum cleaner cord plugged into a hallway outlet that powered a console, and other such common mishaps. As new equipment is brought into service carefully plan how it is powered, if it needs to be on a UPS and/or generator. Check power service panels for tight connections to circuit breakers. Ensure there is nothing that can accidentally fall or be bumped, causing personal injury. Examine driveways for low hanging utility cables that can fall victim to trucks. Guy wires and transmission lines should be free of trees, weeds, and other growths that can come in contact or burn and cause damage. Guy anchors should be protected by fences. If on or near a farm, the anchor fences should be extended to prevent animals from contacting the guy wires.

If your facility(s) qualify for flood insurance, speak with your agent about coverage. Flood insurance covers damages caused by rising waters in certain areas. Water damage caused by broken pipes, overflowed fixtures, and the like are covered by property insurance.

Meet with the Fire Chief in the areas where your facilities are located. Discuss what fire suppression methods they would use in the event of a fire and the best type of extinguishers to have at the ready in your facility. These may vary by locality.

Prepare a "What To Do" manual which clearly outlines procedures to follow in the event of a failure or emergency. Many times operators or other personnel waste precious time trying to reach an engineer by telephone, text, and so on when they can do a simple procedure themselves to get back on the air. Practice and test them on a regular basis. Discuss with your consulting engineer options for auxiliary and/or emergency antenna sites in case they are needed.

7.8 CONCLUSION

Down time can be minimized by anticipating interruptions in routines, preparing alternatives, and practicing implementation. Many times disasters cannot be averted but they can be effectively dealt with when the facility and personnel are properly prepared. The best resource for learning about recovery is from someone experienced in the matter.

A disaster preparation checklist is provided in Table 7.4.1.

TABLE 7.4.1 Disaster Preparation Checklist

- **Insurance**—Are the facilities adequately insured for all loss and claim types? Do the policies specify replacement or new-for-old-without-depreciation coverage? Is the station(s) covered for business interruption loss? If applicable, does the facility have flood insurance? Have you discussed the station's specific needs with your insurance agent?

- **Inventory**—Is there an up-to-date room-by-room inventory of all equipment, furniture, fixtures, utilities, and specialized construction such as soundproofing, antistatic flooring, and satellite dish supports? Has the inventory been kept up to date to maintain pace with new equipment? Are model and serial numbers recorded? Are date of purchase and replacement cost recorded?

- **Alternate Studio/Transmitter Site Plans**—Does the station(s) have portable mixers/switchers and program sources should it be necessary to temporarily relocate the studio? Does the station have an alternate transmitter/antenna site? Do the facilities have generators and adequate fuel supplies?

- **Contacts List**—Is there an up-to-date list of contact information for all employees, public safety officials, contractors, and suppliers? This list should include work, home and cell phone numbers, email and text addresses. Is there a current manual of procedures to employ in the event of an equipment failure, disaster, or other anomaly?

- **The Go-To Person**—Has a person been appointed to act in this capacity in the event of an emergency? Is he/she qualified, informed, and prepared to take charge of all matters and has the staff been directed to follow their instruction implicitly? Have emergency procedures been practiced to identify and rectify weaknesses? Is training an ongoing part of the routine?

- **Data Backup/Facility Security**—Is your computer data routinely backed up? Is the backup data accessible upon demand? Is the security of your computer network continuously monitored and maintained? Do you have replacement equipment that can serve to back up critical computers such as file servers, data storage, etc.? Are your facilities secure? Is access restricted solely to authorized persons? Have the facilities been evaluated by a security and fire prevention professional?

- **Tower Integrity**—Is the tower(s) maintained by a professional? Have all loads been accounted for in a structural integrity analysis? Are guy anchors and tower bases properly fenced and locked? Are critical areas such as transmission lines, ground systems, guy anchors, and towers protected from vehicular contact?

- **Generator**—Does the facility have a generator? UPS? What loads are powered by each device? Is the generator routinely serviced? Are the UPS batteries tested? Does the UPS tolerate generator power quality? For what period is the fuel supply adequate? Have provisions been made for refueling during extended power outages?

- **Documentation**—Does the facility have current documentation on wiring, automation system configuration, satellite receiver interface, IP address for each computer, master list of usernames, and passwords for all company-related sites and equipment? Are all company telephone numbers documented with their uses (business, news, request lines, FAX, etc.)?

CHAPTER 7.5
SAFETY CONSIDERATIONS

Jerry C. Whitaker, Editor-in-Chief

7.1 INTRODUCTION

Electrical safety is important when working with any type of electronic hardware. Because transmission equipment operates at high voltages and currents, safety is doubly important. The primary areas of concern, from a safety standpoint, include:

- Electric shock
- Nonionizing radiation
- Beryllium oxide (BeO) ceramic dust
- Hot surfaces of vacuum tube devices
- Polychlorinated biphenyls (PCBs)

7.2 ELECTRIC SHOCK

Surprisingly little current is required to injure a person. Studies at Underwriters Laboratories (UL) show that the electrical resistance of the human body varies with the amount of moisture on the skin, the muscular structure of the body, and the applied voltage. The typical hand-to-hand resistance ranges between 500 Ω and 600 kΩ, depending on the conditions. Higher voltages have the capability to break down the outer layers of the skin, which can reduce the overall resistance value. UL uses the lower value, 500 Ω as the standard resistance between major extremities, such as from the hand to the foot. This value is generally considered the minimum that would be encountered and, in fact, may not be unusual because wet conditions or a cut or other break in the skin significantly reduces human body resistance.

7.2.1 Effects on the Human Body

Table 7.5.1 lists some effects that typically result when a person is connected across a current source with a hand-to-hand resistance of 2.4 kΩ. The table shows that a current of approximately 50 mA will flow between the hands, if one hand is in contact with a 120 V ac source and the other hand is grounded. The table indicates that even the relatively small current of 50 mA can produce *ventricular fibrillation* of the heart, and perhaps death. Medical literature describes ventricular

TABLE 7.5.1 The Effects of Current on the Human Body

Current	Effect
1 mA or less	No sensation, not felt
More than 3 mA	Painful shock
More than 10 mA	Local muscle contractions, sufficient to cause "freezing" to the circuit for 2.5 percent of the population
More than 15 mA	Local muscle contractions, sufficient to cause "freezing" to the circuit for 50 percent of the population
More than 30 mA	Breathing is difficult, can cause unconsciousness
50–100 mA	Possible ventricular fibrillation
100–200 mA	Certain ventricular fibrillation
More than 200 mA	Severe burns and muscular contractions; heart more apt to stop than to go into fibrillation
More than a few amperes	Irreparable damage to body tissue

fibrillation as rapid, uncoordinated contractions of the ventricles of the heart, resulting in loss of synchronization between heartbeat and pulse beat. The electrocardiograms shown in Fig. 7.5.1 compare a healthy heart rhythm with one in ventricular fibrillation. Unfortunately, once ventricular fibrillation occurs, it will continue. Barring resuscitation techniques, death will ensue within a few minutes.

The route taken by the current through the body has a significant effect on the degree of injury. Even a small current, passing from one extremity through the heart to another extremity, is dangerous and capable of causing severe injury or electrocution. There are cases where a person has contacted extremely high current levels and lived to tell about it. However, usually when this happens, the current passes only through a single limb and not through the body. In these instances, the limb is often lost, but the person survives.

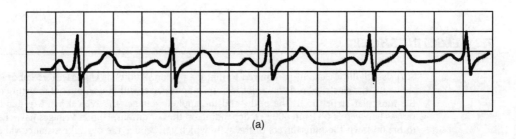

(a)

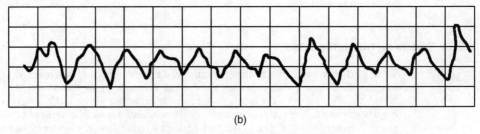

(b)

FIGURE 7.5.1 Electrocardiogram of a human heartbeat: (*a*) healthy rhythm, (*b*) ventricular fibrillation.

Current is not the only factor in electrocution. Figure 7.5.2 summarizes the relationship between current and time on the human body. The graph shows that 100 mA flowing through a human adult body for 2 s will cause death by electrocution. An important factor in electrocution, the *let-go range*, also is shown on the graph. This range is described as the amount of current that causes "freezing," or the inability to let go of the conductor. At 10 mA, 2.5 percent of the population will be unable

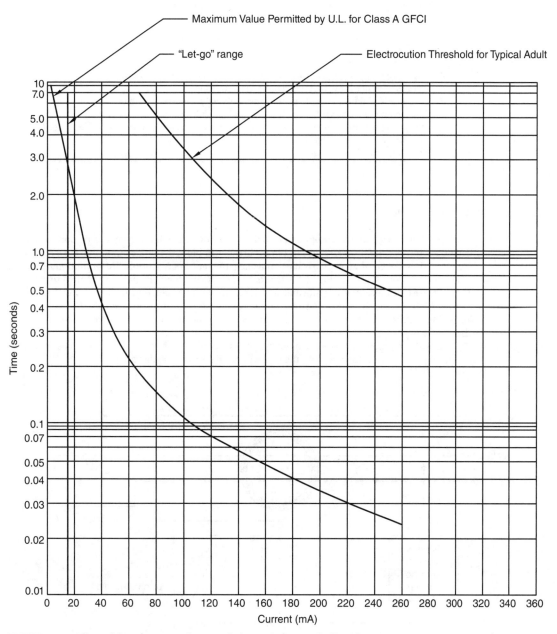

FIGURE 7.5.2 Effects of electric current and time on the human body. Note the "let-go" range.

7.44 BROADCAST MANAGEMENT

to let go of a "live" conductor. At 15 mA, 50 percent of the population will be unable to let go of an energized conductor. It is apparent from the graph that even a small amount of current can "freeze" someone to a conductor. The objective for those who must work around electric equipment is how to protect themselves from electric shock. Table 7.5.2 lists required precautions for personnel working around high voltages.

TABLE 7.5.2 Required Safety Practices for Engineers Working Around High-Voltage Equipment

High-Voltage Precautions
✓ Remove all ac power from the equipment. Do not rely on internal contactors or SCRs to remove dangerous ac.
✓ Trip the appropriate power distribution circuit breakers at the main breaker panel.
✓ Place signs as needed to indicate that the circuit is being serviced.
✓ Switch the equipment being serviced to the *local control* mode as provided.
✓ Discharge all capacitors using the discharge stick provided by the manufacturer.
✓ Do not remove, short circuit, or tamper with interlock switches on access covers, doors, enclosures, gates, panels, or shields.
✓ Keep away from live circuits.

7.2.2 Circuit Protection Hardware

The typical primary panel or equipment circuit breaker or fuse will not protect a person from electrocution. In the time it takes a fuse or circuit breaker to blow, someone could die. However, there are protection devices that, properly used, may help prevent electrocution. The *ground-fault current interrupter* (GFCI), shown in Fig. 7.5.3, works by monitoring the current being applied to the load. The GFI uses a differential transformer and looks for an imbalance in load current. If a current (5 mA, ±1 mA) begins to flow between the neutral and ground or between the hot and ground leads, the differential transformer detects the leakage and opens up the primary circuit within about 2.5 ms.

GFIs will not protect a person from every type of electrocution. If the victim becomes connected to both the neutral and the hot wire, the GFI will not detect an imbalance.

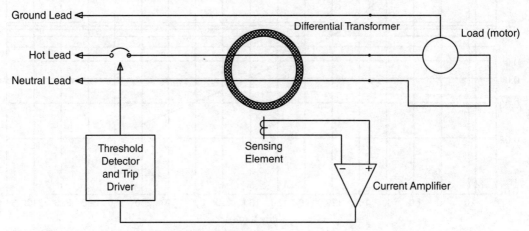

FIGURE 7.5.3 Basic design of a ground-fault interrupter (GFI).

7.2.3 Working with High Voltage

Rubber gloves are commonly used by engineers working on high-voltage equipment. These gloves are designed to provide protection from hazardous voltages or RF when the wearer is working on "hot" ac or RF circuits. Although the gloves may provide some protection from these hazards, placing too much reliance on them can have disastrous consequences. There are several reasons why gloves should be used with a great deal of caution and respect. A common mistake made by engineers is to assume that the gloves always provide complete protection. The gloves found in many facilities may be old or untested. Some may show signs of user repair, perhaps with electrical tape. Few tools could be more hazardous than such a pair of gloves.

Another mistake is not knowing the voltage rating of the gloves. Gloves are rated differently for both ac and dc voltages. For example, a *class 0* glove has a minimum dc breakdown voltage of 35 kV; the minimum ac breakdown voltage, however, is only 6 kV. Furthermore, high-voltage rubber gloves are not usually tested at RF frequencies, and RF can burn a hole in the best of them. It is possible to develop dangerous working habits by assuming that gloves will offer the required protection.

Gloves alone may not be enough to protect an individual in certain situations. Recall the axiom of keeping one hand in a pocket while working around a device with current flowing? That advice is actually based on simple electricity. It is not the "hot" connection that causes the problem, but the ground connection that lets the current begin to flow. Studies have shown that more than 90 percent of electric equipment fatalities occurred when the grounded person contacted a live conductor. Line-to-line electrocution accounted for less than 10 percent of the deaths.

When working around high voltages, always look for grounded surfaces. Keep hands, feet, and other parts of the body away from any grounded surface. Even concrete can act as a ground if the voltage is sufficiently high. If work must be performed in "live" cabinets, then consider using, in addition to rubber gloves, a rubber floor mat, rubber vest, and rubber sleeves. Although this may seem to be a lot of trouble, consider the consequences of making a mistake. Of course, the best troubleshooting methodology is never to work on any circuit without being certain that no hazardous voltages are present. In addition, any circuits or contactors that normally contain hazardous voltages should be firmly grounded before work begins.

Most importantly, defer any work on high voltage equipment to a qualified contractor or consultant. Do not attempt to work on high voltage equipment unless you have the required training and proper tools. Never work alone.

Most companies and organizations have established safety procedures for working with various types of equipment. Know the local rules and follow them.

7.2.3.1 RF Considerations. Engineers sometimes rely on electrical gloves when making adjustments to live RF circuits. This practice, however, can be extremely dangerous. Consider the typical load matching unit shown in Fig. 7.5.4. In this configuration, disconnecting the coil from either L2 or L3 places the full RF output literally at the engineer's fingertips. Depending on the impedances involved, the voltages can become quite high, even in a circuit that normally is relatively tame.

In Fig. 7.5.4 example, assume that the load impedance is approximately $106 + j202$ Ω. With 1 kW feeding into the load, the rms voltage at the matching output will be approximately 700 V. The peak voltage (which determines insulating requirements) will be close to 1 kV, and perhaps more than twice that if the carrier is being amplitude-modulated. At the instant the output coil clip is disconnected, the current in the shunt leg will increase rapidly, and the voltage easily could more than double.

7.2.4 First Aid Procedures

All engineers working around high-voltage equipment should be familiar with first aid treatment for electric shock and burns. Always keep a first aid kit on hand at the facility. Figure 7.5.5 illustrates the basic treatment for victims of electric shock. Copy the information, and post it in a prominent location. Better yet, obtain more detailed information from the local heart association or Red Cross chapter. Personalized instruction on first aid usually is available locally.

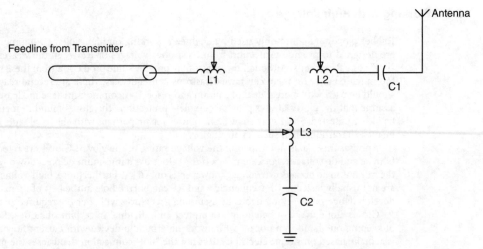

FIGURE 7.5.4 Example of how high voltages can be generated in an RF load matching unit.

7.3 OPERATING HAZARDS

A number of potential hazards exist in the operation and maintenance of high-power transmission equipment. Maintenance personnel must exercise extreme care around such hardware. Consider the following guidelines:

- Use caution around the high-voltage stages of the equipment. Many devices operate at voltages high enough to kill through electrocution. Always break the primary ac circuit of the power supply, and discharge all high-voltage capacitors.

- Minimize exposure to RF radiation. Do not permit personnel to be in the vicinity of open, energized RF generating circuits, RF transmission systems (waveguides, cables, or connectors), or energized antennas. High levels of radiation can result in severe bodily injury, including blindness. Cardiac pacemakers may also be affected.

- Avoid contact with beryllium oxide (BeO) ceramic dust and fumes. BeO ceramic material may be used as a thermal link to carry heat from a tube or power semiconductor to the heat sink. Do not perform any operation on any BeO ceramic that might produce dust or fumes, such as grinding, grit blasting, or acid cleaning. Beryllium oxide dust and fumes are highly toxic, and breathing them can result in serious injury or death. BeO ceramics must be disposed of as prescribed by the device manufacturer.

- Avoid contact with hot surfaces within the equipment. The anode portion of many power tubes are air-cooled. The external surface normally operates at a high temperature (up to 250°C). Other portions of the tube also may reach high temperatures, especially the cathode insulator and the cathode/heater surfaces. All hot surfaces may remain hot for an extended time after the tube is shut off. To prevent serious burns, avoid bodily contact with these surfaces during tube operation and for a reasonable cool-down period afterward. Table 7.5.3 lists basic first aid procedures for burns.

7.3.1 OSHA Safety Considerations

The U.S. government has taken a number of steps to help improve safety within the workplace under the auspices of the Occupational Safety and Health Administration (OSHA). The agency helps industries monitor and correct safety practices. OSHA has developed a number of guidelines designed to

SAFETY CONSIDERATIONS **7.47**

If the victim is not responsive, follow the A-B-Cs of basic life support.

A) AIRWAY: If the victim is unconscious, open airway.

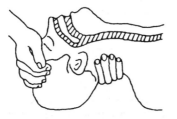

1. Lift up neck
2. Push forehead back
3. Clear out mouth if necessary
4. Observe for breathing

B) BREATHING: If the victim is not breathing, begin artificial breathing.

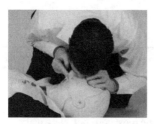
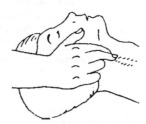

1. Tilt head
2. Pinch nostrils
3. Make airtight seal
4. Provide four quick full breaths

Check carotid pulse. If pulse is absent, begin artificial circulation. Remember that mouth-to-mouth resuscitation must be commenced as soon as possible.

C) CIRCULATION: Depress the sternum 1.2 to 2 inches.

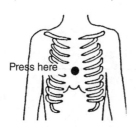
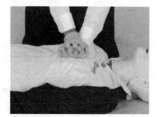

For situations in which there is one rescuer, provide 15 compressions and then 2 quick breaths. The approximate rate of compressions should be 80 per minute.

For situations in which there are two rescuers, provide 5 compressions and then 1 breath. The approximate rate of compressions should be 60 per minute.

Do not interrupt the rhythm of compressions when a second person is giving breaths.

If the victim is responsive, keep warm and quiet, loosen clothing, and place in a reclining position. Call for medical assistance as soon as possible.

FIGURE 7.5.5 Basic first aid treatment for electric shock. (Courtesy Red Cross). Detailed information can be obtained from any Red Cross office. See http://www.redcross.org/ux/take-a-class.

TABLE 7.5.3 Basic First Aid Procedures for Burns

	Extensively Burned and Broken Skin
✓	Loosely cover affected area with a clean sheet or cloth.
✓	Do not break blisters, remove tissue, remove adhered particles of clothing, or apply any salve or ointment.
✓	Treat victim for shock as required.
✓	Arrange for transportation to a hospital as quickly as possible.
✓	If arms or legs are affected, keep them elevated.
	Less Severe Burns (First and Second Degree)
✓	Apply cool (not ice-cold) compresses using the cleanest available cloth article.
✓	Do not break blisters, remove tissue, remove adhered particles of clothing, or apply salve or ointment.
✓	Apply clean, dry dressing if necessary.
✓	Treat victim for shock as required.
✓	Arrange for transportation to a hospital as quickly as possible.
✓	If arms or legs are affected, keep them elevated.

Detailed information can be obtained from any Red Cross office. See: http://www.redcross.org/ux/take-a-class.

help prevent accidents. OSHA records show that electrical standards are among the most frequently violated of all safety standards. Table 7.5.4 lists 16 of the most common electrical violations, including exposure of live conductors, improperly labeled equipment, and faulty grounding.

7.3.1.1 Protective Covers. Exposure of live conductors is a common safety violation. All potentially dangerous electric conductors should be covered with protective panels. The danger is that someone may come into contact with the exposed current-carrying conductors. It is also possible for metallic objects such as ladders, cable, or tools to contact a hazardous voltage, creating a life-threatening condition. Open panels also present a fire hazard.

TABLE 7.5.4 Sixteen Common OSHA Violations (*After* [1].)

Fact Sheet	Subject	NEC[a] Reference
1	Guarding of live parts	110-117
2	Identification	110-122
3	Uses allowed for flexible cord	400-417
4	Prohibited uses of flexible cord	400-418
5	Pull at joints and terminals must be prevented	400-410
6.1	Effective grounding, Part 1	250-251
6.2	Effective grounding, Part 2	250-251
7	Grounding of fixed equipment, general	250-242
8	Grounding of fixed equipment, specific	250-243
9	Grounding of equipment connected by cord and plug	250-245
10	Methods of grounding, cord and plug-connected equipment	250-259
11	AC circuits and systems to be grounded	250-255
12	Location of overcurrent devices	240-224
13	Splices in flexible cords	400-409
14	Electrical connections	110-114
15	Marking equipment	110-121
16	Working clearances about electric equipment	110-116

[a]National Electrical Code.

7.3.1.2 Identification and Marking.
Circuit breakers and switch panels should be properly identified and labeled. Labels on breakers and equipment switches may be many years old and may no longer reflect the equipment actually in use. This is a safety hazard. Casualties or unnecessary damage can be the result of an improperly labeled circuit panel if no one who understands the system is available in an emergency. If a number of devices are connected to a single disconnect switch or breaker, a diagram should be provided for clarification. Label with brief phrases, and use clear, permanent, and legible markings.

Equipment marking is a closely related area of concern. This is not the same thing as equipment identification. Marking equipment means labeling the equipment breaker panels and ac disconnect switches according to device rating. Breaker boxes should contain a nameplate showing the manufacturer, rating, and other pertinent electrical factors. The intent is to prevent devices from being subjected to excessive loads or voltages.

7.3.1.3 Grounding.
OSHA regulations describe two types of grounding: *system grounding* and *equipment grounding*. System grounding actually connects one of the current-carrying conductors (such as the terminals of a supply transformer) to ground. (See Fig. 7.5.6.) Equipment grounding connects all of the noncurrent-carrying metal surfaces together and to ground. From a grounding standpoint, the only difference between a grounded electrical system and an ungrounded electrical system is that the *main bonding jumper* from the service equipment ground to a current-carrying conductor is omitted in the ungrounded system. The system ground performs two tasks:

- It provides the final connection from equipment-grounding conductors to the grounded circuit conductor, thus completing the ground-fault loop.
- It solidly ties the electrical system and its enclosures to their surroundings (usually earth, structural steel, and plumbing). This prevents voltages at any source from rising to harmfully high voltage-to-ground levels.

Note that equipment grounding—bonding all electric equipment to ground—is required whether or not the system is grounded. Equipment grounding serves two important tasks:

- It bonds all surfaces together so that there can be no voltage difference among them.
- It provides a ground-fault current path from a fault location back to the electrical source, so that if a fault current develops, it will rise to a level high enough to operate the breaker or fuse.

The National Electrical Code (NEC) is complex and contains numerous requirements concerning electrical safety. The fact sheets listed in Table 7.5.4, and additional information, are available from OSHA.

7.3.2 Beryllium Oxide Ceramics

Some tubes and power semiconductors contain beryllium oxide (BeO) ceramics. Never perform any operations on BeO ceramics that produce dust or fumes, such as grinding, grit blasting, or acid cleaning. Beryllium oxide dust and fumes are highly toxic, and breathing them can result in serious personal injury or death. Because BeO warning labels may be obliterated or missing, maintenance personnel should contact the device manufacturer before performing any work on a damaged device. Some tubes have BeO internal to the vacuum envelope.

7.3.3 Corrosive and Poisonous Compounds

The external output waveguides and cathode high-voltage bushings of microwave tubes are sometimes operated in systems that use a dielectric gas to impede microwave or high-voltage breakdown. If breakdown does occur, the gas may decompose and combine with impurities, such as air or water vapor, to form highly toxic and corrosive compounds. Examples include Freon gas, which may form

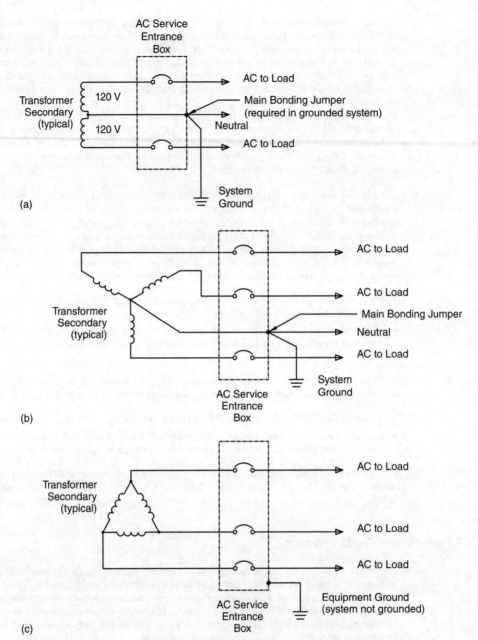

FIGURE 7.5.6 AC service entrance bonding: (*a*) 120 V phase-to-neutral (240 V phase-to-phase), (*b*) 3-phase 208 V wye (120 V phase-to-neutral), (*c*) 3-phase 240 V (or 480 V) delta.

lethal *phosgene*, and sulfur hexafluoride (SF_6) gas, which may form highly toxic and corrosive sulfur or fluorine compounds such as *beryllium fluoride*. When breakdown does occur in the presence of these gases, proceed as follows:

- Ventilate the area to outside air
- Avoid breathing any fumes or touching any liquids that develop
- Take precautions appropriate for beryllium compounds and for other highly toxic and corrosive substances
- Contact the device manufacturer for recommendations on how to proceed

If a coolant other than pure water is used, follow the precautions supplied by the coolant manufacturer.

7.3.4 FC-75 Toxic Vapor

The decomposition products of FC-75 are highly toxic. Decomposition may occur as a result of any of the following:

- Exposure to temperatures above 200°C
- Exposure to liquid fluorine or alkali metals (lithium, potassium, or sodium)
- Exposure to ionizing radiation

Known thermal decomposition products include *perfluoroisobutylene* (PFIB; $[CF_3]_2$ C = CF_2), which is highly toxic in small concentrations.

If FC-75 has been exposed to temperatures above 200°C through fire, electric heating, or prolonged electric arcs, or has been exposed to alkali metals or strong ionizing radiation, take the following steps:

- Strictly avoid breathing any fumes or vapors.
- Thoroughly ventilate the area.
- Strictly avoid any contact with the FC-75.
- Promptly contact an appropriate specialist with experience in such matters.

7.4 NONIONIZING RADIATION

Nonionizing radio frequency radiation (RFR) resulting from high-intensity RF fields is an ongoing concern to engineers who must work around high-power transmission equipment. The principal medical concern regarding nonionizing radiation involves heating of various body tissues, which can have serious effects, particularly if there is no mechanism for heat removal. Research has also noted, in some cases, subtle psychological and physiological changes at radiation levels below the threshold for heat-induced biological effects. However, the consensus is that most effects are thermal in nature.

High levels of RFR can affect one or more body systems or organs. Areas identified as potentially sensitive include the ocular (eye) system, reproductive system, and the immune system. Nonionizing radiation also is thought to be responsible for metabolic effects on the central nervous system and cardiac system.

In spite of these studies, many of which are ongoing, there is still no clear evidence in Western literature that exposure to medium-level nonionizing radiation results in detrimental effects. Russian findings, on the other hand, suggest that occupational exposure to RFR at power densities above 1.0 mW/cm² does result in symptoms, particularly in the central nervous system.

Clearly, the jury is still out as to the ultimate biological effects of RFR. Until the situation is better defined, however, the assumption must be made that potentially serious effects can result from excessive exposure. Compliance with existing standards should be the minimum goal, to protect members of the public as well as facility employees.

7.4.1 NEPA Mandate

The National Environmental Policy Act of 1969 required the Federal Communications Commission to place controls on nonionizing radiation. The purpose was to prevent possible harm to the public at large and to those who must work near sources of the radiation. Action was delayed because no hard and fast evidence existed that low- and medium-level RF energy is harmful to human life. Also, there was no evidence showing that radio waves from radio and TV stations did not constitute a health hazard.

During the delay, many studies were carried out in an attempt to identify those levels of radiation that might be harmful. From the research, suggested limits were developed by the American National Standards Institute (ANSI) and stated in the document known as ANSI C95.1-1982. The protection criteria outlined in the standard are shown in Fig. 7.5.7.

The energy-level criteria were developed by representatives from a number of industries and educational institutions after performing research on the possible effects of nonionizing radiation. The projects focused on absorption of RF energy by the human body, based on simulated human body models. In preparing the document, ANSI attempted to determine those levels of incident radiation that would cause the body to absorb less than 0.4 W/kg of mass (averaged over the whole body) or peak absorption values of 8 W/kg over any 1 g of body tissue.

From the data, the researchers found that energy would be absorbed more readily at some frequencies than at others. The absorption rates were found to be functions of the size of a specific individual and the frequency of the signal being evaluated. It was the result of these absorption rates that culminated in the shape of the *safe curve* shown in the figure. ANSI concluded that no harm would come to individuals exposed to radio energy fields, as long as specific values were not exceeded when averaged over a period of 0.1 hour. It was also concluded that higher values for a brief period would not pose difficulties if the levels shown in the standard document were not exceeded when averaged over the 0.1-hour time period.

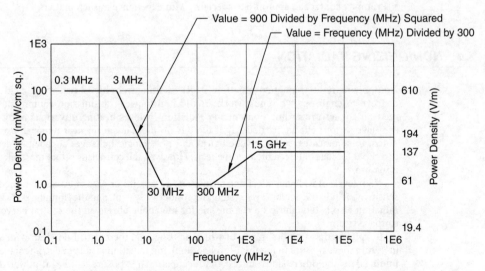

FIGURE 7.5.7 The power density limits for nonionizing radiation exposure for humans.

The FCC adopted ANSI C95.1-1982 as a standard that would ensure adequate protection to the public and to industry personnel who are involved in working around RF equipment and antenna structures.

7.4.1.1 Revised Guidelines. The ANSI C95.1-1982 standard was intended to be reviewed at 5-year intervals. Accordingly, the 1982 standard was due for reaffirmation or revision in 1987. The process was indeed begun by ANSI, but was handed off to the Institute of Electrical and Electronics Engineers (IEEE) for completion. In 1991, the revised document was completed and submitted to ANSI for acceptance as ANSI/IEEE C95.1-1992.

The IEEE standard incorporated changes from the 1982 ANSI document in four major areas:

- An additional safety factor was provided in certain situations. The most significant change was the introduction of new *uncontrolled* (public) exposure guidelines, generally established at one-fifth of the *controlled* (occupational) exposure guidelines. Figure 7.5.8 illustrates the concept for the microwave frequency band.
- For the first time, guidelines were included for body currents; examination of the electric and magnetic fields were determined to be insufficient to determine compliance.
- Minor adjustments were made to occupational guidelines, including relaxation of the guidelines at certain frequencies and the introduction of *breakpoints* at new frequencies.
- Measurement procedures were changed in several aspects, most notably with respect to spatial averaging and to minimum separation from reradiating objects and structures at the site.

IEEE C95.1-2005, "IEEE Standard for Safety Levels with Respect to Human Exposure to Radio Frequency Fields, 3 kHz to 300 GHz," represents the culmination of a 14-year effort. This standard, a major revision of IEEE Standard C95.1-1991, is the fourth revision of the original C95.1 standard. As with the earlier documents, the 2005 revision is designed to "protect against adverse health effects in human beings associated with exposure to electric, magnetic, and electromagnetic fields in the range of 3 kHz to 300 GHz." The basic restrictions for frequencies between 100 kHz and 3 GHz are expressed in terms of a *specific absorption rate* (SAR)—both whole-body-averaged and peak spatial average. From these are derived the *maximum permissible exposure* (MPE), expressed in terms of more easily determined quantities; e.g., incident field strength and power density. The basic rules of C95.1 are presented in graphical form in Fig. 7.5.9.

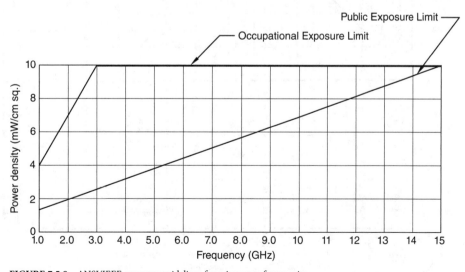

FIGURE 7.5.8 ANSI/IEEE exposure guidelines for microwave frequencies.

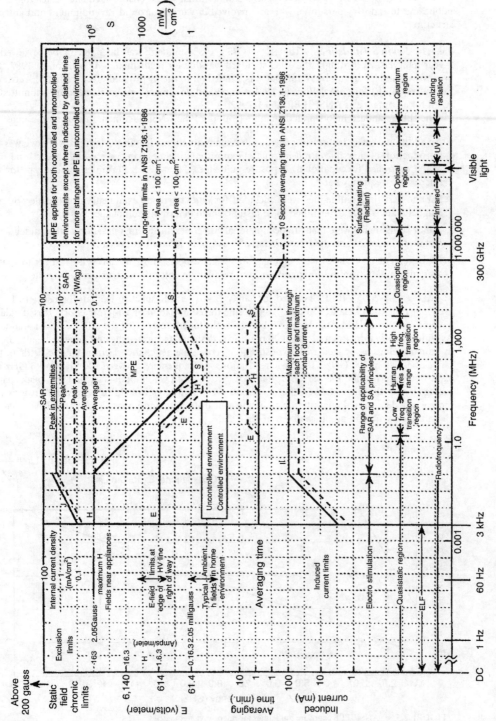

FIGURE 7.5.9 A capsule guide to the IEEE C95.1 standard showing the relevant features. (*From* [2]. *Used with permission.*)

The revised guidelines are complex and beyond the scope of this handbook. Refer to the ANSI/IEEE document for details.

7.4.1.2 Multiple-User Sites. At a multiple-user site, the responsibility for assessing the RFR situation—although officially triggered by either a new user or the license renewal of all site tenants—is, in reality, the joint responsibility of all the site tenants. In a multiple-user environment involving various frequencies, and various protection criteria, compliance is indicated when the fraction of the RFR limit within each pertinent frequency band is established and added to the sum of all the other fractional contributions. The sum must not be greater than 1.0. Evaluating the multiple-user environment is not a simple matter, and corrective actions, if indicated, may be quite complex.

7.4.1.3 Operator Safety Considerations. RF energy must be contained properly by shielding and transmission lines. All input and output RF connections, cables, flanges, and gaskets must be RF leakproof. The following guidelines should be followed at all times:

- Never operate a power tube without a properly matched RF energy absorbing load attached.
- Never look into or expose any part of the body to an antenna or open RF generating tube, circuit, or RF transmission system that is energized.
- Monitor the RF system for radiation leakage at regular intervals and after servicing.

7.4.2 X-Ray Radiation Hazard

The voltages typically used in microwave tubes are capable of producing dangerous X rays. As voltages increase beyond 15 kV, metal-body tubes are capable of producing progressively more dangerous radiation. Adequate X-ray shielding must be provided on all sides of such devices, particularly at the cathode and collector ends, as well as at the modulator and pulse transformer tanks (as appropriate). High-voltage tubes never should be operated without adequate X-ray shielding in place. The X-ray radiation of the device should be checked at regular intervals and after servicing.

7.4.3 Implosion Hazard

Because of the high internal vacuum in power grid and microwave tubes, the glass or ceramic output window or envelope can shatter inward (implode) if struck with sufficient force or exposed to sufficient mechanical shock. Flying debris can result in bodily injury, including cuts and puncture wounds. If the device is made of beryllium oxide ceramic, implosion may produce highly toxic dust or fumes.

In the event of such an implosion, assume that toxic BeO ceramic is involved unless confirmed otherwise.

7.4.4 Hot Coolant and Surfaces

Extreme heat occurs in the electron collector of a microwave tube and the anode of a power grid tube during operation. Coolant channels used for water or vapor cooling also can reach high temperatures (boiling—100°C—and above), and the coolant is typically under pressure (as high as 100 psi). Some devices are cooled by boiling the coolant to form steam.

Contact with hot portions of the tube or its cooling system can scald or burn. Carefully check that all fittings and connections are secure, and monitor back-pressure for changes in cooling system performance. If back pressure is increased above normal operating values, shut the system down and clear the restriction.

For a device whose anode or collector is air-cooled, the external surface normally operates at a temperature of 200–300°C. Other parts of the tube also may reach high temperatures, particularly the cathode insulator and the cathode/heater surfaces. All hot surfaces remain hot for an extended time after the tube is shut off. To prevent serious burns, take care to avoid bodily contact with these surfaces during operation and for a reasonable cool-down period afterward.

7.4.5 Polychlorinated Biphenyls

PCBs belong to a family of organic compounds known as *chlorinated hydrocarbons*. Virtually all PCBs in existence today have been synthetically manufactured. PCBs have a heavy oil-like consistency, high boiling point, a high degree of chemical stability, low flammability, and low electrical conductivity. These characteristics resulted in the widespread use of PCBs in high-voltage capacitors and transformers. Commercial products containing PCBs were widely distributed between 1957 and 1977 under several trade names including:

- Aroclor
- Pyroclor
- Sanotherm
- Pyranol
- Askarel

Askarel is also a generic name used for nonflammable dielectric fluids containing PCBs. Table 7.5.5 lists some common trade names used for Askarel. These trade names typically will be listed on the nameplate of a PCB transformer or capacitor.

TABLE 7.5.5 Commonly Used Names for PCB Insulating Material

PCB Trade Names					
Apirolio	Abestol	Askarel[a]	Aroclor B	Chlorexto	Chlophen
Chlorinol	Clorphon	Diaclor	DK	Dykanol	EEC-18
Elemex	Eucarel	Fenclor	Hyvol	Inclor	Inerteen
Kanechlor	No-Flamol	Phenodlor	Pydraul	Pyralene	Pyranol
Pyroclor	Sal-T-Kuhl	Santothern FR	Santovac	Solvol	Thermin

[a]*Generic* name used for nonflammable dielectric fluids containing PCBs.

PCBs are harmful because once they are released into the environment, they tend not to break apart into other substances. Instead, PCBs persist, taking several decades to slowly decompose. By remaining in the environment, they can be taken up and stored in the fatty tissues of all organisms, from which they are slowly released into the bloodstream. Therefore, because of the storage in fat, the concentration of PCBs in body tissues can increase with time, even though PCB exposure levels may be quite low. This process is called *bioaccumulation*. Furthermore, as PCBs accumulate in the tissues of simple organisms, and as they are consumed by progressively higher organisms, the concentration increases. This process is called *biomagnification*. These two factors are especially significant because PCBs are harmful even at low levels. Specifically, PCBs have been shown to cause chronic (long-term) toxic effects in some species of animals and aquatic life. Well-documented tests on laboratory animals show that various levels of PCBs can cause reproductive effects, gastric disorders, skin lesions, and cancerous tumors.

PCBs may enter the body through the lungs, the gastrointestinal tract, and the skin. After absorption, PCBs are circulated in the blood throughout the body and stored in fatty tissues and a variety of organs, including the liver, kidneys, lungs, adrenal glands, brain, heart, and skin.

The health risk from PCBs lies not only in the PCB itself, but also in the chemicals that develop when PCBs are heated. Laboratory studies have confirmed that PCB by-products, including *polychlorinated dibenzofurans* (PCDFs) and *polychlorinated dibenzo-p-dioxins* (PCDDs), are formed when PCBs or *chlorobenzenes* are heated to temperatures ranging from approximately 900 to 1300°F. Unfortunately, these products are more toxic than PCBs themselves.

7.4.5.1 Governmental Action. The U.S. Congress took action to control PCBs in October 1975 by passing the Toxic Substances Control Act (TSCA). A section of this law specifically directed the EPA to regulate PCBs. Three years later the Environmental Protection Agency (EPA) issued regulations to implement the congressional ban on the manufacture, processing, distribution, and disposal of PCBs. Since that time, several revisions and updates have been issued by the EPA. One of these revisions, issued in 1982, specifically addressed the type of equipment used in industrial plants and transmitting stations. Failure to properly follow the rules regarding the use and disposal of PCBs has resulted in high fines and even jail sentences.

Although PCBs are no longer being produced for electrical products in the United States, there may be PCB transformers or small PCB capacitors still in use or in storage. The threat of widespread contamination from PCB fire-related incidents is one reason behind the EPA's efforts to reduce the number of PCB products in the environment. The users of high-power equipment are affected by the regulations primarily because of the widespread use of PCB transformers and capacitors. These components usually are located in older (pre-1979) systems, so this is the first place to look for them. However, some facilities also maintain their own primary power transformers. Table 7.5.6 lists the primary classifications of PCB devices.

TABLE 7.5.6 Definition of PCB Terms as Identified by the EPA

Term	Definition	Examples
PCB	Any chemical substance that is limited to the biphenyl molecule that has been chlorinated to varying degrees, or any combination of substances that contain such substances.	PCB dielectric fluids, PCB heat-transfer fluids, PCB hydraulic fluids, 2,2¢,4-trichlorobiphenyl
PCB article	Any manufactured article, other than a PCB container, that contains PCBs and whose surface has been in direct contact with PCBs.	Capacitors, transformers, electric motors, pumps, pipes
PCB container	A device used to contain PCBs or PCB articles, and whose surface has been in direct contact with PCBs.	Packages, cans, bottles, bags, barrels, drums, tanks
PCB article container	A device used to contain PCB articles or equipment, and whose surface has not been in direct contact with PCBs.	Packages, cans, bottles, bags, barrels, drums, tanks
PCB equipment	Any manufactured item, other than a PCB container or PCB article container, that contains a PCB article or other PCB equipment.	Microwave systems, fluorescent light ballasts, electronic equipment
PCB item	Any PCB article, PCB article container, PCB container, or PCB equipment that deliberately or unintentionally contains, or has as a part of it, any PCBs.	
PCB transformer	Any transformer that contains PCBs in concentrations of 500 ppm or greater.	
PCB contaminated	Any electric equipment that contains more than 50, but less than 500 ppm of PCBs. (Oil-filled electric equipment other than circuit breakers, reclosers, and cable whose PCB concentration is unknown must be assumed to be PCB-contaminated electric equipment.)	Transformers, capacitors, contaminated circuit breakers, reclosers, voltage regulators, switches, cable, electromagnets

7.4.5.2 PCB Components. The two most common PCB components are transformers and capacitors. A PCB transformer is one containing at least 500 ppm (parts per million) PCBs in the dielectric fluid. An Askarel transformer generally has 600,000 ppm or more. A PCB transformer may be converted to a *PCB-contaminated device* (50–500 ppm) or a *non-PCB device* (less than 50 ppm) by having it drained, refilled, and tested. The testing must not take place until the transformer has been in service for a minimum of 90 days. Note that this is *not* something a maintenance technician can do. It is the exclusive domain of specialized remanufacturing companies.

PCB transformers must be inspected quarterly for leaks. Failed PCB transformers cannot be repaired; they must be properly disposed of.

As of October 1, 1990, the use of PCB transformers (500 ppm or greater) was prohibited in or near commercial buildings when the secondary voltages are 480 V ac or higher.

The EPA regulations also require that the operator notify others of the possible dangers. All PCB transformers (including PCB transformers in storage for reuse) must be registered with the local fire department. The following information must be supplied:

- The location of the PCB transformer(s).
- Address(es) of the building(s) and, for outdoor PCB transformers, the location.
- Principal constituent of the dielectric fluid in the transformer(s).
- Name and telephone number of the contact person in the event of a fire involving the equipment.

Any PCB transformers used in a commercial building must be registered with the building owner. All building owners within 30 meters of such PCB transformers also must be notified. In the event of a fire-related incident involving the release of PCBs, the Coast Guard National Spill Response Center (800-424-8802) must be notified immediately. Appropriate measures also must be taken to contain and control any possible PCB release into water.

Capacitors are divided into two size classes, *large* and *small*. A PCB small capacitor contains less than 1.36 kg (3 lbs) of dielectric fluid. A capacitor having less than 100 inch^3 also is considered to contain less than 3 lb of dielectric fluid. A PCB large capacitor has a volume of more than 200 inch^3 and is considered to contain more than 3 lb of dielectric fluid. Any capacitor having a volume between 100 and 200 inch^3 is considered to contain 3 lb of dielectric, provided the total weight is less than 9 lb. A PCB *large high-voltage capacitor* contains 3 lb or more of dielectric fluid and operates at voltages of 2 kV or greater. A *large low-voltage capacitor* also contains 3 lb or more of dielectric fluid but operates below 2 kV.

Disposal of PCB capacitors must be handled by an EPA-approved PCB disposal facility.

7.4.5.3 PCB Liability Management. Properly managing the PCB risk is not particularly difficult; the keys are understanding the regulations and following them carefully. Any program should include the following steps:

- Locate and identify all PCB devices. Check all stored or spare devices.
- Properly label PCB transformers and capacitors according to EPA requirements.
- Perform the required inspections and maintain an accurate log of PCB items, their location, inspection results, and actions taken.
- Arrange for necessary disposal through a company licensed to handle PCBs. If there are any doubts about the company's license, contact the EPA.
- Report the location of all PCB transformers to the local fire department and to the owners of any nearby buildings.

The importance of following the EPA regulations cannot be overstated. For additional information see the EPA web site: http://www.epa.gov/epawaste/hazard/tsd/pcbs/index.htm.

REFERENCES

[1] National Electrical Code, NFPA #70, National Fire Protection Association, Quincy, MA, http://www.nfpa.org/.

[2] Osepchuk, J. M., and R. C. Peterson, "Safety and Environmental Issues," in *RF and Microwave Applications and Systems*, M. Golio (ed.), CRC Press, Boca Raton, FL, 2008.

BIBLIOGRAPHY

"Current Intelligence Bulletin #45," National Institute for Occupational Safety and Health, Division of Standards Development and Technology Transfer, February 24, 1986.

Code of Federal Regulations, 40, Part 761.

"Electrical Standards Reference Manual," U.S. Department of Labor, Washington, DC.

Hammett, W. F., "Meeting IEEE C95.1-1991 Requirements," *NAB 1993 Broadcast Engineering Conference Proceedings*, National Association of Broadcasters, Washington, DC, pp. 471–476, April 1993.

Hammar, W., *Occupational Safety Management and Engineering*, Prentice Hall, New York, NY.

Markley, D., "Complying with RF Emission Standards," *Broadcast Engineering*, Intertec Publishing, Overland Park, KS, May 1986.

"Occupational Injuries and Illnesses in the United States by Industry," OSHA Bulletin 2278, U.S. Department of Labor, Washington, DC, 1985.

OSHA, "Electrical Hazard Fact Sheets," U.S. Department of Labor, Washington, DC, January 1987.

OSHA, "Handbook for Small Business," U.S. Department of Labor, Washington, DC.

Pfrimmer, J., "Identifying and Managing PCBs in Broadcast Facilities," *1987 NAB Engineering Conference Proceedings*, National Association of Broadcasters, Washington, DC, 1987.

"Safety Precautions," Publication no. 3386A, Varian Associates, Palo Alto, CA, March 1985.

Smith Jr., M. K., "RF Radiation Compliance," *Proceedings of the Broadcast Engineering Conference*, Society of Broadcast Engineers, Indianapolis, IN, 1989.

"Toxics Information Series," Office of Toxic Substances, July 1983.

Whitaker, J. C., *AC Power Systems*, 2nd ed., CRC Press, Boca Raton, FL, 1998.

Whitaker, J. C., G. DeSantis, and C. Paulson, *Interconnecting Electronic Systems*, CRC Press, Boca Raton, FL, 1993.

Whitaker, J. C., *Maintaining Electronic Systems*, CRC Press, Boca Raton, FL, 1991.

Whitaker, J. C., *Power Vacuum Tubes Handbook*, 3rd ed., CRC Press, Boca Raton, FL, 2012.

Whitaker, J. C., *Radio Frequency Transmission Systems: Design and Operation*, McGraw-Hill, New York, NY, 1990.

CHAPTER 7.6
ABOUT THE SOCIETY OF BROADCAST ENGINEERS

Ralph Hogan, Wayne M. Pecena, and John L. Poray
SBE, Indianapolis, IN[1]

7.1 INTRODUCTION

The Society of Broadcast Engineers (SBE) is the only organization devoted to the advancement of all levels and types of broadcast engineering. With more than 5500 members and 114 local chapters, the SBE provides a forum for the exchange of ideas and the sharing of information to help members keep pace with a rapidly changing industry and technology (Fig. 7.6.1). The SBE amplifies the voices of broadcast engineers by validating skills with professional certification, by offering educational opportunities to maintain and expand those skills and by speaking out on technical regulatory issues that affect how our members work.

Our aim: to keep members at the top of their field, enhancing their value to their employers, or if self-employed, preparing them to meet the changing needs of their clients.

7.1.1 Who Are Members of the SBE?

The SBE, a nonprofit professional organization formed in 1964, is committed to serving broadcast engineers no matter where in the field you work. From the studio operator to the maintenance engineer; from the chief engineer to the vice president of engineering, SBE members come from commercial and noncommercial radio and television stations and cable facilities, and from all market sizes. Many members engage the industry on their own as consultants and contractors. Field and sales engineers and engineers from recording studios, schools, production houses, CCTV, corporate audio-visual departments, and other facilities are also members of the SBE.

7.1.2 SBE National Headquarters

The SBE is a professional organization governed by an elected board of directors and professionally managed from its headquarters in Indianapolis, Indiana, USA. For more information, visit the SBE website, www.sbe.org or contact the SBE National Headquarters at 317-846-9000. Office hours are Monday–Friday, 8:30 a.m. to 4:30 p.m., USA Eastern Time.

---

[1]Ralph Hogan, CPBE, DRB, CBNE SBE National Certification Committee Chairman; Wayne Pecena, CPBE, 8-VSB, AMD, DRB, CBNE SBE National Education Committee Chairman; John L. Poray, CAE SBE Executive Director

FIGURE 7.6.1   Logo of the SBE.

7.2 THE SBE CERTIFICATION PROGRAM

In 1975, the Society of Broadcast Engineers established a Certification Program to recognize and raise the professional status of broadcast engineers by providing standards of professional competence (Fig. 7.6.2). It has evolved to become recognized in the industry as the primary method of verifying the attainment of educational standards. It is a service that contributes to the advancement of broadcast engineering for the general benefit of the entire broadcast industry.

FIGURE 7.6.2   SBE certification logo.

With ever-changing technology, the SBE-certified engineer helps the broadcast industry keep pace, helping to deliver to the public entertainment and news utilizing the most up-to-date methods. The program's objectives include:

- To raise the status of broadcast engineers by providing standards of professional competence in the practice of broadcast engineering and related technologies.
- To recognize those individuals who, by fulfilling the requirements of knowledge, experience, responsibility, and conduct, meet those standards of professional competence.
- To encourage broadcast engineers to continue their professional development.

To be eligible for SBE certification, you must have a strong interest in the design, operation, maintenance, or administration of the day-to-day operations associated with a broadcast station or related facility. You must also meet the specific eligibility requirements of the desired certification level. SBE certification is not a license; but rather an achievement that recognizes professional competence by your peers in a professional, independent organization. SBE certification is valid for a period of 5 years. Certification is for individuals only, and may not be used to imply that an organization or firm is certified.

7.2.1 Certification Exams

SBE certification examinations are administered primarily through local SBE chapters under the supervision of proctors approved by the National SBE Certification Committee. Where no chapters exist, the SBE national office will arrange for a suitable exam location and proctor.

7.2.2 SBE Certification and SBE Membership

While membership in the SBE is not a requirement of the SBE Certification Program, because SBE members share in the overhead cost of all SBE activities, nonmember certification fees are higher. Nonmembers who certify at a technologist or engineering level may elect to apply for SBE membership at the time you apply for certification. Of course, we encourage you to become a member of the SBE so you may take advantage of the full range of programs and services available to members.

From the certified operator to the Certified Professional Broadcast Engineer, SBE has a certification for every broadcast engineer and technician. To learn more about SBE certification, visit the "Certification Levels" section of the SBE website at www.sbe.org. There, you may read or download a printable overview of SBE certification, as well as SBE membership.

7.3 SBE EDUCATION PROGRAMS

The Society of Broadcast Engineers provides relevant, affordable educational opportunities to its members and others, using various delivery methods. The Society's programs are designed to help broadcast engineers keep pace with the ever-changing demands of the job and the broadcast industry. They can also help you prepare for SBE certification and qualify for SBE recertification.

Table 7.6.1 lists the major SBE educational opportunities and programs.

TABLE 7.6 SBE Educational Programs

SBE Webinars. The SBE regularly presents original, live programs via the Internet on a wide array of technical, safety, and regulatory broadcast topics, presented by subject matter experts. Each webinar is archived for later viewing, creating a large library of topics from which to view, available 24/7/365.

SBE University. The SBE University is a series of in-depth courses available at the SBE website. Course selections reflect a wide scope of topics, written by experts. They are an affordable way to learn more about broadcast technology from the comfort of your own home or office. Like the webinars, they are available 24/7/365 and you can work through them at your own rate of speed.

SBE Leadership Development Course. A three-day, in-person course that challenges you to refine your leadership skills as you better understand and improve how you interact with others. Taught by a university-level professional instructor, this course helps prepare the individual for engineering management or assist those already holding those responsibilities.

Ennes Workshops. Since 1991, these one-day, in-person seminars have provided thousands of broadcast engineers with valuable information and knowledge to help them carry out their day-to-day engineering responsibilities. They are typically hosted by local SBE chapters. The SBE works in conjunction with the Ennes Educational Foundation Trust, the educational, charitable arm of the SBE, to present these programs.

7.4 SBE'S HISTORY IS PEOPLE

The Society of Broadcast Engineers has had great growth since the organization's founding more than 50 years ago. This growth is largely due to the many members who have contributed greatly to the birth, development, and subsequent success of the organization. SBE's Charter Members were those who were there at the beginning, with a positive attitude and a desire to build. Its past presidents each contributed to the SBE's growth and kept the society on an upward path with their leadership. Those who have held membership over many years and into retirement became Life Members and contribute and provide support in so many ways. A handful of honorary members were recognized for their support of the society or their contributions to the field of broadcasting engineering.

The society's roots can be traced back to 1961 when the Institute of Radio Engineers (IRE) and the American Institute of Electrical Engineers (AIEE) passed a joint resolution calling for a "merger or consolidation… into one organization." The two groups did finally join forces in 1963, forming what is known today as the Institute of Electrical and Electronic Engineers (IEEE). However, the new organization was perceived by some as not addressing the needs of broadcast engineers.

John H. Battison, P.E., CPBE, editor of *Broadcast Engineering* magazine at the time, was one of those people. He wrote an editorial that appeared in his magazine in 1961 that suggested the idea of a new organization just for broadcast engineers. Battison's continued efforts sparked interest across the country. He published a membership application in the April 1963 issue of the magazine and the response was encouraging. With help from family members, Battison mailed membership invitations to almost 5000 radio and TV engineers in the United States and Canada.

Finally, on April 5, 1964, an organizational meeting was held during the National Association of Broadcasters' convention at the Conrad Hilton Hotel in Chicago. Approximately 100 interested broadcast engineers attended. The group formed an organization devoted solely to the needs and interests of broadcast engineers, with Battison as its first president. They at first settled on the name, "Institute of Broadcast Engineers (IBE)." However, because some members feared there might be confusion in the similarity between the names of the IBE and the IBEW (International Brotherhood of Electrical Workers), the name was changed to Society of Broadcast Engineers (SBE) at that very first meeting.

The SBE's first chapter was organized in Binghamton, N.Y. and it is still active today. Its newest chapter is the, Ft. Bragg, NC chapter, having been organized by local broadcast engineers in 2015. Today, there are SBE chapters in every major American city and dozens of others as well.

The SBE Program of Certification was adopted in 1975 and launched in 1977, helping individuals improve their skills and gain recognition for their knowledge and experience through a national testing program. As the industry's preeminent evaluation service for broadcast engineers, SBE certification provides important benefits to both members and employers. Those who have earned SBE certification are proud of their accomplishment, and rightly so.

In recent years, the SBE on a national level has greatly increased its focus on education as keeping up with new technology has become even more important to our members. The society offers several dozen education courses on various topics in traditional in-person settings and via the Internet each year.

The SBE is also known for its involvement in developing in the mid-1970s a nationwide network of more than 200 volunteer frequency coordinators. Through these volunteers, the SBE provides a vital service to stations, networks, and other users of broadcast spectrum who need information about auxiliary broadcast frequencies. Through the efforts of society members, this important spectrum is used efficiently, and interference is minimized.

The society continually monitors regulatory issues and frequently files comments with the FCC and other federal and state agencies. In many cases, the SBE is the only organization representing the technical interests of broadcast engineers on regulatory issues.

Throughout its history, the SBE has strived to advance the field of broadcast engineering and the professional development of the broadcast engineer by being in the forefront as changes in technology have taken place. The society will continue to help its members stay on the leading edge of the broadcast industry by providing leadership, information, certification, education, and opportunities for recognition.

With 114 chapters located throughout the United States and Hong Kong, SBE commands a leadership position in the broadcast industry. Local chapters of the SBE provide services and benefits to members through meetings, technical seminars, certification examinations, leadership opportunities, recognition, and fellowship.

The SBE has entered its next 50 years with an eye toward the future; expanding its reach beyond the traditional station engineer to others with technical roles at broadcast and related facilities. With a strong membership that represents the breadth and scope of the broadcast engineering industry, the SBE—*The Association for Broadcast and Multimedia Technology Professionals*—will surely continue its leadership role in the years to come.

CHAPTER 7.7
LOOKING TOWARD THE FUTURE—ATSC 3.0

Jerry C. Whitaker
ATSC, Washington, D.C.

Dr. Richard Chernock
Triveni Digital, Princeton Junction, NJ

7.1 INTRODUCTION

At this writing, work was underway within the Advanced Television Systems Committee (ATSC) to develop the next-generation digital terrestrial television system, known as ATSC 3.0. The goals of the new system are to improve the television viewing experience with higher audio and video quality, reception on both fixed and mobile devices, and more accessibility, personalization, and interactivity. ATSC is also addressing changing consumer behavior and preferences, providing TV content on a wide variety of devices. Furthermore, ATSC is working to add value to the broadcasting service platform, extending its reach and adding new business models—all without the restriction of backward compatibility with the legacy system.

It is reasonable to ask why this is worth doing. The obvious answer is that technology marches on. The DTV Standard now known as ATSC 1.0 (document A/53 [1]) is 20 years old, and audience expectations are changing rapidly. There are many new competitors. Disruptive forces exist. Wise use of the spectrum is essential. Business opportunities exist for leveraging the power of over-the-air broadcasting and broadband online over-the-top services. So, what is in it for broadcasters?

- Maintaining and building their audience
- Putting content where the viewers are
- Benefiting from new technologies
- Quantitative and qualitative growth
- Developing new revenue streams

During planning for ATSC 3.0, members spent a considerable amount of time identifying "Usage Scenarios," which include the following:

- Flexible use of the spectrum
- Robustness
- Mobile services

7.66 BROADCAST MANAGEMENT

- Ultra HD capabilities
- Hybrid broadcast/broadband services
- Multiview/multiscreen
- 3D video content
- Enhanced and immersive audio
- Advanced accessibility
- Advanced emergency alerting
- Personalization and interactivity
- Advanced advertising and monetization
- Common world standard

Serious planning for ATSC 3.0 began back in 2010, in a Planning Team tasked with looking at what a next-generation system might look like. Requirements were subsequently developed based on the Usage Scenarios. Formal standards development work began in 2012, with the first specification elevated to Candidate Standard status in April 2015.

ATSC 3.0 is being built to last. Technology advances rapidly and so methods to gracefully evolve must be built into the core of the system. An essential capability is to signal when a layer or components of a layer evolve—including the ability to signal minor version changes and updates, and major version changes and updates. The goal of this flexibility is to avoid disruptive technology transitions, and to enable graceful transitions from one technology to another.

7.2 ABOUT THE ATSC 3.0 PROCESS

The overall structure at the Technology Group on ATSC 3.0 (TG3) is illustrated in Fig. 7.7.1. Specialist Groups and ad-hoc groups have made preliminary decisions to select technologies for incorporation into ATSC 3.0. It is important to remember that selections of all technologies are subject to approval

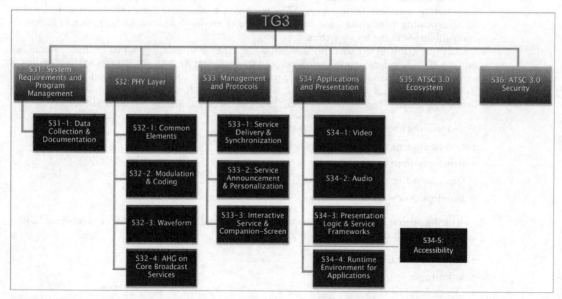

FIGURE 7.7.1 Overall TG3 structure.

of the Technology Group, and ultimately the voting membership in accordance with ATSC due process.

The ATSC 3.0 standard enables system options for the broadcast industry (not just system parameter options) that include high robustness, spectrally efficient operating points, flexible configuration of operating modes, a robust bootstrap process that includes signaling basic system parameters, channel bonding options for spectrum sharing, and many other capabilities.

Work on ATSC 3.0 has been divided into functional layers, as illustrated in Fig. 7.7.2 and described in the following sections.

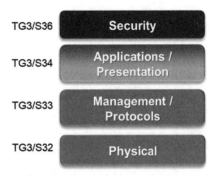

FIGURE 7.7.2 Layers of the ATSC 3.0 system and the specialist group designations.

7.2.1 Physical Layer

The physical layer for ATSC 3.0 embodies the theme of flexibility and future evolution by starting every frame with a bootstrap signal. This bootstrap is extremely robust, able to be received in very challenging RF conditions. Besides providing synchronization, the bootstrap also signals basic information about the technology used in the physical layer itself (major and minor version, which enables graceful evolution of the physical layer in the future) as well as Emergency Alert Service wake up bits, system bandwidth, time to the next frame of a similar service, and the sampling rate of the current frame. The bootstrap is followed by a preamble, which carries the information needed to define the payload framing, including the information needed for the receiver to acquire the data frame. The remainder of the physical layer structure is the data payload itself. This structure is illustrated in Fig. 7.7.3.

By taking advantage of recent advances in modulation, coding, error correction, constellations, and multiplexing, the ATSC 3.0 physical layer offers performance in the BICM chain (bit interleaver, coding, and modulation) that is very close to the Shannon Limit (the theoretical limit for the amount of information that can be carried in a noisy channel). See Fig. 7.7.4. Of particular note is the wide range of operating points available in the new system, especially considering that the current ATSC DTV system (described in document A/53) offers a single operating point—19.39 Mb/s, 15 db C/N.

The ATSC 3.0 physical layer offers broadcasters the capability to operate in a robust/lower bitrate fashion for mobile services (lower left portion of the curve in Fig. 7.7.4) and/or a less robust/higher bitrate fashion for services to large screens in the home (upper right portion of the curve). If desired,

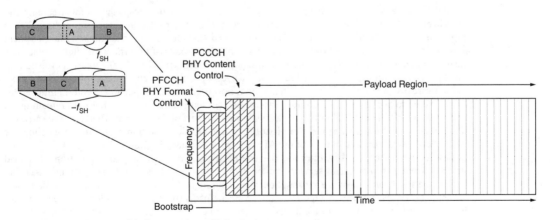

FIGURE 7.7.3 Structure of the bootstrap/preamble/payload.

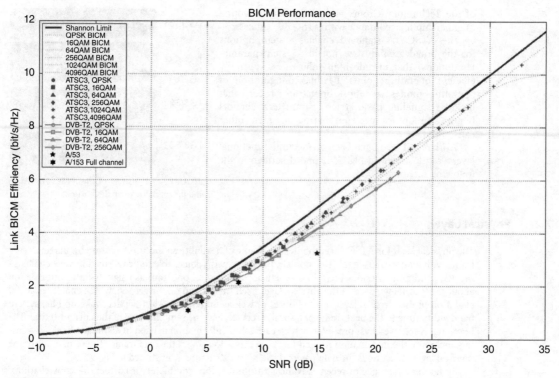

FIGURE 7.7.4 ATSC 3.0 BICM performance.

the broadcaster can also operate with a simultaneous mixture of types of services using either Time Division Multiplexing or Layer Division Multiplexing, or both.

7.2.2 Transport Layer

The ATSC 3.0 transport layer is based on using IP (Internet Protocol) encapsulation for streaming and file delivery, rather than the MPEG-2 Transport Stream (TS) encapsulation as is currently used. When the original DTV system (ATSC 1.0, A/53) was developed, the Internet was in its infancy. Entertainment content was not (yet) delivered over the Internet to homes. Broadcast television and digital cable were the primary entertainment delivery mechanisms, both based on MPEG-2 TS. The Internet evolved very quickly (due to numerous factors, including the lifecycle of computers and the ease of updating codecs and software over a two-way delivery system) and is now used for a large portion of entertainment content delivered to homes (e.g., streaming services such as Netflix and Hulu). Rather than remaining an independent silo, ATSC 3.0 allows broadcasting to become part of the Internet. This allows the creation of new services and business models for broadcasters and enables evolution nearer the pace of how the Internet evolves. A conceptual view of the ATSC 3.0 protocol stack is given in Fig. 7.7.5.

Besides similarities to the Internet, the use of IP transport enables incorporation of hybrid services, where components of services can be delivered by broadcast and broadband in a way that they can be synchronized. Woven throughout the structure of ATSC 3.0 is the basic notion that components of services can be delivered by different mechanisms (broadcast, broadband, or even predelivered by *push* in advance) and combined as needed to create services. This gives the broadcaster a large degree of flexibility and control over the services they offer.

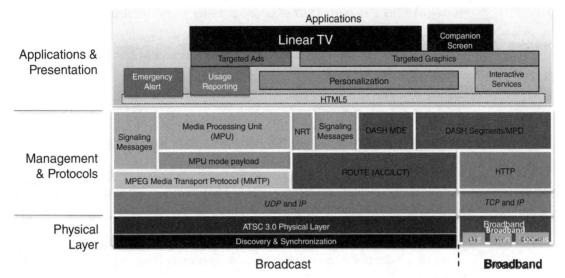

FIGURE 7.7.5 Conceptual protocol stack for ATSC 3.0.

An example use of this hybrid delivery might be a linear television service with multiple audio languages. Imagine a situation where the broadcaster's audience is composed of diverse ethnicities. There might be two languages in widespread use, but enough people speaking other languages to want to include them as alternate audio tracks. With hybrid delivery, it will be possible to include the two more common language audio tracks in the broadcast delivery and make the others available over broadband. The viewer would have the choice of which language to listen to, and it would be synchronized with the video, no matter what the delivery path might be.

The use of IP transport, coupled with another decision—to deliver "streaming" content as chunks of ISOBMFF[1] files (similar to streaming delivery over the Internet) rather than streams of bits—enables another business model that has been difficult to support in the past: the ability to do advertising (or content) insertion at the receiver in a personalized fashion. With MPEG-2 TS, this was a very difficult task, requiring sophisticated buffer manipulation that was beyond the capabilities of an inexpensive receiver. By delivering file segments (along with an associated playlist), insertion in the receiver becomes relatively simple—essentially boiling down to changing the pointer to the next segment to be played.

7.2.3 Applications Layer

The applications layer encompasses video and audio coding, interactive features, accessibility, and other services.

7.2.3.1 Video. The ATSC 1.0 system was a breakthrough when it emerged, offering HDTV video (along with the option of multiple-SD video simulcast). ATSC 3.0 offers the capability to move to Ultra-High Definition Video. With current advances in video coding technology (HEVC), 4K video can fit within the available broadcast bandwidth. Similarly, multiple HD videos can also fit. While not currently in the design, the move to 8K video may be supported in the future (should it be desired).

---

[1] ISO base media file format.

However, besides more pixels, UHD also supports "better" pixels, specifically:

- High Dynamic Range (HRD)—1000-nit color grading rather than 100-nit
- Wider Color Gamut (WCG)—approaching Rec. 2020 rather than Rec. 709
- 10 bits per pixel (rather than 8)
- Higher Frame Rate (HFR)—120 Hz

All of these features provide significant improvements in picture quality, even at HD resolution.

Given the current screen sizes used in the home, as well as the typical viewing distance, many suggest that the other improvements listed above might offer a larger impact to the viewer than simply increasing the pixel count.

7.2.3.2 Audio. While the selection of audio technology for ATSC 3.0 is still under discussion at this writing, the new audio system will provide a number of new capabilities. Interactivity will be an important feature, providing the capability to personalize audio rendering. A broadcaster can choose to give the viewer control over a number of aspects of the audio, including the ability to adjust the dialog level (which can help with certain hearing impairments) or the ability to select which tracks of audio are heard (for example, being able to choose which announcer to listen to during a sporting event—or even to turn the announcer off altogether and simply listen to the crowd).

Television audio has evolved from the original monaural service to stereo (in the analog days), to 5.1 surround (for ATSC 1.0), and now to full immersion in ATSC 3.0, which includes height information (typically 7.1.4 channels). The additional level of immersion adds a significant degree of immediacy to the experience.

A significant departure from the past is that in the ATSC 3.0 system a common audio stream is broadcast and rendered at the receiver appropriately. Besides decreasing the required bitrate, this local rendering allows dynamic range control (DRC) to be applied accurately for the type of receiver (e.g., a home theater as opposed to a portable device with earbuds). Additionally, the receiving device can render correctly for the speaker configuration, including incorrectly placed speakers.

7.2.3.3 Interactivity. Work is underway to develop a robust application runtime environment for ATSC 3.0 that is based on HTML5. Interactivity capabilities are expected to include:

- Targeted ad insertion
- On-demand content launcher
- Flexible use of display real estate (e.g., "screen mash-up")
- Picture-in-Picture, tickers, graphics, etc.
- T-commerce
- Voting and polling
- Games
- News and sports feeds
- Notifications and reminders

ATSC is working with the Consumer Technology Association to bundle features into "profiles." One or more "Universal Profile(s)" are desired. The goal is to promote adoption and use by application developers, and to simplify receiver implementations.

7.2.3.4 Accessibility. Accessibility functions are obviously a key element of any new broadcast system. Work is underway to define the requirements of the new system, and the general architecture. One point of discussion is whether closed captions can be its own essence, rather than always inserted in the video stream. This approach provides for increased flexibility in carriage and presentation.

7.2.4 Security

When ATSC 1.0 was developed, the notion of security for a broadcast television signal was completely off the radar. Since that time, however, the need for security measures has become quite clear. At this writing, the work is in an early stage.

Expected conditional access (CA) functions include the following:

- Subscriptions, Pay-Per-View (PPV), and Video-on-Demand (VOD)
- Content delivered via ISOBMFF, Common Encryption ("CENC"), ISO/IEC 23001-7
- Conditional access support functions, including key security and distribution, entitlement, billing, and customer relationships

Digital rights management (DRM) limits use of content post-reception. It secures the content in the home network, including the primary receiver and second screen communications. It is likely unspecified by the ATSC standard, but likely limited by a CA system.

CA and DRM systems are illustrated in Fig. 7.7.6.

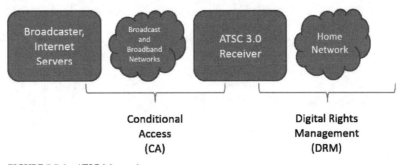

FIGURE 7.7.6 ATSC 3.0 security.

7.3 MOVING FORWARD

The transition to a new technology can be difficult for any industry. For ATSC 3.0, the transition challenges can be broadly divided into two groups:

- The spectrum challenge – In most countries, additional spectrum will not be available for a second service. Stations within a DMA could channel-share to create a dual service during transition.
- The receiver challenge – New devices are needed for consumers. Options include a "stick tuner" (for example, connected via HDMI) for current flat panel TVs, a tuner gateway connected to Wi-Fi, and/or set-top converters. An industry-driven campaign to include tuners in portable devices (tablets, lap tops, smart phones, etc.) will help move the transition forward. In addition, adoption by other countries will foster product development.

The next critical steps for ATSC 3.0 include the following:

(1) Complete the standard.
(2) Develop practical transition scenarios.
(3) Update current business plans and develop new business strategies.
(4) Gain regulatory approval for use of ATSC 3.0.

(5) Commercial launch, which includes the broadcast hardware build-out and cooperative new product development with the CE industry.
(6) Sell the product to consumers. Industry collaboration is essential.

In summary, ATSC 3.0 represents a significant step forward in capabilities for a broadcast television system. It provides a set of flexible capabilities for broadcasters that enable new services and new business cases. The concepts of flexibility, extensibility, and scalability are in the core of the system and will allow graceful evolution over a long period of time.

REFERENCES

[1] ATSC, "ATSC Digital Television System," doc. A/53 Parts 1–6, Advanced Television Systems Committee, Washington, DC, various dates.

[2] ATSC, "Interactive Services Standard," doc. A/105, Advanced Television Systems Committee, Washington, DC, 2015.

ANNEX

Section leader—Jerry C. Whitaker, Editor-in-Chief

This annex is provided as a supplement to the tutorial information contained in the rest of the *SBE Broadcast Engineering Handbook*. The material covers a wide range of subjects, focusing on important, but seldom used or hard-to-find or information of interest to the broadcast engineer.

Engineering is all about data, units, and measurements. To this end, an appendix is included with an extensive compilation of reference data and tables. Of particular note is the units conversion table. While converting from one unit to another may not be required in day-to-day work, a conversion table can be very helpful to have on hand when the need presents itself.

The electromagnetic spectrum is—of course—fundamental to broadcasting. With this in mind, an appendix is provided that puts the subject into perspective, with tables of common frequency band designations. Pointers to detailed information on frequency allocations are also provided.

Because all radio and television broadcast signals are based on specifications cooperatively agreed to by broadcasters, receiver manufacturers, and regulators, an appendix is included that provides an overview of standards organizations and their role in broadcast regulation. Contact information for the standards development organizations (SDOs) working in the broadcast and related fields is included for those who need detailed information on existing and/or emerging standards. One of the challenges in keeping up with the myriad of standards documents applicable to professional audio/video engineering is determining how the various documents relate to the work at hand. It is the goal of this annex to make that task a little easier by providing a one-stop listing of SDOs.

In short, this annex rounds-out the *SBE Broadcast Engineering Handbook* with valuable reference data in an easy-to-use format.

Appendix A: Reference Data and Tables
Appendix B: The Electromagnetic Spectrum
Appendix C: Standards Organizations and SI Units

APPENDIX A
REFERENCE DATA AND TABLES

Compiled by Jerry C. Whitaker, Editor-in-Chief

A.1 INTRODUCTION

The following tables describe standard units, conversion ratios, and other data for engineering work. Data in this table is adapted from [1].

TABLE A.1 Common Standard Units

Name	Symbol	Quantity
ampere	A	electric current
ampere per meter	A/m	magnetic field strength
ampere per square meter	A/m²	current density
becquerel	Bg	activity (of a radionuclide)
candela	cd	luminous intensity
coulomb	C	electric charge
coulomb per kilogram	C/kg	exposure (x and gamma rays)
coulomb per sq. meter	C/m²	electric flux density
cubic meter	m³	volume
cubic meter per kilogram	m³/kg	specific volume
degree Celsius	°C	Celsius temperature
farad	F	capacitance
farad per meter	F/m	permittivity
henry	H	inductance
henry per meter	H/m	permeability
hertz	Hz	frequency
joule	J	energy, work, quantity of heat
joule per cubic meter	J/m³	energy density
joule per kelvin	J/K	heat capacity
joule per kilogram K	J/(kg·K)	specific heat capacity
joule per mole	J/mol	molar energy
kelvin	K	thermodynamic temperature

TABLE A.1 Common Standard Units (*Continued*)

Name	Symbol	Quantity
kilogram	kg	mass
kilogram per cubic meter	kg/m^3	density, mass density
lumen	lm	luminous flux
lux	lx	luminance
meter	m	length
meter per second	m/s	speed, velocity
meter per second sq.	m/s^2	acceleration
mole	mol	amount of substance
newton	N	force
newton per meter	N/m	surface tension
ohm	W	electrical resistance
pascal	Pa	pressure, stress
pascal second	Pa•s	dynamic viscosity
radian	rad	plane angle
radian per second	rad/s	angular velocity
radian per second squared	rad/s^2	angular acceleration
second	s	time
siemens	S	electrical conductance
square meter	m^2	area
steradian	sr	solid angle
tesla	T	magnetic flux density
volt	V	electrical potential
volt per meter	V/m	electric field strength
watt	W	power, radiant flux
watt per meter kelvin	W/(m • K)	thermal conductivity
watt per square meter	W/m^2	heat (power) flux density
weber	Wb	magnetic flux

TABLE A.2 Standard Prefixes

Multiple	Prefix	Symbol
10^{18}	exa	E
10^{15}	peta	P
10^{12}	tera	T
10^{9}	giga	G
10^{6}	mega	M
10^{3}	kilo	k
10^{2}	hecto	h
10	deka	da
10^{-1}	deci	d
10^{-2}	centi	c
10^{-3}	milli	m
10^{-6}	micro	μ
10^{-9}	nano	n
10^{-12}	pico	p
10^{-15}	femto	f
10^{-18}	atto	A

REFERENCE DATA AND TABLES   A.5

TABLE A.3 Common Standard Units for Electrical Work

Unit	Symbol	Unit	Symbol
centimeter	cm	microhenry	µH
cubic centimeter	cm³	microsecond	µs
cubic meter per second	m³/s	microwatt	µW
gigahertz	GHz	milliampere	mA
gram	g	milligram	mg
kilohertz	kHz	millihenry	mH
kilohm	kΩ	millimeter	mm
kilojoule	kJ	millisecond	ms
kilometer	km	millivolt	mV
kilovolt	kV	milliwatt	mW
kilovoltampere	kVA	nanoampere	nA
kilowatt	kW	nanofarad	nF
megahertz	MHz	nanometer	nm
megavolt	MV	nanosecond	ns
megawatt	MW	nanowatt	nW
megaohm	MΩ	picoampere	pA
microampere	µA	picofarad	pF
microfarad	µF	picosecond	ps
microgram	µg	picowatt	pW

TABLE A.4 dB Reference Table

dBm	dBw	Watts	Multiple	Prefix
+150	+120	1,000,000,000,000	10^{12}	1 Terawatt
+140	+110	100,000,000,000	10^{11}	100 Gigawatts
+130	+100	10,000,000,000	10^{10}	10 Gigawatts
+120	+90	1,000,000,000	10^{9}	1 Gigawatt
+110	+80	100,000,000	10^{8}	100 Megawatts
+100	+70	10,000,000	10^{7}	10 Megawatts
+90	+60	1,000,000	10^{6}	1 Megawatt
+80	+50	100,000	10^{5}	100 Kilowatts
+70	+40	10,000	10^{4}	10 Kilowatts
+60	+30	1000	10^{3}	1 Kilowatt
+50	+20	100	10^{2}	1 Hectrowatt
+40	+10	10	10	1 Decawatt
+30	0	1	1	1 Watt
+20	−10	0.1	10^{-1}	1 Deciwatt
+10	−20	0.01	10^{-2}	1 Centiwatt
0	−30	0.001	10^{-3}	1 Milliwatt
−10	−40	0.0001	10^{-4}	100 Microwatts
−20	−50	0.00001	10^{-5}	10 Microwatts
−30	−60	0.000001	10^{-6}	1 Microwatt
−40	−70	0.0000001	10^{-7}	100 Nanowatts
−50	−80	0.00000001	10^{-8}	10 Nanowatts
−60	−90	0.000000001	10^{-9}	1 Nanowatt
−70	−100	0.0000000001	10^{-10}	100 Picowatts
−80	−110	0.00000000001	10^{-11}	10 Picowatts
−90	−120	0.000000000001	10^{-12}	1 Picowatt

TABLE A.5 Standing Wave Ratio Power Conversion

SWR	Reflection Coefficient	Return Loss	Power Ratio	Percent Reflected
1.01:1	0.0050	46.1 dB	0.00002	0.002
1.02:1	0.0099	40.1 dB	0.00010	0.010
1.04:1	0.0196	34.2 dB	0.00038	0.038
1.06:1	0.0291	30.7 dB	0.00085	0.085
1.08:1	0.0385	28.3 dB	0.00148	0.148
1.10:1	0.0476	26.4 dB	0.00227	0.227
1.20:1	0.0909	20.8 dB	0.00826	0.826
1.30:1	0.1304	17.7 dB	0.01701	1.7
1.40:1	0.1667	15.6 dB	0.02778	2.8
1.50:1	0.2000	14.0 dB	0.04000	4.0
1.60:1	0.2308	12.7 dB	0.05325	5.3
1.70:1	0.2593	11.7 dB	0.06722	6.7
1.80:1	0.2857	10.9 dB	0.08163	8.2
1.90:1	0.3103	10.2 dB	0.09631	9.6
2.00:1	0.3333	9.5 dB	0.11111	11.1
2.20:1	0.3750	8.5 dB	0.14063	14.1
2.40:1	0.4118	7.7 dB	0.16955	17.0
2.60:1	0.4444	7.0 dB	0.19753	19.8
2.80:1	0.4737	6.5 dB	0.22438	22.4
3.00:1	0.5000	6.0 dB	0.25000	25.0
3.50:1	0.5556	5.1 dB	0.30864	30.9
4.00:1	0.6000	4.4 dB	0.36000	36.0
4.50:1	0.6364	3.9 dB	0.40496	40.5
5.00:1	0.6667	3.5 dB	0.44444	44.4
6.00:1	0.7143	2.9 dB	0.51020	51.0
7.00:1	0.7500	2.5 dB	0.56250	56.3
8.00:1	0.7778	2.2 dB	0.60494	60.5
9.00:1	0.8000	1.9 dB	0.64000	64.0
10.00:1	0.8182	1.7 dB	0.66942	66.9
15.00:1	0.8750	1.2 dB	0.76563	76.6
20.00:1	0.9048	0.9 dB	0.81859	81.9
30.00:1	0.9355	0.6 dB	0.87513	97.5
40.00:1	0.9512	0.4 dB	0.90482	90.5
50.00:1	0.9608	0.3 dB	0.92311	92.3

TABLE A.6 Specifications of Standard Copper Wire Sizes

Wire Size (AWG)	Diameter (mils)	Circular mil Area	Turns per Linear Inch[1]			Ohms per 100 ft[2]	Current Carrying Capacity[3]	Diameter (mm)
			Enamel	SCE	DCC			
1	289.3	83,810	—	—	—	0.1239	119.6	7.348
2	257.6	05,370	—	—	—	0.1563	94.8	6.544
3	229.4	62,640	—	—	—	0.1970	75.2	5.827
4	204.3	41,740	—	—	—	0.2485	59.6	5.189
5	181.9	33,100	—	—	—	0.3133	47.3	4.621
6	162.0	26,250	—	—	—	0.3951	37.5	4.115
7	144.3	20,820	—	—	—	0.4982	29.7	3.665
8	128.5	16,510	7.6	—	7.1	0.6282	23.6	3.264
9	114.4	13,090	8.6	—	7.8	0.7921	18.7	2.906
10	101.9	10,380	9.6	9.1	8.9	0.9989	14.8	2.588
11	90.7	8234	10.7	—	9.8	1.26	11.8	2.305
12	80.8	6530	12.0	11.3	10.9	1.588	9.33	2.063
13	72.0	5178	13.5	—	12.8	2.003	7.40	1.828
14	64.1	4107	15.0	14.0	13.8	2.525	5.87	1.628
15	57.1	3257	16.8	—	14.7	3.184	4.65	1.450
16	50.8	2583	18.9	17.3	16.4	4.016	3.69	1.291
17	45.3	2048	21.2	—	18.1	5.064	2.93	1.150
18	40.3	1624	23.6	21.2	19.8	6.386	2.32	1.024
19	35.9	1288	26.4	—	21.8	8.051	1.84	0.912
20	32.0	1022	29.4	25.8	23.8	10.15	1.46	0.812
21	28.5	810	33.1	—	26.0	12.8	1.16	0.723
22	25.3	642	37.0	31.3	30.0	16.14	0.918	0.644
23	22.6	510	41.3	—	37.6	20.36	0.728	0.573
24	20.1	404	46.3	37.6	35.6	25.67	0.577	0.511
25	17.9	320	51.7	—	38.6	32.37	0.458	0.455
26	15.9	254	58.0	46.1	41.8	40.81	0.363	0.406
27	14.2	202	64.9	—	45.0	51.47	0.288	0.361
28	12.6	160	72.7	54.6	48.5	64.9	0.228	0.321
29	11.3	127	81.6	—	51.8	81.83	0.181	0.286
30	10.0	101	90.5	64.1	55.5	103.2	0.144	0.255
31	8.9	50	101	—	59.2	130.1	0.114	0.227
32	8.0	63	113	74.1	61.6	164.1	0.090	0.202
33	7.1	50	127	—	66.3	206.9	0.072	0.180
34	6.3	40	143	86.2	70.0	260.9	0.057	0.160
35	5.6	32	158	—	73.5	329.0	0.045	0.143
36	5.0	25	175	103.1	T7.0	414.8	0.036	0.127
37	4.5	20	198	—	80.3	523.1	0.028	0.113
38	4.0	16	224	116.3	83.6	659.6	0.022	0.101
39	3.5	12	248	—	86.6	831.8	0.018	0.090

1. Based on 25.4 mm.
2. Ohms per 1000 feet measured at 20°C.
3. Current-carrying capacity at 700 cm/A.

TABLE A.7 Celsius-to-Fahrenheit Conversion Table

°Celsius	°Fahrenheit	°Celsius	°Fahrenheit
−50	−58	125	257
−45	−49	130	266
−40	−40	135	275
−35	−31	140	284
−30	−22	145	293
−25	−13	150	302
−20	4	155	311
−15	5	160	320
−10	14	165	329
−5	23	170	338
0	32	175	347
5	41	180	356
10	50	185	365
15	59	190	374
20	68	195	383
25	77	200	392
30	86	205	401
35	95	210	410
40	104	215	419
45	113	220	428
50	122	225	437
55	131	230	446
60	140	235	455
65	149	240	464
70	158	245	473
75	167	250	482
80	176	255	491
85	185	260	500
90	194	265	509
95	203	270	518
100	212	275	527
105	221	280	536
110	230	285	545
115	239	290	554
120	248	295	563

TABLE A.8 Inch-to-Millimeter Conversion Table

Inch	0	1/8	1/4	3/8	1/2	5/8	3/4	7/8	Inch
0	0.0	3.18	6.35	9.52	12.70	15.88	19.05	22.22	0
1	25.40	28.58	31.75	34.92	38.10	41.28	44.45	47.62	1
2	50.80	53.98	57.15	60.32	63.50	66.68	69.85	73.02	2
3	76.20	79.38	82.55	85.72	88.90	92.08	95.25	98.42	3
4	101.6	104.8	108.0	111.1	114.3	117.5	120.6	123.8	4
5	127.0	130.2	133.4	136.5	139.7	142.9	146.0	149.2	5
6	152.4	155.6	158.8	161.9	165.1	168.3	171.4	174.6	6
7	177.8	181.0	184.2	187.3	190.5	193.7	196.8	200.0	7
8	203.2	206.4	209.6	212.7	215.9	219.1	222.2	225.4	8
9	228.6	231.8	235.0	238.1	241.3	244.5	247.6	250.8	9
10	254.0	257.2	260.4	263.5	266.7	269.9	273.0	276.2	10
11	279	283	286	289	292	295	298	302	11
12	305	308	311	314	317	321	324	327	12
13	330	333	337	340	343	346	349	352	13
14	356	359	362	365	368	371	375	378	14
15	381	384	387	391	394	397	400	403	15
16	406	410	413	416	419	422	425	429	16
17	432	435	438	441	445	448	451	454	17
18	457	460	464	467	470	473	476	479	18
19	483	486	489	492	495	498	502	505	19
20	508	511	514	518	521	524	527	530	20

TABLE A.9 Conversion of Millimeters to Decimal Inches

mm	Inches	mm	Inches	mm	Inches
1	0.039370	46	1.811020	91	3.582670
2	0.078740	47	1.850390	92	3.622040
3	0.118110	48	1.889760	93	3.661410
4	0.157480	49	1.929130	94	3.700780
5	0.196850	50	1.968500	95	3.740150
6	0.236220	51	2.007870	96	3.779520
7	0.275590	52	2.047240	97	3.818890
8	0.314960	53	2.086610	98	3.858260
9	0.354330	54	2.125980	99	3.897630
10	0.393700	55	2.165350	100	3.937000
11	0.433070	56	2.204720	105	4.133848
12	0.472440	57	2.244090	110	4.330700
13	0.511810	58	2.283460	115	4.527550
14	0.551180	59	2.322830	120	4.724400
15	0.590550	60	2.362200	125	4.921250
16	0.629920	61	2.401570	210	8.267700
17	0.669290	62	2.440940	220	8.661400
18	0.708660	63	2.480310	230	9.055100
19	0.748030	64	2.519680	240	9.448800
20	0.787400	65	2.559050	250	9.842500
21	0.826770	66	2.598420	260	10.236200
22	0.866140	67	2.637790	270	10.629900
23	0.905510	68	2.677160	280	11.032600
24	0.944880	69	2.716530	290	11.417300
25	0.984250	70	2.755900	300	11.811000
26	1.023620	71	2.795270	310	12.204700
27	1.062990	72	2.834640	320	12.598400
28	1.102360	73	2.874010	330	12.992100
29	1.141730	74	2.913380	340	13.385800
30	1.181100	75	2.952750	350	13.779500
31	1.220470	76	2.992120	360	14.173200
32	1.259840	77	3.031490	370	14.566900
33	1.299210	78	3.070860	380	14.960600
34	1.338580	79	3.110230	390	15.354300
35	1.377949	80	3.149600	400	15.748000
36	1.417319	81	3.188970	500	19.685000
37	1.456689	82	3.228340	600	23.622000
38	1.496050	83	3.267710	700	27.559000
39	1.535430	84	3.307080	800	31.496000
40	1.574800	85	3.346450	900	35.433000
41	1.614170	86	3.385820	1000	39.370000
42	1.653540	87	3.425190	2000	78.740000
43	1.692910	88	3.464560	3000	118.110000
44	1.732280	89	3.503903	4000	157.480000
45	1.771650	90	3.543300	5000	196.850000

TABLE A.10 Conversion of Common Fractions to Decimal and Millimeter Units

Common Fractions	Decimal Fractions	mm (approx.)	Common Fractions	Decimal Fractions	mm (appox.)
1/128	0.008	0.20	1/2	0.500	12.70
1/64	0.016	0.40	33/64	0.516	13.10
1/32	0.031	0.79	17/32	0.531	13.49
3/64	0.047	1.19	35/64	0.547	13.89
1/16	0.063	1.59	9/16	0.563	14.29
5/64	0.078	1.98	37/64	0.578	14.68
3/32	0.094	2.38	19/32	0.594	15.08
7/64	0.109	2.78	39/64	0.609	15.48
1/8	0.125	3.18	5/8	0.625	15.88
9/64	0.141	3.57	41/64	0.641	16.27
5/32	0.156	3.97	21/32	0.656	16.67
11/64	0.172	4.37	43/64	0.672	17.07
3/16	0.188	4.76	11/16	0.688	17.46
13/64	0.203	5.16	45/64	0.703	17.86
7/32	0.219	5.56	23/32	0.719	18.26
15/64	0.234	5.95	47/64	0.734	18.65
1/4	0.250	6.35	3/4	0.750	19.05
17/64	0.266	6.75	49/64	0.766	19.45
9/32	0.281	7.14	25/32	0.781	19.84
19/64	0.297	7.54	51/64	0.797	20.24
5/16	0.313	7.94	13/16	0.813	20.64
21/64	0.328	8.33	53/64	0.828	21.03
11/32	0.344	8.73	27/32	0.844	21.43
23/64	0.359	9.13	55/64	0.859	21.83
3/8	0.375	9.53	7/8	0.875	22.23
25/64	0.391	9.92	57/64	0.891	22.62
13/32	0.406	10.32	29/32	0.906	23.02
27/64	0.422	10.72	59/64	0.922	23.42
7/16	0.438	11.11	15/16	0.938	23.81
29/64	0.453	11.51	61/64	0.953	24.21
15/32	0.469	11.91	31/32	0.969	24.61
31/64	0.484	12.30	63/64	0.984	25.00

TABLE A.11 Conversion Ratios for Length

Known Quantity	Multiply by	Quantity to Find
inches (in)	2.54	centimeters (cm)
feet (ft)	30	centimeters (cm)
yards (yd)	0.9	meters (m)
miles (mi)	1.6	kilometers (km)
millimeters (mm)	0.04	inches (in)
centimeters (cm)	0.4	inches (in)
meters (m)	3.3	feet (ft)
meters (m)	1.1	yards (yd)
kilometers (km)	0.6	miles (mi)
centimeters (cm)	10	millimeters (mm)
decimeters (dm)	10	centimeters (cm)
decimeters (dm)	100	millimeters (mm)
meters (m)	10	decimeters (dm)
meters (m)	1000	millimeters (mm)
dekameters (dam)	10	meters (m)
hectometers (hm)	10	dekameters (dam)
hectometers (hm)	100	meters (m)
kilometers (km)	10	hectometers (hm)
kilometers (km)	1000	meters (m)

TABLE A.12 Conversion Ratios for Area

Known Quantity	Multiply by	Quantity to Find
square inches (in^2)	6.5	square centimeters (cm^2)
square feet (ft^2)	0.09	square meters (m^2)
square yards (yd^2)	0.8	square meters (m^2)
square miles (mi^2)	2.6	square kilometers (km^2)
acres	0.4	hectares (ha)
square centimeters (cm^2)	0.16	square inches (in^2)
square meters (m^2)	1.2	square yards (yd^2)
square kilometers (km^2)	0.4	square miles (mi^2)
hectares (ha)	2.5	acres
square centimeters (cm^2)	100	square millimeters (mm^2)
square meters (m^2)	10,000	square centimeters (cm^2)
square meters (m^2)	1,000,000	square millimeters (mm^2)
ares (a)	100	square meters (m^2)
hectares (ha)	100	ares (a)
hectares (ha)	10,000	square meters (m^2)
square kilometers (km^2)	100	hectares (ha)
square kilometers (km^2)	1000	square meters (m^2)

TABLE A.13 Conversion Ratios for Mass

Known Quantity	Multiply by	Quantity to Find
ounces (oz)	28	grams (g)
pounds (lb)	0.45	kilograms (kg)
tons	0.9	tons (t)
grams (g)	0.035	ounces (oz)
kilograms (kg)	2.2	pounds (lb)
tons (t)	100	kilograms (kg)
tons (t)	1.1	tons
centigrams (cg)	10	milligrams (mg)
decigrams (dg)	10	centigrams (cg)
decigrams (dg)	100	milligrams (mg)
grams (g)	10	decigrams (dg)
grams (g)	1000	milligrams (mg)
dekagram (dag)	10	grams (g)
hectogram (hg)	10	dekagrams (dag)
hectogram (hg)	100	grams (g)
kilograms (kg)	10	hectograms (hg)
kilograms (kg)	1000	grams (g)

TABLE A.14 Conversion Ratios for Volume

Known Quantity	Multiply by	Quantity to Find
milliliters (mL)	0.03	fluid ounces (fl oz)
liters (L)	2.1	pints (pt)
liters (L)	1.06	quarts (qt)
liters (L)	0.26	gallons (gal)
gallons (gal)	3.8	liters (L)
quarts (qt)	0.95	liters (L)
pints (pt)	0.47	liters (L)
cups (c)	0.24	liters (L)
fluid ounces (fl oz)	30	milliliters (mL)
teaspoons (tsp)	5	milliliters (mL)
tablespoons (tbsp)	15	milliliters (mL)
liters (L)	100	milliliters (mL)

TABLE A.15 Conversion Ratios for Cubic Measure

Known Quantity	Multiply by	Quantity to Find
cubic meters (m^3)	35	cubic feet (ft^3)
cubic meters (m^3)	1.3	cubic yards (yd^3)
cubic yards (yd^3)	0.76	cubic meters (m^3)
cubic feet (ft^3)	0.028	cubic meters (m^3)
cubic centimeters (cm^3)	1000	cubic millimeters (mm^3)
cubic decimeters (dm^3)	1000	cubic centimeters (cm^3)
cubic decimeters (dm^3)	1,000,000	cubic millimeters (mm^3)
cubic meters (m^3)	1000	cubic decimeters (dm^3)
cubic meters (m^3)	1	steres
cubic feet (ft^3)	1728	cubic inches (in^3)
cubic feet (ft^3)	28.32	liters (L)
cubic inches (in^3)	16.39	cubic centimeters (cm^3)
cubic meters (m^3)	264	gallons (gal)
cubic yards (yd^3)	27	cubic feet (ft^3)
cubic yards (yd^3)	202	gallons (gal)
gallons (gal)	231	cubic inches (in^3)

TABLE A.16 Conversion Reference Data

To Convert	Into	Multiply by
abcoulomb	statcoulombs	2.998×10^{10}
acre	sq. chain (Gunters)	10
acre	rods	160
acre	square links (Gunters)	1×10^5
acre	Hectare or sq. hectometer	0.4047
acre-feet	cubic feet	43,560.0
acre-feet	gallons	3.259×10^5
acres	sq. feet	43,560.0
acres	sq. meters	4047
acres	sq. miles	1.562×10^{-3}
acres	sq. yards	4840
ampere-hours	coulombs	3600.0
ampere-hours	faradays	0.03731
amperes/sq. cm	amps/sq. in	6.452
amperes/sq. cm	amps/sq. meter	10^4
amperes/sq. in	amps/sq. cm	0.1550
amperes/sq. in	amps/sq. meter	1550.0
amperes/sq. meter	amps/sq. cm	10^{-4}
amperes/sq. meter	amps/sq. in	6.452×10^{-4}
ampere-turns	gilberts	1.257
ampere-turns/cm	amp-turns/in	2.540
ampere-turns/cm	amp-turns/meter	100.0
ampere-turns/cm	gilberts/cm	1.257
ampere-turns/in	amp-turns/cm	0.3937
ampere-turns/in	amp-turns/m	39.37
ampere-turns/in	gilberts/cm	0.4950
ampere-turns/meter	amp-turns/cm	0.01
ampere-turns/meter	amp-turns/in	0.0254
ampere-turns/meter	gilberts/cm	0.01257
Angstrom unit	inch	3937×10^{-9}
Angstrom unit	meter	1×10^{-10}
Angstrom unit	micron or (Mu)	1×10^{-4}
are	acre (U.S.)	0.02471
ares	sq. yards	119.60
ares	acres	0.02471
ares	sq. meters	100.0
astronomical unit	kilometers	1.495×10^8
atmospheres	ton/sq. in	0.007348
atmospheres	cm of mercury	76.0
atmospheres	ft of water (at 4°C)	33.90
atmospheres	in of mercury (at 0°C)	29.92
atmospheres	kg/sq. cm	1.0333
atmospheres	kg/sq. m	10,332

TABLE A.16 Conversion Reference Data (*Continued*)

To Convert	Into	Multiply by
atmospheres	pounds/sq. in	14.70
atmospheres	tons/sq. ft	1.058
barrels (U.S., dry)	cubic inches	7056
barrels (U.S., dry)	quarts (dry)	105.0
barrels (U.S., liquid)	gallons	31.5
barrels (oil)	gallons (oil)	42.0
bars	atmospheres	0.9869
bars	dynes/sq. cm	10^4
bars	kg/sq. m	1.020×10^4
bars	pounds/sq. ft	2089
bars	pounds/sq. in	14.50
baryl	dyne/sq. cm	1.000
bolt (U.S. cloth)	meters	36.576
Btu	liter-atmosphere	10.409
Btu	ergs	1.0550×10^{10}
Btu	foot-lb	778.3
Btu	gram-calories	252.0
Btu	horsepower-hour	3.931×10^{-4}
Btu	joules	1054.8
Btu	kilogram-calories	0.2520
Btu	kilogram-meters	107.5
Btu	kilowatt-hour	2.928×10^{-4}
Btu/hour	foot-pounds/s	0.2162
Btu/hour	gram-calories/s	0.0700
Btu/hour	horsepower-hour	3.929×10^{-4}
Btu/hour	watts	0.2931
Btu/min	foot-lbs/s	12.96
Btu/min	horsepower	0.02356
Btu/min	kilowatts	0.01757
Btu/min	watts	17.57
Btu/sq. ft/min	watts/sq. in	0.1221
bucket (br. dry)	cubic cm	1.818×10^4
bushels	cubic ft	1.2445
bushels	cubic in	2150.4
bushels	cubic m	0.03524
bushels	liters	35.24
bushels	pecks	4.0
bushels	pints (dry)	64.0
bushels	quarts (dry)	32.0
calories, gram (mean)	Btu (mean)	3.9685×10^{-3}
candle/sq. cm	Lamberts	3.142
candle/sq. in	Lamberts	0.4870

(*Continued*)

TABLE A.16 Conversion Reference Data (*Continued*)

To Convert	Into	Multiply by
centares (centiares)	sq. meters	1.0
Centigrade	Fahrenheit	(C° × 9/5) + 32
centigrams	grams	0.01
centiliter	ounce fluid (U.S.)	0.3382
centiliter	cubic inch	0.6103
centiliter	drams	2.705
centiliter	liters	0.01
centimeter	feet	3.281×10^{-2}
centimeter	inches	0.3937
centimeter	kilometers	10^{-5}
centimeter	meters	0.01
centimeter	miles	6.214×10^{-6}
centimeter	millimeters	10.0
centimeter	mils	393.7
centimeter	yards	1.094×10^{-2}
centimeter-dynes	cm-grams	1.020×10^{-3}
centimeter-dynes	meter-kg	1.020×10^{-8}
centimeter-dynes	pound-ft	7.376×10^{-8}
centimeter-grams	cm-dynes	980.7
centimeter-grams	meter-kg	10^{-5}
centimeter-grams	pound-ft	7.233×10^{-5}
centimeters of mercury	atmospheres	0.01316
centimeters of mercury	feet of water	0.4461
centimeters of mercury	kg/sq. meter	136.0
centimeters of mercury	pounds/sq. ft	27.85
centimeters of mercury	pounds/sq. in	0.1934
centimeters/second	feet/min	1.9686
centimeters/second	feet/second	0.03281
centimeters/second	kilometers/hour	0.036
centimeters/second	knots	0.1943
centimeters/second	meters/min	0.6
centimeters/second	miles/hour	0.02237
centimeters/second	miles/min	3.728×10^{-4}
centimeters/second/second	feet/second/second	0.03281
centimeters/second/second	km/hour/second	0.036
centimeters/second/second	meters/second/second	0.01
centimeters/second/second	miles/hour/second	0.02237
chain	inches	792.00
chain	meters	20.12
chains (surveyor's or Gunter's)	yards	22.00
circular mils	sq. cm	5.067×10^{-6}
circular mils	sq. mils	0.7854

TABLE A.16 Conversion Reference Data (*Continued*)

To Convert	Into	Multiply by
circular mils	sq. inches	7.854×10^{-7}
circumference	Radians	6.283
cord feet	cubic feet	16
cords	cord feet	8
coulomb	statcoulombs	2.998×10^{9}
coulombs	faradays	1.036×10^{-5}
coulombs/sq. cm	coulombs/sq. in	64.52
coulombs/sq. cm	coulombs/sq. meter	10^{4}
coulombs/sq. in	coulombs/sq. cm	0.1550
coulombs/sq. in	coulombs/sq. meter	1550
coulombs/sq. meter	coulombs/sq. cm	10^{-4}
coulombs/sq. meter	coulombs/sq. in	6.452×10^{-4}
cubic centimeters	cubic feet	3.531×10^{-5}
cubic centimeters	cubic inches	0.06102
cubic centimeters	cubic meters	10^{-6}
cubic centimeters	cubic yards	1.308×10^{-6}
cubic centimeters	gallons (U.S. liq.)	2.642×10^{-4}
cubic centimeters	liters	0.001
cubic centimeters	pints (U.S. liq.)	2.113×10^{-3}
cubic centimeters	quarts (U.S. liq.)	1.057×10^{-3}
cubic feet	bushels (dry)	0.8036
cubic feet	cubic cm	28,320.0
cubic feet	cubic inches	1728.0
cubic feet	cubic meters	0.02832
cubic feet	cubic yards	0.03704
cubic feet	gallons (U.S. liq.)	7.48052
cubic feet	liters	28.32
cubic feet	pints (U.S. liq.)	59.84
cubic feet	quarts (U.S. liq.)	29.92
cubic feet/min	cubic cm/second	472.0
cubic feet/min	gallons/second	0.1247
cubic feet/min	liters/second	0.4720
cubic feet/min	pounds of water/min	62.43
cubic feet/second	million gal/day	0.646317
cubic feet/second	gallons/min	448.831
cubic inches	cubic cm	16.39
cubic inches	cubic feet	5.787×10^{-4}
cubic inches	cubic meters	1.639×10^{-5}
cubic inches	cubic yards	2.143×10^{-5}
cubic inches	gallons	4.329×10^{-3}
cubic inches	liters	0.01639
cubic inches	mil-feet	1.061×10^{5}

(*Continued*)

TABLE A.16 Conversion Reference Data (*Continued*)

To Convert	Into	Multiply by
cubic inches	pints (U.S. liq.)	0.03463
cubic inches	quarts (U.S. liq.)	0.01732
cubic meters	bushels (dry)	28.38
cubic meters	cubic cm	10^6
cubic meters	cubic feet	35.31
cubic meters	cubic inches	61,023.0
cubic meters	cubic yards	1.308
cubic meters	gallons (U.S. liq.)	264.2
cubic meters	liters	1000.0
cubic meters	pints (U.S. liq.)	2113.0
cubic meters	quarts (U.S. liq.)	1057.
cubic yards	cubic cm	7.646×10^5
cubic yards	cubic feet	27.0
cubic yards	cubic inches	46,656.0
cubic yards	cubic meters	0.7646
cubic yards	gallons (U.S. liq.)	202.0
cubic yards	liters	764.6
cubic yards	pints (U.S. liq.)	1615.9
cubic yards	quarts (U.S. liq.)	807.9
cubic yards/min	cubic ft/second	0.45
cubic yards/min	gallons/second	3.367
cubic yards/min	liters/second	12.74
Dalton	gram	1.650×10^{-24}
days	seconds	86,400.0
decigrams	grams	0.1
deciliters	liters	0.1
decimeters	meters	0.1
degrees (angle)	quadrants	0.01111
degrees (angle)	radians	0.01745
degrees (angle)	seconds	3600.0
degrees/second	radians/second	0.01745
degrees/second	revolutions/min	0.1667
degrees/second	revolutions/second	2.778×10^{-3}
dekagrams	grams	10.0
dekaliters	liters	10.0
dekameters	meters	10.0
drams (apothecaries or troy)	ounces (avoirdupois)	0.1371429
drams (apothecaries or troy)	ounces (troy)	0.125
drams (U.S., fluid or apothecaries)	cubic cm	3.6967
drams	grams	1.7718
drams	grains	27.3437
drams	ounces	0.0625
dyne/cm	erg/sq. millimeter	0.01

TABLE A.16 Conversion Reference Data (*Continued*)

To Convert	Into	Multiply by
dyne/sq. cm	atmospheres	9.869×10^{-7}
dyne/sq. cm	inch of mercury at 0°C	2.953×10^{-5}
dyne/sq. cm	inch of water at 4°C	4.015×10^{-4}
dynes	grams	1.020×10^{-3}
dynes	joules/cm	10^{-7}
dynes	joules/meter (newtons)	10^{-5}
dynes	kilograms	1.020×10^{-6}
dynes	poundals	7.233×10^{-5}
dynes	pounds	2.248×10^{-6}
dynes/sq. cm	bars	10^{-6}
ell	cm	114.30
ell	inches	45
em, pica	inch	0.167
em, pica	cm	0.4233
erg/second	Dyne-cm/second	1.000
ergs	Btu	9.480×10^{-11}
ergs	dyne-centimeters	1.0
ergs	foot-pounds	7.367×10^{-8}
ergs	gram-calories	0.2389×10^{-7}
ergs	gram-cm	1.020×10^{-3}
ergs	horsepower-hour	3.7250×10^{-14}
ergs	joules	10^{-7}
ergs	kg-calories	2.389×10^{-11}
ergs	kg-meters	1.020×10^{-8}
ergs	kilowatt-hour	0.2778×10^{-13}
ergs	watt-hours	0.2778×10^{-10}
ergs/second	Btu/min	5688×10^{-9}
ergs/second	ft-lb/min	4.427×10^{-6}
ergs/second	ft-lb/second	7.3756×10^{-8}
ergs/second	horsepower	1.341×10^{-10}
ergs/second	kg-calories/min	1.433×10^{-9}
ergs/second	kilowatts	10^{-10}
farad	microfarads	10^{6}
Faraday/second	ampere (absolute)	9.6500×10^{4}
faradays	ampere-hours	26.80
faradays	coulombs	9.649×10^{4}
fathom	meter	1.828804
fathoms	feet	6.0
feet	centimeters	30.48
feet	kilometers	3.048×10^{-4}
feet	meters	0.3048
feet	miles (naut.)	1.645×10^{-4}

(*Continued*)

TABLE A.16 Conversion Reference Data (*Continued*)

To Convert	Into	Multiply by
feet	miles (stat.)	1.894×10^{-4}
feet	millimeters	304.8
feet	mils	1.2×10^4
feet of water	atmospheres	0.02950
feet of water	in of mercury	0.8826
feet of water	kg/sq. cm	0.03048
feet of water	kg/sq. meter	304.8
feet of water	pounds/sq. ft	62.43
feet of water	pounds/sq. in	0.4335
feet/min	cm/second	0.5080
feet/min	feet/second	0.01667
feet/min	km/hour	0.01829
feet/min	meters/min	0.3048
feet/min	miles/hour	0.01136
feet/second	cm/second	30.48
feet/second	km/hour	1.097
feet/second	knots	0.5921
feet/second	meters/min	18.29
feet/second	miles/hour	0.6818
feet/second	miles/min	0.01136
feet/second/second	cm/second/second	30.48
feet/second/second	km/hour/second	1.097
feet/second/second	meters/second/second	0.3048
feet/second/second	miles/hour/second	0.6818
feet/100 feet	per centigrade	1.0
foot-candle	lumen/sq. meter	10.764
foot-pounds	Btu	1.286×10^{-3}
foot-pounds	ergs	1.356×10^7
foot-pounds	gram-calories	0.3238
foot-pounds	hp-hour	5.050×10^{-7}
foot-pounds	joules	1.356
foot-pounds	kg-calories	3.24×10^{-4}
foot-pounds	kg-meters	0.1383
foot-pounds	kilowatt-hour	3.766×10^{-7}
foot-pounds/min	Btu/min	1.286×10^{-3}
foot-pounds/min	foot-pounds/second	0.01667
foot-pounds/min	horsepower	3.030×10^{-5}
foot-pounds/min	kg-calories/min	3.24×10^{-4}
foot-pounds/min	kilowatts	2.260×10^{-5}
foot-pounds/second	Btu/hour	4.6263
foot-pounds/second	Btu/min	0.07717
foot-pounds/second	horsepower	1.818×10^{-3}
foot-pounds/second	kg-calories/min	0.01945

TABLE A.16 Conversion Reference Data (*Continued*)

To Convert	Into	Multiply by
foot-pounds/second	kilowatts	1.356×10^{-3}
Furlongs	miles (U.S.)	0.125
furlongs	rods	40.0
furlongs	feet	660.0
gallons	cubic cm	3785.0
gallons	cubic feet	0.1337
gallons	cubic inches	231.0
gallons	cubic meters	3.785×10^{-3}
gallons	cubic yards	4.951×10^{-3}
gallons	liters	3.785
gallons (liq. Br. Imp.)	gallons (U.S. liq.)	1.20095
gallons (U.S.)	gallons (Imp.)	0.83267
gallons of water	pounds of water	8.3453
gallons/min	cubic ft/second	2.228×10^{-3}
gallons/min	liters/second	0.06308
gallons/min	cubic ft/hour	8.0208
gausses	lines/sq. in	6.452
gausses	webers/sq. cm	10^{-8}
gausses	webers/sq. in	6.452×10^{-8}
gausses	webers/sq. meter	10^{-4}
gilberts	ampere-turns	0.7958
gilberts/cm	amp-turns/cm	0.7958
gilberts/cm	amp-turns/in	2.021
gilberts/cm	amp-turns/meter	79.58
gills	liters	0.1183
gills	pints (liq.)	0.25
gills (British)	cubic cm	142.07
grade	radian	0.01571
grains	drams (avoirdupois)	0.03657143
grains (troy)	grains (avdp.)	1.0
grains (troy)	grams	0.06480
grains (troy)	ounces (avdp.)	2.0833×10^{-3}
grains (troy)	pennyweight (troy)	0.04167
grains/Imp. gal	parts/million	14.286
grains/U.S. gal	parts/million	17.118
grains/U.S. gal	pounds/million gal	142.86
gram-calories	Btu	3.9683×10^{-3}
gram-calories	ergs	4.1868×10^{7}
gram-calories	foot-pounds	3.0880
gram-calories	horsepower-hour	1.5596×10^{-6}
gram-calories	kilowatt-hour	1.1630×10^{-6}
gram-calories	watt-hour	1.1630×10^{-3}

(*Continued*)

TABLE A.16 Conversion Reference Data (*Continued*)

To Convert	Into	Multiply by
gram-calories/second	Btu/hour	14.286
gram-centimeters	Btu	9.297×10^{-8}
gram-centimeters	ergs	980.7
gram-centimeters	joules	9.807×10^{-5}
gram-centimeters	kg-calories	2.343×10^{-8}
gram-centimeters	kg-meters	10^{-5}
grams	dynes	980.7
grams	grains	15.43
grams	joules/cm	9.807×10^{-5}
grams	joules/meter (newtons)	9.807×10^{-3}
grams	kilograms	0.001
grams	milligrams	1000
grams	ounces (avdp.)	0.03527
grams	ounces (troy)	0.03215
grams	poundals	0.07093
grams	pounds	2.205×10^{-3}
grams/cm	pounds/inch	5.600×10^{-3}
grams/cubic cm	pounds/cubic ft	62.43
grams/cubic cm	pounds/cubic in	0.03613
grams/cubic cm	pounds/mil-foot	3.405×10^{-7}
grams/liter	grains/gal	58.417
grams/liter	pounds/1000 gal	8.345
grams/liter	pounds/cubic ft	0.062427
grams/liter	parts/million	1000.0
grams/sq. cm	pounds/sq. ft	2.0481
hand	cm	10.16
hectares	acres	2.471
hectares	sq. feet	1.076×10^5
hectograms	grams	100.0
hectoliters	liters	100.0
hectometers	meters	100.0
hectowatts	watts	100.0
henries	millihenries	1000.0
horsepower	Btu/min	42.44
horsepower	foot-lb/min	33,000
horsepower	foot-lb/second	550.0
horsepower	kg-calories/min	10.68
horsepower	kilowatts	0.7457
horsepower	watts	745.7
horsepower (boiler)	Btu/hour	33.479
horsepower (boiler)	kilowatts	9.803
horsepower, metric (542.5 ft lb./second)	horsepower (550 ft lb./second)	0.9863

TABLE A.16 Conversion Reference Data (*Continued*)

To Convert	Into	Multiply by
horsepower (550 ft lb./second)	horsepower, metric (542.5 ft lb./second)	1.014
horsepower-hour	Btu	2547
horsepower-hour	ergs	2.6845×10^{13}
horsepower-hour	foot-lb	1.98×10^6
horsepower-hour	gram-calories	641,190
horsepower-hour	joules	2.684×10^6
horsepower-hour	kg-calories	641.1
horsepower-hour	kg-meters	2.737×10^5
horsepower-hour	kilowatt-hour	0.7457
hours	days	4.167×10^{-2}
hours	weeks	5.952×10^{-3}
hundredweights (long)	pounds	112
hundredweights (long)	tons (long)	0.05
hundredweights (short)	ounces (avoirdupois)	1600
hundredweights (short)	pounds	100
hundredweights (short)	tons (metric)	0.0453592
hundredweights (short)	tons (long)	0.0446429
inches	centimeters	2.540
inches	meters	2.540×10^{-2}
inches	miles	1.578×10^{-5}
inches	millimeters	25.40
inches	mils	1000.0
inches	yards	2.778×10^{-2}
inches of mercury	atmospheres	0.03342
inches of mercury	feet of water	1.133
inches of mercury	kg/sq. cm	0.03453
inches of mercury	kg/sq. meter	345.3
inches of mercury	pounds/sq. ft	70.73
inches of mercury	pounds/sq. in	0.4912
inches of water (at 4°C)	atmospheres	2.458×10^{-3}
inches of water (at 4°C)	inches of mercury	0.07355
inches of water (at 4°C)	kg/sq. cm	2.540×10^{-3}
inches of water (at 4°C)	ounces/sq. in	0.5781
inches of water (at 4°C)	pounds/sq. ft	5.204
inches of water (at 4°C)	pounds/sq. in	0.03613
international ampere	ampere (absolute)	0.9998
international Volt	volts (absolute)	1.0003
international volt	joules (absolute)	1.593×10^{-19}
international volt	joules	9.654×10^4
joules	Btu	9.480×10^{-4}
joules	ergs	10^7

(*Continued*)

TABLE A.16 Conversion Reference Data (*Continued*)

To Convert	Into	Multiply by
joules	foot-pounds	0.7376
joules	kg-calories	2.389×10^{-4}
joules	kg-meters	0.1020
joules	watt-hour	2.778×10^{-4}
joules/cm	grams	1.020×10^{4}
joules/cm	dynes	10^{7}
joules/cm	joules/meter (newtons)	100.0
joules/cm	poundals	723.3
joules/cm	pounds	22.48
kilogram-calories	Btu	3.968
kilogram-calories	foot-pounds	3088
kilogram-calories	hp-hour	1.560×10^{-3}
kilogram-calories	joules	4186
kilogram-calories	kg-meters	426.9
kilogram-calories	kilojoules	4.186
kilogram-calories	kilowatt-hour	1.163×10^{-3}
kilogram meters	Btu	9.294×10^{-3}
kilogram meters	ergs	9.804×10^{7}
kilogram meters	foot-pounds	7.233
kilogram meters	joules	9.804
kilogram meters	kg-calories	2.342×10^{-3}
kilogram meters	kilowatt-hour	2.723×10^{-6}
kilograms	dynes	980,665
kilograms	grams	1000.0
kilograms	joules/cm	0.09807
kilograms	joules/meter (newtons)	9.807
kilograms	poundals	70.93
kilograms	pounds	2.205
kilograms	tons (long)	9.842×10^{-4}
kilograms	tons (short)	1.102×10^{-3}
kilograms/cubic meter	grams/cubic cm	0.001
kilograms/cubic meter	pounds/cubic ft	0.06243
kilograms/cubic meter	pounds/cubic in	3.613×10^{-5}
kilograms/cubic meter	pounds/mil-foot	3.405×10^{-10}
kilograms/meter	pounds/ft	0.6720
kilograms/sq. cm	dynes	980,665
kilograms/sq. cm	atmospheres	0.9678
kilograms/sq. cm	feet of water	32.81
kilograms/sq. cm	inches of mercury	28.96
kilograms/sq. cm	pounds/sq. ft	2048
kilograms/sq. cm	pounds/sq. in	14.22
kilograms/sq. meter	atmospheres	9.678×10^{-5}

TABLE A.16 Conversion Reference Data (*Continued*)

To Convert	Into	Multiply by
kilograms/sq. meter	bars	98.07×10^{-6}
kilograms/sq. meter	feet of water	3.281×10^{-3}
kilograms/sq. meter	inches of mercury	2.896×10^{-3}
kilograms/sq. meter	pounds/sq. ft	0.2048
kilograms/sq. meter	pounds/sq. in	1.422×10^{-3}
kilograms/sq. mm	kg/sq. meter	10^6
kilolines	maxwells	1000.0
kiloliters	liters	1000.0
kilometers	centimeters	10^5
kilometers	feet	3281
kilometers	inches	3.937×10^4
kilometers	meters	1000.0
kilometers	miles	0.6214
kilometers	millimeters	10^4
kilometers	yards	1094
kilometers/hour	cm/second	27.78
kilometers/hour	feet/min	54.68
kilometers/hour	feet/second	0.9113
kilometers/hour	knots	0.5396
kilometers/hour	meters/min	16.67
kilometers/hour	miles/hour	0.6214
kilometers/hour/second	cm/second/second	27.78
kilometers/hour/second	feet/second/second	0.9113
kilometers/hour/second	meters/second/second	0.2778
kilometers/hour/second	miles/hour/second	0.6214
kilowatt-hour	Btu	3413
kilowatt-hour	ergs	3.600×10^{13}
kilowatt-hour	foot-lb	2.655×10^6
kilowatt-hour	gram-calories	859,850
kilowatt-hour	horsepower-hour	1.341
kilowatt-hour	joules	3.6×10^6
kilowatt-hour	kg-calories	860.5
kilowatt-hour	kg-meters	3.671×10^5
kilowatt-hour	pounds of water raised from 62° to 212°F	22.75
kilowatts	Btu/min	56.92
kilowatts	foot-lb/min	4.426×10^4
kilowatts	foot-lb/second	737.6
kilowatts	horsepower	1.341
kilowatts	kg-calories/min	14.34
kilowatts	watts	1000.0
knots	feet/hour	6080

(*Continued*)

TABLE A.16 Conversion Reference Data (*Continued*)

To Convert	Into	Multiply by
knots	kilometers/hour	1.8532
knots	nautical miles/hour	1.0
knots	statute miles/hour	1.151
knots	yards/hour	2027
knots	feet/second	1.689
league	miles (approx.)	3.0
light year	miles	5.9×10^{12}
light year	kilometers	9.4637×10^{12}
lines/sq. cm	gausses	1.0
lines/sq. in	gausses	0.1550
lines/sq. in	webers/sq. cm	1.550×10^{-9}
lines/sq. in	webers/sq. in	10^{-8}
lines/sq. in	webers/sq. meter	1.550×10^{-5}
links (engineer's)	inches	12.0
links (surveyor's)	inches	7.92
liters	bushels (U.S. dry)	0.02838
liters	cubic cm	1000.0
liters	cubic feet	0.03531
liters	cubic inches	61.02
liters	cubic meters	0.001
liters	cubic yards	1.308×10^{-3}
liters	gallons (U.S. liq.)	0.2642
liters	pints (U.S. liq.)	2.113
liters	quarts (U.S. liq.)	1.057
liters/min	cubic ft/second	5.886×10^{-4}
liters/min	gal/second	4.403×10^{-3}
lumen	spherical candle power	0.07958
lumen	watt	0.001496
lumens/sq. ft	foot-candles	1.0
lumens/sq. ft	lumen/sq. meter	10.76
lux	foot-candles	0.0929
maxwells	kilolines	0.001
maxwells	webers	10^{-8}
megalines	maxwells	10^6
megohms	microhms	10^{12}
megohms	ohms	10^6
meter-kilograms	cm-dynes	9.807×10^7
meter-kilograms	cm-grams	10^5
meter-kilograms	pound-feet	7.233
meters	centimeters	100.0
meters	feet	3.281
meters	inches	39.37
meters	kilometers	0.001

TABLE A.16 Conversion Reference Data (*Continued*)

To Convert	Into	Multiply by
meters	miles (naut.)	5.396×10^{-4}
meters	miles (stat.)	6.214×10^{-4}
meters	millimeters	1000.0
meters	yards	1.094
meters	varas	1.179
meters/min	cm/second	1667
meters/min	feet/min	3.281
meters/min	feet/second	0.05468
meters/min	km/hour	0.06
meters/min	knots	0.03238
meters/min	miles/hour	0.03728
meters/second	feet/min	196.8
meters/second	feet/second	3.281
meters/second	kilometers/hour	3.6
meters/second	kilometers/min	0.06
meters/second	miles/hour	2.237
meters/second	miles/min	0.03728
meters/second/second	cm/second/second	100.0
meters/second/second	ft/second/second	3.281
meters/second/second	km/hour/second	3.6
meters/second/second	miles/hour/second	2.237
microfarad	farads	10^{-6}
micrograms	grams	10^{-6}
microhms	megohms	10^{-12}
microhms	ohms	10^{-6}
microliters	liters	10^{-6}
microns	meters	1×10^{-6}
miles (naut.)	feet	6080.27
miles (naut.)	kilometers	1.853
miles (naut.)	meters	1853
miles (naut.)	miles (statute)	1.1516
miles (naut.)	yards	2027
miles (statute)	centimeters	1.609×10^{5}
miles (statute)	feet	5280
miles (statute)	inches	6.336×10^{4}
miles (statute)	kilometers	1.609
miles (statute)	meters	1609
miles (statute)	miles (naut.)	0.8684
miles (statute)	yards	1760
miles/hour	cm/second	44.70
miles/hour	feet/min	88
miles/hour	feet/second	1.467

(*Continued*)

TABLE A.16 Conversion Reference Data (*Continued*)

To Convert	Into	Multiply by
miles/hour	km/hour	1.609
miles/hour	km/min	0.02682
miles/hour	knots	0.8684
miles/hour	meters/min	26.82
miles/hour	miles/min	0.1667
miles/hour/second	cm/second/second	44.70
miles/hour/second	feet/second/second	1.467
miles/hour/second	km/hour/second	1.609
miles/hour/second	meters/second/second	0.4470
miles/min	cm/second	2682
miles/min	feet/second	88
miles/min	km/min	1.609
miles/min	knots/min	0.8684
miles/min	miles/hour	60
mil-feet	cubic inches	9.425×10^{-6}
milliers	kilograms	1000
milligrams	grains	0.01543236
milligrams	grams	0.001
milligrams/liter	parts/million	1.0
millihenries	henries	0.001
milliliters	liters	0.001
millimeters	centimeters	0.1
millimeters	feet	3.281×10^{-3}
millimeters	inches	0.03937
millimeters	kilometers	10^{-6}
millimeters	meters	0.001
millimeters	miles	6.214×10^{-7}
millimeters	mils	39.37
millimeters	yards	1.094×10^{-3}
millimicrons	meters	1×10^{-9}
million gal/day	cubic ft/second	1.54723
mils	centimeters	2.540×10^{-3}
mils	feet	8.333×10^{-5}
mils	inches	0.001
mils	kilometers	2.540×10^{-8}
mils	yards	2.778×10^{-5}
miner's inches	cubic ft/min	1.5
minims (British)	cubic cm	0.059192
minims (U.S., fluid)	cubic cm	0.061612
minutes (angles)	degrees	0.01667
minutes (angles)	quadrants	1.852×10^{-4}
minutes (angles)	radians	2.909×10^{-4}
minutes (angles)	seconds	60.0

TABLE A.16 Conversion Reference Data (*Continued*)

To Convert	Into	Multiply by
myriagrams	kilograms	10.0
myriameters	kilometers	10.0
myriawatts	kilowatts	10.0
nepers	decibels	8.686
Newton	dynes	1×10^5
ohm (international)	ohm (absolute)	1.0005
ohms	megohms	10^{-6}
ohms	microhms	10^6
ounces	drams	16.0
ounces	grains	437.5
ounces	grams	28.349527
ounces	pounds	0.0625
ounces	ounces (troy)	0.9115
ounces	tons (long)	2.790×10^{-5}
ounces	tons (metric)	2.835×10^{-5}
ounces (fluid)	cubic inches	1.805
ounces (fluid)	liters	0.02957
ounces (troy)	grains	480.0
ounces (troy)	grams	31.103481
ounces (troy)	ounces (avdp.)	1.09714
ounces (troy)	pennyweights (troy)	20.0
ounces (troy)	pounds (troy)	0.08333
ounces/sq. inch	dynes/sq. cm	4309
ounces/sq. in	pounds/sq. in	0.0625
parsec	miles	19×10^{12}
parsec	kilometers	3.084×10^{13}
parts/million	grains/U.S. gal	0.0584
parts/million	grains/Imp. gal	0.07016
parts/million	pounds/million gal	8.345
pecks (British)	cubic inches	554.6
pecks (British)	liters	9.091901
pecks (U.S.)	bushels	0.25
pecks (U.S.)	cubic inches	537.605
pecks (U.S.)	liters	8.809582
pecks (U.S.)	quarts (dry)	8
pennyweights (troy)	grains	24.0
pennyweights (troy)	ounces (troy)	0.05
pennyweights (troy)	grams	1.55517
pennyweights (troy)	pounds (troy)	4.1667×10^{-3}
pints (dry)	cubic inches	33.60
pints (liq.)	cubic cm	473.2
pints (liq.)	cubic feet	0.01671

(*Continued*)

TABLE A.16 Conversion Reference Data (*Continued*)

To Convert	Into	Multiply by
pints (liq.)	cubic inches	28.87
pints (liq.)	cubic meters	4.732×10^{-4}
pints (liq.)	cubic yards	6.189×10^{-4}
pints (liq.)	gallons	0.125
pints (liq.)	liters	0.4732
pints (liq.)	quarts (liq.)	0.5
Planck's quantum	erg-second	6.624×10^{-27}
poise	gram/cm second	1.00
poundals	dynes	13,826
poundals	grams	14.10
poundals	joules/cm	1.383×10^{-3}
poundals	joules/meter (newtons)	0.1383
poundals	kilograms	0.01410
poundals	pounds	0.03108
pound-feet	cm-dynes	1.356×10^{7}
pound-feet	cm-grams	13,825
pound-feet	meter-kg	0.1383
pounds	drams	256
pounds	dynes	44.4823×10^{4}
pounds	grains	7000
pounds	grams	453.5924
pounds	joules/cm	0.04448
pounds	joules/meter (newtons)	4.448
pounds	kilograms	0.4536
pounds	ounces	16.0
pounds	ounces (troy)	14.5833
pounds	poundals	32.17
pounds	pounds (troy)	1.21528
pounds	tons (short)	0.0005
pounds (avoirdupois)	ounces (troy)	14.5833
pounds (troy)	grains	5760
pounds (troy)	grams	373.24177
pounds (troy)	ounces (avdp.)	13.1657
pounds (troy)	ounces (troy)	12.0
pounds (troy)	pennyweights (troy)	240.0
pounds (troy)	pounds (avdp.)	0.822857
pounds (troy)	tons (long)	3.6735×10^{-4}
pounds (troy)	tons (metric)	3.7324×10^{-4}
pounds (troy)	tons (short)	4.1143×10^{-4}
pounds of water	cubic ft	0.01602
pounds of water	cubic inches	27.68
pounds of water	gallons	0.1198
pounds of water/min	cubic ft/second	2.670×10^{-4}

TABLE A.16 Conversion Reference Data (*Continued*)

To Convert	Into	Multiply by
pounds/cubic ft	grams/cubic cm	0.01602
pounds/cubic ft	kg/cubic meter	16.02
pounds/cubic ft	pounds/cubic in	5.787×10^{-4}
pounds/cubic ft	pounds/mil-foot	5.456×10^{-9}
pounds/cubic in	gm/cubic cm	27.68
pounds/cubic in	kg/cubic meter	2.768×10^{4}
pounds/cubic in	pounds/cubic ft	1728
pounds/cubic in	pounds/mil-foot	9.425×10^{-6}
pounds/ft	kg/meter	1.488
pounds/in	gm/cm	178.6
pounds/mil-foot	gm/cubic cm	2.306×10^{6}
pounds/sq. ft	atmospheres	4.725×10^{-4}
pounds/sq. ft	feet of water	0.01602
pounds/sq. ft	inches of mercury	0.01414
pounds/sq. ft	kg/sq. meter	4.882
pounds/sq. ft	pounds/sq. in	6.944×10^{-3}
pounds/sq. in	atmospheres	0.06804
pounds/sq. in	feet of water	2.307
pounds/sq. in	inches of mercury	2.036
pounds/sq. in	kg/sq. meter	703.1
pounds/sq. in	pounds/sq. ft	144.0
quadrants (angle)	degrees	90.0
quadrants (angle)	minutes	5400.0
quadrants (angle)	radians	1.571
quadrants (angle)	seconds	3.24×10^{5}
quarts (dry)	cubic inches	67.20
quarts (liq.)	cubic cm	946.4
quarts (liq.)	cubic feet	0.03342
quarts (liq.)	cubic inches	57.75
quarts (liq.)	cubic meters	9.464×10^{-4}
quarts (liq.)	cubic yards	1.238×10^{-3}
quarts (liq.)	gallons	0.25
quarts (liq.)	liters	0.9463
radians	degrees	57.30
radians	minutes	3438
radians	quadrants	0.6366
radians	seconds	2.063×10^{5}
radians/second	degrees/second	57.30
radians/second	revolutions/min	9.549
radians/second	revolutions/second	0.1592
radians/second/second	revolutions/min/min	573.0
radians/second/second	revolutions/min/second	9.549

(*Continued*)

TABLE A.16 Conversion Reference Data (*Continued*)

To Convert	Into	Multiply by
radians/second/second	revolutions/second/second	0.1592
revolutions	degrees	360.0
revolutions	quadrants	4.0
revolutions	radians	6.283
revolutions/min	degrees/second	6.0
revolutions/min	radians/second	0.1047
revolutions/min	revolutions/second	0.01667
revolutions/min/min	radians/second/second	1.745×10^{-3}
revolutions/min/min	revolutions/min/second	0.01667
revolutions/min/min	revolutions/second/second	2.778×10^{-4}
revolutions/second	degrees/second	360.0
revolutions/second	radians/second	6.283
revolutions/second	revolutions/min	60.0
revolutions/second/second	radians/second/second	6.283
revolutions/second/second	revolutions/min/min	3600.0
revolutions/second/second	revolutions/min/second	60.0
rod	chain (Gunters)	0.25
rod	meters	5.029
rods	feet	16.5
rods (surveyors' meas.)	yards	5.5
scruples	grains	20
seconds (angle)	degrees	2.778×10^{-4}
seconds (angle)	minutes	0.01667
seconds (angle)	quadrants	3.087×10^{-6}
seconds (angle)	radians	4.848×10^{-6}
slug	kilogram	14.59
slug	pounds	32.17
sphere	steradians	12.57
square centimeters	circular mils	1.973×10^{5}
square centimeters	sq. feet	1.076×10^{-3}
square centimeters	sq. inches	0.1550
square centimeters	sq. meters	0.0001
square centimeters	sq. miles	3.861×10^{-11}
square centimeters	sq. millimeters	100.0
square centimeters	sq. yards	1.196×10^{-4}
square feet	acres	2.296×10^{-5}
square feet	circular mils	1.833×10^{8}
square feet	sq. cm	929.0
square feet	sq. inches	144.0
square feet	sq. meters	0.09290
square feet	sq. miles	3.587×10^{-8}
square feet	sq. millimeters	9.290×10^{4}
square feet	sq. yards	0.1111

TABLE A.16 Conversion Reference Data (*Continued*)

To Convert	Into	Multiply by
square inches	circular mils	1.273×10^6
square inches	sq. cm	6.452
square inches	sq. feet	6.944×10^{-3}
square inches	sq. millimeters	645.2
square inches	sq. mils	10^6
square inches	sq. yards	7.716×10^{-4}
square kilometers	acres	247.1
square kilometers	sq. cm	10^{10}
square kilometers	sq. ft	10.76×10^6
square kilometers	sq. inches	1.550×10^9
square kilometers	sq. meters	10^6
square kilometers	sq. miles	0.3861
square kilometers	sq. yards	1.196×10^6
square meters	acres	2.471×10^{-4}
square meters	sq. cm	10^4
square meters	sq. feet	10.76
square meters	sq. inches	1550
square meters	sq. miles	3.861×10^{-7}
square meters	sq. millimeters	10^6
square meters	sq. yards	1.196
square miles	acres	640.0
square miles	sq. feet	27.88×10^6
square miles	sq. km	2.590
square miles	sq. meters	2.590×10^6
square miles	sq. yards	3.098×10^6
square millimeters	circular mils	1973
square millimeters	sq. cm	0.01
square millimeters	sq. feet	1.076×10^{-5}
square millimeters	sq. inches	1.550×10^{-3}
square mils	circular mils	1.273
square mils	sq. cm	6.452×10^{-6}
square mils	sq. inches	10^{-6}
square yards	acres	2.066×10^{-4}
square yards	sq. cm	8361
square yards	sq. feet	9.0
square yards	sq. inches	1296
square yards	sq. meters	0.8361
square yards	sq. miles	3.228×10^{-7}
square yards	sq. millimeters	8.361×10^5
temperature (°C)+273	absolute temperature (°C)	1.0
temperature (°C)+17.78	temperature (°F)	1.8

(*Continued*)

TABLE A.16 Conversion Reference Data (*Continued*)

To Convert	Into	Multiply by
temperature (°F)+460	absolute temperature (°F)	1.0
temperature (°F)−32	temperature (°C)	5/9
tons (long)	kilograms	1016
tons (long)	pounds	2240
tons (long)	tons (short)	1.120
tons (metric)	kilograms	1000
tons (metric)	pounds	2205
tons (short)	kilograms	907.1848
tons (short)	ounces	32,000
tons (short)	ounces (troy)	29,166.66
tons (short)	pounds	2000
tons (short)	pounds (troy)	2430.56
tons (short)	tons (long)	0.89287
tons (short)	tons (metric)	0.9078
tons (short)/sq. ft	kg/sq. meter	9765
tons (short)/sq. ft	pounds/sq. in	2000
tons of water/24 hour	pounds of water/hour	83.333
tons of water/24 hour	gallons/min	0.16643
tons of water/24 hour	cubic ft/hour	1.3349
volt (absolute)	statvolts	0.003336
volt/inch	volt/cm	0.39370
watt-hours	Btu	3.413
watt-hours	ergs	3.60×10^{10}
watt-hours	foot-pounds	2656
watt-hours	gram-calories	859.85
watt-hours	horsepower-hour	1.341×10^{-3}
watt-hours	kilogram-calories	0.8605
watt-hours	kilogram-meters	367.2
watt-hours	kilowatt-hour	0.001
watt (international)	watt (absolute)	1.0002
watts	Btu/hour	3.4129
watts	Btu/min	0.05688
watts	ergs/second	10^7
watts	foot-lb/min	44.27
watts	foot-lb/second	0.7378
watts	horsepower	1.341×10^{-3}
watts	horsepower (metric)	1.360×10^{-3}
watts	kg-calories/min	0.01433
watts	kilowatts	0.001
watts (Abs.)	Btu (mean)/min	0.056884
watts (Abs.)	joules/second	1
webers	maxwells	10^8
webers	kilolines	10^5

TABLE A.16 Conversion Reference Data (*Continued*)

To Convert	Into	Multiply by
webers/sq. in	gausses	1.550×10^7
webers/sq. in	lines/sq. in	10^8
webers/sq. in	webers/sq. cm	0.1550
webers/sq. in	webers/sq. meter	1550
webers/sq. meter	gausses	10^4
webers/sq. meter	lines/sq. in	6.452×10^4
webers/sq. meter	webers/sq. cm	10^{-4}
webers/sq. meter	webers/sq. in	6.452×10^{-4}
Yards	centimeters	91.44
Yards	kilometers	9.144×10^{-4}
Yards	meters	0.9144
Yards	miles (naut.)	4.934×10^{-4}
Yards	miles (stat.)	5.682×10^{-4}
Yards	millimeters	914.4

REFERENCE

Whitaker, Jerry C. (ed.), *The Electronics Handbook*, 2nd ed., CRC Press, Boca Raton, FL, 2005.

APPENDIX B
THE ELECTROMAGNETIC SPECTRUM

John Norgard

B.1 INTRODUCTION[1]

The electromagnetic (EM) spectrum consists of all forms of EM radiation—EM waves (radiant energy) propagating through space, from dc to light to gamma rays. The EM spectrum can be arranged in order of frequency and/or wavelength into a number of regions, usually wide in extent, within which the EM waves have some specified common characteristics, such as characteristics relating to the production or detection of the radiation. A common example is the spectrum of the radiant energy in white light, as dispersed by a prism, to produce a "rainbow" of its constituent colors. Specific frequency ranges are often called *bands*; several contiguous frequency bands are usually called *spectrums*; and subfrequency ranges within a band are sometimes called *segments*.

The EM spectrum can be displayed as a function of frequency (or wavelength). In air, frequency and wavelength are inversely proportional, $f = c/\lambda$ (where $c \approx 3 \times 10^8$ m/s, the speed of light in a vacuum). The MKS unit of frequency is the Hertz and the MKS unit of wavelength is the meter. Frequency is also measured in the following subunits:

- Kilohertz, 1 kHz = 10^3 Hz
- Megahertz, 1 MHz = 10^6 Hz
- Gigahertz, 1 GHz = 10^9 Hz
- Terahertz, 1 THz = 10^{12} Hz
- Petahertz, 1 PHz = 10^{15} Hz
- Exahertz, 1 EHz = 10^{18} Hz

Or for very high frequencies, *electron volts*, 1 eV ~ 2.41×10^{14} Hz. Wavelength is also measured in the following subunits:

- Centimeters, 1 cm = 10^{-2} meter
- Millimeters, 1 mm = 10^{-3} meter
- Micrometers, 1 μm = 10^{-6} meter (microns)

---

[1]This chapter is adapted from: Norgard, John, "The Electromagnetic Spectrum," in Standard Handbook of Video and Television Engineering, Jerry C. Whitaker and K. Blair Benson (eds.), McGraw-Hill, New York, NY, 2000. Used with permission.

- Nanometers, 1 nm = 10^{-9} meter
- Ångstroms, 1 Å = 10^{-10} meter
- Picometers, 1 pm = 10^{-12} meter
- Femtometers, 1 fm = 10^{-15} meter
- Attometers, 1 am = 10^{-18} meter

B.2 SPECTRAL SUBREGIONS

For convenience, the overall EM spectrum can be divided into three main subregions:

- Optical spectrum
- DC to light spectrum
- Light to gamma ray spectrum

These main subregions of the EM spectrum are next discussed. Note that the boundaries between some of the spectral regions are somewhat arbitrary. Certain spectral bands have no sharp edges and merge into each other, while other spectral segments overlap each other slightly.

B.2.1 Optical Spectrum

The optical spectrum is the "middle" frequency/wavelength region of the EM spectrum. It is defined here as the visible and near-visible regions of the EM spectrum and includes:

- The *infrared (IR)* band, circa 300–0.7 μm (circa 1–429 THz)
- The *visible light* band, 0.7–0.4 μm (429–750 THz)
- The *ultraviolet (UV)* band, 0.4 μm–circa 10 nm (750 THz–circa 30 PHz), approximately 100 eV

These regions of the EM spectrum are usually described in terms of their wavelengths.

Atomic and molecular radiation produce radiant light energy. Molecular radiation and radiation from hot bodies produce EM waves in the IR band. Atomic radiation (outer shell electrons) and radiation from arcs and sparks produce EM waves in the UV band.

B.2.1.1 Visible Light Band.
In the "middle" of the optical spectrum is the visible light band, extending approximately from 0.4 μm (violet) up to 0.7 μm (red); i.e., from 750 THz (violet) down to 429 THz (red). EM radiation in this region of the EM spectrum, when entering the eye, gives rise to visual sensations (colors), according to the spectral response of the eye, which responds only to radiant energy in the visible light band extending from the extreme long wavelength edge of red to the extreme short wavelength edge of violet. The spectral response of the eye is sometimes quoted as extending from 0.38 μm (violet) up to 0.75 or 0.78 μm (red); i.e., from 789 THz down to 400 or 385 THz. This visible light band is further subdivided into the various colors of the rainbow, in decreasing wavelength/increasing frequency as follows:

- Red, a primary color, peak intensity at 700.0 nm (429 THz)
- Orange
- Yellow
- Green, a primary color, peak intensity at 546.1 nm (549 THz)
- Cyan
- Blue, a primary color, peak intensity at 435.8 nm (688 THz)
- Indigo
- Violet

B.2.1.2 IR Band. The IR band is the region of the EM spectrum lying immediately below the visible light band. The IR band consists of EM radiation with wavelengths extending between the longest visible red (circa 0.7 μm) and the shortest microwaves (300 μm–1 mm); i.e., from circa 429 THz down to 1 THz–300 GHz.

The IR band is further subdivided into the "near" (shortwave), "intermediate" (midwave), and "far" (longwave) IR segments as follows[2]:

- *Near* IR segment, 0.7 μm up to 3 μm (429 THz down to 100 THz)
- *Intermediate* IR segment, 3 μm up to 7 μm (100 THz down to 42.9 THz)
- *Far* IR segment, 7 μm up to 300 μm (42.9 THz down to 1 THz)
- Submillimeter band, 100 mm up to 1 mm (3 THz down to 300 GHz). Note that the submillimeter region of wavelengths is sometimes included in the very far region of the IR band.

EM radiation is produced by oscillating and rotating molecules and atoms. Therefore, all objects at temperatures above absolute zero emit EM radiation by virtue of their thermal motion (warmth) alone. Objects near room temperature emit most of their radiation in the IR band. However, even relatively cool objects emit some IR radiation; hot objects, such as incandescent filaments, emit strong IR radiation.

IR radiation is sometimes incorrectly called "radiant heat" because warm bodies emit IR radiation and bodies that absorb IR radiation are warmed. However, IR radiation is not itself "heat." This radiant energy is called "black body" radiation. Such waves are emitted by all material objects. For example, the background cosmic radiation (2.7 K) emits microwaves; room temperature objects (293 K) emit IR rays; the Sun (6000 K) emits yellow light; the Solar Corona (1 million K) emits X rays.

IR astronomy uses the 1 μm to 1 mm part of the IR band to study celestial objects by their IR emissions. IR detectors are used in night vision systems, intruder alarm systems, weather forecasting, and missile guidance systems. IR photography uses multilayered color film, with an IR sensitive emulsion in the wavelengths between 700 and 900 nm, for medical and forensic applications, and for aerial surveying.

B.2.1.3 UV Band. The UV band is the region of the EM spectrum lying immediately above the visible light band. The UV band consists of EM radiation with wavelengths extending between the shortest visible violet (circa 0.4 mm) and the longest X rays (circa 10 nm); i.e., from 750 THz—approximately 3 eV—up to circa 30 PHz—approximately 100 eV.[3]

The UV band is further subdivided into the "near" and the "far" UV segments as follows:

- *Near* UV segment, circa 0.4 μm down to 100 nm (circa 750 THz up to 3 PHz, approximately 3 eV up to 10 eV)
- *Far* UV segment, 100 nm down to circa 10 nm (3 PHz up to circa 30 PHz, approximately 10 eV up to 100 eV)

The far UV band is also referred to as the *vacuum UV band*, since air is opaque to all UV radiation in this region.

UV radiation is produced by electron transitions in atoms and molecules, as in a mercury discharge lamp. Radiation in the UV range is easily detected and can cause florescence in some substances, and can produce photographic and ionizing effects.

In UV astronomy, the emissions of celestial bodies in the wavelength band between 50 and 320 nm are detected and analyzed to study the heavens. The hottest stars emit most of their radiation in the UV band.

---

[2]Some reference texts use 2.5 mm (120 THz) as the breakpoint between the near and the intermediate IR bands, and 10 mm (30 THz) as the breakpoint between the intermediate and the far IR bands. Also, 15 mm (20 Thz) is sometimes considered as the long wavelength end of the far IR band.

[3]Some references use 4, 5, or 6 nm as the upper edge of the UV band.

B.2.2 DC to Light

Below the IR band are the lower frequency (longer wavelength) regions of the EM spectrum, subdivided generally into the following spectral bands (by frequency/wavelength):

- *Microwave* band, 300 GHz down to 300 MHz (1 mm up to 1 meter). Some reference works define the lower edge of the microwave spectrum at 1 GHz.
- *Radio frequency* (RF) band, 300 MHz down to 10 kHz (1 meter up to 30 km)
- *Power/telephony* band, 10 kHz down to dc (30 km up to ∞)

These regions of the EM spectrum are usually described in terms of their frequencies.

Radiations whose wavelengths are of the order of millimeters and centimeters are called microwaves, and those still longer are called RF waves (or *Hertzian waves*).

Radiation from electronic devices produces EM waves in both the microwave and RF bands. Power frequency energy is generated by rotating machinery. Direct current (dc) is produced by batteries or rectified alternating current (ac).

B.2.2.1 Microwave Band.
The microwave band is the region of wavelengths lying between the far IR/submillimeter region and the conventional RF region. The boundaries of the microwave band have not been definitely fixed, but it is commonly regarded as the region of the EM spectrum extending from about 1 mm up to 1 meter in wavelengths; i.e., from 300 GHz down to 300 MHz. The microwave band is further subdivided into the following segments:

- *Millimeter* waves, 300 GHz down to 30 GHz (1 mm up to 1 cm); the Extremely High Frequency band. (Some references consider the top edge of the millimeter region to stop at 100 GHz.)
- *Centimeter* waves, 30 GHz down to 3 GHz (1 cm up to 10 cm); the Super High Frequency band.

The microwave band usually includes the Ultra High Frequency band from 3 GHz down to 300 MHz (from 10 cm up to 1 meter). Microwaves are used in radar, space communication, terrestrial links spanning moderate distances, as radio carrier waves in television broadcasting, for mechanical heating, and cooking in microwave ovens.

B.2.2.2 Radio Frequency (RF) Band.
The RF range of the EM spectrum is the wavelength band suitable for utilization in radio communications extending from 10 kHz up to 300 MHz (from 30 km down to 1 meter). (Some references consider the RF band as extending from 10 kHz to 300 GHz, with the microwave band as a subset of the RF band from 300 MHz to 300 GHz.)

Some of the radio waves in this band serve as the carriers of low-frequency audio signals; other radio waves are modulated by digital information. The *amplitude modulated* (AM) broadcasting band uses waves with frequencies between 535 and 1705 kHz; the *frequency modulated* (FM) broadcasting band uses waves with frequencies between 88 and 108 MHz.

In the United States, the Federal Communications Commission (FCC) is responsible for assigning a range of frequencies to specific services. The International Telecommunications Union (ITU) coordinates frequency band allocation and cooperation on a worldwide basis.

Radio astronomy uses radio telescopes to receive and study radio waves naturally emitted by objects in space. Radio waves are emitted from hot gases (*thermal radiation*), from charged particles spiraling in magnetic fields (*synchrotron radiation*), and from excited atoms and molecules in space (*spectral lines*), such as the 21 cm line emitted by hydrogen gas.

B.2.2.3 Power Frequency (PF)/Telephone Band.
The PF range of the EM spectrum is the wavelength band suitable for generating, transmitting, and consuming low frequency power, extending from 10 kHz down to dc (zero frequency); i.e., from 30 km up in wavelength. In the United States, most power is generated at 60 Hz (some military and computer applications use 400 Hz); in other countries, including Europe, power is generated at 50 Hz.

B.2.3 Frequency Band Designations

The combined microwave, RF (Hertzian Waves), and power/telephone spectra are subdivided into the specific bands given in Table B.1, which lists the international radio frequency band designations and the numerical designations. Note that the band designated (12) has no commonly used name or abbreviation.

The radar band often is considered to extend from the middle of the High Frequency (7) band to the end of the EHF (11) band. The current U.S. Tri-Service radar band designations are listed in Table B.2. An alternate and more detailed subdivision of the UHF (9), SHF (10), and EHF (11) bands is given in Table B.3. Several other frequency bands of interest (not exclusive) are listed in Tables B.4, B.5, and B.6.

TABLE B.1 Frequency Band Designations

Description	Band Designation	Frequency	Wavelength
Extremely Low Frequency	ELF (1) Band	3 Hz up to 30 Hz	100 Mm down to 10 Mm
Super Low Frequency	SLF (2) Band	30 Hz up to 300 Hz	10 Mm down to 1 Mm
Ultra Low Frequency	ULF (3) Band	300 Hz up to 3 kHz	1 Mm down to 100 km
Very Low Frequency	VLF (4) Band	3 kHz up to 30 kHz	100 km down to 10 km
Low Frequency	LF (5) Band	30 kHz up to 300 kHz	10 km down to 1 km
Medium Frequency	MF (6) Band	300 kHz up to 3 MHz	1 km down to 100 m
High Frequency	HF (7) Band	3 MHz up to 30 MHz	100 m down to 10 m
Very High Frequency	VHF (8) Band	30 MHz up to 300 MHz	10 m down to 1 m
Ultra High Frequency	UHF (9) Band	300 MHz up to 3 GHz	1 m down to 10 cm
Super High Frequency	SHF (10) Band	3 GHz up to 30 GHz	10 cm down to 1 cm
Extremely High Frequency	EHF (11) Band	30 GHz up to 300 GHz	1 cm down to 1 mm
—	(12) Band	300 GHz up to 3 THz	1 mm down to 100 mm

B.2.4 Light to Gamma Rays

Above the UV spectrum are the higher frequency (shorter wavelength) regions of the EM spectrum, subdivided generally into the following spectral bands (by frequency/wavelength):

- X ray band, approximately 10 eV up to 1 MeV (circa 10 nm down to circa 1 pm), circa 3 PHz up to circa 300 EHz
- *Gamma ray* band, approximately 1 keV up to ∞ (circa 300 pm down to 0 meter), circa 1 EHz up to ∞

These regions of the EM spectrum are usually described in terms of their photon energies in electron volts. Note that the bottom of the gamma ray band overlaps the top of the X ray band.

It should be pointed out that *cosmic* "rays" (from astronomical sources) are not EM waves (rays) and, therefore, are not part of the EM spectrum. Cosmic rays are high energy charged particles (electrons, protons, and ions) of extraterrestrial origin moving through space, which may have energies as high as 10^{20} eV. Cosmic rays have been traced to cataclysmic astrophysical/cosmological events, such as exploding stars and black holes. Cosmic rays are emitted by supernova remnants, pulsars, quasars, and radio galaxies. Cosmic rays that collide with molecules in the Earth's upper atmosphere produce secondary cosmic rays and gamma rays of high energy that also contribute to natural background radiation. These gamma rays are sometimes called cosmic or *secondary* gamma rays. Cosmic rays are a useful source of high-energy particles for certain scientific experiments.

TABLE B.2 Radar Band Designations

Band	Frequency	Wavelength
A Band	0 Hz up to 250 MHz	∞ down to 1.2 m
B Band	250 MHz up to 500 MHz	1.2 m down to 60 cm
C Band	500 MHz up to 1 GHz	60 cm down to 30 cm
D Band	1 GHz up to 2 GHz	30 cm down to 15 cm
E Band	2 GHz up to 3 GHz	15 cm down to 10 cm
F Band	3 GHz up to 4 GHz	10 cm down to 7.5 cm
G Band	4 GHz up to 6 GHz	7.5 cm down to 5 cm
H Band	6 GHz up to 8 GHz	5 cm down to 3.75 cm
I Band	8 GHz up to 10 GHz	3.75 cm down to 3 cm
J Band	10 GHz up to 20 GHz	3 cm down to 1.5 cm
K Band	20 GHz up to 40 GHz	1.5 cm down to 7.5 mm
L Band	40 GHz up to 60 GHz	7.5 mm down to 5 mm
M Band	60 GHz up to 100 GHz	5 mm down to 3 mm
N Band	100 GHz up to 200 GHz	3 mm down to 1.5 mm
O Band	200 GHz up to 300 GHz	1.5 mm down to 1 mm

TABLE B.3 Detail of UHF, SHF, and EHF Band Designations

Band	Frequency	Wavelength
L Band	1.12 GHz up to 1.7 GHz	26.8 cm down to 17.6 cm
LS Band	1.7 GHz up to 2.6 GHz	17.6 cm down to 11.5 cm
S Band	2.6 GHz up to 3.95 GHz	11.5 cm down to 7.59 cm
C(G) Band	3.95 GHz up to 5.85 GHz	7.59 cm down to 5.13 cm
XN(J, XC) Band	5.85 GHz up to 8.2 GHz	5.13 cm down to 3.66 cm
XB(H, BL) Band	7.05 GHz up to 10 GHz	4.26 cm down to 3 cm
X Band	8.2 GHz up to 12.4 GHz	3.66 cm down to 2.42 cm
Ku(P) Band	12.4 GHz up to 18 GHz	2.42 cm down to 1.67 cm
K Band	18 GHz up to 26.5 GHz	1.67 cm down to 1.13 cm
V(R, Ka) Band	26.5 GHz up to 40 GHz	1.13 cm down to 7.5 mm
Q(V) Band	33 GHz up to 50 GHz	9.09 mm down to 6 mm
M(W) Band	50 GHz up to 75 GHz	6 mm down to 4 mm
E(Y) Band	60 GHz up to 90 GHz	5 mm down to 3.33 mm
F(N) Band	90 GHz up to 140 GHz	3.33 mm down to 2.14 mm
G(A)	140 GHz p to 220 GHz	2.14 mm down to 1.36 mm
R Band	220 GHz up to 325 GHz	1.36 mm down to 0.923 mm

TABLE B.4 Low Frequency Bands of Interest

Band	Frequency
Subsonic band	0–10 Hz
Audio band	10 Hz–10 kHz
Ultrasonic band	10 kHz and up

TABLE B.5 Applications of Interest in the RF Band

Band	Frequency
Longwave broadcasting band	150–290 kHz
AM broadcasting band	535–1705 kHz (1.640 MHz), 107 channels, 10 kHz separation
International broadcasting band	3–30 MHz
Shortwave broadcasting band	5.95–26.1 MHz (eight bands)
VHF TV (Channels 2–4)	54–72 MHz
VHF TV (Channels 5–6)	76–88 MHz
FM broadcasting band	88–108 MHz
VHF TV (Channels 7–13)	174–216 MHz
UHF TV (Channels 14–69)	512–806 MHz[1]

Notes:
[1] Upper TV channels in this band have been reassigned in the US; additional reassignments were under consideration at this writing.

Radiation from atomic inner shell excitations produces EM waves in the X ray band. Radiation from naturally radioactive nuclei produces EM waves in the gamma ray band.

B.2.4.1 X Ray Band.
The X ray band is further subdivided into the following segments:

- *Soft* X rays, approximately 10 eV up to 10 keV (circa 10 nm down to 100 pm), circa 3 PHz up to 3 EHz
- *Hard* X rays, approximately 10 keV up to 1 Mev (100 pm down to circa 1 pm), 3 EHz up to circa 300 EHz

Because the physical nature of these rays was at first unknown, this radiation was called "X rays." The designation continues to this day. The more powerful X rays are called hard X rays and are of high frequencies and, therefore, are more energetic; less powerful X rays are called soft X rays and have lower energies.

X rays are produced by transitions of electrons in the inner levels of excited atoms or by rapid deceleration of charged particles (Brehmsstrahlung or breaking radiation). An important source of X rays is synchrotron radiation. X rays can also be produced when high energy electrons from a heated filament cathode strike the surface of a target anode (usually tungsten) between which a high alternating voltage (approximately 100 kV) is applied.

X rays are a highly penetrating form of EM radiation and applications of X rays are based on their short wavelengths and their ability to easily pass through matter. X rays are very useful in crystallography for determining crystalline structure and in medicine for photographing the body. Because different parts of the body absorb X rays to a different extent, X rays passing through the body provide a visual image of its interior structure when striking a photographic plate. X rays are dangerous and can destroy living tissue. They can also cause severe skin burns. X rays are useful in the diagnosis and nondestructive testing of products for defects.

B.2.4.2 Gamma Ray Band.
The gamma ray band is subdivided into the following segments:

- *Primary* gamma rays, approximately 1 keV up to 1 MeV (circa 300 pm down to 300 fm), circa 1 EHz up to 1000 EHz.
- *Secondary* gamma rays, approximately 1 MeV up to ∞ (300 fm down to 0 meter), 1000 EHz up to ∞.

Secondary gamma rays are created from collisions of high energy cosmic rays with particles in the Earth's upper atmosphere.

TABLE B.6 Applications of Interest in the Microwave Band (Note: applications subject to change.)

Application	Frequency
Aero Navigation	0.96–1.215 GHz
GPS Down Link	1.2276 GHz
Military COM/Radar	1.35–1.40 GHz
Miscellaneous COM/Radar	1.40–1.71 GHz
L-Band Telemetry	1.435–1.535 GHz
GPS Down Link	1.57 GHz
Military COM (Troposcatter/Telemetry)	1.71–1.85 GHz
Commercial COM & Private LOS	1.85–2.20 GHz
Microwave Ovens	2.45 GHz
Commercial COM/Radar	2.45–2.69 GHz
Military Radar (Airport Surveillance)	2.70–2.90 GHz
Maritime Navigation Radar	2.90–3.10 GHz
Miscellaneous Radars	2.90–3.70 GHz
Commercial C-Band SAT COM Down Link	3.70–4.20 GHz
Radar Altimeter	4.20–4.40 GHz
Military COM (Troposcatter)	4.40–4.99 GHz
Commercial Microwave Landing System	5.00–5.25 GHz
Miscellaneous Radars	5.25–5.925 GHz
C-Band Weather Radar	5.35–5.47 GHz
Commercial C-Band SAT COM Up Link	5.925–6.425 GHz
Commercial COM	6.425–7.125 GHz
Military LOS COM	7.125–7.25 GHz
Military SAT COM Down Link	7.25–7.75 GHz
Military LOS COM	7.75–7.9 GHz
Military SAT COM Up Link	7.90–8.40 GHz
Miscellaneous Radars	8.50–10.55 GHz
Precision Approach Radar	9.00–9.20 GHz
X-Band Weather Radar (& Maritime Navigation Radar)	9.30–9.50 GHz
Police Radar	10.525 GHz
Commercial Mobile COM (LOS & ENG)	10.55–10.68 GHz
Common Carrier LOS COM	10.70–11.70 GHz
Commercial COM	10.70–13.25 GHz
Commercial Ku-Band SAT COM Down Link	11.70–12.20 GHz
DBS Down Link & Private LOS COM	12.20–12.70 GHz
ENG & LOS COM	12.75–13.25 GHz
Miscellaneous Radars & SAT COM	13.25–14.00 GHz
Commercial Ku-Band SAT COM Up Link	14.00–14.50 GHz
Military COM (LOS, Mobile, & Tactical)	14.50–15.35 GHz
Aero Navigation	15.40–15.70 GHz
Miscellaneous Radars	15.70–17.70 GHz
DBS Up Link	17.30–17.80 GHz
Common Carrier LOS COM	17.70–19.70 GHz
Commercial COM (SAT COM & LOS)	17.70–20.20 GHz
Private LOS COM	18.36–19.04 GHz
Military SAT COM	20.20–21.20 GHz
Miscellaneous COM	21.20–24.00 GHz
Police Radar	24.15 GHz
Navigation Radar	24.25–25.25 GHz
Military COM	25.25–27.50 GHz
Commercial COM	27.50–30.00 GHz
Military SAT COM	30.00–31.00 GHz
Commercial COM	31.00–31.20 GHz

The primary gamma rays are further subdivided into the following segments:

- *Soft* gamma rays, approximately 1 keV up to circa 300 keV (circa 300 pm down to circa 3 pm), circa 1 EHz up to circa 100 EHz.
- *Hard* gamma rays, approximately 300 keV up to 1 MeV (circa 3 pm down to 300 fm), circa 100 EHz up to 1000 EHz.

Gamma rays are essentially very energetic X rays. The distinction between the two is based on their origin. X rays are emitted during atomic processes involving energetic electrons; gamma rays are emitted by excited nuclei or other processes involving subatomic particles.

Gamma rays are emitted by the nucleus of radioactive material during the process of natural radioactive decay as a result of transitions from high energy excited states to low-energy states in atomic nuclei. Cobalt 90 is a common gamma ray source (with a half-life of 5.26 years). Gamma rays are also produced by the interaction of high energy electrons with matter. "Cosmic" gamma rays cannot penetrate the Earth's atmosphere.

Applications of gamma rays are found both in medicine and in industry. In medicine, gamma rays are used for cancer treatment and diagnoses. Gamma ray emitting radioisotopes are used as tracers. In industry, gamma rays are used in the inspection of castings, seams, and welds.

B.3 FREQUENCY ASSIGNMENT AND ALLOCATIONS

The Communications Act of 1934, as amended, provides for the regulation of interstate and foreign commerce in communication by wire or radio in the United States.[4] This Act is printed in Title 47 of the U.S. Code, beginning with Section 151. The primary treaties and other international agreements in force relating to radiocommunication and to which the United States is a party are as follows:

- The International Telecommunication Convention, signed at Nairobi on November 6, 1982. The United States deposited its instrument of ratification on January 7, 1986.
- The Radio Regulations annexed to the International Telecommunication Convention, signed at Geneva on December 6, 1979 and entered into force with respect to the United States on January 1, 1982.
- The United States-Canada Agreement relating to the Coordination and Use of Radio Frequencies above 30 MHz, effected by an exchange of notes at Ottawa on October 24, 1962, with subsequent revisions and amendments.

B.3.1 The International Telecommunication Union

The International Telecommunication Union (ITU) is the international body responsible for international frequency allocations, worldwide telecommunications standards, and telecommunication development activities. The broad functions of the ITU are the regulation, coordination, and development of international telecommunications. The United States is an active member of the ITU and its work is considered critical to the interest of the United States. The ITU is the oldest of the intergovernmental organizations that have become specialized agencies within the United Nations.

When, at the end of the nineteenth century, wireless (radiotelegraphy) became practicable, it was seen at once to be an invaluable complement of telegraphy by wire and cable, since radio alone could

---

[4]This section is based on: *NTIA Manual of Regulations and Procedures for Federal Radio Frequency Management*, May 2013 Edition, May 2014 Revision, National Telecommunications and Information Administration, U.S. Department of Commerce, Washington, D.C., 2014.

provide telecommunication between land and ships at sea. The first International Radiotelegraph Convention was signed in Berlin in 1906 by 29 countries. Nearly two decades later, in 1924 and 1925, at Conferences in Paris, the International Telephone Consultative Committee (CCIF) and the International Telegraph Consultative Committee (CCIT) were established. This was followed by the 1927 International Radiotelegraph Conference in Washington, D.C. in 1927, which was attended by 80 countries. It was a historical milestone in the development of radio because it was at this Conference that the Table of Frequency Allocations was first devised and the International Radio Consultative Committee (CCIR) was formed.

In 1932, two Plenipotentiary Conferences were held in Madrid: a Telegraph and Telephone Conference and a Radiotelegraph Conference. On that occasion, the two existing Conventions were amalgamated in a single International Telecommunication Convention, and the countries that signed and acceded to it renamed the Union the International Telecommunication Union (ITU) to indicate its broader scope. Four sets of Regulations were annexed to the Convention: telegraph, telephone, radio, and additional radio regulations.

A Plenipotentiary Conference met in Atlantic City, N.J., in 1947 to revise the Madrid Convention. It introduced important changes in the organization of the Union. The International Frequency Registration Board (IFRB) and the Administrative Council were created. Also, the ITU became the specialized agency within the United Nations in the sphere of telecommunications, and its headquarters was transferred from Berne to Geneva.

The Union remained essentially unchanged until 1992, when an Additional Plenipotentiary Conference in Geneva extensively restructured the ITU. The Nice Constitution and Convention of 1989, which had not been ratified, was used as the general model for the 1992 Conference. The CCIR, IFRB, and World Administrative Radio Conference (WARC) functions were incorporated into the Radiocommunication Sector (ITU-R); the CCITT and Telecommunication Conference functions were incorporated into the Telecommunication Standardization Sector (ITU-T); development activities were incorporated into the Telecommunication Development Sector (ITU-D); and the Secretariats were combined into one General Secretariat.

The purposes of the ITU are as follows:

- To promote the development and efficient operation of telecommunication facilities, in order to improve the efficiency of telecommunication services, their usefulness, and their general availability to the public.
- Promote and offer technical assistance to developing countries in the field of telecommunications, to promote the mobilization of the human and financial resources needed to develop telecommunications, and to promote the extension of the benefits of new telecommunications technologies to people everywhere.
- Promote, at the international level, the adoption of a broader approach to the issues of telecommunications in the global information economy and society.

The principal facilities of the ITU are in Geneva.

B.3.2 The Federal Communications Commission

Congress, through adoption of the Communications Act of 1934, created the Federal Communications Commission (FCC) as an independent regulatory agency. Section I of the Act specifies that the FCC was created, "For the purpose of regulation of interstate and foreign commerce in communication by wire and radio so as to make available, so far as possible, to all the people of the United States a rapid, efficient, nationwide, and worldwide wire and radio communication service with adequate facilities at reasonable charges, for the purpose of the national defense, for the purpose of promoting the safety of life and property through the use of wire and radio communication, and for the purpose of securing a more effective execution of this policy by centralizing authority heretofore granted by law to several agencies and by granting additional authority with respect to interstate and foreign commerce in wire and radio communication."

The staff of the FCC performs day-to-day functions of the agency, including license and application processing, drafting of rulemaking items, enforcing rules and regulations, and formulating policy.

B.3.3 National Table of Frequency Allocations

The National Table of Frequency Allocations is comprised of the U.S. Government Table of Frequency Allocations and the FCC Table of Frequency Allocations. The National Table indicates the normal national frequency allocation planning and the degree of conformity with the ITU table. When required in the national interest and consistent with national rights, as well as obligations undertaken by the United States to other countries that may be affected, additional uses of frequencies in any band may be authorized to meet service needs other than those provided for in the National Table. See: http://www.ntia.doc.gov/page/2011/manual-regulations-and-procedures-federal-radio-frequency-management-redbook

BIBLIOGRAPHY

Collocott, T. C., A. B. Dobson, and W. R. Chambers (eds.), *Dictionary of Science & Technology.*

Handbook of Physics, McGraw-Hill, New York, NY, 1958.

Judd, D. B., and G. Wyszecki, *Color in Business, Science and Industry*, 3rd ed., John Wiley and Sons, New York, NY.

Kaufman, Ed, *IES Illumination Handbook*, Illumination Engineering Society.

Lapedes, D. N. (ed.), *The McGraw-Hill Encyclopedia of Science & Technology*, 2nd ed., McGraw-Hill, New York, NY.

Norgard, J., "Electromagnetic Spectrum," *NAB Engineering Handbook*, 9th ed., J. C. Whitaker (ed.), National Association of Broadcasters, Washington, D.C., 1999.

Norgard, J., "Electromagnetic Spectrum," *The Electronics Handbook*, J. C. Whitaker (ed.), CRC Press, Boca Raton, FL, 1996.

NTIA Manual of Regulations and Procedures for Federal Radio Frequency Management, May 2013 Edition, May 2014 Revision, National Telecommunications and Information Administration, U.S. Department of Commerce, Washington, D.C., 2014.

Stemson, A., *Photometry and Radiometry for Engineers*, John Wiley and Sons, New York, N.Y.

The Cambridge Encyclopedia, Cambridge University Press, 1990.

The Columbia Encyclopedia, Columbia University Press, 1993.

Webster's New World Encyclopedia, Prentice Hall, 1992.

Wyszecki, G., and W. S. Stiles, *Color Science, Concepts and Methods, Quantitative Data and Formulae*, 2nd ed., John Wiley and Sons, New York, NY.

APPENDIX C
STANDARDS ORGANIZATIONS AND SI UNITS

Jerry C. Whitaker, Editor-in-Chief

Robert Mancini
Mancini Enterprises, Whittier, CA

C.1 INTRODUCTION

Standardization usually starts within a company as a way to reduce costs associated with parts stocking, design drawings, training, and retraining of personnel. The next level might be a cooperative agreement between firms making similar equipment to use standardized dimensions, parts, and components. Competition, trade secrets, and the NIH factor ("not invented here") often generate an atmosphere that prevents such an understanding. Enter the professional engineering society, which offers a forum for discussion between users and engineers while downplaying the commercial and business aspects.

To those outside the standardization process, the wheels of progress may appear to turn very slowly. This is the inevitable result of considering all sides of an issue. Work on any standard usually begins with one or more proposals from one or more organizations. To the proponent, their submission is usually believed to be complete and sufficient to address all observed needs. To a potential user of the standard, however, another view may exist. When multiple proposals are offered to address a single need (or collection of needs) the process can become quite involved, and time-consuming.

While nearly everyone involved in standards work would like to see projects move swiftly through the process, they realize that no single organization or group has all the answers to a particular problem—or has even thought of all the questions. A great deal of creativity emerges from the competition of ideas. Finding solutions to complex problems takes time and requires participants to occasionally give up on their favored approach and agree that someone else's approach is better. It is this focus on developing the best ideas that makes the process work.

It is clear that no process involving humans is perfect, and critics can always find examples of standardization efforts that failed to meet the requirements of the user and were therefore never implemented, or implemented and then quickly faded away. This situation may be the result of rushing the process and not considering all of the sides of a particular issue. It may also be the result of timing. Standardization work, like most any other product, can be adversely impacted by timing. A standard may be developed and finalized too early in the technology lifecycle. If so, it may be outdated by the time it is issued because the underlying technology has continued to move forward, leaving it of little value in the marketplace. On the other hand, a standard developed too late in the process may remain unused because a proprietary solution reached the market first and has become a de facto standard.

It is difficult to address timing when it comes to standardization work. Technologists do their best to issue standards at the point the underlying technology is stable and the user base wants the product. As a practical matter, each standards organization offers its work to the marketplace. Many times they are successful; sometimes they are not. Taken on the whole, however, the work advances the state of the art.

C.2 THE STANDARDS DEVELOPMENT ORGANIZATION

Membership in a Standards Development Organization (SDO) may be international, regional, national, industry, or "open source." Participation may be limited to one member/one country, one member/one company, or individualized membership. As illustrated in Fig. C.1, there are typically two types of SDOs:

- Fundamental standards organization, which sets core standards for specialized applications, generally from a clean sheet of paper.
- Applications standards organization, which develops standards used for specific market purposes. This type of group may use a mixture of fundamental standards from several SDOs and also create "glue" standards that tie different standards together for a particular application.

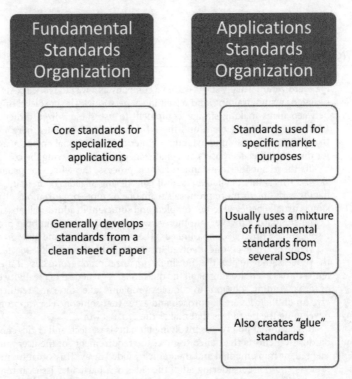

FIGURE C.1 Basic types of SDOs.

There are a number of considerations for various SDO that complicate the standards development process, including the relationship of standards to regulatory bodies and intellectual property provisions. In the latter case, SDOs nearly always have clearly defined patent policies and IPR (intellectual property rights) provisions to facilitate the exchange of ideas in an open environment. The SDO, for example, may require disclosure of IPR that might be included in a draft specification, and furthermore require the participant to agree to license the IPR on Reasonable and Non-Discriminatory (RAND) terms.

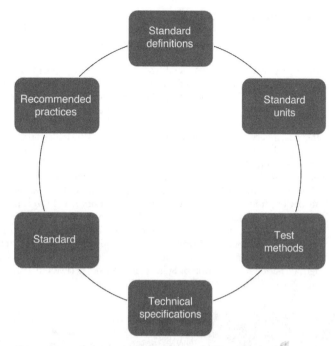

FIGURE C.2 Common types of documents developed by an SDO.

Each SDO has a particular mix of document types that it develops; Fig. C.2 shows the most common ones. Generally speaking, they can be grouped into the following broad categories:

- Standard definitions (nomenclature)
- Standard units (e.g., English or MKS/CGS)
- Test methods
- Technical specifications for interoperability
- Standard
- Recommended Practice

Standards are seldom available at no cost; however, there are exceptions. The number of workhours that go into the development of a typical standard is considerable, and while the participant's time and expenses may be "donated" by their employers, the SDO has associated administrative expenses that must be recovered.

C.2.1 Professional Society Engineering Committees

The engineering groups that collate and coordinate activities that are eventually presented to standardization bodies encourage participation from all concerned parties. Meetings are often scheduled in connection with technical conferences to promote greater participation. Other necessary meetings are usually scheduled in geographical locations of the greatest activity in the field. There are usually no charges to attend the meetings. An interest in these activities can also be served by reading the reports from these groups in the appropriate professional journals. These wheels may seem to grind exceedingly slowly at times, but the adoption of standards that may be used for decades should not be taken lightly.

C.2.2 The History of Modern Standards

Key dates in the development of modern standards are given in Fig. C.3. As shown, in 1836, the U.S. Congress authorized the Office of Weights and Measures (OWM) for the primary purpose of ensuring uniformity in custom house dealings. The Treasury Department was charged with its operation. As advancements in science and technology fueled the industrial revolution, it was apparent that standardization of hardware and test methods was necessary to promote commercial development and to compete successfully with the rest of the world. The industrial revolution in the 1830s introduced the need for interchangeable parts and hardware. Economical manufacture of transportation equipment, tools, and other machinery was possible only with mechanical standardization.

By the late 1800s, professional organizations of mechanical, electrical, chemical, and other engineers were founded with this aim in mind. The Institute of Electrical Engineers developed standards between 1890 and 1910 based on the practices of the major electrical manufacturers of the time. Such activities were not within the purview of the OWM, so there was no government involvement during this period. It took the pressures of war production in 1918 to cause the formation of the American Engineering Standards Committee (AESC) to coordinate the activities of various industry and engineering societies. This group became the American Standards Association (ASA) in 1928.

Parallel developments would occur worldwide. The International Bureau of Weights and Measures was founded in 1875, the International Electrotechnical Commission (IEC) in 1904, and the International Federation of Standardizing Bodies (ISA) in 1926. Following World War II (1946)

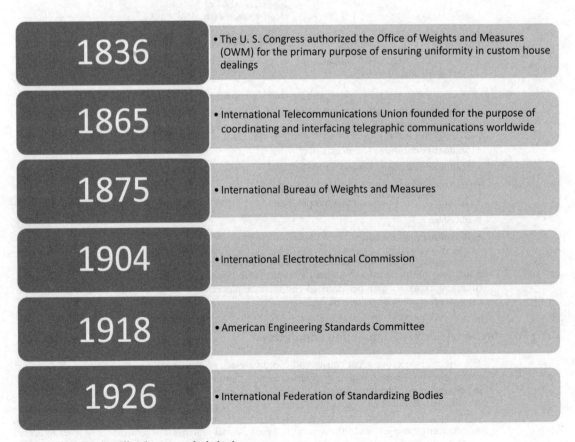

FIGURE C.3 Timeline of key dates in standards development.

this group was reorganized as the International Organization for Standardization (ISO) comprised of the ASA and the standardizing bodies of 25 other countries.

The International Telecommunications Union (ITU) was founded in 1865 for the purpose of coordinating and interfacing telegraphic communications worldwide. Today, its member nations develop regulations and voluntary recommendations, and provide coordination of telecommunications development. A subgroup, the International Radio Consultative Committee (CCIR) was concerned with certain transmission standards and the compatible use of the frequency spectrum, including geostationary satellite orbit assignments.

Standardized transmission formats to allow interchange of communications over national boundaries are the purview of the ITU-R committee. Because these standards involve international treaties, negotiations are channeled through national administrations.

C.3 PRINCIPAL STANDARDS ORGANIZATIONS

Hundreds of SDOs exist worldwide, each targeted at a specific segment of technology and markets. In the following sections some of the major organizations working in the broadcast area are described in brief.

C.3.1 International Organization for Standardization (ISO)

ISO is the world's largest developer and publisher of international standards. ISO is a network of the national standards institutes of more than 150 countries, one member per country, with a central Secretariat in Geneva, Switzerland, that coordinates the system.

ISO is a nongovernmental organization that forms a bridge between the public and private sectors. On the one hand, many of its member institutes are part of the governmental structure of their countries, or are mandated by their government. On the other hand, other members have their roots uniquely in the private sector, having been set up by national partnerships of industry associations. As such, ISO enables a consensus to be reached on solutions that meet both the requirements of business and the broader needs of society.

The ISO is one of three global sister organizations (ISO, IEC, and ITU) that develop international standards for the world. When appropriate, these groups cooperate to ensure that international standards fit together seamlessly and complement each other. Joint committees ensure that international standards combine all relevant knowledge of experts working in related areas.

C.3.2 International Electrotechnical Commission (IEC)

The International Electrotechnical Commission is a global organization that publishes consensus-based international standards and manages conformity assessment systems for electric and electronic products, systems, and services, collectively known as electrotechnology. IEC publications serve as a basis for national standardization and as references when drafting international tenders and contracts. Over 10,000 experts from industry, commerce, government, test and research labs, academia, and consumer groups participate in IEC standardization work.

Founded in 1906, the IEC provides a platform to companies, industries, and governments for meeting, discussing, and developing the international standards they require. All IEC international standards are consensus-based and represent the needs of key stakeholders of the nations participating in IEC work. Every member country has one vote and a say in what goes into an IEC international standard.

The IEC promotes world trade and economic growth and encourages the development of products, systems, and services that are safe, efficient, and environmentally friendly.

C.3.3 International Telecommunication Union (ITU)

The International Telecommunication Union (ITU) is the United Nations specialized agency for information and communication technologies. ITU allocates global radio spectrum and satellite orbits, develops the technical standards that ensure networks and technologies seamlessly interconnect, and strives to improve access to communications technologies to underserved communities worldwide.

Founded on the principle of international cooperation between governments (Member States) and the private sector (Sector Members, Associates, and Academia), ITU is a global forum through which parties work toward consensus on a wide range of issues affecting the future direction of the communications industry. ITU has been based on public/private partnership since its inception. It currently has a membership of more than 190 countries and over 700 private sector entities and academic institutions. ITU is headquartered in Geneva, Switzerland, and has 12 regional and area offices around the world. ITU membership represents a cross-section of the global communications technology sector, from the world's largest manufacturers and carriers to small players working with new and emerging technologies, along with leading research and development institutions.

C.3.4 American National Standards Institute (ANSI)

The American National Standards Institute (ANSI) has served in its capacity as administrator and coordinator of the United States private sector voluntary standardization system for more than 90 years. Founded in 1918 by five engineering societies and three government agencies, the Institute remains a private, nonprofit membership organization supported by a diverse constituency of private and public sector organizations.

ANSI coordinates policies to promote procedures, guidelines, and the consistency of standards development. Due process procedures ensure that participation is open to all persons who are materially affected by the activities without domination by a particular group. Written procedures are available to ensure that consistent methods are used for standards developments and appeals. Various levels of membership are available that support the U.S. voluntary standardization system as members of the ANSI federation.

The functions of ANSI include: (1) serving as a clearinghouse on standards development and supplying standards-related publications and information, and (2) the following business development issues:

- Provides national and international standards information necessary to market products worldwide.
- Offers American National Standards that assist companies in reducing operating and purchasing costs, thereby assuring product quality and safety.
- Offers an opportunity to voice opinion through representation on numerous technical advisory groups, councils, and boards.
- Furnishes national and international recognition of standards for credibility and force in domestic commerce and world trade.
- Provides a path to influence and comment on the development of standards in the international arena.

Prospective standards must be submitted by an ANSI accredited standards developer. There are three basic methods that may be used:

- Accredited organization method. This approach is most often used by associations and societies having an interest in developing standards. Participation is open to all interested parties as well as members of the association or society. The standards developer must fashion its own operating procedures, which must meet the general requirements of the ANSI procedures.
- Accredited standards committee method. Standing committees of directly and materially affected interests develop documents and establish consensus in support of the document. This method is most often used when a standard affects a broad range of diverse interests or where multiple associations or societies with similar interests exist. These committees are administered by a secretariat, an organization that assumes the responsibility for providing compliance with the pertinent operating procedures. The committee can develop its own operating procedures consistent with ANSI requirements, or it can adopt standard ANSI procedures.
- Accredited canvass method. This approach is used by smaller trade associations or societies that have documented current industry practices and desire that these standards be recognized

nationally. Generally, these developers are responsible for a small number of standards. The developer identifies those who are directly and materially affected by the activity in question and conducts a letter ballot canvass of those interests to determine consensus. Developers must use standard ANSI procedures.

Note that all methods must fulfill the basic requirements of public review, voting, consideration, and disposition of all views and objections, and an appeals mechanism.

In order to maintain ANSI accreditation, standards developers are required to consistently adhere to a set of requirements or procedures known as the "ANSI Essential Requirements," which govern the consensus development process. Due process is the key to ensuring that standards are developed in an environment that is equitable, accessible, and responsive to the requirements of various stakeholders. The open and fair process ensures that all interested and affected parties have an opportunity to participate in a standard's development. It also serves and protects the public interest since standards developers accredited by ANSI must meet the Institute's requirements for openness, balance, consensus, and other due process safeguards.

The introduction of new technologies or changes in the direction of industry groups or engineering societies may require a mediating body to assign responsibility for a developing standard to the proper group. The Joint Committee for Intersociety Coordination (JCIC), for example, operates under ANSI to fulfill this need.

ANSI promotes the use of U.S. standards internationally, advocates U.S. policy and technical positions in international and regional standards organizations, and encourages the adoption of international standards as national standards where they meet the needs of the user community.

C.3.5 Advanced Television Systems Committee

The Advanced Television Systems Committee (ATSC) is an international, nonprofit organization developing voluntary standards for digital television. The ATSC has approximately 130 member organizations representing the broadcast, broadcast equipment, motion picture, consumer electronics, computer, cable, satellite, and semiconductor industries. ATSC is the organization that standardized the digital television system for the United States and elsewhere. Current work is focused on ATSC 3.0, the next generation television service. Additional information is available at www.atsc.org.

C.3.6 Audio Engineering Society

The AES (www.aes.org) was organized in 1948 primarily to serve the needs of the high quality audio recording and reproduction community. The Society maintains a standards committee (AESSC) that supervises the work of subcommittees and working groups. Drafts of proposed standards are published in the *Journal of the Audio Engineering Society* (*JAES*) for review and comment by all interested parties. Current AES standards address measurement methods, commercial loudspeaker specifications, and digital audio recording/transmission systems.

C.3.7 Consumer Technology Association

The Consumer Technology Association (CTA, http://www.cta.tech), formerly the Consumer Electronics Association (CEA), unites companies within the consumer technology sector to advance the industry. CTA is the recognized authority on market research and forecasts, consumer surveys, legislative and regulatory news, engineering standards, and training resources. CTA makes an ongoing effort to grow the CE industry by developing essential industry standards to enable interoperability between new products hitting the market and existing devices. CTA has more than 70 committees, subcommittees, and working groups within its Technology and Standards program. CTA is accredited by the American National Standards Institute (ANSI).

C.3.8 Electronics Industries Alliance

Many of the early standards relating to radio and television broadcasting were developed by equipment manufacturers under the banner of the Radio Manufacturers Association (RMA), later the RETMA (add Electronics and Television), then the Electronics Industries Association (EIA), and now the Electronic Industries Alliance (www.eia.org).

The EIA is a national trade organization made up of a number of product divisions. Some of the best known EIA standards activities are in the areas of data communications, instrumentation, broadcast transmitters, video transmission, video monitors, and RF interference.

With the proliferation and expansion of electronics, the EIA is now divided into many sectors and groups. Of particular interest to the broadcaster are the Consumer Technology Association (CTA, see above); Electronic Components, Assemblies, Equipment, and Supplies Association (ECA); and Telecommunications Industry Association (TIA). For a comprehensive, searchable listing of all electronic international standards, visit http://global.ihs.com/.

C.3.9 Institute of Electrical and Electronic Engineers

The IEEE has many branches (professional groups) that serve the standardization needs of the electrical, electronic, and computer industries. Presently available standards relate to definitions, measurement techniques, and test methods. The Institute of Radio Engineers (long since joined the IEEE) was responsible for measurement standards and techniques in the early years of radio and television. The Standards Coordinating Committees publish books and documents covering definitions of electric and electronics terms, graphic symbols and reference designations for engineering drawings, and letter symbols for measurement units. See www.ieee.org.

C.3.10 Society of Cable Telecommunications Engineers

The SCTE standards program provides an ANSI-accredited forum for the development of technical specifications supporting the cable telecommunications industry. This work program includes:

- Data and telephony over cable
- Application platform development
- Digital video signal performance
- Emergency alert systems
- Network management systems
- Cables, connectors and amplifiers
- Construction and maintenance practices

C.3.11 Society of Motion Picture and Televisions Engineers

Organizations such as the SMPTE (www.smpte.org), composed primarily of users of equipment and processes, are able to accomplish what is nearly impossible in the manufacturing community. Namely, to provide a forum where users and manufacturers can distill the best of current technology to promote basic interchangeability in hardware and software. A chronology of the development of this engineering society provides insights as to how such organizations adapt to the needs of advancing technologies.

Around 1915, it became obvious that the rapidly expanding motion picture industry must standardize basic dimensions and tolerances of film stocks and transport mechanisms. After two unsuccessful attempts to form industry-based standardizing committees, the Society of Motion Picture Engineers was formed. The founding goals were to standardize the nomenclature, equipment, and processes of the industry; to promote research and development in the industry's science and technology; and to remain independent of, while cooperating with, its business partners. It is this independent quality of a professional society that makes it possible to mediate strongly held opinions of business competitors.

By the late 1940s, it was apparent that the future of motion pictures and television would involve sharing technologies, techniques, and the market for visual education and entertainment. SMPE became SMPTE. In comparatively recent times the Society has been assigned more responsibility for television standards. The recording and reproduction of television signals has become the province of SMPTE standardization efforts. An index of the work of the engineering committees is published yearly. The basic SMPTE documents issued as a part of the organization's standardization efforts are:

- **Engineering Guidelines**—documents for the implementation of test materials and equipment operation.
- **Recommended Practices**—these include specifications for test materials, generic equipment setup and operating techniques, and mechanical dimensions involving operational procedures.
- **Standards**—mechanical specifications for film, tape, cassettes, and transport mechanisms; electrical recording and reproduction characteristics; and protocol and software for digital video systems.

C.4 ACQUIRING REFERENCE DOCUMENTS

Standards are seldom available at no cost. The number of man-hours that go into the development of a typical standard is mind-boggling. The participant's time and expenses are donated, but the standards developing organization has associated administrative expenses that must be recovered. The cost of a printed standard reflect these costs as distributed over the anticipated demand. Standards issuing organizations will supply lists of available standards on request. Internet web sites of the organizations also contain up-to-date listings, descriptions, and ordering information. Many organizations now issue sets of standards on a CD-ROM or via download at considerable savings over the cost of printed versions. Proposed new and revised standards are published for comment in professional society journals, such as those published by SMPTE and AES.

ATSC Standards—Advanced Television Systems Committee (ATSC), 1776 K Street N.W., Suite 1200 Washington, D.C. 20006, USA; Phone 202-872-9160; Fax 202-872-9161; Internet http://www.atsc.org/.

CEA Standards—Consumer Technology Association (CTA), 1919 Eads Street, Arlington, VA 22202, USA; Phone 703-907-7634; Fax 703-907-7693; Internet http://www.cta.tech.

IEEE Standards—Institute of Electrical and Electronics Engineers, Inc. 445 Hoes Lane, PO Box 1331, Piscataway, N.J. 08855-1331, USA; Phone 800-678-IEEE (4333); Outside USA and Canada 732-981-9667; Internet: http://www.ieee.org; e-mail: customer-service@ieee.org.

IETF Standards—Internet Engineering Task Force (IETF), c/o Corporation for National Research Initiatives, 1895 Preston White Drive, Suite 100, Reston, VA 20191-5434, USA; Phone 703-620-8990; Fax 703-758-5913; Internet: http://www.ietf.org/rfc.

ISO Central Secretariat—1, rud de Varembe, Case postale 56, CH-1211 Geneve 20, Switzerland; Phone +41-22-749-01 11; Fax +41-22-733-34-30; Internet: http://www.iso.ch; e-mail: central@iso.ch.

ISO Standards—Global Engineering Documents, World Headquarters, 15 Inverness Way East, Englewood, CO 80112-5776, USA; Phone 800-854-7179; Fax 303-397-2750; Internet: http://global.ihs.com.

SCTE Standards—Society of Cable Telecommunications Engineers Inc., 140 Philips Road, Exton, PA 19341; USA; Phone 610-363-6888; Fax 610-363-5898; Internet: http://scte.org e-mail: scte@scte.org.

SMPTE Standards—Society of Motion Picture and Television Engineers, 595 W. Hartsdale Avenue, White Plains, NY 10607-1824, USA; Phone 914-761-1100; Fax: 914-761-3115; Internet: http://www.smpte.org.

C.5 TABULAR DATA

Whenever possible, specifications and other engineering documents should use standard values that have industry-wide acceptance. This helps to eliminate confusion on the part of readers and tends to facilitate usability outside of specific industries.

Many of the following tables refer to "SI Units." The International System of Units (French, Système International d'Unités, SI) is the modern form of the metric system, and is the most widely used system of units of measurement built on seven base units. It defines 22 named units, and includes many more unnamed coherent derived units. The system also establishes a set of 20 prefixes to the unit names and unit symbols that may be used when specifying multiples and fractions of the units.

Data in the following tables are adapted from [1].

TABLE C.1 Names and Symbols for the SI Base Units

Physical Quantity	Name of SI Unit	Symbol for SI Unit
Length	meter	M
Mass	kilogram	kg
Time	second	s
electric current	ampere	A
thermodynamic temperature	kelvin	K
amount of substance	mole	mol
luminous intensity	candela	Cd

TABLE C.2 Units in Use Together with the SI (These units are not part of the SI, but it is recognized that they will continue to be used in appropriate contexts.)

Physical Quantity	Name of Unit	Symbol for Unit	Value in SI Units
time	minute	min	60 s
time	hour	h	3600 s
time	day	d	86,400 s
plane angle	degree	°	$(\pi/180)$ rad
plane angle	minute	′	$(\pi/10,800)$ rad
plane angle	second	″	$(\pi/648,000)$ rad
length	ångstrom	Å	10^{-10} m
area	barn	b	10^{-28} m^2
volume	liter	l, L	dm^3 = 10^{-3} m^3
mass	ton	t	Mg = 10^3 kg
pressure	bar	bar	10^5 Pa = 10^5 N m^{-2}
energy	electronvolt[1]	eV (= e × V)	$\approx 1.60218 \times 10^{-19}$ J
mass	unified atomic mass unit[1,2]	u (= m_a (^{12}C)/12)	$\approx 1.66054 \times 10^{-27}$ kg

Notes:

1 The values of these units in terms of the corresponding SI units are not exact as they depend on the values of the physical constants e (for the electronvolt) and N_A (for the unified atomic mass unit), which are determined by experiment.

2 The unified atomic mass unit is also sometimes called the Dalton, with the symbol Da.

TABLE C.3 Derived Units with Special Names and Symbols

Physical Quantity	Name of SI Unit	Symbol for SI Unit	Expression in Terms of SI Base Units
frequency[1]	hertz	Hz	s^{-1}
force	newton	N	$m\ kg\ s^{-2}$
pressure, stress	pascal	Pa	$N\ m^{-2} = m^{-1}\ kg\ s^{-2}$
energy, work, heat	joule	J	$N\ m = m^2\ kg\ s^{-2}$
power, radiant flux	watt	W	$J\ s^{-1} = m^2\ kg\ s^{-3}$
electric charge	coulomb	C	$A\ s$
electric potential, electromotive force	volt	V	$J\ C^{-1} = m^2\ kg\ s^{-3}\ A^{-1}$
electrical resistance	ohm	Ω	$V\ A^{-1} = m^2\ kg\ s^{-3}\ A^{-2}$
electric conductance	siemens	S	$\Omega^{-1} = m^{-2}\ kg^{-1}\ s^3\ A^2$
electric capacitance	farad	F	$C\ V^{-1} = m^{-2}\ kg^{-1}\ s^4\ A^2$
magnetic flux density	tesla	T	$V\ s\ m^{-2} = kg\ s^{-2}\ A^{-1}$
magnetic flux	weber	Wb	$V\ s = M^2\ kg\ s^{-2}\ A^{-1}$
inductance	henry	H	$V\ A^{-1}\ s = m^2\ kg\ s^{-2}\ A^{-2}$
Celsius temperature[2]	degree Celsius	°C	K
luminous flux	lumen	lm	cd sr
illuminance	lux	lx	cd sr m^{-2}
activity (radioactive)	becquerel	Bq	s^{-1}
absorbed done (of radiation)	gray	Gy	$J\ kg^{-1} = m^2\ s^{-2}$
dose equivalent (dose equivalent index)	Sievert	Sv	$J\ kg^{-1} = m^2\ s^{-2}$
phase angle	radian	rad	$1 = m\ m^{-1}$
solid angle	sterdian	sr	$1 = m^2\ m^{-2}$

Notes:

1 For radial (circular) frequency and for angular velocity the unit rad s^{-1}, or simply s^-, should be used, and this may not be simplified to Hz. The unit Hz should be used only for frequency in the sense of cycles per second.

2 The Celsius temperature θ is defined by the equation $\theta/°C = T/K - 273.15$. The SI unit of Celsius temperature interval is the degree Celsius, °C, which is equal to the kelvin, K.

TABLE C.4 The Greek Alphabet

Greek Letter			Greek Name	English Equivalent
Α	α		Alpha	a
Β	β		Beta	b
Γ	γ		Gamma	g
Δ	δ		Delta	d
Ε	ε		Epsilon	ĕ
Ζ	ζ		Zeta	z
Η	η		Eta	ē
Θ	θ	ϑ	Theta	th
Ι	ι		Iota	i
Κ	κ		Kappa	k
Λ	λ		Lambda	l
Μ	μ		Mu	m
Ν	ν		Nu	n
Ξ	ξ		Xi	x
Ο	ο		Omicron	ŏ
Π	π		Pi	p
Ρ	ρ		Rho	r
Σ	σ	ς	Sigma	s
Τ	τ		Tau	t
Υ	υ		Upsilon	u
Φ	φ	ϕ	Phi	ph
Χ	χ		Chi	ch
Ψ	ψ		Psi	ps
Ω	ω		Omega	ō

TABLE C.5   Constants

	π Constants									
π	3.14159	26535	89793	23846	26433	83279	50288	41971	69399	37511
$1/\pi$	0.31830	98861	83790	67153	77675	26745	02872	40689	19291	48091
π^2	9.8690	44010	89358	61883	44909	99876	15113	53136	99407	24079
$\log_e \pi$	1.14472	98858	49400	17414	34273	51353	05871	16472	94812	91531
$\log_{10} \pi$	0.49714	98726	94133	85435	12682	88290	89887	36516	78324	38044
$\log_{10} \sqrt{2\pi}$	0.39908	99341	79057	52478	25035	91507	69595	02099	34102	92128
	Constants Involving e									
e	2.71828	18284	59045	23536	02874	71352	66249	77572	47093	69996
$1/e$	0.36787	94411	71442	32159	55237	70161	46086	74458	11131	03177
e^2	7.38905	60989	30650	22723	04274	60575	00781	31803	15570	55185
$M = \log_{10} e$	0.43429	44819	03251	82765	11289	18916	60508	22943	97005	80367
$1/M = \log_e 10$	2.30258	50929	94045	68401	79914	54684	36420	76011	01488	62877
$\log_{10} M$	9.63778	43113	00536	78912	29674	98645	-10			
	Numerical Constants									
$\sqrt{2}$	1.41421	35623	73095	04880	16887	24209	69807	85696	71875	37695
$\sqrt[3]{2}$	1.25992	10498	94873	16476	72106	07278	22835	05702	51464	70151
$\log_e 2$	0.69314	71805	59945	30941	72321	21458	17656	80755	00134	36026
$\log_{10} 2$	0.30102	99956	63981	19521	37388	94724	49302	67881	89881	46211
$\sqrt{3}$	1.73205	08075	68877	29352	74463	41505	87236	69428	05253	81039
$\sqrt[3]{3}$	1.44224	95703	07408	38232	16383	10780	10958	83918	69253	49935
$\log_e 3$	1.09861	22886	68109	69139	52452	36922	52570	46474	90557	82275
$\log_{10} 3$	0.47712	12547	19662	43729	50279	03255	11530	92001	28864	19070

REFERENCE

[1] Whitaker, Jerry C. (ed.), *The Electronics Handbook*, CRC Press, Boca Raton, FL, 1996.

BIBLIOGRAPHY

"About ISO," International Organization for Standardization, Geneva, Switzerland, http://www.iso.org/iso/about.htm.

"About the IEC, Vision and Mission, International Electrotechnical Commission, Geneva, Switzerland, http://www.iec.ch/about/.

"About the ITU," International Telecommunication Union, Geneva, Switzerland, http://www.itu.int/en/about/Pages/default.aspx.

"An Introduction to ANSI," American National Standards Institute, Washington, D.C., http://ansi.org/about_ansi/introduction/introduction.aspx?menuid=1.

Lide, D. R. (ed.), CRC Handbook of Chemistry and Physics, 76th ed., CRC Press, Boca Raton, FL, 1992.

Whitaker, J. C., and K. Blair Benson (eds.), *Standard Handbook of Video and Television Engineering*, McGraw-Hill, New York, NY, 2000.

Whitaker, J. C., and R. K. Mancini, *Technical Documentation and Process*, CRC Press, Boca Raton, FL, 2013.

Zwillinger, D. (Ed.), CRC Standard Mathematical Tables and Formulae, 30th ed., CRC Press, Boca Raton, FL, 1996.

INDEX

2D effects, video, 5.65
2W/4W, interfaces for wireless intercom, 5.94–5.95
2.4 GHz, 5.94
3D effects, video, 5.65
3D models, of design mockups, 6.6
3-way handshake, TCP, 4.43–4.44
4G LTE technology, 2.236
8-VSB. *See also* DTV (digital television)
   channel occupancy, 2.136
   filters, 2.137–2.140
   I-signal, 2.136–2.137
   pilot insertion, 2.137
   power meter for measuring signal power, 2.211
   power requirements, 2.146–2.148
   Q-signal, 2.140
   RF spectrum description, 2.172–2.179
   RF upconverter, 2.140–2.141
   signal, 2.133–2.134
   specialist certification, 2.151
   spectral display, 2.142–2.143
   spectral mask for, 2.141–2.142
   transmission monitoring, 2.143–2.146
   transmission system characteristics, 2.180
   VSB baseband, 2.166–2.169
   VSB baseband spectral description, 2.169–2.172
32-bit addresses, IPv6, 4.40
128-bit addresses, IPv6, 4.40
802.xx standard, 4.6. *See also* Ethernet

A/53:
   A/153 replacing legacy A/53 standard, 3.57–3.58
   mobile DTV transmission infrastructure, 3.67
   terrestrial DTV transmission standard, 2.154–2.155
A/110, mobile DTV transmission infrastructure, 3.67
A/153. *See also* Mobile DTV
   mobile DTV transmission infrastructure, 3.67
   replacing legacy A/53 standard, 3.57–3.58
ABCs (airway, breathing, circulation), of life support, 7.49
ABIP (Alternative Broadcast Inspection Program):
   introduction, 1.17–1.18
   scope of, 1.18–1.20
   summary, 1.20

AC power:
   advanced system protection, 6.77–6.78
   batteries in UPS systems and, 6.83–6.85
   blackout effects, 6.74
   choosing a generator, 6.78–6.81
   critical system bus, 6.71–6.72
   designing for fault-tolerance, 6.69
   dual feeder system for power loss protection, 6.76–6.77
   introduction, 6.69
   noise produced by standby power systems, 6.82
   peak power shaving, 6.77
   plant configuration, 6.70–6.71
   plant maintenance, 6.72–6.73
   power distribution options, 6.70
   references, 6.85
   standby power systems, 6.73–6.75
   transmitter best practices and, 2.30–2.31
   transmitter power supplies, 2.15
   UPS systems, 6.82
AC-3 Descriptor, PSIP descriptors, 3.37
ACATS (Advisory Committee for Advanced Television Service):
   in design of new DTV system, 2.153
   elevation patterns for television broadcasts, 2.120
   receive antenna pattern, 2.124
ACC (Advanced Audio Coding), 5.149
Access control, assessment aspect of disaster planning, 7.38
Accessibility requirements:
   captioning, 1.24–1.28
   introduction, 1.21–1.22
   new directions in, 1.29
   radio reading services for print handicapped, 1.22–1.23
   summary, 1.30
   TV services for visually handicapped (video description), 1.23–1.24
Acoustics, in facility design:
   handling background noise, 5.11–5.12
   handling noise of mechanical systems and ducts, 5.12–5.15
   performance criteria, 5.10–5.11
   sound isolation techniques, 5.15–5.17

ACPR (adjacent channel power), measuring, 2.220–2.223
ADA (Americans with Disabilities Act), 1.24
`adaptation_field`, MPEG-2 stream packets, 3.7–3.9
Adaptive precorrection, of transmitters, 2.20–2.21
Address masks, IPv4, 4.29
Addressing schemes, network systems, 4.58–4.59. *See also* IP (Internet Protocol)
Adjacent channel power (ACPR), measuring, 2.220–2.223
Advanced Audio Coding (ACC), 5.149
Advanced Emergency Alerting (AEA), in ATSC 3.0, 3.66–3.67
Advanced system protection, for AC power, 6.77–6.78
Advisory Committee for Advanced Television Service (ACATS):
  in design of new DTV system, 2.153
  elevation patterns for television broadcasts, 2.120
  receive antenna pattern, 2.124
AEA (Advanced Emergency Alerting), in ATSC 3.0, 3.66–3.67
AES (Audio Engineering Society):
  client cards, 5.99–5.100
  overview of, 5.19, A.55
  packet delay or latency and, 5.23–5.24
AES 67-2013 standard. *See* AoIP (audio over IP)
AES3 standard, 5.19
AES10 standard. *See* MADI (Multichannel Audio Digital Interface)
AESC (American Engineering Standards Committee), A.52
Aggregation techniques:
  cellular aggregation technology, 2.238–2.240
  hybrid aggregation, 2.241–2.243
Aging time, MAC addresses, 4.20
Air conditioning. *See* HVAC (heating, ventilation, and air conditioning)
Air cooling, for transmitters, 2.199, 6.46–6.48
Air filters, in transmitter maintenance, 6.95
Alternative Broadcast Inspection Program. *See* ABIP (Alternative Broadcast Inspection Program)
Altitude, operating parameters for HVAC systems, 6.49
AM (amplitude modulation):
  exciters, 2.16
  introduction, 2.87
  power amplifiers, 2.17–2.18
  theory, 2.3–2.6
  transmitters. *See* Transmitters (AM and FM)
AM antennas:
  feeder lines and coupling networks, 2.89–2.90
  final adjustment of complex conjugate impedance, 2.95–2.96
  ground system design, 2.88–2.89
  impedance requirements for digital transmission, 2.24–2.26
  introduction, 2.87–2.88
  monitoring directional systems, 2.96–2.97
  preliminary network setup and adjustment, 2.94–2.95

AM antennas (*Cont.*):
  RF networks and, 2.91–2.94
  system components, 2.91
  technical principles, 2.90
  troubleshooting directional arrays, 2.98–2.99
  troubleshooting single radiator problem, 2.97–2.98
AM stations:
  ABIP review of, 1.19–1.20
  Chief Operator making field measurements, 1.14
  equipment inspection, 6.105
  maintaining transmission system, 6.111–6.115
  shortwave compared with, 2.102
AM-AM (amplitude modulation to amplitude modulation):
  conversion, 2.191
  precorrection and, 2.22
American Engineering Standards Committee (AESC), A.52
American National Standards Institute. *See* ANSI (American National Standards Institute)
American Standards Association (ASA), A.52
American Wire Gauge (AWG), 4.21
Americans with Disabilities Act (ADA), 1.24
Amplifiers:
  AM power amplifiers, 2.17–2.18
  components of wired party-line intercom systems, 5.84–5.85
  directional couplers, 2.18–2.19
  FM power amplifiers, 2.18
  high power amplifiers, 2.148–2.151
  linearity of amplification effecting performance, 2.21–2.23
  loss-pass filters in, 2.18
Amplifiers (talent boxes), Interrupted Fold-Back and, 5.96–5.97
Amplifiers, television transmitters:
  amplification chain, 2.185
  in broadcast transmitter diagram, 2.184
  high efficiency amplification techniques, 2.193–2.195
  overview of, 2.184
  power amplifiers, 2.189–2.193
  RF amplification technologies, 2.185–2.189
Amplitude:
  measuring head to shoulder amplitude ratio, 2.219–2.220
  measuring signal amplitude, 2.215–2.216
Amplitude modulation. *See* AM (amplitude modulation)
AM-PM (amplitude modulation to phase modulation):
  conversion, 2.192
  precorrection and, 2.22
Analog partyline intercom systems:
  advantages of, 5.90–5.92
  components, 5.84–5.90
  overview of, 5.84
Analog services:
  client cards (analog), 5.99–5.100
  computing service and interfering field strengths for television, 2.120–2.124

Analog services (*Cont.*):
    impedance requirements for analog transmission, 2.25–2.26
    migrating from analog to digital FM modulation, 2.231–2.233
Announcements, Mobile DTV, 3.47
ANSI (American National Standards Institute):
    controls on nonionizing radiation, 7.54–7.55
    EIA-568A/EIA-568B wiring schemes, 4.11–4.12
    industry standards for equipment racks, 6.25
    overview of, A.54–A.55
Antenna tuning unit (ATU), maintaining AM stations, 6.11
Antennas:
    ABIP review of registration of antenna structure, 1.19
    AM. *See* AM antennas
    assessment aspect of disaster planning, 7.38
    corporate feed systems, 2.85
    designing ENG system, 5.150
    environmental management and cooling, 6.48
    factory testing, 2.75–2.76
    gain, 2.71–2.75
    impedance requirements for digital transmission, 2.24–2.26
    including receive antenna pattern for television, 2.124–2.129
    installing, 2.77–2.80
    maintaining, 2.80–2.83
    maintaining AM stations, 6.114–6.115
    optimum take-off angle for shortwave, 2.111
    placing for wireless intercom, 5.95–5.96
    radiator types, 2.70–2.71
    radomes and slot covers, 2.84–2.85
    for shortwave broadcasts, 2.107–2.108
    signal noise as function of antenna VSWR, 2.226
    tower maintenance, 6.119–6.122
    transmitting antennas. *See* Transmitting antennas
    VSWR and pressurization requirements, 2.85
AoIP (audio over IP):
    audio monitoring, 5.42
    audio quality, 5.22–5.23
    building broadcast studio, 5.26–5.27
    comparing AES to Ethernet- based systems, 5.21
    controls, 5.24
    delay (latency) and, 5.23–5.24
    designing security for, 5.29–5.30
    frequently asked questions, 5.33–5.35
    multicast IP communications, 4.36–4.37
    in outside broadcast truck, 5.47–5.48
    overview of, 5.19–5.20
    planning network for, 5.27–5.29
    QoS (Quality of Service), 5.31
    Source Advertising, 5.24
    surround sound support, 5.22
    Switch options for, 5.31–5.32
    synchronization clock in, 5.24
    use with PCs, 5.22
    uses of, 5.25–5.26
AOTS (Automatic Overlap Transfer Switch), switching to/from standby power, 6.75

Application Framework toolkit, ATSC Mobile DTV system, 3.63
Application layer (Layer 7):
    of OSI model, 4.7
    overview of, 4.58
Applications, network systems, 4.64–4.65
Applications layer, ATSC 3.0, 7.71–7.72
Arab States Broadcasting Union (ASBU), 2.103
Archival MAMs, 5.79
Archiving, uses of media asset management, 5.78
Area conversion ratios, A.12
Area Model, predictive models in Longley-Rice, 2.131
ARPANET, 4.79
Arrays, curtain arrays, 2.114
ASA (American Standards Association), A.52
ASBU (Arab States Broadcasting Union), 2.103
Ascension numbering, documentation in wire management, 6.12–6.13
ASI (asynchronous serial interface), MPEG, 2.156
Asia-Pacific Broadcasting Union (ABU), 2.103
Assessment, in disaster planning and recovery, 7.38–7.39
Asynchronous serial interface (ASI), MPEG, 2.156
ATM (Asynchronous Transport Mode), shifting from circuits to packets, 4.80
ATOM feeds, sources of dynamic graphics, 5.51
ATSC (Advanced Television System):
    8-VSB signal and, 2.133
    documentation of mobile DTV standard, 3.58–3.59
    DTV standard, 2.153–2.154, 7.67
    DTV transmission system. *See* DTV (digital television), RF transmission system
    introduction, 3.57
    Mobile DTV. *See* Mobile DTV
    NRT (non-real-time) content, 3.49–3.50
    overview of, A.55
    reference documents, A.57
ATSC 3.0:
    applications layer, 7.71–7.72
    challenges in, 7.73
    emergency alerting, 3.53–3.55
    introduction, 7.67–7.68
    next steps for, 7.73–7.74
    overall TG3 structure, 7.68
    physical layer, 7.69–7.70
    references, 7.74
    security, 7.73
    transport layer, 7.70–7.71
Attached storage, storage options in file-based workflow, 4.65
Attenuation:
    coaxial transmission lines, 2.36–2.38
    de-rating factors, 2.46–2.47
Attenuator, features of spectrum analyzers, 2.214
ATU (antenna tuning unit), maintaining AM stations, 6.11
Audio:
    applications layer of ATSC 3.0 and, 7.72
    ATSC Mobile DTV content files, 3.61
    CIAB (Channel in a Box), 5.52
    compensating for latency in video display, 5.43

Audio (*Cont.*):
  file-based workflow, 4.63
  mobile services for streaming, 3.44
  MPEG-2 specification, 3.3
  MPEG-4 audio coding, 3.65
  off air monitoring, 5.55
Audio control room, 5.9–5.10
Audio Engineering Society (AES). *See* AES (Audio Engineering Society)
Audio engineers, 5.32–5.33
Audio meter, 5.37–5.38
Audio mixer, in outside broadcast truck, 5.47
Audio monitoring systems:
  broadcasts and, 5.38
  connectivity and signals types, 5.42–5.43
  functionality and operation of, 5.38–5.42
  future of, 5.43–5.44
  introduction, 5.37–5.38
  physical form of, 5.38
Audio over IP. *See* AoIP (audio over IP)
Audio PGM/AUX, Interfaces for wireless intercom, 5.95
Audio routers/routing, 5.21, 5.25
Audio system interconnection:
  AoIP and, 5.19–5.20
  audio routing, 5.21
  benefits of Ethernet, 5.20–5.21
  building broadcast studios, 5.26–5.27
  comparing AES audio to Ethernet- based systems, 5.21
  controls, 5.24
  delay (latency) and, 5.23–5.24
  designing security for AoIP, 5.29–5.30
  frequently asked questions regarding AoIP, 5.33–5.35
  introduction, 5.19
  network engineering for audio engineers, 5.32–5.33
  network requirements, 5.31
  planning network for AoIP, 5.27–5.29
  QoS (Quality of Service), 5.31
  quality of AoIP, 5.22–5.23
  references, 5.35–5.36
  Source Advertising, 5.24
  surround sound support, 5.22
  switch options for AoIP, 5.31–5.32
  synchronization clock in AoIP, 5.24
  uses of AoIP, 5.25–5.26
  using AoIP with PCs, 5.22
Audio-over sources, baseband master control and, 5.52
Auto down-mix, special and extended audio functions, 5.43
Automatic Overlap Transfer Switch (AOTS), switching to/from standby power, 6.75
Automation systems:
  case study, 5.73
  current evolution of, 5.73–5.74
  introduction and history, 5.71–5.73
  next generation, 5.74–5.75
  summary, 5.75
Average power:
  de-rating factors, 2.46
  ratings for coaxial lines, 2.39–2.41

AWG (American Wire Gauge), 4.21
Azimuth pattern:
  antenna characteristics, 2.69
  antenna gain and, 2.71–2.72

Back light, in three point lighting, 5.106
Backups:
  assessment in disaster planning, 7.39
  disaster preparation checklist, 7.41
Background noise, handling in facility design, 5.11–5.12
Balanced filters. *See* CIF (constant impedance filter)
Band. *See* Frequency band
Bandpass filters:
  for DTV receivers, 2.161–2.162
  SSB (single sideband), 2.136
  use with junction and directional combiners, 2.61–2.62
Bandwidth, requirements for RF transmission, 2.8–2.10
BAS (Broadcast Auxiliary System), 2.231
Baseband:
  spectral description, 2.169–2.172
  VSB baseband, 2.166–2.169
Batteries:
  characteristics of, 6.83–6.84
  maintaining HVAC system, 6.117
  SLA (sealed lead acid), 6.84–6.85
  types of, 6.83–6.85
Battison, John H., 7.66
BCIM (bit interleaver, coding, and modulation), physical layer of ATSC 3.0, 7.69–7.70
BCROs (Broadcast Rights Objects), 3.63
BCs (Boundary Clocks), 4.74
Belt pack user stations, components of wired party-line intercom systems, 5.85
BeO (Beryllium oxide) ceramics:
  operating hazards, 7.48
  safety considerations, 7.51
Best Master Clock Algorithm (BMCA), in PTP, 4.70
Best practices, RF transmission, 2.28–2.30
BGP, routing protocol options, 4.34–4.35
Bit interleaver, coding, and modulation (BCIM), physical layer of ATSC 3.0, 7.69–7.70
Black boxes, wire management, 6.21
Blackout effects, AC power and, 6.74
BMCA (Best Master Clock Algorithm), in PTP, 4.70
Bonded modem system, configuring modems for IP-based ENG, 5.152–5.153
Bonding, elements of grounding system, 6.60–6.61
Boundary Clocks (BCs), 4.74
BPM (Business Process Management), 5.75
Breaking news, field implementation of ENG, 5.157–5.158
Break-out cables, wire management, 6.21
Broadcast accessibility. *See* Accessibility requirements
Broadcast Auxiliary System (BAS), 2.231
Broadcast center, audio monitoring and, 5.39
Broadcast Exchange Format (BXF), 5.56
Broadcast facility design:
  component selection and installation, 6.4–6.6
  construction considerations, 6.3–6.4

Broadcast facility design (*Cont.*):
  design mockups, 6.6
  introduction, 6.3
  references, 6.9
  technical documentation, 6.7–6.9
  working with contractors, 6.9
Broadcast Rights Objects (BCROs), 3.63
Broadcast studios. *See* Studios
The Broadcaster's BIGBOOK Project, 1.8
Broadcasts:
  Broadcast plant, 4.3–4.4
  centralized. *See* Centralized broadcasting:
  managing, 7.1–7.2
  operating hazards, 4.5
  remote. *See* Remote audio broadcasting:
  types of, 2.69
  use of IP networks. *See* IP networks
Broads, lighting instruments, 5.136
Brown, George H., 2.88
Budgets:
  analyzing requirements for system engineering, 7.26–7.29
  role of management and leadership in, 7.7–7.9
Buffer and timing model, Mobile DTV, 3.46
Bulkhead interconnects, in wire management, 6.16
Bulkhead panel, grounding and, 6.63–6.66
Burns, first aid procedures, 7.50
Business managers, role in interdepartmental cooperation, 7.14
Business Process Management (BPM), 5.75
BXF (Broadcast Exchange Format), 5.56

CA (conditional access), ATSC security, 7.73
Cables. *See also* Wire management
  cable lengths for medium-wave radiators, 2.89–2.90
  converting to VLANs, 4.61
  layout in detailed design, 7.26
Calibrations, duties of Chief Operator, 1.13
Cameras, components of wired party-line intercom systems, 5.88
CAP (Common Alerting Protocol). *See also* EAS/CAP (Emergency Alert System/Common Alerting Protocol)
  distribution of alert or emergency messages, 1.34
  inspection of EAS/CAP equipment, 1.18
  Mobile Emergency Alert System and, 3.49
  Mobile Emergency Alert System interface with, 3.53
Capacitive reactance, antenna components, 2.91
Capacitors, antenna components, 2.91
Capital budgets:
  analyzing budget requirements, 7.27
  role of management and leadership in budgeting, 7.7–7.9
Caption Service Descriptor, PSIP, 3.37–3.38
Captioning services:
  challenges, 1.24–1.25
  new directions in, 1.29
  summary of requirements, 1.25–1.29
Carrier networks:
  history of TV transport over, 4.79–4.80
  shift from circuits to packets in, 4.80–4.81

Carrier-to-noise ratio (C/N):
  8-VSB signal and, 2.133
  COFDM and, 2.234–2.235
Carson's bandwidth rule (CBR), 2.232
CAT (Conditional Access Table), program specific information in MPEG-2, 3.15
CAT 5, 5E, and 6:
  Ethernet cable, 4.11
  wiring standards, 4.21
Catwalks, lighting suspension and rigging systems, 5.113
CBR (Carson's bandwidth rule), 2.232
CBTE (Certified Broadcast Television Engineer), 2.151
CD (compact disc), audio standards, 2.154
CEA (Consumer Electronics Association), 1.25
Ceilings, sound isolation techniques, 5.16
Celsius-to-Fahrenheit conversion table, A.8
Cell Information Table (CIT), Mobile DTV, 3.45
Cellular:
  designing ENG system, 5.151
  electronic newsgathering and, 5.146–5.147
  hybrid aggregation, 2.241–2.243
  transmission systems for ENG, 5.154
Cellular newsgathering. *See* CNG (cellular newsgathering)
Central processing unit (CPU), 5.98
Centralized broadcasting:
  distributed content and control, 5.59
  fully centralized, 5.58–5.59
  interconnection, 5.60
  overview of, 5.57
  remote control and remote monitoring, 5.59–5.60
  remote monitoring, 5.61
  topology of, 5.58
  traffic and related considerations, 5.60–5.61
Certification program, SBE (Society of Broadcasters), 7.64–7.65
Certified Broadcast Television Engineer (CBTE), 2.151
CFLs (compact fluorescent lamps), light sources, 5.107, 5.129
CFR (Code of Federal Regulations), television standard, 1.11, 2.153
CG (character generator), superimposing titles over video, 5.67
Chairperson, managing meetings, 7.20–7.21
Change orders, project tracking and control, 7.28–7.29
Channel in a Box (CIAB), 5.50, 5.52–5.53
Channel type grids, lighting suspension and rigging systems, 5.112
Channels, spacing and rejection requirement for FM channel combiners, 2.63–2.65
Character generator (CG), superimposing titles over video, 5.67
Chebyshev filters:
  types of mask filters, 2.203
  use with junction and directional combiners, 2.62
Chemical grounding rods, in grounding system, 6.56–6.58

Chief Operator (CO):
    duties of, 1.12–1.14
    introduction, 1.11
    selection and designation, 1.11–1.12
    summary, 1.15
China Radio International, 2.106
Chroma key:
    hard cyc and, 5.118
    video keying, 5.66–5.67
CIAB (Channel in a Box), 5.50, 5.52–5.53
CIDR (classless interdomain routing), 4.41
CIF (constant impedance filter):
    installing, 2.209
    RF upconverter and. *See* Mask filters
    types of mask filters, 2.204
CIRAF (Conferencia Internacional de Radiodifusión por Altas Frecuencias):
    defining target zones, 2.114
    shortwave zones, 2.104
Circuit breakers, OSHA guidelines, 7.51
Circuits:
    components of matrix intercom systems, 5.98
    history of TV transport over carrier networks, 4.79–4.80
Cisco switches:
    AoIP capable, 5.32
    VLAN configuration examples, 4.48–4.50
CIT (Cell Information Table), Mobile DTV, 3.45
Classes, IPv4, 4.26–4.27
Classful vs. classless IP addressing, 4.28
Classless interdomain routing (CIDR), 4.41
Cleaning, in transmitter maintenance, 6.95
Client cards (analog), 5.99–5.100
Clocks:
    STC (System Time Clock) in MPEG-2, 3.23–3.25
    timing for DTV receivers, 2.162
Closed captions, 5.55. *See also* Captioning services
Cloud:
    on-premise MAM system vs., 5.80
    storage options in file-based workflow, 4.65
C/N (carrier-to-noise ratio):
    8-VSB signal and, 2.133
    COFDM and, 2.234–2.235
CNG (cellular newsgathering):
    4G LTE technology, 2.236
    cellular aggregation technology and CNG workflow, 2.238–2.240
    overview of, 2.236
    UE (user equipment), 2.236–2.238
Coaxial cavity filters:
    accounting for frequency drift, 2.209
    retuning, 2.210
    types of mask filters, 2.206
Coaxial transmission lines:
    attenuation, 2.36–2.38
    average power ratings, 2.39–2.41
    cable lengths for medium-wave radiators, 2.89–2.90
    characteristic impedance, 2.36
    connector power ratings, 2.43
    cutoff frequency, 2.35–2.36
    de-rating factors, 2.45–2.48

Coaxial transmission lines (*Cont.*):
    differential expansion, 2.44–2.45
    inspecting and maintaining, 2.57–2.58
    installing rigid lines, 2.53–2.56
    installing semiflexible lines, 2.50–2.52
    introduction, 2.33
    peak power ratings, 2.41–2.43
    pressurization of, 2.56–2.57
    rigid systems, 2.25
    selecting rigid lines, 2.52–2.53
    selecting semiflexible lines, 2.50
    semiflexible systems, 2.48–2.49
    types of, 2.33–2.35
    velocity factor, 2.44
    voltage standing wave ratio, 2.43–2.44
Code of Federal Regulations (CFR), television standard, 1.11, 2.153
Coefficient of thermal expansion (CTE), thermal stability of mask filters, 2.208–2.209
COFDM (Coded orthogonal frequency-division multiplexing):
    benefits of, 2.233–2.234
    DRM system using, 2.109
    overview of, 2.234–2.235
    peak power ratings for coaxial lines, 2.41–2.43
Coils, antenna components, 2.91
Collaboration, uses of media asset management, 5.78
Collisions:
    defined, 2.114
    network systems, 4.60
    QoS (Quality of Service), 4.60
    shortwave, 2.106
Collocated wireless intercoms, 5.95
Color coding, wire management and, 6.16–6.17
Color Quality Scale (CQS), in evaluation of light sources, 5.109
Color Rendering Index (CRI), in evaluation of light sources, 5.108–5.109
Color temperature, 5.108
Colors, characteristics of (frequency/wavelength), 5.108
Combiners, FM channels. *See* FM channel combiners
Combining methods, for HD radio, 2.11–2.13
Comb-line filters:
    accounting for frequency drift in mask filters, 2.209
    retuning, 2.210
    types of mask filters, 2.205, 2.208
Common Alerting Protocol. *See* CAP (Common Alerting Protocol)
Communications Act (1934), A.45
Communications and Video Accessibility Act (CVAA), 1.25
Compact disc (CD), audio standards, 2.154
Compact fluorescent lamps (CFLs), light sources, 5.107, 5.129
Complex conjugate impedance, final adjustment of AM antenna, 2.95–2.96
Composite baseband, requirements for RF transmission, 2.10

Computer floors, equipment rack enclosures, 6.37–6.38
Computer Room Air Conditioning (CRAC), 6.42–6.43
Concert halls. *See* Studios
Conditional access (CA), ATSC security, 7.73
Conditional Access Table (CAT), program specific information in MPEG-2, 3.15
Conditional switching, special and extended audio functions, 5.43
Conductors:
  inductance of grounding system and, 6.61
  protective covers for, 7.50–7.51
Conduit, wire management and, 6.20
CONELRAD (Control of Electromagnetic Radiation), 1.31
Conferencia Internacional de Radiodifusión por Altas Frecuencias (CIRAF):
  defining target zones, 2.114
  shortwave zones, 2.104
Connectivity:
  audio monitoring systems, 5.42–5.43
  matrix intercom systems and, 5.101–5.102
  network components, 4.5
Connectors:
  Ethernet, 4.13
  intercom systems, 5.99
  power ratings for coaxial connectors, 2.43
Consoles, light controls on, 5.126
Constant flow cooling systems, 6.49
Constant impedance filter. *See* CIF (constant impedance filter)
Consumer Electronics Association (CEA), 1.25
Consumer Technology Association (CTA), A.55
Content, Mobile DTV file types, 3.61
Content Advisory Descriptor, PSIP, 3.37
Content delivery framework, for Mobile DTV, 3.43–3.44
Content Manager, for Mobile Emergency Alert System, 3.52–3.53
Contractors, working with in designing broadcast facilities, 6.9
Control, master. *See* Master control
Control key panels, user/subscriber, 5.102–5.104
Control of Electromagnetic Radiation (CONELRAD), 1.31
Control panels, video, 5.68–5.69
Control rooms:
  illustration of, 5.8
  production facility design, 5.9–5.10
Control/monitoring system:
  in fault diagnostics, 2.27
  operating parameters for HVAC systems, 6.52
  for television transmitters, 2.197–2.198
  for transmitters, 2.19
Controls:
  AoIP, 5.24
  lighting, 5.125–5.126, 5.139–5.140
  project, 7.28–7.29
  transmitter, 5.61

Conversion reference table, A.14–A.35
Coolant, safety of hot coolant, 7.57–7.58
Cooling. *See also* Environmental management
  designing cooling systems, 6.41
  equipment rack layout and, 6.35–6.36
  for equipment rooms, 6.42–6.43
  heater exchangers, 6.44
  high availability design, 6.44–6.45
  operating parameters and, 6.49–6.52
  satellite antennas and, 6.48
  site heat and, 6.45–6.46
  for studios, 6.42
  television transmitters, 2.199–2.200
  transmitters and, 6.46–6.48
Cooperation, role of management and leadership in interdepartmental cooperation, 7.14–7.15
Coordinated Universal Time (UTC), PSIP, 3.31
Copper wire:
  specifications of copper wire sizes, A.7
  in Ufer grounding system, 6.59
Corporate feed systems (superturnstile, panels, FM), 2.85
Corrosive compounds, safety considerations, 7.51–7.53
Coupling networks, for AM antenna, 2.89–2.90
CPU (central processing unit), 5.98
CQS (Color Quality Scale), in evaluation of light sources, 5.109
CRAC (Computer Room Air Conditioning), 6.42–6.43
CRI (Color Rendering Index), in evaluation of light sources, 5.108–5.109
Critical system bus, AC power and, 6.71–6.72
Cross-coupled filters, 2.66–2.68, 2.203
Cross-over cables, Ethernet, 4.12
Crosstalk, on party-line intercoms, 5.91
CRPD (Rights of Persons with Disabilities), 1.21
CTA (Consumer Technology Association), A.55
CTE (coefficient of thermal expansion), thermal stability of mask filters, 2.208–2.209
Cubic conversion ratios, A.12
Cueing, use in automation systems, 5.73
Curtain arrays, 2.114
Curtain track, 5.116–5.118
Curtains, 5.116–5.118
Cutoff frequency, 2.35–2.36
CVAA (Communications and Video Accessibility Act), 1.25
Cyclorama (curtains):
  hard cyc, 5.118–5.119
  lighting and, 5.116–5.118
Cylindrical waveguide filters, 2.207

Data:
  backing up, 7.39
  equalizers. *See* Equalizers
Data interleaver:
  applying to MPEG data packets, 2.157–2.158
  data deinterleaver for DTV receiver, 2.165
Data processing, DTV transmission system, 2.156–2.157

Data randomizer:
  in ATSC data processing, 2.157
  for DTV receiver, 2.165
Data synchronization, for DTV receivers, 2.162
Databases:
  in media asset management, 5.77
  wiring database, 6.12–6.13
Data-flow layers, of OSI model, 4.8–4.9
Data-Link Layer (Layer 2):
  Ethernet II frame, 4.16–4.17
  Ethernet switches, 4.19–4.23
  LLC sublayer of, 4.15
  MAC sublayer of, 4.15–4.16
  overview of, 4.8, 4.14, 4.57
  switching and routing and, 4.59
  VLANs, 4.17–4.18
dB (decibels), reference table, A.5
dBFs (decibels full scale), audio meters using, 5.38
DC power:
  power supplies, 2.197
  transmitter power supplies, 2.15
DCC (Dynamic Carrier Control), power consumption for AM transmitters, 2.26–2.27
DE (differential expansion):
  de-rating factors, 2.45–2.48
  overview of, 2.44–2.45
Deaf-blind, new directions in accessibility, 1.29
decibels full scale (dBFs), audio meters using, 5.38
Decoding Time Stamp (DTS), in MPEG-2 timing model, 3.26
DECT, regulation of wireless intercoms, 5.94
De-emphasis function, for improving RF performance, 2.19–2.20
Delay (latency). *See* Latency
Delay matching, for AM transmitters, 2.24
Delay/loss corrections, for FM channel combiners, 2.68
Demolition, facility upgrades and, 6.3
Department of Defense (DoD) model, 4.9–4.11
Department of Homeland Security (DHS), role in emergency management, 1.34
Descriptors:
  MPEG-2, 3.11, 3.18–3.19
  PSIP, 3.36–3.38
  SMT, 3.46
Design:
  of broadcast facility. *See* Broadcast facility design
  electronic systems, 7.24–7.26
  mockups, 6.6
  planning for success, 7.18
  of production facility. *See* Production facility design
Design development phase, 7.23
Development, role of management and leadership in, 7.9–7.10
DHCP (Dynamic Host Configuration Protocol), 4.56
DHS (Department of Homeland Security), role in emergency management, 1.34
Diagnostics, fault diagnostics, 2.27
Diagrams, use in detailed design, 7.26–7.27
Dial pads, control panels and, 5.103

Differential expansion (DE):
  de-rating factors, 2.45–2.48
  overview of, 2.44–2.45
Diffraction, propagation theory, 2.129–2.131
Digital lighting approach, 5.121–5.122
Digital linear tape (DLT), storage options in file-based workflow, 4.65
Digital mobile radio (DMR), interfacing to analog PL intercom, 5.90
Digital partyline intercom systems, 5.92–5.93
Digital radio mondale (DRM), shortwave broadcasts, 2.109
Digital Rights Management. *See* DRM (Digital Rights Management)
Digital services:
  computing service and interfering field strengths for television, 2.120–2.124
  migrating from analog to digital FM modulation, 2.231–2.233
Digital signal processing (DSP), 2.16, 5.146
Digital signal processors (DSPs), 2.16
Digital television. *See* DTV (digital television)
Digital transmission, AM antenna impedance requirements for, 2.24–2.25
Digital Versatile Disk (DVD):
  MPEG-2 specification, 3.3
  surround sound support, 5.22
Dimming:
  centralized and distributed, 5.124–5.125
  ELV (electronic low-voltage dimmers), 5.127
  light controls and, 5.125–5.126
  VAC dimmers, 5.120
DIN (German Industrial Standard), equipment rack standard, 6.25
Dipole antenna arrays, 2.107–2.108
Direct beam fluorescent, lighting instruments, 5.134–5.135
Direct beam incandescent scoops and broads, lighting instruments, 5.136
Direct beam LED panels, lighting instruments, 5.135–5.136
Directional antennas:
  monitoring, 2.96–2.97
  troubleshooting directional arrays, 2.98–2.99
Directional couplers, amplifiers, 2.18–2.19
Directional filter combining system, 2.60–2.61
Disaster planning and recovery:
  anticipation, 7.37
  assessment, 7.38–7.39
  checklist, 7.41
  common precautions, 7.40
  introduction, 7.37
  preparation, 7.38
  who is in charge and personnel matters, 7.40
Discontinuities (program transitions), MPEG-2 timing model and, 3.26
Dissolves, video, 5.65
Distributed content and control, centralized broadcasting, 5.59
Distributed content, store and forward methods and, 5.49

Distributed television service (DTS):
  model for television coverage and interference, 2.117
  receive antenna pattern, 2.124
Distribution workflow, network systems, 4.66
DLT (digital linear tape), storage options in file-based workflow, 4.65
DMR (Digital mobile radio), interfacing to analog PL intercom, 5.90
DMX:
  distribution, 5.122–5.124
  light controls, 5.139
Document control, for wiring database, 6.12–6.13
Documentation:
  disaster preparation checklist, 7.41
  of project plans, 7.17–7.19
  self-documenting wire, 6.14
  of system design, 7.24
  technical documentation for new facility, 6.7–6.9
  in wire management, 6.12–6.14
Documents, ABIP, 1.18
DoD (Department of Defense) model, 4.9–4.11
Doherty:
  high-efficiency amplification techniques for transmitters, 2.193–2.195
  solid state amplifiers and, 2.150–2.151
  television transmitters and, 2.185
Doherty, William H., 2.150–2.151
Dolby ATMOS, 5.44
Dolby decoding, 5.43
Doors, sound isolation techniques, 5.16
Downstream keys, video effects, 5.66
Dramatic production (Theater) studio, 5.5. *See also* Production facility design; Studios
Driving frequency, sound isolation techniques, 5.17
DRM (digital radio mondale), shortwave broadcasts, 2.109
DRM (Digital Rights Management):
  ATSC security, 7.73
  building broadcast studio and, 5.26
  Mobile DTV Service Protection System, 3.63, 3.65
DSP (Digital Signal Processing), 2.16, 5.146
ΔT, operating parameters for HVAC systems, 6.49
DTS (Decoding Time Stamp), in MPEG-2 timing model, 3.26
DTS (distributed television service):
  model for television coverage and interference, 2.117
  receive antenna pattern, 2.124
DTV (digital television):
  ATSC 1 standard, 7.67
  captioning challenges, 1.24–1.25
  FCC regulations, 2.117
  role of PSIP in, 3.29
DTV (digital television), RF considerations:
  8-VSB channel occupancy, 2.136
  8-VSB power, 2.146–2.148
  8-VSB signal and, 2.133–2.134
  8-VSB specialist certification, 2.151
  8-VSB transmission monitoring, 2.143–2.146
  FCC spectral mask, 2.141–2.143
  high-power amplifiers, 2.148–2.151
  introduction, 2.133

DTV (digital television), RF considerations (*Cont.*):
  I-signal, 2.136–2.140
  Q-signal, 2.140
  references, 2.151
  RF upconverter, 2.140–2.141
  single sideband transmissions, 2.134–2.136
  terminology, 2.151
DTV (digital television), RF measurement:
  ACPR (adjacent channel power), 2.220–2.223
  digital signal overload, 2.213–2.215
  dynamic range and intermodulation and, 2.213
  of emissions mask compliance, 2.219–2.220
  EVM (error vector magnitude), 2.224
  introduction, 2.211
  MER (modulation error ratio), 2.223–2.224
  of pilot frequency, 2.216–2.218
  with power meter, 2.211–2.212
  references, 2.230
  relationship of response errors to signal noise, 2.224–2.226
  of shoulder amplitude ratio, 2.219
  of signal amplitude, 2.215–2.216
  of signal flatness, 2.218–2.219
  signal quality, 2.223
  signal quality measurements, 2.227–2.229
  signal terminology and, 2.212
  signal to noise budget, 2.226–2.227
  SNR (signal-to-noise ratio), 2.223
  with spectrum analyzer, 2.212
  spectrum analyzer specifications, 2.215
  with vector analyzer or TV analyzer, 2.229–2.230
DTV (digital television), RF transmission system:
  description of, 2.154–2.155
  introduction, 2.151, 2.153–2.154
  receiver for, 2.160–2.165
  system parameters, 2.179–2.182
  transmitter for, 2.155–2.160
  VSB baseband, 2.166–2.169
  VSB baseband spectral description, 2.169–2.172
  VSB RF spectrum description, 2.172–2.179
Dual feeder system, for power loss protection, 6.76–6.77
Ducts, acoustics in facility design, 5.13–5.15
Due process, FCC licensing, 1.5
DVD (Digital Versatile Disk):
  MPEG-2 specification, 3.3
  surround sound support, 5.22
Dynamic Carrier Control (DCC), power consumption for AM transmitters, 2.26–2.27
Dynamic Host Configuration Protocol (DHCP), 4.56
Dynamic range, of spectrum analyzer, 2.213
Dynamic routing, 4.32–4.33

EAN (Emergency Alert Notification), 1.33–1.34
Earthing, 6.53. *See also* Grounding
Earthquakes, disaster planning and, 7.38
EAS (Emergency Alert System). *See also* M-EAS (Mobile Emergency Alert System)
  Chief Operator's responsibility for field logs, 1.14
  improvements to, 1.33–1.35
  introduction, 1.31

EAS (Emergency Alert System) (Cont.):
  Mobile Emergency Alert System and, 3.49
  Mobile Emergency Alert System interface with, 3.53
  resources for, 1.35
EAS/CAP (Emergency Alert System/Common Alerting Protocol):
  inspection of EAS/CAP equipment, 1.18
  interface of Mobile Emergency Alert System with EAS/CAP, 3.53
  rules for all broadcast and cable operations, 1.34–1.35
EAT (Emergency Alert Table):
  Mobile DTV signaling, 3.45
  Mobile Emergency Alert System and, 3.50
EBU (European Broadcasting Union), 5.109–5.110
Edit Decision List (EDL), use in automation systems, 5.74
Editing suites, in production facility, 5.9–5.10
EDL (Edit Decision List), use in automation systems, 5.74
Education:
  role of management and leadership in, 7.13–7.14
  SBE programs, 7.65
Effects, video, 5.65–5.66
Efficiency:
  amplification techniques for transmitters, 2.193–2.195
  antenna characteristics, 2.69
  of transmitters, 2.26–2.27
EIA (Electronic Industry Alliance):
  EIA-568A/EIA-568B wiring schemes, 4.11–4.12
  EIA/TIA-485 standard for DMX 512, 5.143
  industry standards for equipment racks, 6.27
  overview of, A.56
  references for studio lighting, 5.143
EIGRP, routing protocol options, 4.34–4.35
EIT (Event Information Table):
  overview of, 3.34–3.35
  PSIP descriptors, 3.37
  PSIP table intervals, 3.40–3.41
Electric shock:
  circuit protection hardware, 7.46
  effects on human body, 7.43–7.46
  overview of, 7.43–7.46
  working with high voltage, 7.47
Electrical systems, maintaining, 6.117–6.119
Electrical work, standard units of measurements, A.5
Electrode, grounding electrode, 6.53–6.54
Electromagnetic (EM) spectrum:
  assigning and allocating frequencies, A.45–A.47
  DC to light, A.40
  designating frequency bands, A.41
  introduction, A.37–A.38
  light to gamma rays, A.41–A.45
  optical spectrum, A.38–A.39
  references, A.45–A.47
  subregions, A.38
  of visible light, 5.106
Electronic Industry Alliance. See EIA (Electronic Industry Alliance)
Electronic low-voltage dimmers (ELV), 5.127

Electronic newsgathering. See ENG (electronic newsgathering)
Electronic Newsroom Technique (ENT), 1.28
Electronic Program Guide (EPG):
  PSIP support for, 3.29
  showing relationship of PSIP tables, 3.38–3.39
Electronic Service Guide (ESG):
  Mobile DTV support, 3.66
  Mobile Emergency Alert System and, 3.52
Electrostatic discharge (ESD), 6.38
Elevation pattern:
  antenna characteristics, 2.69
  antenna gain and, 2.71–2.72
  computing service and interfering field strengths for television, 2.120–2.121
Ellipsoidal Reflector Spotlight (ERS), lighting instruments, 5.131–5.133
ELV (electronic low-voltage dimmers), 5.127
EM (electromagnetic) spectrum. See Electromagnetic (EM) spectrum
Emergency Alert Notification (EAN), 1.33–1.34
Emergency Alert System. See EAS (Emergency Alert System)
Emergency Alert System/Common Alerting Protocol (EAS/CAP). See EAS/CAP (Emergency Alert System/Common Alerting Protocol)
Emergency Alert Table (EAT):
  Mobile DTV signaling, 3.45
  Mobile Emergency Alert System and, 3.50
Emergency power sources, 6.74
Emissions limits, for television stations, 2.201–2.203
Emissions mask compliance, 2.219–2.220
Encapsulation/de-encapsulation:
  IP networks, 4.9–4.11
  in Layer 2 and Layer 3 conceptual network, 4.48
  MPEG-2 sections, 3.13
  MPEG-2 transport, 3.4–3.5
Encryption standards, Mobile DTV, 3.63
ENG (electronic newsgathering):
  audio system in van, 5.46
  benefits of COFDM, 2.233–2.234
  cellular aggregation technology and CNG workflow, 2.238–2.240
  cellular newsgathering (CNG), 2.236–2.238
  cellular revolution and, 5.146–5.147
  cellular transmission systems, 5.154
  changes in news coverage and, 5.161
  economics of portable systems, 5.156–5.157
  features of ideal ENG system, 5.147–5.149
  field implementation (getting the story), 5.157–5.158
  field operations, 5.46–5.47
  hybrid aggregation, 2.241–2.243
  introduction, 2.231
  introduction and history, 5.145–5.146
  IP-based ENG systems, 5.149–5.150
  migrating from analog to digital FM modulation, 2.231–2.233
  migrating from KU to KA band IP satellite systems, 2.240–2.241

ENG (electronic newsgathering) (*Cont.*):
  migrating to hybrid microwave solutions, 2.241
  modem configuration options, 5.151–5.154
  overview of COFDM, 2.234–2.235
  references, 2.246
  robust approach to getting story while maintaining uninterrupted transmission, 5.158–5.161
  summary, 2.243–2.245
  system design, 5.150–5.151
  vans, 5.46
  Wi-Fi and, 5.154–5.155
Engineering groups, A.51
Engineering Managers. *See* Management and leadership
Ennes Workshops, SBE education programs, 7.65
Ensembles, of ATSC data, 3.59
ENT (Electronic Newsroom Technique), 1.28
Enterprise Service Bus (ESB), next generation in automation systems, 5.75
Envelope Delay, signal noise as function of, 2.225
Envelope Tracking (drain modulation), transmitter amplification techniques, 2.194–2.195
Environmental management. *See also* Cooling
  designing cooling systems, 6.41
  for equipment rooms, 6.42–6.43
  heater exchangers, 6.44
  high availability design, 6.44–6.45
  introduction, 6.41
  maintaining antennas and, 2.80–2.81
  operating parameters and, 6.49–6.52
  satellite antennas and, 6.48
  site heat and, 6.45–6.46
  for studios, 6.42
  transmitter best practices and, 2.29–2.30
  transmitters and, 6.46–6.48
EPG (Electronic Program Guide):
  PSIP support for, 3.29
  showing relationship of PSIP tables, 3.38–3.39
Equalizers:
  delay/loss corrections in FM combiners, 2.68
  linear equalizer for DTV receivers, 2.162–2.163
Equipment:
  analyzing budget requirements, 7.27
  maintenance inspections, 6.105–6.106
  selecting and installing for new facility, 6.4–6.6
  selecting in system design process, 7.25
Equipment rack enclosures:
  computer floors, 6.37
  configuration options, 6.32
  grounding, 6.36–6.37
  industry standard for, 6.25–6.28
  introduction, 6.25
  layout, 6.25, 6.35–6.36
  references, 6.39
  selecting, 6.32–6.33
  types of, 6.28–6.32
Equipment rooms, cooling for, 6.42–6.43
Error Indicator bit, Mobile DTV content delivery and, 3.43–3.44

Error vector magnitude. *See* EVM (error vector magnitude)
ERS (Ellipsoidal Reflector Spotlight), lighting instruments, 5.131–5.133
ESB (Enterprise Service Bus), next generation in automation systems, 5.75
ESD (electrostatic discharge), 6.38
ESG (Electronic Service Guide):
  Mobile DTV support, 3.66
  Mobile Emergency Alert System and, 3.52
Ethernet:
  802.xx standard, 4.6
  AoIP based on, 5.19–5.20
  benefits for use with AoIP, 5.20–5.21
  cables, 4.11
  clock synchronization methods, 4.69
  comparing AES audio to Ethernet-based systems, 5.21
  connectors, 4.13
  cross-over cables, 4.12
  delay (latency) and audio quality, 5.23–5.24
  Ethernet II (DIX) frame, 4.16–4.17
  IP networks based on. *See* IP networks
  Layer 2 Ethernet switches, 4.19–4.23
  limitations on clock synchronization in Ethernet networks, 4.72–4.73
  planning AoIP network, 5.27–5.29
  surround sound support, 5.22
  synchronous Ethernet (SyncE), 4.77
  types of media for Ethernet cables, 4.13–4.14
  VLAN configuration examples, 4.48–4.50
ETMs (Extended Text Messages), 3.36
ETT (Extended Text Table):
  overview of, 3.36
  PSIP descriptors, 3.37
European Broadcasting Union (EBU), 5.109–5.110
European International Electrotechnical Commission. *See* IEC (European International Electrotechnical Commission)
Event coverage, news broadcasts, 5.47–5.48
Event Information Table. *See* EIT (Event Information Table)
EVM (error vector magnitude):
  defined, 2.151
  measuring signal quality, 2.224
  monitoring 8-VSB transmissions, 2.143–2.144
Exams, SBE certification, 7.65
Exciters:
  applying RF transport stream to, 2.183
  overview of, 2.16–2.17
  VSB exciter, 2.159–2.160
Executive manager, role in program management, 7.29–7.30
Exothermic bonding, bonding elements of grounding system, 6.60–6.61
Expertise, planning for success and, 7.19
Extended Text Messages (ETMs), 3.36
Extended Text Table (ETT):
  overview of, 3.36
  PSIP descriptors, 3.37
Extensible Markup Language (XML), 5.80

Facility:
　designing broadcast facility. See Broadcast facility design
　designing production facility. See Production facility design
　physical. See Physical facility
Factory tests:
　for antenna pattern and gain specification, 2.73
　for antennas, 2.75–2.76
Faders, controlling lighting, 5.126
Fades, video, 5.65
Fahrenheit, Celsius-to-Fahrenheit conversion table, A.8
Fast Information Channel (FIC):
　mechanisms for mobile service signaling, 3.45
　Mobile DTV data channels, 3.60
Fault diagnostics, 2.27
Fault-tolerance, designing AC power system for, 6.69
FC-75 toxic vapor, 7.53
FCC (Federal Communications Commission):
　Advisory Committee for Advanced Television Service, 2.120, 2.124
　Electronic Newsroom Technique, 1.28
　emissions limits for TV stations, 2.201–2.203
　frequency assignments and allocations, A.46–A.47
　shortwave regulations, 2.101
　spectral mask, 2.141–2.143
　television regulations, 2.115–2.117, 2.153
Feasibility studies, systems engineering and, 7.28
Feature news, field implementation of ENG, 5.157–5.158
FEC (forward error correction):
　8-VSB signal and, 2.133
　defined, 2.151
　in designing ENG system, 5.151
　mobile DTV content delivery and, 3.43
　in ST2022-1, 4.83–4.85
Federal Communications Commission. See FCC (Federal Communications Commission)
Federal Emergency Management Agency (FEMA):
　administering Emergency Alert Notification system, 1.33–1.34
　interface of Mobile Emergency Alert System with EAS/CAP, 3.53
Feeder lines, for AM antenna, 2.89–2.90
FEMA (Federal Emergency Management Agency):
　administering Emergency Alert Notification system, 1.33–1.34
　interface of Mobile Emergency Alert System with EAS/CAP, 3.53
FETs (field-effect transistors), RF amplification technologies, 2.185
Fiber optic links, planning for AoIP, 5.29
FIC (Fast Information Channel):
　mechanisms for mobile service signaling, 3.45
　Mobile DTV data channels, 3.60
Field Engineering Check List, for antenna installation, 2.81–2.82
Field measurements, duties of Chief Operator, 1.13–1.14
Field operations, ENG, 5.46–5.47
Field programmable gate arrays (FPGAs), 2.16
Field strength, computing interfering field strengths for television, 2.119–2.124
Field-effect transistors (FETs), RF amplification technologies, 2.185
File Delivery over Unidirectional Transport (FLUTE), 3.44, 3.46
File roll-in, video, 5.67–5.68
File-based workflow:
　fundamental technologies, 4.63–4.67
　overview of, 4.62–4.63
Fill light, in three point lighting, 5.105
"Film at Eleven!," 5.145, 5.161
Filters, 8-VSB, 2.137–2.140
Filters, FM:
　cross-coupled filters, 2.66–2.68
　directional filter combining system, 2.60–2.61
　order and loss in FM channel combiners, 2.65–2.66
FIMS (Framework for Interoperability of Media Services), 5.74–5.75
Final stage power amplifiers, 2.17–2.19
FIR (finite impulse response) filters, 8-VSB, 2.139
Fire:
　maintaining transmission system, 6.119
　operating parameters for HVAC systems, 6.52
　precautions in disaster planning, 7.40
Firewalls:
　NAT and, 4.56
　overview of, 4.59
First aid:
　equipment at transmitter sites, 6.88
　procedures, 7.47–7.48, 7.50
Fixed pipe grid, lighting suspension and rigging systems, 5.111–5.112
Flat panel displays, 5.55
Flatness, of signal, 2.225
FLIRs (Forward Looking Infrared) cameras, in tuning HVAC systems, 6.51
Flood, precautions in disaster planning, 7.40
Flood lights, 5.133–5.136
Flooded cables, 6.18
Floor plans, use in detailed design, 7.26
Floor-mounted open frame equipment racks, 6.28
Floors, sound isolation techniques, 5.16
Flow diagrams, use in detailed design, 7.26–7.27
Fluorescent, light sources, 5.129
FLUTE (File Delivery over Unidirectional Transport), 3.44, 3.46
Flypack systems, in remote news broadcasts, 5.48
FM (frequency modulation):
　bands for FM channels, 2.70
　corporate feed system, 2.85
　exciters, 2.16–2.17
　migrating from analog to digital FM, 2.231–2.233
　power amplifiers, 2.18
　theory, 2.6–2.10
　transmitters. See Transmitters (AM and FM)
FM channel combiners:
　channel spacing and rejection, 2.63–2.65
　cross-coupled filters, 2.66–2.68
　delay/loss corrections, 2.68
　directional filter combining system, 2.60–2.61

INDEX **I.13**

FM channel combiners (*Cont.*):
  filter order and group delay, 2.65–2.66
  filter order and loss, 2.66
  frequency response, 2.61–2.62
  IM suppression, 2.62–2.63
  introduction, 2.59
  junction combiners, 2.59–2.60
  references, 2.68
  types of, 2.59
FM directional filter combining system, 2.60–2.61
FM stations:
  Chief Operator making field measurements, 1.14
  corporate feed systems, 2.85
  equipment inspection, 6.105
  maintaining transmission system, 6.106–6.107
FMO (Frequency Management Organization), 2.104–2.105
Formats:
  graphics files, 5.64
  HD cameras, 5.63–5.64
  video files, 5.67
Forward error correction. *See* FEC (forward error correction)
Forward Looking Infrared (FLIRs) cameras, in tuning HVAC systems, 6.51
FOT (Frequency of Optimum Traffic):
  defined, 2.114
  shortwave and, 2.111
FPGAs (field programmable gate arrays), 2.16
Fragmentation, IP packets, 3.44
Frames/subframes, Mobile DTV, 3.59
Framework for Interoperability of Media Services (FIMS), 5.74–5.75
Free space propagation, propagation theory, 2.129–2.131
Free-standing equipment racks, 6.30
Frequencies, shortwave:
  international bands, 2.102–2.104
  managing, 2.104–2.105
  for optimum traffic, 2.111
  requesting, 2.106
Frequency band:
  accounting for frequency drift in mask filters, 2.209
  assigning and allocating, A.45–A.47
  designating, A.41
  international shortwave bands, 2.102–2.104
  mask filters and, 2.207–2.208
Frequency Management Organization (FMO), 2.104–2.105
Frequency modulation. *See* FM (frequency modulation)
Frequency of Optimum Traffic (FOT):
  defined, 2.114
  shortwave and, 2.111
Frequency range, of spectrum analyzer, 2.213
Frequency response:
  effecting performance, 2.23–2.24
  in selection of FM channel combiner, 2.61–2.62
Fresnel lens, lighting instruments, 5.130–5.131
Fuel lines/tanks, maintaining generators and, 6.102–6.104

Fuel tanks, assessment aspect of disaster planning, 7.39
Fully centralized broadcasting, 5.58–5.59. *See also* Centralized broadcasting

Gain, antenna, 2.71–2.75
Gamma ray band, in EM spectrum, A.41, A.43, A.45
GAT (Guide Access Table):
  Mobile DTV system, 3.63
  tables in mobile service signaling, 3.45
GBIC (Gigabit Ethernet Interface Converter), 4.13
General Manager (GM), role in interdepartmental cooperation, 7.14
General-Purpose Input/Output (GPIO):
  matrix intercom systems and, 5.101
  use in automation systems, 5.73–5.74
Generators:
  choosing, 6.78–6.80
  in disaster planning and recovery, 7.38
  disaster preparation checklist, 7.41
  maintaining, 6.99–6.105
  types of, 6.80–6.81
Genlock over IP, 4.77
German Industrial Standard (DINJ), equipment rack standard, 6.25
GFCI (ground-fault current interrupter), 7.46
Gigabit Ethernet, 4.13–4.14
Gigabit Ethernet Interface Converter (GBIC), 4.13
Glass-to-Glass latency, IP-based ENG systems, 5.149–5.150
Global Positioning System (GPS):
  clock synchronization methods, 4.69
  representation of System Time Table in GPS seconds, 3.31
GM (General Manager), role in interdepartmental cooperation, 7.14
Goals, planning for success, 7.18
Go-to person, disaster planning and, 7.40–7.41
GPIO (General-Purpose Input/Output):
  matrix intercom systems and, 5.101
  use in automation systems, 5.73–5.74
GPS (Global Positioning System):
  clock synchronization methods, 4.69
  representation of System Time Table in GPS seconds, 3.31
Graphics:
  CIAB (Channel in a Box), 5.52
  file-based workflow, 4.63
  sources of dynamic graphics, 5.51
  types of video input, 5.64
Green screen (hard cyc), 5.118–5.119
Green screen lights, lighting instruments, 5.137–5.138
Grids, lighting suspension and rigging systems, 5.111–5.113
Ground rods:
  chemical grounding rods, 6.56–6.58
  ground system options, 6.62–6.63
  overview of, 6.54–6.55
Ground-fault current interrupter (GFCI), 7.46

Grounding:
  bonding elements of grounding system, 6.60–6.61
  bulkhead panel, 6.63–6.66
  checklist for, 6.66–6.67
  chemical grounding rods, 6.56–6.58
  designing ground system for AM antenna, 2.88–2.89
  equipment racks, 6.35–6.37
  grounding electrode in, 6.53–6.54
  grounding interface, 6.54–6.56
  inductance of grounding system, 6.61
  introduction, 6.53
  OSHA guidelines, 7.51
  references, 6.67
  system options, 6.62–6.63
  tower maintenance, 6.120
  Ufer ground system, 6.58–6.60
Grounding electrode, 6.53–6.54
Grounding interface, 6.54–6.56
Group delay, FM channel combiners, 2.65–2.66
Groups, Mobile DTV, 3.60
Guide Access Table (GAT):
  Mobile DTV system, 3.63
  tables in mobile service signaling, 3.45

H.264 standard, 5.149–5.150
HA (high availability), 4.86, 6.44–6.45
HAAT (height above average terrain), FCC regulation of television, 2.116
Hard cyc, 5.118–5.119
Hardware, for master control:
  Baseband master control, 5.51–5.52
  CIAB (Channel in a Box), 5.52–5.53
  overview of, 5.50–5.51
Hardware, limitations and precorrection, 2.20–2.21
HD (high-definition):
  camera formats, 5.63–5.64
  newsgathering, 2.231
  television, 2.153
  video formats, 5.63
HD radio:
  building broadcast studio and, 5.26
  maintaining transmission system, 6.107
  system, 2.31–2.32
  theory, 2.10–2.15
HDSDI:
  baseband master control, 5.51
  IP-based ENG systems, 5.149
HDTV:
  applications layer of ATSC 3.0 and, 7.71
  DTV transmission system, 2.154
  surround sound support, 5.22
Head:
  head to shoulder amplitude ratio, 2.219–2.220
  of signal, 2.212
Headsets:
  components of wired party-line intercom systems, 5.86–5.87
  control panels and, 5.103
Heating. *See* HVAC (heating, ventilation, and air conditioning)

Height above average terrain (HAAT), FCC regulation of television, 2.116
HF (high frequency) broadcasts. *See* Shortwave broadcasts
HFCC (High Frequency Coordination Conference):
  Frequency Management Organizations coordinating with, 2.104
  overview of, 2.103
  shortwave frequency requests, 2.106
High availability (HA), 4.86, 6.44–6.45
High Frequency Coordination Conference. *See* HFCC (High Frequency Coordination Conference)
High frequency (HF) broadcasts. *See* Shortwave broadcasts
High power amplifiers, DTV, 2.148–2.151
High voltage:
  operating hazards, 7.48
  precautions, 7.46
  working with, 7.47
High-definition. *See* HD (high-definition)
Hilbert transform:
  defined, 2.151
  phase shift required for 8-VSB modulation, 2.136
Hop count, routing protocols and, 4.34–4.35
Horizontal pattern, 2.71
Host Address component, IP addressing, 4.26
Hot spots, Wi-Fi, 5.147
Hot surfaces:
  operating hazards, 7.48
  safety considerations, 7.57–7.58
HVAC (heating, ventilation, and air conditioning):
  cooling television transmitters, 2.200
  designing cooling systems, 6.41
  in detailed design, 7.26
  environmental management, 6.41
  for equipment rooms, 6.42–6.43
  heater exchangers, 6.44
  high availability design, 6.44–6.45
  maintaining, 6.115–6.117
  maintaining electrical systems, 6.115–6.117
  operating parameters, 6.49–6.52
  for studios, 6.42
Hybrid aggregation, 2.241–2.243
Hybrid microwave solutions, 2.241

I and Q constellation, monitoring 8-VSB transmissions, 2.143
IAAIS (International Association of Audio Information Services), 1.22
IANA (Internet Assigned Numbers Authority):
  IPv6 and, 4.39–4.40
  network topology and, 4.56
IBB (International Broadcasting Union), in frequency management, 2.104
ICANN (Internet Corporation for Assigned Names and Numbers), management of IP addresses, 4.56
ICMP (Internet Control Message Protocol), 4.35–4.36
ICPM (incidental carrier phase modulation), 2.146
IDFs (Intermediate Distribution Frames), 6.20

IEC (European International Electrotechnical Commission):
　history of standardization, A.52
　industry standards for equipment racks, 6.25–6.27
　overview of, A.53
IEEE (Institute of Electrical and Electronic Engineers):
　controls on nonionizing radiation, 7.55–7.56
　Ethernet standards, 4.11
　information technology systems and, 4.1
　networking standards, 4.5–4.6
　overview of, A.56
　reference documents, A.57
　STP standard (802.1d), 4.20–4.21
　tagged Ethernet frames (802.1Q), 4.18
IEEE 1588 standard. *See also* PTP (Precision Time Protocol)
　accuracy of PTP, 4.71–4.72
　clock synchronization methods, 4.70
　message flow, 4.71
　NTP compared with PTP, 4.76
　specifications of, 4.74
　TCs (Transparent Clocks), 4.73
IETF (Internet Engineering Task Force):
　information technology systems and, 4.1
　networking standards, 4.5–4.6
　PTP standards, 4.74
　reference documents, A.57
　RFCs (Request-for-Comments), 4.53
IF filter, for DTV receivers, 2.161–2.162
IFB (Interrupted Fold-Back):
　communication circuit, 5.96–5.97
　IP-based ENG systems, 5.149–5.150
IGMP (Internet Group Management Protocol):
　Layer 3 protocols, 4.35–4.36
　MPEG-2 transport streams over IP networks, 4.81
　as multicast protocol, 4.38
IGP (interior gateway protocols), 4.34–4.35
IM (intermodulation):
　in selection of FM channel combiner, 2.62–2.63
　signal noise as function of transmitter IM, 2.225–2.226
　of spectrum analyzer, 2.213–2.215
　transmitter limitations, 2.20
Impedance:
　AM antenna impedance requirements for digital transmission, 2.24–2.26
　coaxial transmission lines, 2.36
　complex conjugate impedance in AM antenna, 2.95–2.96
　methods for representing complex impedance, 2.91
Implosion hazard, safety considerations, 7.57
Incandescent cyclorama lights, lighting instruments, 5.136–5.137
Incandescent groundrow cyc lights, lighting instruments, 5.137
Incandescent light sources, 5.127
Inches, converting to/from millimeters, A.9–A.10
Incidental carrier phase modulation (ICPM), 2.146
Indoor heater exchangers, in environmental management, 6.44
Inductance, of grounding system, 6.61

Inductive output tube. *See* IOT (inductive output tube)
Inductive reactance, antenna system components, 2.91
Inductors, antenna system components, 2.91
Infrared (IR) thermometers, in tuning HVAC systems, 6.51
Infrastructure, of network systems, 4.55–4.56, 4.60
Ingest department, audio monitoring and, 5.41
Input sources, for Mobile Emergency Alert System, 3.50–3.52
Input/output (I/O):
　client cards and, 5.99–5.100
　devices in OB truck, 5.47–5.48
Inspections, duties of Chief Operator, 1.13
Installation:
　of antennas, 2.77–2.80
　of mask filters, 2.209–2.210
　of rigid coaxial systems, 2.53–2.56
　of semiflexible coaxial systems, 2.50–2.52
Institute for Telecommunications Sciences (ITS). *See* Longley-Rice model
Institute of Electrical and Electronic Engineers. *See* IEEE (Institute of Electrical and Electronic Engineers)
Insurance:
　assessment aspect of disaster planning, 7.39
　common precautions in disaster planning, 7.40
　disaster preparation checklist, 7.41
Integrated Public Alert and Warning System. *See* IPAWS (Integrated Public Alert and Warning System)
Interactivity, applications layer of ATSC 3.0 and, 7.72
Intercom (talkback) system:
　analog systems, see Analog partyline intercom systems
　in outside broadcast truck, 5.47
　overview of, 5.83
　for production. *See* Production intercom systems
Interdepartmental cooperation, role of management and leadership in, 7.14–7.15
Interdigital filters:
　accounting for frequency drift in mask filters, 2.209
　retuning, 2.210
　types of mask filters, 2.205
Interduct, wire management and, 6.20
Interference. *See also* Television coverage and interference
　allowable percentages, 2.128
　computing, 2.119–2.124
　maximum site to cell interference evaluation, 2.127
Interior gateway protocols (IGP), 4.34–4.35
Intermediate Distribution Frames (IDFs), 6.20
Intermediate power amplifier. *See* IPA (intermediate power amplifier)
Intermodulation. *See* IM (intermodulation)
Internal circuitry, components of matrix intercom systems, 5.98
International Association of Audio Information Services (IAAIS), 1.22
International bands, shortwave, 2.102–2.104
International Broadcasting Union (IBB), in frequency management, 2.104

International Standards Organization. *See* ISO (International Standards Organization)
International Telecommunications Union (ITU), 4.74
Internet Assigned Numbers Authority. *See* IANA (Internet Assigned Numbers Authority)
Internet Control Message Protocol (ICMP), 4.35–4.36
Internet Corporation for Assigned Names and Numbers (ICANN), management of IP addresses, 4.58
Internet Engineering Task Force. *See* IETF (Internet Engineering Task Force)
Internet Group Management Protocol. *See* IGMP (Internet Group Management Protocol)
Internet Protocol. *See* IP (Internet Protocol)
Internet Protocol Security (IPSec), 4.35–4.36
Internetworking (routing):
 comparing interior gateway protocols, 4.35
 dynamic routing, 4.32–4.33
 hop count and, 4.34–4.35
 Layer 3 switches, 4.35
 overview of, 4.32
 routing protocols and routing tables, 4.33–4.34
 static routing, 4.32
Interrupted Fold-Back (IFB):
 communication circuit, 5.96–5.97
 IP-based ENG systems, 5.149–5.150
Inter-symbol interference (ISI), 8-VSB filters and, 2.138
Interval signals:
 defined, 2.114
 shortwave broadcasts, 2.112–2.113
Inventory system:
 disaster preparation checklist, 7.41
 information in, 7.5
I/O (input/output):
 client cards and, 5.99–5.100
 devices in OB truck, 5.47–5.48
Ionization, and recombination in ionosphere, 2.109
Ionosphere, shortwave broadcasts and, 2.109–2.111
IOT (inductive output tube):
 high power amplifiers, 2.148–2.149
 multi-stage depressed collector (MSDC-IOT), 2.150
 RF amplification technologies, 2.188–2.189
 RF upconverter and, 2.140
 television transmitters and, 2.184
 use on UPS system, 2.149–2.150
IP (Internet Protocol):
 IP routers at Layer 3, 4.59
 IP-based ENG systems, 5.149–5.150
 matrix intercom systems communicating over, 5.100–5.101
 overview of, 4.58
 providing transport for Mobile DTV system, 3.66–3.67
 reverse engineering IPv4 address, 4.50–4.53
 use in automation systems, 5.74–5.76
IP addressing:
 classful vs. classless addressing, 4.28
 IPv4 address masks, 4.29
 IPv4 Classes, 4.26–4.27
 IPv4 Network and Host Address components, 4.26
 IPv4 packet header, 4.25

IP addressing (*Cont.*):
 IPv4 private and public addresses, 4.30–4.31
 IPv4 reserved and special addresses, 4.32
 IPv6, 4.39–4.42
 mobile DTV content delivery and, 3.44
 NAT (Network Address Translation), 4.30–4.31
 network topology and, 4.56
 PAT (Port Address Translation), 4.31
 reverse engineering IPv4 address, 4.50–4.53
 subnetting and supernetting, 4.30
IP networks:
 building the segmented network, 4.44–4.46
 Data-Link Layer (Layer 2), 4.14
 encapsulation/de-encapsulation, 4.9–4.11
 encapsulation/de-encapsulation and, 4.24
 Ethernet frames and, 4.15–4.17
 frame flow, 4.24
 Layer 2 Ethernet switches, 4.19–4.23
 Layer 3 internetworking (routing), 4.32–4.35
 Layer 3 IP addressing, 4.23–4.32
 Layer 3 protocols, 4.35–4.36
 Layer 3 summary, 4.42–4.43
 layers of OSI model, 4.8–4.9
 LLC sublayer of Data-Link Layer, 4.15
 MAC sublayer of Data-Link Layer, 4.15–4.16
 multicast IP communications, 4.36–4.39
 Network Layer (Layer 3), 4.23
 networking standards, 4.5–4.6
 OSI model, 4.6–4.8
 PHY (physical layer), 4.11–4.14
 references, 4.53
 reverse engineering IPv4 address, 4.50–4.53
 review of IP network, 4.4–4.5
 summary, 4.46–4.48
 technology review, 4.4–4.5
 Transport Layer (Layer 4), 4.43–4.44
 VLAN configuration examples, 4.48–4.50
 VLANs, 4.17–4.18
IP transport, for Mobile DTV:
 announcements, 3.47
 content delivery framework, 3.43–3.44
 introduction, 3.43
 references, 3.47
 services, 3.44–3.45
 signaling, 3.45–3.46
 terminology, 3.47
 timing and buffer model, 3.46
IPA (intermediate power amplifier):
 overview of, 2.17
 RF upconverter and, 2.140
 as special case, 2.192–2.193
 television transmitters and, 2.184
IPAWS (Integrated Public Alert and Warning System):
 as input to Mobile Emergency Alert System, 3.50–3.51
 interface of Mobile Emergency Alert System with EAS/CAP, 3.53
 role in emergency management, 1.33–1.34
IP-based ENG systems:
 cellular transmission and, 5.154
 economics of portable systems, 5.156–5.157

IP-based ENG systems (*Cont.*):
  general considerations, 5.156
  modem configuration options, 5.151–5.154
  overview of, 5.149–5.150
  system design, 5.150–5.151
  Wi-Fi and, 5.154–5.155
IPSec (Internet Protocol Security), 4.35–4.36
IPv4:
  address masks, 4.29
  classes, 4.26–4.27
  multicast applications, 4.37–4.38
  Network and Host Address components, 4.26
  as outgrowth of ARPANET, 4.79
  overcoming limits in IPv4 address space, 4.39
  overview of, 4.58
  packet header, 4.25
  private and public addresses, 4.30–4.31
  reserved and special addresses, 4.32
  reverse engineering IPv4 address, 4.50–4.53
  subnetting and supernetting, 4.30
IPv6:
  32-bit and 128-bit addresses, 4.40
  autoconfiguration feature, 4.41–4.42
  classification of addresses, 4.41
  IANA (Internet Assigned Numbers Authority) and, 4.39–4.40
  overcoming limits in IPv4 address space, 4.39
  overview of, 4.58
IR (infrared), in optical spectrum, A.38–A.39
IR (infrared) thermometers, in tuning HVAC systems, 6.51
ISI (inter-symbol interference), 8-VSB filters and, 2.138
I-signal:
  defined, 2.151
  monitoring 8-VSB transmissions, 2.143
  overview of, 2.136–2.137
  phase tracking for DTV receivers, 2.163–2.164
  pilot insertion and, 2.137
  Q-signal orthogonal to, 2.140
  VSB filters and, 2.137–2.140
ISO (International Standards Organization):
  history of standardization, A.52
  network topology and, 4.56
  networking standards, 4.5–4.6
  OSI model. *See* OSI (Open System Interconnection) model
  overview of, A.53
  reference documents, A.57
Isochronous governor, generator types, 6.81
IT (information technology) systems, 4.1–4.2
ITU (International Telecommunications Union):
  frequency assignments and allocations, A.45–A.46
  history of standardization, A.52
  overview of, A.53–A.54
  PTP standards, 4.74

Job description, 7.5
Junction combiners, in selection of FM channel combiner, 2.59–2.60

KA band:
  hybrid aggregation, 2.241–2.243
  migrating from KU to KA band IP satellite systems, 2.240–2.241
Kean, J.C., 1.22
Key effects:
  chroma key, 5.66–5.67
  video effects, 5.65–5.66
Key light, in three-point lighting, 5.105
Keys, user/subscriber control key panels, 5.103
Kirchhoff's Law, 2.90
Klystrons, high power amplifiers, 2.148
KU band, 2.240–2.241

Labels, in wire management, 6.12, 6.14
Land mobile radio (LMR), 5.90
Latency:
  AoIP and, 5.23–5.24
  Audio delay compensating for latency in video display, 5.43
  Glass-to-Glass latency in IP-based ENG, 5.149–5.150
  network systems, 4.60
  off air monitoring and, 5.55
  QoS (Quality of Service), 4.60
Laterally diffused metal oxide silicon (LDMOS):
  RF amplification technologies, 2.185
  solid state amplifiers and, 2.150–2.151
Layer 1. *See* Physical layer (Layer 1)
Layer 2. *See* Data-Link Layer (Layer 2)
Layer 3. *See* Network Layer (Layer 3)
Layer 3 switches, 4.35
Layer 4 (Transport Layer):
  overview of, 4.8, 4.57
  protocols, 4.43–4.44
Layer 5 (Session Layer), 4.57
Layer 6 (Presentation Layer), 4.58
Layer 7 (Application layer):
  of OSI model, 4.7
  overview of, 4.58
Layout, equipment rack enclosures, 6.35–6.36
LDMOS (laterally diffused metal oxide silicon):
  RF amplification technologies, 2.185
  solid state amplifiers and, 2.150–2.151
Leadership. *See* Management and leadership
LED cyclorama lights, lighting instruments, 5.137
LED tape, lighting instruments, 5.138
LEDs (light emitting diodes):
  digital lighting approach, 5.121–5.122
  solid state light source, 5.127
  status LEDs on control panels, 5.104
  studio light sources, 5.107
Length conversion ratios, A.12
LEPs (Light Emitting Plasma):
  light sources, 5.128–5.129
  studio light sources, 5.107
LeSEA Broadcasting, shortwave frequency requests, 2.106
Licensing, FCC:
  basics, 1.4–1.5
  requirements, 1.3–1.4
  resources for, 1.5–1.9
Light control devices, 5.140

Light emitting diodes. *See* LEDs (light emitting diodes)
Light Emitting Plasma (LEPs):
   light sources, 5.128–5.129
   studio light sources, 5.107
Light source information, 5.106–5.108
Lighting:
   components of lighting system, 5.110
   controls, 5.125–5.126
   CQS (Color Quality Scale), 5.109
   CRI (Color Rendering Index), 5.108–5.109
   cyclorama, curtains, and curtain track, 5.116–5.118
   digital lighting approach, 5.121–5.122
   dimming, 5.124–5.125
   electrical power distribution and capacity, 5.119–5.121
   flood lights, 5.133–5.136
   hard cyc, 5.118–5.119
   introduction, 5.105
   light control devices, 5.139–5.140
   light source information, 5.106–5.108
   mounting, 5.138–5.139
   network data and DMX distribution and, 5.122–5.124
   other factors in, 5.140
   references, 5.141–5.143
   sources, 5.126–5.129
   spotlights, 5.130–5.133
   summary, 5.141
   supplementary lights, 5.136–5.138
   suspension and rigging systems, 5.110–5.116
   three point, 5.105–5.106
   TLCI (Television Lighting Consistency Index), 5.109–5.110
Lightning, protecting equipment against, 2.29
Li-ion (lithium ion) batteries, 6.83
Linear effects, amplifier predistortion/precorrection, 2.196
Linear equalization, for DTV receivers, 2.162–2.163
Linear Tape-Open (LTO), 4.65
Linearity of amplification:
   effecting performance, 2.21–2.23
   power amplifiers, 2.189–2.191
Link aggregation (port trunking), redundancy options for Ethernet switches, 5.28
Li-polymer (lithium polymer) batteries, 6.83
Liquid cooling, for transmitters, 2.199–2.200, 6.46–6.48
Lithium ion (Li-ion) batteries, 6.83
Lithium polymer (Li-polymer) batteries, 6.83
LLC (Logical Link Control), sublayer of Data-Link Layer, 4.15
LMR (land mobile radio), 5.90
Log books, in transmitter maintenance, 6.88–6.90
Logical (physical) addresses, MAC sublayer responsible for, 4.15
Logical Link Control (LLC), sublayer of Data-Link Layer, 4.15
Log-periodic antennas (LPA), 2.107
Longley-Rice model:
   computing service and interfering field strengths for television, 2.119–2.124
   overview of, 2.129–2.132

Long-Term Key Message (LTKM), 3.63
Loss-pass filters, in final stage power amplifiers, 2.18
Loudness measurement, special and extended audio functions, 5.43
Loudspeakers, 5.103. *See also* Speakers
Low ceiling method, lighting suspension and rigging systems, 5.113
Low power television stations. *See* LPTV (Low power television stations)
Lowest usable frequency (LUF):
   defined, 2.114
   frequency of optimum shortwave traffic, 2.111
LPA (Log-periodic antennas), 2.107
LPTV (low power television stations):
   elevation patterns for television broadcasts, 2.120
   emissions limits for TV stations, 2.201
   establishing protected service areas for television, 2.119
   model for television coverage and interference, 2.117
LTKM (Long-Term Key Message), 3.63
LTO (Linear Tape-Open), 4.65
LUF (Lowest usable frequency):
   defined, 2.114
   frequency of optimum shortwave traffic, 2.111

M&C (monitoring and control), for high availability, 6.44–6.45
MAC (Media Access Control):
   Ethernet switches learning MAC addresses, 4.19–4.20
   overview of MAC addressing, 4.58
   sublayer of Data-Link Layer, 4.15–4.16
MADI (Multichannel Audio Digital Interface):
   audio monitoring and, 5.42
   client cards, 5.100
   in outside broadcast truck, 5.47
   overview of, 5.19
Main Distribution Frames (MDFs), in wire management, 6.20
Maintenance:
   AC power plant, 6.72–6.73
   AM stations, 6.111–6.115
   antennas, 2.80–2.83
   coaxial transmission lines, 2.57–2.58
   of electrical systems, 6.117–6.119
   equipment inspection, 6.105–6.106
   of essential services, 7.10–7.12
   FM stations, 6.106–6.107
   generators, 6.99–6.105
   of HVAC systems, 6.115–6.117
   introduction, 6.87
   of mask filters, 2.210
   role of management and leadership in budgeting, 7.8
   site visits, 6.87–6.90
   of towers, 6.119–6.122
   transmission lines, 6.96–6.99
   transmitters, 2.31–2.32, 6.90–6.96
   TV stations, 6.107–6.110
Maintenance logs, ABIP review of, 1.19

MAM (media asset management):
  categorizing types of, 5.79–5.80
  history of, 5.78–5.79
  introduction, 5.77
  summary, 5.81
  uses of, 5.78
  XML and, 5.80
Managed switches, 4.59
Management and leadership:
  budgeting and, 7.7–7.9
  interdepartmental cooperation, 7.14–7.15
  introduction, 7.3
  maintaining essential services, 7.10–7.12
  planning and oversight of physical facility, 7.12
  scope of, 7.3–7.7
  staffing oversight and personnel management, 7.10
  strategic building development, 7.13
  strategic planning and development, 7.9–7.10
  summary, 7.14–7.15
  training and education, 7.13–7.14
Manifold junction combiner, 2.29
Manufacturer's recommendations, RF transmission, 2.28–2.30
Mask filters:
  band considerations, 2.207–2.208
  defined, 2.151
  emissions limits and, 2.201–2.203
  installing, 2.209–2.210
  introduction, 2.201
  maintaining and retuning, 2.210
  reflective and constant impedance filters, 2.204
  RF upconverter and, 2.141
  thermal stability of, 2.208–2.209
  transmitters and, 2.198–2.199
  types of, 2.204–2.207
Mass conversion ratios, A.12
Master control:
  baseband master control, 5.51–5.52
  BXF (Broadcast Exchange Format), 5.56
  CIAB (Channel in a Box), 5.52–5.53
  closed captions and metadata content, 5.55
  evolution of master control room for automation, 5.73
  function of, 5.49–5.50
  hardware, 5.50–5.51
  holy grail of, 5.53–5.54
  introduction, 5.49
  off air monitoring, 5.55
  quality control, 5.56
  relationship to traffic, 5.56
Master control room (MCR), audio monitoring and, 5.41
Master Guide Table (MGT), PSIP, 3.32–3.33, 3.40–3.41
Master stations, components of wired party-line intercom systems, 5.86
Matrix intercom systems:
  client cards and, 5.99–5.100
  communicating over IP, 5.100–5.101
  components, 5.98–5.99
  connectivity, 5.101–5.102

Matrix intercom systems (*Cont.*):
  GPIO (General-Purpose Input/Output), 5.101
  overview of, 5.97–5.98
  relay outputs, 5.101
  user/subscriber control key panels, 5.102–5.104
MAX RC (Maximal Ratio Combining), COFDM and, 2.234–2.235
Maximum Transmission Unit (MTU), 3.44
Maximum usable frequency (MUF):
  defined, 2.114
  frequency of optimum shortwave traffic, 2.111
MCR (master control room), audio monitoring and, 5.41
MDCL (Modulation Dependent Carrier Level), 2.26–2.27
MDFs (Main Distribution Frames), in wire management, 6.20
ME (Mix-Effects), video, 5.65
Mean-time between failure (MTBF):
  reducing, 6.47
  redundancy options and, 2.28
M-EAS (Mobile Emergency Alert System):
  adding to mobile DTV, 3.49–3.50
  advantages of, 3.52
  ATSC 3.0 standard and, 3.53–3.55
  content manager and transmission software, 3.52–3.53
  input sources, 3.50–3.52
  interface with current EAS/CAP equipment, 3.53
  introduction, 3.49
  references, 3.53–3.55
  relationship to national alerting infrastructure, 3.50
  scenario for use of, 3.52
Measurement:
  fractions to decimal conversion, A.11
  inch to/from millimeter conversion tables, A.9–A.10
  international system of, A.58
  length and area conversion ratios, A.12
  mass and volume conversion ratios, A.13
  RF. *See* DTV (digital television), RF measurement:
  standard units of, A.3–A.5
Mechanical systems, handling acoustic in facility design, 5.12–5.15
Media, file-based workflow, 4.63
Media Access Control. *See* MAC (Media Access Control)
Media asset management. *See* MAM (media asset management)
Media management:
  architecture, 4.66–4.67
  in file-based workflow, 4.66
Medium wave (MW), 2.88–2.89. *See also* AM (amplitude modulation)
Meetings:
  attributes of good, 7.21–7.22
  structure of, 7.20–7.21
MER (Modulation error ratio):
  measuring, 2.223–2.224
  monitoring 8-VSB transmissions, 2.144
MER (modulation error ratio):
  measuring, 2.223–2.224
  monitoring 8-VSB transmissions, 2.144
Message (data), network components, 4.5

Metadata:
  in file-based workflow, 4.63, 4.65
  master control and, 5.55
  network systems, 4.65–4.66
  PSI (program specific information) in MPEG-2, 3.15–3.16
Metal oxide silicon field effect transistor (MOSFET):
  RF amplification technologies, 2.184
  solid state amplifiers and, 2.150–2.151
Metcalf, Robert, 5.21
Method of moments (MOM), monitoring directional systems, 2.96–2.97
MGT (Master Guide Table), PSIP, 3.32–3.33, 3.40–3.41
M/H (Mobile/Handheld). *See* Mobile DTV
Microphones:
  components of wired party-line intercom systems, 5.86
  control panels and, 5.103
Microwave:
  FM microwave links supporting SD, 2.231
  hybrid aggregation, 2.241–2.243
  migrating to hybrid microwave solutions, 2.241
Microwave band:
  applications of, A.44
  in EM spectrum, A.40
Middleware, servers and, 4.64–4.65
Millimeters, converting to/from inches, A.9–A.10
Mix-Effects (ME), video, 5.65
Mixing console, in outside broadcast truck, 5.47
Mobile DTV:
  adding Mobile Emergency Alert System to, 3.49–3.50
  content delivery, 3.43–3.44
  diagram of transmission and transport systems, 3.62
  documentation of ATSC DTV standard, 3.58–3.59
  evolution of, 3.65–3.66
  foundations of, 3.58
  introduction, 3.57–3.58
  IP transport for. *See* IP transport, for Mobile DTV
  Recommended Practice, 3.66
  references, 3.67
  scalable full-channel option, 3.65
  services, 3.44–3.45
  signaling, 3.45–3.46
  system overview, 3.59–3.65
  timing and buffer model for, 3.46–3.47
  transmission infrastructure, 3.66–3.67
Mobile Emergency Alert System. *See* M-EAS (Mobile Emergency Alert System)
Mobile/Handheld (M/H). *See* Mobile DTV
Modems:
  bonded modem system, 5.152–5.153
  configuration settings for ENG system, 5.151
  single modem system, 5.153–5.154
Modulation. *See also* FM (frequency modulation)
  de-rating factors, 2.45
  migrating from analog to digital FM modulation, 2.231–2.233
Modulation Dependent Carrier Level (MDCL), 2.26–2.27

Modulation error ratio (MER):
  measuring, 2.223–2.224
  monitoring 8-VSB transmissions, 2.144
Modulator, in processing digital signals, 2.183
MOM (method of moments), monitoring directional systems, 2.96–2.97
Monitoring. *See also* Control/monitoring system
  directional antennas, 2.96–2.97
  off air monitoring and, 5.55
  quality control, 5.56
  remote monitoring with centralized broadcasting, 5.59–5.61
  wireless ITE (In-the-Ear), 5.96
Monitoring and control (M&C), for high availability, 6.44–6.45
Moore, Gordon, 4.3–4.4
Moore's law, 4.3–4.4
MOSFET (metal oxide silicon field effect transistor)
  RF amplification technologies, 2.184
  solid state amplifiers and, 2.150–2.151
Motorized lighting fixtures, 5.138
Motorized rigging options, 5.114–5.115
MPEG (Moving Picture Experts Group):
  COFDM and, 2.233
  data processing, 2.156–2.158
MPEG-2:
  `adaptation_field`, 3.8–3.9
  ATSC transport layer and, 7.70
  descriptors, 3.18–3.19
  discontinuities (program transitions), 3.26
  DTS (Decoding Time Stamp), 3.26
  history of TV transport over carrier networks, 4.80
  introduction, 3.3
  IP transport replacing for ATSC Mobile DTV, 3.57
  multiplex concepts, 3.9–3.10
  network composition, 3.5–3.6
  `NULL` stream packet, 3.9
  packetization of sections into stream packets, 3.13–3.15
  PAT (Program Association Table), 3.16–3.17
  PCR (Program Clock Reference), 3.25
  PES (Packetized Elementary Stream) packets, 3.19–3.20
  PES packet syntax, 3.20–3.21
  PMT (Program Map Table), 3.16, 3.18
  `private_section` syntax, 3.11–3.13
  PSI (program specific information), 3.15–3.16
  PTS (Presentation Time Stamp), 3.25
  references, 3.27
  section segmentation and encapsulation, 3.13
  segmentation, encapsulation, and packetization, 3.4–3.5
  shift from circuits to packets in carrier networks, 4.80–4.81
  specification, 3.3–3.4
  STC (System Time Clock), 3.23–3.25
  stream packet, 3.6–3.8
  system timing, 3.22
  tables, sections and descriptors, 3.11
  timing model, 3.22–3.23
  transport, 3.1

MPEG-2 (*Cont.*):
  transport streams, 4.81
  `TS_program_map_section` syntax, 3.18–3.19
  TS-Over-RTP method, 4.83–4.84
  TS-Over-UDP method, 4.82
MPEG-4:
  IP-based ENG systems, 5.149
  use by ATSC Mobile DTV, 3.64–3.65
MPX (Multiplex), 2.10
MSDC (multi-stage depressed collector):
  MSDC-IOT, 2.150
  RF amplification technologies, 2.189
  television transmitters and, 2.184
MTBF (mean-time between failure):
  reducing, 6.47
  redundancy options and, 2.28
MTU (Maximum Transmission Unit), 3.44
MUF (Maximum usable frequency):
  defined, 2.114
  frequency of optimum shortwave traffic, 2.111
Multicast:
  communication types, 4.5
  introduction to, 4.36–4.39
Multicast addresses:
  overview of, 4.58–4.59
  UDP and, 4.81
Multichannel Audio Digital Interface. *See* MADI (Multichannel Audio Digital Interface)
Multichannel program origination center, 5.73
Multiplex (MPX), 2.10
Multiplexing, MPEG-2, 3.9–3.10
Multipurpose studio, 5.5. *See also* Production facility design
Multi-stage depressed collector. *See* MSDC (multi-stage depressed collector)
Music:
  acoustical performance criteria, 5.10
  Mobile DTV content files, 3.61
MW (medium wave), 2.88–2.89. *See also* AM (amplitude modulation)

NaCL (salt), reducing soil resistivity to grounding, 6.56
NASB (National Association of Shortwave Broadcasters, Inc.), 2.103
NAT (Network Address Translation), 4.30–4.31, 4.56
The National Association of Broadcasters Legal Guide to Broadcast Law and Regulation, 1.9
National Electrical Code (NEC):
  installing power cables and, 6.3–6.4
  standard for generator installation, 6.81
National Environmental Policy Act (NEPA), 7.54–7.55
National Institute of Standards and Technology (NIST), 5.109
National Table of Frequency Allocations, A.47
National Television System Committee. *See* NTSC (National Television System Committee)
NBW (Necessary bandwidth), requirements for RF transmission, 2.8
NC (noise-criterion) curves, acoustical performance criteria, 5.11
NEC (National Electrical Code):
  installing power cables and, 6.3–6.4
  standard for generator installation, 6.81
Necessary bandwidth (NBW), requirements for RF transmission, 2.8
NEMA, industry standards for equipment racks, 6.26–6.27
NEPA (National Environmental Policy Act), 7.54–7.55
Network Address component, IP addressing, 4.26
Network Address Translation (NAT), 4.30–4.31, 4.56
Network Attached Storage, storage options in file-based workflow, 4.65
network data, lighting and, 5.122–5.124
Network engineering, for audio engineers, 5.32–5.33
Network Layer (Layer 3):
  IP networks, 4.23
  Layer 3 internetworking (routing), 4.32–4.35
  Layer 3 IP addressing, 4.23–4.32
  Layer 3 protocols, 4.35–4.36
  overview of, 4.8, 4.57
  summary, 4.42–4.43
Network operations centers. *See* NOCs (network operations centers)
Network switches, at Layer 2, 4.59
Network Time Protocol. *See* NTP (Network Time Protocol)
Networks:
  applications and servers, 4.64–4.65
  collisions and latency, 4.60
  diagram of hybrid network, 4.56
  distribution workflow, 4.66
  example of broadcast network infrastructure, 4.60
  file-based workflow, 4.62–4.64
  infrastructure of, 4.55–4.56, 4.60
  introduction, 4.55
  IP networks. *See* IP networks
  managed and unmanaged switches, 4.59
  media management architecture, 4.66–4.67
  metadata, 4.65–4.66
  MPEG-2, 3.5–3.6
  network topology and, 4.56
  OSI layers and, 4.56–4.58
  planning for AoIP, 5.27–5.29
  protocols and addressing schemes, 4.58–4.59
  public network for ideal ENG system, 5.148
  requirements for audio system, 5.31
  segment options, 4.59–4.61
  standards, 4.5–4.6
  storage options, 4.65
  system demands and, 4.61–4.62
  topology, 4.56
News broadcasts. *See also* ENG (electronic newsgathering):
  changes in news coverage, 5.161
  ENG and SNG vans, 5.46
  ENG field operations, 5.46–5.47
  flypack systems, 5.48
  overview of, 5.46
  sports and event coverage, 5.47–5.48
NiCd (nickel cadmium) batteries, 6.83
NiMH (Nickel metal hydride) batteries, 6.83

NIST (National Institute of Standards and Technology), 5.109
Nitrogen generator, maintaining transmitters, 6.99
NLE (nonlinear editing), 5.74
NOCs (network operations centers):
 distributed content and control, 5.59
 fully centralized broadcasting and, 5.58
 interconnection, 5.60
 next generation in automation systems, 5.76
 remote monitoring and control, 5.59–5.60
 remote monitoring of centralized operations, 5.61
Noise-criterion (NC) curves, acoustical performance criteria, 5.11
Noise/sound, 6.51. *See also* Acoustics, in facility design
Nonionizing radiation. *See* RFR (radio frequency radiation)
Nonlinear editing (NLE), 5.74
Nonlinear effects, amplifier predistortion/precorrection, 2.195
Non-real-time (NRT) content, 3.49–3.50
NPR radio, radio reading services for print handicapped, 1.22–1.23
NRT (non-real-time) content, 3.49–3.50
NTP (Network Time Protocol):
 clock synchronization methods, 4.69
 overview of, 4.75–4.77
 timing and buffer model for mobile DTV, 3.46–3.47
NTSC (National Television System Committee):
 history of TV transport over carrier networks, 4.80
 television standard, 2.153
 transition from analog to digital television, 3.57
NULL transport stream packet, MPEG-2, 3.9
Nylon wire ties, 6.15
Nyquist filter:
 8-VSB filters, 2.137–2.138
 defined, 2.151

OB (outside broadcast) truck, 5.39, 5.47–5.48
Object-based audio, 5.44
OET-69:
 Longley-Rice model and, 2.129
 model for television coverage and interference, 2.116–2.117
OFDM (Orthogonal Frequency Division Multiplexing), 2.10
Off air monitoring, 5.55
Office of Weights and Measures (OWM), A.51–A.52
Ohm's law, 2.90
OMA BCAST:
 Digital Rights, 3.63
 Service Guide (SG), 3.47
OMA-RME (OMA Rich Media Environment), 3.63
Open System Interconnection model. *See* OSI (Open System Interconnection) model
Operating hazards, 7.48
Operational budgets, 7.7
Optical spectrum, A.38–A.39
Orchestration:
 in file-based workflow, 4.67
 next generation in automation systems, 5.75–5.76

Orthogonal Frequency Division Multiplexing (OFDM), 2.10
OSHA guidelines and violations, 7.48, 7.50–7.51
OSI (Open System Interconnection) model:
 building the segmented network, 4.44–4.46
 Data-Link Layer (Layer 2), 4.14
 Ethernet frame and, 4.16–4.17
 IP networks, 4.6–4.8
 IPv6 and, 4.39–4.42
 Layer 2 Ethernet switches, 4.19–4.23
 Layer 3 internetworking (routing), 4.32–4.35
 Layer 3 IP addressing, 4.23–4.32
 Layer 3 protocols, 4.35–4.36
 Layer 3 summary, 4.42–4.43
 layers of, 4.8–4.9, 4.56–4.58
 LLC sublayer of Data-Link Layer, 4.15
 MAC sublayer of Data-Link Layer, 4.15–4.16
 multicast IP communications, 4.36–4.39
 Network Layer (Layer 3), 4.23
 overview of, 4.56–4.58
 PHY (physical layer), 4.11–4.14
 TCP/IP model and DoD model compared with, 4.9–4.11
 Transport Layer (Layer 4), 4.43–4.44
 VLANs, 4.17–4.18
OSPF, routing protocol options, 4.34–4.35
OTT (over the top) channels, 5.53
Outdoor heater exchangers, 6.44
Outside broadcast (OB) truck, 5.39, 5.47–5.48
Over the top (OTT) channels, 5.53
Overengineering (overdesign), 7.25
Overlays, video effects, 5.66
OWM (Office of Weights and Measures), A.51–A.52

PA (plate current), in transmitter maintenance, 6.92
PA (power amplifier):
 AM–AM conversion, 2.191
 AM–PM conversion, 2.192
 design considerations, 2.192
 IPA (intermediate power amplifier), 2.192–2.193
 linearity of, 2.189–2.191
 RF upconverter and, 2.140
 television transmitters and, 2.184, 2.189
Packet header, iPv4, 4.25
Packet identifiers. *See* PIDs (packet identifiers)
Packetization:
 MPEG-2 transport, 3.4–3.5
 of sections into stream packets, 3.13–3.15
Packetized Elementary Stream packets. *See* PES (Packetized Elementary Stream) packets
Panels:
 corporate feed systems, 2.85
 user/subscriber control key panels, 5.102–5.104
PAPR (Peak average power ratio), for HD radio, 2.13
PAR (Parabolic Aluminized Reflector), lighting instruments, 5.132–5.133
Parades, Mobile DTV, 3.60
Party-line. *See* PL (party-line)
Pass band drift, 2.209

PAT (Port Address Translation), 4.31
PAT (Program Association Table):
  overview of, 3.16–3.17
  PSI (program specific information) in MPEG-2, 3.15
Pay-Per-View (PPV), subscription security, 7.71
PCBs (polychlorinated biphenyls):
  components of, 7.60
  governmental controls, 7.59
  liability management, 7.60–7.61
  safety considerations, 7.58–7.59
PCR (Program Clock Reference), in MPEG-2 timing model, 3.25
PCs (personal computers):
  AoIP use with, 5.22
  controlling lighting, 5.126
  delay (latency) and audio quality, 5.23–5.24
  virtual intercoms and, 5.104
Peak average power ratio (PAPR), for HD radio, 2.13
Peak power:
  de-rating factors, 2.45–2.46, 2.48
  ratings for coaxial lines, 2.41–2.43
  shaving, 6.77
Peak-envelope power (PEP):
  8-VSB power, 2.146
  SSB (single sideband), 2.134–2.136
People/personnel, planning for success, 7.18
PEP (peak-envelope power):
  8-VSB power, 2.146
  SSB (single sideband), 2.134–2.136
PEP (Primary Entry Point) stations, in Emergency Alert Notification, 1.33–1.34
Performance:
  AM antenna impedance requirements for digital transmission, 2.24–2.26
  frequency response affecting performance, 2.23–2.24
  hardware limitations and precorrection, 2.20–2.21
  linearity of amplification effecting performance, 2.21–2.23
  Pre-emphasis function for improving performance, 2.19–2.20
  system design and, 7.24
Personal computers. *See* PCs (personal computers)
Personnel:
  disaster planning and recovery, 7.40
  management, 7.10
PES (Packetized Elementary Stream) packets:
  MPEG-2, 3.19–3.20
  overview of, 3.4
  syntax, 3.20–3.21
Phase monitor, monitoring directional systems, 2.96–2.97
Phase noise, tracking for DTV receivers, 2.163–2.164
Phase-locked loop (PLL), in SDI, 4.69
Phasing method, single sideband, 2.134
Phasors, maintaining AM stations, 6.113
Physical (logical) addresses, MAC sublayer responsible for, 4.15

Physical facility:
  assessment aspect of disaster planning, 7.38, 7.39
  common precautions in disaster planning, 7.40
  role of management and leadership in planning and developing, 7.12
  strategic building development, 7.13
Physical layer, ATSC 3.0, 7.69–7.70
Physical layer (Layer 1):
  cross-over cables, 4.12
  Ethernet cable and, 4.11
  overview of, 4.8, 4.57
  types of media for Ethernet cables, 4.13–4.14
  wire management and, 6.15–6.16
PIDs (packet identifiers):
  labeling PSIP tables, 3.30
  MPEG-2 stream packets, 3.7–3.8
  program elements in MPEG-2, 3.18
Pilot carrier:
  in 8-VSB signal, 2.212
  defined, 2.151
  pilot inserter for ATSC signal processing, 2.159
Pilot frequency, spectrum analyzer for measuring, 2.216–2.218
Pilot insertion, 8-VSB, 2.137
PIM (Protocol Independent Multicast), 4.38–4.39
PL (party-line):
  analog systems, see Analog partyline intercom systems
  digital systems. *See* Digital partyline intercom systems:
  overview of, 5.84
Planning:
  documented plans in successful projects, 7.17–7.19
  role of management and leadership in, 7.9–7.10
Plate current (PA), in transmitter maintenance, 6.92
Plate voltage (PV), in transmitter maintenance, 6.92
Playout and transmission, audio monitoring and, 5.42
Plenum-rated cable, 6.17
PLL (phase-locked loop), in SDI, 4.69
PMT (Program Map Table):
  overview of, 3.18
  PSI (program specific information) in MPEG-2, 3.16
  PSIP descriptors, 3.37
POE (Power-Over-Ethernet), 4.21
Point-to-Point model, predictive models in Longley-Rice, 2.131
Poisonous compounds, safety considerations, 7.51–7.53
Polarization, antenna characteristics, 2.69
Pole-operated spotlights, lighting instruments, 5.131
Pole-placed filters, mask filters and, 2.203
Polychlorinated biphenyls. *See* PCBs (polychlorinated biphenyls)
Port Address Translation (PAT), 4.31
Port trunking (Link aggregation), redundancy options for Ethernet switches, 5.28
Post production department, audio monitoring and, 5.41–5.42
Power:
  8-VSB, 2.146–2.148
  AC. *See* AC power

Power (*Cont.*):
  DC. *See* DC power
  distribution and capacity, 5.119–5.121
  distribution options, 6.70
  installing power cables, 6.3–6.4
  managing wiring. *See* Wire management
  operating parameters for HVAC systems, 6.49
  requirements for FM transmitters, 2.14–2.15
  standing wave ratio power conversion, A.6
Power, coaxial lines:
  average power ratings, 2.39–2.41
  connector power ratings, 2.43
  peak power ratings, 2.41–2.43
Power amplifier. *See* PA (power amplifier)
Power loads, in detailed design, 7.26
Power meters, 2.211–2.212
Power-Over-Ethernet (POE), 4.21
Power sourcing devices (PSD), 4.21
Power sourcing equipment (PSE), 4.21
Power supplies:
  components of matrix intercom systems, 5.98
  components of wired party-line intercom systems, 5.84
  television transmitters, 2.197
  transmitter, 2.15–2.16
Power/telephony band, in EM spectrum, A.40
PPV (Pay-Per-View), subscription security, 7.73
Precision Time Protocol. *See* PTP (Precision Time Protocol)
Precorrection:
  television transmitter amplifiers and, 2.195–2.197
  transmitters (AM and FM), 2.22–2.23
Predistortion, television transmitter amplifiers and, 2.195–2.197
Pre-emphasis function, for improving RF performance, 2.19–2.20
Prefixes, for standards units of measurement, A.4
Preparation checklist, disaster planning and recovery, 7.41
Presentation Layer (Layer 6), 4.58
Presentation Time Stamp (PTS), in MPEG-2 timing model, 3.25
Pressurization:
  of coaxial transmission lines, 2.56–2.57
  maintaining transmission lines, 6.97–6.98
  requirements for transmitting antennas, 2.85
Primary Entry Point (PEP) stations, in Emergency Alert Notification, 1.33–1.34
Print handicapped, radio reading services for, 1.22–1.23
Private addresses:
  iPv4, 4.30–4.31
  network topology and, 4.56
`private_section`, MPEG-2:
  overview of, 3.4
  syntax, 3.11–3.13
Process elements, project and systems management, 7.19–7.20
Production, in file-based workflow, 4.65

Production facility design:
  acoustical performance criteria, 5.10–5.11
  audio monitoring and, 5.41
  control room considerations, 5.9–5.10
  examples of studio designs, 5.4–5.9
  handling background noise, 5.11–5.12
  handling noise of mechanical systems and ducts, 5.12–5.15
  introduction, 5.3
  references, 5.17–5.18
  sound isolation techniques, 5.15–5.17
  studio design considerations, 5.3
Production intercom systems:
  advantages of party-line systems, 5.90–5.92
  analog client cards, 5.99–5.100
  analog partyline systems, 5.84
  communicating over IP, 5.100–5.101
  digital partyline systems, 5.92–5.93
  GPIO (General-Purpose Input/Output), 5.101
  IFB (Interrupted Fold-Back), 5.96–5.97
  introduction, 5.83–5.84
  matrix systems, 5.97–5.99
  other connectivity options, 5.101–5.102
  relay outputs, 5.101
  user/subscriber control key panels, 5.102–5.104
  virtual intercoms, 5.104
  wired party-line system components, 5.84–5.90
  wireless system interfaces, 5.94–5.96
  wireless system regulations, 5.93–5.94
Production systems, 5.1–5.2
Production team, 5.39–5.41
Program and System Information Protocol. *See* PSIP (Program and System Information Protocol)
Program Association Table (PAT):
  overview of, 3.16–3.17
  PSI (program specific information) in MPEG-2, 3.15
Program Clock Reference (PCR), in MPEG-2 timing model, 3.25
Program elements, MPEG-2, 3.18
Program Map Table. *See* PMT (Program Map Table)
Program specific information (PSI), MPEG-2, 3.15–3.16
Program transitions (discontinuities), MPEG-2 timing model and, 3.26
`program_number`, MPEG-2, 3.5
Programs, managing, 7.29
Project and systems management:
  introduction, 7.17
  meetings, 7.20–7.22
  planning for success, 7.17–7.19
  process elements, 7.19–7.20
  summary and references, 7.22
Project manager, 7.31–7.33
Projects, tracking and controlling, 7.28–7.29
Propagation theory, 2.129–2.131
Protocol Independent Multicast (PIM), 4.38–4.39
Protocols:
  Layer 3 protocols, 4.35–4.36
  LLC sublayer of Data-Link Layer, 4.15
  network components, 4.5

Protocols (*Cont.*):
  network systems, 4.58–4.59
  transport Layer (Layer 4), 4.43–4.44
PSD (power sourcing devices), 4.21
PSE (power sourcing equipment), 4.21
PSI (program specific information), MPEG-2, 3.15–3.16
PSIP (Program and System Information Protocol):
  assigning channel numbers, 3.45
  descriptors, 3.36–3.38
  Event Information Table, 3.34–3.35
  Extended Text Table, 3.36
  graphic illustration of, 3.38
  information sources for, 3.39–3.40
  introduction, 3.29–3.30
  Master Guide Table, 3.32–3.33
  Rating Region Table, 3.33–3.34
  System Time Table, 3.31
  table intervals, 3.40–3.41
  tables, 3.30–3.31
  Virtual Channel Table, 3.33
  virtual channels, 3.30
PTP (Precision Time Protocol):
  accuracy of, 4.71–4.72
  BCs (Boundary Clocks)d, 4.74
  clock synchronization methods, 4.69
  drawbacks/limits of network components, 4.72–4.73
  IEEE 1588 standard, 4.74
  NTP compared with, 4.76–4.77
  overview of, 4.69–4.70
  selecting master node, 4.70–4.71
  SMPTE ST2059-2 standard, 4.75
  TCs (Transparent Clocks), 4.73–4.74
PTS (Presentation Time Stamp), in MPEG-2 timing model, 3.25
Public addresses:
  iPv4, 4.30–4.31
  network topology and, 4.56
The Public and Broadcasting, 1.5–1.7
Public file, of ABIP documents, 1.18
Public network, for ideal ENG system, 5.148
Purchases, categorizing, 7.9
PV (plate voltage), in transmitter maintenance, 6.92

QAM (Quadrature Amplitude Modulation), 2.234
QC (quality control), 5.41, 5.56
QoS (Quality of Service):
  COFDM and, 2.233
  configuration settings, 4.59–4.60
QPSK (Quadrature Phase Shift Keying), 2.234
Q-signal:
  defined, 2.151
  DTV, 2.140
  monitoring 8-VSB transmissions, 2.143
  phase tracking for DTV receivers, 2.163–2.164
QSL cards, shortwave listening and, 2.113

Race way, wire management, 6.18–6.20
Racks, in OB (outside broadcast) truck, 5.39. *See also* Equipment rack enclosures
Radial ground screen, designing ground system for AM antenna, 2.89
Radiation, EM spectrum. *See* Electromagnetic (EM) spectrum
Radiation center above mean sea level (RCAMSL), 2.116
Radiator antennas:
  cable lengths for medium-wave radiators, 2.89–2.90
  overview of, 2.70–2.71
  troubleshooting, 2.97–2.98
Radio frequency. *See* RF (radio frequency)
Radio frequency radiation. *See* RFR (radio frequency radiation)
Radio links, planning for AoIP, 5.29
Radio Miami International, 2.106
Radio reading services, for print handicapped, 1.22–1.23
Radio stations. *See also* Production facility design:
  maintaining transmission lines, 6.96
  studio design, 5.8–5.9
Radomes, maintaining antennas, 2.84–2.85
Raised cosine filter, 2.151
Rapid Spanning Tree (RST), 5.28
Rating Region Table. *See* RRT (Rating Region Table)
RCAMSL (radiation center above mean sea level), 2.116
RDM (Remote Device Management), 5.123
Real-time Transport Control Protocol (RTCP), 3.46–3.47
Real-time Transport Protocol. *See* RTP (Real-time Transport Protocol)
Receive antenna pattern, in model for television coverage, 2.124–2.129
Receive host, network components, 4.5
Receivers, for DTV:
  data deinterleaver, 2.165
  data derandomizer, 2.165
  data synchronization and clock timing, 2.162
  IF filter and synchronous detector, 2.161–2.162
  linear equalization, 2.162–2.163
  overview of, 2.160–2.161
  phase tracking, 2.163–2.164
  RS (Reed-Solomon) decoder, 2.165
  TCM (trellis-coded modulation), 2.164–2.165
  tuner, 2.161
Reception reports, shortwave broadcasts, 2.113–2.114
Recombination, and ionization in ionosphere, 2.109
Recommended Practices (RP), ATSC Mobile DTV system, 3.66
Recording studios (music studio), 5.4. *See also* Production facility design; Studios
Rectangular waveguide filters, 2.206–2.207
Redundancy:
  benefits of, 2.28
  Ethernet switches and, 5.28
Reed-Solomon. *See* RS (Reed-Solomon)
Reflection, propagation theory, 2.129–2.131
Reflective filters, 2.204
Refraction, propagation theory, 2.129–2.131
Regulations, wireless intercom system, 5.93–5.94
Relay outputs, matrix intercom systems, 5.101

I.26 INDEX

Remote audio broadcasting:
  ENG and SNG vans, 5.46
  ENG field operations, 5.46–5.47
  flypack systems, 5.48
  future of, 5.48
  introduction, 5.45
  monitoring, 5.39
  news broadcasts, 5.46
  sports and event coverage, 5.47–5.48
Remote control:
  ABIP review of, 1.19
  with centralized broadcasting, 5.59–5.60
Remote Device Management (RDM), 5.123
Remote master stations, PL intercom systems, 5.86
Remote mic-kill, PL intercom systems, 5.86
Remote monitoring, with centralized broadcasting, 5.59–5.61
Remote phosphor LED, light sources, 5.129
Replacement parts, maintaining essential services, 7.11–7.12
Representational State Transfer (REST), 5.79
Request-for-Comments. *See* RFCs (Request-for-Comments)
Required Monthly Test (RMT), EAS/CAP rules, 1.34
Required Weekly Test (RWT), EAS/CAP rules, 1.34
Reserved addresses, IPv4, 4.32
Resolution bandwidth, measuring DTV, 2.213–2.215
Resources, planning for success, 7.18
REST (Representational State Transfer), 5.79
Reusable alkaline batteries, 6.83
Revenue budgets, 7.7
Revolving field alternator, generator types, 6.80
RF (radio frequency):
  AM and FM transmitters. *See* Transmitters (AM and FM)
  channel for ATSC Mobile DTV, 3.59
  coaxial lines. *See* Coaxial transmission lines
  DTV consideration. *See* DTV (digital television), RF considerations
  DTV measurement. *See* DTV (digital television), RF measurement
  DTV transmission. *See* DTV (digital television), RF transmission system
  in EM spectrum, A.40
  FM channel combiners. *See* FM channel combiners
  radiation hazards, 7.48
  spectrum description for DTV, 2.172–2.179
  transmission, 2.1–2.2
  upconverter, 2.140–2.141
  working with high voltage, 7.47
RF (radio frequency) networks:
  final adjustment of complex conjugate impedance, 2.95–2.96
  "L" networks, 2.91–2.94
  preliminary setup and adjustment of "T" network, 2.94–2.95
RFCs (Request-for-Comments):
  IETF, 4.53
  networking standards and, 4.6
  TS-over-RTP method, 4.83–4.84

RFR (radio frequency radiation):
  NEPA mandate, 7.54–7.55
  overview of, 7.53–7.54
  revised guidelines, 7.55–7.57
Rhombic antennas, 2.107
Rigging systems, for lighting, 5.110–5.116
Rights of Persons with Disabilities (CRPD), 1.21
Rigid coaxial systems:
  attenuation, 2.36–2.38
  average power ratings, 2.39–2.41
  characteristic impedance, 2.36
  construction of, 2.35
  cutoff frequencies, 2.36
  installing, 2.53–2.56
  overview of, 2.25
  peak power ratings, 2.41–2.43
  pressurization of, 2.56–2.57
  selecting, 2.52–2.53
RIP (Routing Information Protocol), 4.34–4.35
RJ-45 connectors, for twisted-pair cable, 4.13
RMT (Required Monthly Test), EAS/CAP rules, 1.34
Robusticity, maintaining uninterrupted transmission, 5.158–5.161
Roll-in, video file, 5.67
Root dispersion, NTP, 4.76
Routing. *See also* Internetworking (routing)
  audio routing, 5.21
  IP router vs. SDI router, 4.55
Routing Information Protocol (RIP), 4.34–4.35
Routing protocols:
  comparing interior gateway protocols, 4.35
  hop count and, 4.34–4.35
  overview of, 4.33–4.34
Routing tables, 4.33–4.34
RP (Recommended Practices), ATSC Mobile DTV system, 3.66
RRT (Rating Region Table):
  in mobile DTV service signaling, 3.45
  PSIP, 3.33–3.34, 3.41
RS (Reed-Solomon):
  applying RS encoder to MPEG data packets, 2.157
  ATSC Mobile DTV frames, 3.59–3.60
  decoder for DTV receiver, 2.165
  mobile DTV content delivery and RS frames, 3.43–3.44
RS-232, in automation systems, 5.73–5.74
RS-422, in automation systems, 5.73–5.74
RS-485, in automation systems, 5.74
RSS feeds, sources of dynamic graphics, 5.51
RST (Rapid Spanning Tree), 5.28
RTCP (Real-time Transport Control Protocol), 3.46–3.47
RTP (Real-time Transport Protocol):
  MPEG-2 transport streams over IP networks, 4.81
  timing and buffer model for mobile DTV, 3.46–3.47
  TS-over-RTP method, 4.83–4.84
RWT (Required Weekly Test), EAS/CAP rules, 1.34

Sabotage, assessment aspect of disaster planning, 7.38
Safety considerations:
  ABCs of life support, 7.49

Safety considerations (*Cont.*):
   beryllium oxide (BeO) ceramics, 7.51
   circuit protection hardware, 7.46
   corrosive and poisonous compounds, 7.51–7.53
   electric shock, 7.43–7.46
   FC-75 toxic vapor, 7.53
   first aid equipment at transmitter sites, 6.88
   first aid procedures, 7.47–7.48, 7.50
   hot coolant and surfaces, 7.57–7.58
   implosion hazard, 7.57
   introduction, 7.43
   nonionizing radiation, 7.53–7.57
   operating hazards, 7.48
   operating parameters for HVAC systems, 6.51
   OSHA guidelines and violations, 7.48, 7.50–7.51
   PCBs (polychlorinated biphenyls), 7.58–7.61
   references, 7.61
   working with high voltage, 7.47
   x-ray radiation hazard, 7.57
Salt (NaCL), reducing soil resistivity to grounding, 6.56
Satellite:
   hybrid aggregation, 2.241–2.243
   migrating from KU to KA band IP satellite systems, 2.240–2.241
satellite antennas, 6.48
Satellite news gathering. *See* SNG (satellite news gathering)
SAW (surface-acoustical wave) filters, 2.140, 2.162
SBE (Society of Broadcasters):
   8-VSB specialist certification, 2.151
   certification program, 7.64–7.65
   education programs, 7.65
   educational courses, 7.10, 7.14
   introduction, 7.63
   people in history of, 7.65–7.66
SBE Leadership Development Course, 7.65
SBE National Headquarters, 7.63
SBE University, 7.65
SC (Standard Connector), 4.13
Scalable Full Channel Mobile Mode (SFCMM), 3.65
Scalable full-channel option, ATSC Mobile DTV system, 3.65
Schematics, use in detailed design, 7.26
Scoops, lighting instruments, 5.136
SCR (silicon controlled rectifier), for fixture dimming, 5.124
SCTE (Society of Cable Telecommunications Engineers):
   overview of, A.56
   reference documents, A.57
SD (standard definition):
   FM microwave links supporting, 2.231
   standards, 2.154
   video formats, 5.63
SDI (serial digital interface):
   baseband master control, 5.51
   evolution of, 4.61–4.62
   IP-based ENG systems, 5.149
   moving to IP-based workflow, 4.69
   network infrastructure and, 4.55

SDNs (Software Defined Networks), 4.62
SDO (Standards Development Organization), A.50–A.51
SDP (Session Description Protocol), 3.61
SDTV, 2.154
Sealed lead acid (SLA) batteries, 6.83–6.85
Searches, media asset management and, 5.78
Seasonal schedules, for shortwave broadcasts, 2.103
Sections, MPEG-2:
   overview of, 3.11
   packetization into stream packets, 3.13–3.15
   segmentation and encapsulation, 3.13
Secure Sockets Layer (SSL), 5.29
Security, ATSC Mobile DTV file types, 3.61
Segmentation:
   MPEG-2 sections, 3.13
   MPEG-2 transport, 3.4–3.5
Segments:
   defined, A.37
   network systems, 4.59–4.61
   VLANs and subnets, 4.59
Semiflexible coaxial systems:
   attenuation, 2.36–2.38
   average power ratings, 2.39–2.41
   characteristic impedance, 2.36
   construction of, 2.35
   installing, 2.50–2.52
   overview of, 2.48–2.49
   peak power ratings, 2.41–2.43
   pressurization of air dielectric lines, 2.56–2.57
   selecting, 2.50
Send host, network components, 4.5
Serial digital interface. *See* SDI (serial digital interface)
Servers, network systems, 4.64–4.65
Service contracts, 7.11–7.12
Service Guides, ATSC Mobile DTV system, 3.63
Service ID, Mobile DTV service, 3.44
Service Labeling Table (SLT), in mobile service signaling, 3.45
Service Location Descriptor (SLD), PSIP:
   descriptors, 3.37
   table intervals, 3.40
Service loops, in wire management, 6.17
Service Map Table (SMT), in mobile DTV service signaling, 3.45–3.46
Service-Oriented Architecture (SOA), in media asset management, 5.79
Services:
   Mobile DTV, 3.44–3.45, 3.59, 3.61–3.62
   role of management and leadership in maintaining, 7.10–7.12
Services area:
   computing service and interfering field strengths for television, 2.119–2.124
   establishing protected service areas for television, 2.117–2.119
Session Description Protocol (SDP), 3.61
Session Layer (Layer 5), 4.57
SFCMM (Scalable Full Channel Mobile Mode), 3.65
SFD (Start Frame Delimiter), in Ethernet frame, 4.16

SFP (Small Form Factor Pluggable), Gigabit Ethernet, 4.13–4.14
Short-Term Messages (STKM), 3.63
Shortwave broadcasts:
    antenna types for, 2.107–2.108
    compared with domestic AM stations, 2.102
    digital radio mondale, 2.109
    FCC regulation of, 2.101
    frequency management, 2.104–2.105
    frequency of optimum traffic, 2.111
    frequency requests, 2.106
    international bands, 2.102–2.104
    interval signals, 2.112–2.113
    introduction, 2.101
    ionosphere and, 2.109–2.111
    reception reports, 2.113–2.114
    references, 2.114
    shortwave transmitters, 2.106–2.107
    single sideband transmissions, 2.109
    smoothed sunspot number and, 2.111–2.112
    terminology, 2.114
Shortwave Listening (SWL), 2.113
Shoulder amplitude ratio, measuring head to shoulder amplitude ratio, 2.219–2.220
SI units, international system of measurement, A.58
Signal:
    audio monitoring systems, 5.42–5.43
    processing, 2.156
    spectrum analyzer for measuring amplitude, 2.215–2.216
    spectrum analyzer for measuring flatness, 2.218–2.219
    terminology, 2.212
Signal quality:
    determining, 2.223
    EVM (error vector magnitude), 2.224
    measurements, 2.227–2.229
    measuring digital signal overload, 2.213–2.215
    MER (modulation error ratio), 2.223–2.224
    relationship of response errors to signal noise, 2.224–2.226
    SNR (signal-to-noise ratio), 2.223
Signaling:
    components of wired party-line intercom systems, 5.86
    Mobile DTV, 3.45–3.46, 3.61, 3.65–3.66
Signal-to-noise ratio. *See* SNR (signal-to-noise ratio)
Silicon controlled rectifier (SCR), for fixture dimming, 5.124
Single modem system, configuring for IP-based ENG, 5.153–5.154
Single sideband (SSB):
    experimentation with, 2.109
    fundamentals of, 2.134–2.136
Single-sideband suppressed carrier (SSSC):
    8-VSB as SSSC transmission, 2.136
    overview of, 2.134
SINPO code, shortwave listening and, 2.113
Site heat, cooling and, 6.45–6.46
Site visits, maintenance and, 6.87–6.90
Skin depth, coaxial lines, 2.37
Skin effect, ground-system inductance and, 6.61–6.62
Skips, shortwave in ionosphere, 2.110
SLA (sealed lead acid) batteries, 6.83–6.85
Slack, in wire management, 6.17
SLD (Service Location Descriptor), PSIP:
    descriptors, 3.37
    table intervals, 3.40
Slot covers, maintaining antennas, 2.84–2.85
SLT (Service Labeling Table), in mobile service signaling, 3.45
Small Form Factor Pluggable (SFP), Gigabit Ethernet, 4.13–4.14
Smartphones, 5.104
Smoothed sunspot number (SSN), shortwave broadcasts and, 2.111–2.112
SMPS (Switch mode power supply), 2.15
SMPTE (Society of Motion Picture Engineers):
    overview of, A.56–A.57
    reference documents, A.57
    ST2020 standard for Dolby metadata, 5.43
    ST2022-1 (forward error correction), 4.83–4.85
    ST2022-2 (TS-over-RTP method), 4.83–4.84
    ST2022-6 (uncompressed video on IP networks), 4.86–4.88
    ST2022-7 (seamless reconstruction of signal paths for high availability), 4.86
    ST2059-2 (PTP standards), 4.74–4.475
SMT (Service Map Table), in mobile DTV service signaling, 3.45–3.46
SNG (satellite news gathering). *See also* ENG (electronic newsgathering)
    audio monitoring and, 5.39
    audio system in van, 5.46
    cellular revolution and, 5.147
    history of, 5.145–5.146
    overview of, 2.236
SNR (signal-to-noise ratio):
    budget, 2.226–2.227
    computing service and interfering field strengths for television, 2.120
    in designing ENG system, 5.150–5.151
    DTV transmission system and, 2.155
    FCC regulation of television, 2.116
    formula for, 2.123–2.124
    improving RF performance, 2.19–2.20
    measuring signal quality, 2.223
    monitoring 8-VSB transmissions, 2.143–2.144
    relationship of response errors to signal noise, 2.224–2.226
    signal to noise budget, 2.226–2.227
SOA (Service-Oriented Architecture), in media asset management, 5.79
Society of Broadcasters. *See* SBE (Society of Broadcasters)
Society of Motion Picture Engineers. *See* SMPTE (Society of Motion Picture Engineers)
Softlights (direct beam), lighting instruments, 5.134
Softlights (indirect beam), lighting instruments, 5.133–5.134
Software, for Mobile Emergency Alert System, 3.52–3.53
Software Defined Networks (SDNs), 4.62

Soil types, grounding and, 6.55–6.56
Solar radiation, de-rating factors, 2.46–2.48
Solid state relay (SSR), for fixture dimming, 5.124
Solid State (SSD), storage options, 4.65
Solid-state devices:
   high power amplifiers, 2.150–2.151
   RF amplification technologies, 2.185–2.187
Sound control, in OB (outside broadcast) truck, 5.39–5.41
Sound isolation techniques, in facility design, 5.15–5.17
Source Advertising, AoIP and, 5.24
Source assignment panels, PL intercom systems, 5.87
Source/destination numbering, in wire management, 6.13–6.14
Sources (inputs), switch operation, 5.63–5.64
Sources, lighting, 5.126–5.129
Spare parts/equipment, benefits of redundancy options, 2.28
Speakers:
   audio monitors and, 5.43
   components of wired party-line intercom systems, 5.85
   control panels and, 5.103
Special addresses, IPv4, 4.32
Specialist certification, 8-VSB, 2.151
Specifications, system design and, 7.24
Spectral description, VSB baseband, 2.169–2.172
Spectral display, 8-VSB, 2.142–2.143
Spectral mask, 8-VSB, 2.141–2.143
Spectrum analyzer:
   measuring adjacent channel power, 2.220–2.223
   measuring digital signal overload, 2.213–2.215
   measuring dynamic range and intermodulation, 2.213
   measuring emissions mask compliance, 2.219–2.220
   measuring pilot frequency, 2.216–2.218
   measuring shoulder amplitude ratio, 2.219
   measuring signal amplitude, 2.215–2.216
   measuring signal flatness, 2.218–2.219
   specifications, 2.215
   test equipment, 2.212
Spectrums, A.37
Speech, acoustical performance criteria, 5.10
Speech-to-text, in accessibility, 1.29
Sports, news broadcasts, 5.47–5.48
Spotlights, 5.130–5.133
SSB (single sideband):
   experimentation with, 2.109
   fundamentals of, 2.134–2.136
SSD (Solid State), storage options, 4.65
SSI (synchronous serial interface), MPEG, 2.156
SSL (Secure Sockets Layer), 5.29
SSN (smoothed sunspot number), shortwave broadcasts and, 2.111–2.112
SSR (solid state relay), for fixture dimming, 5.124
SSSC (single-sideband suppressed carrier):
   8-VSB as SSSC transmission, 2.136
   overview of, 2.134
ST2020 standard for Dolby metadata, 5.43
ST2022-1 (forward error correction), 4.83–4.85
ST2022-2 (TS-over-RTP method), 4.83–4.84
ST2022-6 (uncompressed video on IP networks), 4.86–4.88
ST2022-7, seamless reconstruction of signal paths for high availability, 4.86
ST2059-2, PTP standards, 4.74–4.475
Staffing:
   management and leadership in oversight, 7.10
   requirements for RF transmission, 2.27
Stage Announce, interfaces for wireless intercom, 5.95
Standard Connector (SC), 4.13
Standard definition. *See* SD (standard definition)
Standard units of measurement:
   for electrical work, A.5
   overview of, A.3–A.4
   prefixes for, A.4
Standards:
   history of, A.51–A.53
   introduction, A.49
   principal organizations in, A.53–A.57
   professional society engineering committees, A.51
   references, A.57
   SDO (Standards Development Organization), A.50–A.51
   tabular data and, A.58–A.60
Standby power systems:
   advanced system protection, 6.77–6.78
   blackout effects, 6.74
   choosing a generator, 6.78–6.81
   dual feeder system, 6.76–6.77
   generators in disaster planning and recovery, 7.38
   noise produced by, 6.82
   overview of, 6.73–6.75
   peak power shaving, 6.77
Standing wave ratio (SWR), power conversion, A.6
Star point junction combiner, 2.29
Start Frame Delimiter (SFD), in Ethernet frame, 4.16
Static IP addresses, 4.56
Static routing, 4.32
Station logs:
   ABIP review of, 1.19
   duties of Chief Operator, 1.13–1.14
STC (System Time Clock):
   discontinuities (program transitions) and, 3.26
   in MPEG-2 timing model, 3.23–3.25
STKM (Short-Term Messages), 3.63
STL (studio-to-transmitter) links, 7.39
Storage, options in file-based workflow, 4.65
Storage Area Networks, 4.65
Store and forward methods, distributed content and, 5.49
STP (Spanning Tree Protocol), 4.20–4.21
Straight cuts, video transitions, 5.65
Strategic planning, management and leadership in, 7.9–7.10
Stream packets, packetization of sections into, 3.13–3.15
Streaming:
   audio and video, 3.44
   IP-based ENG systems, 5.150
Stress, operating parameters for HVAC systems, 6.52
STT (System Time Table), PSIP, 3.31, 3.41

Studios. *See also* Production facility design
  audio monitoring and, 5.41
  building, 5.26–5.27
  cooling for, 6.42
  cost of studio space, 5.105
  design examples, 5.4–5.9
  equipment racks. *See* Equipment rack enclosures
  facility issues, 6.1
  grounding system. *See* Grounding
  licensing basics, 1.4
  lighting. *See* Lighting
  overview of, 5.6
  recording studios (music studio), 5.4
  television studio, 5.6–5.7
  wiring. *See* Wire management
Studio-to-transmitter (STL) links, 7.39
Stuffing Indicator field, in mobile DTV content delivery, 3.43–3.44
Subnets:
  IPv4, 4.30
  IPv6, 4.41
  mobile DTV content delivery and, 3.44
  segment options, 4.59
Sunspot activity:
  CIRAF target zones and, 2.104
  smoothed sunspot number and, 2.111–2.112
Supernetting, IPv4, 4.30
Superturnstiles, corporate feed system, 2.85
Supplementary lights, 5.136–5.138
Surface-acoustical wave (SAW) filters, 2.140, 2.162
Surround sound:
  AoIP support, 5.22
  future of audio, 5.44
Suspension systems, for lighting, 5.110–5.116
Switch mode power supply (SMPS), 2.15
Switch panels, OSHA guidelines, 7.51
Switches. *See also* VLANs (Virtual LANs)
  AoIP capable, 5.32
  building network for AoIP, 5.27–5.28
  CIAB (Channel in a Box), 5.52
  at Layer 2, 4.59
  Layer 2 Ethernet switches, 4.19–4.23
  Layer 3 switches, 4.35
  managed and unmanaged, 4.59
  video switchers. *See* Video switchers
Switchgear, maintaining AC power equipment, 6.73
SWL (Shortwave Listening), 2.113
SWR (standing wave ratio), power conversion, A.6
Symbols, in technical documentation, 6.8
sync_message, limitations on clock synchronization in Ethernet networks, 4.72–4.73
SyncE, 4.77
Synchronization clock, in AoIP, 5.24
Synchronization codes, in ATSC signal processing, 2.159
Synchronous detector (VSB), for DTV receivers, 2.161–2.162
Synchronous serial interface (SSI), MPEG, 2.156
System demands, network systems and, 4.61–4.62
System design. *See* Systems engineering
System parameters, for DTV, 2.179–2.182

System Time Clock (STC):
  discontinuities (program transitions) and, 3.26
  in MPEG-2 timing model, 3.23–3.25
System Time Table (STT), PSIP, 3.31, 3.41
System timing, MPEG-2:
  discontinuities (program transitions), 3.26
  DTS (Decoding Time Stamp), 3.26
  overview of, 3.22
  PCR (Program Clock Reference), 3.25
  PTS (Presentation Time Stamp), 3.25
  STC (System Time Clock), 3.23–3.25
  timing model, 3.22–3.23
Systems engineer, role in program management, 7.33–7.34
Systems engineering:
  analyzing budget requirements, 7.26–7.28
  designing electronic systems, 7.24–7.25
  detailed design, 7.25–7.26
  ENG systems, 5.150–5.151
  executive manager in, 7.29–7.30
  introduction, 7.23–7.24
  managing programs, 7.29
  members of project teams, 7.35
  project manager in, 7.31–7.33
  references, 7.35
  systems engineer in, 7.33–7.34
  tracking and controlling projects, 7.28–7.29
Systems management. *See* Project and systems management
Systems specification, MPEG-2, 3.3–3.4

Tables:
  in mobile service signaling, 3.45
  MPEG-2, 3.11
Tables, PSIP:
  Event Information Table, 3.34–3.35
  Extended Text Table, 3.36
  graphic showing relationship of, 3.38
  intervals, 3.40–3.41
  Master Guide Table, 3.32–3.33
  overview of, 3.30–3.31
  Rating Region Table, 3.33–3.34
  System Time Table, 3.31
  Virtual Channel Table, 3.33
Tablets, virtual intercoms and, 5.104
Tagging media, media asset management, 5.78
Take-off angle, of antenna:
  defined, 2.114
  shortwave and, 2.111
Talent boxes (amplifiers), Interrupted Fold-Back and, 5.96–5.97
Talkback (intercom) system:
  analog systems. *See* Analog partyline intercom systems
  in outside broadcast truck, 5.47
  overview of, 5.83
  for production. *See* Production intercom systems
TCM (trellis-coded modulation):
  in ATSC data processing, 2.158–2.159
  decoder for DTV receiver, 2.164–2.165

INDEX I.31

TCP (Transport Control Protocol):
  comparing with UDP, 4.45
  data transfer using, 4.43–4.44
  overview of, 4.58
TCP/IP model, OSI model compared with, 4.9–4.11
TCs (Transparent Clocks), PTP, 4.73–4.74
TDM (time division multiplexing), 5.97–5.98
Teams:
  building, 7.19
  members of system engineering project teams, 7.35
Technical documentation:
  ABIP, 1.18
  broadcast facility design, 6.7–6.9
Technology assessment, systems engineering and, 7.28
Technology Group (TG), ATSC, 7.68–7.69
Telecommunications:
  under *Code of Federal Regulations*, Title 47, 1.11
  history of TV transport over carrier networks, 4.79–4.80
  shift from circuits to packets in carrier networks, 4.80–4.81
Telecommunications Industry Association. *See* TIA (Telecommunications Industry Association)
Television coverage and interference:
  computing service and interfering field strength, 2.119–2.124
  establishing protected service areas, 2.117–2.119
  including receive antenna pattern, 2.124–2.129
  introduction, 2.115–2.116
  Longley-Rice model for, 2.129–2.132
  model needed for, 2.116–2.117
  summary, 2.132
Television Decoder Circuitry Act, 1.24
Television Lighting Consistency Index (TLCI), 5.109–5.110
Television outside broadcast. *See* Remote audio broadcasting
Television Parental Guidelines (TVPG), 3.33
Television production control room, 5.9–5.10
Television stations. *See* TV stations
Television studio, 5.6–5.7. *See also* Production facility design; Studios
Television transmitters:
  amplification chain, 2.185
  amplifiers, 2.184
  control and metering, 2.197–2.198
  cooling, 2.199–2.200
  high efficiency amplification techniques, 2.193–2.195
  introduction, 2.183–2.184
  mask filters, 2.198–2.199
  power amplifiers, 2.189–2.193
  power supplies, 2.197
  predistortion/precorrection, 2.195–2.197
  references, 2.200
  RF amplification technologies, 2.185–2.189
  troubleshooting, 2.198
TEM (Transverse electromagnetic):
  in coaxial transmission, 2.33–2.34
  cutoff frequency and, 2.35

Temperature:
  Celsius-to-Fahrenheit conversion, A.8
  de-rating factors and ambient temperature, 2.46–2.47
  grounding and, 6.55
  in transmitter maintenance, 6.95
Terrain Integrated Rough Earth Model (TIREM), 2.129
terrestrial Virtual Channel Table (TVCT), PSIP, 3.33, 3.41
Test equipment:
  power meter, 2.211–2.212
  spectrum analyzer, 2.212
Tetrode, high power amplifiers, 2.148
Text, Extended Text Table, 3.36
TG (Technology Group), ATSC, 7.68–7.69
T-H (tungsten-halogen):
  incandescent light source, 5.127
  studio light sources, 5.106
Theater (dramatic production) studio, 5.5. *See also* Production facility design; Studios
Thermal stability, of mask filters, 2.208–2.209
Three point lighting, 5.105–5.106
TIA (Telecommunications Industry Association):
  EIA-568A/EIA-568B wiring schemes, 4.11–4.12
  EIA/TIA-485 standard for DMX 512, 5.143
  references for studio lighting, 5.143
Time allocation, planning for success, 7.18
Time division multiplexing (TDM), 5.97–5.98
Time information transfer:
  accuracy of PTP, 4.71–4.72
  BCs (Boundary Clocks)d, 4.74
  drawbacks/limits of network components, 4.72–4.73
  Genlock over IP, 4.77
  IEEE 1588 standard, 4.74
  introduction, 4.69
  NTP (Network Time Protocol), 4.75–4.77
  PTP (Precision Time Protocol), 4.69–4.70
  references, 4.78
  selecting PTP master node, 4.70–4.71
  SMPTE ST2059-2 standard, 4.75
  summary, 4.77–4.78
  SyncE, 4.77
  TCs (Transparent Clocks), 4.73–4.74
Timestamps, timing and buffer model for mobile DTV, 3.46–3.47
Timing, MPEG-2. *See* System timing, MPEG-2
Timing and buffer model, Mobile DTV, 3.46
TIREM (Terrain Integrated Rough Earth Model), 2.129
Titling, video, 5.67–5.68
TLCI (Television Lighting Consistency Index), 5.109–5.110
Topology, of centralized broadcasting, 5.58
Towers:
  ABIP reviewing inspection of, 1.19
  assessment aspect of disaster planning, 7.39
  disaster preparation checklist, 7.41
  installing antennas, 2.77
  maintaining, 6.119–6.122

TPC (Transport Parameter Channel), 3.45
TPO (Transmitter power output), 1.19
TPO (transmitter power output), 2.185
Track and Trolley, lighting suspension and rigging systems, 5.114
Tracking projects, 7.28–7.29
Traffic:
    centralized broadcasting and, 5.60–5.61
    relationship of master control, 5.56
Training, role of management and leadership in, 7.13–7.14
Transitions, video, 5.65
Transmission infrastructure, Mobile DTV, 3.66–3.67
Transmission lines:
    coaxial. *See* Coaxial transmission lines
    maintaining, 6.96–6.99
    operating parameters for HVAC systems, 6.51
Transmission monitoring, 8-VSB, 2.143–2.146
Transmission software, for Mobile Emergency Alert System, 3.52–3.53
Transmitter output power (TPO), 2.120
Transmitter power output (TPO), 1.19, 2.185
Transmitters:
    assessment aspect of disaster planning, 7.39
    controls, 5.61
    cooling, 6.46–6.48
    disaster preparation checklist, 7.41
    for DTV, 2.155–2.160
    maintaining, 6.90–6.96
    shortwave, 2.106–2.107
    signal to noise budget, 2.226–2.227
    site visits, 6.87–6.90
    television. *See* Television transmitters
Transmitters (AM and FM):
    8-VSB signal and, 2.133–2.134
    AM antenna impedance requirements, 2.24–2.26
    AM shortwave transmitters, 2.106–2.107
    AM theory, 2.3–2.6
    benefits of having redundancy options and spares, 2.28
    control and monitoring, 2.19
    control and monitoring systems in fault diagnostics, 2.27
    efficiency in reducing cost of ownership, 2.26–2.27
    exciters, 2.16–2.17
    final stage power amplifiers, 2.17–2.19
    FM theory, 2.6–2.10
    frequency response effecting performance, 2.23–2.24
    hardware limitations and precorrection, 2.20–2.21
    HD radio and, 2.10–2.15
    HD radio system, 2.31–2.32
    intermediate power amplifiers, 2.17
    introduction, 2.3
    linearity of amplification effecting performance, 2.21–2.23
    manufacturer's recommendations and industry best practices, 2.28–2.30
    overview, 2.15
    overview of RF transmission, 2.1–2.2
Transmitters (AM and FM) (*Cont.*):
    power supplies, 2.15–2.16
    pre-emphasis function for improving performance, 2.19–2.20
    references, 2.32
    staffing requirements, 2.27
    transmitter overview, 2.15
Transmitting antennas:
    corporate feed systems (superturnstile, panels, FM), 2.85
    factory testing, 2.75–2.76
    gain, 2.71–2.75
    installing, 2.77–2.80
    introduction, 2.69–2.70
    maintaining, 2.80–2.83
    radiator types, 2.70–2.71
    radomes and slot covers, 2.84–2.85
    summary and references, 2.86
    VSWR and pressurization requirements, 2.85
Transparent Clocks (TCs), PTP, 4.73–4.74
Transport Control Protocol. *See* TCP (Transport Control Protocol)
Transport layer, ATSC 3.0, 7.70–7.71
Transport Layer (Layer 4):
    overview of, 4.8, 4.57
    protocols, 4.43–4.44
Transport Parameter Channel (TPC), 3.45
Transport streams, MPEG-2:
    `adaptation_field`, 3.8–3.9
    ATSC transport layer and, 7.70
    multiplexing and, 3.9–3.10
    NULL stream packet, 3.9
    over IP networks, 4.81
    overview of, 3.6–3.8
    Transport Stream packet, 3.6–3.8
    TS-over-RTP method, 4.83–4.84
    TS-over-UDP method, 4.82
Transverse electromagnetic (TEM):
    in coaxial transmission, 2.33–2.34
    cutoff frequency and, 2.35
Trellis-coded modulation (TCM):
    in ATSC data processing, 2.158–2.159
    decoder for DTV receiver, 2.164–2.165
Troubleshooting:
    directional arrays, 2.98–2.99
    television transmitters, 2.198
Trucks, outside broadcast truck, 5.47–5.48
Trunking, matrix intercom systems and, 5.102
`TS_program_map_section` syntax, MPEG-2, 3.18–3.19
TS-over-RTP method, 4.83–4.84
TS-over-UDP method, 4.82
Tubular fluorescent, light sources, 5.129
Tuner, for DTV receivers, 2.161
Tungsten-halogen (T-H):
    incandescent light source, 5.127
    studio light sources, 5.106
TV analyzer, measurement with, 2.229–2.230
TV service, for visually handicapped, 1.23–1.24
TV stations:

TV stations (*Cont.*):
  bands for TV channels, 2.70
  captioning requirements, 1.25–1.29
  equipment inspection, 6.105
  maintaining transmission lines, 6.96
  maintaining transmission system, 6.107–6.110
TVCT (terrestrial Virtual Channel Table), PSIP, 3.33, 3.41
TVPG (Television Parental Guidelines), 3.33
TW (two-wire) systems, 5.83. *See also* Analog partyline intercom systems
Twisted-pair cable:
  Ethernet, 4.11
  RJ-45 connectors for, 4.13
Two-way radio, interfacing to analog PL intercom, 5.89–5.90
Two-wire (TW) systems, 5.83. *See also* Analog partyline intercom systems

UDP (User Datagram Protocol):
  comparing with TCP, 4.45
  data transfer using, 4.43–4.44
  MPEG-2 transport streams over IP networks, 4.81
  in multicast communications, 4.37
  overview of, 4.58
  TS-over-UDP method, 4.82
UE (user equipment), CNG, 2.236–2.238
Ufer ground system, 6.58–6.60
UHD (ultra-high definition) video, 7.71–7.72
UHF (ultra-high frequency):
  band considerations for mask filters, 2.208
  elevation patterns for television broadcasts, 2.120–2.121
  establishing protected service areas for television, 2.118
  radomes and slot covers for UHF antennas, 2.84
  regulation of wireless intercoms, 5.94
  retuning mask filters, 2.210
  RF amplification technologies, 2.188
  spectrum compatibility with DTV, 2.154
  TV broadcast bands, 2.70
  for wireless partyline intercom systems, 5.83
UL (Underwriters Laboratory), 6.26
Ultra-high definition (UHD) video, 7.71–7.72
Ultra-high frequency. *See* UHF (ultra-high frequency)
Ultraviolet, in optical spectrum, A.38–A.39
Uncompressed video, Video over IP, 4.86–4.88
Underwriters Laboratory (UL), 6.26
Unicast:
  communication types, 4.5
  unicast addresses, 4.58
Uninterruptible power supply. *See* UPS (uninterruptible power supply)
Unmanaged switches, 4.59
Upconverters. *See* Exciters
UPS (uninterruptible power supply):
  anticipation in disaster planning, 7.37
  assessment aspect of disaster planning, 7.39
  batteries in, 6.83–6.85

UPS (uninterruptible power supply) (*Cont.*):
  IOT use on, 2.149–2.150
  maintaining transmission system, 6.117–6.118
  overview of, 6.82
Upstream keys, video effects, 5.66
User Datagram Protocol. *See* UDP (User Datagram Protocol)
User equipment (UE), CNG, 2.236–2.238
User/subscriber control key panels, 5.102–5.104
UTC (Coordinated Universal Time), PSIP, 3.31

Vans:
  ENG (electronic newsgathering), 5.46
  SNG (satellite news gathering), 5.46
Vapor phase cooling, for transmitters, 6.48
Variable flow system, cooling, 6.49
VCT (Virtual Channel Table), PSIP, 3.33, 3.40–3.41
Vector analyzer, measurement with, 2.229–2.230
Vector signal analyzer (VSA), monitoring 8-VSB transmissions, 2.143
Velcro wire bundling, 6.15
Velocity factor, coaxial transmission lines, 2.44
Ventilation. *See* HVAC (heating, ventilation, and air conditioning)
Ventricular fibrillation, of heart, 7.43–7.44
Versioning, in ATSC 3.0, 3.65–3.67
Very-high frequency. *See* VHF (very-high frequency)
Vestigial sideband (VSB) modulation. *See also* 8-VSB:
  ATSC Mobile DTV based on, 3.58
  frames/subframes, 3.59
VHF (very-high frequency):
  band considerations for mask filters, 2.207–2.208
  elevation patterns for television broadcasts, 2.120–2.121
  radomes and slot covers for VHF antennas, 2.84
  retuning mask filters, 2.210
  RF amplification technologies, 2.187
  spectrum compatibility with DTV, 2.154
  TV broadcast bands, 2.70
Vibration, sound isolation techniques, 5.16–5.17
Video:
  applications layer of ATSC 3.0 and, 7.71–7.72
  ATSC Mobile DTV content files, 3.61
  CIAB (Channel in a Box), 5.52
  file-based workflow, 4.63
  mobile services for streaming, 3.44
  MPEG-2, 3.3
  MPEG-4, 3.64
  off air monitoring, 5.55
Video control room, 5.9–5.10
Video file roll-in, 5.67–5.68
Video over IP:
  multicast IP communications, 4.36–4.37
  standards, 4.88
  uncompressed video, 4.86–4.88
Video programmers, captioning requirements, 1.27
Video servers, baseband master control and, 5.51

Video switchers:
    chroma key, 5.66–5.67
    control panels, 5.68–5.69
    effects, 5.65–5.66
    features of, 5.63–5.64
    introduction, 5.63
    summary (big picture), 5.69
    tilting, 5.67–5.68
    transitions, 5.65
    video file roll-in, 5.67–5.68
Video transport, over IP networks:
    forward error correction in ST2022-1, 4.83–4.85
    history of TV transport over carrier networks, 4.79–4.80
    introduction, 4.79
    MPEG-2 transport streams, 4.81
    seamless reconstruction for high availability, 4.86
    shift from circuits to packets in carrier networks, 4.80–4.81
    TS-over-RTP method, 4.83–4.84
    TS-over-UDP method, 4.82
    uncompressed video, 4.86–4.88
Video-On-Demand (VOD), subscription security, 7.73
Virtual Channel Table (VCT), PSIP, 3.33, 3.40–3.41
Virtual channels, PSIP, 3.30
Virtual intercoms, 5.104
Virtual IP addresses, 4.23
visible light, electromagnetic spectrum of, 5.106
Visible light, in optical spectrum, A.38
Visually handicapped (video description), TV service for, 1.23–1.24
VLANs (Virtual LANs):
    conceptual diagram of implementation, 4.47
    configuration examples, 4.48–4.50
    designing security for AoIP, 5.29
    port assignments, 4.17
    segment options, 4.59–4.61
    tagged Ethernet frame, 4.18
    virtual view of, 4.18
VOA (Voice of America), 2.101
VOACAP (Voice of America Coverage Analysis Program), 2.112
VOD (Video-On-Demand), subscription security, 7.73
Voice of America Coverage Analysis Program (VOACAP), 2.112
Voice of America (VOA), 2.101
Voice over IP (VoIP):
    Ethernet switches used with, 5.19
    multicast IP communications, 4.36–4.37
VoIP (Voice over IP):
    Ethernet switches used with, 5.19
    multicast IP communications, 4.36–4.37
Voltage, working with high voltage, 7.47
Voltage standing wave ratio. *See* VSWR (voltage standing wave ratio)

Volume conversion ratios, A.12
VSA (vector signal analyzer), monitoring 8-VSB transmissions, 2.143
VSB (vestigial sideband) modulation. *See also* 8-VSB
    ATSC Mobile DTV based on, 3.58
    frames/subframes, 3.59
VSWR (voltage standing wave ratio):
    de-rating factors, 2.45–2.46
    directional couplers and, 2.18–2.19
    overview of, 2.43–2.44
    peak power ratings for coaxial lines, 2.41
    requirements for transmitting antennas, 2.85
    signal noise as function of antenna VSWR, 2.226
Waivers, from FCC licensing requirements, 1.5
Wakefield, Doug, 1.23
Walls/barriers, sound isolation techniques, 5.15–5.16
Waveform:
    AM, 2.3–2.4
    FM, 2.7–2.8
    HD radio using hybrid waveform, 2.11–2.12
Web browsers, native client vs. browser based MAMs, 5.80
Web sites:
    ABIP review of station web site, 1.19
    FCC, 1.9
Webinars, SBE education programs, 7.65
Wi-Fi:
    cellular revolution and, 5.147
    electronic newsgathering and, 5.154–5.155
Winch and counterweight, lighting suspension and rigging systems, 5.114
Windows, sound isolation techniques, 5.16
Wipes, video transitions, 5.65
Wire management:
    challenges in, 6.21
    cleaning and removing wiring, 6.21–6.23
    color coding, 6.16–6.17
    conduit and interduct, 6.20
    documentation, 6.12–6.14
    installation, 6.17
    introduction, 6.11
    labels, 6.12
    physical layer and, 6.15–6.16
    references, 6.24
    rules/guidelines for, 6.23–6.24
    service loops and slack, 6.17
    special types of wire, 6.17–6.18
    specifications of copper wire sizes, A.7
    third party verification, 6.20–6.21
    wire tray and race way, 6.18–6.20
    wiring closets, 6.20
Wire tray, in wire management, 6.18–6.20
Wired partyline intercom systems (analog):
    advantages of, 5.90–5.92
    components, 5.84–5.90
    overview of, 5.84
Wired partyline intercom systems (digital), 5.92–5.93

Wireless intercom system:
  interfaces, 5.94–5.96
  regulations, 5.93–5.94
Wireless ISO/Talk-around, interfaces for wireless intercom, 5.95
Wireless ITE (In-the-Ear) monitoring, 5.96
Wiring closets, 6.20
Wood:
  collisions and unauthorized frequencies, 2.106
  defined, 2.114

Work in progress MAMs, 5.79
Work order, 7.6

X ray band, in EM spectrum, A.41, A.43
XML (Extensible Markup Language), 5.80
X-ray radiation hazard, 7.57